AF886572

MULTICOLOUR EDITION

# HEAT THERMODYNAMICS AND STATISTICAL PHYSICS

For B. Sc. Students

Revised Edition

**BRIJ LAL**
M.Sc.
*Former, Reader in Physics*
*Hindu College, University of Delhi*

**Dr. N. SUBRAHMANYAM**
M.Sc., Ph.D.
*Former, Reader in Physics*
*Kirori Mal College, University of Delhi*

*Revised by*

**P.S. HEMNE**
M.Sc.
*Post Graduate Deptt. of Physics*
*Nevjabai Hitkarini College*
*Brahmapuri (M.S.)*
*(Nagpur University)*

S. CHAND
PUBLISHING

S Chand And Company Limited
(ISO 9001 Certified Company)

**S Chand And Company Limited**

**(ISO 9001 Certified Company)**

***Head Office:*** D-92, Sector–2, Noida – 201301, U.P. (India), Ph. 91-120-4682700

***Registered Office:*** A-27, 2nd Floor, Mohan Co-operative Industrial Estate, New Delhi – 110 044, Phone: 011-49731800

www.**schandpublishing.com**; e-mail: **info@schandpublishing.com**

**Branches**

| | | |
|---|---|---|
| Chennai | : | Ph: 23632120; chennai@schandpublishing.com |
| Guwahati | : | Ph: 2738811, 2735640; guwahati@schandpublishing.com |
| Hyderabad | : | Ph: 40186018; hyderabad@schandpublishing.com |
| Jalandhar | : | Ph: 4645630; jalandhar@schandpublishing.com |
| Kolkata | : | Ph: 23357458, 23353914; kolkata@schandpublishing.com |
| Lucknow | : | Ph: 4003633; lucknow@schandpublishing.com |
| Mumbai | : | Ph: 25000297; mumbai@schandpublishing.com |
| Patna | : | Ph: 4011400; patna@schandpublishing.com |




*First Edition 1968*
*Subsequent Editions and Reprints 1994, 95 (Twice), 97, 98, 99, 2000, 2001, 2006, 2007 (Twice)*
*Revised Edition 2010; Reprints 2010, 2012 (Twice), 2014 (Twice), 2015, 2016 (Twice), 2017, 2018 (Twice) 2019 (Twice), 2020, 2021 (Twice), 2022*

***Reprint 2025***

**ISBN:** 978-81-219-2813-7 **Product Code:** H6TSP68THRM10ENZX10O

PRINTED IN INDIA

By Shri Shyamlal Printing Press Private Limited, Plot 20/4, Site-IV, Industrial Area Sahibabad, Ghaziabad – 201 010 and Published by S. Chand And Company Limited, A-27, 2nd Floor, Mohan Co-operative Industrial Estate, New Delhi – 110 044.

# PREFACE TO THE REVISED EDITION

I express my sincere thanks to the readers for showing their overwhelming response to the first colored edition of Heat, Thermodynamics and Statistical Physics, by Brij Lal and N. Subrahmanyam and Revised by me. I feel proud to present this modified second colored edition, which is thoroughly revised. In this edition, Exercise at the end of each chapter now contains questions asked by the various Indian Universities upto summer 2009. Numerical Examples are also modified by incorporating the solved numerical problems asked up to summer 2009, by various universities.

In order to fulfill the demand from the students and teachers community, at large, to elaborate the article 1.27, *"Applications to the atmospheric Physics"* and article 8.36, *"Greenhouse Effect"*, I am glad to add one more chapter **Global Warming** at the end of the present book as 18th chapter. Global Warming is one of the most serious challenges facing us today. Leading U.S.A. -based prominent physicists, including 23 Nobel Prize winners and Economists of the entire globe are worried about this problem. To protect the health and economic well-being of current and future generations, they have unanimously appealed to reduce our emissions of heat-trapping gases by using the technology, know-how, and practical solutions already at our disposal. Every one of us can contribute to reduce the severity of this problem. In view of need of the day, this chapter on Global Warming may be included as a separate chapter in the new syllabus. Hope the readers will appreciate this new venture.

NAGPUR

**P.S. HEMNE**
(Reviser)

# ACKNOWLEDGEMENT

While preparing the present book, large number of books of different authors were referred. The author does not claim originality in content or presentation of the material included in this book. The author records here his indebtedness to the authors of those original works from where much of the information is liberally drawn.

I (PSH) am greatly indebted to Dr. N.S. Kokode, Principal, Dr. S.V. Moharil, Reader and Mr. Shekhar Bannore, Librarian, N. H. college, Brahmapuri, and my other colleagues for their valuable help while writing the book. My special thanks are due for my wife Vaishali, daughter Dr. Vishakha and son Vishal (Mechanical Engineer) for their constant encouragement.

I express my sincere thanks to Mrs. Nirmala Gupta Ji, CMD, Shri Navin Joshi, Vice-President (Publishing) and Shri Bhagirath Kaushik, General Manager (S & M) of S. Chand & Company Ltd. for giving the opportunity to revise the book ***"Heat Thermodynamics and Statistical Physics"*** by Prof. Brij Lal & Prof. N. Subrahmanyam. I also thank the editorial staff of S. Chand & Company Ltd. especially Mr. Shishir Bhatnagar (Pre-Press Manager), Mr. Rupesh Gupta, Subject Editor and Mr. Benny K. Abraham for designing and layouting of this book in the present form.

NAGPUR

**P. S. HEMNE**
(Reviser)

# UGC SYLLABUS

## KINETIC THEORY OF MATTER, THERMODYNAMICS AND STATISTICAL PHYSICS

### 4.1 KINETIC THEORY OF MATTER (25)

Ideal Gas : Kinetic model, deduction of Boyle's law, interpretation of temperature, estimation of rms speeds of molecules, Brownian motion, Estimate of Avogadrow number, Equipartition of energy, Specific heats of monoatomic gas, Extension to - di and triatomic gases, Behaviour at low temperatures, Adiabatic expansion of an ideal gas, Application to atmospheric physics. (7)

Real Gas : Van der Waals' gas, Equation of state, Nature of Van der Waals' forces, Comparison with experimental PV curves, The critical constants, gas, and vapour, Joule expansion of ideal gas, and of a Van der Waals' gas, Joule's coefficient, Estimates of J-T cooling. (6)

Liquefaction of Gases : Boyle temperature and Inversion temperature, Principle of regenerative cooling, and cascade cooling, Liquefaction of hydrogen and helium, Refrigeration cycle, meaning of efficiency. (6)

Transport Phenomena in Gases : Molecular collisions, mean free path and collision cross-section, estimates of molecular diameter and mean free path, Transport of mass, momentum and energy and interrelationship, Dependence on temperature and pressure. (6)

### 4.2 THERMODYNAMICS (25)

The Laws of Thermodynamics : The Zeroth law, various indicator diagrams, work done by and on the system, First law of thermodynamics, Internal energy as a state function and other applications, Reversible and irreversible changes, Carnot's cycle and its efficiency, Carnot's theorem and the Second law of thermodynamics, Different versions of second law, Practical cycles used in internal combustion engines, Entropy, principle of increase of entropy. The thermodynamic scale of temperature, its identity with perfect gas scale, Impossibility of attaining the absolute zero, Third law of thermodynamics. (8)

Thermodynamic Relationships : Thermodynamic variables, extensive and intensive, Maxwell's general relationships, application to Joule – Thomson cooling and adiabatic cooling in a general system, Van der Waals' gas, Clausius - Clapeyron heat equation, Thermodynamic potentials and equilibrium of thermodynamical systems, Relation with thermodynamical variables, Cooling due to adiabatic demagnetisation, Production and measurement of very low temperatures. (10)

Black Body Radiation : Pure temperature dependence, Stefan - Boltzmann law, pressure of radiation, spectral distribution of energy in black body radiation, Wien's displacement law, Rayleigh – Jean's law, the Ultraviolet catastrophy, Planck's quantum postulates, Planck's radiation law, complete fit with experiment, Interpretation of behaviour of specific heats of gases at low temperatures. (7)

### 4.3 STATISTICAL PHYSICS (25)

The Statistical Basis of Thermodynamics : Probability and thermodynamic probability, Principle of equal a priori probability, Probability distribution and its narrowing with increase in number of particles, The expression for average properties, constraints, accessible and inaccessible states, Distribution of particles with a given total energy into particles with a given total energy into a discrete set of energy states. (6)

Some universal laws : The mu-Space representation, division of mu-space into energy sheets and into phase cells of arbitrary size, Applications to one dimensional Harmonic Oscillator and free particles, Equilibrium between two Systems in thermal contact, Bridge with macroscopic Physics, Entropy and Probability, boltzmann entropy relation, Statistical interpretation of Second law of thermodynamics, Boltzmann's canonical distribution law and its applications, rigorour from of equipartition of energy. (9)

Maxwellian Distribution of Speeds in an Ideal gas : Distribution of speeds and of velocities experimental verification, distinction between mean, rms and most probable speed values, Doppler broadening of Spectral lines. (4)

Transition to quantum statistics : 'h' as a natural constant and its implications, Cases of Particle in a one dimensional box and one dimensional harmonic oscillator, Indistinguishability of particles and its consequences, Bose–Einstein and Fermi–Dirac conditions, Application to liquid helium, free electrons in a metal, and photons in blackbody chamber, Fermi Level and Fermi Energy. (6)

# CONTENTS

## SECTION I : KINETIC THEORY OF MATTER

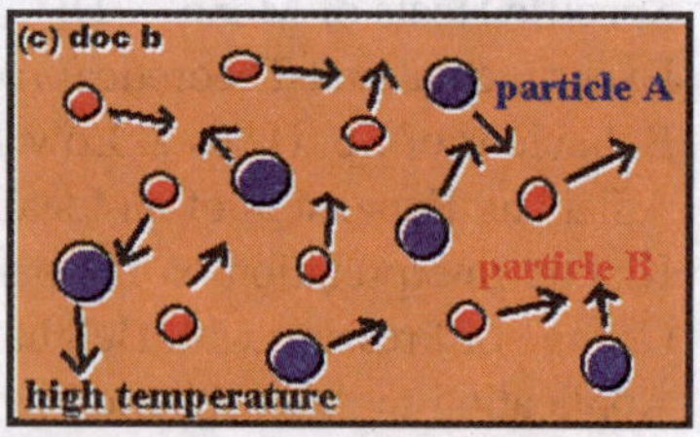
(c) doc b
particle A
particle B
high temperature

## SECTION II : THERMODYNAMICS

## 5. ENTROPY

## 6. THERMODYNAMICAL RELATIONSHIPS



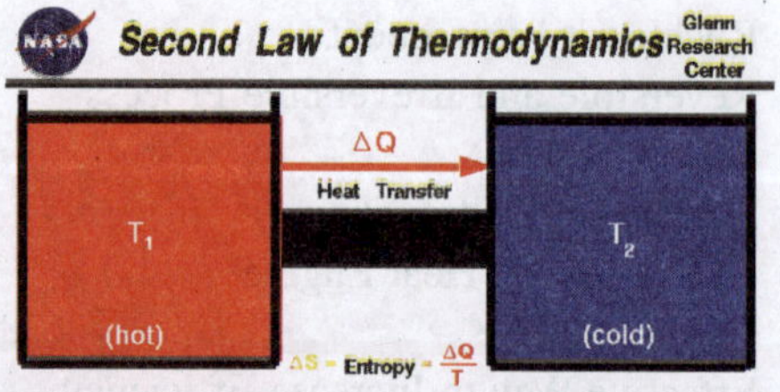

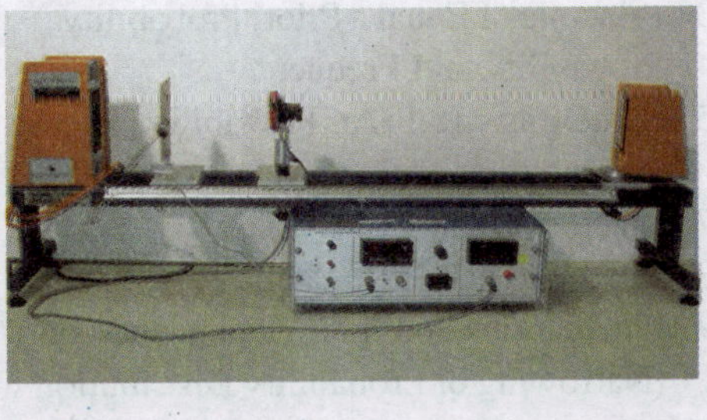

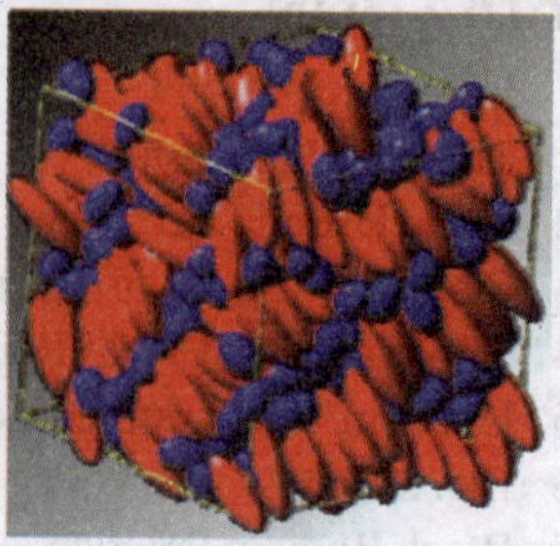

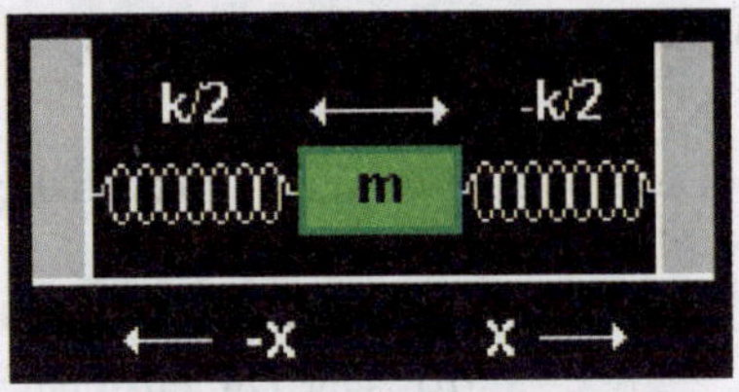
k/2
-k/2
m
-X
X

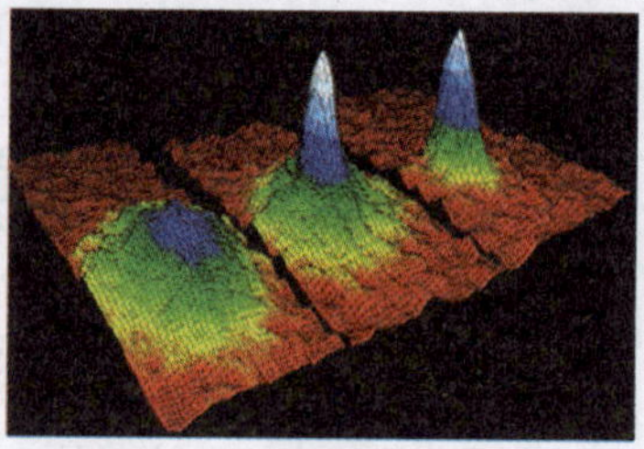

## SECTION IV : HEAT FLOW AND AIR CONDITIONING

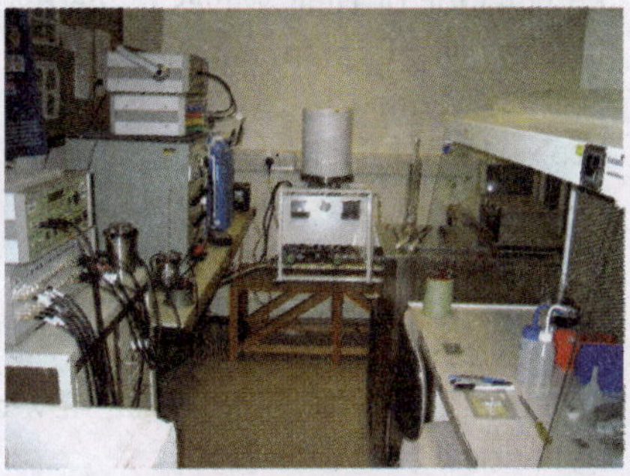



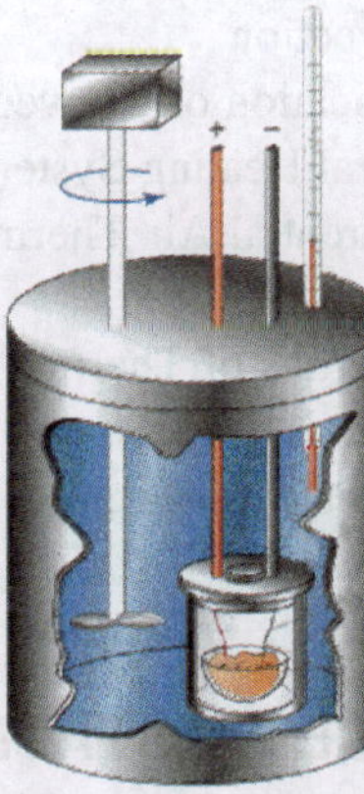

## 17. AIR CONDITIONING SYSTEMS 620 – 627

## 18. GLOBAL WARMING 628 – 640

# SECTION – I

# Kinetic Theory of Matter

1. Ideal Gas
2. Behaviour of Real Gases
3. Transport Phenomena in Gases

# KINETIC THEORY OF MATTER

The kinetic theory of matter rests upon two distinct but closely related hypotheses: the hypothesis of the molecular structure of matter and the hypothesis that heat is a manifestation of the random to-and-fro motion of the molecules.

The hypothesis of the molecular structure of matter merely asserted that all matter is composed of very small particles, known as molecules. In case of compound molecules, motions of the atoms within the molecules must also be taken into account in considering the heat energy of matter. The heat motion of the molecules is especially simple in matter that is in the gaseous state. The average distance between the molecules in the gaseous state is roughly ten times as great as in the solid or liquid state. The total energy of the gas is very approximately equal to the sum of the separate energies of its different molecules. However the potential energy arising from intermolecular forces between pairs of molecules may be treated as negligible in the gaseous state.

These two simplifying facts bring the properties of the gaseous state of matter within the range of mathematical treatment. The kinetic theory of gases attempts to account, in terms of the molecular structure of matter, for all the non-chemical and non-electrical properties of gases.

One of the earliest triumphs of the kinetic theory was its successful explanation of the laws that had been found to connect the density, pressure and temperature of the gas. A quantitative relation between the pressure and the molecular velocities is easily obtained, using the kinetic theory.

# 1 Ideal Gas

## At a Glance

• Introduction • Three States of Matter • Concept of Ideal or Perfect Gas • Kinetic Model • Expression for the Pressure Exerted by a Gas • Estimation of r.m.s. Speeds of Molecules • Deduction of Boyle's Law • Kinetic Energy per unit Volume of a Gas • Derivation of Gas Equation • Derivation of Gas Laws • Avogadrow's Hypothesis • Graham's Law of Diffusion • RMS Velocity as a Function of Absolute Temperature • Brownian Motion • Langevin's Theory of Brownian Motion • Einstein's Theory of Brownian Motion • Distribution of Brownian Particles in a Vertical Column • Brownian Motion of a Galvanometer • Degrees of Freedom • Maxwell's Law of Equipartition of Energy • Relation Between Molar Specific Heats and Degrees of Freedom • Specific Heats of Mono-, Di-, and Polyatomic Gas • Experimental and Theoretical Agreement of $c_p$, $c_v$ and $\gamma$ • Behaviour of Sp. Heats at Low Temperature (Freezing of Degrees of Freedom) • Adiabatic Expansion of an Ideal Gas • Kinetic Interpretation of Temperature • Change of Pressure with Height • Application to Atmospheric Physics

## INTRODUCTION

The experiments on the conversion of work into heat have shown clearly that heat is a form of energy. This energy is connected with the motion of molecules of which the matter is made of. Kinetic theory of matter gives an explanation to the nature of this motion and the nature of heat energy.

According to Kinetic theory of matter, every substance (in the form of solid, liquid or gas) consists of a very large number of very small particles called the molecules. The molecules are the smallest particles of a substance that can exist in free state. The molecules possess the characteristic properties of the parent substance. The molecules are in a state of continuous motion with all possible velocities. The velocity of the molecules increases with rise in temperature.

The energy possessed by the molecules can be of two forms, kinetic or potential. When the stop-cock of an evacuated flask is opened, air rushes in quickly to fill in the space. This shows that the molecules possess rapid motion and hence they possess kinetic energy. When a solid expands on heating, the molecules are pulled apart against the forces of intermolecular attraction. The amount of work done in separating the molecules to larger distances manifests itself as the potential energy of the molecules. Thus the amount of heat given to a solid substance increases the energy of the molecules and this increase in energy is partly kinetic and partly potential.

Thus, the kinetic theory of matter is based on the following three points: (*i*) matter is made up of molecules, (*ii*) molecules are in rapid motion, and (*iii*) molecules experience forces of attraction between one another.

The validity and the soundness of any theory depends on its capability to explain the observed facts, *viz*., the three sates of matter, expansion, change of state and conduction.

## 1·1 Three States of Matter

### Solid

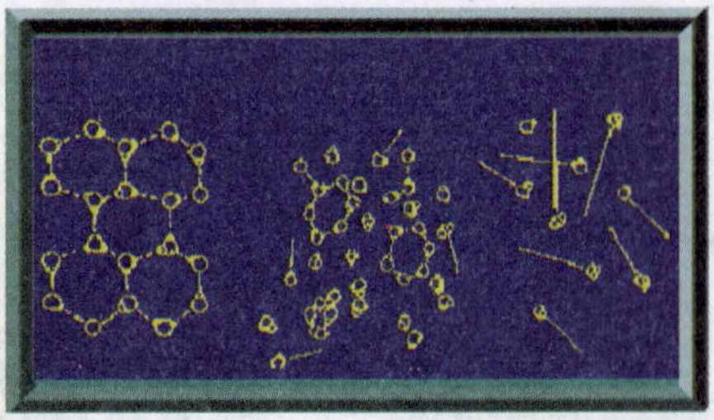

In the case of a solid, the intermolecular distances are small and hence the forces of intermolecular attraction are large. The only motion permissible to a molecule is oscillation or vibration about its mean position which is fixed. Hence a solid has definite size and shape. Heating a solid means giving more energy to the molecules. This will enable the molecules to vibrate more violently about their mean position. The molecules move further apart and this results in the expansion of the solid. With the supply of more heat to the solid, the amplitude of vibration of the molecules increases to such an extent that the molecules will be free to leave their mean position because the kinetic energy of the molecules is greater than the potential energy due to the forces of intermolecular attraction. Now the molecules are free to move about within the body of the substance and this corresponds to the liquid state of the substance. The amount of heat supplied, which is equivalent to the work done in pulling the molecules apart against the forces of intermolecular attraction, determines the latent heat of fusion of the solid.

SOLID

### Liquid

In a liquid, the molecules are farther apart than the solid. The molecules are not confined to any fixed mean position. They are free to move about within the volume of the liquid. Hence the liquids have their own size but no shape.

The definite size of the liquid is due to the mutual forces of attraction between the molecules. The molecules present on the surface of the liquid experience a resultant downward force due to the molecules present within the sphere of influence and hence the surface of a liquid behaves as a stretched membrane. This is the phenomenon of surface tension. The volume of the liquid is fixed at a given temperature but the molecular distances are comparatively larger than that of a solid. Therefore the molecules experience less force of attraction and they can occupy any position within the liquid. Hence the liquid occupies the shape of the containing vessel.

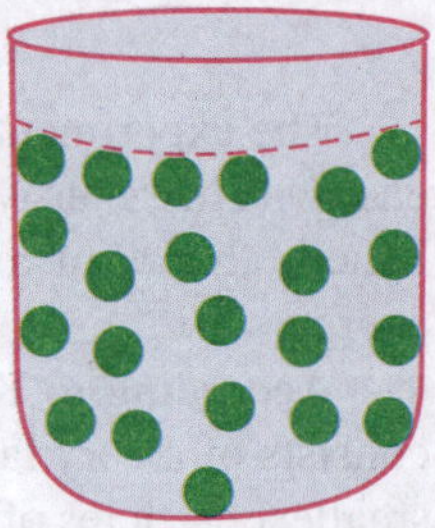

LIQUID

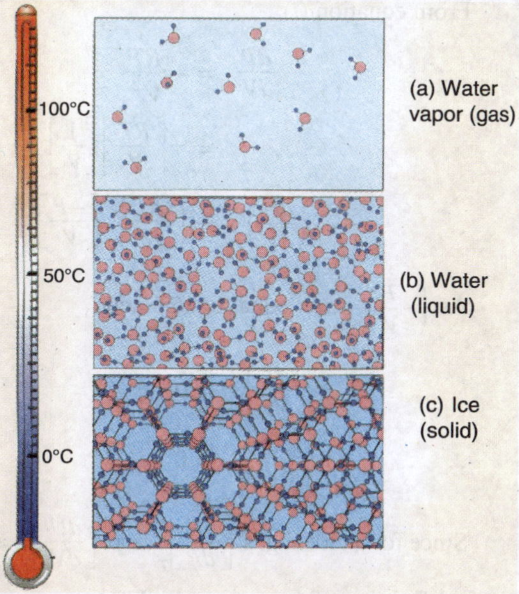


The molecules are always in a state of random motion. They collide with each other and they may or may not change their respective kinetic energies. As already explained, every molecule on the surface of the liquid is held in its position as a result of the forces of intermolecular attraction and surface tension. Due to the collisions a certain molecule may acquire more energy than the energy with which it is held. As soon as this happens, the molecule flies off, *i.e.*, it leaves the liquid surface. This is called *evaporation*. If no external heat is supplied it is obvious that the mean kinetic energy of the molecules decreases and the temperature of the liquid falls. This explains cooling caused by evaporation. If the moving vapour molecules come near to each other at sufficiently low velocities, they experience a greater magnitude of force of attraction. The molecules coalesce and this results in *condensation*. Rate of evaporation increases with increase in temperature of the liquid and decreases in pressure over the liquid surface.

## Gas

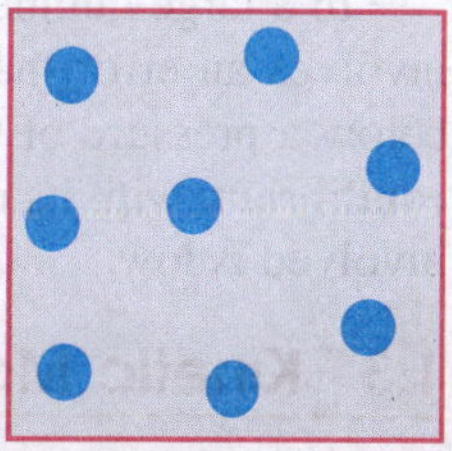
GAS

In the case of a gas, the inter-molecular distances are much larger than that of a solid or a liquid and the molecules of a gas are free to move about in the entire space available to them. Hence a gas has no shape or size. If the space occupied by a gas is increased, the gas will occupy the whole of the new space uniformly. This is due to the fact that the molecules possess rapid random motion and the force of inter-molecular attraction is negligibly small. The molecules move about independently in straight lines and the only restriction to this movement is collision with another molecule or collision with the walls of the containing vessel. In a gas the more predominant factor is the movement of the molecules than the force of attraction between the molecules.

## 1.2 Concept of Ideal or Perfect Gas

At very low pressures, the forces of inter-molecular attraction are negligible. Therefore, at low pressures only, a real gas obeys the equation $PV = RT$. An ideal gas can be defined as a real gas at low pressure. For a real gas the internal energy ($U$) is a function of pressure and temperature. By definition, the ideal gas should satisfy the equation

$$PV = RT \qquad ...(i)$$

and

$$\left[\frac{dU}{dP}\right]_T = 0 \qquad ...(ii)$$

where $dU$ is a small change in internal energy of an ideal gas with pressure $dP$ at constant temperature $T$.

$$\left(\frac{dU}{dV}\right)_T = \left[\frac{dU}{dP}\right]_T \left[\frac{dP}{dV}\right]_T$$

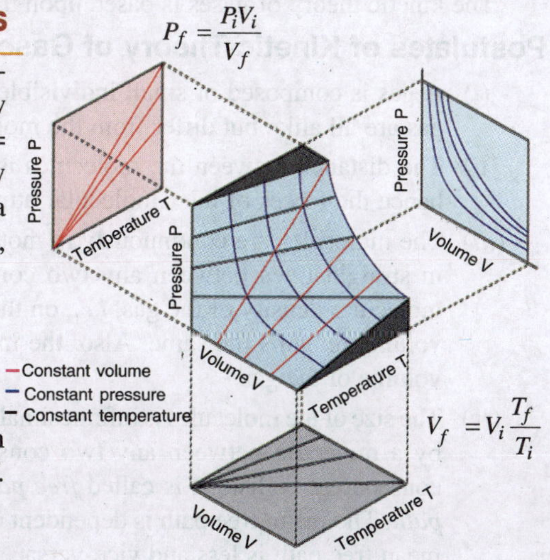

From equation (*i*)

$$\frac{dP}{dV} = -\frac{RT}{V^2}$$

$$= -\left[\frac{RT}{V}\right]\left[\frac{1}{V}\right] = -\frac{P}{V}$$

$$\therefore \quad \left(\frac{dU}{dV}\right)_T = \left[\frac{dU}{dP}\right]_T \left[\frac{-P}{V}\right]_T \qquad ...(iii)$$

As $\left(-\frac{P}{V}\right)$ is not equal to zero and

$$\left[\frac{dU}{dP}\right]_T = 0$$

$$\therefore \quad \left[\frac{dU}{dV}\right]_T = 0 \qquad ...(iv)$$

Since for an ideal gas $\left[\frac{dU}{dP}\right]_T$ and $\left[\frac{dU}{dV}\right]_T$ are both equal to zero, the internal energy of an ideal gas is a function of temperature only,

$$\therefore \quad U = f(T) \qquad ...(v)$$

In a large number of calculations, a real gas can be taken approximately as an ideal gas. This involves an error that can be tolerated in the calculations. For all practical purposes, real gases below a pressure of two atmospheres can be considered as an ideal gases. A saturated vapour in equilibrium with its own liquid can be considered to have the properties of an ideal gas. The error involved is low.

## 1.3 Kinetic Model

According to the kinetic theory of gases, which takes into account of the microscopic structure of matter, a gas enclosed in a vessel consists of very large number of atoms and molecules moving in all directions with all possible velocities in a haphazard manner, thereby colliding with one another. The well-known laws of classical mechanics are applicable to the motion of these molecules.

The kinetic theory of gases is based upon the following important postulates :

### Postulates of Kinetic Theory of Gases

(*i*) A gas is composed of small indivisible particles called molecules. The molecules of same gas are all alike but differ from the molecules of other gases.

(*ii*) The distance between the molecules is large as compared to that of a solid or liquid and hence the forces of inter-molecular attraction are negligible.

(*iii*) The molecules are continuously in motion with varying velocities and the molecules move in straight lines between any two consecutive collisions. The collision do not alter the molecular density of the gas, *i.e.*, on the average the number of molecules present in a unit volume remains the same. Also, the molecules do not accumulate at any place within the volume of the gas.

(*iv*) The size of the molecules is infinitesimally small as compared to the average distance traversed by a molecule between any two consecutive collisions. The distance between any two consecutive collisions is called *free path* and the average distance is called the *mean free path*. The mean free path is dependent on the pressure of the gas. If the pressure is high, the mean free path is less and vice-versa.

(*v*) The time of impact is negligible in comparison to the time taken to traverse the free path.

(*vi*) The molecules are perfectly hard elastic spheres and the whole of their energy is kinetic.

## 1.4 Expression for the Pressure Exerted by a Gas

When a certain mass of a perfect gas is enclosed in a vessel, the molecules frequently collide with the walls of the vessel. This continuous impact of the molecules on the walls of the vessel accounts for the pressure of the gas.

Consider the cubical vessel ABCDEFGH of side $l$ cm containing the gas (Fig. 1.1). The volume of the vessel and hence that of the gas is $l^3$ *cc*. Let $m$ and $n$ represents the mass of each gas molecule and number of molecules present in the vessel respectively.

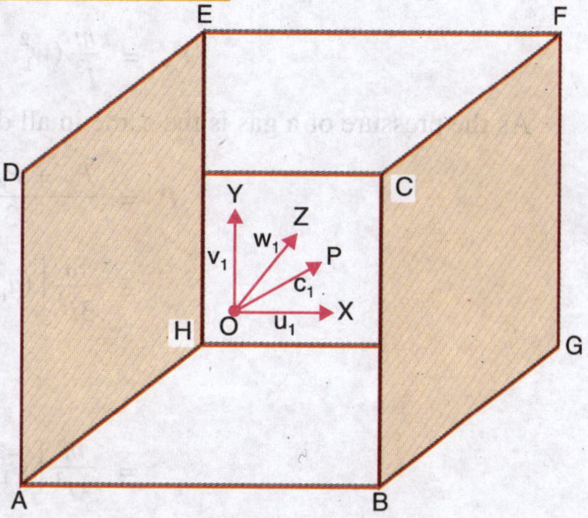


**Fig. 1.1**

Consider a molecule $P$ moving in a random direction with a velocity $C_1$. The velocity can be resolved in to three mutually perpendicular components $u_1$, $v_1$ and $w_1$ along the $X$, $Y$ and $Z$ axes respectively.

Therefore, $$C_1^2 = u_1^2 + v_1^2 + w_1^2$$

The component of the velocity with which the molecule $P$ will strike the opposite face $BCFG$ is $u_1$ and the momentum of the molecule is $mu_1$. This molecule is reflected back with the same momentum $mu_1$ in an opposite direction and after traversing a distance $l$ will strike the opposite face $ADEH$.

The change in momentum produced due to the impact is

$$mu_1 - (-mu_1) = 2\,mu_1$$

As the velocity of the molecule is $u_1$, the time interval between two successive impacts on the wall $BCFG$ is

$$\frac{2l}{u_1} \text{ seconds}$$

∴ No. of impacts per second

$$= \frac{1}{\frac{2l}{u_1}}$$

$$= \frac{u_1}{2l}$$

Change in momentum produced in one second due to the impact of this molecule is

$$2\,mu_1 \times \frac{u_1}{2l} = \frac{mu_1^2}{l}$$

The force $F_X$ due to the impact of all the $n$ molecules in one second

$$= \frac{m}{l}\left[u_1^2 + u_2^2 + \ldots + u_n^2\right]$$

Force per unit area on the wall $BCFG$ of $ADEH$ is equal to the pressure $P_X$

$$P_X = \frac{m}{l \times l^2}(u_1^2 + u_2^2 + u_3^2 + \ldots\ldots + u_n^2)$$

Similarly the pressure $P_Y$ on the walls *CDEF* and *ABGH* is given by

$$P_Y = \frac{m}{l^3}(v_1^2 + v_2^2 + ...... + v_n^2)$$

and the pressure $P_Z$ on the walls *ABCD* and *EFGH* is given by

$$P_Z = \frac{m}{l^3}(w_1^2 + w_2^2 + ...... + w_n^2)$$

As the pressure of a gas is the same in all directions, the mear pressure *P* is given by

$$P = \frac{P_X + P_Y + P_Z}{3}$$

$$= \frac{m}{3l^3}\Big[(u_1^2 + v_1^2 + w_1^2) + (u_2^2 + v_2^2 + w_2^2) + (u_3^2 + v_3^2 + w_3^2) + ......$$

$$+ (u_n^2 + v_n^2 + w_n^2)\Big]$$

$$= \frac{m}{3l^3}\left[C_1^2 + C_2^2 + C_3^2 + ...... C_n^2\right] \quad ...(i)$$

But volume, $V = l^3$. Let *C* be the root-mean-square velocity of the molecules (R.M.S. velocity).

Then
$$C^2 = \frac{C_1^2 + C_2^2 + C_3^2 + ...... C_n^2}{n}$$

or
$$nC^2 = C_1^2 + C_2^2 + C_3^2 + ...... C_n^2$$

Substituting this value in equation (*i*), we get

$$P = \frac{mnC^2}{3V} \quad ...(1.1)$$

But $M = mn$ where *M* is the mass of the gas of volume *V*, *m* is the mass of each molecule and *n* is the number of molecules in a volume *V*.

$$\therefore \quad P = \frac{MC^2}{3V}$$

or
$$P = \frac{1}{3}\rho C^2 \quad ...(1.2)$$

$$\because \quad \frac{M}{V} = \rho \text{ the density of the gas.}$$

From equation (1.2)

$$C^2 = \frac{3P}{\rho}$$

or
$$C = \sqrt{\frac{3P}{\rho}} \quad ...(1.3)$$

[**Note.** R.M.S. velocity *C* is the square root of the mean of the squares of the individual velocities and it is not equal to the mean velocity of the molecules.]

**TABLE 1.1 Molecular r.m.s. Velocities at 0°C**

| Gas | Molecular Weight | R.M.S. Velocity in cm/s |
|---|---|---|
| Hydrogen | 2.016 | $18.4 \times 10^4$ |
| Helium | 4 | $13.1 \times 10^4$ |
| Nitrogen | 28 | $4.95 \times 10^4$ |
| Oxygen | 32 | $4.61 \times 10^4$ |
| Argon | 40 | $4.14 \times 10^4$ |
| Carbon dioxide | 44 | $3.95 \times 10^4$ |
| Chlorine | 71 | $3.11 \times 10^4$ |

## 1.5 Estimation of rms Speeds of Molecules

(*a*) For hydrogen.

The density of hydrogen at N.T.P. is 0.000089 g/cc. Therefore, $C$ for hydrogen can be calculated as follows :

$$C = \sqrt{\frac{3P}{\rho}}$$

$$= \sqrt{\frac{3 \times 76 \times 13.6 \times 981}{0.000089}}$$

$$= 18.4 \times 10^4 \text{ cm/s}$$

(*b*) For oxygen, the density at N.T.P. is $16 \times 0.000089$ g/cc.

$$\therefore \quad C \text{ for oxygen} = \sqrt{\frac{3 \times 76 \times 13.6 \times 981}{16 \times 0.000089}}$$

$$= 4.61 \times 10^4 \text{ cm/s}$$

(*c*) For air, the density at N.T.P. is 0.001293 g/cc.

$$\therefore \quad C \text{ for air} = \sqrt{\frac{3 \times 76 \times 13.6 \times 981}{0.001293}}$$

$$= 4.85 \times 10^4 \text{ cm/s.}$$

## 1.6 Deduction of Boyle's Law

According to the kinetic theory,

$$P = \frac{1}{3}\rho C^2$$

$$P = \frac{1}{3}\frac{M}{V}C^2$$

$$PV = \frac{1}{3}MC^2 \qquad \text{...(1.4)}$$

At constant temperature $T$, $C^2$ is constant. Therefore, at a constant temperature

$$\frac{1}{3}MC^2 = \text{constant}$$

Robert Boyle

Hence, $PV$ = constant, at constant temperature. This is Boyle's Law.

## 1.7 Kinetic Energy per unit volume of a gas

According to kinetic theory,

$$P = \frac{1}{3}\rho C^2$$

$$= \frac{2}{3}\cdot\frac{1}{2}\rho C^2$$

$$= \frac{2}{3}\cdot E \qquad ...(1.5)$$

where $E = \frac{1}{2}\rho C^2$ and is equal to the kinetic energy per unit volume of the gas. ρ is the mass per unit volume. Hence, the pressure of a gas is numerically equal to two-thirds of the mean kinetic energy of translation of a unit volume of the molecules.

## 1.8 Derivation of Gas Equation

From kinetic theory,

$$P = \frac{1}{3}\rho C^2$$

$$P = \frac{1}{3}\frac{M}{V}C^2$$

$$PV = \frac{1}{3}MC^2$$

Consider one gram molecule of a gas at an absolute temperature $T$. The mean energy of the molecules

$$= \frac{1}{2}MC^2$$

$$= \frac{1}{2}NmC^2$$

$$\therefore \quad PV = \frac{1}{3}NmC^2$$

$$= \frac{2}{3}N\cdot\frac{1}{2}\cdot mC^2$$

Mean kinetic energy of a molecule

$$= \frac{1}{2}mC^2$$

$$= \frac{3}{2}kT$$

$$\therefore \quad PV = \frac{2}{3}N\cdot\frac{3}{2}kT$$

$$PV = NkT$$

But $$N\times k = R$$

$$\therefore \quad PV = RT \qquad ...(1.6)$$

***Note on the Gas Equation.***

In the gas equation

$$PV = RT$$

$P$ is in dynes/sq cm

$$R = 8.31 \times 10^7 \text{ ergs/g mol-K}$$

$T$ is in $K$

$V$ is the volume in *cc* per gram molecule.

## 1.9 Derivation of Gas Laws

**(*i*) Boyle's Law**

According to the kinetic theory,

$$P = \frac{1}{3}\rho C^2$$

$$P = \frac{1}{3}\frac{M}{V}C^2$$

$$PV = \frac{1}{3}MC^2 \qquad \text{...}(i)$$

At a constant temperature $T$, $C^2$ is constant. Therefore at a constant temperature

$$\frac{1}{3}MC^2 = \text{constant}$$

Hence $\qquad PV = \text{constant} - \text{at constant temperature.}$

**(*ii*) Charles' Law**

According to the kinetic theory

$$P = \frac{1}{3}\rho C^2$$

$$P = \frac{1}{3}\frac{M}{V}C^2$$

$$PV = \frac{1}{3}MC^2$$

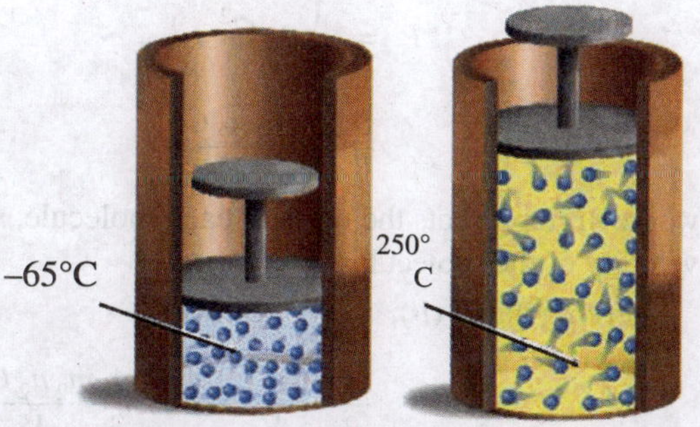

Charles' Law

Consider one gram molecule of a gas at absolute temperature $T$.

$$M = mN$$

$$PV = \frac{1}{3}NmC^2 \qquad \text{...}(i)$$

The mean kinetic energy of a molecule

$$\frac{1}{2}mC^2 = \frac{3}{2}kT$$

$$mC^2 = 3\,kT$$

Substituting this value in equation (*i*),

$$\therefore \quad PV = NkT \qquad \text{...}(ii)$$

where $N$ is the Avogadro's number and $k$ is the Boltzmann's constant.

If $\qquad P$ is constant,

$$V \propto T$$

It means that for a given mass of gas, the volume is directly proportional to the absolute temperature provided the pressure remains constant. This is Charles' Law

***(iii)* Regnault's Law**

From equation (*ii*)

$$PV = NkT$$

When $V$ is constant,

$$P \propto T$$

It means for a given mass of gas, the pressure is directly proportional to its absolute temperature provided the volume remains constant. This is Regnault's law.

## 1.10 Avogadro's Hypothesis

Consider two gases $A$ and $B$ at a pressure $P$ and each having a volume $V$.

Mass of each molecule of the first gas = $m_1$

Number of molecules of the first gas = $n_1$

Mean square velocity of the molecules of the first gas = $C_1^2$

For the first gas

$$P = \frac{1}{3}\rho_1 C_1^2$$

$$= \frac{1}{3}\frac{m_1 n_1 C_1^2}{V} \qquad ...(i)$$

Similarly for the second gas

$$P = \frac{1}{3}\rho_2 C_2^2$$

$$= \frac{1}{3}\frac{m_2 n_2 C_2^2}{V} \qquad ...(ii)$$

where $m_2$ represents the mass of each molecule, $n_2$ the number of molecules and $C_2^2$ the mean square velocity of the molecules.

From (*i*) and (*ii*),

$$\frac{1}{3}\frac{m_1 n_1 C_1^2}{V} = \frac{1}{3}\frac{m_2 n_2 C_2^2}{V}$$

**LORENZO ROMANO AMEDEO AVOGADRO,**

**Conte di Quaregna e di Cerreto (1776–1856)**

- An Avogadro's number of standard soft drink cans would cover the surface of the earth to a depth of over 200 miles.
- If you had Avogadro's number of unpopped popcorn kernels and spread them across the United States of America, the country would be covered in popcorn to a depth of over 9 miles.
- If we were able to count atoms at the rate of 10 million per second; it would take about 200 billion years to count the atoms in one mole.

or $$m_1 n_1 C_1^2 = m_2 n_2 C_2^2 \qquad ...(iii)$$

If the two gases are at the same temperature $T$, the mean kinetic energy of the molecules of both the gases is the same

$$\frac{1}{2} m_1 C_1^2 = \frac{1}{2} m_2 C_2^2$$

$$\therefore \quad m_1 C_1^2 = m_2 C_2^2 \qquad ...(iv)$$

From (*iii*) and (*iv*),

$$n_1 = n_2$$

Hence, equal volumes of all gases under similar conditions of temperature and pressure have the same number of molecules. This represents Avogadro's hypothesis.

## 1.11 Graham's Law of Diffusion of Gases

According to the kinetic theory

$$P = \frac{1}{3} \rho C^2$$

or $$C = \sqrt{\frac{3P}{\rho}}$$

$$C \propto \frac{1}{\sqrt{\rho}}$$

It means that the root mean square velocity of the molecules of a gas is inversely proportional to the square root of its density. Consider two gases whose root mean square velocities are $C_1$ and $C_2$.

$$\frac{C_1}{C_2} = \sqrt{\frac{\rho_2}{\rho_1}}$$

or $$\frac{r_1}{r_2} = \sqrt{\frac{\rho_2}{\rho_1}} \qquad ...(1.7)$$

Here, $r_1$ and $r_2$ represent the rates of diffusion of the two gases.

## 1.12 RMS Velocity as a Function of Absolute Temperature

The pressure of a gas, according to kinetic theory, is

$$P = \frac{1}{3} \rho C^2$$

$$P = \frac{1}{3} \frac{MC^2}{V}$$

$$PV = \frac{1}{3} MC^2$$

For 1 mole of an ideal gas at temperature $T$ K,

$$PV = RT$$

$$\therefore \quad \frac{1}{3} MC^2 = RT$$

$$\frac{1}{2} MC^2 = \frac{3}{2} RT$$

Let m be the mass of each gas molecule and *N* be Avogadro's number.

$$M = m \times N$$

$$\frac{1}{2} mNC^2 = \frac{3}{2} RT$$

$$\frac{1}{2} mC^2 = \frac{3}{2} \frac{R}{N} T$$

$$= \frac{3}{2} kT \qquad ...(1.8)$$

where $k$ is called Boltzmann's constant.

Thus, from equation (1.8), the mean kinetic energy of a molecule is directly proportional to the absolute temperature ($T$) of a gas. When the temperature of the gas is increased, the mean kinetic energy of the molecules increases. When heat is withdrawn from a gas, the mean kinetic energy of the molecules decreases. Thus, at absolute zero temperature, the kinetic energy should be zero. It means, at $T = 0$ $K$, the molecules are in a perfect state of rest and have no kinetic energy. But, before the absolute zero temperature is reached, all gases change their states to liquids and solids.

Also from equation (1.8), $C^2 \propto T$ or $C \propto \sqrt{T}$ . It means that the root mean square (r.m.s) velocity of the molecules is also directly proportional to the square root of the absolute temperature.

## 1.13 Brownian Motion

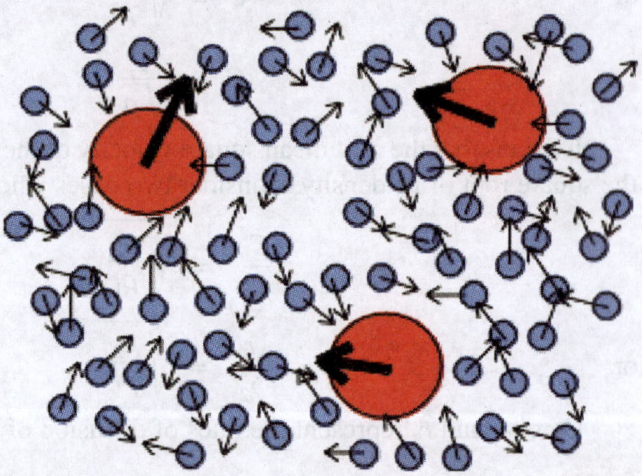

Brownian motion was first observed in 1827 by Brown, a botanist. The phenomenon of Brownian movement can be easily observed in a laboratory if a colloidal solution is examined under an ultramicroscope. As the direction of illumination is perpendicular to the axis of the microscope, the suspended particles in the solution look like bright illuminated spots. These illuminated particles continuously move to and fro in a random haphazard way. The particles spin, rise, sink and rise again. The movement of the particles is continuous and spontaneous. This non-stop random and haphazard motion of the particles is called *Brownian motion*.

The motion of the particles becomes more violent with increase in temperature and more conspicuous in a liquid of lower viscosity. Particles of smaller size appear to be more agitated than the bigger particles. Further, no two particles execute the same type of motion.

The phenomenon of Brownian movement gives a clear picture of the gaseous state of matter. The motion of the molecules of a gas is similar in nature to the Brownian motion of suspended particles in colloidal solution. It may be noted that the laws of kinetic theory of gases are applicable to Brownian particles also.

### Essential Features :

(*i*) The motion of each particle is completely irregular and random. No two particles are found to execute the same motion.

(*ii*) The motion is continuous and takes place for ever.

(*iii*) The smaller particles appear to be more agitated than the larger ones.

(*iv*) The motion is independent of the nature of the suspended particles.

(v) The motion becomes more violent on increasing the temperature.
(vi) The motion is not modified due to the shaking of colloidal solution.
(vii) The Brownian motion cannot be observed with particles of large size.
(viii) The motion is more conspicuous in a liquid of lower viscosity.
(ix) The laws of kinetic theory of gases are applicable to Brownian motion too.

## 1.14. Langevin's Theory of Brownian Motion

According to Langevin, the force experienced by a suspended particle is of two kinds :

(i) Frictional force proportional to the velocity

$$f\left(\frac{dx}{dt}\right) = 6\pi\eta av$$

(ii) Force due to all external influences of the surrounding fluid.

Consider the motion of a particle in a specified direction say $x$ – direction. The force on the particle is given by the equation.

$$m\left(\frac{d^2x}{dt^2}\right) = -f\left(\frac{dx}{dt}\right) + F \qquad ...(i)$$

Here, $-f\left(\frac{dx}{dt}\right)$ represents the component of the frictional force in the $x$ – direction. $F$ is the combined force due to other influences. Multiply equation (i) by $x$,

$$mx\left(\frac{d^2x}{dt^2}\right) = -fx\left(\frac{dx}{dt}\right) + Fx \qquad ...(ii)$$

Also $\quad \frac{d}{dt}(x^2) = 2x\left(\frac{dx}{dt}\right)$ and $x\left(\frac{dx}{dt}\right) = \frac{1}{2}\frac{d}{dt}(x^2)$

$$\frac{d}{dt}\left[\frac{d}{dt}(x^2)\right] = 2\left[x\left(\frac{d^2x}{dt^2}\right) + \left(\frac{dx}{dt}\right)^2\right]$$

$$\therefore \quad x\left(\frac{d^2x}{dt^2}\right) = \frac{1}{2}\frac{d}{dt}\left[\frac{d}{dt}(x^2)\right] - \left(\frac{dx}{dt}\right)^2 \qquad ...(iii)$$

Substituting these values in equation (ii)

$$\frac{m}{2}\frac{d}{dt}\left[\frac{d}{dt}(x^2)\right] - m\left(\frac{dx}{dt}\right)^2 = -\frac{f}{2}\frac{d}{dt}(x^2) + Fx \qquad ...(iv)$$

Equation (iv) represents the motion of a single particle. For all the particles, the mean value can be written as

$$\frac{m}{2}\frac{d}{dt}\overline{\left[\frac{d}{dt}(x^2)\right]} - x\overline{\left(\frac{dx}{dt}\right)^2} = -\frac{f}{2}\frac{d}{dt}\overline{(x^2)} + \overline{Fx} \qquad ...(v)$$

As the force $F$ varies completely in an irregular manner, it can be assumed that

$$\overline{Fx} = 0$$

Further, from the law of equipartition of velocities,

$$m\overline{\left(\frac{dx}{dt}\right)^2} = kT$$

Also take $\frac{d}{dt}(x^2) = \frac{d}{dt}\left(\overline{x^2}\right) = U$

Substituting these values in equation (*v*)

$$\frac{m}{2}\left(\frac{dU}{dt}\right) + \frac{fU}{2} = kT$$

or
$$\frac{dU}{dt} + \left(\frac{f}{m}\right)U = \frac{2kT}{m} \qquad ...(vi)$$

The general, solution of equation (*vi*) is

$$U = \frac{2kT}{f} + A\left[e^{-\left(\frac{f}{m}\right)t}\right] \qquad ...(vii)$$

As the value of $m$ is very small, the value of $(f/m)$ is very large and hence $e^{\left(\frac{-f}{m}\right)t}$ is negligibly small

∴
$$u = \frac{2kT}{f} = \frac{d\,\overline{x^2}}{dt} \qquad ...(viii)$$

For a time interval $t = 0$ to $t = \tau$, integrating equation (*viii*), we get

$$\overline{x^2} - \overline{x_0^2} = \left(\frac{2kT}{f}\right)\tau$$

At $t = 0$, $x_0 = 0$ and for small values, $x^2$ can be written as $\Delta x^2$

∴
$$\overline{\Delta x^2} = \left(\frac{2kT}{f}\right)\tau \qquad ...(ix)$$

For each interval of time $\tau$, the displacement $\Delta x$ of the particle in the $x$ – direction is determined. The mean square value $\overline{\Delta x^2}$ is calculated. In actual practice the particle makes millions of collisions and moves along zig-zag paths. The value $\overline{\Delta x^2}$ is only loosely related to the actual path.

According to Stokes's formula,

$$f\left(\frac{dx}{dt}\right) = 6\pi\eta a v = 6\,\pi\eta a\left(\frac{dx}{dt}\right)$$

∵
$$f = 6\pi\eta a$$

∴
$$\overline{\Delta x^2} = \frac{2kT\tau}{6\pi\eta a} = \frac{kT\tau}{3\,\pi\eta a} \qquad ...(x)$$

or
$$\left[\overline{\Delta x^2}\right]^{\frac{1}{2}} = \frac{(kT)^{\frac{1}{2}}\,(\tau)^{\frac{1}{2}}}{(3\pi\eta a)^{\frac{1}{2}}} \qquad ...(xi)$$

This theory indicates that $\overline{\Delta x^2}$ is not dependent on the mass of the particle. In the experiments of Perrin, the masses of the particles varied in the ratio 1 to 15000. However, within the limits of experimental error, the value of Boltzmann's constant $k$ is the same.

Further,

$$\left(\overline{\Delta x^2}\right)^{\frac{1}{2}} \propto \tau^{\frac{1}{2}}$$

and
$$\left(\overline{\Delta x^2}\right)^{\frac{1}{2}} \propto \frac{1}{\eta^{\frac{1}{2}}} \qquad ...(1.9)$$

The effect of temperature is not very large because $\left(\overline{\Delta x^2}\right)^{\frac{1}{2}} \propto T^{\frac{1}{2}}$. However, viscosity decreases rapidly with increase in temperature. Thus, the pure temperature effect is negligibly small in comparison to the effect of viscosity.

Diffusion, fluctuations in concentration and Brownian motion represent a single phenomenon. Diffusion is a macroscopic phenomenon whereas Brownian motion is a microscopic phenomenon.

## 1.15. Einstein's Theory of Brownian Motion

According to Einstein's theory of transitional Brownian motion, the particles tend to diffuse into the medium in course of time. Consequently, the diffusion coefficient must be related to the Brownian movement.

Albert Einstein

The diffusion coefficient can be calculated in two different ways :

(1) From the irregular motion of the suspended particles.

(2) From the difference in osmotic pressure caused by the differences in concentration of the suspended particles.

Let $D$ be the diffusion coefficient. Consider an imaginary cylinder with its axis along the $x$-axis. The end faces $P$ and $Q$ are separated by a distance $\Delta$. Let $n_1$ and $n_2$ be molecular concentration at the end faces of the cylinder and $A$ the area of cross-section.

The number of particles crossing the surface $P$ to the right in time $T = \frac{1}{2} n_1 A\Delta$

Similarly, the number of particles crossing the surface $Q$ in the opposite direction $= \frac{1}{2} n_2 A\Delta$

It should be noted that half the particles contained in the imaginary cylinder move towards right whereas the other half move towards left.

The excess number of particles crossing a middle layer to the right $= \frac{1}{2}(n_1 - n_2) A\Delta$

From the definition of diffusion coefficient, the number,

$$\frac{1}{2}(n_1 - n_2) A\Delta = -D\frac{dn}{dx}\tau A$$

Here, $\left(\frac{dn}{dx}\right)$ is the concentration gradient.

But
$$(n_1 - n_2) = -\Delta\frac{dn}{dx}$$

$$\therefore \quad -\frac{1}{2}\Delta^2\left(\frac{dn}{dx}\right) = -D\left(\frac{dn}{dx}\right)\tau$$

$$\therefore \quad \Delta^2 = 2D\tau$$

or
$$D = \frac{\Delta^2}{2\tau} \qquad ...(i)$$

Now, $D$ is to be calculated from the concept of osmotic pressure. If $p_1$ and $p_2$ are the osmotic pressures at the end $P$ and $Q$, then from the gas laws

$$p_1 = n_1 kT$$
$$p_2 = n_2 kT$$

Thus the cylinder experiences a resultant force, $(p_1 - p_2) A = (n_1 - n_2) kTA$ along the +ve $x$ – direction. This force is experienced by the particles contained in the cylinder.

The number of particles in the cylinder

$$= n A \Delta$$

where $n$ is the mean concentration.

Therefore, the force acting on a single particle

$$f' = \frac{(n_1 - n_2)\, kTA}{nA\,\Delta} \quad \text{...}(ii)$$

Substituting the value of $(n_1 - n_2) = -\Delta \dfrac{dn}{dx}$ in equation $(ii)$

$$f' = -\left(\frac{dn}{dx}\right)\left(\frac{kT}{n}\right)$$

$$= 6\pi\eta av$$

$$\therefore \quad = -\left(\frac{kT}{n}\right)\left(\frac{dn}{dx}\right)$$

$$\text{or} \quad nv = -\left(\frac{kT}{6\,\pi\eta a}\right)\left(\frac{dn}{dx}\right) \quad \text{...}(iii)$$

Here, $nv$ is the number of particles moving to the right per unit area per second

$$\therefore \quad nv = -D\left(\frac{dn}{dx}\right) \quad \text{...}(iv)$$

Equating $(iii)$ and $(iv)$

$$-D\left(\frac{dn}{dx}\right) = -\left(\frac{kT}{6\pi\eta a}\right)\left(\frac{dn}{dx}\right)$$

$$\text{or} \quad D = \left(\frac{kT}{6\pi\eta a}\right) = \left(\frac{RT}{N}\right)\left(\frac{1}{6\pi\eta a}\right) \quad \text{...}(v)$$

From equations $(i)$ and $(v)$

$$\frac{\Delta^2}{2\tau} = \left(\frac{RT}{N}\right)\left(\frac{1}{6\pi\eta a}\right)$$

$$\Delta^2 = \frac{RT}{N}\left(\frac{1}{3\pi\eta a}\right)\tau \quad \text{...}(vi)$$

Equations $(i)$ and $(v)$ have been verified experimentally. Also, with the help of equation $(v)$, the value of Avogadro's number is calculated.

## 1.16 Distribution of Brownian Particles in a Vertical Column

The Brownian particles form a gas in equilibrium under the action of earth's gravitation. The concentration of the particles in a vertical column decreases with height due to gravity. Consider two layers of the particles at heights $h$ and $(h + dh)$. Let $P$ and $(P + dp)$ be the pressures respectively and $\rho$ the density of the gas at a height $h$. Consider a unit area of the layer. The force due to gravity acting on the particles between two layers

$$= (1 \times dh)\,\rho g = \rho g dh$$

The net force on the layer

$$= (P + dP) - P + [\rho g dh]$$

In the equilibrium state, the net force must be equal to zero. Therefore,

$$(P + dP) - P + [\rho g dh] = 0$$

or $$dP + \rho g dh = 0$$

or $$dP = -\rho g dh \qquad ...(i)$$

For a perfect gas,

$$P = nkT$$

and $$dP = kT\,dn$$

and $$\rho = mn$$

Here, $n$ is the number of molecules per unit volume and $m$ is the mass of each molecule.

Substituting, we get

$$kTdn = -mngdh$$

$$\frac{dn}{n} = -\left(\frac{mg}{kT}\right)dh$$

Also $k = \frac{R}{N}$, here $N$ is Avogadro's number.

$$\therefore \quad \frac{dn}{n} = -\left(\frac{Nmg}{RT}\right)dh \qquad ...(ii)$$

Integrating

$$\log n = -\left(\frac{Nmg}{RT}\right)h + K \qquad ...(iii)$$

Here $K$ is a constant.
At $h = h_0$, $n = n_0$

$$\therefore \quad \log n_0 = -\left(\frac{Nmg}{RT}\right)h_0 + K$$

or $$K = \log n_0 + \left(\frac{Nmg}{RT}\right)h_0$$

Substituting this value of $K$ in equation (*ii*),

$$\log n = -\left(\frac{Nmg}{RT}\right)h + \left(\frac{Nmg}{RT}\right)h_0 + \log n_0$$

$$\log\left(\frac{n}{n_0}\right) = -\left(\frac{Nmg}{RT}\right)(h - h_0) \qquad ...(iv)$$

$$\therefore \quad n = n_0 \cdot e^{-\left(\frac{Nmg}{RT}\right)(h - h_0)} \qquad ...(v)$$

## Determination of Avogadro's number

From equation (*iv*),

$$N = \frac{RT \log\left(\frac{n_0}{n}\right)}{mg\,(h - h_0)} \qquad ...(vi)$$

The effective mass of the suspended particle,

$$m = \frac{4}{3}\pi a^3 (d_1 - d_2)$$

where $a$ is the radius of the particle, $d_1$ is the density of the particle and $d_2$ is the density of the intervening fluid.

$$\therefore \quad N = \frac{3RT \log\left(\frac{n_0}{n}\right)}{4\pi\, a^3 g\,(d_1 - d_2)\,(h - h_0)} \qquad ...(vii)$$

Thus, from equation (*vii*), the value of Avogadro's number ($N$) can be calculated if the values of $a$ and $d_1$ are known.

**Determination of $d_1$ :** Perrin used the method of finding density ($d_1$) with a specific gravity bottle. Let $m_1$ and $m_2$ be the masses of water and emulsion filling the specific gravity bottle (volume $V$) and $m_3$ the mass of the particles left over after desiccation in an oven. Taking $d$ as the density of water,

Volume of granules,

$$= \left(\frac{m_1}{d}\right) - \left(\frac{m_2 - m_3}{d}\right)$$

$\therefore$ Density of granules,

$$d_1 = \frac{m_3}{\left(\frac{m_1}{d}\right) - \left(\frac{m_2 - m_3}{d}\right)}$$

or

$$d_1 = \frac{m_3\, d}{(m_1 + m_3 - m_2)} \qquad \text{...(viii)}$$

**Determination of $a$ :** The radius of the particles is determined by Stoke's formula.

The terminal velocity $v$ of the particles is determined by experiment.

$$\therefore \quad 6\,\pi\eta\, a\, v = \frac{4}{3}\,\pi\, a^3\,(d_1 - d_2)\, g$$

$$\therefore \quad a = \left[\frac{9\,\eta v}{2(d_1 - d_2)\, g}\right]^{\frac{1}{2}} \qquad \text{...(ix)}$$

Substituting these values of $d_1$ and $a$ in equation (*vii*), the value of Avogadro's number can be calculated.

To measure the values of $n$ and $n_0$ at heights $h$ and $h_0$, photographs of the suspended particles are taken and the number of particles in these photographs is counted. The value of Avogadro's number determined by Perrin was found to be $6.82 \times 10^{26}$ moles/kg–mole which is slightly higher than the present accepted value of $6.023 \times 10^{26}$ molecules/kg–mole.

## 1.17 Brownian Motion of a Galvanometer

In the case of a moving coil galvanometer with a lamp and scale arrangement, the initial position is adjusted so that the spot of light is at the zero mark on the scale. Even when no current passes through the galvanometer coil, it is sometimes noticed that the spot of light has shifted from zero. This is due to the random zig-zag collision of the air molecules with the suspended system. When these collisions produce unequal torques in opposite directions, the suspended system continuously fluctuates and results in unsteady, zero setting. This shift of zero setting is due to Brownian motion of a galvanometer.

For a suspension fibre, the elastic potential energy is $\frac{1}{2}C\theta^2$ where $C$ is the torsional rigidity of the fibre. Similarly, the rotational kinetic energy of the suspended system is $\frac{1}{2}I\omega^2$. The mean elastic potential energy = $\frac{1}{2}C\overline{\theta^2}$ and mean rotational kinetic energy = $\frac{1}{2}I\overline{\omega^2}$. Each of the quantities are equal to $\frac{1}{2}kT$. Here k is the Boltzmann's constant, $T$ is the temperature in degree kelvin.

$$\therefore \quad \frac{1}{2}C\,\overline{\theta^2} = \frac{1}{2}kT \qquad \text{...(i)}$$

and $$\frac{1}{2} k \overline{\omega^2} = \frac{1}{2} kT \quad ...(ii)$$

The root mean square angular displacement of the suspended system is given by

$$\theta_{rms} = \sqrt{\theta^2} = \sqrt{\frac{kT}{C}} \quad ...(iii)$$

For a fine quartz suspension fibre $C$ is approximately equal to $10^{-13}$ N-m/radian.

and $$k = 1.38 \times 10^{-23}$$

For a room temperature of 300 K.

$$\theta_{rms} = \sqrt{\frac{kT}{C}}$$

$$\theta_{rms} = \left[\frac{1.38 \times 10^{-23} \times 300}{10^{-13}}\right]^{\frac{1}{2}}$$

$$\theta_{rms} = 2 \times 10^{-4} \text{ radian}$$

If the scale is kept at a distance of one metre from the mirror of the galvanometer, then the displacement of the spot of light from zero is given by

$$2(\theta_{rms}) = \frac{dx}{1}$$

$$dx = 4 \times 10^{-4} \text{ m} = 0.4 \text{ mm}. \quad ...(1.10)$$

For a large value of $C$, the value of $dx$ increases.

## 1.18 Degree of Freedom

### Definition

The total number of independent variables required to describe completely the state of motion of a body are called its degrees of freedom.

An ant moving along a straight line *i.e.*, a wire, it has only one degree of freedom ($x$). If it moves on a plane, then it has two degrees of freedom ($x$, $y$). A flying insect in air, say a mosquito has 3 degrees of freedom ($x$, $y$, $z$). A rigid body, however, not only moves but also rotates about any axis passing through itself. Hence, it has 3 degrees of freedom due to rotational motion. Thus, it has in all 6 degrees of freedom.

### Monoatomic molecule

A Monoatomic molecule, *e.g.*, Neon (Ne), Helium (He) etc., has 3 degrees of freedom (all are translational) as shown in Fig. 1.2.

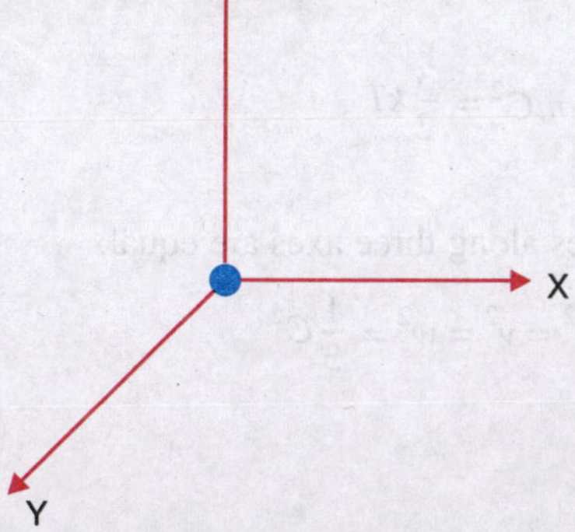


**Fig. 1.2** Monoatomic molecule.

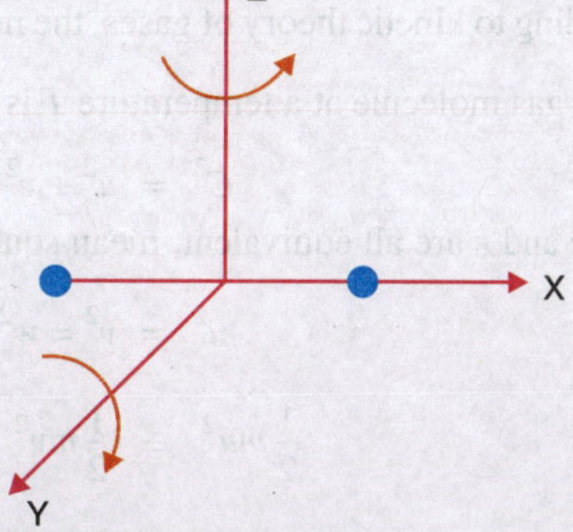


**Fig. 1.3** Diatomic molecule.

### Diatomic molecule

A diatomic molecule having a dumb-bell shape *e.g.*, HCl, $Cl_2$, $O_2$, CO etc. has 5 degrees of freedom (3 translational + 2 rotational), as shown in Fig. 1.3.

### At high temperature

At high temperatures, in addition to the translational and rotational degrees of freedom, the vibrational degrees of freedom are also excited and in that case a diatomic molecule possesses 7 degrees of freedom in all.

### Triatomic molecule

A triatomic molecule, if it is non-linear *e.g.*, $H_2S$, $SO_2$ etc. has 6 degrees of freedom (3 translational + 3 rotational). Fig. 1.4 shows a triatomic non-linear molecule.

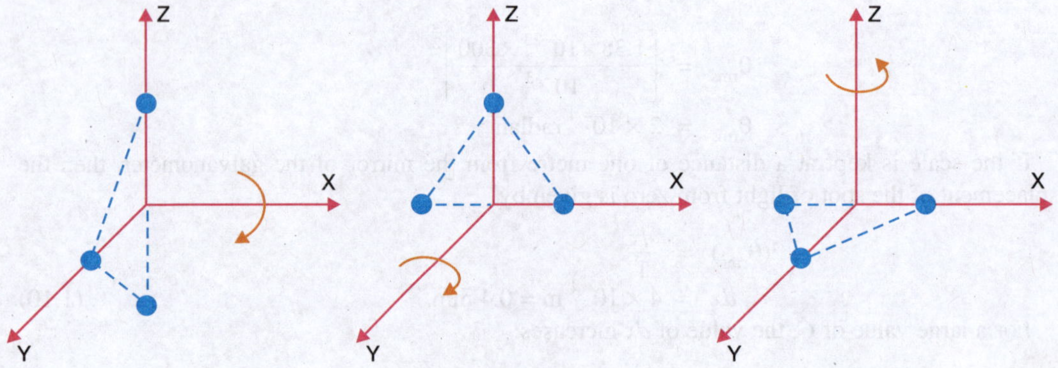


**Fig. 1.4** Triatomic non-linear molecule.

A triatomic molecule, if it is linear *e.g.*, $CO_2$, $CH_2$ etc., has 5 degrees of freedom, similar to diatomic molecule (3 translational + 2 rotational). Fig. 1.5 shows a triatomic linear molecule.

In addition to translational and rotational motion as discussed above, the molecule also possesses vibratory motion, giving rise to vibrational degrees of freedom, particularly at high temperature.

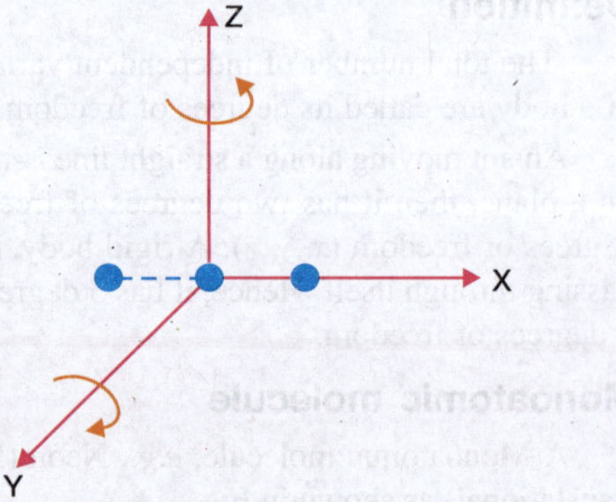


**Fig. 1.5** Triatomic Linear molecule.

## 1.19 Maxwell's Law of Equipartition of Energy

According to kinetic theory of gases, the mean kinetic energy of a gas molecule at a temperature $T$ is given by $\frac{1}{2}mC^2 = \frac{3}{2}kT$.

But $$C^2 = u^2 + v^2 + w^2$$

As $x$, $y$ and $z$ are all equivalent, mean square velocities along three axes are equal.

$$\therefore \quad u^2 = v^2 = w^2 \quad \text{Hence, } u^2 = v^2 = w^2 = \frac{1}{3}C^2$$

$$\text{or} \quad \frac{1}{2}mu^2 = \frac{1}{2}mv^2 = \frac{1}{2}mw^2$$

$$\therefore \quad \frac{1}{2}mC^2 = 3\left[\frac{1}{2}mu^2\right] = 3\left[\frac{1}{2}mv^2\right] = 3\left[\frac{1}{2}mw^2\right]$$

$$= \frac{3}{2}kT$$

$$\therefore \quad \frac{1}{2}mu^2 = \frac{1}{2}kT$$

$$\frac{1}{2}mv^2 = \frac{1}{2}kT$$

$$\frac{1}{2}mw^2 = \frac{1}{2}kT \qquad \text{...(1.11)}$$

Therefore, the average kinetic energy associated with each degree of freedom = $\frac{1}{2}kT$.

Thus, the energy associated with each degree of freedom (whether translatory or rotatory) is $\frac{1}{2}kT$. This represents the theorem of equipartition of energy.

**Statement.** It states that for any dynamical system in thermal equilibrium, the total energy is divided equally among all the degrees of freedom and the energy associated per molecule per degree of freedom is $\frac{1}{2}kT$; where $k$ is Boltzmann's constant and $T$ is absolute temperature of the gas.

## 1.20 Relation between Molar Specific Heats and Degrees of Freedom

For an ideal gas, there is no force of attraction between the molecules, hence the internal energy $U$ of the gas is wholly kinetic. If $f$ is the number of degrees of freedom, then according to the law of equipartition of energy,

Average energy per molecule = $\frac{1}{2}fkT$

$\therefore$ Total energy per gm. molecule $U = N \times \frac{1}{2}fkt$

where $N$ is Avogadro's number.

$$\therefore \quad U = \frac{1}{2}fRT \qquad \left(\because \frac{R}{N} = k\right)$$

The molar specific heat of a gas at constant volume,

$$C_V = \frac{dU}{dT} = \frac{d}{dT}\left(\frac{1}{2}fRT\right) = \frac{1}{2}fR = R\left(\frac{f}{2}\right)$$

and molar specific heat of the gas at constant pressure,

$$C_P = C_V + R = \frac{1}{2}fR + R = R\left(\frac{f}{2}+1\right)$$

$\therefore$ Ratio of two specific heats,

$$\gamma = \frac{C_P}{C_V} = \frac{R\left(\frac{f}{2}+1\right)}{R\left(\frac{f}{2}\right)}$$

$$\therefore \quad \gamma = 1 + \frac{2}{f} \qquad \text{...(1.12)}$$

## 1.21 Specific Heats of Mono–, Di– and Polyatomic Gas

**(i) Mono-atomic gas.** A mono-atomic gas molecule has one atom. Each molecule has three degrees of freedom due to translatory motion only.

Energy associated with each degree of freedom = $\frac{1}{2}kT$. Therefore energy associated with three degrees of freedom

$$= \frac{3}{2}kT$$

Consider one gram molecule of a gas.

Energy associated with one gram molecule of a gas

$$= N \times \frac{3}{2}kT$$

$$= \frac{3}{2}(N \times k)T$$

$$\therefore \quad U = \frac{3}{2}RT \qquad (\because N \times k = R) \qquad \text{...(1.13)}$$

This energy of the gas is due to the energy of its molecules. It is called internal energy $U$. For an ideal gas, it depends upon temperature only.

∴ Specific heat at constant volume, $C_v$ is given by

$$C_V = \frac{dU}{dT} = \frac{3}{2}R \qquad \text{...(1.14)}$$

($\frac{dU}{dT}$ is the increase in internal energy per unit degree rise of temperature.)

But

$$C_P - C_V = R$$

$$C_P = C_V + R$$

$$= \frac{3}{2}R + R = \frac{5}{2}R$$

For a mono-atomic gas, ratio of specific heats

$$\gamma = \frac{C_P}{C_V}$$

$$= \frac{\frac{5}{2}R}{\frac{3}{2}R} = \frac{5}{3} = 1.67 \qquad \text{...(1.15)}$$

The value of $\gamma$ is found to be true experimentally for mono-atomic gases like argon and helium.

**(ii) Diatomic gas.** A diatomic gas molecule has two atoms. Such a molecule at ordinary temperatures has three degrees of freedom of translation and two degrees of freedom of rotation.

Energy associated with each degree of freedom = $\frac{1}{2}kT$.

∴ Energy associated with 5 degrees of freedom = $\frac{5}{2}kT$.

Consider one gram molecule of gas.

Energy associated with 1 gram molecule of a diatomic gas

$$= N \times \frac{5}{2}kT = \frac{5}{2}N\,kT$$

$$U = \frac{5}{2}RT$$

$$C_V = \frac{dU}{dT}$$

$$= \frac{5}{2}R$$

But $$C_P - C_V = R \qquad \text{...(1.16)}$$

or $$C_P = C_V + R$$

$$= \frac{5}{2}R + R = \frac{7}{2}R$$

Therefore, the ratio of specific heats

$$\gamma = \frac{C_P}{C_V}$$

$$= \frac{\frac{7}{2}R}{\frac{5}{2}R} = \frac{7}{5} = 1.40 \qquad \text{...(1.17)}$$

The value of $\gamma = 1.40$ has been found to be true experimentally for diatomic gases like hydrogen, oxygen, nitrogen etc.

### Variation of $\gamma$ with temperature

At high temperature, the vibrational degree of freedom also becomes prominent and hence taking this into account, the total degrees of freedom in a diatomic molecule possesses (3 translational + 2 rotational + 1 vibrational) 7 degrees of freedom.

$$\therefore \qquad \gamma = \frac{C_P}{C_V} = 1 + \frac{2}{f} = 1 + \frac{2}{7} = 1.29$$

***(a)* At low temperatures.** In case of **hydrogen molecule ($H_2$)** only *translatory* motion is possible at low temperatures of the order of 70 K and has only three degrees of freedom like that of a monoatomic gas. Hence,

$$C_V = \frac{3}{2}R \quad \text{and} \quad \gamma = 1.67$$

***(b)* At ordinary temperatures.** At ordinary temperatures of the order of 250 K, the hydrogen molecule which is like a dumb-bell shape has a rotational motion in addition to translational motion. It has, therefore, (3 translational + 2 rotational), five degrees of freedom and, hence

$$C_V = \frac{5}{2}R \quad \text{and} \quad \gamma = 1.40$$

***(c)* At high temperatures.** At high temperatures beyond 750 K, the collisions in the molecules cause the atoms to vibrate. Therefore, in addition to translational and rotational degrees of freedom, the hydrogen molecule also possesses vibrational degree of freedom. The hydrogen molecule in all possesses 7 degrees of freedom. Hence

$$C_V = \frac{7}{2}R \quad \text{and} \quad \gamma = 1.29$$

It should be noted that different gases show these effects at different temperatures.

***(iii)* Triatomic gas**

***(a)* Non-linear molecule.** A triatomic non-linear gas molecule has 6 degrees of freedom (3 translational + 3 rotational) neglecting the vibrational energy. Energy associated with 1 gram molecule

$$U = N \times \frac{6}{2} kT = 3RT$$

$$C_V = \frac{dU}{dT} = 3R \qquad \text{...(1.18)}$$

But

$$C_P - C_V = R$$

$$C_P = C_V + R$$

$$= 3R + R = 4R$$

The ratio of two specific heats

$$\gamma = \frac{C_P}{C_V}$$

$$= \frac{4R}{3R} = 1.33 \qquad \text{...(1.19)}$$

In addition to 3 translational and 3 rotational degrees of freedom, it also possesses energy due to vibrational motion at high temperature. The total number of effective degrees of freedom in such a molecule are *seven*, Hence,

$$C_v = \frac{7}{2} R \quad \text{and} \quad \gamma = 1 + \frac{2}{7} = 1.29$$

***(b)* Linear molecule.** A triatomic linear gas molecule has 5 degrees of freedom. This is similar to diatomic molecule and possess $\gamma = 1.40$

Taking into consideration vibrational degrees of freedom, a triatomic linear molecule possesses in all 6 degrees of freedom, Hence

$$C_v = \frac{6}{2} R \quad \text{and} \quad \gamma = 1 + \frac{2}{6} = 1.33$$

Thus, the values of $C_P$, $C_V$ and $\gamma$ can be calculated depending upon the degrees of freedom of a gas molecule.

## 1.22 Experimental and Theoretical Agreement of $C_P$, $C_V$ and $\gamma$

The experimental values of $C_P$, $C_V$ and $\gamma$ for monoatomic gases agree with the theoretical values. But the values of $C_P$ and $C_V$ for some diatomic gases and for most of the polyatomic gases are larger than the theoretical values. In addition, experimental values are found to increase with the temperature.

### Explanation

In finding the theoretical values for diatomic and polyatomic gases, we have not considered the energy due to vibrations of atom. At high temperatures the atoms in each molecule can vibrate relative to each other. This increases the number of degrees of freedom.

## 1.23 Behaviour of Sp. Heats at Low Temperature (Freezing of Degrees of Freedom)

**Monoatomic gas.** In a monoatomic gas, the molecule (having only one atom) behaves like a point mass. It has only kinetic energy of translation. Hence total energy of such a molecule is divided among only 3 degrees of freedom. Thus

$$C_V = \frac{3}{2} R = 1.5\, R$$

$$C_P = C_V + R = \frac{3}{2}R + R = \frac{5}{2}R = 2.5\,R$$

and
$$\gamma = \frac{C_P}{C_V} = \frac{5}{3} = 1.67$$

This is in good agreement with the experimental values of $C_V$ and $\gamma$. Furthermore, the specific heats of mono-atomic gases are found to be practically independent of temperature, in agreement with the theory.

**Diatomic gas.** A diatomic gas molecule has a dumb-bell structure. For example, a hydrogen molecule.

***(i)* At high temperature.** Beyond 750 K, the collisions causes the atoms in the molecule to vibrate. In addition to translational and rotational motion, vibrational motion also takes place and the molecule has in all 7 degrees of freedom (3 translational + 2 rotational + 2 vibrational). The theory predicts

$$C_V = \frac{7}{2}R = 3.5\,R$$

and
$$\gamma = \frac{9}{7} = 1.29$$

These values are in good agreement with experimental values.

***(ii)* At room temperature.** The molecule has only rotational motion in addition to translatory motion. Thus, it has 5 degrees of freedom (3 translational + 2 rotational). The theory predicts

$$C_V = \frac{5}{2}R = 2.5\,R$$

and
$$\gamma = \frac{7}{5} = 1.40$$

These values are also in good agreement with experimental values.

***(iii)* At low temperature.** At low temperature, say about 20 K, the specific heat capacities of hydrogen (the only diatomic gas that remains a gas at very low temperatures) decreases to $\frac{3}{2}R$; the value predicted by theory for a monoatomic gas.

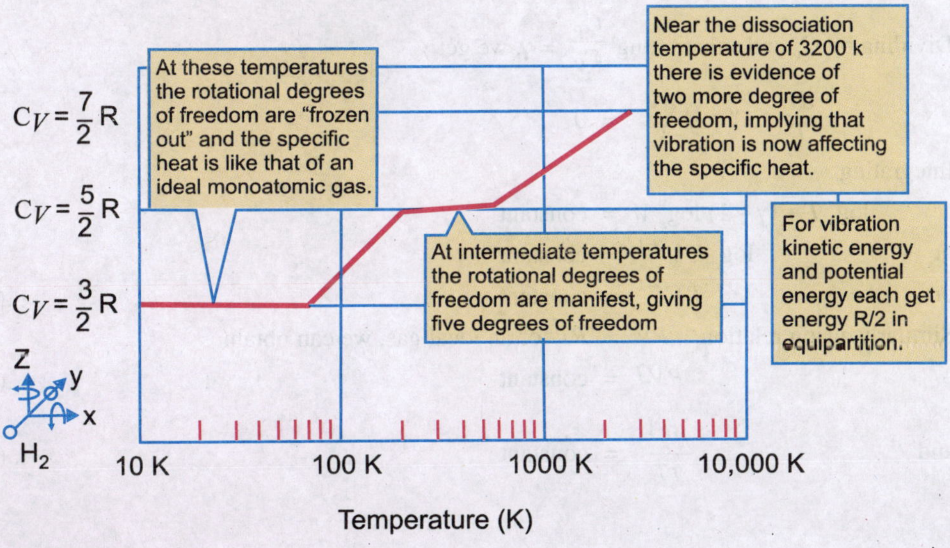

Thus, at low temperature, neither the rotational nor the vibrational degrees of freedom contribute. In other words, the degrees of freedom go on freezing as the temperature lowers down and ultimately at 0K; the translational degree of freedom reduces to zero and the molecule completely comes to rest. This is known as **freezing of degrees of freedom.**

## 1.24 Adiabatic Expansion of an Ideal Gas

We consider 1 mole of an ideal gas undergoing an adiabatic change in its state. In an adiabatic process neither the heat is allowed to enter nor to leave the system. Hence $\delta Q = 0$. For an ideal gas, the internal energy $U$ depends only on the temperature.

$$\therefore \quad dU = C_V\, dT \qquad \left(\because C_V = \frac{dU}{dT}\right)$$

For a small change in the state of a system

$$\delta Q = dU + PdV$$

For adiabatic change, $0 = C_V dT + PdV$

$$\therefore \quad dT = -\frac{PdV}{C_V}$$

By definition, the ideal gas should satisfy the equation

$$PV = RT$$

or

$$P = \frac{RT}{V}$$

Substituting for $P$, we get

$$C_V dT + \frac{RT}{V} dV = 0$$

or

$$C_V \frac{dT}{T} + R\frac{dV}{V} = 0$$

or

$$C_V \frac{dT}{T} + (C_P - C_V)\frac{dV}{V} = 0 \qquad \because C_P - C_V = R$$

Dividing by $C_V$ and substituting $\frac{C_P}{C_V} = \gamma$, we get

$$\frac{dT}{T} + (\gamma - 1)\frac{dV}{V} = 0$$

Integrating

$$\log_e T + (\gamma - 1)\log_e V = \text{constant}$$

or

$$\log_e TV^{\gamma-1} = \text{constant}$$

or

$$TV^{\gamma-1} = \text{constant} \qquad \text{...(1.20)}$$

Similarly, using relation $PV = RT$ for an ideal gas, we can obtain

$$PV^{\gamma} = \text{constant} \qquad \text{...(1.21)}$$

and

$$\frac{P^{\gamma-1}}{T^{\gamma}} = \text{constant} \qquad \text{...(1.22)}$$

## 1.25 Kinetic Interpretation of Temperature

Consider a system of $N$ molecules of monoatomic gas. If $E$ is the total energy of the system, then some of the energy will be in the form of translational energy and the remaining may be rotational, vibrational potential energies of the molecules. Let $E_t$ be the translational energy, then according to the law of equipartition of energy

$$\frac{E_t}{N} = \frac{3}{2}kT \quad \text{...}(i)$$

where $k$ is Boltzmann's constant and $T$ is temperature of the system in Kelvin. But $\frac{E_t}{N}$ is nothing but the average translational energy per molecule, denoted by $\bar{\varepsilon}_i$.

$$\therefore \quad \bar{\varepsilon}_i = \frac{E_t}{N} = \frac{3}{2}kT \quad \text{...}(ii)$$

or

$$\bar{\varepsilon}_i \propto T \quad \text{...}(iii)$$

Thus, the average kinetic energy of translation per molecule of an ideal monoatomic gas in a given mass is directly proportional to absolute temperature of the gas.

Hence, the average kinetic energy of translation of each molecule of the gas is zero, if $T = 0$.

### Conclusion :

The absolute zero temperature is that temperature at which the average kinetic energy of translation of each molecule of the gas is zero. *i.e.*, the temperature at which the molecular translational velocities of a gas are reduced to zero.

Thus, we may say that large, small and medium temperature of a system means large, small and medium average kinetic energy of translation of each of its molecules. Such an interpretation of temperature has been supported by Maxwell, who established theoretically that in a mixture of gases, there is an exchange of energy between different kinds of molecules untill the average kinetic energy of translation of each molecule is the same. Thus, if a number of gases having the same kinetic energy of translation are mixed together, there will be no net exchange of energy. Thus, the gases having same kinetic energies of translation of the molecules are at the same temperature. As a result, we may say that the average kinetic energy of translation of the molecules is proportional to the absolute temperature.

This is kinetic interpretation of temperature.

## 1.26 Change of Pressure with Height

The pressure of the atmosphere decreases with increase in height from sea level. The density of air is not the same at all levels. If the density of air were the same at all levels, the atmospheric pressure at sea level would correspond to a vertical column of 8 km of air only. But is has been found that air exists even at a height of 200 km.

Consider that the pressure of air at a height $h = P$. For an increase in height $dh$, the decrease in pressure is $dP$ (Fig. 1.6).

$$dP = -(dh)\,\rho g \quad \text{...}(i)$$

[–*ve* sign shows that the pressure decreases with increase in height]

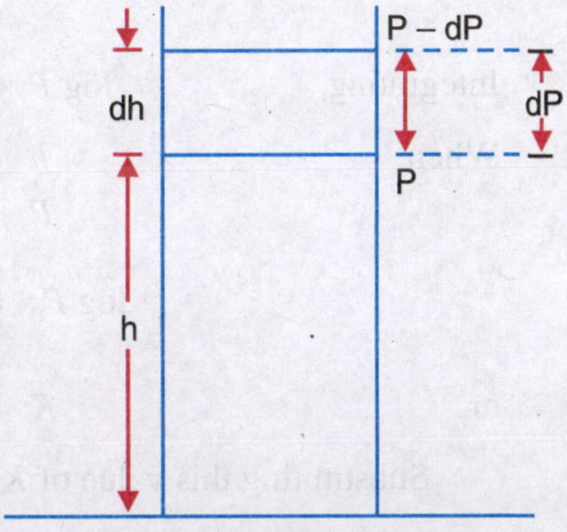


Fig. 1.6

Let $M$ be the molecular weight of the gas and $V$ its volume.

Then, $$\rho = \frac{M}{V}$$

and $$dP = -(dh)\frac{M}{V}g$$

But for a perfect dry air

$$PV = RT$$

$$V = \frac{RT}{P}$$

$$\therefore \quad dP = -(dh)\frac{M.Pg}{RT}$$

$$\frac{dP}{P} = -\frac{Mg}{RT}(dh) \qquad ...(ii)$$

Integrating

$$\log P = -\frac{Mg}{RT}h + C \quad \text{[where } C \text{ is a constant]}$$

When $$h = 0,$$

$$\therefore \quad P = P_o$$

$$C = \log P_0$$

Substituting, we get $$\log P = -\frac{Mg}{RT}h + \log P_0$$

$$\log\frac{P}{P_0} = -\frac{Mg}{RT}h \qquad ...(iii)$$

$$P = P_o e^{-\frac{Mg}{RT}h} \qquad ...(iv)$$

Equation (*ii*) will be true only under isothermal conditions. But air temperature decreases uniformly with height. Assuming that the lapse rate is $\alpha$ in the lower portion of the atmosphere (troposphere),

$$T = T_0 - \alpha h$$

where $T_0$ is the temperature of the ground.

Substituting this value in equation (*ii*)

$$\therefore \quad \frac{dP}{P} = -\frac{Mg}{R}\frac{dh}{(T_0 - \alpha h)}$$

Integrating, $$\log P = \frac{Mg}{R\alpha}\log(T_0 - \alpha h) + K \qquad ...(v)$$

When $$h = 0,$$

$$P = P_0$$

$$\log P_0 = \frac{Mg}{R\alpha}\log T_0 + K$$

or $$K = \log P_0 - \frac{Mg}{R\alpha}\log T_0$$

Substituting this value of $K$ in equation (*v*)

$$\log P = \frac{Mg}{R\alpha} \log (T_0 - \alpha h) + \log P_0 - \frac{Mg}{R\alpha} \log T_0$$

$$\log \frac{P}{P_0} = \frac{Mg}{R\alpha} \log \left(\frac{T_0 - \alpha h}{T_0}\right) \quad ...(vi)$$

$$P = P_o e^{\frac{Mg}{R\alpha} \log\left(\frac{T_o - \alpha h}{T_o}\right)} \quad ...(vii)$$

## 1.27 Application to Atmospheric Physics (Convective Equilibrium of the Atmosphere)

The atmospheric temperature decreases with altitude upto a height of about 20 km. The rate of fall of temperature is about 5°$C$ per Km height and is known as the lapse rate. The atmosphere is divided into two regions troposphere and stratosphere.

1. **Troposphere or the convective zone.** It is the region in which the temperature falls with increase in height from the ground.

2. **Stratosphere or advective zone.** It is the region in which the decrease in temperature does not take place with increase in height.

Troposphere

The layer that separates these two regions is known as *tropopause.* The height of tropopause is different at different places on the earth. It is about 14 km at the equator and about 10 km at the poles.

Stratosphere

The fall in temperature with altitude in the troposphere is due to convection. When the heat radiations from the sun fall on the surface of the earth, it gets heated. The atmospheric air surrounding the earth gets heated by conduction and radiation. The heated air moves up and convection currents are set up. The heated air that moves up from the regions of higher to lower pressure is adiabatically cooled. Similarly when the colder air moves down from lower to higher pressure regions it is adiabatically heated. Thus a *convective equilibrium is set up* and there is a gradual fall of temperature with increase in height.

The earth gets heated up due to the heat radiations from the sun. The layers of the atmosphere close to the ground get heated up and convection currents start in the atmosphere. As the warm air rises up into region of low pressure, it **expands adiabatically.** For an atmosphere, in convective equilibrium, there is a vertical fall of temperature from the earth's surface. The **dry adiabatic lapse rate** is calculated as follows :

The relation between pressure and temperature for an adiabatic process is

$$\frac{P^{\gamma-1}}{T^{\gamma}} = \text{const.}$$

or

$$P^{\gamma-1} T^{-\gamma} = \text{const.}$$

Differentiating

$$(\gamma - 1) P^{\gamma-2} \, dP . T^{-\gamma} + P^{\gamma-1} (-\gamma) T^{-\gamma-1} dT = 0$$

Dividing by $P^{\gamma-1} T^{-\gamma}$

$$(\gamma - 1) \frac{dP}{P} + (-\gamma) \frac{dT}{T} = 0$$

$$\frac{dP}{P} = \frac{\gamma}{\gamma - 1} \frac{dT}{T} \quad ...(i)$$

For an increase in height $dh$, the decrease in pressure is $dP$

$$\therefore \quad dP = \rho g\, dh = -\frac{M}{V}\, g\, dh$$

But $\quad PV = RT$

or $\quad V = \dfrac{RT}{P}$

$$\therefore \quad dP = -\frac{MP}{RT}\, g\, dh$$

or $$\frac{dP}{P} = -\frac{Mg}{RT}\, dh \qquad ...(ii)$$

Equating (*i*) and (*ii*)

$$\left(\frac{\gamma}{\gamma-1}\right)\frac{dT}{T} = -\frac{Mg}{RT}\, dh$$

$$\frac{dT}{dh} = -\frac{Mg}{R}\left(\frac{\gamma-1}{\gamma}\right) \qquad ...(iii)$$

This equation holds good only for perfect dry air. The lapse rate $\dfrac{dT}{dh}$ can be calculated from equation (*iii*).

Substituting the values,

$$M = 29\ g$$
$$g = 981\ \text{cm/s}^2$$
$$R = 8.31\times 10^7 \text{ ergs per gram mol per } K,$$

and $\quad \gamma = 1{\cdot}4$

$$\frac{dT}{dh} = -\frac{29\times 981}{8{\cdot}31\times 10^7}\left(\frac{1{\cdot}40-1}{1{\cdot}40}\right)$$
$$= 10^{-4}\ K \text{ per cm}$$
$$= 10^{-4}\ {}^\circ C \text{ per cm}$$

Thus, the lapse rate for a height of 1 km = $10^{-4} \times 10^5 = 10°C$. But it has been found that the average lapse rate is lower than this value. Under average conditions the lapse rate has been found to be between 5°C and 6·5°C per km.

However at night, when the earth is radiating heat to the atmosphere, the layers of air nearer the surface may cool to a temperature lower than that of the upper layers. Thus a *negative lapse rate is set up.*

## NUMERICAL EXAMPLES

### EXAMPLE 1.1

*Calculate the number of molecules in one c.c. of oxygen at N.T.P. using the following data :*

*Density of mercury* = *13.6 g/cm³*

*r.m.s. velocity of oxygen molecules at 0°C*

= $4.62 \times 10^4$ *cm/s*

*Mass of one molecule of oxygen* = $52.8 \times 10^{-24}$ *g.*

**SOLUTION.** The pressure is given by

$$P = \frac{1}{3}\rho C^2$$

Let the mass of each molecule be $m$ and the number of molecules in one cc = $n$

$$\therefore \quad P = \frac{1}{3} mnC^2$$

or $$n = \frac{3P}{mC^2}$$

Here, $$P = 76 \times 13.6 \times 980 \text{ dynes/cm}^2$$
$$m = 52.8 \times 10^{-24} \text{ g}$$
$$C = 4.62 \times 10^4 \text{ cm/s}$$

∴ $$n = \frac{3 \times 76 \times 13.6 \times 980}{52.8 \times 10^{-24} \times (4.62 \times 10^4)^2}$$
$$\mathbf{n = 2.697 \times 10^{19}}.$$

### EXAMPLE 1.2

*At what celsius temperature will oxygen molecules have the same root mean square velocity as that of hydrogen molecules at – 100°C ?*

**Solution.** The energy of a gas molecule is,

$$\frac{1}{2}mC^2 = \frac{3}{2}kT$$

For hydrogen molecules

$$\frac{1}{2}m_1C_1^2 = \frac{3}{2}kT_1 \quad ...(i)$$

For oxygen molecules

$$\frac{1}{2}m_2C_2^2 = \frac{3}{2}kT_2 \quad ...(ii)$$

Dividing (*i*) by (*ii*),

$$\frac{m_1C_1^2}{m_2C_2^2} = \frac{T_1}{T_2} \quad ...(iii)$$

Here, $C_1 = C_2$ and $T_1 = -100°\text{C} = -100 + 273 = 173$ K
$$T_2 = ?$$
$$\frac{m_2}{m_1} = 16$$

From equation (*iii*),

$$T_2 = \frac{m_2C_2^2T_1}{m_1C_1^2}$$
$$= 16 \times 173$$
$$= 2768 \text{ K}$$

or $$\mathbf{T_2} = (2768 - 273) = \mathbf{2495°C.}$$

### EXAMPLE 1.3

*Calculate r.m.s. velocity of the oxygen molecules at 27°C.*

**SOLUTION.** First calculate the r.m.s. velocity of oxygen at N.T.P.

$$C = \sqrt{\frac{3P}{\rho}}$$

Here, $$P = 76 \times 13.6 \times 980 \text{ dynes/cm}^2$$
$$\rho = 16 \times 0.000089$$

$$\therefore \qquad C = \sqrt{\frac{3\times76\times13.6\times980}{16\times0.000089}}$$

$$C = 4.6 \times 10^4 \text{ cm/s}$$

Let the r.m.s. velocity of oxygen molecules at 27°C be $C_1$

$$\therefore \qquad \frac{C_1}{C} = \sqrt{\frac{T_1}{T}}$$

$$C_1 = C\times\sqrt{\frac{T_1}{T}}$$

Here,

$$C = 4.6 \times 10^4 \text{ cm/s}$$

$$T = 273 \text{ K}$$

$$T_1 = 27°\text{C} = (27 + 273) \text{ K}$$

$$= 300 \text{ K}$$

$$C_1 = 4.6 \times 10^4 \sqrt{\frac{300}{273}}$$

$$\mathbf{C_1 = 4.84 \times 10^4 \text{ cm/s.}}$$

**EXAMPLE 1.4**

*Calculate the volume occupied by 3.2 grams of oxygen at 76 cm of Hg and 27°C.*

**SOLUTION.**

$$P = 76 \times 13.6 \times 980 \text{ dynes/sq.cm}$$

$$T = (27 + 273) = 300 \text{ K}$$

$$R = 8.31 \times 10^7 \text{ erg/g mol–K}$$

For an ideal gas,

$$PV = RT$$

$$V = \frac{RT}{P} = \frac{8.31\times10^7\times300}{76\times13.6\times980}$$

$$= 24610 \text{ cc per g mol}$$

Volume for 3.2 g of oxygen

$$= \frac{24610\times3.2}{32}$$

$$= \mathbf{2461 \text{ cc.}}$$

**EXAMPLE 1.5**

*Calculate the number of molecules in one cubic metre of an ideal gas at N.T.P.*

**(Purvanchal 2000)**

**Solution.** For one gram molecule of a gas,

$$PV = RT$$

But

$$R = Nk$$

$$PV = NkT$$

Let $n$ be the number of molecules per cc. Then

$$P = nkT \qquad [\because \text{ when } V = 1,\ N = n]$$

or

$$n = \frac{P}{kT}$$

But $$k = \frac{R}{N}$$

$$\therefore \quad n = \frac{PN}{RT} \qquad \text{...(1.23)}$$

Number of molecules in one cubic metre volume

$$x = n \times 10^6$$

$$x = \frac{PN \times 10^6}{RT}$$

Here,

$$P = 76 \times 13.6 \times 980 \text{ dynes/cm}^2$$
$$N = 6.023 \times 10^{23}$$
$$R = 8.31 \times 10^7 \text{ erg/g mol–K}$$
$$T = 273 \text{ K}$$

$$\therefore \quad x = \frac{76 \times 13.6 \times 980 \times 6.023 \times 10^{23} \times 10^6}{8.31 \times 10^7 \times 273}$$

$$\mathbf{x = 2.688 \times 10^{25}}.$$

### EXAMPLE 1.6

*The number of molecules per cc of a gas is $2.7 \times 10^{19}$ at N.T.P. Calculate the number of molecules per cc of the gas.*

(*i*) *at 0°C and $10^{-6}$ mm pressure of mercury and*

(*ii*) *at 39°C and $10^{-6}$ mm pressure of mercury.*

**Solution.** For a unit volume of a gas

$$P = \frac{1}{3} mnC^2$$

(*i*) At 0°C, r.m.s. velocity is euqal to $C$.
At N.T.P.

$$P_1 = \frac{1}{3} mn_1C^2 \qquad \text{...(i)}$$

At 0°C and $10^{-6}$ pressure

$$P_2 = \frac{1}{3} mn_2C^2 \qquad \text{...(ii)}$$

From (*i*) and (*ii*)

$$\frac{P_1}{P_2} = \frac{n_1}{n_2}$$

or

$$n_2 = \frac{n_1 \times P_2}{P_1} \qquad \text{...(iii)}$$

Here

$$n_1 = 2.7 \times 10^{19}$$
$$P_1 = 76 \times 13.6 \times 980 \text{ dynes/cm}^2$$
$$P_2 = 10^{-6} \text{ mm of Hg}$$
$$= 10^{-7} \text{ cm of Hg}$$
$$= 10^{-7} \times 13.6 \times 980 \text{ dynes/cm}^2$$

Using (*iii*),

$$n_2 = \frac{2.7 \times 10^{19} \times 10^{-7} \times 13.6 \times 980}{76 \times 13.6 \times 980}$$

$$\mathbf{n_2 = 3.553 \times 10^{10}}$$

(*ii*) Let the r.m.s. velocity at 0°C be $C_2$ and at 39°C be $C_3$. Number of molecules per cc at 0°C and $10^{-6}$ mm of Hg pressure

$$= n_2$$

and at 39°C and $10^{-6}$ mm of Hg pressure

$$= n_3$$

For unit volume, $$P = \frac{1}{3} mnC^2$$

Here pressure is the same in both the cases

$$\therefore \quad \frac{1}{3} mn_2C_2^2 = \frac{1}{3} mn_3C_3^2$$

$$\therefore \quad n_2C_2^2 = n_3C_3^2$$

$$n_3 = n_2 \times \frac{C_2^2}{C_3^2}$$

But $$C \propto \sqrt{T}$$

$$\therefore \quad \frac{C_2^2}{C_3^2} = \frac{T_2}{T_3}$$

$$\therefore \quad n_3 = \frac{n_2 \times T_2}{T_3}$$

Here
$$n_2 = 3.553 \times 10^{10}$$
$$T_2 = 0°\text{C} = 273 \text{ K}$$
$$T_3 = 39°\text{C} = (273 + 39) = 312 \text{ K}$$

$$n_3 = \frac{3.553 \times 10^{10} \times 273}{312}$$

$$\mathbf{n_3 = 3.109 \times 10^{10}.}$$

**EXAMPLE 1.7**

*Calculate the mean translational kinetic energy per molecule of a gas at 727°C, given $R = 8.32$ joules/mole–K and Avogadro's number $N = 6.06 \times 10^{23}$.* **(Ranchi 2000)**

**SOLUTION.** Given :

$$T = 627 + 273 = 1000 \text{ K}$$
$$R = 8.320 \text{ joules/mole–K}$$
$$N = 6.06 \times 10^{23}$$

$$k = \frac{R}{N} = \left(\frac{8.32}{6.06 \times 10^{23}}\right) \text{ joules/mole–K}$$

Mean translational K.E. per molecule,

$$= \frac{3}{2} kT$$

$$= \frac{3 \times 8 \cdot 32 \times 1000}{2 \times 6.06 \times 10^{23}}$$

$$= \mathbf{2.059 \times 10^{-20} \text{ joule.}}$$

**EXAMPLE 1.8**

*Calculate the root mean square velocity of a molecule of mercury vapour at 300 K. (Given : molecular weight of mercury = 221).* **(Rohilkhand 2000)**

**SOLUTION.** Mean kinetic energy of one molecule of mercury

$$= \frac{1}{2}mC^2 = \frac{3}{2}kT$$

Let $N$ be the Avogadro's number.

Then
$$\frac{1}{2}mNC^2 = \frac{3}{2}kNT$$
$$\frac{1}{2}MC^2 = \frac{3}{2}RT$$
$$C = \sqrt{\frac{3RT}{M}}$$

Here the molecular weight of mercury, $M = 221$.

$$\therefore \quad C = \sqrt{\frac{3\times 8.3\times 10^7\times 300}{221}} = \mathbf{1.93 \times 10^4 \text{ cm/s.}}$$

**EXAMPLE 1.9**

*Calculate the temperature at which the r.m.s. velocity of a hydrogen molecule will be equal to the speed of the earth's first satellite (i.e., v = 8 km/s)* **(Kanpur 2001)**

**SOLUTION.** Energy for 1 gram molecule of hydrogen

$$= \frac{1}{2}Mv^2 = \frac{3}{2}RT$$
$$T = \frac{Mv^2}{3R}$$
$$= \frac{2\times(8\times10^5)^2}{3\times 8.3\times 10^7}$$
$$T = \mathbf{5.14 \times 10^3 \text{ K.}}$$

**EXAMPLE 1.10**

*At what temperature, pressure remaining constant, will the r.m.s. velocity of a gas be half its value at 0°C ?*

**SOLUTION.**
$$\frac{C_2}{C_1} = \sqrt{\frac{T_2}{T_1}}$$

Here
$$C_2 = \frac{C_1}{2}$$
$$T_2 = ?$$
$$T_1 = 273 \text{ K}$$

$$\therefore \quad \frac{1}{2} = \sqrt{\frac{T_2}{273}}$$
$$\frac{1}{4} = \frac{T_2}{273}$$
$$T_2 = \frac{273}{4} = \mathbf{68.25 \text{ K}}$$

or
$$T_2 = \mathbf{-204.75°C}$$

## EXAMPLE 1.11

*At what temperature, pressure remaining unchanged, will the r.m.s. velocity of hydrogen is double of its value at N.T.P.?* **(Nagpur 2008)**

**SOLUTION.** The r.m.s. velocity is given by

$$C = \sqrt{\frac{3RT}{M}}$$

$$\therefore \quad C \propto \sqrt{T}$$

Let the r.m.s. velocity at $T_0$ (*i.e.,* at 273 K) be $C$ and at $T$ be $2C$, then

$$\frac{2C}{C} = \sqrt{\frac{T}{T_0}}$$

or $$\frac{T}{T_0} = 4$$

$$T = 4\,T_0 = 4 \times 273 = 1092 \text{ K}$$

or $$t = (1092 - 273) = \mathbf{819°C.}$$

## EXAMPLE 1.12

*Calculate the total random kinetic energy of one gram of nitrogen at 300 K.* **(Kerala 2002)**

**SOLUTION.** Total random energy for one molecule of nitrogen

$$= \frac{3}{2}RT$$

$\therefore$ total random energy for one gram of nitrogen

$$E = \frac{3}{2}\frac{RT}{M} \qquad \text{where } M = 28 \text{ g}$$

$$= \frac{3 \times 8.3 \times 10^7 \times 300}{2 \times 28}$$

$$= 133.4 \times 10^7 \text{ erg}$$

$$E = \mathbf{133.4 \text{ joule.}}$$

## EXAMPLE 1.13

*Calculate the total random kinetic energy of 2g of helium at 200 K. (Given R = 8.3 × $10^7$ erg/g mol – K)* **(Patna 2000)**

**SOLUTION.** The total random energy for 1*g* of helium

$$= \frac{3RT}{2M}$$

$\therefore$ The total random energy of 2*g* of helium

$$= \frac{2 \times 3RT}{2M}$$

$$= \frac{3RT}{M}$$

$$= \frac{3 \times 8.3 \times 10^7 \times 200}{4}$$

$$= 1245 \times 10^7 \text{ ergs}$$

$$= \mathbf{1245 \text{ joules.}}$$

**EXAMPLE 1.14**

*Show that for a gas possessing f degrees of freedom, the ratio of two specific heats is* $1+\frac{2}{f}$ *and the atomicity of a molecule is given by* $f = \frac{2}{\gamma - 1}$.

**(Garhwal 2000, Utkal 2003, Calcutta 2003, Guwahati 2000, Nagpur 2001, 2003)**

**SOLUTION.** Let us consider 1 mole of a gas. Let $f$ be the number of degrees of freedom of each molecule. The gas contains the number of molecules equal to the Avogadro's number $N$. According to the law of equipartition of energy, the K.E. associated with each degree of freedom is $= \frac{1}{2}kT$.

$\therefore$ the total internal energy of 1 mole of the gas

$$U = N\left(f \times \frac{1}{2}kT\right)$$

$$= \frac{1}{2}Nf\frac{R}{N}T \qquad \left(\because \quad k = \frac{R}{N}\right)$$

$$= \frac{1}{2}f\,RT$$

Now the molar specific heat at constant volume $C_V$ is

$$C_V = \left(\frac{\partial U}{\partial T}\right)_V = \frac{\partial}{\partial T}\left(\frac{1}{2}f\,RT\right)$$

$$= \frac{fR}{2}$$

For a perfect gas, $C_P - C_V = R$

$$\therefore \quad C_P = C_V + R = \frac{fR}{2} + R$$

$$= \frac{(f+2)R}{2}$$

Hence, $\gamma = \frac{C_P}{C_V} = \frac{(f+2)R}{fR} = 1 + \frac{2}{f}$

or **atomicity** $= f = \frac{2}{\gamma - 1}$.

**EXAMPLE 1.15.**

*Calculate the values of the molar heat capacities $C_V$ and $C_P$ of a gas, if the ratio of the heat capacities is 1.33. What is the atomicity of the gas? Given R = 8.31 J/mol – K.*

**(K.U. 2001)**

**SOLUTION.** We have $C_P - C_V = R$ and $\frac{C_P}{C_V} = \gamma$

or $C_P = \gamma C_V$

Substituting, $\gamma C_V - C_V = R$

$$C_V = \frac{R}{\gamma - 1}$$

(*i*) $C_V = \frac{R}{\gamma - 1} = \frac{8.31}{1.33 - 1} =$ **25.18 J/mol – K**

(*ii*) $C_P = \gamma C_V = 1.33 \times 25.18 =$ **33.49 J/mol – K** (*iii*) Atomicity : The number of degrees of freedom $f$ of each molecule is

$$f = \frac{2}{\gamma - 1} = \frac{2}{1.33-1} = \frac{2}{0.33}$$

$$f = \mathbf{6.}$$

The number of degrees of freedom is 6.

∴ the gas is polyatomic.

**EXAMPLE 1.16**

*Given Avogadro's number = 6.02 × $10^{23}$ and Boltzmann's constant 1.37 × $10^{-23}$ $JK^{-1}$, calculate (a) kinetic energy of translation of an oxygen molecule at 27°C, (b) the total kinetic energy of an oxygen molecule at 27°C and (c) the total K.E. of a gram molecule of oxygen at 27°C.* **(Calcutta 2003)**

**SOLUTION.** Oxygen is a diatomic molecule. At room temperature *i.e.,* at 27°C, it has 5 degrees of freedom (3 translational + 2 rotational). Here $T = (273 + 27) = 300$ K.

(*a*) K.E. of translation per molecule

$$= \frac{3}{2}kT$$

$$= \frac{3}{2} \times 1.37 \times 10^{-23} \times 300$$

$$= \mathbf{6.165 \times 10^{-21}\ joules.}$$

(*b*) Total K.E. per molecule

$$= \frac{5}{2}kT$$

$$= \frac{5}{2} \times 1.37 \times 10^{-23} \times 300$$

$$= \mathbf{10.275 \times 10^{-21}\ joules.}$$

(*c*) Total K.E. per gram mole

$$= N \times \frac{5}{2}kT$$

$$= 6.02 \times 10^{23} \times 10.275 \times 10^{-21}$$

$$= \mathbf{6185.55\ joules.}$$

**EXAMPLE 1.17**

*Particles of mass 6.2 × $10^{-17}$ kg are suspended in a liquid at 27°C and the rms speed is found to be 1.4 × $10^{-2}$ m/s. Find Avogadro's number from the equipartition theorem. (R = 8.314 J/mole K).* **(Nagpur 2005/W)**

**Solution.** According to the law of equipartition of energy, the average K.E. of a particle per degree of freedom = $\frac{1}{2}$ $kT$. But a particle has 3 dregrees of freedom (translational only).

∴ Average K.E. of theparticle = $\frac{3}{2}kT$

But from kinctic theory of gases, average K.E. of the particle = $\frac{1}{2}$ $mC^2$

$$\therefore \quad \frac{3}{2}kT = \frac{1}{2}mC^2 \quad \text{or} \quad \frac{3RT}{2N} = \frac{1}{2}mC^2 \qquad \left(\because k = \frac{R}{N}\right)$$

or $$N = \frac{3RT}{mC^2} = \frac{3 \times 8.31 \times 300}{6.2 \times 10^{-17} \times (1.4 \times 10^{-2})\,2} = \mathbf{6.155 \times 10^{23}/mol.}$$

**EXAMPLE 1.18**

*Calculate the total random kinetic energy of one gram-mole of oxygen at 330 K.*

*[R = 8.31 J/mole-K]* **(Madurai 2002)**

**Solution.** Total random kinetic energy of one gram mole of oxygen.

$$U = \frac{3}{2}RT$$

$$U = \frac{3}{2}\times 8.31 \times 300$$

$$U = \mathbf{3.739 \times 10^3\ J.}$$

**EXAMPLE 1.19**

*Calculate the number of molecules in one litre of an ideal gas at 136.5°C temperature and 3 atmospheric pressure.*

**Solution.** Let the number of molecules per cc = $n$

$$n = \frac{PN}{RT}$$

Number of molecules in one litre,

$$x = n \times 10^3$$

$$x = \frac{PN \times 10^3}{RT}$$

Here

$$P = 3 \times 76 \times 13.6 \times 980 \text{ dynes/cm}^2$$
$$N = 6.023 \times 10^{23}$$
$$R = 8.31 \times 10^7 \text{ erg/g mol–K}$$
$$T = (273 + 136.5)$$
$$= 409.5 \text{ K}$$

$$x = \frac{3\times 76\times 13.6\times 980\times 6.023\times 10^{23}\times 10^3}{8.31\times 10^7\times 409.5}$$

$$x = \mathbf{5.376 \times 10^{22}.}$$

**EXAMPLE 1.20**

*Calculate the total random kinetic energy of one gm-molecule of oxygen at 300 K.*

**(Madurai 2003)**

**Solution.** Total random kinetic energy per gram-molecule of oxygen

$$= \frac{1}{2}mC^2 \times N$$

$$= \frac{3}{2}k\,T.N$$

$$= \frac{3}{2}\cdot\frac{R}{N}T.N$$

$$= \frac{3}{2}RT$$

$$= \frac{3}{2}\times 8.3\times 10^7 \times 300$$

$$= 3.735 \times 10^{10} \text{ ergs}$$

$$= \mathbf{3735\ joules.}$$

**Note.** The total random kinetic energy for one gram-molecule of any gas *i.e.*, hydrogen, oxygen, helium, nitrogen, air etc. is the same at the same temperature.

### EXAMPLE 1.21

*Calculate the average kinetic energy of molecule of a gas at a temperature of 300 K.* **(Meerut 2004)**

**SOLUTION.** Average K.E. of a molecule

$$= \frac{1}{2}mC^2 = \frac{3}{2}kT$$

Here $k$ is Boltzmann's constant.

$$k = 1.38 \times 10^{-16} \text{ erg/molecule-deg}$$

$$\text{Average K.E. of a molecule} = \frac{3}{2} \times 1.38 \times 10^{-16} \times 300$$

$$= \mathbf{6.21 \times 10^{-14} \text{ ergs.}}$$

**Note.** The average kinetic energy of a molecule of any gas *i.e.,* hydrogen, oxygen, helium, nitrogen, air, etc., is the same at the same temperature.

### EXAMPLE 1.22

*Explain on the basis of kinetic theory, why upper atmosphere is cooled while its kinetic temperature is of the order of 1000 K.*

**(Purvanchal 2004)**

**SOLUTION.**

The quantity of heat (*i.e.,* heat content) of a individual molecule in the upper atmosphere is increased about 3.5 times than that at the earths's surface. Kinetic temperature on earth's surface is about 300 K., while in the upper atmosphere it is of the order of 1000 K. But the pressure and hence the density is very low in the upper atmosphere, *i.e.,* in the upper atmosphere the number of molecules per cc is very small. Thus obviously the total heat content in the upper atmosphere is very small as compared to the heat content on the earth in the same volume of the gas. This results **the temperature of the upper atmosphere very low.**

### EXAMPLE 1.23

*The mean kinetic energy of molecules of hydrogen at 0°C is 5.64 × $10^{-21}$ J and molar gas constant is 8.32 J/mole–K. Calculate Avogadro's number.*

**SOLUTION.** K.E. per molecule $E = \frac{3}{2}kT$

$$E = \frac{3}{2}\left(\frac{R}{N}\right)T$$

or

$$N = \frac{3RT}{2E}$$

Here

$$E = 5.64 \times 10^{-21} \text{ J}$$

$$R = 8.32 \text{ J/mole–K}$$

$$T = 273 \text{ K}$$

$$\therefore \quad N = \frac{3 \times 8.32 \times 273}{2 \times 5.64 \times 10^{-21}}$$

$$N = \mathbf{6.182 \times 10^{23}.}$$

### EXAMPLE 1.24

*Calculate the energy of 5 gram molecules of ozone ($O_3$) at 27°C. Given R = 8.3 J/g-mole/K.*

**(Agra 2008, Lucknow 2000)**

**SOLUTION.** Here

$$U = n\left[\frac{3}{2}RT\right]$$

$$n = 5$$

$$R = 8.3 \text{ J/g–mole/K}$$

$$T = 27 + 273 = 300 \text{ K}$$

$$U = 5\times\left(\frac{3}{2}\right)\times 8.3\times 300$$

$$\mathbf{U = 1.8675 \times 10^4 \text{ J.}}$$

### EXAMPLE 1.25

*Prove that*

$$\frac{K}{\eta C_v} = 1 \textit{ for a gas.}$$

**SOLUTION.** The expression for coefficient of viscosity

$$\eta = \frac{mC}{3\lambda} \quad ...(i)$$

The coefficient of thermal conductivity

$$K = \frac{mC\,C_v}{3\lambda} \quad ...(ii)$$

Dividing (*ii*) by (*i*)

$$\frac{K}{\eta} = C_V$$

$$\therefore \quad \frac{K}{\eta C_v} = \mathbf{1.}$$

### EXAMPLE 1.26

*If the RMS speed of molecules of hydrogen at NTP is 1.84 km/s, calculate the RMS speed of oxygen at NTP. The molecular weight of hydrogen and oxygen are 2 and 32 respectively.*

**(IIT Entrance 2000)**

**SOLUTION.** Here

$$C_1 = 1.84 \text{ km/s}$$

$$C_2 = ?$$

$$\frac{\rho_1}{\rho_2} = \frac{2}{32} = \frac{1}{16}$$

$$\frac{C_2}{C_1} = \sqrt{\frac{\rho_1}{\rho_2}}$$

$$\frac{C_2}{1.84} = \sqrt{\frac{1}{16}}$$

$$C_2 = \frac{1.84}{4}$$

$$\therefore \quad C_2 = 0.46 \text{ km/s} = \mathbf{460 \text{ m/s.}}$$

**EXAMPLE 1.27**

*Calculate RMS speed of molecule of a gas which has density of 1.24 × $10^{-5}$ gm/cm$^3$ at a temperature of 273 K and pressure of $10^{-2}$ atm. Identify the gas.*

**(Nagpur 2008)**

**SOLUTION.** The RMS speed at 273 K is

$$C = \sqrt{\frac{3P}{\rho}}$$

where $P = 10^{-2} \text{ atm} = 76 \times 10^{-2} \times 13.6 \times 980 \text{ dynes/cm}^2$

and $\rho = 1.24 \times 10^{-5} \text{ gm/cm}^3.$

$$\therefore \quad C = \sqrt{\frac{3 \times 76 \times 10^{-2} \times 13.6 \times 9.80}{1.24 \times 10^{-5}}} = 4.949 \times 10^4 \text{ cm/s}$$

The gas is nitrogen whose RMS value at NTP is 4.95 × $10^4$ cm/s.

## EXERCISE–I

**1.1** Describe kinetic model of a gas with basic postulates. Write down the expression for pressure exerted by a gas contained in a vessel. **(M.S. Uni. Tamil Nadu 2007, Lucknow 2004, Guwahati 2000, Calcutta; 2003, Nagpur 2008, 2004, Patna 2004)**

**1.2** Deduce laws of ideal gas on the basis of kinetic theory of gases. Derive the perfect ges equation. **(Agra 2007, G.N.D.U. 2002, Patna 2004)**

**1.3** Give kinetic interpretation of temperature on the basis of kinetic theory of gases and define absolute zero temperature. **(Calcutta 2000, Nagpur 2007)**

**1.4** On the basis of kinetic theory of gases, derive an expression for average translational kinetic energy per molecule. **(Nagpur 2000)**

**1.5** Give any four postulates of kinetic theory of gases. **(Madurai 2003)**

**1.6** Explain the basic principles of the kinetic theory of gases and show that the pressure of an ideal gas is proportional to its density. **(Nagpur 2003, Patna 2001, Ranchi 2002)**

**1.7** In terms of kinetic theory of gases, obtain an expression for the temperature dependence of pressure of a gas in closed container. **(M.D.U. 2000)**

**1.8** What are essential features of Brownian motion? Explain its basis. Obtain the expression for the distribution of Brownian particles in a vertical isothermal column. Describe how its can be used for experimental determination Avogadro's number. **(Garhwal 2000, Meerut 2002, Calcutta 2003)**

**1.9** Show that for an ideal gas, the internal energy is a function of temperature alone. Derive the relation for adiabatic expansion of an ideal gas. **(Calcutta 2003, 2004)**

**1.10** Explain Brownian motion. Discuss Langevin's theory for Brownian motion. **(M.S. Uni. Tamil Nadu 2007, Patna 2005)**

**1.11** What is Brownian motion? Derive Einstein's relation for Brownian motion. **(Agra 2007, Calcutta 2004, M.S. Uni. Tamil Nadu 2007)**

**1.12** Prove that Avogadro's number

$$N = \frac{3RT}{4\pi a^3 g(d_1 - d_2)(h - h_0)} \log\left(\frac{n_0}{n}\right)$$

and Einstein's equation

$$\Delta^2 = \frac{RT}{N}\left(\frac{1}{3\pi\eta a}\right)$$ **(Garhwal 2002)**

**1.13** Explain the concept of degrees of freedom by taking suitable examples. **(Nagpur 2004/s, Meerut 2004)**

**1.14** Give essential features of Brownian motion.

**1.15** Give the theory of estimation of Avogadro's number. **(Agra 2002, 2000)**

**1.16** Explain the term : "degrees of freedom". How many the degrees of freedom will monoatomic, diatomic and triatomic molecule have in the absence of vibratory motion? **(Alagappa 2004, Kuvempa 2005, Nagpur Uni. 2008, 2007, 2006)**

**1.17** Show that the average kinetic energy per molecule of a perfect gas per degree of freedom is $\frac{1}{2}$ kT. **(Purvanchal 2002, Meerut 2005, Nagpur 2004, Alagappa 2004)**

**1.18** Show that root mean square speed of the molecule is inversely proportional to square root of molecular weight. **(Nagpur 2008, 2004)**

**1.19** Explain how many degrees of freedom of non-linear tri-atomic molecule will have, including vibratory motion. **(Nagpur 2005/W)**

**1.20** State and prove the Maxwell's law of equipartition of energy. **(Meerut 2003, 2004, Nagpur 2009, 2005/w, 2005/s, 2004/s, Alagappa 2004, Agra 2004, Bangalore 2004)**

**1.21** Using the law of equipartition of energy, show that for a gas possessing $f$ degrees of freedom, the ratio of two specific heats ($\gamma$) is $\left(1+\frac{2}{f}\right)$. Estimate the value of $\gamma$ for monoatomic, diatomic and polyatomic gases. Compare the values with the experimental results. **(Alagappa 2004/s, Meerut 2005, 2003, 2001, Nagpur 2004/5, Garhwal 2000, Rohilkhand 2000)**

**1.22** What do you mean by freezing of degrees of freedom? Explain behaviour of specific heats at low temperature.

**1.23** Explain the term atomicity of gases. Obtain the atomicity for mono–, di– and triatomic molecules.

**1.24** State the law of equipartition of energy. Apply this law to find an expression for the total internal energy of one gm-mole of a monoatomic ideal gas at an absolute temperature T. Find also the expression for the molar specific heat at constant volume. **(Calcutta 2005)**

**1.25** Explain low temperatures behaviour of specific heats of gases. **(Nagpur 2009, 2005/s)**

**1.26** Do the experimental values of $C_p$, $C_v$ and $\gamma$ agree with the theoretical values ? Explain. **(Nagpur 2004)**

**1.27** State the law of equipartition of energy. Prove the law for a perfect gas, whose molecules have $n$ degrees of freedom. What are the expected values of $\gamma$ for Neon and nitrogen. **(Nagpur 2006, Purvanchal 2004, Ranchi 2005, Mysore 2004, Kuvempa 2005)**

**1.28** Show from the kinetic theory of gases that the mean kinetic energy of translation of a molecule of a gas is $\frac{3}{2}kT$, where $k$ is the Boltzmann's constant. **(Meerut 2001, Ranchi 2001)**

**1.29** Explain what do you mean by degrees of freedom. State the law of equipartition of energy. Prove that for a perfect gas whose molecules have $f$ degrees of freedom

$$\frac{C_P}{C_V} = 1+\frac{2}{f}$$

Hence show that for a mono-atomic gas $\gamma = 1.67$, for a diatomic gas $\gamma = 1.4$ and for triatomic gas $\gamma = 1.33$. **(Purvanchal 2005, 2003, Nagpur 2006, Garhwal 2002)**

**1.30** Describe the 'stability of the atmosphere'. What is mean by lapse rate? **(Alagappa 2004)**

**1.31** Hot air rises up; Why is it cooler at the top of a mountain than near sea level. **(Nagpur 2009)**

**1.32** Assuming that air behaves like a perfect gas and its temperature is constant throughout, derive a relation between pressure of air at ground level and pressure at height $h$. **(Agra 2005)**

**1.33** Write short notes on :

(*i*) Degrees of freedom. **(Alagappa 2004, Nagpur 2005)**

(*ii*) Law of equipartition of energy. **(Patna 2004)**

(*iii*) Avogadro's number. **(Lucknow 2004, 2003)**

(*iv*) Brownian motion. **(M.S. Uni. Tamil Nadu 2007, Agra 2003, Lucknow 2003)**

(*v*) Kinetic interpretation of temperature. **(Madras 2001 Calcutta 2000, Madurai 2001)**

(*vi*) Freezing of degrees of freedom. **(Nagpur 2001)**

(*vii*) Adiabatic expansion of a gas. **(Nagpur 2005)**

**1.34** With what speed would 1 gram molecule of hydrogen at 27°C be moving in order that the translational kinetic energy of its centre of mass is equal to the total random kinetic energy of all its molecules ? (Molecular weight of hydrogen = 2) **[Ans.** $1.93 \times 10^5$ cm/s] **(Nagpur 2000)**

**1.35** Calculate the temperature at which r.m.s. velocity of a helium molecule will be equal to the speed of the earth's first satellite, *i.e.*, $v$ = 8 km/s. **[Ans.** $10.28 \times 10^3$ K] **(Madras 2001)**

**1.36** Calculate the mean kinetic energy of a molecule of a gas at 1000°C. Given:

$R = 8.31 \times 10^7$ ergs/gram mol–K

$N = 6.02 \times 10^{23}$ **[Ans.** $2.07 \times 10^{-13}$ ergs] **(Kuvempa 2002)**

**1.37** Calculate the r.m.s. velocity of the molecules of hydrogen at 0°C. Molecular weight of hydrogen = 2.016 and $R = 8.31 \times 10^7$ ergs/ gram mole °C. **[Ans.** $18.4 \times 10^4$ cm/s]

**1.38** Calculate the values of molar heat capacity $c_v$ and $c_p$ of a diatomic gas. Given : R = 8.31 J/mol K. Calculate also the value of $\gamma$. **[Ans.** $c_p$ = 2.9074, $c_v$ = 2.077] **(Nagpur 2009, 2007, 2005/s)**

**[Hint:** - $\frac{C_P}{C_V} = 1 + \frac{2}{f}$ , Here $f$ = 5 for diatomic gas $C_p - C_V = R$]

**1.39** A particle under Brownian motion at 27°C has r.m.s speed of $10^{-2}$ m/s. Find the mass of the partide. Boltzmann canstant k = $1.38 \times 10^{-23}$ J/K. **[Ans.** $1.24 \times 10^{-15}$ kg] **(Kerala 2003)**

**1.40** Calculate the mean kinetic energy of a molecule of a gas at 1,000°C. Given,

$R = 8.31 \times 10^7$ ergs/gram mol–K

$N = 6.02 \times 10^{23}$ **[Ans.** $2.07 \times 10^{-13}$ ergs] **(Meerut, 2001)**

**1.41** If the density of nitrogen is 1.25 g/litre at N.T.P., calculate the R.M.S. velocity of its molecules. **[Ans.** $4.95 \times 10^4$ cm/s]

**1.42** At what temperature is the R.M.S. speed of oxygen molecules twice their R.M.S. Speed at 27°C ? **[Ans.** 927°C] **(Kerala, 2000)**

**1.43** Calculate the R.M.S. velocity of the molecules of hydrogen at 0°C. Molecular weight of hydrogen = 2.016 and $R = 8.31 \times 10^7$ ergs/gram mole °C. **[Ans.** $18.4 \times 10^4$ cm/s] **(Alagappa, 2001)**

**1.44** Calculate the R.M.S. velocity of the hydrogen molecules at room temperature, given that one litre of the gas at room temperature and normal pressure weigh's 0.086 g. **[Ans.** $1.88 \times 10^5$ cm/s)]

**1.45** At what temperature the molecular velocity (r.m.s) of oxygen will become half as that of hydrogen at NTP? Molecular wt. of oxygen and hydrogen are 32 and 2 respectively. **[Ans.** 1092 K] **(Nagpur 2008, 2004)**

**1.46** State the law of equipartition of energy. Calculate the total molecular translational and rotational energy of 1 gm of hydrogen gas at 100 K and 300 K. assuming the molecules to be dumb-bell shaped. Will the specific heat at constant volume be the same at these two temperatures? **(Calcutta 2003)**

**1.47** If $c_p$ = 29.08 J mole$^{-1}$ K$^{-1}$ and $c_v$ = 20.77 J mole$^{-1}$ K$^{-1}$, find the number of degrees of freedom. What is the atomicity of gas molecule? **(Nagpur 2003)**

**1.48** Show that the pressure of an ideal gas is equal to 2/3 of the translational kinetic energy of the molecules in a unit volume. **(Nagpur 2009)**

## OBJECTIVE QUESTIONS

**1.1** The molecules of an ideal gas have:
(*a*) only kinetic energy (*b*) only potential energy
(*c*) Both K.E. and P.E. (*d*) None of them

**1.2** The value of Boltzmann's constant is
(*a*) $1.38 \times 10^{-16}$ $JK^{-1}$ (*b*) 1.38 $JK^{-1}$
(*c*) $1.38 \times 10^{-23}$ $JK^{-1}$ (*d*) 8.314 $JK^{-1}$

**1.3** According to kinetic theory of gases the relation between pressure *P* density *P* and mean square velocity *C* is
(*a*) $P = \frac{1}{2} \rho C^2$ (*b*) $P = \frac{1}{3} \rho C^2$
(*c*) $P = \frac{1}{2} \rho C$ (*d*) $P = \frac{1}{3} \rho C$

**1.4** The temperature of a gas is doubled, then its root mean square speed becomes
(*a*) double (*b*) $\sqrt{2}$ times
(*c*) half (*d*) $\frac{1}{\sqrt{2}}$ times **(Garhwal, 2001)**

**1.5** The average velocity of the molecules in a gas in equilibrium is:
(*a*) proportional $\sqrt{T}$ (*b*) proportional to $T^2$
(*c*) proportional to T (*d*) equal to zero

**1.6** The average energy of the molecules of monoatomic gas at temperature *T* is
(*a*) $\frac{1}{2} kT$ (*b*) $\frac{3}{2} kT$
(*c*) $kT$ (*d*) $\frac{5}{2} kT$

**1.7** A diatomic gas molecule has translational, rotational and vibrational degrees of freedom. Then the ratio $\frac{c_p}{c_v}$ is :
(*a*) 1.29 (*b*) 1.33
(*c*) 1.4 (*d*) 1.67

**1.8** The ratio of two specific heats $\left(\frac{c_p}{c_v}\right)$ of a diatomic gas is:
(*a*) 1.67 (*b*) 1.40
(*c*) 1.00 (*d*) 2 **(Rohilkhand, 2003)**

**1.9** The kinetic energy per unit volume of a perfect gas is equal to:
(*a*) $\frac{2}{3} P$ (*b*) $\frac{3}{2} P$
(*c*) $\frac{P}{3}$ (*d*) $\frac{1}{2} P$ where *P* is pressure **(Garhwal, 2005)**

**1.10** For a gas the r.m.s speed at 800 K is:
(*a*) four times the value of 200 K (*b*) twice the value of 200 K
(*c*) half the value of 200 K (*d*) same as at 200 K

## ANSWERS

| | | | |
|---|---|---|---|
| **1.1** (*a*) | **1.2** (*c*) | **1.3** (*b*) | **1.4** (*b*) |
| **1.5** (*d*) | **1.6** (*b*) | **1.7** (*a*) | **1.8** (*b*) |
| **1.9** (*b*) | **1.10** (*b*) | | |

# Behaviour of Real Gases

# 2

## At a Glance



## INTRODUCTION

An ideal or perfect gas is that which strictly obeys gas laws like Boyle's law, Charles' law as well as gas equation $PV = RT$ under all conditions. But all real gases viz., hydrogen, helium, nitrogen, oxygen etc., show some deviation from the gas laws particularly at low temperatures and high pressures. At very low pressure, the inter-molecular forces are negligible. Therefore, at low pressures, all real gases obey the gas equation $PV = RT$. The internal energy of an ideal gas is a function of temperature only but for a real gas, internal energy ($U$) is a function of pressure and temperature. In fact, no gas is perfect or ideal in the true sense.

## 2.1 Change of State

The molecules in a liquid have random motion. The average kinetic energy of the molecules depends upon the temperature. The molecules are held together to occupy a particular volume due to intermolecular attraction.

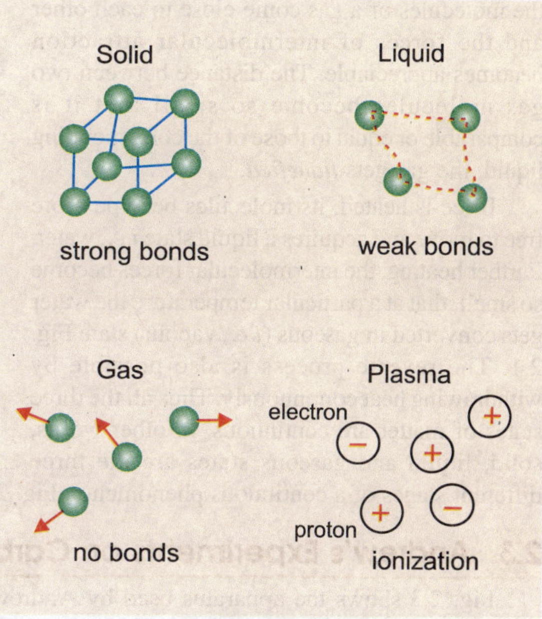


Fig. 2.1

Consider water kept in a tray and exposed to the atmosphere. Generally, the atmosphere is not saturated with water vapours. The molecules of the liquid which are in a state of constant random motion collide with each other resulting in exchange of kinetic energies. Thus, there exists the possibility of a particular molecule in the vicinity of water air interface, acquiring sufficiently high kinetic energy. This energy of the molecule may be high enough to enable the molecule to break away from the forces of attraction of the surrounding molecules. In fact, a molecule on the interface experiences minimum force of attraction. Such a molecule flies off into vapour phase. When this molecule leaves the liquid, the mean kinetic energy of the remaining liquid molecules decreases and there is fall in temperature. This explains cooling caused by evaporation. Some vapour molecules in the atmosphere, also coalesce and get absorbed by the water surface. This is called *condensation*. Evaporation and condensation continue simultaneously but rate of evaporation is higher than the rate of condensation.

An essential pre-condition for cooling due to evaporation is the capacity of air to absorb more moisture.

If the liquid is heated, and if the rise in temperature of the liquid is more than the cooling caused by evaporation, the temperature of the liquid increases in addition to the increase in the rate of evaporation. A stage will come when the rate of heat supply is exactly equal to the energy needed for evaporation. At this stage, evaporation takes place without change in temperature. This is *boiling*. At the boiling point, the molecules are pulled further apart and their positions are no longer restricted to the space occupied by the liquid. This explains the change of state from liquid to gas or vapour. The amount of work done in pulling the molecules apart measures the **latent heat of vaporisation.** At the boiling point, the heat supplied to the liquid is used up only in pulling the molecules apart against the forces of intermolecular attraction and hence no rise in temperature of the liquid is noticed till the whole of the liquid is changed to the gaseous state. 1 cc of water when converted to steam occupies a volume of 1600 cc. This means that the mean intermolecular distance is approximately 11.7 times more ( $\sqrt[3]{1600} = 11.7$ approximately) in a gas than the corresponding liquid.

## 2·2 Continuity of State

A substance, usually exists in three states *viz.*, solid, liquid and gas. For a long time, it was supposed that gases like hydrogen, helium, nitrogen, oxygen, carbon dioxide etc., exist only in gaseous form. Moreover, it was also supposed that the liquid and gaseous states are non-continuous and are distinctly separate from one another. But at high pressure and low temperature, there are marked deviations from Boyle's law. At high pressure, the molecules of a gas come close to each other and the forces of intermolecular attraction becomes appreciable. The distance between two gas molecules become so small that it is comparable or equal to those of the corresponding liquid, the gas gets *liquefied*.

If ice is heated, its molecules become more free to move and acquires a liquid state *i.e.*, water. Further heating, the intermolecular forces become so small, that at a particular temperature the water gets converted in gaseous (*i.e.*, vapour) state Fig. 2.1. The reverse process is also possible by withdrawing heat continuously. Thus all the three states of matter are continuous. In other words, solid, liquid and gaseous states are the three different stages of a continuous phenomenon Fig. 2.2.

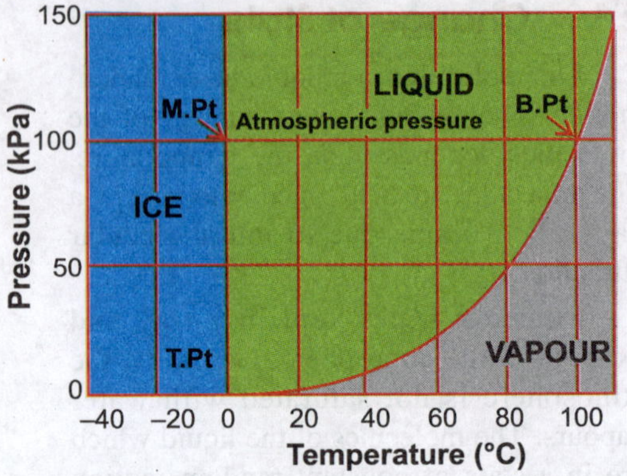


**Fig. 2.2**

## 2.3 Andrew's Experiments on Carbon Dioxide

Fig. 2.3 shows the apparatus used by Andrew in 1869, to study the isothermals of carbon dioxide at different temperatures.

The apparatus consists of two similar glass tubes *A* and *B*, having thick capillary tubes at the top and bulbs in the middle. Initially in tube *A* pure dry air is passed for a long time and then sealed at both the ends. Similarly, in tube *B*, carbon dioxide is passed for a long time (nearly 24 hours) and then both the ends are sealed. The lower ends of both the tubes are immersed in mercury. A small pellet of mercury is drawn in both the tubes by alternately heating and cooling the tubes. Both the tubes are fixed in *H* shaped copper vessel having steel stoppers *S* and *S*′. The vessel is filled with water. By screwing in the plunger, the water is compressed and the pressure is transmitted to the enclosed gases. Pressures of the order of 400 atmospheres can thus be applied.

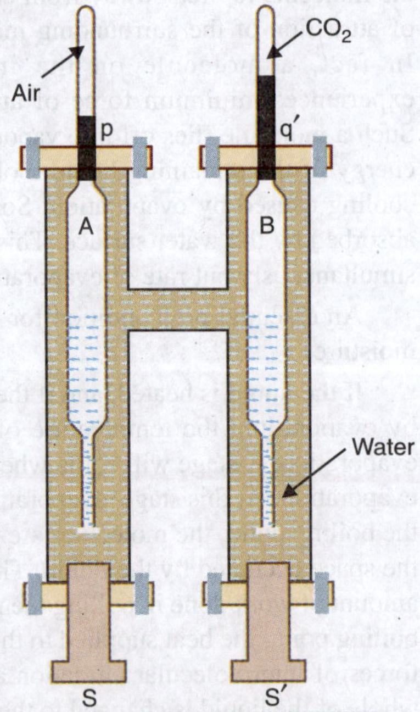


**Fig 2.3**

Since the pressures of air in *A* and $CO_2$ in *B* are the same, from the volume of air in *A*, pressure of $CO_2$ can be calculated assuming air to obey Boyle's law. The volume of $CO_2$ is read from tube *B*, as it is calibrated to read the volume directly. Both the tubes are surrounded by water baths (not shown). Air is kept at constant temperature. The temperature of $CO_2$ could be maintained at any desired temperature between 0°C to 100°C. In Andrew's actual experiments, isothermals were drawn at temperatures 13.1°C, 21.5°C, 31.1°C, 32.5°C, 35.5°C and 48.1°C as shown in Fig. 2.4.

### Discussion of Results

From the isothermals of $CO_2$ at various temperatures (Fig 2.4), following conclusions can be drawn.

**1.** At 13.1°C, the portion *AB* represents the gaseous states of $CO_2$ up to the point *B*, it obeys Boyle's law from *B* to *C*, volume decreases suddenly (enormous decrease) with slight increase in pressure. The portion *BC* represents the change of $CO_2$ from the gaseous to liquid state. At *C*, the gas has been liquefied completely. The portion *CD* represents the liquid state of $CO_2$.

**2.** At 21.5°C, the curve is similar. The horizontal portion *BC* has decreased further to *FG*.

**3.** At 31.1°C, the horizontal portion is just vanished. This is called critical isothermal for $CO_2$. Above this temperature *i.e.*, 35.5°C and 48.1°C the horizontal portion does not appear at all and the gas obeys Boyle's law from one end to the other. This isothermals are similar to the isothermals for air as shown in Fig. 2.4.

**4.** Thus, for $CO_2$, the critical isothermal divides into two regions. Above the critical isothermal, the $CO_2$ gas cannot be liquefied howsoever large the applied pressure may be. Below the critical isothermal, $CO_2$ gas can be liquefied by the application of pressure alone. The dotted curve whose highest point *P* is called the critical point and the temperature 31.1°C of the isothermal is called the critical temperature.

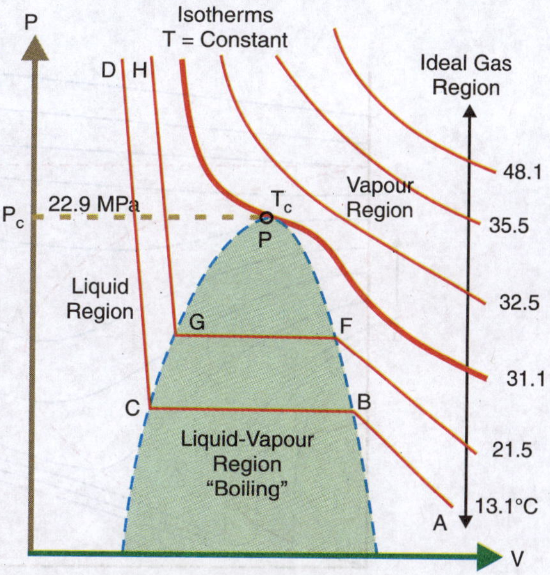


**Fig 2.4** Isothermals of $CO_2$ and air

## 2.4 Critical Constants

**1. Critical temperature ($T_c$):** It is defined as the maximum temperature at which a gas can be liquefied by the increase of pressure alone. Above this temperature, the gas cannot be liquefied howsoever large the applied pressure may be.

**2. Critical pressure ($P_c$):** The pressure necessary to liquefy a gas at critical temperature is called the critical pressure.

**3. Critical volume ($V_c$):** It is the volume which unit mass of a gas occupies at the critical temperature and critical pressure.

These three quantities, $T_c$, $P_c$ and $V_c$ are called critical constants of a gas.

**4. Critical point:** It is that point on the isothermal for the critical temperature at which the change of state from the gaseous state to the liquid state takes place under constant values of $P_c$ and $V_c$.

## 2.5 Behaviour of Gases at High Pressure

The isothermals between *PV* and *P* for various gases have been drawn and the general nature of the curves is as shown in Fig. 2.5.

The general nature of the curves is the same for all gases.

**1.** At high temperature, the value of *PV* increases with increase in *P*.

**2.** At lower temperatures, the value of *PV* decreases initially with increase in *P*. It becomes minimum at a particular pressure and then increases with increase in pressure. The locus of these minima is shown by the dotted curve *A*.

**3.** At temperatures below the critical temperature, there is a sudden decrease in the value of *PV* with increase in pressure. This corresponds to the change of state from gas to liquid. When the liquefaction is complete, *PV* gradually increases with increase in *P*. The shaded area represents the region of liquefaction.

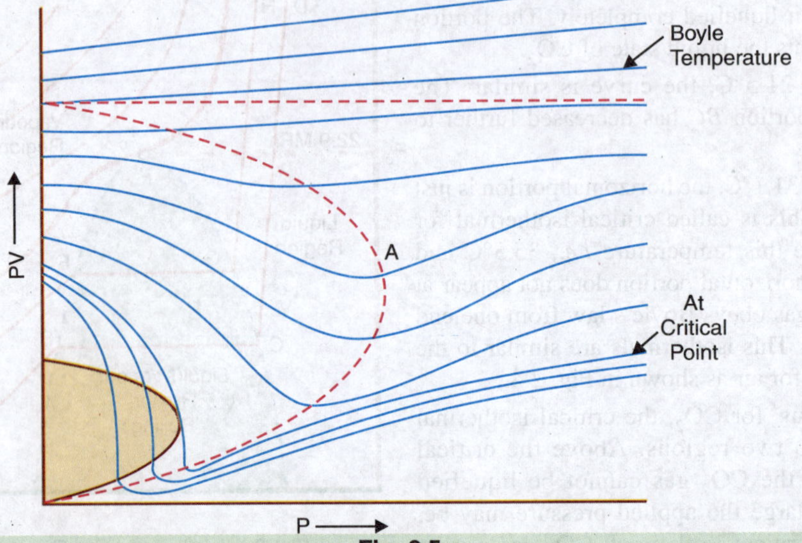


**Fig. 2.5**

Thus, the Boyle's law $PV = RT$ is obeyed only under ideal conditions, *i.e.*, at high temperature and low pressure.

Attempts were made to modify Boyle's law. Based on the experimental results of Andrew, Amagat and Holborn, K. Onnes suggested for all real gases at high pressure and at constant temperature, an empirical relation

$$PV = A + BP + CP^2 + DP^3 + \ldots\ldots \quad (2.1)$$

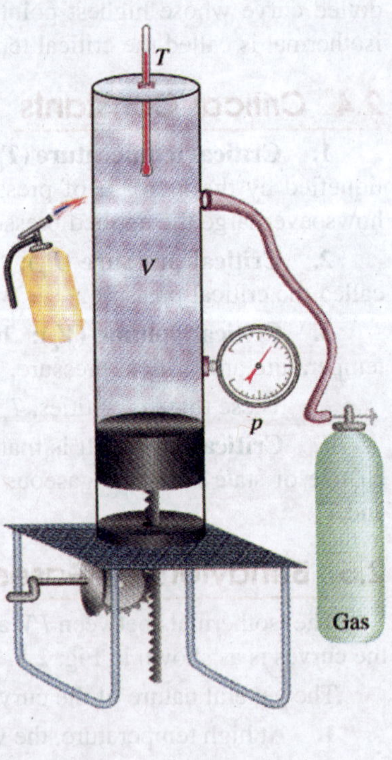


where *A*, *B*, *C*, *D*, etc. are constants for a given temperature depending on the nature of the gas and are called **virial coefficients**. *A* is called the first virial coefficient, *B* the second virial coefficient, *C* the third virial coefficient, and so on. These coefficients are in decreasing order of magnitude, *i.e.*, $A > B > C > D$ and so on. The ratio $B/A$ is of the order of $10^{-3}$ while $C/A$ is of the order of $10^{-5}$. A is simply equal to $RT$ because as $P \to 0$, equation (*ii*) reduces to $PV = RT$. The second, third and fourth terms become negligible in comparison to first at very low pressure and then gases obey Boyle's law accurately– lower the pressure, higher is the degree of accuracy.

## 2.6 Boyle Temperature

The coefficient *B* has a special importance. For all gases, it varies in a similar manner. Its value is negative at low temperatures, but gradually increases to zero and becomes positive as the temperature is increased. The temperature at which coefficient *B* is zero is called the **Boyle temperature**, represented by $T_B$. At this temperature Boyle's law holds good over a wide range of pressure provided the constants *C*, *B*, .... are all negligible. Thus, at Boyle temperature.

$$B = \frac{d}{dP}(PV) = 0 \quad ...(i)$$

and $$PV = A = \text{constant.} \quad (ii)$$

From Van der Waals' equation of state for a real gas

$$\left(P + \frac{a}{V^2}\right)(V - b) = RT \quad ...(iii)$$

where $a$ and $b$ are Van der Waals' constants.

Solving eq. (*iii*), we get

$$PV - Pb + \frac{a}{V} - \frac{ab}{V^2} = RT$$

or $$PV = RT + Pb - \frac{a}{V} + \frac{ab}{V^2} \quad ...(iv)$$

we can write approximately,

$$V = \frac{RT}{P} \quad ...(v)$$

Putting this value of V in equation (*iv*), we get

$$PV = RT + Pb - \frac{a}{RT}P + \frac{ab}{R^2T^2}P^2$$

$$= RT + \left(b - \frac{a}{RT}\right)P + \frac{ab}{R^2T^2}P^2 \quad ...(vi)$$

Comparing this with equation (2.1), we have

$$A = RT, B = b - \frac{a}{RT} \quad \text{and} \quad C = \frac{ab}{R^2T^2} \quad ...(vii)$$

At Boyle Temperature $T = T_B$ and $B = 0$, therefore, we have

$$0 = b - \frac{a}{RT_B}$$

or $$T_B = \frac{a}{Rb} \quad ...(2.2)$$

This equation gives the value of Boyle's temperature for all gases obeying Van der Waals' equation of state. Clearly Boyle temperature ($T_B$) depends upon $a$ and $b$ which are different for different gases. Hence **Boyle temperature is different for different gases**. For nitrogen $T_B$ is 50°C, for hydrogen it is –169°C while for helium it is – 254°C.

The behaviour of gases above, below and at Boyle temperature is expressed by plotting the product PV against $P$ for different temperatures (Fig. 2.6). The general nature of the curves is the same for all gases. Following conclusions can be drawn:

**1. Below Boyle Temperature, ($T < T_B$) :** The value of $PV$ first decreases (as $B$ is –ve) with increase in $P$, reaches a minimum at a particular pressure and then begins to increase. This is because at high pressures the terms $CP^2$, $DP^3$ etc. of equation (2.1)

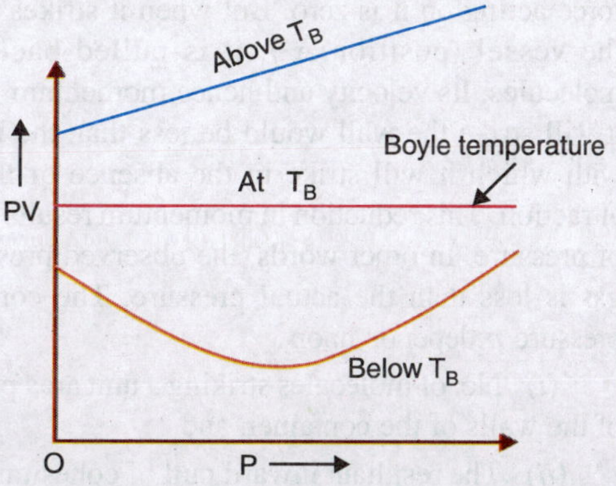


Fig. 2.6

become more effective and are positive and thus $PV$ increases with increase in $P$ after reaching a minimum. This suggests the existence of intermolecular attraction because below $T_B$, the gases are highly compressible.

**2. Above Boyle Temperature ($T > T_B$) :** Above Boyle temperature, $PV$ continuously increases with increase in $P$. This is because, the coefficient $B$ is +ve above Boyle temperature and thus the terms on R.H.S. of equation (2.1) are positive. It now indicates that the intermolecular attraction is now less significant.

**3. At Boyle temperature ($T = T_B$) :** The product PV is almost constant for a long range of pressure. Thus, Boyle's law is obeyed over a wide range of pressures.

Thus, below the Boyle temperature, the gases are highly compressible and this suggests the existence of intermolecular attraction. Beyond Boyle temperature, Boyle's law is obeyed and intermolecular attractions are less significant.

**Definition:** Boyle temperature is that temperature above which a real gas behaves like an ideal gas and obeys Boyle's law.

## 2.7 Reasons for Modification of Gas Equation

The perfect gas equation $PV = RT$ is not obeyed by real gases, particularly at high pressure and low temperature. At ordinary temperatures and pressures, the deviation is small but at high pressures and low temperatures the deviations are very large. In the derivation of the perfect gas equation, on the basis of kinetic theory of gases, the following assumptions have been made, which do not hold true in case of real gases. These are (*i*) the size of the molecule of the gas is negligible, and (*ii*) the forces of intermolecular attraction are absent. But in actual practice, at high pressure the size of the molecule of the gas becomes significant and cannot be neglected, in comparison with the volume of the gas. Also at high pressure, the molecules come closer and the forces of molecular attraction are appreciable. Hence, a perfect gas equation needs modification.

## 2.8 Van der Waals Equation of State

In 1879 Van der Waals modified the perfect gas equation $PV = RT$ by applying corrections for

1. Intermolecular force of attraction and
2. Finite size of molecules.

### 1. Correction for Pressure:

Consider a molecule A of the gas, well inside the vessel as shown in Fig. 2.7. It is attracted by other molecules in all directions with the same force and the net force acting on it is zero. But when it strikes the wall of the vessel (position A′), it is pulled back by other molecules. Its velocity and hence momentum with which it will strike the wall would be less than the momentum with which it will strike in the absence of the force of attraction. This reduction in momentum results in decrease of pressure. In other words, the observed pressure of the gas is less than the actual pressure. The correction for pressure $p$ depends upon

(*i*) No. of molecules striking a unit area per unit time of the walls of the container, and

(*ii*) The resultant inward pull of cohesion on each of the striking molecules.

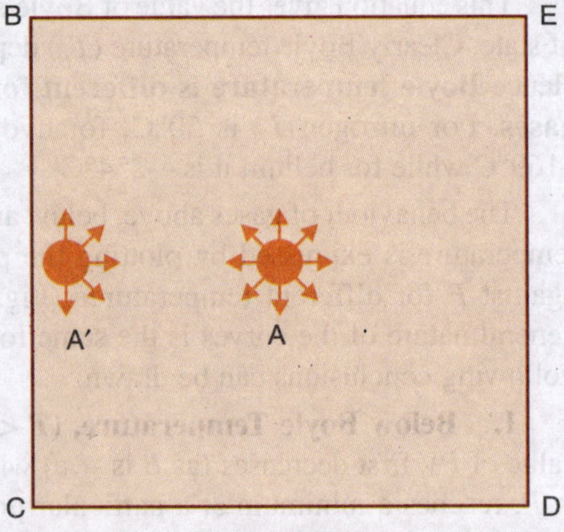


Fig. 2.7

Each of these factors is proportional to the density of the gas.

∴ Correction for pressure $p \propto \rho^2 \propto \dfrac{1}{V^2}$

∴ $$p = \frac{a}{V^2}$$

Here $a$ is a constant and $V$ is the volume of the gas.

Therefore, the corrected or real pressure

$$= P + p = \left(P + \frac{a}{V^2}\right) \qquad ...(2.3)$$

where $P$ is the observed pressure.

## 2. Correction for Volume :

Due to the finite size of gas molecules, the actual space available for the movement of the molecules is less than the volume of the vessel. The molecules have the sphere of influence (Fig. 2.8) around them of radius ($2r$), within which no other molecule can penetrate.

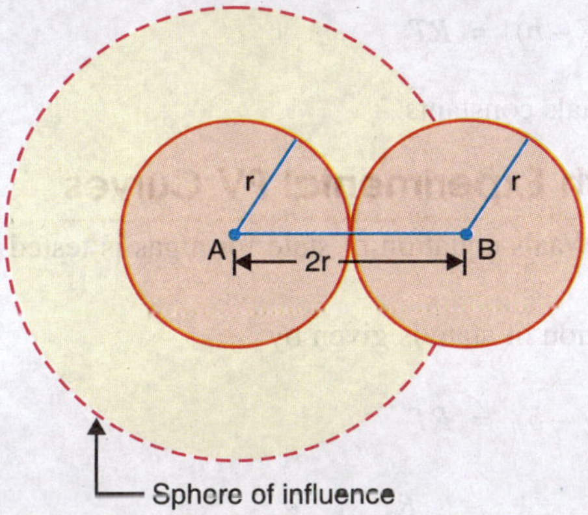


**Fig. 2.8**

Here $r$ is the radius of each gas molecule.

Volume of the molecule $= x = \dfrac{4}{3}\pi r^3$

The centre of any two molecules can approach each other only by a minimum distance of $2r$. The volume of sphere of influence of each molecule

$$s = \frac{4}{3}\pi\,(2r)^3 = 8 \times \frac{4}{3}\pi r^3 = 8\,x$$

Let us fill the whole space of the volume $V$ with $n$ molecules one by one.

The volume available for the first molecule $= V$

The volume available for the second molecule $= V - 8x$

$= V - s$

The volume available for *nth* molecule $= [V - (n - 1)\,s]$.

∴ Average space available for each molecule

$$= \frac{V + (V - s) + (V - 2s) + ........ + [V - (n - 1)\,s]}{n}$$

$$= \frac{n\,V}{n} - \frac{s}{n}\,\{1 + 2 + 3 + ........ + (n - 1)\}$$

$$= V - \frac{s}{n} \cdot \frac{(n-1)n}{2}$$

$$= V - \frac{ns}{2} + \frac{s}{2}$$

As the number of molecules is very large $\frac{s}{2}$ can be neglected.

∴ Average space available of each molecule

$$= V - \frac{ns}{2}$$

$$= V - \frac{n\,(8x)}{2} \qquad (\because s = 8x)$$

$$= V - 4\,(n\,x)$$

$$= V - b \qquad ...(2.4)$$

Where $b = 4\,(n\,x)$ = four times the actual volume of the molecules.

Thus, the Van der waals equation of state for a gas is

$$\left(P + \frac{a}{V^2}\right)(V - b) = RT \qquad ...(2.5)$$

where $a$ and $b$ are Van der waals constants.

## 2.9 Comparison with Experimental PV Curves

The validity of Van der Waals equation of state for a gas is tested by comparing its *PV* curves with the experimental results.

The Van der Waals equation of state is given by

$$\left(P + \frac{a}{V^2}\right)(V - b) = RT$$

or
$$P = \frac{RT}{(V-b)} - \frac{a}{V^2} \qquad ...(2.6)$$

we see that if $V \to \infty, P \to 0$. Again if $V$ is very small and $V \to b$, then $P \to \infty$.

A family of plots *i.e.*, curves between pressure and volume at various temperatures are drawn using equation 2.5. The curves are as shown in Fig. 2.9. In the graph, the horizontal portion is absent. But in place of it, the curve *ABCDE* is obtained. This does not agree with the experimental isothermals for $CO_2$ as obtained by Andrew. However, the portion *AB* has been explained as due to super cooling of the vapours and the portion *ED* is due to super-heating of the liquid. But the portion *BCD* cannot be explained because it shows decrease in volume with decrease in pressure, which is not true in actual practice. At higher temperatures, however, both theoretical and practical *PV* curves are similar.

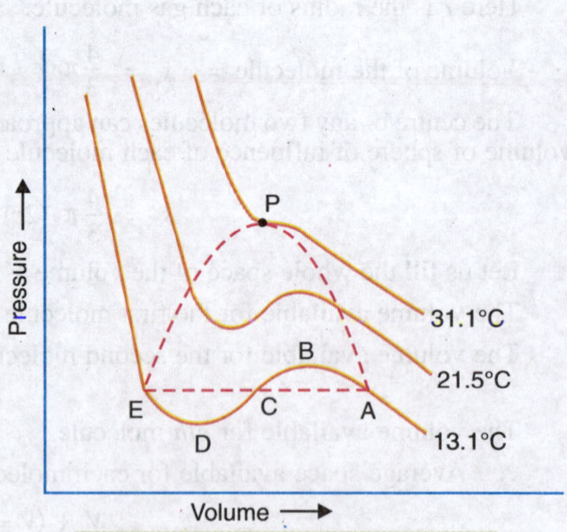


Fig. 2.9

## 2.10 Estimation of Critical Constants

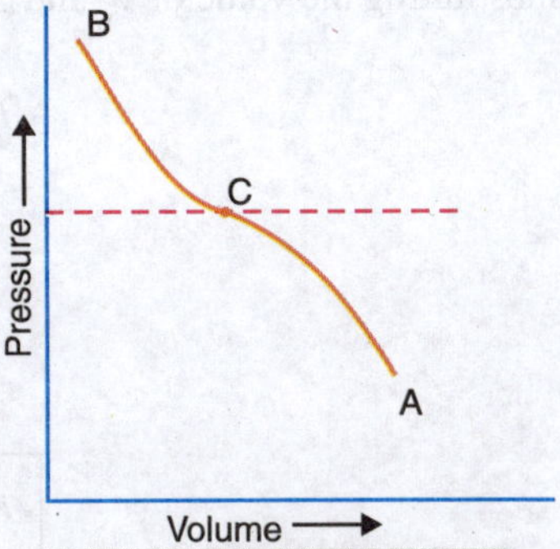


Fig. 2.10

Consider the critical isothermal *ACB* as shown in Fig. 2.10. At the critical point *C*, the curve is horizontal. Therefore, at point *C*, $\left(\frac{dP}{dV}\right) = 0$. At this point, the tangent also crosses the curve. Therefore, the tangent at such a point is said to be stationary and the point is called **'the point of inflexion'.** At the point of inflexion, we have

$$\frac{d^2P}{dV^2} = 0$$

Hence, at critical point *C*, we have $\frac{dP}{dV} = 0$ and $\frac{d^2P}{dV^2} = 0$.

The Van der Waals equation is

$$P = \frac{RT}{(V-b)} - \frac{a}{V^2} \quad \text{...(i)}$$

$$\therefore \quad \frac{dP}{dV} = -\frac{RT}{(V-b)^2} + \frac{2a}{V^3} \quad \text{...(ii)}$$

and

$$\frac{d^2P}{dV^2} = +\frac{2RT}{(V-b)^3} - \frac{6a}{V^4} \quad \text{...(iii)}$$

At the critical point *C*, we have

$$T = T_c,\ P = P_c,\ V = V_c,\ \frac{dP}{dV} = 0 \text{ and } \frac{d^2P}{dV^2} = 0$$

Substituting in the above equations,

$$P_c = \frac{RT_c}{(V_c - b)} - \frac{a}{V_c^2} \quad \text{...(iv)}$$

$$\frac{RT_c}{(V_c - b)^2} = \frac{2a}{V_c^3} \quad \text{...(v)}$$

and

$$\frac{2RT_c}{(V_c - b)^3} = \frac{6a}{V_c^4} \quad \text{...(vi)}$$

Dividing (*v*) by (*vi*),

$$\frac{V_c - b}{2} = \frac{V_c}{3} \quad \text{or} \quad 3V_c - 3b = 2V_c$$

$$\therefore \quad \boxed{V_c = 3b} \quad \text{...(2.7)}$$

Substituting the value of $V_c$ in equation (*v*), we get

$$\frac{RT_c}{(3b-b)^2} = \frac{2a}{(3b)^3}$$

$$\frac{RT_c}{4b^2} = \frac{2a}{27b^3}$$

$$\therefore \quad \boxed{T_c = \frac{8a}{27bR}} \quad \text{...(2.8)}$$

Substituting the value of $V_c$ and $T_c$ in equation (*iv*), we get

$$P_c = \frac{8aR}{27bR(3b-b)} - \frac{a}{(3b)^2}$$

$$= \frac{8a}{27 \times 2b^2} - \frac{a}{9b^2}$$

$$= \frac{a}{9b^2}\left[\frac{8}{6} - 1\right] = \frac{a}{9b^2} \times \frac{1}{3}$$

$$\boxed{P_c = \frac{a}{27b^2}} \quad \text{...(2.9)}$$

**TABLE 2.1 Critical Temperature and Pressure of Common Gases**

| *Substance* | *Critical Temperature °C* | *Critical Pressure (atmosphere)* |
|---|---|---|
| Helium | – 268°C | 2.26 |
| Hydrogen | – 240°C | 12.80 |
| Nitrogen | – 146°C | 33.50 |
| Air | – 140°C | 39.00 |
| Argon | – 122°C | 48.00 |
| Oxygen | – 118°C | 50.00 |
| Carbon dioxide | 31°C | 73.00 |
| Ammonia | 130°C | 15.00 |
| Chlorine | 146°C | 76.00 |
| Water | 374°C | 218.00 |

## 2.11 Constants of Van der Waals Equation

From equation (2.8),

$$T_c = \frac{8a}{27bR}$$

$$\therefore \quad \frac{a}{b} = \frac{27RT_c}{8} \quad \text{...(vii)}$$

But

$$P_c = \frac{a}{27b^2}$$

$$\therefore \quad \frac{a}{b^2} = 27P_c \quad \text{...(viii)}$$

Dividing (*vii*) by (*viii*),

$$\frac{a}{b} \times \frac{b^2}{a} = \frac{27RT_c}{8} \times \frac{1}{27P_c}$$

$$\therefore \qquad b = \frac{RT_c}{8P_c} \qquad \text{...(2.10)}$$

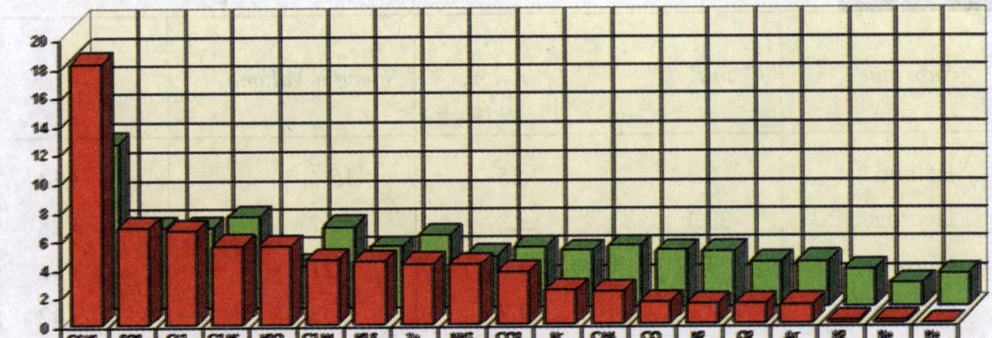


| | C6H6 | SO2 | Cl2 | C2H6 | H2O | C2H4 | H2S | Xe | NH3 | CO2 | Kr | CH4 | CO | N2 | O2 | Ar | H2 | Ne | He |
|---|---|---|---|---|---|---|---|---|---|---|---|---|---|---|---|---|---|---|---|
| a [atm ltr²/mol²] | 18.24 | 6.803 | 6.579 | 5.562 | 5.536 | 4.53 | 4.49 | 4.25 | 4.225 | 3.64 | 2.349 | 2.283 | 1.505 | 1.408 | 1.378 | 1.363 | 0.2476 | 0.2135 | 0.03457 |
| b [0.01 ltr/mol] | 11.54 | 5.636 | 5.622 | 6.38 | 3.049 | 5.714 | 4.287 | 5.105 | 3.707 | 4.267 | 3.978 | 4.278 | 3.985 | 3.913 | 3.183 | 3.219 | 2.661 | 1.709 | 2.37 |

**Fig. 2.11.** Vander Waals coefficients

Substituting for $b$ in equation (*vii*), we get

$$\frac{a}{\frac{RT_c}{8P_c}} = \frac{27\,RT_c}{8}$$

$$\therefore \qquad a = \frac{RT_c}{8\,P_c} \times \frac{27\,RT_c}{8}$$

$$\text{or} \qquad a = \frac{27\,R^2 T_c^2}{64\,P_c} \qquad \text{...(2.11)}$$

Equation (2.10) and (2.11) gives the value of constants $a$ and $b$ respectively.

## 2.12 Critical Coefficient

The quantity $\frac{RT_c}{P_c V_c}$ is called the critical coefficient. Its value is obtained by substituting the values of $P_c$, $V_c$ and $T_c$.

$$\therefore \qquad \text{Critical coefficient} = \frac{RT_c}{P_c V_c} = \frac{R \times \frac{8a}{27\,bR}}{\frac{a}{27\,b^2} \times 3b}$$

$$= \frac{8\,a}{27\,b} \times \frac{27\,b^2}{3\,ab} = \frac{8}{3}$$

$$= 2.667 \qquad \text{...(2.12)}$$

The calculated value of critical coefficient $\frac{RT_c}{P_c V_c}$ is $\frac{8}{3}$ and it is the same for all gases.

The experimental values of critical coefficient for different gases are given in Table 2.2.

**TABLE 2.2 Experimental values of Critical Coefficient $\frac{RT_c}{P_c V_c}$**

| *Substance* | $T_c$ *in K* | $P_c$ *in atm.* in cm$^3$/g | *Specific Volume* | $\frac{RT_c}{P_c V_c}$ |
|---|---|---|---|---|
| Helium | 5.1 | 2.25 | 15.4 | 3.13 |
| Hydrogen | 33.1 | 12.8 | 32.2 | 3.28 |
| Nitrogen | 125.9 | 33.5 | 3.21 | 3.42 |
| Oxygen | 154.2 | 49.7 | 2.32 | 3.42 |
| Carbon dioxide | 304 | 72.8 | 2.17 | 3.48 |
| Water | 647 | 218 | 3.181 | 4.30 |

**NOTE:** Here $V_c$ = molecular wt. × specific volume.

The experimental value of the critical coefficient of all gases is greater than the theoretical value of 2.667.

## 2.13 Limitations of Van der Waals Equation

**1.** The values of constants '$a$' and '$b$' obtained by different methods differ considerably. Although Van der Waals theory assumes them to be constant. The value of $a$ is found to depend on the temperature. At very high temperatures, it tends to zero.

**2.** According to theory, $V_c = 3b$, but it is found to depend on the nature of gas. Experimentally, it is found that $V_c \cong 2b$.

**3.** The theoretical value of critical coefficient $\frac{RT_c}{P_c V_c}$ is 2.667 for all gases. However, it varies from gas to gas with an average value of 3.7 for most of the gases. This constant appears to depend on the molecular structure of the gas.

Thus no gas obeys Van der Waals equation closely in the vicinity of the critical point.

## 2.14 Reduced Equation of State

When the actual pressure, volume and temperature of a fixed mass of a gas are divided by the critical values of these quantities, the resulting ratios are called the reduced pressure ($P_r$), reduced volume ($V_r$) and reduced temperature ($T_r$) respectively.

Thus, we have

$$P_r = \frac{P}{P_c}, \quad V_r = \frac{V}{V_c} \quad \text{and} \quad T_r = \frac{T}{T_c}$$

$$\therefore \qquad P = P_c P_r = \frac{a}{27b^2} P_r$$

$$V = V_c V_r = 3b\, V_r$$

$$T = T_c T_r = \frac{8a}{27\,Rb} T_r$$

Substituting these values of $P$, $V$ and $T$ in equation

$$\left(P + \frac{a}{V^2}\right)(V - b) = RT$$

we get

$$\left(\frac{a}{27b^2} P_r + \frac{a}{9b^2 V_r^2}\right)(3b\, V_r - b) = R\, \frac{8a}{27\, Rb} T_r$$

or

$$\frac{a}{27b^2}\left(P_r + \frac{3}{V_r^2}\right) 3b\left(V_r - \frac{1}{3}\right) = \frac{8a}{27b} T_r$$

or

$$\left(P_r + \frac{3}{V_r^2}\right)\left(V_r - \frac{1}{3}\right) = \frac{8}{3} T_r \qquad ...(2.13)$$

This equation does not contain the constants $a$ and $b$; which are characteristic of a particular gas. The equation is called Van der Waals reduced equation of State.

## 2.15 Properties of Matter Near Critical Point

Based on experiments of Andrews, Amagat and others, the state of matter near the critical point can be summarised as follows:

1. The densities of the vapour and the liquid generally approach each other and their densities are equal at the critical point.
2. At the critical point or just near the critical point, the line of demarcation between the liquid and the vapour disappears. Consequently, there must exist mutual diffusion between the two phases and the surface tension must be zero. This also means that the forces of intermolecular attraction in the liquid and vapour states must be equal at the critical point.
3. The whole mass of the substance exhibits a flickering experience which indicates that there may be variations of density inside mass. This has been experimentally verified.
4. The compressibility of the vapour becomes infinite at the critical temperature.

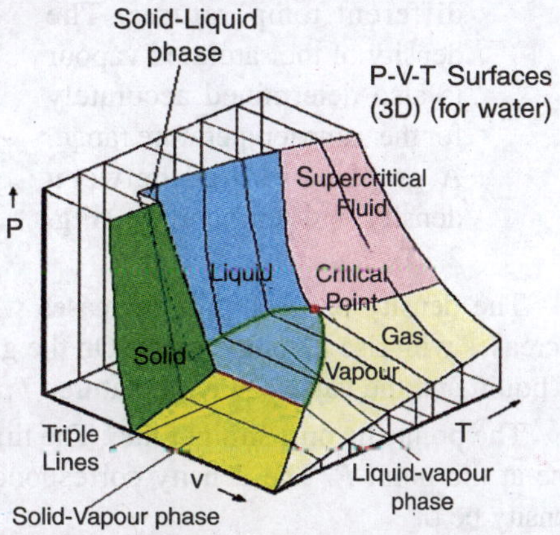


**Fig. 2.12**

The experiments performed by Callendar and others have shown that the densities of the liquid and vapour are not equal at the critical temperature. The densities are equal at a temperature slightly higher than the critical temperature. The critical temperature of water is 374°C. It was found that the densities are equal at 380°C. At 374°C, the latent heat of water was found to be 72.4 calories per gram. At 380°C, the latent heat of water is zero.

To conclude, the critical point for any substance is not a particular temperature but it is a narrow region (critical region) (fig. 2.12). **The critical point is that point at which the properties of the two phases become identical.**

## 2.16 Experimental Determination of Critical Constants

1. The critical constants can be determined by plotting the isothermals between $P$ and $V$ at different temperatures of the gas. The critical point is found on the graph, where the horizontal portion of the curve (change from gaseous to the liquid state) just vanishes. The pressure, volume and temperature corresponding to the critical point give the *critical constants* of the gas.

2. With the help of Cagniard de la Tour's apparatus, the temperatures at which the liquid meniscus disappears and reappears are carefully noted. The mean of these two temperatures gives the critical temperature of the gas. The bulb is surrounded by a bath whose temperature is controlled by a sensitive thermostat and the temperature is varied by small amounts (say 0.1°C). The pressure is read with the help of a manometer. The pressure corresponding to the critical temperature gives the critical pressure. However, the measurement of critical volume offers a practical difficulty because even if the temperature is slightly below the critical temperature, the change in volume is enormous even for a small change in pressure.

3. Cailetet and Mathias have suggested a method for the determination of critical volume. Density of the liquid is determined accurately at different temperatures. The density of the saturated vapour is also determined accurately for the same temperature range. A graph is plotted between density and temperature (Fig. 2.13).

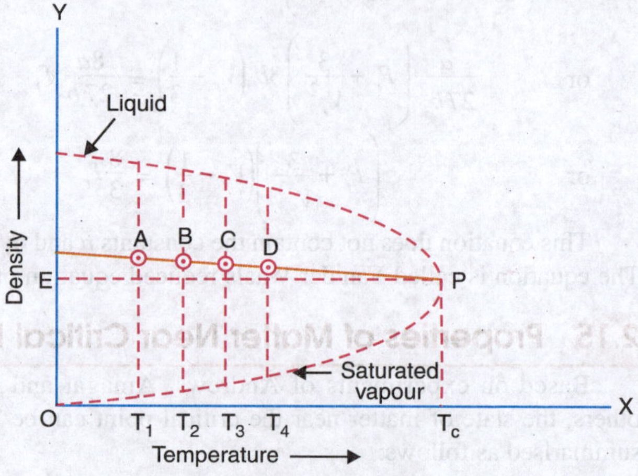


Fig. 2.13

The density of the liquid decreases with rise in temperature and that of the saturated vapour increases with rise in temperature. On the graph the points *A, B, C* and *D* represent the mean density of liquid and the vapour at temperatures $T_1$, $T_2$, $T_3$, $T_4$ etc. respectively.

The points lie on a straight line. The line is produced and the curves are extrapolated to meet the line at the point *P*. The density corresponding to the point *P* is obtained from the graph. Let this density be ρ.

The specific volume *i.e.* volume per unit mass

$$V_c = \frac{1}{\rho}$$

This represents the critical volume. The temperature corresponding to the point *P* gives the critical temperature.

## 2.17 Joule's Law for a perfect Gas (Intermolecular Attraction)

If a fixed mass of a gas is allowed to expand without doing any external work under adiabatic condition *i.e.*, no heat allowed to enter or leave the gas. In other words, the gas is allowed to undergo what is called free expansion, then if the gas molecules attract one another, internal work will have to be done by the gas in pulling them apart and this will take place at the expense of internal energy (heat energy) or the gas. This will result in *cooling effect i.e.,* the gas cools on expansion. On the other hand, if the molecules repel one another, the expansion would be accompanied by a rise of temperature. If there are no intermolecular attractions, there would be no change in temperature. Joule's experiment provides the answer to this curiosity.

## Joule's Experiment

Joule performed an experiment to study the intermolecular attraction between the molecules of a gas. He designed an apparatus to measure the change in energy of a gas when its pressure and volume are changed.

*A* and *B* are two flasks connected by a tube and fitted with a stop-cock *S*. A was filled with the gas at high pressure and *B* was perfectly evacuated (Fig. 2.14). The bulbs were surrounded by a water bath and its temperature was noted with a sensitive thermometer. The stop-cock was opened and the gas rushed from *A* to *B*. If work was done during expansion, cooling was expected.

But in this experiment no cooling was observed. This was puzzling as the work had been done by the gas in overcoming the forces of inter-molecular attractions. Joule was able to explain the drawback in his experiment by assuming that the gas has expanded into vacuum and so no cooling was observed.

In this experiment only the pressure and volume of the gas changed and as the bath did not show any change in temperature, it is evident that the temperature of the gas remained constant.

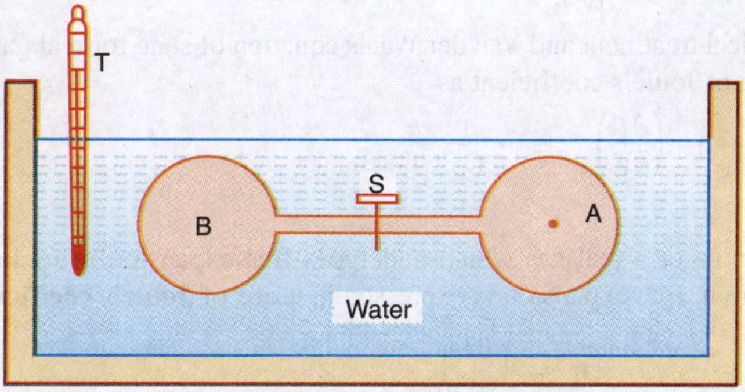


**Fig. 2.14**

According to Joule, this experiment also shows that the internal energy of a gas is a function of temperature only, irrespective of the changes in pressure and volume.

Thus, so long as the temperature of the gas remains constant, its internal energy remains unchanged. This experimental result is often referred as **Joule's law** and may be stated that the internal energy ($U$) of a gas depends only on its temperature ($T$) and is quite independent of its volume ($V$) and pressure ($P$). Mathematically, internal energy ($U$) is a function of temperature ($T$) alone, i.e.

$$U = f(T)$$

$$\therefore \qquad \left(\frac{\partial U}{\partial V}\right)_T = 0 \quad \text{and} \quad \left(\frac{\partial U}{\partial P}\right)_T = 0 \qquad ...(i)$$

$$\text{and} \qquad \left(\frac{\partial U}{\partial T}\right)_P = \left(\frac{\partial U}{\partial T}\right)_V \qquad ...(ii)$$

The equation (*i*) implies that the internal energy of a system does not change with volume or pressure so long as the temperature remains constant. The equation (*ii*) indicates that the increase of internal energy per degree of temperature is the same, no matter whether the pressure or the volume is kept constant during the process.

It should be noted that the Joule's law is obeyed strictly by a perfect gas only.

The porous plug experiment designed by Joule and Kelvin, however, showed cooling due to the work done by a gas in over-coming the inter-molecular attraction when it was passed from the high pressure to the low pressure side.

## 2.18 Expression for Joule's Coefficient (η)

Let us consider the free expansion of a real gas in Joule's experiment. In a real gas, there exist intermolecular forces and hence the gas is allowed to expand freely. In Joule's expermiment, the intermolecular separation changes and the gas has to do some internal work against the intermolecular forces in separating the molecules further away from each other. This increases the potential energy of the gas and since no heat is allowed to enter or leave the gas (due to rapid expansion), it will happen at the cost of its kinetic energy. Thus the K.E. of the gas decreases resulting in a fall of its temperature. Thus, if the intermolecular forces are *attractive*, a real gas will show a *cooling effect* when allowed to suffer free expansion. On the similar reasoning, we may say that if intermolecular forces are *repulsive*, the expansion will result on the heating of the gas. But experimentally it is seen that Joule's expansion of a real gas always produces cooling and hence the intermolecular forces in a gas are attractive in nature.

The result of a free expansion is measured in terms of Joule's coefficient, defined by

$$\text{Joule's coefficient} = \eta = \left(\frac{\partial T}{\partial V}\right)_U$$

Using thermodynamical treatment and Van der Waals equation of state for real gas (or Van der Waals gas) we get the value of Joule's coefficient as

$$\eta = \left(\frac{\partial T}{\partial V}\right)_U = -\frac{1}{C_V}\frac{a}{V^2}$$

### Proof

Consider 1 kg mole of a real gas which undergoes free expansion in Joule's experiment. The change in temperature in free expansion is expressed in terms of **Joule's coefficient** defined by

$$\eta = \left(\frac{\partial T}{\partial V}\right)_U \qquad ...(i)$$

Here $\partial T$ is the change in temperature due to change in volume $\partial V$ in free expansion of the gas.

In Van der Waals gas, due to intermolecular forces of attraction, an internal pressure developed is given by $p = \frac{a}{V^2}$. Therefore, the internal work done against this internal pressure in Joule's free expansion is

$$dW = p.dV = \frac{a}{V^2}\,dV \qquad ...(ii)$$

where a is Van der Waals' constant and $V$ is the initial volume of the gas. This internal work is done at the expense of its internal kinctic energy and consequently its internal potential energy increases. Thus,

$$\begin{aligned} \text{Decrease in internal K.E.} &= \text{Increase in internal P.E.} \\ &= -dW \\ &= -\frac{a}{V^2}dV \qquad ...(iii) \end{aligned}$$

The decrease in internal K.E. results in decrease in temperature (cooling effect). Let $dT$ be the decrease in temperature due to free expansion of the real gas and $C_v$ be its molar specific heat at constant volume, then

$$\text{Decrease in internal K.E.} = C_v\,dT \qquad ...(iv)$$

Equating (*iii*) and (*iv*), we get

$$C_v\,dT = -\frac{a}{V^2}dV$$

or

$$\frac{dT}{dV} = -\frac{1}{C_V}\cdot\frac{a}{V^2}$$

Using equation (*i*), the Joule's coefficient (η) will be

$$\eta = \left(\frac{\partial T}{\partial V}\right)_U$$

or $$\eta = -\frac{1}{C_V}\cdot\frac{a}{V^2} \qquad ...(v)$$

## Alternative Method

Consider 1 kg mole of a real gas which underdoges free expansion in Joule's experiment. Let $\partial T$ be the change in temperature, corresponding to the change in volume $\partial V$, then the Joule's coefficient is defined as

$$\eta = \left(\frac{\partial T}{\partial V}\right)_U \qquad ...(i)$$

The variation of internal energy $U$ with its volume $V$ at any temperature $T$ is obtained by using laws of thermodynamics. According to first law of thermodynamics

$$dQ = dU + dW = dU + PdV \qquad ...(ii)$$

From 2nd law of thermodynamics, in terms of entropy

$$dQ = TdS \qquad ...(iii)$$

Equating (*ii*) and (*iii*), we have

$$dU + PdV = TdS$$

$$dU = TdS - PdV$$

For isothermal change, we can write

$$\left(\frac{\partial U}{\partial V}\right)_T = T\left(\frac{\partial S}{\partial V}\right)_T - P \qquad ...(iv)$$

Using Maxwell's second thermodynamic relation

$$\left(\frac{\partial S}{\partial V}\right)_T = \left(\frac{\partial P}{\partial T}\right)_V \qquad ...(v)$$

Substituting in equation (iv), we get

$$\left(\frac{\partial U}{\partial V}\right)_T = T\left(\frac{\partial P}{\partial T}\right)_V - P \qquad ...(vi)$$

According to Van der Waals' equation of state for a real gas (also called as Van der Waals' gas),

$$\left(P + \frac{a}{V^2}\right)(V - b) = RT$$

or $$P = \frac{RT}{V-b} - \frac{a}{V^2}$$

Differentiating *w.r.* to $T$, we get

$$\left(\frac{\partial P}{\partial T}\right)_V = \frac{R}{V-b} \qquad (\because \text{ Assuming } V \text{ constant})$$

Substituting, in equation (*vi*), we get

$$\left(\frac{\partial U}{\partial V}\right)_T = T\cdot\frac{R}{V-b} - \frac{RT}{V-b} + \frac{a}{V^2} = \frac{a}{V^2} \qquad ...(vii)$$

The internal energy $U$ is also a function of volume in addition to temperature.

Hence, for a real gas, $U = f(T, V)$ ...(*viii*)

Therefore, $$dU = \left(\frac{\partial U}{\partial V}\right)_T dV + \left(\frac{\partial U}{\partial T}\right)_V dT$$

But during free expansion of a real gas in Joule's experiment, the internal energy of gas remains constant *i.e.* $dU = 0$

$$\therefore \qquad 0 = \left(\frac{\partial U}{\partial V}\right)_T dV + \left(\frac{\partial U}{\partial T}\right)_V dT$$

or $$\left(\frac{\partial T}{\partial V}\right)_U = -\frac{\left(\frac{\partial U}{\partial V}\right)_T}{\left(\frac{\partial U}{\partial T}\right)_V} \qquad \text{...(ix)}$$

But the quantity $\left(\frac{\partial U}{\partial T}\right)_V$ is defined as $C_v$, the specific heat at constant volume.

$$\therefore \qquad \left(\frac{\partial T}{\partial V}\right)_U = -\frac{1}{C_V}\cdot\left(\frac{\partial U}{\partial V}\right)_T$$

$$= -\frac{1}{C_V}\frac{a}{V^2} \qquad [\because \text{Eq. } (vii)]$$

Therefore, the Joule's coefficient defined by *eq.* (*i*),

$$\eta = \left(\frac{\partial T}{\partial V}\right)_U = -\frac{1}{C_V}\frac{a}{V^2} \qquad \text{...(2.14)}$$

**Discussion.** Since the Van der Waals' constant $a$ is a positive quantity, the Joule's coefficient ($\eta$) from equation (2.14) for a real gas is a negative quantity. Thus, *when a real gas is allowed to suffer Joule's free expansion i.e. a throttle, a cooling effect is produced which is in quite agreement with the experimental results.*

## 2.19 Differentiation Between Gas and Vapour

The dotted curve (blue) shown in Fig. 2.4, joining the horizontal part of the isothermals is called the **border curve.** The border curve clearly indicates the different states of substance. To its right, the substance is in gaseous state, within it a mixture of saturated vapour and liquid and to its left, it is entirely in the liquid state.

At critical temperature, the volume and density of the saturated vapour becomes equal to the volume and density of the liquid. It is not possible to distinguish or differentiate the vapour (or gas) from the liquid.

Below the critical temperature, the gas is termed as vapour and above the critical temperature it is called a gas.

## 2.20 Joule-Thomson Effect

If a gas initially at a constant high pressure is allowed to suffer throttle expansion through the porous plug of silk wool or cotton wool having a number of fine pores, to a region of constant lower pressure adiabatically, a change of temperature of the gas (either cooling or heating) is observed. This effect is called Joule-Thomson or Joule-Kelvin effect.

Joule in collaboration with William Thomson [Lord Kelvin] devised a very sensitive technique known as **porous plug experiment** and performed number of experiments from 1852 to 1862 and established beyond doubt the existence of intermolecular attraction. In this effect, the total heat function $H = U + PV$ remains constant.

At ordinary temperatures, all gases except hydrogen show cooling effect on passing through the porous plug, but hydrogen shows a heating effect. At sufficiently low temperatures all gases show a cooling effect.

## 2.21 Joule-Thomson Porous Plug Experiment

The experimental set up for the porous plug experiment to study the Joule-Thomson effect is shown in Fig. 2.15. It consists of the following main parts:

**1.** The apparatus consists of a porous plug having two perforated brass discs *D, D*.

**2.** The space between *D, D* is packed with cotton wool or silk fibers.

**3.** The porous plug is fitted in a cylindrical box-wood *W* which is surrounded by a vessel containing cotton wool (Fig. 2.15). This is done to avoid loss or gain of heat from the surroundings.

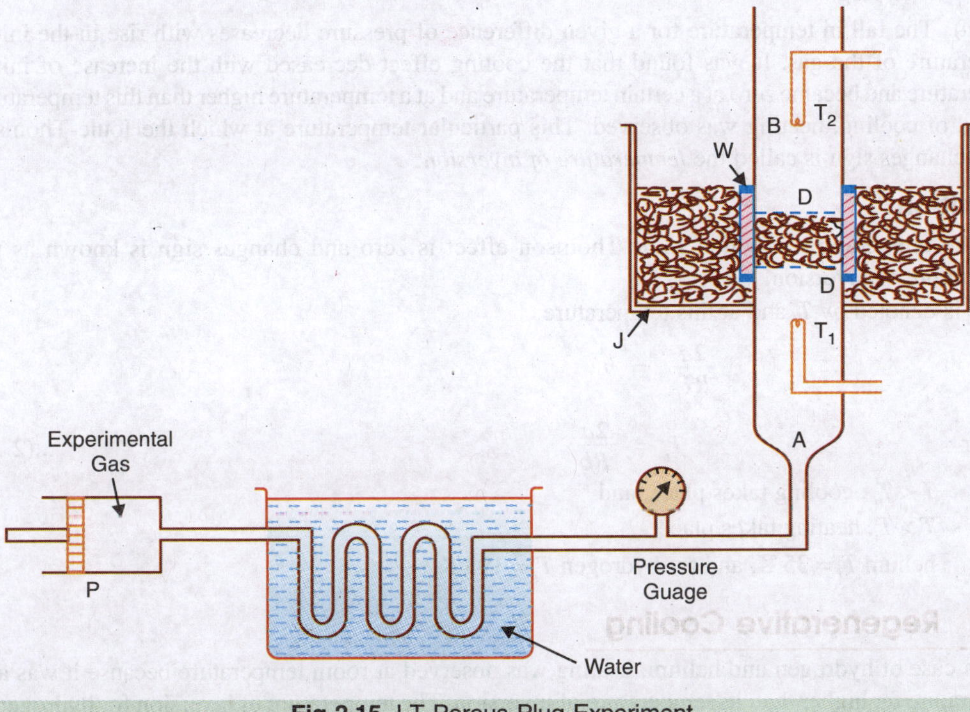


**Fig 2.15** J-T Porous Plug Experiment

**4.** $T_1$ and $T_2$ are two sensitive platinum resistance thermometers and they measure the temperatures of the incoming and outgoing gas.

**5.** The gas is compressed to a high pressure with the help of piston *P* and it is passed through a spiral tube immersed in water bath maintained at a constant temperature. If there is any heating of the gas due compression, this heat is absorbed by the circulating water in the water bath.

### Experimental procedure

By means of the compression pump *P*, the experimental gas is passed slowly and uniformly through the porous plug keeping the high pressure constant, read by pressure gauge. During its passage through the porous plug, the gas is throttled, *i.e.* the separation between the molecules increases. On passing through the porous plug, the volume of the gas increases against the atmospheric pressure. As there is no loss or gain of heat during the whole process, the expansion of the gas takes place adiabatically.

The initial and final temperatures are noted by means of the platinum resistance thermometers $T_1$ and $T_2$.

## Experimental Results

### Inversion Temperature

The behaviour of a large number of gases was studied at various inlet temperatures of the gas and the results obtained are as follows:

(*i*) At sufficiently low temperatures, all gases show a cooling effect.

(*ii*) At ordinary temperatures, all gases except hydrogen and helium show cooling effect. Hydrogen and helium show heating instead of cooling at room temperature.

(*iii*) The fall in temperature is directly proportional to the difference in pressure on the two sides of the porous plug.

(*iv*) The fall in temperature for a given difference of pressure decreases with rise in the initial temperature of the gas. It was found that the cooling effect decreased with the increase of initial temperature and became zero at a certain temperature and at a temperature higher than this temperature, instead of cooling, heating was observed. This particular temperature at which the Joule-Thomson effect changes sign is called the *temperature of inversion.*

### Definition

The temperature at which Joule-Thomson effect is zero and changes sign is known as the temperature of inversion.

It is denoted by $T_i$ and at this temperature

$$\frac{2a}{RT_i} = b$$

or

$$T_i = \frac{2a}{Rb} \qquad \text{...(2.15)}$$

when $T < T_i$, cooling takes place, and

$T > T_i$, heating takes place.

for helium $T_i = 35$ K, and for hydrogen $T_i = 193$ K.

## 2.22 Regenerative Cooling

In case of hydrogen and helium, heating was observed at room temperature because it was at a temperature far higher than its temperature of inversion. The temperature of inversion for hydrogen is – 80°C and for helium is –258°C. If helium is passed through the porous plug at a temperature lower than –258°C, it will also show cooling effect. It means any gas below its temperature of invession shows a cooling effect when it is passed through the porous plug or a throttle valve. This is called regenerative cooling or Joule-Kelvin cooling. This principle is used in the liquefaction of the so called permanent gases like nitrogen, oxygen, hydrogen and helium.

## 2.23 Estimates of *J-T* Cooling (Theory of Porous Plug Experiment)

### Theory

The simple arrangement of the Joule-Thomson porous plug experiment is shown in Fig. 2.16. The gas is allowed to pass through the porous plug from the high pressure side to the low pressure side. Due to this large difference of pressure, the gas flows through the fine pores and becomes throttled or wire drawn, *i.e.* the molecules of the gas are further drawn apart from one another.

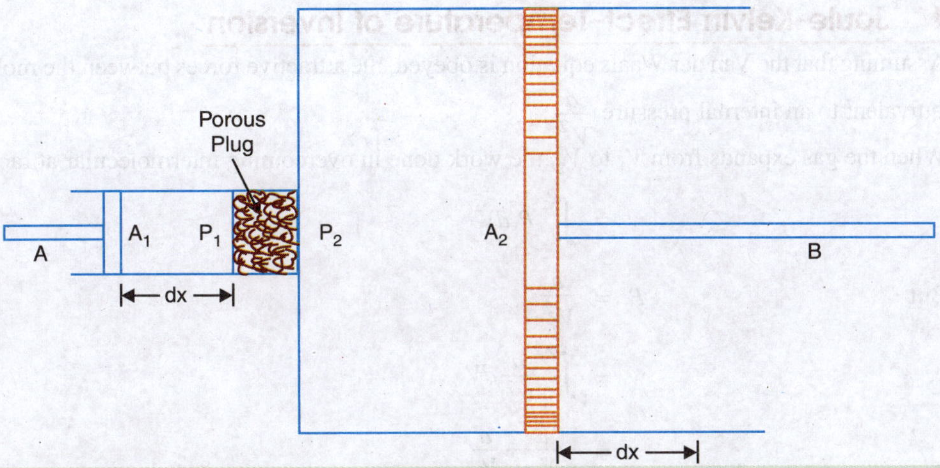


Fig 2.16

Consider one gram molecule of a gas to the left and to right of the porous plug. Let $P_1$, $V_1$ and $P_2$, $V_2$ represent the pressure and volume on the two sides of the porous plug. When the piston A is moved through a certain distance *dx*, the piston *B* also moves through the same distance *dx*. The work done on the gas by the piston $A = P_1A_1dx = P_1V_1$. The work done by the gas on the piston $B = P_2A_2dx = P_2V_2$. Thus, the net external work done by the gas is $P_2V_2 - P_1V_1$.

If *w* is the work done by the gas in separating the molecules against their intermolecular attractions, the total amount of work done by the gas is

$$(P_2V_2 - P_1V_1) + w$$

No heat is gained or lost to the surroundings. There are three possible cases :

(*i*) Below the Boyle temperature,

$$P_1V_1 < P_2V_2$$

and $P_2V_2 - P_1V_1$ is + *ve*. *w* must be either positive or zero. Thus a net +*ve* work is done *by the gas* and there must be cooling when the gas passes though the porous plug.

(*ii*) At the Boyle temperature if $P_1$ is not very high

$$P_1V_1 = P_2V_2$$

and

$$P_2V_2 - P_1V_1 = 0$$

The total work done by the gas in this case is *w*. Therefore colling effect at this temperature is only due to the work done by the gas in overcoming intermolecular attractions.

(*iii*) Above the Boyle temperature

$$P_1V_1 > P_2V_2$$

or

$$P_2V_2 - P_1V_1 \text{ is } -ve$$

Thus, the observed effect will depend upon whether $(P_2V_2 - P_1V_1)$ is greater or less than *w*.

If $w > (P_1V_1 - P_2V_2)$, cooling will be observed.

If $w < (P_1V_1 - P_2V_2)$, heating will be observed.

Thus, the cooling or heating of a gas due to free expansion through a porous plug from a high pressure to a low pressure side will depend on (*i*) the deviation from Boyle's law and (*ii*) work done in overcoming intermolecular attractions.

## 2.24 Joule-Kelvin Effect–Temperature of Inversion

Assuming that the Van der Waals equation is obeyed, the attractive forces between the molecules are equivalent to an internal pressure $\frac{a}{V^2}$.

When the gas expands from $V_1$ to $V_2$, the work done in overcoming intermolecular attractions

$$w = \int_{V_1}^{V_2} P.dV$$

But $$P = \frac{a}{V^2}$$

$\therefore$ $$w = \int_{V_1}^{V_2} \frac{a}{V^2} dV$$

$$= -\frac{a}{V_2} + \frac{a}{V_1}$$

If $V_1$ and $V_2$ represent the gram molecular volumes on the high and the low pressure sides respectively, the external work done by the gas is

$$(P_2V_2 - P_1V_1)$$

Hence the total work done by the gas

$$W = (P_2V_2 - P_1V_1) + w$$

$$= (P_2V_2 - P_1V_1) - \frac{a}{V_2} + \frac{a}{V_1}$$

Van der Waals equation of state for a gas is

$$\left(P + \frac{a}{V^2}\right)(V - b) = RT$$

or $$PV + \frac{a}{V} - bP - \frac{ab}{V^2} = RT$$

or $$PV = RT + bP - \frac{a}{V}$$

$$\left[\frac{ab}{V^2} \text{ is negligible}\right]$$

$$W = \left[RT + bP_2 - \frac{a}{V_2}\right] - \left[RT + bP_1 - \frac{a}{V_1}\right] - \frac{a}{V_2} + \frac{a}{V_1}$$

$$= b[P_2 - P_1] + 2a\left[\frac{1}{V_1} - \frac{1}{V_2}\right]$$

But $$V_1 = \frac{RT}{P_1} \text{ and } V_2 = \frac{RT}{P_2}$$

$\therefore$ $$W = b[P_2 - P_1] + 2a\left[\frac{P_1}{RT} - \frac{P_2}{RT}\right]$$

or $$W = -b[P_1 - P_2] + \frac{2a}{RT}[P_1 - P_2]$$

or $$W = (P_1 - P_2)\left(\frac{2a}{RT} - b\right) \quad ...(i)$$

Suppose the fall in temperature is $\delta T$

$$W = JH$$
$$= J\,[MC_P\,\delta T]$$

where $M$ is the gram-molecular weight of the gas

$$\therefore \qquad JMC_P\,\delta T = (P_1 - P_2)\left(\frac{2a}{RT} - b\right)$$

or

$$\delta T = \left[\frac{P_1 - P_2}{JMC_P}\right] \times \left(\frac{2a}{RT} - b\right) \qquad ...(ii)$$

(*i*) Since $P_1 - P_2$ is + ve

$\delta T$ will be + ve if $\left(\frac{2a}{RT} - b\right)$ is + ve

*i.e.*,

$$\frac{2a}{RT} > b \quad \text{or} \quad T < \frac{2a}{Rb} \qquad ...(iii)$$

Therefore, cooling will take place if the temperature of the gas is less than $\frac{2a}{Rb}$.

(*ii*) For $\delta T$ to be zero, from equation (*ii*),

$$\frac{2a}{RT} - b = 0$$

or

$$T = \frac{2a}{Rb}$$

This temperature is called the temperature of inversion and is represented by $T_i$

$$T_i = \frac{2a}{Rb} \qquad ...(iv)$$

(*iii*) $\delta T$ will be negative, if

$$\left(\frac{2a}{RT} - b\right) \text{ is } - \text{ve}$$

*i.e.*,

$$b > \frac{2a}{RT}$$

or

$$T > \frac{2a}{Rb}$$
$$T > T_i$$

Therefore, heating will take place if the temperature of the gas is more than the temperature of inversion.

## Results

(*i*) If the gas is at the temperature of inversion, then no cooling or heating is observed when it is passed through the porous plug.

(*ii*) If the gas is at a temperature lower than the temperature of inversion, cooling will take place when it is passed through the porous plug. This is called *regenerative cooling* or *Joule-Kelvin cooling*. This principle has been used in the liquefaction of the so-called permanent gases like nitrogen, oxygen, hydrogen and helium.

(*iii*) If the gas is at a temperature higher than the temperature of inversion, instead of cooling, heating is observed when the gas is passed through the porous plug.

## 2.25 Relation between Boyle Temperature, Temperature of Inversion and Critical Temperature

Temperature of inversion,

$$T_i = \frac{2a}{Rb} \qquad ...(2.16)$$

Boyle temperature,

$$T_B = \frac{a}{Rb} \qquad ...(2.17)$$

Critical temperature,

$$T_c = \frac{8a}{27\,Rb} \qquad ...(2.18)$$

From (2.16) and (2.17), we have

$$T_i = 2\,T_B$$

From (2.16) and (2.18), we have

$$\frac{T_i}{T_c} = \frac{2a}{Rb} \cdot \frac{27\,Rb}{8a}$$

$$= \frac{27}{4} = 6.75 \qquad ...(2.19)$$

The experimental value of $\frac{T_i}{T_c}$ for actual gases is just less than 6.

It means that the temperature of inversion is very much higher than the critical temperature. For hydrogen $T_i = 190$ K and $T_c = 33$ K. As $T_i \geqslant T_c$, the methods employing regenerative cooling Joule-Kelvin cooling) are preferred to those employing the initial cooling of the gas below the critical temperature.

Since ratio $\frac{T_i}{T_c}$ is a number (*i.e.*, 6.75) from equation 2.19, the ratio of temperature of inversion and critical temperature does not depend upon the nature of the gas.

## 2.26 Distinction between Joule Expansion, Joule-Thomson Expansion and Adiabatic Expansion

In all the three processes *i.e.*, Joule expansion, Joule-Thomson expansion and adiabatic expansion, the system under consideration is thermally isolated from external systems, *i.e.*, $dQ = 0$. In other words, the heat is neither allowed to enter nor to leave the system. Hence, according to definition, all the three processes are adiabatic. But in **Joule expansion**, the system is also isolated *mechanically* from the external system and hence the net external work done by or on the gas is zero, *i.e.*, $W = 0$.

In **Joule-Thomson expansion**, on the contrary, the system is not isolated mechanically from the external systems and both the pistons can move keeping the pressures $P_1$ and $P_2$ constant. The net external work done by the gas in passing through the porous-plug

$$W = P_2V_2 - P_1V_1$$

In **adiabatic expansion**, the system is not mechanically isolated from the external systems and also there is no any attempt to keep the pressure of the gas constant. The expanding gas performs external work against the pressure of the atmosphere, which is given by

$$W = \int P.dV$$

Here, the energy required for doing this work against the constant external pressure is drawn from the internal energy of the gas. Hence there is a decrease in internal energy, resulting in fall of temperature of the gas.

Applying the first law of thermodynamics

$dQ = dU + dW$ to all these three processes, keeping in mind that $dQ = 0$, we have

(*i*) in Joule expansion, $dU = 0$

(*ii*) in Joule-Thomson expansion, $dU = - dW = P_1V_1 - P_2V_2$

(*iii*) in adiabatic expansion, $dU = - \int P.dv$

In the first case *i.e.*, Joule expansion, the cooling effect is produced. Here the gas has to perform internal work against intermolecular forces of attraction during *free expansion*, at the expense of its internal kinetic energy. In Joule expansion, where $dU = 0$, a real gas will always show *some cooling effect.*

In the second case *i.e.*, Joule-Thomson expansion, the net external work done by the gas passing through porous, plug may produce a cooling effect or a heating effect depending upon whether $P_2V_2$ is greater or smaller than $P_1V_1$.

However, in the third case *i.e.*, adiabatic expansion, external work is done at the expense of internal energy of the gas itself. Hence, there will always be large fall of temperature. In other words, in adiabatic expansion cooling is mainly due to external work at the cost of its internal energy ($U$).

## 2.27 Nature of Van der Waals Forces

The matter is made up of atoms and molecules and has the tendency to occupy minimum space at low temperatures through transforming into liquid and solid state. From this we conclude that the atoms or molecules must exert a force of attraction on each other. As the atoms or molecules do not lose their individuality inspite of this attractive force, we may further conclude that there must be repulsive forces at short intermolecular distances. The simplest model giving rise to such forces was assumed by Van der Waal, according to which, "The molecules are hard and impenetrable surrounded by attractive field".

This model does not explain the observed compressibility of solids and does not agree with the modern concept of electronic structure of atom suggested by Bohr and Rutherford according to which, "The atom consists of a positively charged nucleus surrounded by negatively charged electrons in shells". Hence, it may be assumed that the forces between atoms and molecules are due to interaction between the electrons and the nuclei of the two systems. Hence, we are led to the conclusion that intermolecular forces are essentially *electrostatic*. However, there are also magnetic forces which are ordinarily small and may be neglected. These forces of attraction are referred to Van der Waals forces.

Except for atoms having closed shells viz. Helium (He), Argon (Ar), Neon (Ne) etc., almost all other atoms are electrically unsymmetrical giving rise to electric dipole. The molecules formed of such atoms will also behave as dipoles. The dipole moment of such atoms continuously varies in magnitude and direction. Therefore, the average dipole moment is always zero. However, in the case of molecules, this average is not necessarily zero. Hence, it posseses permanent dipole moment. Such molecules are called polar molecules. On the other hand, if the average dipole moment of a molecule vanishes, the molecule is said to be non-polar molecule.

If such non-polar molecules are placed in an electric field, we get induced dipole. Thus, the Van der Waals forces between the molecules of a matter (gas), more conveniently the potential energy of the system, may be due to :

(*i*) Dipole - dipole interaction.

(*ii*) Dipole - induced dipole interaction.

(*iii*) Dispersion interaction.

Thus, there are three types of Van der Waal's forces. These are

**1.** Orientation forces between polar molecules because of their permanent electric dipole moment.

**2.** Polarization forces between polar and non-polar molecules; these forces arise because of polarization of a non-polar molecule in the electric field of a polar molecule, and

**3.** Dispersion forces between non-polar molecules; these forces arise because of the occurrence of momentary electric dipole moments due to the rotational motion of the electrons. The mutual potential energy of two molecules, at a distance of seperation $r$, due to Van der Waals' forces can be shown to be $-\frac{A}{r^6}$, where $A$ is a constant.

If two molecules, are very close to each other so that their charge distributions overlap, the intermolecular force becomes repulsive. The mutual potential energy of two molecules due to the repulsive interaction is represented by the empirical expression $\frac{B}{r^{12}}$. Hence, the total potential energy of two molecules is expressed by

$$U = \frac{B}{r^{12}} - \frac{A}{r^6}$$

The exponential form $ce^{-r/\rho}$ where $\rho$ is a measure of the range of interaction is also used for the repulsive interaction.

If we plot the potential energies as function of the separation $r$ between two molecules, we get curves as shown in Fig. 2.17. The curves for the force

$$F = -\frac{dU}{dr}$$

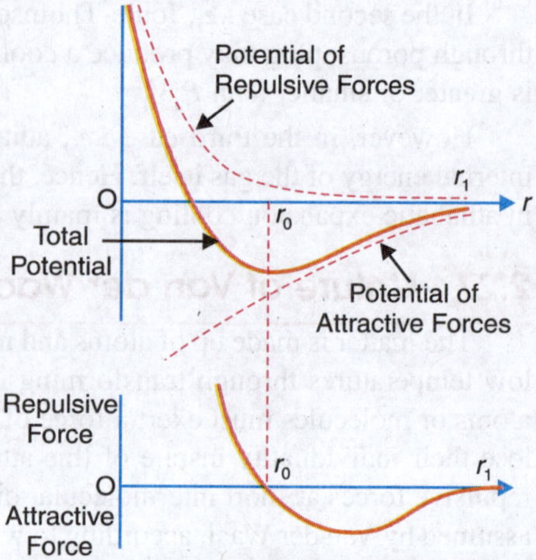

Fig 2.17

is shown in Fig. 2.17. It is seen that the attractive force between two molecules is effective within the distance of separation from $r_0$ at which the total potential energies is minimum to $r_1$ at which the force becomes negligible. In the case of simple molecules of diameter of about $2.5 \times 10^{-10}$ m, the value of the equilibrium distance $r_0$ is about $3.5 \times 10^{-10}$ m and that of $r_1$ is about $10^{-9}$ m which is about 4 times the molecular diameter.

Van der Waals' forces are much weaker than those which exist in ionic and covalent bonds. Consequently molecular crystals and crystals of inert gases have low melting and boiling points and little mechanical strength. These forces account for many familiar phenomena such as friction, surface tension, viscosity, cohesion, adhesion, liquefaction of gases, solidification liquids etc.

## 2.28 Origin of Van der Waals' Forces

### 1. In Polar Molecules

In a molecule, the protons can be considered as equivalent to a positive charge at some point in the molecule. This point is called the "centre of gravity" of the protons. Similarly the electrons are quivalent to an equal negative charge at the "centre of gravity" of the electrons. In a polar molecule, these points are separated by a short distance and therefore these molecules have electric dipole moments. All diatomic molecules, such as HCl, CO, etc. consisting of different atoms are polar.

In HCl molecule [Fig. 2.18 (*a*)], the dipole moment is due to some excess of positive charge at the hydrogen end and a corresponding excess of negative charge at the chlorine end. In some molecules, the dipole moment is due to unsymmetrical geometrical structure. For example, in $H_2O$ molecule [Fig. 2.18 (*b*)], the dipole moment is due to the unsymmetric structure; the O-H axes are inclined to each other at about 105°.

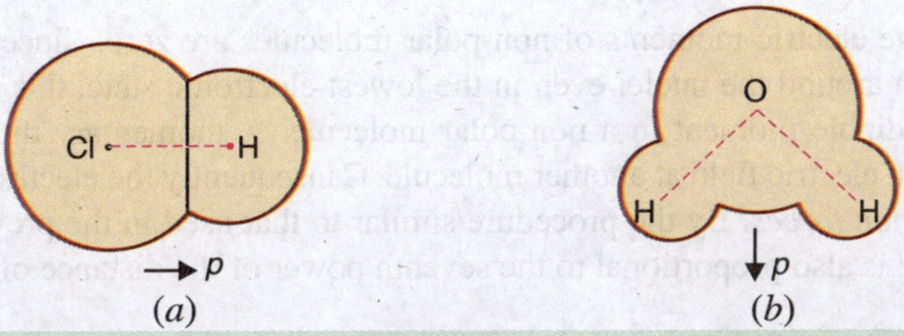


**Fig. 2.18**

In a substance consisting of polar molecules, there are orientation forces between molecular electric dipoles, which tend to bring the interacting system into a position of minimum potential energy. It can be shown that the mean value of the mutual potential energy of two molecular dipoles in the substance is given by (at high temperatures)

$$\bar{U} = -\frac{1}{4\pi\varepsilon_0} \cdot \frac{2}{3} \cdot \frac{p^4}{kT\, r^6} \quad \text{(In S.I. system)} \qquad ...(2.21)$$

where $p$ is the electric dipole moment, $\varepsilon_o$ is the permittivity of free space, $k$ is Boltzmann's constant, $T$ is the absolute temperature and $r$ is the distance of seperation.

The –ve sign indicates that the force is attractive. The force is given by

$$F = -\frac{d\bar{U}}{\partial r} = \frac{1}{4\pi\varepsilon_0} \cdot \frac{12}{3} \cdot \frac{p^4}{kT\, r^7} \qquad ...(2.22)$$

Thus, the force is proportional to $\frac{1}{r^7}$.

## 2. Interaction between a polar and a nonpolar molecules

The electric field $E$ at a distance $r$ from a molecular dipole of moment $p$ is given by

$$E = \frac{1}{4\pi\varepsilon_0} \cdot \frac{p}{r^3} \sqrt{1 + 3\cos^2\theta} \qquad ...(2.23)$$

where $\theta$ is the angle between the direction of $r$ and the axis of the dipole.

The field $E$ induces an electric dipole moment $P_1$ in the non-polar molecule. The induced dipole moment is given by

$$p_1 = \alpha E = \frac{1}{4\pi\varepsilon_0} \cdot \frac{\alpha\, p}{r^3} \sqrt{1 + 3\cos^2\theta} \qquad ...(2.24)$$

where $\alpha$ is the polarizability of the molecule.

The potential energy of the induced dipole in the field $E$ is given by

$$U = -p_1 E = -(\alpha E) E = -\alpha E^2$$

or

$$U = -\left(\frac{1}{4\pi\varepsilon_0}\right)^2 \cdot \frac{\alpha\, p^2}{r^6} (1 + 3\cos^2\theta) \qquad ...(2.25)$$

This is the expression for the mutual potential energy of the two molecules. The negative sign shows the force between them is attractive. The force is given by

$$F = -\frac{\partial\bar{U}}{\partial r} = \left(\frac{1}{4\pi\varepsilon_0}\right)^2 \cdot \frac{6\alpha\, p^2}{r^7} (1 + 3\cos^2\theta) \qquad ...(2.26)$$

Thus, the force is proportional to $\frac{1}{r^7}$.

### 3. Interaction between non-polar molecules

The time average electric moments of non-polar molecules are zero. Since the electrons are in the rotational motion around the nuclei even in the lowest electronic state, this motion gives rise to momentary electric dipole moment in a non-polar molecule. A momentary dipole moment of one molecule produces an electric field at another molecule. Cansequently the electric dipoles give rise to the so called *dispersion forces*. By the procedure similar to that used in the previous case, it can be shown that this force is also proportional to the seventh power of the distance of separation between two molecules.

## NUMERICAL EXAMPLES

**EXAMPLE 2.1**

*Calculate the Van der Waals constants for dry air, given that $T_c = 132$ K, $P_c = 37.2$ atmospheres and R per mole = 82.07 $cm^3$ atmos $K^{-1}$.* **(Nagpur 2005, 2009)**

**SOLUTION.** Given:

$$P_c = 37.2 \text{ atmospheres}$$
$$T_c = 132 \text{ K}$$
$$R = 82.07 \text{ cm}^3 \text{ atmos K}^{-1}$$

(*i*)
$$a = \frac{27}{64} \times \frac{R^2 T_c^2}{P_c}$$
$$= \left(\frac{27}{24}\right) \frac{(82.07)^2\,(132)^2}{37.2}$$
$$\mathbf{a = 13.31 \times 10^6 \text{ atmos cm}^6}$$

(*ii*)
$$b = \frac{RT_c}{8P_c}$$
$$= \frac{82.07 \times 132}{8 \times 37.2}$$
$$\mathbf{b = 36.41 \text{ cm}^3}$$

**EXAMPLE 2.2**

*Calculate critical temperature for $CO_2$, given that a = 0.00874 atmos, $cm^6$ and b = 0.0023 $cm^3$.* **(Nagpur 2007 Similar)**

**SOLUTION.** Critical temperature is given by

$$T_c = \frac{8a}{27\,Rb}$$

To evaluate the value of gram molecular gas constant $R$ at a pressure of 1 atmosphere for a gram molecular volume at O°C or 273 K from Van der Waals equation

$$\left(P + \frac{a}{V^2}\right)(V - b) = RT$$

Substituting the values,

$$\left(1 + \frac{0.0874}{1}\right)(1 - 0.0023) = R \times 273$$

$$\therefore \quad R = \frac{1.00874 \times 0.9977}{273}$$

Thus $$T_c = \frac{8a}{27\,Rb} = \frac{8 \times 0.0874 \times 273}{27 \times 1.00874 \times 0.9977 \times 0.023}$$

$$= 305.7 \text{ K}$$

$\therefore$ $$\mathbf{T_c = 32.7°C}$$

**EXAMPLE 2.3**

*The Van der Waals constants a and b for 1 gram molecule of hydrogen are a = 0.245 atms-litre$^2$-mole$^{-2}$ and b = 2.67 × 10$^{-2}$ litre-mole$^{-1}$. Calculate the critical temperature.*

**(Kerala 2001)**

**SOLUTION.** Critical temperature

$$T_c = \frac{8a}{27\,Rb}$$

Here,

$$a = 0.245 \times 76 \times 13.6 \times 980 \times 10^6 \text{ dynes-cm}^4\text{-mole}^{-2}$$
$$b = 2.67 \times 10^{-2} \times 10^3 \text{ cm}^3 \text{ mole}^{-1}$$
$$R = 8.31 \times 10^7 \text{ ergs/mole–K}$$

$\therefore$ $$T_c = \frac{8 \times 0.245 \times 76 \times 13.6 \times 980 \times 10^6}{27 \times 8.31 \times 10^7 \times 2.67 \times 10^{-2} \times 10^3}$$

$$\mathbf{T_c = 33.14 \text{ K} = -239.86°C}$$

**EXAMPLE 2.4**

*Van der Waals' constants of a gas are a = 0.245 liter$^2$ atms mole$^{-2}$, b = 2.67 × 10$^{-10}$ litre. mole$^{-1}$. The initial volume and pressure of the gas are 2 litre and 4 atmosphere respectively. The gas is made to suffer a free expansion to tripple its volume. Find the change in temperature of the gas. Given $C_v$ = 5 cals. deg$^{-1}$ mole$^{-1}$.* **(Nagpur 2009)**

**SOLUTION.** The Joule's coefficient for a real gas is given by

$$\eta = \left(\frac{\partial T}{\partial V}\right)_U = -\frac{1}{C_V} \cdot \frac{a}{V^2}$$

or $$dT = -\frac{1}{C_V} \cdot \frac{a}{V^2} dV$$

Here,

$$a = 0.245 \text{ litre}^2 \text{ . atm . mole}^{-2}$$
$$= 0.245 \times (10^3)^2 \times 10^{-12} \times 10^5 \text{ N m}^4 \text{ mole}^{-2}$$
$$= 0.245 \times 10^{-1} \text{ Nm}^4 \text{ mole}^{-2}$$
$$= 0.0245 \text{ Nm}^4 \text{ mole}^{-2}$$
$$C_V = 5 \text{ cals . deg}^{-1} \text{ . mole}^{-1}$$
$$V = 2 \text{ litre} = 2 \times 10^3 \text{ cm}^3 = 2 \times 10^{-3} \text{ m}^3$$
$$dV = 2 \times 3 - 2 = 4 \text{ litre} = 4 \times 10^{-3} \text{ cm}^3 = 4 \times 10^{-3} \text{ m}^3$$

Substituting, the change in temperature will be

$$dT = -\frac{1}{5 \times 4.2} \times \frac{0.0245}{(2 \times 10^{-3})^2} \times 4 \times 10^{-3}$$

$$= -\frac{0.0245 \times 4 \times 10^3}{5 \times 4.2 \times 4}$$

$$= \mathbf{-1.17 \text{ K} = -1.17°C}$$

Thus, the temperature will fall by 1.17°C.

## EXAMPLE 2.5

*Van der Waal's constants for a gas are $a = 6.9 \times 10^{-2}\ J\,m^3\ mole^{-2}$ and $b = 2.9 \times 10^{-5}\ m^3\ mole^{-1}$. The universal gas constant $R = 8.31\ J\ mole^{-1}\ K^{-1}$. Calculate the critical temperature of the gas.*

**(Nagpur 2004, Kerala 2001)**

**SOLUTION.** Critical temperature

$$T_c = \frac{8a}{27Rb}$$

$$= \frac{8\times 6.9\times 10^{-2}\ J\ \text{m}^3\ \text{mole}^{-2}}{27\times 8.31\ J\ \text{mole}^{-1}\ K^{-1}\times 2.9\times 10^{-5}\ \text{m}^3\ \text{mole}^{-1}}$$

$$= \mathbf{84.83\ K}$$

## EXAMPLE 2.6

*Calculate the values of Van der Waal constants a and b in Van der Waals' equation for He, when critical pressure is $0.23 \times 10^6\ N/m^2$ and critical volume is $58 \times 10^{-8}\ m^3/mol$, $P_c = 37.2$ atm, R per mole = $82.07\ cm^3$ atoms $K^{-1}$.*

**(Nagpur 2006)**

**SOLUTION.** Given :

$$P_c = 0.23\times 10^6\ \text{N/m}^2$$

$$V_c = 58\times 10^{-8}\ \text{m}^3/\text{mol}$$

Critical volume

$$V_c = 3b$$

$$b = \frac{V_c}{3} = \frac{58\times 10^{-8}}{3} = 19.33\times 10^{-8}\text{m}^3/\text{mole} \quad ...(i)$$

and

$$\frac{a}{b^2} = 27P_c = 27\times 0.23\times 10^6$$

∴

$$a = b^2\times 27\times 0.23\times 10^6$$

$$= (19.33\times 10^{-8})^2\times 27\times 0.23\times 10^6$$

or,

$$a = 2.32\times 10^{-7}\ \text{Jm}^3\ \text{mole}^{-2} \quad ...(ii)$$

## EXERCISE—II

**2.1** In what way a real gas differs from an ideal gas. **(Meerut 2000, G.N.D.U. 2000)**

**2.2** Describe Andrew's experiments on carbon dioxide. Discuss the results obtained. Show that the liquid and the gaseous states are two distant stages of a continuous phenomenon. **(Kerala 2001, Meerut 2000, Bangalore 2000)**

**2.3** Differentiate between gas and vapour using Andrew's experimental isothermals for $CO_2$.

**2.4** What are critical constants of a gas? **(Alagappa 2004)**

**Define :** (*i*) Critical temperature $T_c$

(*ii*) Critical pressure $P_c$

(*iii*) Critical volume $V_c$

**2.5** Define Boyle temperature and inversion temperature. Explain the behavior of real gases above and below the Boyle temperature. **(Nagpur 2009, 2007, Calcutta 2004)**

**2.6** Is the Boyle temperature different for different gases? Explain why.

**2.7** Explain the reasons for modification of perfect gas equation. **(Nagpur 2005, 2004)**

**2.8** Derive and discuss the Van der Waals equation of state of a gas. Compare the results with Andrew's experimental *PV* curves. **(Purvanchal 2006, Patna 2005, Nagpur 2005, 2004)**

**2.9** What do you understand by Virial coefficients? What is the value of first Virial coefficient? How does the second Virial coefficient vary with temperature ? **(Calcutta 2001)**

**2.10** Express the Van der Waals' equation in the following Virial form:

$$PV = R+\left[1+\frac{B}{V}+\frac{C}{V^3}+.....\right]$$

where *B*, *C*, ... are Virial coefficients. From above equation find out the expression of the second virial coefficient *B* and the Boyle temperature. What is the significance of Boyle temperature?
**(Calcutta 2004)**

**2.11** Distinguish between a perfect gas and a real gas. Derive Van der Waals' equation of state and use it to obtain the expressions for the critical constants in terms of the constants of the Van der Waals' equation.
**(Calcutta 2005, Ranchi 2004, M.D.U. 2003, 2001, 2000, Kerala 2001, K.U. 2001, Utkal 2003, Nagpur 2003, 2001)**

**2.12** What are the critical constants of a gas? Calculate the values of these constants in terms of the constants of the Van der Waals equation. **(Nagpur 2005, Alagappa 2004, Calcutta 2003)**

**2.13** Define critical coefficient of a gas. Is it same for all gases? Does experimental values agree with the theoretical value? **(Nagpur 2005)**

**2.14** Mention limitations of Van der Waals equation. **(Nagpur 2004, Garhwal 2001)**

**2.15** Derive Van der Waals reduced equation of state of a gas. **(Kerala 2001, Garhwal 2000)**

**2.16** What is Joule-Thomson effect? How is it experimentally established? How will you interpret the effect? **(Nagpur 2009, 2008, 2007, 2006, 2004, Bangalore 2005, Mysore 2005, Patna 2005, Madras 2004, Agra 2008, 2006)**

**2.17** Describe the porous plug experiment. What conclusions have been drawn from it? What is inversion temperature? **(Bangalore 2004, 2005, Alagappa 2004, Ranchi 2004, Madras 2005)**

**2.18** What is temperature of inversion ? Derive the expression for inversion temperature for Van der Waals gas, $T_i = \frac{2a}{Rb}$. **(Purvanchal 2006, 2003)**

**2.19** Define Jule-Thomson coefficient for the real gas.
**(Garhwal 2005, Nagpur, 2007, 2006, 2004, Purvanchal 2006)**

**2.20** Give the theory of Joule-Thomson effect. Hence explain why hydrogen and helium show heating effect at ordinary temperatures. **(Nagpur 2004, Garhwal 2000, 2002, 2005, Bangalore 2002)**

**2.21** Give the theory of the porous plug experiment. Obtain conditions for heating and cooling of a gas.
**(Gujrat 2005, Alagappa 2005, Bangalore 2002, 2004, Nagpur 2009, 2007)**

**2.22** Show that in Porous-plug experiment the total heat function (enthalpy) remains constant.
**(Nagpur 2009, 2007)**

**2.23** What are the critical constants of a gas? State and explain Van der Waals equation. Calculate the Van der Waals constants *a* and *b* in terms of $V_c$, $P_c$ and $T_c$. **(Nagpur 2007)**

**2.24** What is Joule-Thomson effect? Distinguish between adiabatic expansion and Joule-Thomson effect.
**(Nagpur, 2008, 2009, Kuvempa Uni. 2005)**

**2.25** How is the ideal gas equation modified when mutual attraction between molecules and the finite size of the molecules are taken into consideration? Obtain relations between critical constants and the constants appearing in the Van der Waals equation.

**2.26** Derive the reduced equation of state for a gas starting from Van der Waals equation of state. Show that if the two gases have the same reduced pressure and volume, they also have the same reduced temperature?
**(Kerala 2001, Garhwal 2000)**

**2.27** Starting from elementary ideas, set up Van der Waals equation of state and show how the constants of the equation can be expressed in terms of the critical constants. What is the experimental bearing of this equation? **(Meerut 2000)**

**2.28** Explain the corrections introduced by Van der Waals in the gas equation. Show that for a gas obeying Van der Waals equation

$$\frac{RT_c}{P_cV_c} = \frac{8}{3}$$

**(Utkal 2003, K.U. 2001, Nagpur 2003, 2004, Meerut 2002, 2000, Kerala 2001, Calcutta 2003)**

**2.29** Draw Andrew's curves for $CO_2$ at different temperatures and obtain expression for critical constants from them. Show that

$$\frac{T_i}{T_c} = \frac{27}{4}$$

**2.30** Derive and discuss the Van der Waals' equation of state of a gas. Mention its defects. **(Calcutta 2003)**

**2.31** Explain the principle of regenerative cooling. **(Nagpur 2008, Garhwal 2005, Bangalore 2004, Kuvempa 2005)**

**2.32** Obtain equation of Joule-Kelvin coefficient $\eta$ by discussing Joule-Thomson porous plug experiment in detail. **(Calcutta 2004, Nagpur 2004, Gujrath 2005, 2004, Mumbai 2005)**

**2.33** Deduce Van dar Waals equation of state. How far does it conform to Andrews' experimental results on carbon dioxide? What is the importance of Andrews' experiments in the problem of liquefaction of gases? **(Punjab 2003)**

**2.34** Define Boyle point, temperature of inversion and critical temperature of a gas. How Boyle temperature is related to critical temperature. **(Nagpur 2009, 2005, winter)**

**2.35** Prove that the ratio of temperature of inversion and critical temperature does not depend upon the nature of the gas. **(Ranchi 2000)**

**2.36** Distinguish between ideal and real gases. **(Bangalore 2004, Mysore 2004)**

**2.37** Describe briefly the porous plug experiment. Deduce a theoretical expression for the Joule-Thomson cooling for 1 gram molecule of a gas obeying Van der Waals equation of state. Why does hydrogen show a negative Joule-Thomson effect? **(Mysore 2004, Madias 2004)**

**2.38** Describe experiment to find the value of the critical constants of a gas.

**2.39** State limitations of the ideal gas equation. **(Nagpur 2005)**

**2.40** Explain Joule's law of expansion. Discuss the results of Joule's experiment after describing the experimental details of the experiment. How this expansion differs from adiabatic and Joule-Thomson expansion? **(Purvanchal 2007, Rohilkhand 2000)**

**2.41** Derive an expression for Joule's coefficient for real gas. **(Nagpur 2009, 2004)**

**2.42** Write short notes on:

(*i*) Van der Waals' gas equation **(Kanpur 1995, Rohilkhand 2000)**
(*ii*) Behavior of gases at high pressure.
(*iii*) J-T cooling **(Nagpur 2005)**
(*iv*) Critical point.
(*v*) Temperature of inversion. **(Nagpur 2004, 2005, Mumbai 2005, Mysore 2005)**
(*vi*) Reduced equation of state for a gas.
(*vii*) Porous plug experiment. **(Rohilkhand 2001)**
(*viii*) Boyle's temperature **(Nagpur 2004, 2005)**
(ix) Adiabatic cooling and Joule-Kelvin cooling. **(Bangalore 2002)**

**2.43** The Van der Waals constants for carbon dioxide are $a = 1.32 \times 10^4$ N m$^4$ mole$^{-2}$ and $b = 3.64 \times 10^{-5}$ m$^3$ mole$^{-1}$. Calculate critical pressure and temperature. **(Kuvempa Uni. 2005)**

**2.44** Calculate the fall in temperature produced by adiabatic throttling process of a gm. mole of oxygen when a pressure is reduced by 50 atmospheres. The initial temperature of gas is 27°C. Given for oxygen $C_p$ = 7.03 cal/mole $K$, $a$ = 1.32 litre$^2$ × atms mole$^{-2}$, $b = 3.12 \times 10^{-2}$ litre mole$^{-1}$ and $R$ = 8.3 Joule/K mole.

**[Ans.** 12.7°C] **(Nagpur 2008, Panjab 2005)**

**2.45** Calculate the change in temperature when helium gas suffers Joule-Thomson expansion at – 173°C, the pressure difference on the two sides of the porous plug being 20 atmospheres. Does the gas show a heating effect or a cooling effect in this expansion at – 173°C. Given $R$ = 8.3 Joule/K mole and for helium Van der Waal's constant $a$ = 0.0341 litre$^2$ atms mole$^{-2}$ and $b$ = 0.0237 litre mole$^{-1}$.

**[Ans.** *dT = – 1.5 K, Since Ti = – 240°C, heating effect*]

## OBJECTIVE QUESTIONS

**2.1** The Van der Waals' equation of states for a real gas is

(*a*) $\left(P+\frac{a}{V^2}\right)(V-b)=RT$ (*b*) $PV=RT$

(*c*) $\left(P-\frac{a}{V^2}\right)(V-b)=RT$ (*d*) $\left(P-\frac{a}{V^2}\right)(V+b)=RT$ **(Kanpur 2001)**

**2.2** Choose the wrong statement for Joule expansion

(*a*) Joule expansion is a free expansion. (*b*) In Joule-expansion, gas expands into vacuum.

(*c*) No external work is done in Joule expansion. (*d*) No internal work is done in Joule expansion.

**2.3** Choose the wrong statement from the following

(*a*) Joule expansion always produces a cooling effect

(*b*) Joule-Thomson effect is always a cooling effect

(*c*) In Joule expansion, only internal work is done

(*d*) In Joule-Thomson expansion, external work is done on the gas as well as by the gas.

**2.4** In general gas equation $PV=RT$, Van der Waal introduced a correction factor $\frac{a}{V^2}$ in pressure. This term represents

(*a*) Effective area of molecules (*b*) Mean velocity of gas molecules

(*c*) Volume occupied by molecules (*d*) Attractive force between molecules

**2.5** The dimensions of the constant $b$ in Van der Waals' gas equation are that of

(*a*) Pressure (*b*) Volume

(*c*) Pressure/Volume (*d*) Volume/Pressure

**2.6** The value of critical volume $V_c$ according to Van der Waals' gas equation is

(*a*) $V_c=b$ (*b*) $V_c=3b$

(*c*) $V_c=2b$ (*d*) none of these

**2.7** According to Van der Waals' gas equation, critical coefficient $\frac{RT_c}{P_c V_c}$ is equal to

(*a*) 8 (*b*) $\frac{8}{3}$

(*c*) 8.3 (*d*) 1

**2.8** Two gases have equal reduced pressure and reduced temperature. The gas having higher molecular weight

(*a*) shall have larger reduced volume

(*b*) shall have smaller reduced volume

(*c*) their reduced volume is the same

(*d*) no relationship between them, sometimes decreases

**2.9** In Joule-Thomson effect: **(Garhwal 2005)**

(*a*) Temperature decreases (*b*) Temperature increases

(*c*) some times temperature increases (*d*) None of these

**2.10** For an ideal gas Joule's law is:

(*a*) internal energy of an ideal gas does not depend on its pressure and volume and is a function of temperature only

(*b*) internal energy of an ideal gas does not depend on its temperature and volume and is a function of pressure only.

(*c*) internal energy of an ideal gas does not depend on its pressure and temperature and is a function of volume only.

(*d*) internal energy of an ideal gas depends upon pressure, volume and temperature.

## ANSWERS

| | | | |
|---|---|---|---|
| **2.1** (*a*) | **2.2** (*d*) | **2.3** (*b*) | **2.4** (*d*) |
| **2.5** (*b*) | **2.6** (*b*) | **2.7** (*b*) | **2.8** (*c*) |
| **2.9** (*a*) | **2.10** (*a*) | | |

# Transport Phenomena in Gases

# 3

### At a Glance

• Introduction • Molecular Collision • Mean Free Path (λ) • Sphere of Influence • Collision Cross-section • Expression for Mean Free Path • Variation of λ with Temperature and Pressure • Transport Phenomena • Viscosity (Coefficient η) • Effect of Temperature on η • Effect of Pressure on η • Thermal Conductivity (Coefficient K) • Relation Between η and K • Effect of Temperature on K • Effect of Pressure on K • Largest Thermal Conductivity of Hydrogen • Self Diffusion (Coefficient D) • Effect of Temperature and Pressure on D • Relation Between η and D

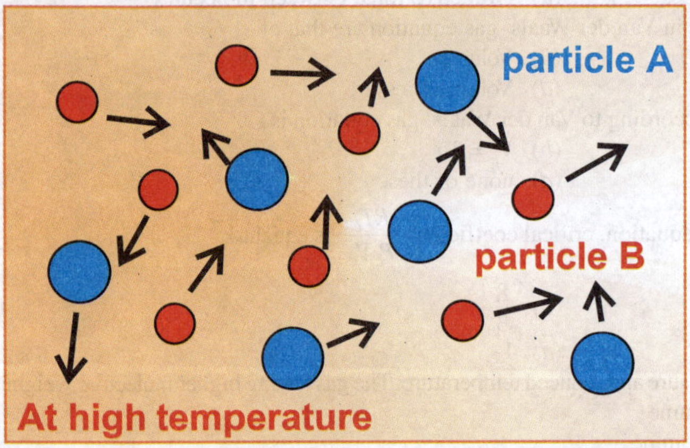


## INTRODUCTION

In the kinetic theory of gases, it is assumed that the molecules of a gas are all indentical and perfectly elastic spheres of negligible but finite size. Due to haphazard motion, the molecules constantly collide on each other. At room temperature, an air molecule suffers about $4 \times 10^9$ collisions per second and travels a mean free path about $10^{-7}$ m. When the gas is not in equilibrium, the random *i.e.*, haphazard motion of gas molecules give rise to viscosity, conductivity and diffusion, when it tries to attain an equilibrium state by transporting momentum, heat and mass respectively.

## 3.1 Molecular Collisions

According to kinetic theory of gases, the molecules are of finite size. During the random motion of gas molecules in all possible directions and with all possible velocities, they collide with each other. Between two successive collisions, a molecule moves with a constant speed, along a straight line as shown in Fig. 3.1. These collisions are known as molecular collisions.

## 3.2 Mean Free Path (λ)

At a given temperature, the molecules suffer continuous collisions with one another. Between any two collisions, the molecule travels freely a certain distance in a straight line. This distance is known as **free path.** Thus, the path covered by a gas molecule between any two consecutive collisions in a straight line is called the free path. The direction of the molecule is changed after every collision. After a number of collisions, the total path appears to be zig-zag and free path is not constant as shown in Fig. 3.1. Therefore, a term mean free path is used and is defined as the average distance travelled by a gas molecule between two successive collisions. It is denoted by λ. If the total distance travelled after $N$ collisions is $S$, then mean free path (λ) is given by

$$\lambda = \frac{S}{N}$$

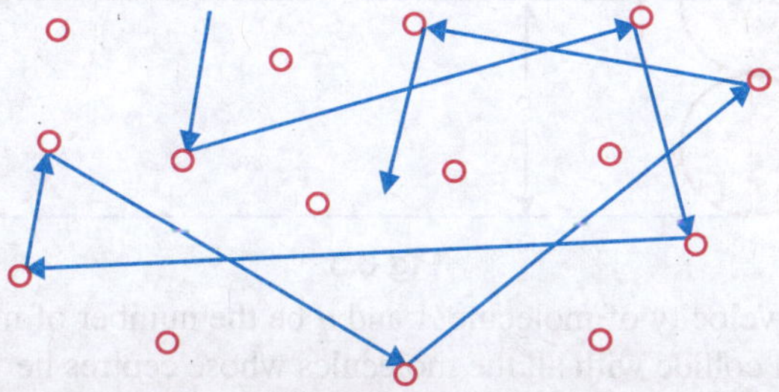

**Fig. 3.1**

**Mean free time (τ):** The average time taken by a gas molecule between two successive collisions is called mean free time (τ) and is given by

$$\tau = \frac{\lambda}{C}$$

where $C$ is the average velocity of a gas molecule.

## 3.3 Sphere of Influence

In the kinetic theory for an ideal gas, it is assumed that all the gas molecules are identical and perfectly elastic spheres moving randomly in all possible directions. In doing so, they are continuously colliding against each other. For the sake of simplicity of calculations, it is assumed that the molecule under consideration say A, is in motion, all other molecules are at rest Fig. (3.2). Let σ be the diameter of each gas molecule. Taking the centre of molecule A as a centre, if we draw a sphere of radius σ, then molecule A will collide with all those molecules whose centres lie within this sphere. Such a sphere is called 'sphere of influence'.

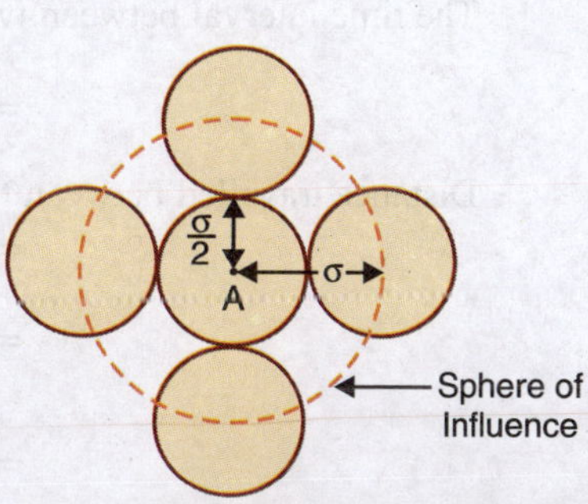


**Fig. 3.2**

## 3.4 Collision Cross-section

If we take a cross-section along the diameter of a sphere of influence, we get a collision cross-section. The area = $\pi\sigma^2$ is called the collision cross-section as shown by dotted lines in Fig. 3.2. It is the effective area which determines the probability that a molecule of diameter $\sigma$ will collide on another molecule of the same diameter.

## 3.5 Expression for Mean Free Path (Clausius' Expression)

Suppose the molecules of a gas are all identical perfectly elastic spheres, each of diameter $\sigma$. It is assumed that the molecules do not exert any force on each other. To simplify the calculations, we further assume that (1) only the molecule under consideration is in motion while all other molecules are at rest. (2) The sphere of influence of the molecule has a radius $\sigma$ i.e. equal to the diameter of the molecule Fig. 3.3.

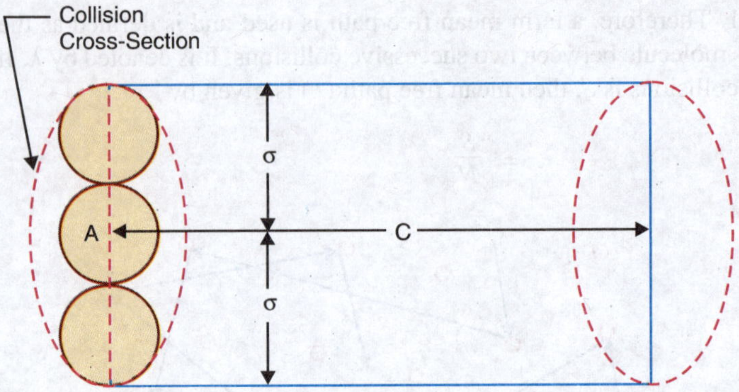


Fig 3.3

Let $c$ be the average velocity of molecule $A$ and $n$ be the number of molecules per unit volume, then in one second, $A$ will collide with all the molecules whose centres lie within a cylinder of radius $\sigma$ and length $c$.

No. of molecules in the cylinder

$$= \pi\sigma^2 cn$$

No. of collisions made by molecule A in 1 second

$$= \pi\sigma^2 cn$$

$\therefore$ 1 Collision takes place in $\dfrac{1}{\pi\sigma^2 cn}$ seconds.

The time interval between two successive collisions

$$= \frac{1}{\pi\sigma^2 cn} \text{ seconds}$$

$\therefore$ Distance travelled between two successive collisions

$$= \text{speed} \times \text{time}$$

$$= c \times \frac{1}{\pi\sigma^2 cn}$$

$$= \frac{1}{\pi\sigma^2 n}$$

$\therefore$ Mean free path, $$\lambda = \frac{1}{\pi\sigma^2 n} \qquad ...(3.1)$$

This is Clausius' expression for the mean free path.

### Maxwell's Formula

The above expression (3.1) is not exact because it is based on the assumption that only one molecule under consideration is in motion while all other are at rest. If the motion of all the molecules is taken into account and the Maxwell's law of distribution of molecular velocities is applied, which must hold in steady state, the corrected formula for mean free path is

$$\lambda = \frac{1}{\sqrt{2}\,\pi\sigma^2 n} \qquad \text{...(3.2)}$$

This is Maxwell's formula for the mean free path.

If $\lambda$ and $n$ are known, the molecular diameter $\sigma$ can be calculated. Collision frequency is $f = \frac{c}{\lambda}$ where $c$ is average velocity of the gas molecule.

## 3.6 Variation of λ with Temperature and Pressure

Consider 1 mole of an ideal gas. Then

$$PV = RT$$

Dividing both sides by N, the Avogadro's number, we have,

$$\frac{PV}{N} = \frac{RT}{N}$$

$$P = \frac{N}{V}\cdot\frac{R}{N}.T$$

∴ The pressure $P$ of a gas is given by

$$P = n\,k\,T \qquad \text{...(3.3)}$$

Where, $n = \frac{N}{V}$, the no. of molecules per unit volume and $k = \frac{R}{N}$, the Boltzmann's constant.

Equation (3.3) gives, $n = \frac{P}{kT}$

Substituting for $n$ in equation (3.2), the mean free path

$$\lambda = \frac{1}{\sqrt{2}\,\pi\sigma^2 n}$$

or

$$\lambda = \frac{kT}{\sqrt{2}\,\pi\sigma^2 P} \qquad \text{...(3.4)}$$

Thus, the mean free path varies directly as the absolute temperature ($T$) and inversely as the pressure.

## 3.7 Transport Phenomena

According to the kinetic theory of gases, the molecules of a gas are in a state of thermal agitation. The gas, therefore, attains a steady state or equilibrium state by transporting momentum, heat (thermal energy) and mass from one layer of a gas to another layer, giving rise to the viscosity, conductivity and diffusion respectively and the phenomena is called transport phenomena.

The transport phenomena occur only in the non-equilibrium state of a gas.

## 3.8 Viscosity: Transport of Momentum

The different layers of a non-equilibrium gas may have different velocities. This will result in the relative motion of the different layers with respect to one another. In such a case, the layer moving faster will impart momentum to the layer moving slower to bring about an equilibrium state. Thus, the transport of momentum gives rise to the phenomenon of viscosity.

Consider a gas flowing over a horizontal surface *OX* from left to right. The velocity of the layer in contact with the surface *OX* is zero and goes on increasing along *OZ* at a uniform rate $\frac{dv}{dz}$, known as velocity gradient (Fig 3.4).

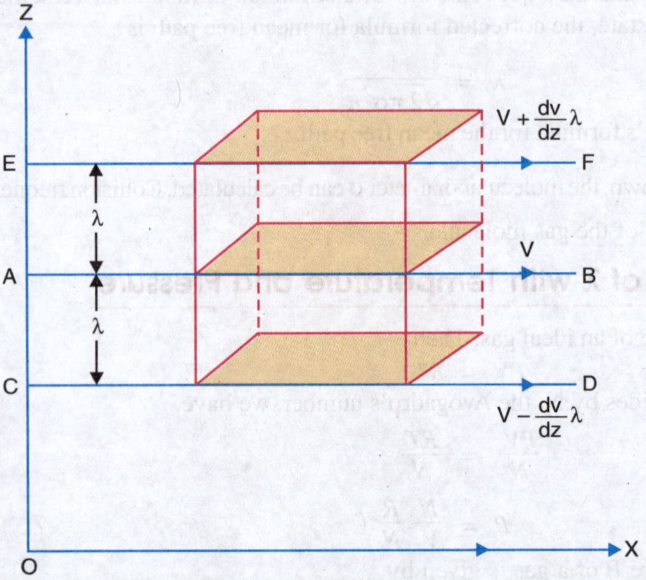


**Fig 3.4**

Let us consider a layer *AB* moving with a drift velocity *V*, situated at a distance *Z* from *O*. All the molecules in this layer have the same drift velocity. Consider two more layers *EF* and *CD* just above and below *AB* respectively at a distance λ equal to the mean free path of the molecules, so that the molecules moving vertically up or down do not suffer any collision while moving between the two layers.

The velocity of the gas in the layer $EF = V + \frac{dv}{dz}\lambda$

and the velocity of the gas in the layer $CD = V - \frac{dv}{dz}\lambda$

Due to thermal agitation, the gas molecules are moving in all directions, thus, the average number of molecules along any one ( + ve or – ve) axis will be $\frac{1}{6}$ th of the total number of molecules in the gas.

Let $n$ be number of molecules per unit volume,

$m$ be the mass of each gas molecule,

$c$ be the average velocity of a molecule at a given temperature of the gas, then

Number of molecules passing downwards from *EF* to *CD* per unit area of the *AB* in one second

$$= \frac{nc}{6}$$

∴ Forward momentum lost per unit area per second by the layer *EF*

$$= m \times \frac{nc}{6} \times \left(V + \frac{dv}{dz}\lambda\right)$$

Similarly, the number of molecules passing upwards from *CD* to *EF* per unit area of the layer *AB* in one second is also

$$= \frac{nc}{6}$$

$\therefore$ Forward momentum gained per unit area per second by the layer *EF*

$$= m \times \frac{nc}{6} \times \left(V - \frac{dv}{dz}\lambda\right)$$

$\therefore$ Net momentum lost by the layer *EF* per unit area per second

$$= \frac{mnc}{6}\left\{\left(V + \frac{dv}{dz}\lambda\right) - \left(V - \frac{dv}{dz}\lambda\right)\right\}$$

$$= \frac{1}{3} mnc\lambda \frac{dv}{dz}$$

The layer *CD* below *AB* gains the same amount of momentum. Hence, the layer *EF* above *AB* tends to accelerate its motion and the layer *CD* below *AB* tends to retard its motion.

The backward dragging force per unit area = gain or loss of momentum per unit area per second.

$$\therefore \quad F = \frac{1}{3} mnc\lambda \frac{dv}{dz} \qquad ...(3.5)$$

This must be equal to the viscous or tangential force $\left(\eta \frac{dv}{dz}\right)$ acting per unit area of the layer *AB*, $\eta$ being the coefficient of viscosity of the gas.

$$\therefore \quad \eta \frac{dv}{dz} = \frac{1}{3} mnc\lambda \frac{dv}{dz}$$

or

$$\eta = \frac{1}{3} mnc\lambda$$

or

$$\eta = \frac{1}{3} \rho c\lambda \qquad ...(3.6)$$

where $\rho = mn$, the density of the gas

## 3.9 Effect of Temperature on η

The density ($\rho$) of the gas decreases with increase in temperature but $\lambda$, the mean free path increases in the same proportion so that $\rho\lambda$ remains constant. Since the average molecular speed *c* is directly proportional to the square root of its absolute temperature $\left(c \propto \sqrt{T}\right)$, the coefficient of viscosity ($\eta$) will also be proportional to $\sqrt{T}$.

Therefore,

$$\eta \propto \sqrt{T} \qquad ...(3.7)$$

In other words, the coefficient of viscosity of a gas increases with increase in temperature.

## 3.10 Effect of Pressure on η

The density ($\rho$) of a gas increases with increase in pressure but $\lambda$, the mean free path decreases in the same proportion, so that $\rho\lambda$ remains constant. Moreover, the average molecular speed *c* is independent of pressure. Thus, using relation

$$\eta = \frac{1}{3}\rho c\lambda$$

$\eta$, the coefficient of viscosity is independent of pressure.

## 3.11 Thermal Conductivity: Transport of Thermal Energy

The different layers of a gas may have different temperatures. The molecules will carry kinetic energy (heat) from regions of higher temperature to the region of lower temperature to bring about an equilibrium state. Thus, the transport of thermal energy (thermal) gives rise to the phenomenon of thermal conductivity.

Consider a fixed mass of a gas at rest as shown in Fig. 3.5. Suppose there is a uniform temperature gradient $\frac{d\theta}{dz}$ along *OZ*. In the steady state, the temperature of each horizontal layer parallel to

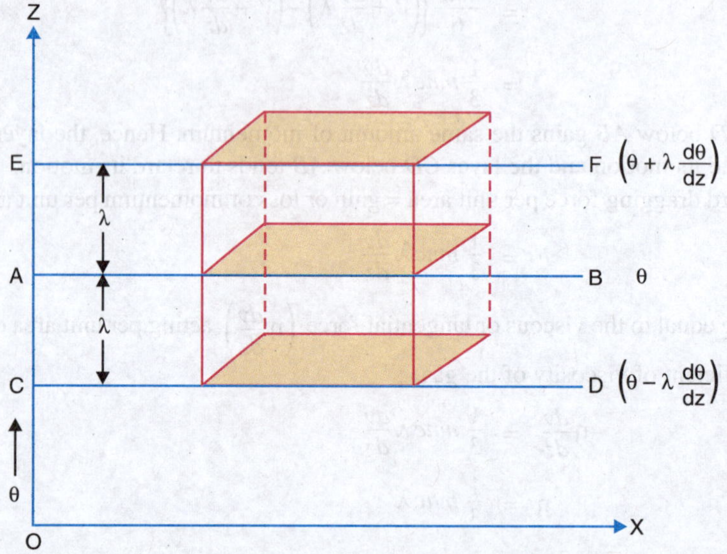


**Fig. 3.5**

*OX* is constant. Let θ be the temperature of layer *AB*. The temperature of the layer *EF* which is at a distance λ (the mean free path) above *AB* is $\left(\theta + \lambda \frac{d\theta}{dz}\right)$ and the layer *CD* at the distance λ below *AB* will be $\left(\theta - \lambda \frac{d\theta}{dz}\right)$. This means the gas above the layer *AB* is at higher temperature than the gas below *AB*. Hence, molecules of gas coming from *EF* and passing downward across *AB* possesses more K.E. than the molecules coming from *CD* and passing upward across *AB*. Thus, the average energy of molecules below *AB* is continuously increasing and that of molecules above *AB* is continuously decreasing, till the equilibrium is reached.

Number of molecules crossing unit area of the layer *AB* upward or downward per second $\frac{nc}{6}$, where *n* is the number of gas molecules per unit volume and *c* is the average molecular velocity.

If *m* is the mass of each gas molecule, then the mass of gas crossing unit area of layer *AB* upward or downward per second = $\frac{mnc}{6}$.

If $C_v$ be the specific heat of the gas at constant volume, then

Heat energy carried by the molecules in crossing unit area of the layer *AB* in the downward direction per second

$$= \text{mass} \times \text{specific heat} \times \text{temperature}$$

$$= \frac{mnc}{6} \times C_V \times \left(\theta + \lambda \frac{d\theta}{dz}\right)$$

Similarly, heat energy carried by the molecules crossing in unit area of the layer *AB* in the upward direction per second

$$= \frac{mnc}{6} \times C_V \times \left(\theta - \lambda \frac{d\theta}{dz}\right)$$

∴ Net transfer of heat energy per unit area of layer *AB* in the downward direction per second

$$Q = \frac{mnc}{6} \times C_V \times \left(\theta + \lambda \frac{d\theta}{dz}\right) - \frac{mnc}{6} \times C_V \times \left(\theta - \lambda \frac{d\theta}{dz}\right)$$

$$= \frac{mnc}{6} \times C_V \times 2\lambda \frac{d\theta}{dz} = \frac{1}{3} mnc \times C_V \lambda \frac{d\theta}{dz}$$

Since $mn = \rho$, the density of the gas,

$$Q = \frac{1}{3} \rho c \lambda \, C_V \frac{d\theta}{dz} \qquad \text{...(3.8)}$$

The coefficient of thermal conductivity (K) of the gas is defined as the quantity of heat that flows per unit area per second per unit temperature gradient.

or
$$Q = K \frac{d\theta}{dz} \qquad \text{...(3.9)}$$

Equating equation (3.8) and (3.9),

$$K \frac{d\theta}{dz} = \frac{1}{3} \rho c \lambda C_V \frac{d\theta}{dz}$$

or,
$$\text{K} = \frac{1}{3} \rho c \lambda C_V \qquad \text{...(3.10)}$$

## 3.12 Relation between η and K

The coefficient of viscosity of a gas is given by

$$\eta = \frac{1}{3} \rho c \lambda$$

and, the coefficient of thermal conductivity is

$$\text{K} = \frac{1}{3} \rho c \lambda \, C_V$$

∴
$$\text{K} = \eta \, C_V \qquad \text{...(3.11)}$$

The relation is, however, does not hold good with the experimental values since it is true only if all the heat supplied is converted into kinetic energy of translation.

According to Chapman and Enskog, the corrected value is

$$\text{K} = \varepsilon \eta \, C_V$$

where $\varepsilon = \frac{1}{4}(9\gamma - 5)$, where $\gamma$ is ratio of two specific heats of the gas.

$$\text{K} = \frac{1}{3} \varepsilon \rho c \lambda \, C_V \qquad \text{...(3.12)}$$

## 3.13 Effect of Temperature on K

The coefficient of thermal conductivity (K) is given by equation (3.10) as

$$\text{K} = \frac{1}{3} \rho c \lambda \, C_V$$

$$= \frac{1}{3} mnc.\, C_V \lambda \qquad (\because \rho = mn)$$

$$\text{K} = \frac{1}{3} mnc.\, C_V \times \frac{1}{\sqrt{2}\, \pi \sigma^2 n}$$

$$K = \frac{mcC_V}{3\sqrt{2}\pi\sigma^2} \quad \text{...(3.13)}$$

But $$C \propto \sqrt{T}$$

∴ $$K \propto \sqrt{T}$$

Thus, the coefficient of thermal conductivity of a gas is directly proportional to the square root of its absolute temperature. In other words, coefficient of thermal conductivity increases with increase in the temperature.

## 3.14 Effect of Pressure on K

In the relation, $K = \varepsilon\eta C_V$, all the quantities $\varepsilon$, $\eta$ and $C_V$ are independent of pressure. Hence the coefficient of conductivity K is also independent of pressure.

## 3.15 Largest Thermal Conductivity of Hydrogen

The coefficient of thermal conductivity of a gas is given by

$$K = \frac{1}{3}\rho c\lambda C_V$$

Substituting for $$\lambda = \frac{1}{\sqrt{2}\,\pi\sigma^2 n}$$

we have, $$K = \frac{1}{3\sqrt{2}}\frac{mcC_V}{\pi\sigma^2} \qquad (\because \rho = mn)$$

If $M$ is the molecular weight and $N$, the Avogadro's number, then $m = \frac{M}{N}$ and $C_V$ its molar specific heat, then

$$K = \frac{1}{3\sqrt{2}}\cdot\frac{c}{\pi\sigma^2}\left(\frac{M}{N}\right)\frac{C_V}{M}$$

$$= \frac{1}{3\sqrt{2}}\cdot\frac{C_V}{\pi N\sigma^2}c \quad \text{...(3.14)}$$

In this relation, N is constant, being Avogadro's number. For a diatomic gas, the value of gram molecular specific heat $C_V$ is the same for all gases being equal to $\frac{5}{2}R$. Assuming equal molecular diameters ($\sigma$) for all gases

$$K \propto c$$

But $$c = \text{average speed of the molecules}$$

$$= \sqrt{\frac{8kT}{\pi m}}$$

Where K is Boltzmann's constant.

$$K \propto \frac{1}{\sqrt{m}}$$

Thus, coefficient of thermal conductivity (K) is inversely proportional to the square root of the mass of the molecule of the gas.

As the mass of molecule of hydrogen is the least among diatomic gases, the thermal conductivity (K) of hydrogen gas will be greatest at a given temperature.

## 3.16 Self Diffusion: Transport of Mass

The phenomenon of self diffusion is due to the transport of mass from a region of higher concentration to a region of lower concentration to bring about an equilibrium.

Consider a gas having a concentration gradient $\frac{dn}{dz}$ along *OZ*. Let *n* be the concentration of layer *AB*.

Consider two more layers *EF* and *CD* above and below that of *AB* at a distance λ equal to the mean free path of the molecules of the gas, so that the molecules moving vertically upward or downward do not collide between these two layers (Fig. 3.6).

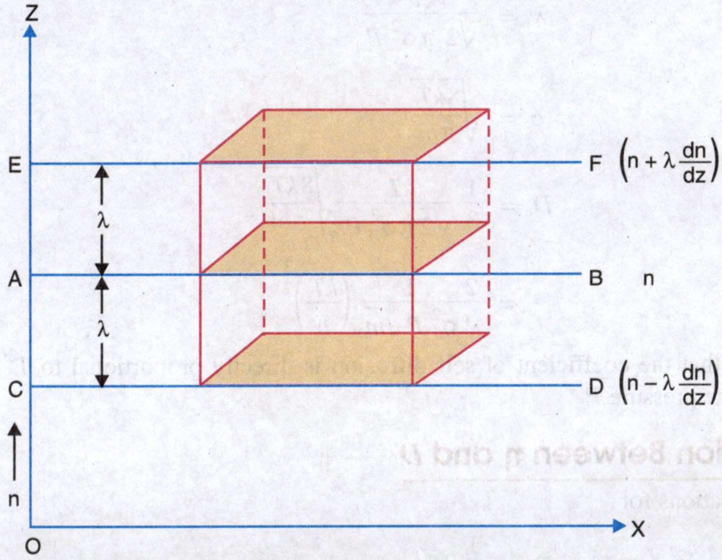


**Fig 3.6**

Concentration at the layer $EF = n + \lambda \frac{dn}{dz}$

and concentration at the layer $CD = n - \lambda \frac{dn}{dz}$

Therefore, the number of molecules coming from layer *EF* and crossing *AB* downward per unit area per second

$$= \frac{1}{6} c \left(n + \lambda \frac{dn}{dz}\right)$$

and number of molecules coming from layer *CD* and crossing *AB* upward per unit area per second

$$= \frac{1}{6} c \left(n - \lambda \frac{dn}{dz}\right)$$

∴ Net number of molecules crossing per unit area per second of layer *AB* in downward direction

$$= \frac{1}{6} c \left(n + \lambda \frac{dn}{dz}\right) - \frac{1}{6} c \left(n - \lambda \frac{dn}{dz}\right)$$

$$= \frac{1}{6} c \lambda \frac{dn}{dz}$$

The coefficient of diffusion is defined as the ratio of the number of molecules crossing per unit area in one second to the rate of change of concentration with distance.

$$\therefore \quad \text{Coefficient of diffusion } D = \frac{\frac{1}{3} c\lambda \frac{dn}{dz}}{\frac{dn}{dz}}$$

$$D = \frac{1}{3} c\lambda \qquad \text{...(3.15)}$$

## 3.17 Effect of Temperature and Pressure

We have $\lambda = \dfrac{kT}{\sqrt{2}\,\pi\sigma^2 P}$ from eq. 34.

and $c = \sqrt{\dfrac{8kT}{\pi m}}$

$$D = \frac{1}{3} \cdot \frac{kT}{\sqrt{2}\pi\sigma^2 P} \cdot \sqrt{\frac{8kT}{\pi m}}$$

$$= \frac{2}{3} \frac{1}{\sigma^2 P \sqrt{m}} \cdot \left(\frac{kT}{\pi}\right)^{\frac{3}{2}} \qquad \text{...(3.16)}$$

This shows that the coefficient of self diffusion is directly proportional to $T^{3/2}$ and inversely proportional to the pressure $P$.

## 3.18 Relation Between η and *D*

From the relations for

$$\eta = \frac{1}{3} \rho c \lambda$$

and $D = \dfrac{1}{3} c\lambda$

we have $D = \dfrac{\eta}{\rho}$ ...(3.17)

This relation between $D$ and $\eta$ does not agree well with the experimental values for a gas. A more detailed analysis yields the relation

$$D = f' \frac{\eta}{\rho}$$

where $f'$ is a numerical factor whose value is nearly 1.4.

The values of Table 3.1 the density ($\rho$), average speed (C), molecular diameter ($\sigma$), mean free path ($\lambda$), coefficient of viscosity ($\eta$), coefficient of viscosity (K) at N.T.P. and specific heat capacity at constant volume ($C_V$) for some gases are given in the following table. All the quantities are in *SI* units.

**TABLE 3.1**

| Gas | $\sigma$ m | $\rho$ kg/m$^3$ | $C$ m/s | $\lambda$ m | $\eta$ Ns/m$^2$ | K W/mk | $C_v$ J/kgK |
|---|---|---|---|---|---|---|---|
| Helium | $2.18 \times 10^{-10}$ | 0.179 | 1202.1 | $28.5 \times 10^{-4}$ | $18.6 \times 10^{-6}$ | $1415 \times 10^{-4}$ | 3157 |
| Hydrogen | $2.47 \times 10^{-10}$ | 0.090 | 1699.9 | $18.3 \times 10^{-8}$ | $8.35 \times 10^{-6}$ | $1684 \times 10^{-4}$ | 10142 |
| Nitrogen | $3.5 \times 10^{-10}$ | 1.25 | 454.4 | $9.44 \times 10^{-8}$ | $16.7 \times 10^{-6}$ | $243 \times 10^{-4}$ | 741 |
| Oxygen | $3.39 \times 10^{-10}$ | 1.429 | 424.9 | $9.95 \times 10^{-8}$ | $19.2 \times 10^{-6}$ | $244 \times 10^{-4}$ | 652 |
| Carbon dioxide | $4.18 \times 10^{-10}$ | 1.977 | 362.4 | $6.3 \times 10^{-8}$ | $14.3 \times 10^{-6}$ | $145 \times 10^{-4}$ | 640 |
| Air | – | 1.293 | – | – | $18.325 \times 10^{-6}$ | $241 \times 10^{-4}$ | 708 |

## NUMERICAL EXAMPLES

### EXAMPLE 3.1

*The viscosity of a gas at NTP was found to be $1.66 \times 10^{-4}$ dynes/cm$^2$ per unit velocity gradient. The r.m.s. velocity of the molecules is $4.5 \times 10^4$ cm/s. The density of gas is 1.25 grams per litre.* **(Nagpur 2009, 2001, Madurai 2003)**

*Calculate* (*i*) *mean free path of the molecules of the gas,*

(*ii*) *frequency of collision.*

(*iii*) *molecular diameter of the gas molecules.*

**SOLUTION.** Here

$$\eta = 1.66 \times 10^{-4} \text{ units}$$
$$c = 4.5 \times 10^{-4}$$
$$\rho = 1.25 \times 10^{-3} \text{ g/cc}$$

(*i*)
$$\lambda = \frac{3\eta}{\rho c}$$
$$= \frac{3 \times 1.66 \times 10^{-4}}{1.25 \times 10^{-3} \times 4.5 \times 10^{4}}$$
$$\boldsymbol{\lambda = 9 \times 10^{-6} \text{ cm}}$$

(*ii*) Frequency of collision = number of collisions per second

$$= \frac{\text{r.m.s. Velocity}}{\text{Mean free path}}$$
$$N = \frac{c}{\lambda}$$
$$= \frac{4.5 \times 10^{4}}{9 \times 10^{-6}}$$
$$\boldsymbol{N = 5 \times 10^{9}}$$

(*iii*) Avogadro's number = $6.023 \times 10^{23}$

$$\text{Number of molecules per } cc = n = \frac{6.023 \times 10^{23}}{22400}$$

According to Maxwell's relation

$$\lambda = \frac{1}{\sqrt{2}.\pi\sigma^2 n} = \frac{1}{1.44 \times \pi\sigma^2 n}$$

$$\sigma = \frac{1}{\sqrt{1.44 \times \pi \times n \times \lambda}}$$

$$\sigma = \frac{1}{\sqrt{1.44 \times 3.142 \times \left(\frac{6.023 \times 10^{23}}{22400}\right) \times 9 \times 10^{-6}}}$$

$$\sigma = \mathbf{3 \times 10^{-8}\ cm}$$

**EXAMPLE 3.2**

*The diameter of nitrogen molecule is $3.2 \times 10^{-10}$ m. The number of molecules at 0°C and 1 atm. pressure is $2.69 \times 10^{25}$ per $m^3$. Calculate mean free path for nitrogen molecules.* **(Nagpur 2004)**

**SOLUTION.** The mean free path $\lambda$ is given by

$$\lambda = \frac{1}{\sqrt{2}\,\pi\sigma^2 n}$$

$$= \frac{1}{\sqrt{2} \times 3.14 \times (3.2 \times 10^{-10})^2 \times 2.69 \times 10^{25}}$$

$$= \frac{1}{122.32} \times 10^{-5}$$

$$= \mathbf{0.008175 \times 10^{-5}\ m}$$

**EXAMPLE 3.3**

*Find the coefficient of viscosity of nitrogen at N.T.P. from the following data:*

$\rho = 1.25$ kg/m$^3$, $\bar{c} = 454.4$ m/s, $\lambda = 9.44 \times 10^{-8}$ m **(Nagpur 2005)**

**SOLUTION.** The mean free path is given by

$$\lambda = \frac{3\eta}{\rho\bar{c}}$$

∴ Coefficient of viscosity, $\eta = \frac{\lambda\rho\bar{c}}{3}$

$$= \frac{9.44 \times 10^{-8} \times 1.25 \times 454.4}{3}$$

$$= \mathbf{1787 \times 10^{-8}\ Nm^{-2}\ \text{per unit velocity gradient}}$$

**EXAMPLE 3.4**

*Calculate the molecular diameter of Nitrogen molecules if $n = 2.7 \times 10^{25}$ molecules per $m^3$ and mean free path $\lambda$ for nitrogen is $8 \times 10^{-8}$ m.* **(Nagpur 2007, 2004)**

**SOLUTION.** The mean free path $\lambda$ is given by

$$\lambda = \frac{1}{\sqrt{2}\,\pi\sigma^2 n}$$

$$\therefore \quad \sigma^2 = \frac{1}{\sqrt{2}\,\pi\lambda n}$$

$$= \frac{1}{\sqrt{2}\times 3.14\times 8\times 10^{-8}\times 2.7\times 10^{25}}$$

$$= \frac{1}{95.918\times 10^{17}} = \frac{1}{959.18\times 10^{16}}$$

$\therefore$ $$\sigma = \mathbf{0.03229 \times 10^{-8}\ m}$$

**EXAMPLE 3.5**

*In an experiment, the viscosity of the gas was found to be 2.25 × $10^{-4}$ CGS units. The r.m.s. velocity of the molecules is 4.5 × $10^4$ cm/s. The density of the gas is 1 gram per litre. Calculate the mean free path of the molecules.* **(Bangalore 2001)**

**SOLUTION.** Here

$$\eta = 2.25 \times 10^{-4} \text{ CGS units}$$

$$c = 4.5 \times 10^4 \text{ cm/s}$$

$$\rho = 1 \text{ g/litre}$$

$$\rho = 10^{-3} \text{ g/cc}$$

and

$$\lambda = \frac{3\eta}{\rho c}$$

$$= \frac{3\times 2.25\times 10^{-4}}{10^{-3}\times 4.5\times 10^4}$$

$$\lambda = \mathbf{15 \times 10^{-6}\ cm}$$

**EXAMPLE 3.6**

*Calculate the mean free path of a gas molecule, given that the molecular diameter is 2 × $10^{-8}$ cm and the number of molecule per cc is 3 × $10^{19}$.* **(Purvanchal 2005, Kuvempa Uni. 2005)**

**SOLUTION.**

$$\lambda = \frac{1}{\sqrt{2}\,\pi\sigma^2 n}$$

$$\lambda = \frac{1}{1.44\times 3.14\times (2\times 10^{-8})^2\times 3\times 10^{19}}$$

$$\lambda = \mathbf{2.12 \times 10^{-5}\ cm}$$

[**Note:** The mean free path is less than the wavelength of light in the visible spectrum.]

**EXAMPLE 3.7**

*Calculate the mean free path of gas molecules in a chamber of $10^{-6}$ mm of mercury pressure, assuming the molecular diameter to be 2A°. One gram molecule of the gas occupies 22.4 liters at N.T.P. Take the temperature of the chamber to be 273 K.* **(Calcutta 2002)**

**SOLUTION.** At 760 mm Hg pressure and 273 K temperature, the number of molecules in 22.4 litres of a gas is

$$= 6.023 \times 10^{22}$$

Therefore, the number of molecules per $cm^3$ in the chamber at $10^{-6}$ mm of pressure and 273 K temperature

$$n = \frac{6.023\times 10^{22}\times 10^{-6}\times 10^{-1}}{22400\times 760\times 10^{-1}}$$

$$n = 3.538 \times 10^{10} \text{ molecules/cm}^3$$

$$\sigma = 2\ \text{Å} = 2\times 10^{-8} \text{ cm}$$

$$\text{Mean free path} = \lambda = \frac{1}{\sqrt{2}\pi\sigma^2 n}$$

$$= \frac{1}{1.44 \times 3.14 \times (2 \times 10^{-8})^2 \times 3 \times 538 \times 10^{10}}$$

$$\lambda = \mathbf{1.97 \times 10^4 \text{ cm}}$$

**EXAMPLE 3.8**

*Calculate the mean free path of nitrogen molecule at 27°C temperature and one atmospheric pressure. The molecular diameter of nitrogen is 3.5 × 10$^{-8}$ cm.*

**SOLUTION.** One atmospheric pressure = 76 cm height of mercury

$$P = 76 \times 10^{-3} \times 13.6 \times 10^3 \times 9.8$$

$$P = 1.013 \times 10^5 \text{ Nm}^{-2}$$

Temperature, $T = 273 + 27 = 300$ K

Boltzmann's constant, $k = 1.38 \times 10^{-23}$ J/k

Molecular diameter, $\sigma = 3.5 \times 10^{-8}$ cm $= 3.5 \times 10^{-10}$ m

$$\text{Mean free path, } \lambda = \frac{kT}{\sqrt{2}\pi P\sigma^2}$$

$$= \frac{kT}{\sqrt{2}\pi P\sigma^2}$$

$$= \frac{1.38 \times 10^{-23} \times 300}{1.414 \times 3.14 \times 1.013 \times 10^5 \times (3.5 \times 10^{-10})^2}$$

$$\lambda = \mathbf{7.5 \times 10^{-8} \text{ m}}$$

**EXAMPLE 3.9**

*The diameter of the molecule of a gas is 2 × 10$^{-8}$ cm and Boltzmann's constant is 1.38 × 10$^{-23}$ J/K. Calculate the mean free path at NTP.* **(Garhwal 2000)**

**SOLUTION.** Here $d = 2 \times 10^{-8}$ cm $= 2 \times 10^{-10}$ m

$k = 1.38 \times 10^{-23}$ J/K

Let $n$ be the number of the molecules per cubic meter.

$$PV = RT$$

$$PV = NkT$$

$$n = \left(\frac{N}{V}\right) = \frac{P}{kT}$$

$$\text{At NTP} \quad n = \frac{0.76 \times 13.6 \times 10^3 \times 9.8}{1.38 \times 10^{-23} \times 273}$$

$$n = 2.688 \times 10^{25}$$

$$\text{Mean free path} \quad \lambda = \frac{1}{\sqrt{2}\pi d^2 n} = \frac{1}{3.14 \times (2 \times 10^{-10})^2 \times 2.688 \times 10^{25} \times 1.414}$$

$$\lambda = \mathbf{2.094 \times 10^{-7} \text{ m}}$$

**EXAMPLE 3.10**

*For oxygen at standard conditions, calculate (a) Collision frequency of the molecules, and (b) the mean free path. Given at standard conditions for oxygen,*

*(i) the number of molecules per cubic metre* $= 3 \times 10^{25}$

*(ii) Diameter of oxygen molecule* $= 3.6 \times 10^{-10}$ *m* **(Pune 2001)**

**SOLUTION.**

Here $k = 1.38 \times 10^{-23}$

and $m = \dfrac{32}{6.023\times10^{26}} = 5.31\times10^{26}$ *kg*

Average speed of oxygen molecule at 0°C (273 K)

$$v = \sqrt{\frac{8\,kT}{\pi\, m}}$$

$$v = \sqrt{\frac{8\times1.38\times10^{-23}\times273}{3.14\times5.31\times10^{-26}}}$$

$$v = \mathbf{426\ m/s}$$

Also $d = 3.6 \times 10^{-10}$ m

$n = 3 \times 10^{25}$ molecules/m$^3$

(*a*) Collision frequency (number of collisions per second)

$$= \pi d^2\, vn$$

$$= 3.14\,(3.6 \times 10^{-10})^2 \times 426 \times 3\ 10^{25}$$

$$= \mathbf{5.2 \times 10^9}$$

(*b*) Mean free path

$$\lambda = \frac{1}{\sqrt{2}\,\pi d^2\, n} = \frac{1}{3.14\times(3.6\times10^{-10})^2\times3\times10^{25}\times1.414}$$

$$\lambda = \mathbf{5.792 \times 10^{-8}\ m.}$$

**EXAMPLE 3.11**

*The molecular diameter of an ideal gas is* $2 \times 10^{-10}$ *m at a temperature of 20°C and pressure 1 atmosphere. Calculate the*

(*a*) *mean free path*

(*b*) *collision frequency*

*Velocity of molecules at*

*20°C = 511 m/s,*

*Take I atmospheric pressure*

$= 1.01 \times 10^5$ *N/m*$^2$

**SOLUTION.**

(*a*) Here $T = 293$ K

As the gas is ideal, the equation

$PV = NkT$ can be applied

$$\therefore \quad n = \frac{N}{V} = \frac{P}{kT} = \frac{1.01\times10^5}{(1.38\times10^{-23})\times293}$$

$$n = 2.5 \times 10^{25}\ \text{molecules/m}^3$$

mean free path

$$\lambda = \frac{1}{1.414 \times \pi d^2 n}$$

$$\lambda = \frac{1}{1.414 \times \pi (2 \times 10^{-10})^2 \times (2.5 \times 10^{25})}$$

$$\boldsymbol{\lambda = 2.25 \times 10^{-7}\ \text{m}}$$

(*b*) collision frequency

$$f = \frac{\bar{v}}{\lambda} = \frac{511}{2.25 \times 10^{-7}}$$

$$\boldsymbol{f = 2.27 \times 10^{9}\ \text{per second}}$$

## EXAMPLE 3.12

*Calculate the number of collisions per second of a molecule of a gas having mean free path $1.876 \times 10^{-7}$ m. Take average speed of the molecule as 511 m/s.* **(Madras 2000)**

**SOLUTION.** Mean free path $\lambda = 1.876 \times 10^{-7}$ m

$$v = 511 \text{ m/s}$$

Collision frequency,

$$f = \frac{\bar{v}}{\lambda} = \frac{511}{1.876 \times 10^{-7}} = 2.72 \times 10^{9} \text{ per second}$$

## EXAMPLE 3.13

*Calculate the radius of an oxygen molecule if its coefficient of thermal conductivity*

$$K = 24 \times 10^{-3} \text{ J/m-s-K at } 0°\text{C}$$

*and* $C_v = 20.9 \times 10^{3}$ *J/kilo-mole-K,*

*Boltzmann's constant* $k = 1.38 \times 10^{23}$ *J/K*

*and mass of an oxygen molecule*

$$m = 5.31 \times 10^{-26} \text{ kg}$$

**(Rajasthan 2001)**

**SOLUTION.**

$$K = \frac{mC\,C_v}{3(\sqrt{2})\pi\sigma^2}$$

But $\frac{1}{2}mC^2 = \frac{3}{2}kT$ or $C = \sqrt{\frac{3kT}{m}}$

$$\sigma^2 = \frac{mC\,C_v}{3 \times 1.414 \times 3.142 \times K}$$

$$\sigma^2 = \frac{(\sqrt{3kTm})\,C_v}{3 \times 1.414 \times 3.142 \times K}$$

$$\sigma^2 = \frac{[3 \times 1.38 \times 10^{-23} \times 273 \times 5.31 \times 10^{-26}]^{1/2} \times 20.9 \times 10^{3}}{3 \times 1.414 \times 3.142 \times 24 \times 10^{-3}}$$

or $\boldsymbol{\sigma = 3.036 \times 10^{11}\ \text{m}}$

## EXAMPLE 3.14

*The average speed of molecules of a gas is $4.5 \times 10^{2}$ m/s. The density of gas is $1.25$ kg/m$^3$ and the coefficient of viscosity of gas is $1.66 \times 10^{-5}$ Ns/m$^2$. Find mean free path and frequency of collision.* **(Nagpur 2005/w)**

**SOLUTION.** Given:

$$\eta = 1.66 \times 10^{-15} \text{ Ns/m}^2$$

$$\rho = 1.25 \text{ kg/m}^3$$

$$\bar{c} = 4.5 \times 10^2$$

Mean free path, $$\lambda = \frac{3\eta}{\rho\bar{c}} = \frac{3\times1.66\times10^{-5}}{1.25\times4.5\times10^2}$$

$$= \mathbf{8.85 \times 10^{-8}\ m}$$

Frequency of collision,

$$f = \frac{\bar{c}}{\lambda} = \frac{4.5\times10^2}{8.85\times10^{-8}}$$

$$= \mathbf{5.085 \times 10^9\ collisions/s}$$

## EXAMPLE 3.15

*Calculate the frequency of sound at which the wavelength of sound wave becomes equal to the mean free path in oxygen at 0°C and 1.0 atmospheric pressure. Take diameter of oxygen molecule as 3 × 10⁻⁸ cm.* **(Calcutta 2000)**

**SOLUTION.** The mean free path is given by

$$\lambda = \frac{1}{\sqrt{2}\,(\pi\sigma^2 n)}$$

At 0°C and 1 atmospheric pressure means at N.T.P.,

$$\eta = \frac{6\times10^{23}}{22400} \quad \text{and} \quad \sigma = 3 \times 10^{-8}\ \text{cm},$$

$$\therefore \quad \lambda = \frac{1}{1.414\times3.14\times(3\times10^{-16})^2\times(6\times10^{23}/22400)}$$

$$= \frac{22400}{1.414\times3.14\times9\times10^{-16}\times6\times10^{23}}$$

$$= \mathbf{9.37 \times 10^{-6}\ cm.}$$

This wavelength, as per given problem, represents the wavelength of sound waves. Therefore, the frequency, corresponding to this wavelength will be

$$\nu = \frac{V}{\lambda} = \frac{332\times100\ \text{cm/sec}}{9.37\times10^{-6}\ \text{cm}}$$

$$= \mathbf{3.54 \times 10^9\ Hertz}$$

## EXAMPLE 3.16

*Calculate the diameter of a molecule of benzene, if $\eta = 2.79 \times 10^{19}$ molecules per c.c. and mean free path $\lambda$ for benzene = $2.2 \times 10^{-6}$ cm.* **(Nagpur 2008)**

**SOLUTION.** The mean free path $\lambda$ is given by

$$\lambda = \frac{1}{\sqrt{2}(\pi\sigma^2 n)}$$

$$\therefore \quad \sigma^2 = \frac{1}{\sqrt{2}(\pi\lambda n)}$$

$$= \frac{1}{1.414\times3.14\times2.2\times10^{-6}\times2.79\times10^{19}}$$

$$= \frac{1}{2.725 \times 10^{14}} = 36.2404 \times 10^{-16}$$

$\therefore$ Molecular diameter, $\sigma$ = **6.02 × 10⁻⁸ cm.**

## EXAMPLE 3.17

*The mean free path of molecules of a certain gas at pressure p and temperature T is $2 \times 10^{-5}$ cm. Deduce the mean free path under the following conditions:*

*(i) pressure $p \times 10^{-6}$, temperature T.*

*(ii) pressure p/2, temperature 2T.*

*Also calculate the probability that under condition, pressure p and temperature T, molecules would traverse $6 \times 10^{-5}$ cm. without collision.*

**SOLUTION.** The mean free path at pressure p and temperature $T$ is given by

$$\lambda = \frac{kT}{(\sqrt{2})\pi\sigma^2 p} \quad \text{...(i)}$$

(*i*) The mean free path at pressure $p \times 10^{-6}$ and temperature $T$ will be

$$\lambda_1 = \frac{kT}{(\sqrt{2})\pi\sigma^2 p \times 10^{-6}} \quad \text{...(ii)}$$

Dividing equation (*ii*) by equation (*i*), we get

$$\frac{\lambda_1}{\lambda} = 10^6 \quad \text{or} \quad \lambda_1 = 10^6 \lambda = 10^6 \times 2 \times 10^{-5} \text{ cm}$$

$$\boldsymbol{\lambda_1 = 20 \text{ cm.}} \quad \text{...(I)}$$

(*ii*) The mean free path at pressure $p/2$ and temperature 2 $T$ will be

$$\lambda_2 = \frac{k.2T}{\sqrt{2}\pi\sigma^2 (p/2)} = \frac{4kT}{\sqrt{2}\pi\sigma^2 p} \quad \text{...(iii)}$$

Dividing eqn. (*iii*) by eqn. (*i*), we get

$$\frac{\lambda_2}{\lambda} = 4 \quad \text{or} \quad \lambda_2 = 4\lambda = 4 \times 2 \times 10^{-5}$$

$$\boldsymbol{\lambda_2 = 8 \times 10^{-5} \text{ cm.}} \quad \text{...(II)}$$

Let the probability that a molecule will collide in traversing a distance $dx$ be $\mu dx$, where $\mu$ is some constant related to $\lambda$. Then the fraction of molecules colliding with a distance $dx$ after traversing a distance $x$ will be

$$\frac{dN}{N} = -\mu\, dx \quad \text{...(iv)}$$

where $N$ is the number of molecules which traverse freely a distance $x$. Negative sign indicates the decrease of probability with increase in distance.

Integrating (*iv*), we get

$$\log_e N = -\mu x + c \qquad \text{where } C \text{ is a constant of integration.}$$

and is evaluated by boundary conditions.

At $\quad x = 0, N = N_o$

Substituting, $\quad \log_e N_o = C$

$\therefore \quad \log_e N = -\mu x + \log_e N_o$

or $$\log_e\left(\frac{N}{N_0}\right) = \mu x$$

$$\frac{N}{N_o} = e^{-\mu x} \qquad ...(v)$$

The required probability is given by

$$P = e^{-\mu x_1} \qquad ...(vi)$$

To deduce the ratio μ and λ, from equation (*v*), which gives

$$dN = N_o\, e^{-\mu x}\,(-\mu\, dx)$$

$$\therefore \quad P\,(x_1 + dx > x > x_1) = -\frac{dN}{N_0} = \mu e^{-\mu x}\,dx$$

$$\therefore \quad \lambda = \int_0^{\infty} x_1\,\mu\, e^{-\mu x}\,dx = \frac{1}{\mu}$$

or $$\mu = \frac{1}{\lambda}$$

Substituting in eqn. (*vi*), we get, the required probability as

$$P = e^{-x_1/\lambda}$$

Here $\lambda = 2 \times 10^{-5}$ cm, and $x_1 = 6 \times 10^{-5}$ cm.

Probability, $$P = e^{(-6 \times 10^{-5}/2 \times 10^{-5})}$$

$$P = e^{-3}$$

## EXERCISE–III

**3.1** Define the terms

(*i*) mean free path,

**(Nagpur 2007, 2005, Bangalore 2002, 2004, Agra 2003, Mysore 2004, Kamraj 2003, Calcutta 2005, M.S. Uni. Tamil Nadu 2007, Kuvempa 2005, Agra 2006, 2004)**

(*ii*) mean free time of a gas molecule. **(Nagpur 2007, 2006/s)**

**3.2** Explain the term sphere of influence and collision cross-section. **(Nagpur 2007, 2004, 05)**

**3.3** Derive the Clausius expression for mean free path (λ) on the basis of kinetic theory of gases. **(M.S. Uni. Tamil Nadu 2007, Patna 2004, Kanpur 2002, Nagpur Uni. 2007)**

**3.4** Obtain Maxwell's expression for mean free path $\lambda = \frac{1}{\sqrt{2}\,\pi\sigma^2 n}$, where σ is the molecular diameter and *n* is the number of molecules per unit volume, on the bais of kinetic theory of gases.

**(Nagpur 2007, 2004, 2003, 2001, Ranchi 2005, Bangalore 2002, 2005, Mysore 2005, Purvanchal 2005, 2003)**

**3.5** Explain the variation of mean free path (λ) with temperature and pressure. **(Bangalore 2002)**

**3.6** What are transport phenomena? Explain in brief the viscosity, conductivity and self diffusion on the basis of kinetic theory of gases. **(Agra 2006, Nagpur 2009, 2004, 05, Ranchi 2005, Patna 2004)**

**3.7** Derive an expression for the viscosity (η) of a gas in terms of mean free path of its molecules. Show that it is independent of pressure but depends upon the temperature of the gas. Discuss the effect of pressure and temperature on coefficient of viscosity.

**(Nagpur 2009, 2007, 2005/w, 2005/s, Ranchi 2005, Osmania 2005, Alagappa 2004, Patna 2004, 2000, 2002, Agra 2002, Delhi 2000, Kanpur 2001, Bangalore 2000, Calcutta 2005, Purvanchal 2006, 2002, M.S. Uni. Tamil Nadu 2007)**

**3.8** Derive an expression for thermal conductivity (K) of a gas on the basis of kinetic theory of gases. Show that coefficient of thermal conductivity of hydrogen should be largest among all diatomic molecules. **(Madras 2004, 05, Ranchi 2005, Alagappa 2004, Agra 2006, 2000, Meerut 2002, Nagpur 2009, 2006, M.S. Uni. Tamil Nadu 2007)**

**3.9** Obtain the relation between coefficient of viscosity ($\eta$) and coefficient thermal conductivity (K). Why the experimental values do not follow this relationship. **(Nagpur 2006)**

**3.10** Derive the relation for coefficient of self-diffusion $D$ and show that it is directly proportional to $T^{3/2}$ and inversely proportional to the pressure. **(Calcutta 2004, Garhwal 2001, 02, Meerut 2001)**

**3.11** Obtain the relation between coefficient of viscosity and coefficient of diffusion.

**3.12** Explain the effect of temperature and pressure on coefficient of viscosity. **(Nagpur 2008, 2004, Purwanchal 2004)**

**3.13** What is the effect of temperature and pressure on thermal conductivity? **(Nagpur 2008, 2006, 2004)**

**3.14** Discuss the effect of temperature and pressure on mean free path. **(Garhwal 2002)**

**3.15** What is meant by the mean free path of the molecules of a gas. Obtain an expression for it and show that it is inversely proportional, to the pressure and directly proportional to the absolute temperature of the gas. **(Agra 2002, Nagpur 2009, Rohilkhand 2001)**

**3.16** Write notes on

(*i*) Transport phenomenon **(Patna 2005, Nagpur Uni. 2008, Purvanchal 2007)**

(*ii*) Relation between thermal and electrical conductivities **(Patna 2004)**

(*iii*) Mean free path ($\lambda$). **(Agra 2006, Kanpur 2000, Rohilkhand 2000, 2002)**

(*iv*) Frequency of collision. **(Agra 2001)**

(*v*) Degrees of freedom **(Kumaun 1997)**

**3.17** Calculate the diameter of a molecule of a gas if the number of molecules per $cm^3$ in a gas is $3 \times 19^{19}$ and mean free path is $2 \times 10^{-8}$ cm. [**Ans.** $7.28 \times 10^{-7}$ cm] **(Utkal 2003)**

**3.18** The diameter of the molecules of a gas is $3 \times 10^{-10}$ m. Find the mean free path $\lambda$ at S.T.P. [**Ans.** $1.03 \times 10^{-7}$ m] **(Garhwal 2000, Purvanchal 2004)**

**3.19** Calculate the radius of an oxygen molecule if its coefficient of viscosity at 0°C is $19.2 \times 10^{-6}$ N-s/m$^2$. Given: mass of an oxygen molecule = $5.31 \times 10^{-26}$ kg. Boltzmann constant $k = 1.38 \times 10^{-23}$ joule/degree. [**Ans.** 1.5 R] **(Purvanchal 2006)**

**3.20** Calculate the mean free path of argon molecules at 25°C and 1 atm. pressure. Given $\sigma = 2.56$ A° and $k = 1.38 \times 10^{-23}$ Joule/K. [**Ans.** $1.4 \times 10^{-7}$ metre]

**3.21** The density of nitrogen at N.T.P. is 1.25 kg/m$^3$ and its coefficient of viscosity is $1.66 \times 10^{-5}$ M.K.S. units. Determine mean free path of a molecule of a gas. [**Ans.** $8.084 \times 10^{-5}$ m]

**3.22** The mean free path of molecules of a gas at pressure $P$ and temperature $T$ is $\lambda$. What will be the mean free path if

(*i*) the pressure in kept constant and the temperature is doubled

(*ii*) the temperature is kept constant and the pressure is doubled.

[**Hint :** the mean free path will be doubled in the first case and will be halved in the second case]

**3.23** The mean free path of the molecules of a gas is $2 \times 10^{-7}$ metre at pressure $p$ and temperature 200 K. Calculate its value of (*i*) *p*, 400 *K*, (*ii*) 2*p*, 200 *K*, (*iii*) $\frac{1}{2}p$, 400 K

[**Ans.** (*i*) $4 \times 10^{-7}$ m (*ii*) $1 \times 10^{-7}$ metre (*iii*) $8 \times 10^{-7}$ metre]

**3.24** The mean free path of a gas molecule at pressure $p$ and temperature $T$ is $5.2 \times 10^{-6}$ metre. What will be the mean free path at (*i*) a pressure $2p$ and temperature $\frac{T}{4}$, (*ii*) pressure $p/2$ and temperature $2T$.

[**Ans.** (*i*) $0.15 \times 10^{-6}$ m (*ii*) $4.8 \times 10^{-6}$ m] **(Nagpur 2001, Bangalore 2001)**

**3.25** Determine the temperature at which mean free path of oxygen at 1 atmospheric pressure is $9.95 \times 10^{-8}$ m. Given: the radius of oxygen molecule is $1.695 \times 10^{-10}$ m. **[Ans.** 100°C]

**3.26** A cross-section for collision of slow neutrons in hydrogen is about $20 \times 10^{-24}$ cm$^2$. What is the mean free path of such neutrons in hydrogen at N.T.P.? Normal pressure is $10^5$ N/m$^2$ and Boltzmann constant $k = 1.38 \times 10^{-23}$ J/k.

[**Hint :** $n = \dfrac{p}{kT}$, collision cross-section = $\pi\sigma^2 = 20 \times 10^{-24}$ cm$^2$] **[Ans.** 13.4 m]

**3.27** Calculate the root mean square velocity of oxygen molecules at 27°C. Density of oxygen at N.T.P. is 1.43 kg.m$^{-3}$. **[Ans.** $4.835 \times 10^2$ ms$^{-1}$]

**3.28** Calculate the diameter of molecule of benzene if $n = 2.79 \times 10^{19}$ molecules per c.c. and mean free path $\lambda$ for benzene is $2.2 \times 10^{-6}$ cm. **[Ans.** $6.06 \times 10^{-8}$ cm]

**3.29** Calculate the mean free path of a gas molecule of diameter 3.2 Å. The number of molecules per unit volume is $2.5 \times 10^{25}$ m$^{-3}$. **[Ans.** $8.79 \times 10^{-8}$ m]

**3.30** The viscosity of a gas (oxygen) at a temperature of 16°C is 169 micropoise. Calculate the diameter of the molecules of the gas. Avogadro's number = $6.02 \times 10^{23}$. Molecular weight of oxygen = 32 and Boltzmann's constant K = $1.38 \times 10^{-23}$ J/K. **[Ans.** $3.97 \times 10^{-10}$ m]

**3.31** Calculate the fraction of molecules which will be travelling undeflected after traversing 0.693 times the mean free path. Calculate the distance after which only one percent of the incident molecules remain undeflected. **(Calcutta 2003)**

## OBJECTIVE QUESTIONS

**3.1** Viscosity of a gas is due to transport of

(*a*) momentum (*b*) energy
(*c*) mass (*d*) none of these

**3.2** Out of the following which has the highest thermal conductivity at the same temperature

(*a*) oxygen (*b*) helium
(*c*) hydrogen (*d*) carbon dioxide

**3.3** In diffusion, the transport of the following occurs:

(*a*) momentum (*b*) energy
(*c*) mass (*d*) none of these

**3.4** At very low temperatures, the coefficient of viscosity of a gas

(*a*) decreases with decrease of prcssure.
(*b*) increases with inercase of pressure.
(*c*) is independent of pressure.
(*d*) is equal to precssure.

**3.5** The mean free path of a gas molecule is inersely proportional to **(Kanpur 2001)**

(*a*) square of the diameter of the molecule.
(*b*) square root of the diameter of the molecule.
(*c*) molecular diameter.
(*d*) fourth power of the molecular diameter.

**3.6** The mean free path of molecules of a certain gas at pressure $p$ and temperature $T$ is $2 \times 10^{-6}$ m. The mean free path at pressure $p \times 10^{-6}$ and temperature $T$ is

(*a*) $2 \times 10^{-1}$ m (*b*) $2 \times 10^{-3}$ m
(*c*) $2 \times 10^{-4}$ m (*d*) $2 \times 10^{-10}$ m **(Rohilkhand 2003)**

**3.7** The molecular density in a gas is $n$ and the diameter of its molecule is $d$, then the mean free path of molecule is **(Kanpur 2002)**

(*a*) $\dfrac{\pi}{nd^2}$ (*b*) $\dfrac{1}{\pi nd}$

(*c*) $\dfrac{1}{\sqrt{2}\,\pi nd^2}$ (*d*) $\dfrac{1}{3\sqrt{2}\,\pi nd^3}$

**3.8** Viscosity of a gas is directly proportional to: **(Garhwal 2002)**

(*a*) Temperature (*b*) Density of gas

(*c*) $\sqrt{T}$ (*d*) $T^2$

**3.9** The mean free path of gas at pressure *p* and temperature *T* is **(Grahwal 2002)**

(*a*) $\dfrac{kT}{(\sqrt{2})\sigma^2 \pi p}$ (*b*) $\dfrac{p}{(\sqrt{2})\pi\sigma^2 kT}$

(*c*) $\dfrac{\sqrt{kT}}{\pi\sigma^2 p}$ (*d*) $\dfrac{1}{\sigma}\sqrt{\dfrac{kT}{2\pi p}}$

## ANSWERS

| | | | |
|---|---|---|---|
| **3.1** (*a*) | **3.2** (*c*) | **3.3** (*c*) | **3.4** (*c*) |
| **3.5** (*a*) | **3.6** (*a*) | **3.7** (*c*) | **3.8** (*b*) |
| **3.9** (*a*) | | | |

# SECTION – II

# Thermodynamics

# THERMODYNAMICS

Thermodynamics is a branch of physics which mainly deals with the transformation of heat into mechanical work. Today, the scope of thermodynamics has very much increased covering all the branches of physics, chemistry and engineering and has innumerable applications. Thermodynamics does not take into account the atomic constitution of matter i.e. structure etc. but it only deals with the large scale response (macroscopic properties) of the system that we can observe and measure in experiments.

The zeroth law of thermodynamics introduces the concept of temperature and the concept of thermodynamic equilibrium when two objects attain the same temperature when brought in thermal contact. During the process of reaching thermal equilibrium, heat is transferred between the objects. The first law of thermodynamics represents the relationship between heat and mechanical work while the second law depicts the manner in which these energy changes take place. In short, thermodymanics is an empirical and experimental science and based upon the general laws of nature which govern the conversion of heat in to mechanical work and vice-versa.

In 1906, Nernst proposed a general principle, known as third law of thermodynamics, which explains the nature of bodies in the neighbourhood of absolute zero temperature. At absolute zero temperature, the entropy tends to zero and the constituents of any matter are in a perfect order.

# Laws of Thermodynamics

# 4

## At a Glance

• Thermodynamic System • Zeroth Law of Thermodynamics • Concept of Heat • Thermodynamic Equilibrium • Work : A Path Dependent Function • Internal Energy (U) • First Law of Thermodynamics • Internal Energy as a State Function • Specific Heats of a Gas • Applications of First Law of Thermodynamics • The Indicator Diagram • Work done during an Isothermal Process • Work done during an Adiabatic Process • Slopes of Adiabatics and Isothermals • Relation between Adiabatic & Isothermal Elasticities • Cooling due to Adiabatic Reversible Process • Clement and Desormes Method-Determination of γ • Partington's Method • Ruchhardt's Experiment • Reversible and Irreversible Process • Heat Engines • Definition of Efficiency • Carnot's Ideal Heat Engine • Carnot's Cycle • Effective way to Increase Efficiency • Carnot's Engine and Refrigerator • Coefficient of Performance • Second Law of Thermodynamics • Carnot's Theorem • Steam Engine • Otto Cycle • Petrol Engine • Diesel Engine • Multicylinder Engine

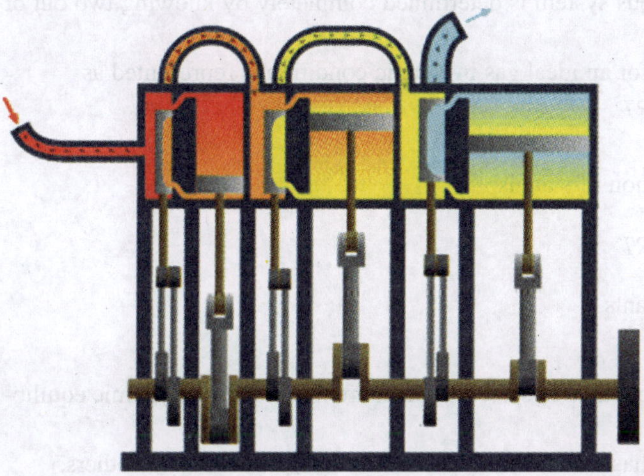

## 4.1 Thermodynamic System

A system may be defined as a definite quantity of matter (solid, liquid or gases) bounded by some closed surface. The simplest example of a system is a gas contained in a cylinder with a movable piston. When a system is completely uniform throughout, such as gas, or mixture of gases, or a pure solid, or a liquid or a solution is called a homogeneous system. When a system consists of two or more phases, which are separated from one another by definite boundary surfaces, it is said to be a heterogeneous system. Examples of heterogeneous system are: a liquid and its vapour, two immiscible or partially miscible liquids. Anything outside this system which can exchange energy with it and has a direct bearing on the behaviour of the system is called as its **surroundings**.

A system may be separated from its surroundings by a real or imaginary boundary through which heat or mechanical energy may pass. The existence of the boundary is, however, essential to visualize the system distinctly from rest of the universe. A thermodynamic system may contain no substance at all, but may consist of radiant energy or electric and magnetic field. The combination of a system and its surroundings is called an *universe*.

## Thermodynamic Variables and equation of state

The thermodynamic state or macroscopic state of a system is determined by four observable properties. These properties are the composition, pressure, volume and temperature, and are called the *variables of state*. For a homogeneous system, consisting of a single substance, the composition is fixed. Hence the state of the system is determined completely by the three variables: pressure ($P$), volume ($V$) and temperature ($T$).

For a system, in general, there is an equation of state

$$f(P, V, T) = 0$$

Therefore, the pressure, volume and temperature are not independent variables. Consequently, the thermodynamic state of a homogeneus system is determined completely by knowing two out of three variables.

For example: the equation of state for an ideal gas in a static condition is represented as

$$PV = RT$$

where R is universal gas constant.

For a Van der Waals' gas, the equation of state is

$$\left(P + \frac{a}{V^2}\right)(V - b) = RT$$

where $a$ and $b$ are Van der Waal's constants.

## Limitations

(*i*) The equation of state is not applicable to systems which are not in thermodynamic equilibrium.

(*ii*) Every thermodynamic system has its own equation of state independent of the others.

(*iii*) An equation of state expresses the peculiar behaviour of one individual system which distinguishes it from the others.

(*iv*) An equation of state is not a theoretical deduction from thermodynamics but is an experimental backing to it.

The thermodynamic systems in engineering are gas, vapour, steam, mixture of gasoline vapour and air, ammonia vapours and its liquid. In physics, thermodynamics includes besides the above, systems like stretched wires, thermocouples, magnetic materials, electrical condenser, electrical reversible cells, solid and surface films etc.

## There are three classes of system

(*i*) **Open system :** A system which can exchange matter and energy with the surroundings is called an *open system*. e.g. Air compressor : Air at low pressure enters and air at high pressure leaves the system i.e. there is an exchange of matter and energy with the surroundings.

(*ii*) **Closed system :** A system which can exchange only energy (and not matter) with the surroundings is called a *closed system*. e.g., Gas enclosed in a cylinder expands when heated and drives the piston outwards. The boundary of the system moves but the matter (here gas) in the system remains constant.

***(iii)* Isolated system:** A system which is thermally insulated and has no communication of heat or work with the surroundings is called *isolated system.*

## 4.2 Zeroth Law of Thermodynamics

When a hot body A is brought in thermal contact with a cold body *B*, heat flows from *A* to *B* and after some time the flow stops. The bodies are then said to be in thermal equilibrium with each other.

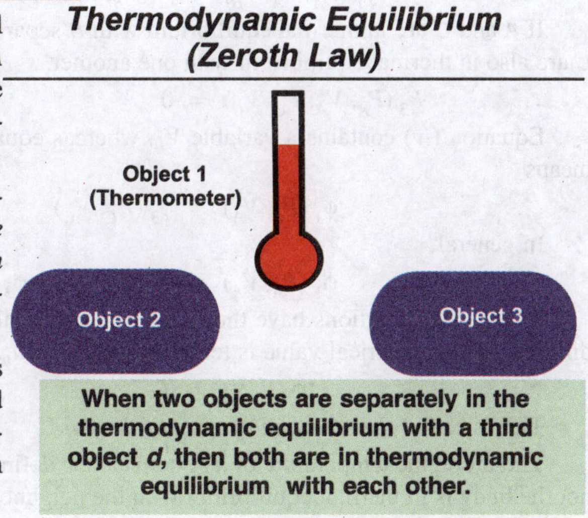


**Statement:** *The zeroth law of thermodynamics states that if two bodies A and B are each separately in thermal equilibrium with a third body C, then A and B are also in thermal equilibrium with each other.*

**Explanation:** We consider two systems *A* and *B* insulated from each other but in good thermal contact with a common system *C*. Systems *A* and *B* will attain thermal equilibrium with third system *C* [Fig. 4.1 (*a*)].

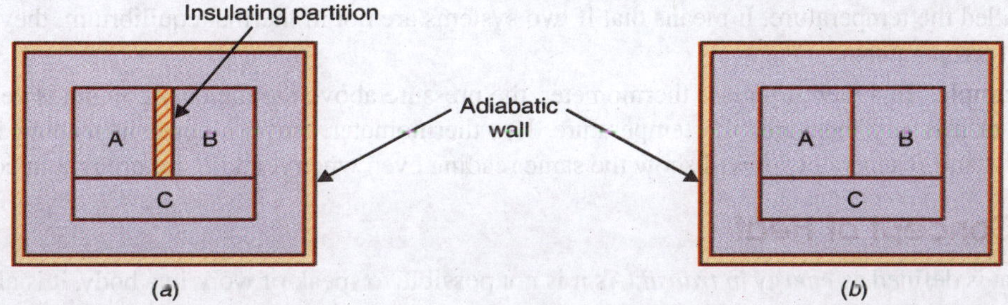


**Fig. 4.1**

Now if the insulating partition is removed and systems *A* and *B* are brought into thermal contact, we find that there is no further change. This means that the systems *A* and *B* are also in thermal equilibrium with another. This is known as zeroth law of thermodynamics.

### Conversely the law can be stated as follows:

If three or more systems are in thermal contact, each to each, by means of diathermal walls and are all in thermal equilibrium together, then any two systems taken separately are in thermal equilibrium with one another.

Consider three fluids *A*, *B* and *C*. Let $P_A$, $V_A$ represent the pressure and volume of *A*, $P_B$, $V_B$, the pressure and volume of *B*, and $P_C$, $V_C$ are the pressure and volume of *C*.

If *A* and *B* are in thermal equilibrium, then

$$\phi_1 (P_A, V_A) = \phi_2 (P_B, V_B)$$

or $$F_1 (P_A, V_A, P_B, V_B) = 0 \qquad ...(i)$$

Expression (*i*) can be solved, and

$$P_B = f_1 (P_A, V_A, V_B) \qquad ...(ii)$$

If *B* and *C* are in thermal equilibrium

$$\phi_2 (P_B, V_B) = \phi_3 (P_C, V_C)$$

or $F_2 (P_B, V_B, P_C, V_C) = 0$

Also $P_B = f_2 [V_B, P_C, V_C]$ ...(*iii*)

From equations (*ii*) and (*iii*) for *A* and *C* to be in thermal equilibrium separately,

$$f_1 (P_A, V_A, V_B) = f_2 (V_B, P_C, V_C) \quad ...(iv)$$

If *A* and *C* are in thermal equilibrium with *B* separately, then according to the zeroth law, *A* and *C* are also in thermal equilibrium with one another.

$$\therefore \quad F_3 (P_A, V_A, P_C, V_C) = 0 \quad ...(v)$$

Equation (*iv*) contains a variable $V_B$, whereas equation (*v*) does not contain the variable $V_B$. It means

$$\phi_1 (P_A, V_A) = \phi_3 (P_C, V_C) \quad ...(vi)$$

In general,

$$\phi_1 (P_A, V_A) = \phi_2 (P_B, V_B) = \phi_3 (P_C, V_C) \quad ...(vii)$$

These three functions have the same numerical value though the parameters (*P*, *V*) of each are different. This numerical value is termed as *temperature* (*T*) of the body.

$$\therefore \quad \phi (P, V) = T \quad ...(viii)$$

This is called *the equation of state of the fluid.*

Therefore, the temperature of a system can be defined as the property that determines whether or not the body is in thermal equilibrium with the neighbouring systems. If a number of systems are in thermal equilibrium, this common property of the system can be represented by a single numerical value called the temperature. It means that if two systems are not in thermal equilibrium, they are at different temperatures.

**Example.** In a mercury glass thermometer, the pressure above the mercury column is zero and volume of mercury measures the temperature. If a thermometer shows a constant reading in two systems *A* and *B* seperately, it will show the same reading even when *A* and *B* are brought in contact.

## 4.3 Concept of Heat

Heat is defined as *energy in transit*. As it is not possible to speak of work in a body, it is also not possible to speak of heat in a body. Work is either done on a body or by a body. Similarly, heat can flow from a body or to a body. If a body is at a constant temperature, it has both mechanical and thermal energies due to the molecular agitations and it is not possible to separate them. So, in this case, we cannot talk of heat energy. It means, if the flow of heat stops, the word heat cannot be used. It is only used when there is transfer of energy between two or more systems.

Consider two systems *A* and *B* in thermal contact with one another and surrounded by adiabatic walls.

For the system *A*,

$$Q' = U_2 - U_1 + W \quad ...(i)$$

where *Q* is the heat energy transferred, $U_1$ is the initial internal energy, $U_2$ is the final internal energy and W is the work done.

Similarly for the system *B*,

$$Q' = U_2' - U_1' + W' \quad ...(ii)$$

Adding (*i*) and (*ii*)

$$\therefore \quad Q + Q' = (U_2 - U_1) + W + (U_2' - U_1') + W'$$

$$Q + Q' = [(U_2 + U_2') - (U_1 + U_1')] + (W + W') \quad ...(iii)$$

The total change in the internal energy of the composite system

$$= [(U_2 + U_2') - (U_1 + U_1')]$$

The work done by the composite system = $W + W'$

It means that the heat transferred by the composite system = $Q + Q'$. But the composite system is surrounded by adiabatic walls and the net heat transferred is zero.

$$\therefore \quad Q + Q' = 0$$

$$\text{or} \quad Q = -Q' \qquad \text{...(iv)}$$

Thus, for two systems $A$ and $B$ in thermal contact with each other, and the composite system surrounded by adiabatic walls, the heat gained by one system is equal to the heat lost by the other system.

## 4.4 Thermodynamic Equilibrium

Any state of homogeneous system in which any two of the three variables $P$, $V$ and $T$ remain constant, as long as the external conditions remain unchanged is said to be in thermodynamic equilibrium.

A system in thermodynamic equilibrium must satisfy the following requirements strictly :

*(i)* **Mechanical Equilibrium.** For a system to be in mechanical equilibrium, there should be no macroscopic movement within the system (*i.e.* no unbalanced forces acting) or of the system with respect to its surroundings.

*(ii)* **Thermal Equilibrium.** For a system to be in thermal equilibrium, there should be no temperature difference between the parts of the system or between the system and the surroundings.

*(iii)* **Chemical Equilibrium.** For a system to be in chemical equilibrium, there should be no chemical reaction within the system and also no movement of any chemical constituent from one part of the system to the other.

### Quasistatic Process

A finite unbalanced force may cause the system to pass through *non-equilibrium* states. A quasistatic process is defined as the process in which the deviation from thermodynamic equilibrium is infinitesimal and all the states through which the system passes during a quasistatic process can be considered as equilibrium states.

A quasistatic process is an ideal concept which can never be satisfied rigorously in practice. However, in actual practice, many processes closely approach a quasistatic process with no significant error.

## 4.5 Work: A path dependent function

A system can be taken from state $A$ to state $B$ by many ways as shown in fig. 4.2. Let the coordinates of state A and B be $A$ ($P_1$ $V_1$) and $B$ ($P_2$, $V_2$) respectively, where $P_1$ and $V_1$ represents pressure and volume of state $A$ and $P_2$ and $V_2$ that of state $B$.

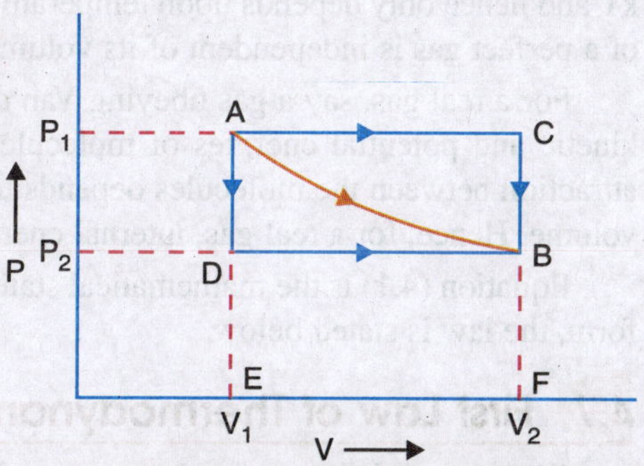


Fig. 4.2

**1.** Starting from point $A$, the pressure is continuously decreased from $P_1$ to $P_2$ along the curve $AB$ so that volume increases from $V_1$ to $V_2$. The work done by the system is given by

$$W_1 = \int_{V_1}^{V_2} PdV = \text{area } ABFE$$

**2.** Starting from point $A$ the volume $V_1$ is kept constant in going from $A$ to $D$, the pressure decreases from $P_1$ to $P_2$ and then $P_2$ is kept constant from $D$ to $B$. The work done in this process *i.e.* along $ADB$ path given by

$$W_2 = P_2 (V_2 - V_1) \;=\; \text{area } DBFE$$

**3.** Again starting from point $A$, the pressure $P_1$ is kept constant in going from $A$ to $C$ and then the volume $V_2$ is kept constant from $C$ to $B$. The work done in this process *i.e.* along path $ACB$ is given by

$$W_3 = P_1 (V_2 - V_1) \;=\; \text{area } ACFE$$

$$\therefore \qquad W_1 \neq W_2 \neq W_3$$

Thus, the work done depends on the path of the process when the system goes from one thermodynamic state to another.

## 4.6 Internal Energy (U)

The energy content of a system is called its internal energy. It is the sum of following forms of energy of the system:

(*i*) kinetic energy due to translational, rotational and vibrational motion of the molecules, all of which depend only on the temperature,

(*ii*) potential energy due to intermolecular forces, which depends on the separation between the molecules, and

(*iii*) the energy of electrons and nuclei.

In practice, it is not possible to measure the total internal energy of a system in any given state. Only change in its value can be measured.

If the state of the system is changed from an initial state 1 to a final state 2, by supplying heat $Q$ to the system and if $W$ is the work done by the system during the change, then increase in the internal energy of the system is given by

$$(U_2 - U_1) \;=\; Q - W \qquad \text{...(4.1)}$$

where $U_1$ is the internal energy in state 1, and $U_2$ the internal energy in state 2.

### Internal Energy of a Real gas and Ideal gas

The internal energy of a perfect (or ideal) gas is only dependent upon temperature where as that of a real gas, it is a function of both temperature and volume.

For a perfect gas, the internal energy is only kinetic energy of its molecules. It is proportional to kT and hence only depends upon temperature ($T$), k being Boltzmann's constant. The internal energy of a perfect gas is independent of its volume.

For a real gas, say a gas obeying Van der Waals' equation, the internal energy is the sum of the kinetic and potential energies of molecules due to their mutual forces of attraction. The force of attraction between the molecules depends upon the intermolecular distance and is thus a function of volume. Hence, for a real gas, internal energy is a function of both, the temperature and volume.

Equation (4.1) is the mathematical statement of the first law of thermodynamics. In a convenient form, the law is stated below.

## 4.7 First Law of Thermodynamics

**Statement:** When a certain amount of heat Q is supplied to a system which does external work W in passing from state 1 to state 2, the amount of heat is equal to sum of the increase in the internal energy of the system and the external work done by the system. In symbols, the law is expressed as

$$Q \;=\; (U_2 - U_1) + W \qquad \text{...(4.2)}$$

For a very small change in the state of the system, the law is expressed in the differential form as

$$\delta Q = dU + \delta W \qquad ...(4.3)$$

where $\delta Q$ and $\delta W$ are not perfect differentials but $dU$ is a perfect differential because $U$ is a function of the state of the system.

It should be remembered that all the quantities in equation (4.2) or (4.3) must be expressed in unit of energy. The amount of heat $Q$ or $\delta Q$ is taken positive if heat is supplied to the system, and negative if heat is removed from it. Similarly, the work $W$ or $\delta W$ is positive when the external work is done by the system in expansion and negative if the work is done on it in compression.

### Significance of the First Law:

The first law of thermodynamics is important because

*(i)* it is applicable to any process by which a system undergoes a physical or chemical change.

*(ii)* it introduces the concept of the internal energy, and

*(iii)* it provides method for determining the change in the internal energy.

***First Law of Thermodynamics***

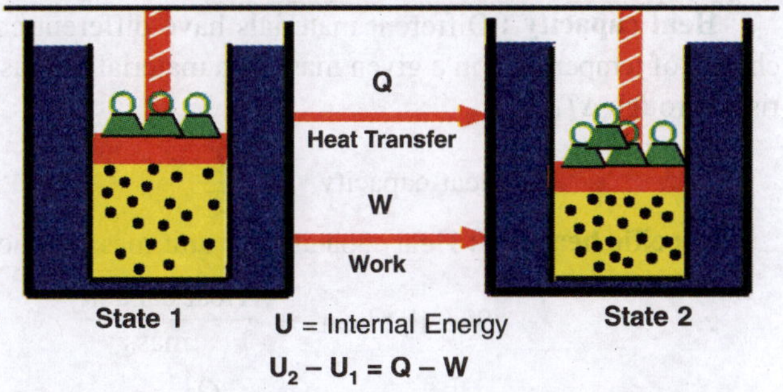


Any thermodynamic system in an equillbrium state possesses state variable called the internal energy (U). Between any two equilibrium states, the change in internal energy is equal to the difference of the heat transfer into the system and work done by the system.

### Limitations of the First Law:

The first law of thermodynamics is based on the principle of conservation of energy of a system. Though it is applicable to every process in nature between the equilibrium states, it does not specify the condition under which a system can use its heat energy to produce a supply of mechanical work. It also does not say how much of the heat energy can be converted into work.

## 4.8 Internal Energy as a State Function

Suppose $Q$ is the heat supplied to a working substance forming a system and $W$ is the work done by it in taking the system from state 1 to state 2, then the increase in the internal energy of the system is given by first law of thermodynamics as

$$(U_2 - U_1) = Q - W$$

where $U_1$ is the initial internal energy of the system in state 1 and $U_2$ the final internal energy in state 2 along path $A$ as shown in fig. 4.3.

If the state of the system is changed from the same initial state 1 to the same final state 2 by different paths, say $B$ or $C$, it is found experimentally that the difference $(Q - W)$ is the same for all paths between the states 1 and 2. Hence we conclude:

When a system changes from one state to another, the change of internal energy of the system has a definite value, independent of the paths between the two states. Hence, internal energy is called as state function or point function and not a path function.

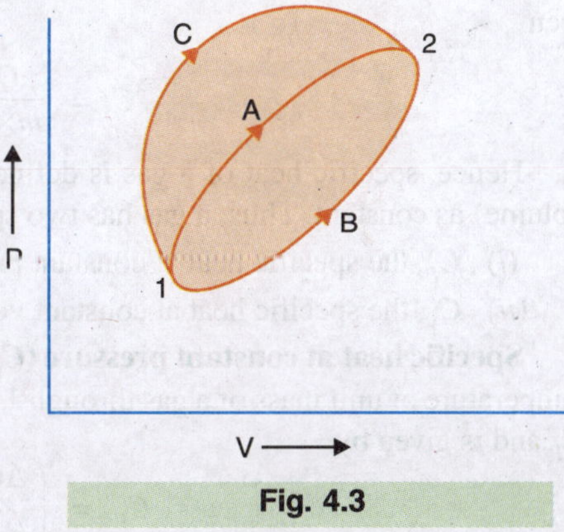


**Fig. 4.3**

In other words, the internal energy is a function of any two of the thermodynamic variables $P$, $V$ and $T$.

Thus, $U = f(P, V)$ or $U = f(V, T)$ or $U = f(T, P)$

Hence, the differential

$$dU = \left(\frac{\partial U}{\partial P}\right)_V dP + \left(\frac{\partial U}{\partial V}\right)_P dV \qquad ...(4.4)$$

is a perfect differential of the thermodynamic variables $P$ and $V$. Internal energy is an extensive variable.

## 4.9 Specific Heats of a Gas

**Heat capacity :** Different materials have different capacity to absorb heat to produce a given change of temperature in a given mass. If a material of mass $m$ absorbs heat $Q$, so that its temperature rises through $\Delta T$, then

$$\text{Heat capacity} = \frac{Q}{\Delta T} \qquad ...(i)$$

**Specific heat ($C$) :** Heat capacity per unit mass is known as specific heat and is denoted by $C$.

$$\therefore \qquad \text{Specific heat} = \frac{\text{Heat capacity}}{\text{mass}}$$

$$C = \frac{Q}{m.\Delta T} \qquad ...(ii)$$

**Definition :** The specific heat of a material is defined as the quantity of heat required to raise the temperature of unit mass of the material through 1 degree.

The unit in C.G.S. is calories per gram per °C. In S.I. it is Joule per kg per °C. For example:

Specific heat of water = 1 cal/gm °C

= 1 Kilo calories/kg °C.

= $4.18 \times 10^3$ Joule/kg °C.

This definition is sufficient for solids and liquids. Since, the gases are compressible, the specific heat of a gas may vary from 0 to infinity. For examples, if a gas is compressed, it's temperature rises, without supplying any heat to it *i.e.* $Q = 0$. Hence, specific heat

$$C = \frac{Q}{m.\Delta T} = 0 \qquad ...(iii)$$

On the contrary, if the gas is allowed to expand freely, without any rise in temperature ($\Delta T = 0$) then

$$C = \frac{Q}{m \times o} = \infty \qquad ...(iv)$$

Hence, specific heat of a gas is defined by considering any one of the two (either pressure or volume) as constant. Thus, a gas has two specific heats:

(*i*) $C_P$, the specific heat at constant pressure, and

(*ii*) $C_V$, the specific heat at constant volume.

**Specific heat at constant pressure ($C_P$) :** It is defined as the amount of heat required to raise the temperature of unit mass of a gas through 1 °C, when its pressure is kept constant. It is represented by $C_P$ and is given by

$$C_P = \left(\frac{\Delta Q}{\Delta T}\right)_P \qquad ...(v)$$

**Specific heat at constant volume ($C_V$) :** It is defined as the amount of heat required to raise the temperature of unit mass of a gas through 1 °C, when its volume is kept constant. It is represented by $C_V$ and is given by

$$C_V = \left(\frac{\Delta Q}{\Delta T}\right)_V \quad ...(vi)$$

**$C_P$ is greater than $C_V$ :** Heat capacity of a gas is different under these two conditions. Suppose the heat is supplied to a gas and is allowed to expand at constant pressure. Then, the heat supplied is used up in doing two things:

(*i*) it raises the temperature of the gas (*i.e.* increase in its internal energy) and

(*ii*) it does work in expanding the gas against the external pressure.

$$\delta Q = dU + \delta W$$
$$= dU + PdV$$

On the other hand, when gas is heated at constant volume, no work is done ($\delta W = P.dV = 0$) and hence whole of the heat supplied is used to raise its temperature. Thus, more heat is required for increasing the temperature of the gas through 1 °C at constant pressure than at constant volume. Hence specific heat of a gas at constant pressure ($C_P$) is greater than the specific heat at constant volume ($C_V$); *i.e.*, $C_P > C_V$.

## 4.10 Applications of First Law of Thermodynamics

### 4.10.1 Specific Heat of a Gas (*V* and *T* independent)

The internal energy of a system is a single valued function of the state variables viz. pressure, volume, temperature etc. In the case of gas, any two of the variables *P, V, T* are sufficient to define completely its state. If *V* and *T* are chosen as the independent variables,

$$U = f(V, T) \quad ...(i)$$

Differentiating equation (*i*)

$$dU = \left(\frac{\partial U}{\partial T}\right)_V dT + \left(\frac{\partial U}{\partial V}\right)_T dV \quad ...(ii)$$

If an amount of heat $\delta Q$ is supplied to a thermodynamical system, say an ideal gas and if the volume increases by $dV$ at a constant pressure $P$, then according to the first law of thermodynamics

$$\delta Q = dU + \delta W$$

Here $\delta W = P.dV$

$\therefore$ $\delta Q = dU + P.dV$

Substituting the value of $dU$ from equation (*ii*),

$$\delta Q = \left(\frac{\partial U}{\partial T}\right)_V dT + \left(\frac{\partial U}{\partial V}\right)_T dV + PdV \quad ...(iii)$$

Dividing both sides by dT,

$$\frac{\delta Q}{dT} = \left(\frac{\partial U}{\partial T}\right)_V + \left(\frac{\partial U}{\partial V}\right)_T \frac{dV}{dT} + \frac{PdV}{dT}$$

or

$$\left(\frac{\delta Q}{dT}\right) = \left(\frac{\partial U}{\partial T}\right)_V + \left[P + \left(\frac{\partial U}{\partial V}\right)_T\right]\frac{dV}{dT} \quad ...(iv)$$

If the gas is heated at constant volume,

$$\left(\frac{\delta Q}{\partial T}\right)_V = C_V$$

and $$\frac{dV}{dT} = 0$$

$$\therefore \quad \left(\frac{\delta Q}{dT}\right)_V = \left(\frac{\partial U}{\partial T}\right)_V = C_V \qquad ...(v)$$

when the gas is heated at constant pressure,

$$\left(\frac{\delta Q}{dT}\right) = C_P$$

$\therefore$ from equation (*iv*)

$$C_P = \left(\frac{\partial U}{\partial T}\right)_V + \left[P + \left(\frac{\partial U}{\partial V}\right)_T\right]\left(\frac{\partial V}{\partial T}\right)_P$$

$$\therefore \quad C_P = C_V + \left[P + \left(\frac{\partial U}{\partial V}\right)_T\right]\left(\frac{\partial V}{\partial T}\right)_P$$

or $$C_P - C_V = \left[P + \left(\frac{\partial U}{\partial V}\right)_T\right]\left(\frac{\partial V}{\partial T}\right)_P \qquad ...(vi)$$

From Joule's experiment, for an ideal gas on opening the stopcock, no work was done and no heat transfer took place. So $\delta Q = 0 = dU + 0$. Therefore, $dU = 0$. Even though the volume changed, while the temperature is constant, there is no change in the internal energy.

$$\left(\frac{\partial U}{\partial V}\right)_T = 0$$

From the ideal gas equation

$$PV = RT$$

or $$P\left(\frac{\partial V}{\partial T}\right)_P = R \qquad ...(vii)$$

$$\therefore \quad C_p - C_V = P\left(\frac{\partial V}{\partial T}\right)_P + \left(\frac{\partial U}{\partial V}\right)_T\left(\frac{\partial V}{\partial T}\right)_P$$

But $$\left(\frac{\partial U}{\partial V}\right)_T = 0$$

$$\therefore \quad C_P - C_V = P\left(\frac{\partial V}{\partial T}\right)_P = R$$

$$\therefore \quad C_P - C_V = R \qquad ...(viii)$$

This formula is known as Mayer's relation.

Here $C_P$, $C_V$ and $R$ are expressed in the same unit. From equation (*iii*),

$$\delta Q = \left(\frac{\partial U}{\partial T}\right)_V dT + \left[P + \left(\frac{\partial U}{\partial V}\right)_T\right] dV \qquad ...(ix)$$

For a process at constant temperature

$$dT = 0$$

$$\therefore \quad (\delta Q)_T = P(dV)_T + \left(\frac{\partial U}{\partial V}\right)_T (dV)_T \qquad ...(x)$$

This equation represents the amount of heat energy supplied to a system in an isothermal reversible process and is equal to the sum of the work done by the system and increase in its internal energy.

### For a Reversible Adiabatic Process:

We write, $\delta Q = 0$

Therefore, from equation (*ix*)

$$0 = C_V dT + \left[P + \left(\frac{\partial U}{\partial V}\right)_T\right] dV$$

or

$$C_V\, dT = -\left[P + \left(\frac{\partial U}{\partial V}\right)_T\right] dV$$

Dividing throughout by $dV$,

$$C_V \left(\frac{\partial T}{\partial V}\right) = -\left[P + \left(\frac{\partial U}{\partial V}\right)_T\right] \quad \text{...(xii)}$$

The isobaric volume coefficient of expansion

$$\alpha = \frac{1}{V}\left(\frac{\partial V}{\partial T}\right)_P$$

$$\therefore \quad \alpha V = \left(\frac{\partial V}{\partial T}\right)_P$$

We have

$$C_P - C_V = P\left(\frac{\partial V}{\partial T}\right)_P$$

$$\therefore \quad \frac{C_P - C_V}{\alpha V} = P$$

But

$$\left(\frac{\partial U}{\partial V}\right)_T = 0 = P - P$$

or

$$\left(\frac{\partial U}{\partial V}\right)_T = \left(\frac{C_P - C_V}{\alpha V}\right) - P \quad \text{...(xiii)}$$

or

$$-\left(\frac{C_P - C_V}{\alpha V}\right) = -\left[P + \left(\frac{\partial U}{\partial V}\right)_T\right] \quad \text{...(xiv)}$$

From equation (*xii*) and (*xiv*),

$$C_V\left(\frac{\partial T}{\partial V}\right) = -\left(\frac{C_P - C_V}{\alpha V}\right)$$

or

$$\left(\frac{\partial T}{\partial V}\right) = \frac{C_V - C_P}{\alpha\, V\, C_V} \quad \text{...(4.5)}$$

This expression holds good for an adiabatic reversible process.

### 4.10.2 Isochoric process (*i.e.* Volume constant)

If a system undergoes a change in which the volume remains constant, the process is called isochoric. At constant volume, no external work is done *i.e.* $\delta W = 0$

$\therefore$ Heat absorbed is given by $\delta Q = dU$

This expression may be used to define the internal energy of a system.

Thus, the increase in the internal energy of a system is equal to the heat absorbed by the system, at constant volume. The work done in isochoric process is zero.

### 4.10.3 Isobaric process (i.e. pressure constant)

If a system undergoes a change in which the pressure is kept constant, the process is called isobaric. Suppose $Q$ is the heat absorbed by a system at constant pressure $P$ and suppose its volume increases from $V_1$ to $V_2$. Then from the first law, we have

$$Q = (U_2 - U_1) + W$$

or
$$Q = (U_2 - U_1) + P(V_2 - V_1)$$

or
$$Q = (U_2 + PV_2) - (U_1 + PV_1)$$
$$= H_2 - H_1 \quad \text{...(4.6)}$$

where
$$H = U + PV.$$

From equation (4.6), we conclude that the heat absorbed at constant pressure is equal to increase in quantity $H$, called as *enthalpy*. Like the internal energy $U$, the quantity $H$ is a function of thermodynamic variables. The quantity $H$ is also called the heat function at constant pressure.

### 4.10.4 Adiabatic process

When a system undergoes from an initial state to a final state in such a way that no heat leaves or enters the system, the process is called adiabatic. In this process.

$$Q = 0$$

$\therefore$
$$U_2 - U_1 = -W$$

For an ideal gas
$$U_2 - U_1 = C_V d_T$$

$\therefore$
$$W = -C_V d_T$$

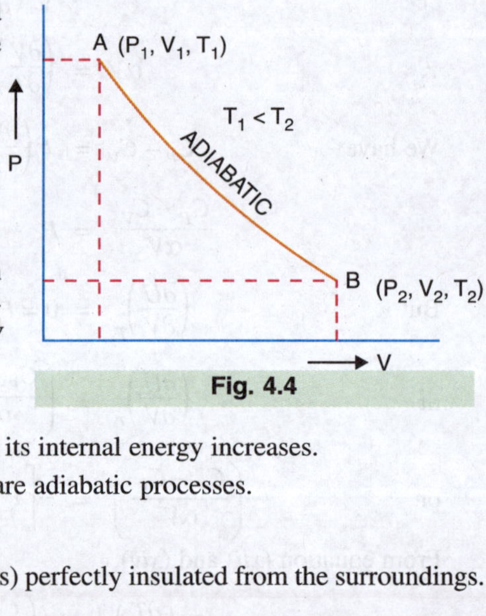

Fig. 4.4

Here $W$ is the work done by the system. Thus, when a system expands adiabatically, its internal energy decreases. If the system is compressed adiabatically then $W$ is negative. Thus, in case of compression

$$U_2 - U_1 = -(-W) = W$$

Therefore, in an adiabatic compression of a system its internal energy increases.

The processes that take place suddenly or quickly are adiabatic processes.

### Adiabatic Equation of a Perfect Gas.

Consider 1 gram of the working substance (ideal gas) perfectly insulated from the surroundings. Let the external work done by the gas be $\delta W$.

Applying the first law of thermodynamics

$$\delta Q = dU + \delta W$$

But
$$\delta Q = 0$$

and
$$\delta W = P.dV$$

where $P$ is the pressure of the gas and $dV$ is the change in volume.

$\therefore$
$$0 = dU + \frac{P.dV}{J} \quad \text{...}(i)$$

where $J$ is mechanical equivalent of heat.

As the external work is done by the gas at the cost of its internal energy, there is fall in temperature by $dT$.

$$dU = 1 \times C_v \times dT \quad (\because m = 1)$$

$$C_v.dT + \frac{P.dV}{J} = 0 \quad \text{...}(ii)$$

For an ideal gas of 1 gram, we have

$$PV = rT \quad \text{...(iii)}$$

where $r$ is ordinary gas constant.

Differentiating,

$$PdV + VdP = rdT$$

Substituting the value of $dT$ in equation (*ii*),

$$C_v\left[\frac{PdV+VdP}{r}\right]+\frac{PdV}{J} = 0$$

$$C_v\,[PdV + VdP] + r\frac{PdV}{J} = 0$$

But, $$\frac{r}{J} = C_p - C_v$$

$$\therefore \quad C_v.PdV + C_vVdP + C_pPdV - C_v\,PdV = 0$$

$$C_pPdV + C_vVdP = 0$$

Dividing by $C_vPV$,

$$\frac{C_p}{C_v}\cdot\frac{dV}{V}+\frac{dP}{P} = 0$$

But $$\frac{C_p}{C_v} = \gamma$$

$$\therefore \quad \frac{dP}{P}+\gamma\frac{dV}{V} = 0$$

Integrating, $\log P + \gamma \log V =$ constant

$$\log PV^{\gamma} = \text{constant}$$

or $$\boxed{PV^{\gamma} = \text{constant}} \quad \text{...(iv)}$$

This is the equation connecting pressure and volume during an adiabatic process and is known as *Poisson's law*.

Taking $$PV = rT$$

or $$P = \frac{rT}{V}$$

$$\left(\frac{rT}{V}\right).V^{\gamma} = \text{constant}$$

But $r$ is constant

$$rTV^{\gamma-1} = \text{constant}$$

$$\therefore \quad TV^{\gamma-1} = \text{constant} \quad \text{...(v)}$$

Also $$V = \frac{rT}{P}$$

$$P\left[\frac{rT}{P}\right]^{\gamma} = \text{constant}$$

or $$\frac{r^{\gamma}T^{\gamma}}{P^{\gamma-1}} = \text{constant}$$

or $$\frac{P^{\gamma-1}}{T^{\gamma}} = \text{constant} \quad \text{or} \quad \boxed{TP^{(1-\gamma)/\gamma} = \text{constant}} \quad \text{...(vi)}$$

Thus, during an adiabatic process

(i) $PV^{\gamma}$ = constant

(ii) $TV^{\gamma-1}$ = constant, and

(iii) $\dfrac{P^{\gamma-1}}{T^{\gamma}}$ = constant

### 4.10.5 Cyclic process

For the cyclic process, the law can be stated as

$$\oint dQ = \oint dU + \oint dW$$

In a cyclic process, the system is restored to the initial state at the end of the cycle.

Since, the internal energy is a state function,

$$\therefore \quad \oint dU = 0$$

$$\text{or} \quad \oint dQ = \oint dW$$

∴ Total work obtained = Net Heat supplied

Thus, no work is obtained if no heat is supplied or work can be obtained only at the cost of energy. From indicator diagram (Fig. 4.5), we write

$$Q = \text{Area } ACBDA$$

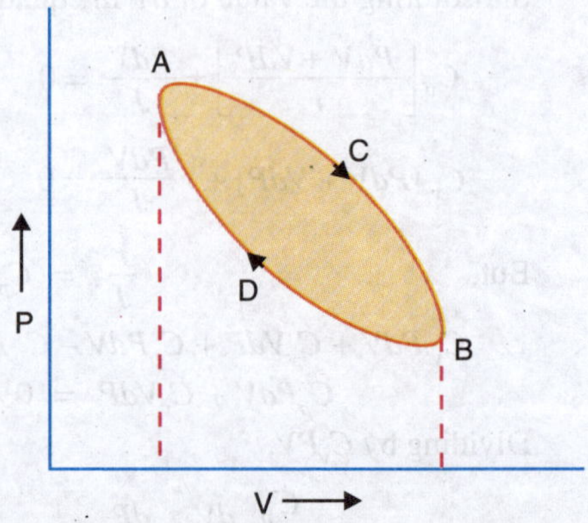


Fig. 4.5

### 4.10.6 Non-Isolated system (Internal energy of the Universe)

System and surroundings together constitutes the universe. In a non-isolated system, heat transfer between the system and the surroundings is allowed. When a heat $\delta Q$ is supplied to the system its internal energy increases by $dU$ and it performs an external work $\delta W$.

$$dQ = dU + \delta W$$

$$\text{or} \quad dU = \delta Q - \delta W$$

This gives the change in internal energy of the system.

Here $\delta Q$ is +ve, since heat enters the system.

$\delta W$ is +ve, since work is done by the system.

On applying the first law of thermodynamic to the surroundings, which loses a heat $\delta Q$ and receives a work $\delta W$ *i.e.* a work $\delta W$ is done on the surroundings; hence both $\delta Q$ and $\delta W$ are negative for the surroundings.

∴ Change in internal energy of the surroundings

$$dU' = -\delta Q - (-\delta W)$$

$$dU' = \delta W - \delta Q \qquad \text{...(4.7)}$$

$$\text{Net change in internal energy} = dU + dU'$$

$$= (\delta Q - \delta W) + (\delta W - \delta Q)$$

$$= 0$$

Thus, the net change in internal energy is always zero.

### 4.10.7 Isothermal process

If a system is perfectly conducting to the surroundings and the temperature remains constant throughout the process, it is called an isothermal process. Consider a working substance at a certain pressure and temperature and having volume represented by the point A (Fig. 4.6).

Pressure is decreased and work is done by the working substance at the cost of its internal energy and suffers a fall in temperature. But the system is perfectly conducting to the surroundings. It absorbs heat from the surroundings and maintains a constant temperature.

Thus, from $A$ to $B$, the temperature remains constant. The curve $AB$ is called the isothermal curve or isothermal. In going from $B$ to $A$ back, the system gives out extra heat to the surroundings and maintains the temperature constant.

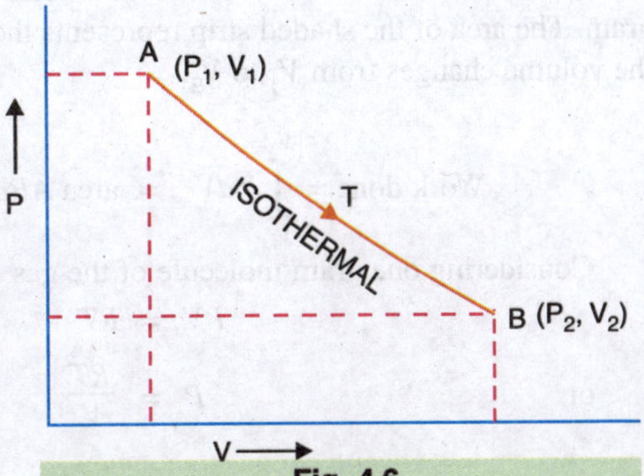

Fig. 4.6

Thus, during isothermal process, the temperature of the working substance remains constant. It can absorb heat or give heat to the surroundings. The equation for isothermal process is

$$PV = RT = \text{constant (for one gram molecule of a gas)}$$

For $n$ gram molecules of a gas,

$$PV = nRT$$

For an ideal gas undergoing isothermal process,

$$(U_2 - U_1) = 0$$

∴ From the first law of thermodynamics, we get

$$Q = W$$

*i.e.* in an isothermal process the heat supplied to an ideal gas is equal to the work done by the gas.

## 4.11 The Indicator Diagram

It is very convenient to represent the behaviour of an engine by an indicator diagram. This helps to understand the performance of heat engines. Suppose that a system is taken from an initial equilibrium state $A$ to the final equilibrium state $B$ along path $AB$ and the process is quasistatic (Fig 4.7).

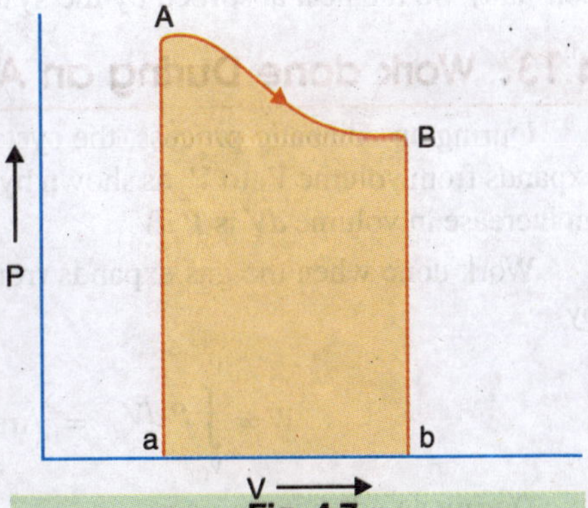

Fig. 4.7

Each point on the path traced represents the state of the system in terms of coordinates ($P$, $V$). Such a trace obtained in an actual engine is called indicator diagram or $P - V$ diagram.

The work done by the gas in expanding against pressure $P$ is $PdV$, where $dV$ is the infinitesimal change in volume. For finite process $AB$, the work done

$$= \int PdV = \text{area } ABba$$

Thus, the indicator diagram directly indicates the work done by the engine during each cycle of operation, the work being equal to the area enclosed by the indicator diagram. The work done is +ve if the indicator diagram is traced in clock-wise direction and –ve if the direction is anticlock wise.

## 4.12 Work done During an Isothermal Process

When a gas is allowed to expand isothermally, work is done by it.

Let the initial and final volumes be $V_1$ and $V_2$ respectively. Fig 4.8 represents the indicator diagram. The area of the shaded strip represents the work done for a small change in volume dV. When the volume changes from $V_1$ to $V_2$

$$\text{Work done} = \int_{V_1}^{V_2} P\,dV = \text{area } ABba$$

Considering one gram molecule of the gas

$$PV = RT$$

or

$$P = \frac{RT}{V}$$

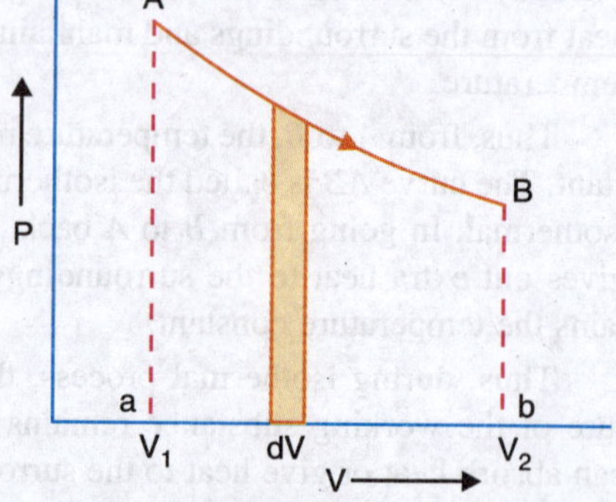


Fig. 4.8

$$\therefore \quad W = RT\int_{V_1}^{V_2} \frac{dV}{V}$$

$$W = RT \times 2.303 \log_{10} \frac{V_2}{V_1} \quad \text{...(4.8)}$$

Also

$$P_1 V_1 = P_2 V_2$$

$$\frac{V_2}{V_1} = \frac{P_1}{P_2}$$

$$\therefore \quad W = RT \times 2.303 \log_{10} \frac{P_1}{P_2} \quad \text{...(4.9)}$$

Here, the change in the internal energy of the system is zero (because the temperature remains constant). So the heat absorbed by the system is equal to the work done by it.

## 4.13 Work done During an Adiabatic Process

During an adiabatic process, the system is thermally insulated from the surroundings. The gas expands from volume $V_1$ to $V_2$ as shown by indicator diagram [Fig. 4.9]. The work done by the gas for an increase in volume $dV$ is $P\,dV$.

Work done when the gas expands from $V_1$ to $V_2$ is given by

$$W = \int_{V_1}^{V_2} P\,dV = \text{Area } ABba$$

During an adiabatic process

$$PV^{\gamma} = \text{constant} = \text{K}$$

or

$$P = \frac{K}{V^{\gamma}}$$

$$\therefore \quad W = K\int_{V_1}^{V_2} \frac{dV}{V^{\gamma}}$$

$$= \frac{K}{1-\gamma}\left[\frac{1}{V_2^{\gamma-1}} - \frac{1}{V_1^{\gamma-1}}\right]$$

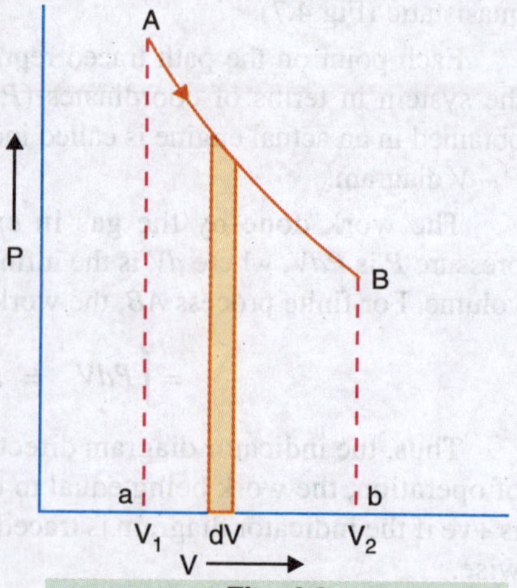


Fig. 4.9

Since $A$ and $B$ lie on the same adiabatic,

$$P_1 V_1^{\gamma} = P_2 V_2^{\gamma} = K$$

$$W = \frac{1}{1-\gamma}\left[\frac{K}{V_2^{\gamma-1}} - \frac{K}{V_1^{\gamma-1}}\right]$$

$$= \frac{1}{1-\gamma}\left[\frac{P_2 V_2^{\gamma}}{V_2^{\gamma-1}} - \frac{P_1 V_1^{\gamma}}{V_1^{\gamma-1}}\right]$$

$$W = \frac{1}{1-\gamma}[P_2 V_2 - P_1 V_1] \quad \text{...(4.10)}$$

Taking $T_1$ and $T_2$ as the temperatures at points $A$ and $B$ respectively and considering one gram molecule of the gas

$$P_1 V_1 = RT_1$$

$$P_2 V_2 = RT_2$$

Substituting these values in equation (4.10)

$$W = \frac{1}{1-\gamma}[RT_2 - RT_1]$$

$$W = \frac{R}{1-\gamma}[T_2 - T_1]$$

$$= \frac{R}{\gamma-1}[T_1 - T_2] \quad \text{...(4.11)}$$

Hence the work done in adiabatic process depends only upon the initial and final temperatures $T_1$ and $T_2$. Thus, *the work done along any adiabatic between two isothermals is independent of the particular adiabatic*.

## 4.14 Slopes of Adiabatics and Isothermals

In an isothermal process

$$PV = \text{constant}$$

Differentiating

$$P\,dV + V\,dP = 0$$

or

$$\frac{dP}{dV} = -\frac{P}{V} \quad \text{...(4.12)}$$

In an adiabatic process

$$P V^{\gamma} = \text{constant}$$

Differentiating,

$$P \gamma V^{\gamma-1}\,dV + V^{\gamma}\,dP = 0$$

$$\frac{dP}{dV} = -\frac{\gamma P}{V} \quad \text{...(4.13)}$$

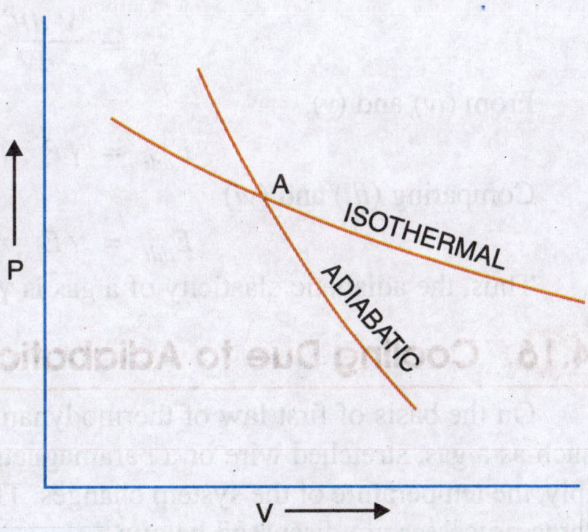


Fig. 4.10

Therefore, the slope $\left(\frac{dP}{dV}\right)$ of an adiabatic is $\gamma$ times the slope of the isothermal.

Hence, the adiabatic curve is steeper than the isothermal curve (Fig 4.10) at a point $A$ where the two curves intersect each other.

## 4.15 Relation Between Adiabatic and Isothermal Elasticities

### 1. Isothermal Elasticity

During an isothermal process

$$PV = \text{constant}$$

Differentiating,

$$PdV + VdP = 0$$

or $$\frac{V.dP}{-dV} = P \quad \text{...(i)}$$

From the definition of elasticity of a gas

$$E_{iso} = \frac{dP}{-dV/V}$$

$$= \frac{VdP}{-dV} \quad \text{...(ii)}$$

From (*i*) and (*ii*)

$$E_{iso} = P \quad \text{...(iii)}$$

### 2. Adiabatic Elasticity

During an adiabatic process

$$PV^{\gamma} = \text{constant}$$

Differentiating, $P\gamma V^{\gamma-1}\, dV + V^{\gamma}\, dP = 0$

or $$\frac{VdP}{-dV} = \gamma P \quad \text{...(iv)}$$

From the definition of elasticity of a gas

$$E_{adi} = \frac{dP}{-dV/V}$$

$$= \frac{V.dP}{-dV} \quad \text{...(v)}$$

From (*iv*) and (*v*),

$$E_{adi} = \gamma P \quad \text{...(vi)}$$

Comparing (*iii*) and (*vi*)

$$E_{adi} = \gamma E_{iso}$$

Thus, the adiabatic elasticity of a gas is $\gamma$ times the isothermal elasticity.

## 4.16 Cooling Due to Adiabatic Reversible Processes

On the basis of first law of thermodynamics, we can show that when a thermodynamic system such as a gas, stretched wire or a paramagnetic salt is made to undergo an adiabatic process reversibly, the temperature of the system changes. The thermodynamic coordinates taking part in the above three processes are discussed below:

**1. Gas expansion:** A system of gas is a function of three parameters, the pressure ($P$), the volume ($V$) and the temperature ($T$), *i.e.*, $f(P, V, T)$.

**2. Stretched wire:** System of stretched wire is a function of stretching force ($F$), the length of the stretching wire ($l$) and the temperature of the wire ($T$), *i.e.*, $f(F, l, T)$. Here pressure and volume are considered to be constant.

**3. Paramagnetic salt :** A system of paramagnetic salt is a function of the magnetic field ($H$), the intensity of magnetisation ($I$) and the temperature of the salt ($T$), *i.e.*, $f(H, I, T)$.

In all these cases, we can calculate the work done ($dW$) in a reversible infinitesimal change *i.e.*, the corresponding changes in above parameters. Let us write down the work done in each case.

**1.** In gas expansion, the work done by the system is

$$dW = P \cdot dV$$

**2.** In stretching a wire, the work done is

$$dW = -F \cdot dl$$

**3.** The paramagnetic salt in a magnetic field,

$$dW = -H\,dl$$

We presume here that the work done by the system is positive. For example: in case of gas expansion, the gas does work against a pressure, in stretching a wire, the work is done by tension $F$ to produce a change in length through $dl$, because the wire by iteself wants to contract due to its elasticirity and not to expand, as is the case in a gas. Hence –ve sign is used with $Fdl$. The same is true for the case of a paramagnetic salt in magnetic field, *i.e.*, $dW = -Hdl$.

A general formula, which will describe the above three systems can be of the form

$$dW = X\,dx \qquad ...(i)$$

where X is called as generalised force and dx as generalised extension. The generalised force ($P, -F, -H$) is a unique function of the extensions ($V, l, I$) respectively, at a fixed temperature. The generalised parameters for each case will be

**1.** for Gas expansion, $X = P$ and $dx = dV$

**2.** for Stretching a wire, $X = -F$ and $dx = dl$

**3.** for paramagnetic salt in field, $X = -H$ and $dx = dI$

From the first law of thermodynamics

$$\delta Q = dU + \delta W$$

In adiabatic process, $\delta Q = 0$

$$\therefore \quad dU = -\delta W$$

In terms of generalised parameters

$$dU = -X\,dx$$

But in adiabatic process,

$$dU = C_v\,dT$$

$$\therefore \quad C_v\,dT = -X\,dx$$

or

$$dT = -\frac{1}{C_v} X\,dx \qquad ...(ii)$$

In adiabatic process, $P$ and $V$ are related by

$$PV^\gamma = \text{constant}$$

Differentiating, we get

$$PV^{\gamma-1}\,dV + V^\gamma\,dP = 0$$

or

$$PdV = -\frac{V}{\gamma}dP \qquad ...(iii)$$

for $n$ gm. moles of a gas, the equation of state is

$$PV = nRT$$

At constant pressure, differenting *w.r.* to $T$, we get

$$P\left(\frac{\partial V}{\partial T}\right)_P = nR$$

Multiplying both sides by $T$, we get

$$PT\left(\frac{\partial V}{\partial T}\right)_P = nRT = PV$$

or

$$T\left(\frac{\partial V}{\partial T}\right)_P = V$$

Substituting for $V$ in eqn. (*iii*), we have

$$PdV = -\frac{T}{\gamma}\left(\frac{\partial V}{\partial T}\right)_P dP$$

Substituting for $PdV$ in eqn. (*ii*), we get

$$dT = \frac{T}{\gamma C_V}\left(\frac{\partial V}{\partial T}\right)_P dP$$

But

$$\gamma = \frac{C_P}{C_V} \quad \text{or} \quad \gamma C_V = C_P$$

$$\therefore \quad dT = \frac{T}{C_P}\left(\frac{\partial V}{\partial T}\right)_P dP \qquad \text{...(iv)}$$

In terms of generalised coordinates, the change in temperature during adiabatic process may be written as

$$dT = \frac{T}{C_X}\left(\frac{\partial x}{\partial T}\right)_X dx \qquad \text{...(v)}$$

This is the equation of state of the system in terms of generalised parameters. Substituting the corresponding values of parameters, the change in temperature of the system undergoing adiabatic process can be evaluated. Let us examine the same for the three cases:

## 1. Adiabatic expansion of gas

In case of a gas

$$X = P \quad \text{and} \quad x = V$$

substituting in eqn. (*v*), the change

$$dT = \frac{T}{C_P}\left(\frac{\partial V}{\partial T}\right)_P dP \qquad \text{...(vi)}$$

During expansion, $dP$ decreases *i.e.* falls. It means the temperature of the gas also decreases. Thus, if a highly campressed gas is suddenly allowed to expand adiabatically, the gas cools considerably.

If $\alpha$ is the coefficient of volume expansion, given by $\alpha = \frac{1}{V}\left(\frac{\partial V}{\partial T}\right)_P$, then

$$dT = \frac{TV\alpha}{C_P} dP \qquad \text{...(vii)}$$

## 2. Stretching of a wire

In case of adiabatic stretching of a wire,

$$X = -F \quad \text{and} \quad x = l$$

Substituting in eqn. (*v*), the change in temperature

$$dT = \frac{-T}{C_F}\left(\frac{\partial l}{\partial T}\right)_F dF \quad \text{...(viii)}$$

Since length increasess with increase in temperature, $\left(\frac{\partial l}{\partial T}\right)$ is positive, from eqn (*viii*), therefore, we conclude that wire of those substances which expand on heating, show a cooling effect when stretched adiabatically.

If β is coefficient of linear expansion, *m* the mass per unit tength, then $\beta = \frac{1}{l}\left(\frac{\partial l}{\partial T}\right)_F$ or $\left(\frac{\partial l}{\partial T}\right)_F = \beta l$ and $C_F = mlC_P$

$$\therefore \quad dT = -\frac{T}{mlC_P}\beta l d_F = \frac{-T\beta}{mC_P}dF \quad \text{...(ix)}$$

This gives fall in temperature due to adiabatic stretching of a wire.

### 3. Adiabatic demagnetisation of paramagnetic salt

In this case, $X = -H$ and $x = I$

Substituting in eqn. (*v*), the change in temperature

$$dT = -\frac{T}{C_H}\left(\frac{\partial I}{\partial T}\right)_H dH \quad \text{...(x)}$$

where $C_H$ is the specific heat of the substance at constant field.

Here $\left(\frac{\partial I}{\partial T}\right)_H$ is always negative. Hence R.H.S. is always positive. If *H* decreases, *dH* is – *ve* and hence *dT* is –*ve*. Hence, a cooling is produced during adiabatic demagnetisation of a paramagnetic substance.

## 4.17 Clement and Desormes Method–Determination of γ

Clement and Desormes in 1819 designed an experiment to find γ, the ratio between the two specific heats of a gas.

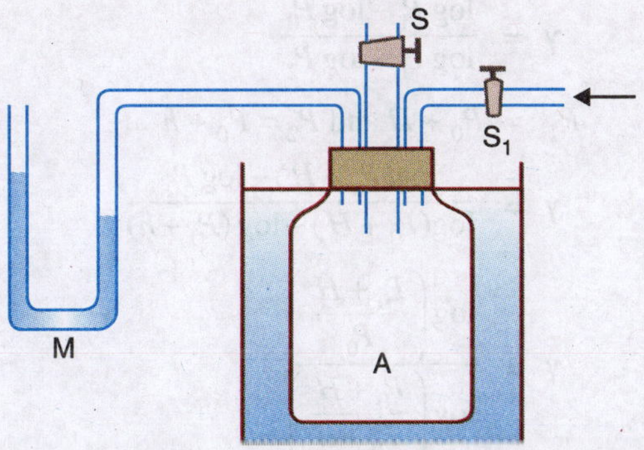


**Fig. 4.11**

The vessel *A* has a capacity of 20 to 30 litres and is fitted in a box containing cotton and wool. At the top end, three tubes are fitted as shown in Fig. 4.11. Through $S_1$, dry air is forced into the vessel *A*. The stop cock $S_1$ is closed when the pressure inside *A* is slightly greater than the atmospheric pressure. Let the difference in level on the two sides of the manometer be H and the atmospheric pressure be $P_0$. The pressure of air inside the vessel is $P_1$.

The stop-cock $S$ is suddenly opened and closed just at the moment when the levels of the liquid on the two sides of the manometer are the same. Some quantity of air escapes to the atmosphere. The air inside the vessel expands adiabatically. The temperature of air inside the vessel falls due to adiabatic expansion. The air inside the vessel is allowed to gain heat from the surroundings and it finally attains the temperature of the surroundings. Let the pressure at the end be $P_2$ and the difference in levels on the two sides of the manometer be $h$.

## Theory

Consider a fixed mass of air left in the vessel in the end. This mass of air has expanded from volume $V_1$ (less than the volume of the vessel) at pressure $P_1$ to volume $V_2$ at pressure $P_0$. The process is adiabatic as shown by the curve $AB$ (Fig. 4.12).

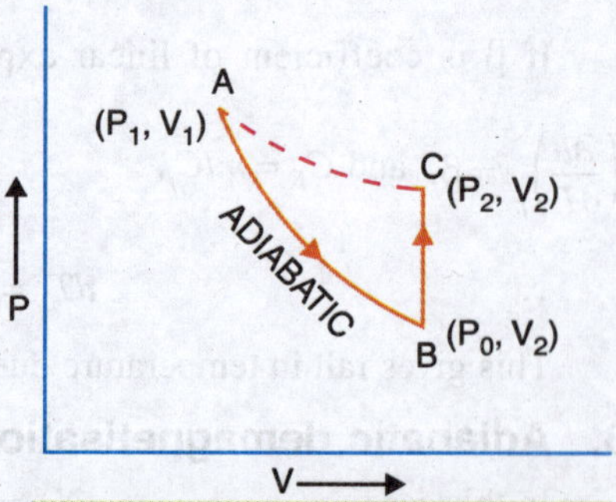


Fig. 4.12

$$P_1 V_1^{\gamma} = P_0 V_2^{\gamma}$$

$$\frac{P_1}{P_0} = \left(\frac{V_2}{V_1}\right)^{\gamma} \qquad \text{...}(i)$$

Finally the point $C$ is reached. The points $A$ and $C$ are at the room temperature. Therefore $AC$ can be considered as an isothermal.

$$P_1 V_1 = P_2 V_2$$

$$\frac{V_2}{V_1} = \frac{P_1}{P_2} \qquad \text{...}(ii)$$

Substituting the value of $\frac{V_2}{V_1}$ in equation ($i$),

$$\frac{P_1}{P_0} = \left(\frac{P_1}{P_2}\right)^{\gamma}$$

Taking logarithms,

$$\log P_1 - \log P_0 = \gamma [\log P_1 - \log P_2]$$

$$\gamma = \frac{\log P_1 - \log P_0}{\log P_1 - \log P_2} \qquad \text{...}(iii)$$

But $\quad P_1 = P_0 + H$ and $P_2 = P_0 + h$

$$\therefore \quad \gamma = \frac{\log(P_0 + H) - \log P_0}{\log(P_0 + H) - \log(P_0 + h)}$$

$$\gamma = \frac{\log\left(\frac{P_0 + H}{P_0}\right)}{\log\left(\frac{P_0 + H}{P_0 + h}\right)}$$

$$= \frac{\log\left(1 + \frac{H}{P_0}\right)}{\log\left(1 + \frac{H - h}{P_0 + h}\right)}$$

Approximately, $$\gamma = \frac{\frac{H}{P_0}}{\frac{H-h}{P_0}} = \frac{H}{H-h}$$

Hence $$\gamma = \frac{H}{H-h} \qquad ...(iv)$$

Similarly, γ for any gas can be determined by this method.

### Drawbacks

When the stop-cock is opened, a series of oscillations are set up. This is shown by the up and down movement of the liquid in the manometer. Therefore, the exact moment when the stopcock should be closed is not known. The pressure may not be equal to the atmospheric pressure when the stop-cock is closed. It may be higher or less than the atmospheric pressure. Thus the result obtained will not be accurate.

## 4.18 Partington's Method

Lummer, Pringsheim and Partington designed an apparatus to determine the value of γ. In this method, the pressure and temperature are measured accurately before and after the adiabatic expansion.

The apparatus consists of a vessel *A* having a capacity between 130 and 150 litres. The valve *L* can be opened and closed suddenly.

It is controlled by a spring arrangement (Fig. 4.13). Dry air (or gas) at a pressure higher than the atmospheric pressure is forced into the vessel *A* and the stop-cock *S* is closed. The oil manometer *M* is used to measure the pressure of air inside the vessel *A*. The bolometer *B* (a platinum wire) and a sensitive galvanometer are used in the Wheatstone's bridge arrangement.

The vessel is surrounded by a constant temperature bath. Let the initial pressure and temperature be $P_1$ and $T_1$ (room temperature). The bridge is kept slightly disturbed from the balanced position.

The valve *L* is suddenly opened and closed. The Wheatstone's bridge is at once adjusted for balanced position. The temperature of air inside *A* has decreased due to adiabatic expansion of air. Let the temperature inside be $T_0$ and the atmospheric pressure $P_0$. If the apparatus is allowed to remain as such for some time, it will gain heat from the surroundings and the balance point gets disturbed. In order that the balance point remains undisturbed, some pieces of ice are added into the water surrounding the vessel *A*. When the temperature of water bath is the same as that of air just after adiabatic expansion, the bridge will remain balanced.

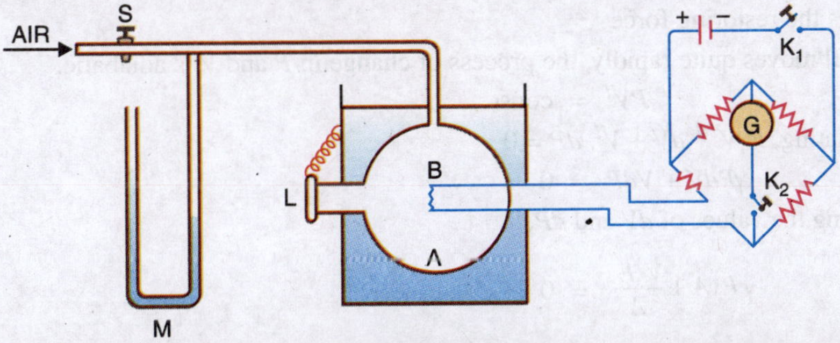


Fig 4.13

The temperature $T_0$ of the bath represents the temperature of air after the adiabatic expansion.

$$\frac{P_1^{\gamma-1}}{T_1^{\gamma}} = \frac{P_0^{\gamma-1}}{T_0^{\gamma}}$$

$$\left(\frac{P_1}{P_0}\right)^{\gamma-1} = \left(\frac{T_1}{T_0}\right)^{\gamma}$$

$$(\gamma-1)(\log P_1 - \log P_0) = \gamma[\log T_1 - \log T_0]$$

$$\gamma = \frac{\log P_1 - \log P_0}{(\log P_1 - \log P_0) - (\log T_1 - \log T_0)}$$

As $P_1$, $P_0$, $T_1$ and $T_0$ are known, $\gamma$ can be calculated. The value of $\gamma$ for air at 17°C is found to be 1.4034.

### Advantages.

1. Due to the large volume of the vessel, the expansion is adiabatic.
2. The temperatures are measured accurately just before and after the adiabatic expansion.

### Drawbacks.

This method cannot be used to find the value of $\gamma$ at higher temperatures because it is not possible to determine the cooling correction accurately.

## 4.19 Ruchhardt's Experiment

In 1929, Ruchhardt designed an apparatus to find the value of $\gamma$. It is based on the principle of mechanics. Air (or gas) is enclosed in a big jar (Fig. 4.14). A tube of uniform area of cross section is fitted and a ball of mass $m$ fits in the tube just like a piston. In the equilibrium position, the ball is at the point A. The pressure P of air inside the vessel, is given by

$$P = P_0 + \frac{mg}{A}$$

where $P_0$ is the atmospheric pressure and $A$ is the area of cross-section of the tube.

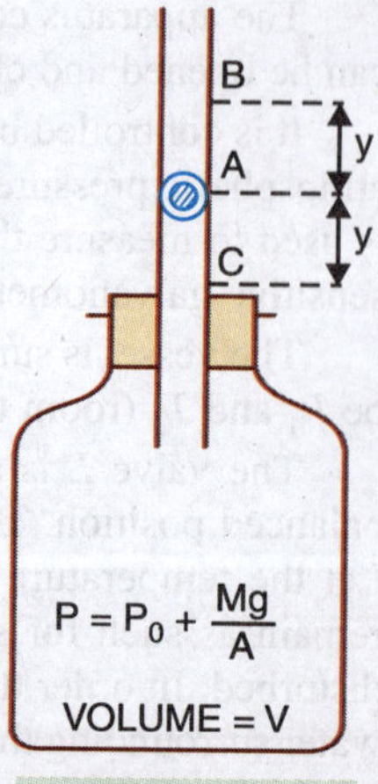

Fig. 4.14

The ball is given a small downward movement through a distance $y$. It moves up and down and executes simple harmonic motion. Let the change in volume be $dV$ and change in pressure be $dP$.

$$dV = yA$$

and

$$dP = \frac{F}{A}$$

where $F$ is the restoring force

As the ball moves quite rapidly, the process of change in $P$ and $V$ is adiabatic.

$$\therefore \quad PV^{\gamma} = \text{const}$$

Differentiating, $P\gamma V^{\gamma-1}\, dV + V^{\gamma}.\, dP = 0$

$$\gamma P dV + V dP = 0$$

Substituting the values of $dV$ and $dP$,

$$\gamma PyA + \frac{V.F}{A} = 0$$

But

$$F = M.\frac{d^2y}{dt^2}$$

or

$$F = -\frac{\gamma PA^2}{V}.y \qquad ...(i)$$

$$\therefore \qquad \frac{d^2 y}{dt^2} + \frac{\gamma PA^2}{VM} \cdot y = 0 \qquad ...(ii)$$

Equation (*ii*) represents S.H.M.

Let the time period be $t$

$$t = 2\pi\sqrt{\frac{\text{displacement}}{\text{acceleration}}}$$

$$= 2\pi\sqrt{\frac{y}{-d^2 y / dt^2}}$$

$$t = 2\pi\sqrt{\frac{VM}{\gamma PA^2}}$$

$$\gamma = \frac{4\pi^2 VM}{PA^2 t^2} \qquad ...(iii)$$

As $V, M, P$ and $A$ are already known, only the time period $t$ is to be determined. Hence $\gamma$ can be calculated.

This experiment is not very accurate because the ball comes to rest in a short time due to friction and the time period cannot be determined accurately.

## Rinkle's Modification

Rinkle suggested a modification in Ruchhardt's experiment. The ball is kept in the equilibrium position where the pressure of air (or gas) is exactly equal to the atmospheric pressure. The ball is allowed to drop and the distance L through which it drops before starting to move up again is noted. This distance is found accurately by taking a series of photographs of the moving ball.

$$\text{Work done} = \int_0^L F.dy$$

But

$$F = \frac{\gamma P_0 A^2 y}{V}$$

$$W = \frac{\gamma P_0 A^2}{V}\int_0^L y dy$$

$$= \frac{\gamma P_0 A^2 L^2}{2V} \qquad ...(iv)$$

$$\text{Work done} = MgL \qquad ...(v)$$

Equating (*iv*) and (*v*)

$$\frac{\gamma P_0 A^2 L^2}{2V} = MgL$$

$$\gamma = \frac{2MgV}{P_0 A^2 L} \qquad ...(vi)$$

As $M, g, V, P_0$ and $A$ are known, $L$ is measured from the photographic film. Hence $\gamma$ can be calculated. The error in the result is due to the following assumptions: (*i*) that there is no friction, (*ii*) that the gas is ideal and (*iii*) the process is completely adiabatic.

## 4.20 Reversible and Irreversible Process

### Reversible Process

A reversible process is one which can be retracted in opposite direction by changing the external conditions infinitesimally.

The working substance in the reverse process passes through all the stages as in the direct process in such a way that all changes occuring in the direct process are exactly repeated in the opposite order and inverse sense and no changes are left in any of the bodies participating in the process or in the surroundings. If heat is absorbed by the substance in the direct process, the same quantity of heat will be given out by it in the reverse process, and if work is done by the substance in the direct process, an equal amount of work will be done on the substance in the reverse process. Thus, there is no wastage or loss of energy at all in the reversible process.

Consider a gas enclosed in a cylinder, made of a perfectly conducting material and immersed in a large tank of water of a constant temperature. Suppose the gas is compressed very slowly such that its temperature remains unchanged throughout. Obviously to do so, small pauses will have to be given in between various small compressions to enable the heat generated by compression to pass out in to the surrounding water. If now after reducing considerably the volume of the gas, it is now allowed to expand isothermally and the expansion be just as infinitely small as done in compression with similar pauses in between, heat will flow in from enclosing water and the temperature of the gas remains unchanged. Exactly the same amount of heat will be received during the expansion as was given up during compression. Thus, all the stages of the direct process are retracked in the opposite direction and inverse order. Hence the infinitesimally slow isothermal expansion and compression of a gas is a **reversible process**. Infact **all isothermal and adiabatic operations are reversible when carried out very slowly.**

The process will not be reversible if there is any loss of heat due to friction, radiation or conduction. If the changes take place rapidly, the process will not be reversibly. The energy used in over coming friction can not be retraced.

The conditions of reversibility for any heat engine or process can be stated as follows:

**1.** The pressure and temperature of the working substance must not differ appreciably from those of the surrounding at any stage of the cycle of operation.

**2.** All the processes taking place in the cycle of operation must be infinitely slow.

**3.** The working parts of the engine must be completely free from friction.

**4.** There should not be any loss of energy due to conduction, convection or radiation during the cycle of operation.

It should be remembered that the complete reversible process or cycle of operation is only an ideal case. In an actual process, there is always loss of heat due to friction, conduction, convection or radiation.

### Irreversible Process

In nature all changes are irreversible because of the following reasons : (1) The conditions for thermodynamic equilibrium *i.e.* mechanical, thermal or chemical equilibrium are not satisfied because a natural process does not take place quasi-statically. (2) Dissipative effects, such as friction, viscosity, inelasticity, electric resistance, eddy-formation etc. are always present.

Examples of some natural irreversible processes are: spontaneous expansion of a gas into an evacuated space, spontaneous conduction of heat along a metal bar which is hot at one end and cold at the other, transfer of heat by radiation from a hotter to a colder body, transfer of electricity through a resistor. Nonetheless reversible processes are most important in thermodynamics because they can be handled analytically.

## 4.21 Heat Engines

Heat Engine

Any practical machine which converts heat into mechanical work is called a heat engine. Heat engines in their operation absorb heat at a higher temperature, converts part of it into mechanical work, and rejects the remaining heat at a low temperature. In this process, a working substance is used. In steam engines, the working substance is water vapour, and in all gas engines the working substance is a combustible mixture of gases.

In any heat engine, the working substance goes through certain changes of pressure, volume and temperature, and then returns to the initial state. The complete changes through which the working substance undergoes from its initial state and back to its starting state constitute one cycle of operation.

## 4.22 Definition of Efficiency

The efficiency, η, of a heat engine is defined as the ratio of the mechanical work done by the engine in one cycle to the heat absorbed from the high temperature source. Thus

$$\eta = \frac{Q_1 - Q_2}{Q_1} \qquad ...(4.14)$$

where $Q_1$ is the heat absorbed from the source at high temperature, $Q_2$, is the heat rejected to a sink at low temperature and $(Q_1 - Q_2)$ is the mechanical work done by the engine in one cycle.

Since $(Q_1 - Q_2) < Q_1$, the efficiency can never be 100%.

## 4.23 Carnot's Ideal Heat Engine

In 1824 the French engineer Sadi Carnot conceived a theoretical engine which is free from all practical imperfections. Such an engine can not be realised in practice. It has maximum efficiency and it is an ideal heat engine. Sadi Cornot's heat engine requires the following important parts (Fig. 4.15):

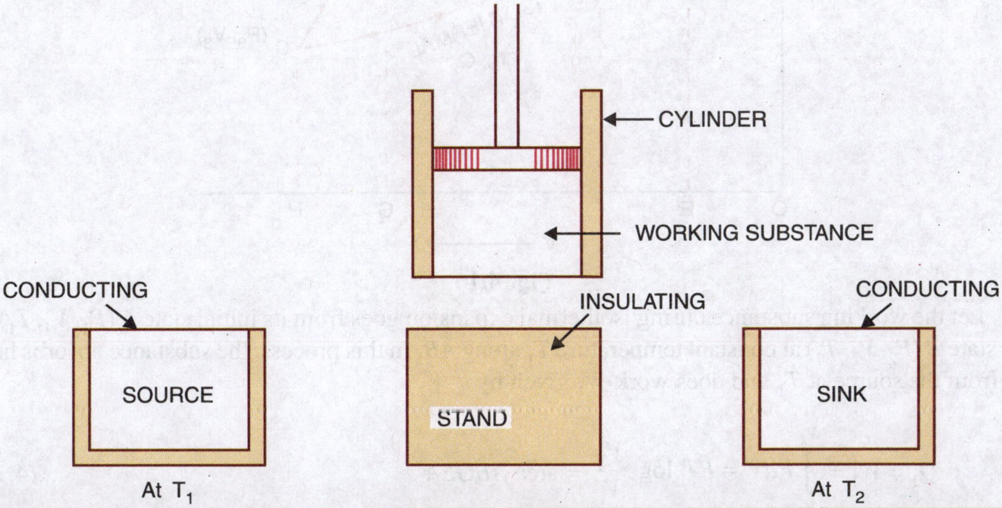


Fig. 4.15

**1.** A **cylinder** having perfectly non-conducting walls, a perfectly conducting base and is provided with a perfectly non-conducting piston which moves without friction in the cylinder. The cylinder contains one mole of perfect gas as the working substance.

**2. Source.** A reservoir maintained at a constant temperature $T_1$ from which the engine can draw heat by perfect conduction. It has infinite thermal capacity and any amount of heat can be drawn from it at constant temperature $T_1$.

**3. Heat insulating stand.** A perfectly non-conducting platform acts as a stand for adiabatic processes.

**4. Sink.** A reservoir maintained at a constant lower temperature $T_2$ ($T_2 < T_1$) to which the heat engine can reject any amount of heat. The thermal capacity of sink is infinite so that its temperature remains constant at $T_2$, no matter how much heat is given to it.

## 4.24 Carnot's Cycle

In order to obtain a continuous supply of work, the working substance is subjected to the following cycle of quasi-static operations known an Carnot's cycle (Fig. 4.16).

**1. Isothermal expansion.** The cylinder is first placed on the source, so that the gas acquires the temperature $T_1$ of the source. It is then allowed to undergo quasi-static expansion. As the gas expands, its temperature tends to fall. Heat passes into the cylinder through the perfectly conducting base which is in contact with the source. The gas therefore, undergoes slow isothermal expansion at the constant temperature $T_1$.

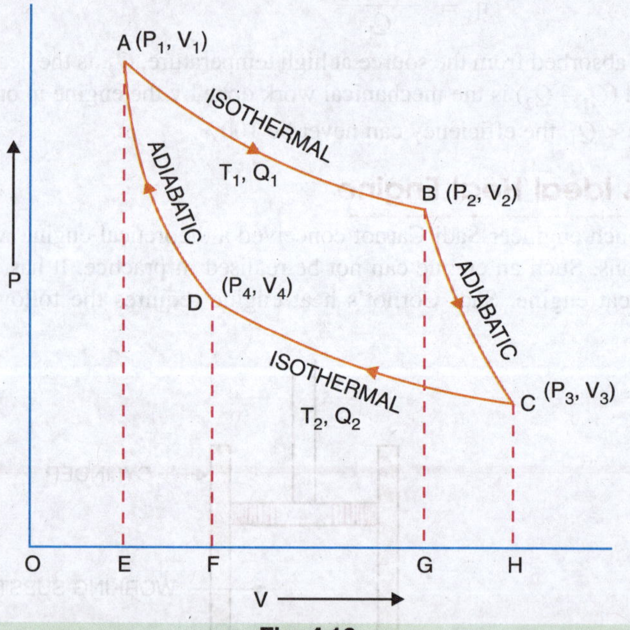


**Fig. 4.16**

Let the working substance during isothermal expansion goes from its initial state $A$ ($P_1$, $V_1$, $T_1$) to the state $B$ ($P_2$, $V_2$, $T_1$) at constant temperature $T_1$ along $AB$. In this process, the substance absorbs heat $Q_1$ from the source at $T_1$ and does work $W_1$ given by

$$Q_1 = W_1 = \int_{V_1}^{V_2} PdV = RT_1 \log_e \frac{V_2}{V_1} = \text{area } ABGEA \qquad ...(4.15)$$

**2. Adiabatic expansion.** The cylinder is now removed from the source and is placed on the insulating stand. The gas is allowed to undergo slow adiabatic expansion, performing external work at the expense of its inernal energy, until its temperature falls to $T_2$, the same as that of the sink.

This operation is represented by the adiabatic *BC*, starting from the state $B\ (P_2, V_2, T_1)$ to the state $C\ (P_3, V_3, T_2)$. In this process, there is no transfer of heat, the temperature of the substance falls to $T_2$ and it does some external work $W_2$ given by

$$W_2 = \int_{V_2}^{V_3} P.dV = K\int_{V_2}^{V_3} \frac{dV}{V^\gamma}$$

($\because$ During adiabatic process, $PV^\gamma$ = constant = $K$)

$$= \frac{KV_3^{1-\gamma} - KV_2^{1-\gamma}}{1-\gamma}$$

$$= \frac{P_3 V_3 - P_2 V_2}{1-\gamma} \qquad (\because P_2 V_2^\gamma = P_3 V_3^\gamma = K)$$

$$= \frac{RT_2 - RT_1}{1-\gamma}$$

$$= \frac{R(T_1 - T_2)}{\gamma - 1} = \text{Area } BCHGB \qquad ...(4.16)$$

**3. Isothermal Compression.** The cylinder is now removed from the insulating stand and is placed on the sink which is at a temperature $T_2$. The piston is now very slowly moved inwards so that the work is done on the gas. The temperature tends to increase due to heat produced by compression since the conducting base of the cylinder is in contact with the sink, the heat developed passes to the sink and the temperature of the gas remains constant at $T_2$. Thus the gas undergoes isothermal compression at a constant temperature $T_2$ and gives up some heat to the sink.

This operation is represented by the isothermal *CD*, starting from the state $C\ (P_3, V_3, T_2)$ to the state $D\ (P_4, V_4, T_2)$. In this process, the substance rejects heat $Q_2$ to the sink at $T_2$ and work $W_3$ is done on the substance given by

$$Q_2 = W_3 = \int_{V_3}^{V_4} PdV = RT_2 \log_e \frac{V_4}{V_3} \qquad ...(4.17)$$

$$= -RT_2 \log_e \frac{V_3}{V_4} = \text{area } CHFDC$$

(– *ve* signe indicates that work is done on the working substance)

**4. Adiabatic Compression.** The cylinder is now removed from sink and again placed on the insulating stand. The piston is slowly moved inwards so that the gas is adiabatically compressed and the temperature rises. The adiabatic compression is continued till the gas comes back to its original condition *i.e.* state $A\ (P_1, V_1, T_1)$, thus completing one full cycle.

This operation is represented by adiabatic *DA*, starting from $D\ (P_4, V_4, T_2)$ to the final state $A\ (P_1, V_1, T_1)$. In this process, work $W_4$ is done on the substance and is given by

$$W_4 = \int_{V_4}^{V_1} P.dV$$

$$= -\frac{R(T_1 - T_2)}{\gamma - 1} = \text{Area } DFEAD \qquad ...(4.18)$$

(– *ve* sign idicates that work is done on the working substance. Since $W_2$ and $W_4$ are equal and opposite, they cancel each other.)

## Work done by the engine per cycle

During the above cycle, the working substance absorbs an amount of heat $Q_1$ from the source and rejects $Q_2$ to the sink.

Hence, the net amount of heat absorbed by the gas per cycle

$$= Q_1 - Q_2$$

The net work done by the engine per cycle

$$= W_1 + W_2 + W_3 + W_4$$

$$= W_1 + W_3 \qquad (\because W_2 = -W_4)$$

From the graph, the net work done per cycle

$$= \text{area } ABGEA + \text{area } BCHGB - \text{area } CHFDC - \text{area } DFEAD$$

$$= \text{area } ABCDA$$

Thus, the area enclosed by the Carnot's cycle consisting of two isothermals and two adiabatics gives the net amount of work done per cycle.

In the cyclic process,

$$\text{Net heat absorbed} = \text{Net work done per cycle.}$$

$$Q_1 - Q_2 = W_1 + W_3$$

$$= RT_1 \log_e \frac{V_2}{V_1} - RT_2 \log_e \frac{V_3}{V_4} \qquad ...(4.19)$$

Since the points $A$ and $D$ lie on the same adiabatic $DA$

$$T_1 V_1^{\gamma-1} = T_2 V_4^{\gamma-1}$$

$$\frac{T_2}{T_1} = \left(\frac{V_1}{V_4}\right)^{\gamma-1} \qquad ...(4.20)$$

Similarly, points $B$ and $C$ lie on the same adiabatic $BC$

$$T_1 V_2^{\gamma-1} = T_2 V_3^{\gamma-1}$$

$$\therefore \quad \frac{T_2}{T_1} = \left(\frac{V_2}{V_3}\right)^{\gamma-1} \qquad ...(4.21)$$

From equations (5.20) and (5.21),

$$\left(\frac{V_1}{V_4}\right)^{\gamma-1} = \left(\frac{V_2}{V_3}\right)^{\gamma-1}$$

or

$$\frac{V_1}{V_4} = \frac{V_2}{V_3}$$

or

$$\frac{V_2}{V_1} = \frac{V_3}{V_4}$$

Substituting in equation (5.19), we get

$$\text{Net work done} = Q_1 - Q_2 = RT_1 \log_e \frac{V_2}{V_1} - RT_2 \log_e \frac{V_2}{V_1}$$

$$W = (Q_1 - Q_2) = R(T_1 - T_2) \log_e \frac{V_2}{V_1} \qquad ...(4.22)$$

## Efficiency

The efficiency of the heat engine is the rate of quantity of heat converted into work (Useful output) per cycle to the total amount of heat absorbed per cycle.

Efficiency, $$\eta = \frac{\text{Useful Output}}{\text{Input}} = \frac{W}{Q_1}$$

$$= \frac{(Q_1 - Q_2)}{Q_1}$$

$$= \frac{R(T_1 - T_2)\log_e \frac{V_2}{V_1}}{RT_1 \log_e \frac{V_2}{V_1}}$$

$$\eta = \frac{T_1 - T_2}{T_1} \qquad \text{...(4.23)}$$

or $$\eta = 1 - \frac{T_2}{T_1} \qquad \text{...(4.24)}$$

From equation (4.24), we conclude that the efficiency depends only upon the temperature of the source and sink and is always less than unity. The efficiency is independent of the nature of working substance. From equation, $\eta = 1 - \frac{T_2}{T_1}$, we get $\eta = 1$, if $T_2 = 0$K *i.e.* the temperature of the sink is at absolute zero degrees. In practice, it is never possible to reach absolute zero and hence 100% conversion of heat energy into mechanical work is not possible.

Again, the efficiency is minimum or zero when $T_1 = T_2$ *i.e.* the temperature of the source is equal to the temperature of sink, then $\eta = 0$ *i.e.* the engine does not work.

The Carnot's heat engine is perfectly reversible. It can be operated in the reverse direction also. Then it works as a refrigerator. The heat $Q_2$ is taken from the sink and external work is done on the working substance and heat $Q_2$ is rejected to the source at a higher temperature (principle of a refrigerator).

Moreover in the Carnot's heat engine, the process of isothermal and adiabatic expansions and compressions are carried out very-very slowly *i.e.* quasi-static. This is an ideal case. Any practical engine can not satisfy these conditions. Therefore, all practical engines have an efficiency less than the Carnot's engine.

## 4.25 Effective Way to Increase Efficiency

The expression for the efficiency of a Carnot's engine is

$$\eta = 1 - \frac{T_2}{T_1}$$

The efficiency $\eta$ can be increased by one of the following ways:

(*i*) Using a heat-source at constant temperature $T_1$ and heat-sink at temperature as low as possible, or

(*ii*) Using a heat-sink at constant temperature $T_2$ and a heat-source at temperature as high as possible.

In practice, it is not convenient to use heat sink at a temperature below that of the atmosphere. Therefore, the more effective way to increase the efficiency is to use a heat-source at temperature as high as possible.

## 4.26 Carnot's Engine and Refrigerator

Carnot's cycle is perfectly reversible. It can work as a heat engine and also as a refrigerator.

When it is operated as a heat engine, it absorbs heat $Q_1$ from the source at a temperature $T_1$, does an amount of work W and rejects heat $Q_2$ to the sink at temperature $T_2$, ($T_2 < T_1$) as shown in [Fig. 4.17 (*a*)].

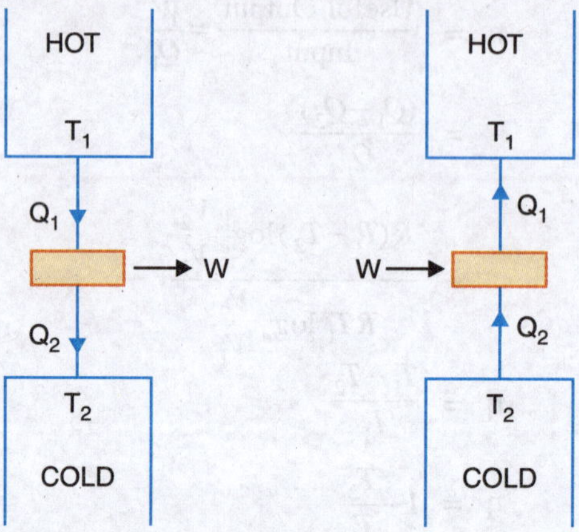


(*a*) Heat Engine (*b*) Refrigerator

**Fig. 4.17**

When it is operated as a refrigerator, it absorbs heat $Q_2$ from the sink at temperature $T_2$. $W$ amount of work is done on it by some external means and rejects heat $Q_1$ to the source at temperature $T_1$, $(T_1 > T_2)$ as shown in [Fig. 4.17 (*b*)].

In the second case [Fig. 4.17 (*b*)], heat flows from a body at a lower temperature $T_2$ to a body at a higher temperature $T_1$ with the help of external work done on the working substance. This action is that of a refrigerator. In every cycle, heat $Q_2$ is extracted from the cold body. This will not be possible if the cycle is not completely reversible.

## 4.27 Coefficient of Performance

The amount of heat absorbed at the lower temperature is $Q_2$. The amount of work done on the working substance by the external agency (input energy) = $W$ and the amount of heat rejected = $Q_1$. Here $Q_2$ is the desired refrigerating effect in each cycle.

$$\therefore \quad \text{Coefficient of performance, } P = \frac{Q_2}{W} = \frac{Q_2}{Q_1 - Q_2} \quad \text{...(4.25)}$$

Suppose 200 joules of energy is absorbed at the lower temperature and 100 joules of work is done on it by external help. Then, 200 + 100 = 300 joules are rejected at the higher temperature.

The coefficient of performance

$$P = \frac{Q_2}{W}$$

$$= \frac{Q_2}{Q_1 - Q_2}$$

$$= \frac{200}{300 - 200} = 2$$

Therefore, the coefficient of performance of a refrigerator is 200%.

In the case of heat engine, the efficiency cannot be more than 100% but in case of a refrigerator, the coefficient of performance can be much higher than 100%.

If the working substance is an ideal gas, then

$$\frac{Q_1}{T_1} = \frac{Q_2}{T_2} = \frac{Q_1 - Q_2}{T_1 - T_2}$$

or $$\frac{Q_2}{Q_1 - Q_2} = \frac{T_2}{T_1 - T_2}$$

Thus, equation (4.25) can also be expressed as

$$\text{coefficient of performance, } P = \frac{T_2}{T_1 - T_2} \quad ...(4.26)$$

## 4.28 Second Law of Thermodynamics

The first law of thermodynamics states the equivalence of heat and energy. It simply tells that whatever work is obtained, an equivalent amount of heat is used up, or vice versa. It does not say anything about the direction in which the change might occur or about the range or limit to which it can be possible. The first law shows that perpetual motion of the first kind is impossible *i.e.* energy can not be created out of nothing or production of energy without disappearance of an equivalent energy of another form is not possible.

The first law has no answer why heat always flows from a body at high temperature to a body at lower temperature and does not flow in the reverse direction. It can not explain why vast amount of a vailable heat connot be converted into mechanical work. It we could control and make use of the limitless store of heat viz-solar energy, we could have an inexhaustible supply of useful inergy. Thus, we could have a perpetual motion machine not forbidden by the first law. This is called perpetual motion of the second kind. In practice, however there is no engine which can convert the heat from the single source in to useful work without rejecting some heat to a heat sink at a lower temperature *i.e.* perpetual motion of the second kind *i.e.* production of useful energy from the internal energy of one body is impossible.

It was the quest of several such questions which led to the formulation of second law of thermodynamics. The second law is a generalization of certain experiences and observations and is concerned with the direction in which energy transfer takes place. The law has been stated in a number of ways, which means the same thing.

### Lord Kelvin's Statement

In a heat engine the working substance extracts heat from the source, converts a part of it into work and rejects the rest to a sink at a lower temperature. The temperature of the source must be higher than the surroundings and engine will not work when the temperatures of source and sink are the same. No engine has ever been constructed which converts all the heat absorbed from the source into work without rejecting a part of it to the cold body. As the engine absorbs more and more heat from the hot body, the latter suffers a continuous fall in temperature and if a continuous supply of work is desired, the hot body will in the long run become as cold as its surroundings. Then no heat flow will be possible, the engine will stop working and hence no mechanical work will be obtained. It means that we can not obtain a continuous supply of work from a single supply of it *i.e.* the presence of colder body is a must for the continuous conversion of heat into work. Such considerations led Lord Kelvin to state that

"It is impossible to get a continuous supply of work from a body by cooling it to a temperature lower than that of its surroundings."

### Planck's Statement

"It is impossible to construct an engine which, working in a complete cycle, will produce no effect other than the raising of a weight and the cooling of a heat reservoir."

Thus, it is impossible to construct an engine which working in a complete cycle, will produce no effect other than the absorption of heat from a reservoir and its conversion into an equivalent amount of work i.e. perpetual motion of the second kind is impossible. The engine must reject a part of the heat absorbed to a sink at lower temperature.

These two statements can be combined into one equivalent statement, known as the **Kelivn-Planck's statement** of the second law of thermodynamics.

### Kelvin-Planck Statement

"It is impossible to construct an engine which, operating in a cycle, has the sole effect of extracting heat from a reservoir and performing an *equivalent* amount of work."

### Clausius's Statement

According to Clausius "It is impossible for a self-acting machine working in a cyclic process, unaided by external agency, to transfer heat from a body at a lower temperature to a body at a higher temperature."

In other words it may be stated as "Heat cannot flow of itself from a colder body to a hotter body."

This statement is based upon the performance of refrigerator – a heat engine working in the backward direction. This statement means that natural flow of heat is always from a hot body to a cold body. If heat is to be transferred from cold body to hot body, work will have to be done by external agency. A refrigerator is a device which transfers heat from a colder body to hotter body by doing external work on the working substance. The compression is brought about by an external agency i.e. 'electricity' by performing work on the working substance.

## 4.29 Carnot's Theorem

### Statement

From the second law of thermodynamics two important results are derived; these conclusions are taken together to constitute Carnot's theorem which may be stated in the following forms.

(*a*) 'No engine can be more efficient than a perfectly reversible engine working between the same two temperatures.'

(*b*) 'The efficiency of all reversible engines, working between the same two temperatures is the same, whatever the working substance.'

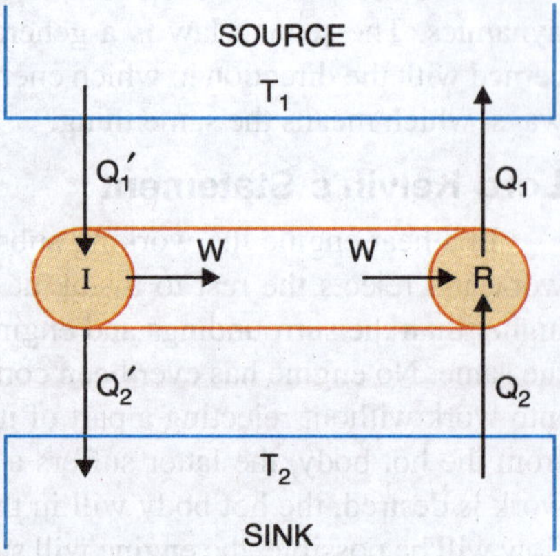


Fig. 4.18

### Proof:

**First Part:** To prove the first part of the theorem, we consider two engines *R* and *I* working between the temperatures $T_1$ and $T_2$ where $T_1 > T_2$ [Fig. 4.18]. Of these two engines *R* is reversible and *I* is irreversible.

Suppose *I* is more efficient than *R*. Suppose in each cycle, R absorbs the quantity of heat $Q_1$ from the source at $T_1$ and rejects the quantity of heat $Q_2$ to the sink at $T_2$. Suppose in each cycle *I* absorbs the quantity of heat $Q_1'$ from the source at $T_1$ and gives up the quantity of heat $Q_2'$ to the sink at $T_2$. Let the two engines do the same amount of work *W* in each cycle. According to the assumption *I* is more efficient than *R*.

$$\therefore \qquad \frac{Q_1' - Q_2'}{Q_1'} > \frac{Q_1 - Q_2}{Q_1}$$

or $\frac{W}{Q_1'} > \frac{W}{Q_1}$

or $Q_1 > Q_1'$

and $Q_1' - Q_2' = Q_1 - Q_2$

or $Q_2 - Q_2' = Q_1 - Q_1'$

Now because $Q_1 > Q_1'$

$\therefore$ $Q_2 > Q_2'$

Now suppose the two engines are coupled together so that $I$ drives $R$ backwards and suppose they use the same source and sink. The combination forms a self-acting machine in which $I$ supplies external work $W$ and $R$ absorbs this amount of work in its reverse cycle. $I$ in its cycle absorbs heat $Q_1'$ from the source and gives up heat $Q_2'$ to the sink. $R$ in its reverse cycle, absorbs heat $Q_2$ from the sink and gives up heat $Q_1$ to the source.

The net result of the complete cycle of the coupled engines is given by

Gain of heat by the source at $T_1 = Q_1 - Q_1'$

Loss of heat by the sink at $T_2 = Q_2 - Q_2'$

External work done on the system = 0

Thus, the coupled engines forming a self-acting machine unaided by any external agency transfer heat continuously from a body at low temperature to a body at a higher temperature.

This conclusion is contrary to the second law of thermodynamics, according to which in a cyclic process heat cannot be transferred from one body to another at a higher temperature by a self-acting machine. Hence our assumption is incorrect and we conclude that no engine can be more efficient than a perfectly reversible engine working between the same temperatures.

**Second Part:** The second part of the theorem may be proved by the same arguments as before. For this purpose, we consider two reversible engines $R_1$ and $R_2$ and assume that $R_2$ is more efficient than $R_1$. Proceeding in the same way we can show that $R_2$ cannot be more efficient than $R_1$. Therefore, all reversible engines working between the same two temperatures have the same efficiency.

Thus, the efficiency of a perfectly reversible engine depends only on the temperatures between which the engine work, and is independent of the nature of the working substance.

## 4.30 Steam Engine

It consists of a steam chest and a cylinder. These are connected by the steam ports $P_1$ and $P_2$. $P$ is a piston that moves inside the cylinder. $V$ is a sliding valve which can also move to and fro such that the port $P_1$ or $P_2$ is open to the steam chest. The piston and the sliding valve are connected to the main shaft eccentrically at diametrically opposite points such that when the shaft rotates they move in opposite directions. $W$ is a flywheel attached to the main shaft. The coupling between the main rod and the piston rod is such that the linear motion of the piston rod is converted to circular motion of the shaft. The machinery to be operated is connected to the main shaft.

### Working

1. Steam at high pressure from the boiler enters the steam chest. The steam enters the cylinder through the port $P_1$. The piston $P$ is pushed down due to the high pressure of steam (Fig. 4.19) The sliding valve moves up. Both the ports $P_1$ and $P_2$ are closed. This steam inside the cylinder expands and moves the piston further in the downward direction. At this position, the port $P_1$ is connected to the exhaust valve $E$ and $P_2$ is opened.

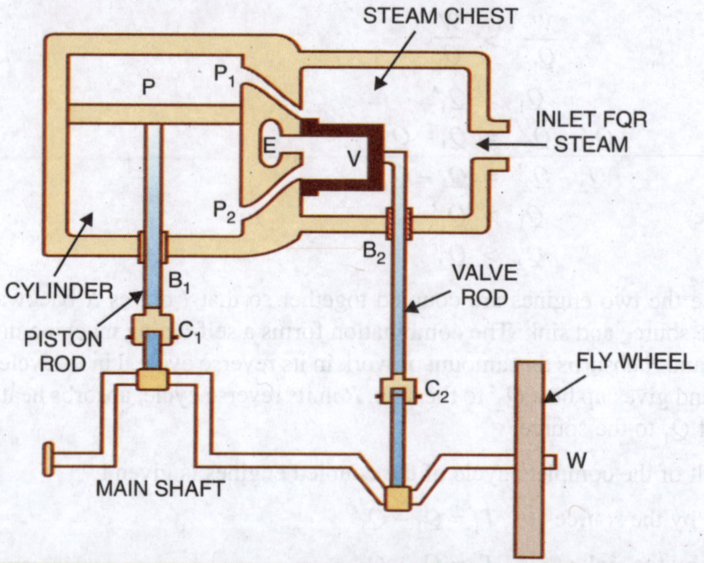


Fig. 4.19

**2.** When steam enters through the port $P_2$, the piston is pushed up and the sliding valve moves down.

The dead steam in the cylinder is exhausted out to the atmosphere through the exhaust valve *E*. Both the ports $P_1$ and $P_2$ are closed and the steam in the cylinder expands. The piston is pushed further to complete the stroke. Finally the port $P_2$ is closed and $P_1$ opens. The process is continuously repeated.

Steam Engine

$B_1$ and $B_2$ are the stuffing boxes. These boxes allow for the free and smooth sliding of the valve rod and the piston rod. $C_1$ and $C_2$ are the cross heads. When the piston moves to and fro, the shaft is rotated and finally the machinery rotates.

## Dead centres

Twice in the rotation of the shaft, the force acts at right angles to the shaft. Theoretically a force does not produce any effect when it acts at right angles to the shaft. For this purpose ***the flywheel*** is used so that the momentum gained by the flywheel enables the shaft to rotate beyond the dead centre in each half rotation. Thus, the flywheel makes the motion of the shaft smooth and not jerky.

**Power.** Suppose the average pressure of steam = $P$ newtons/metre$^2$. The length of the stroke = $L$ metres and the area of cross section of the piston = $A$ sq metres.

Total force on the piston $= (PA)$ newtons.

Distance moved by the piston in a double acting steam engine

$$= 2L$$

$\therefore$ Work done for one rotation of the shaft

$$= (PA) \times 2L \text{ joules.}$$

Let the shaft rotate $N$ times per second.

Work done per second $= (PA) \times 2\,L \times N$ watts

$\therefore$ Power of the engine $= \dfrac{2\,PLAN}{1000}$ kilowatts

**Efficiency.** Most of the heat energy in the steam engine is wasted. Therefore, its efficiency is very low. It is of the order of 8 to 10%. Efficiency is the ratio of the useful mechanical work ($W$) done by the engine to the amount heat ($Q$) supplied by the fuel for the same time.

**Example:** In a double acting steam engine the average pressure of steam is $10^5$ newton/m$^2$. The length of the stroke is 1 metre and the area of the piston is 0.15 sq. metre. Find the power of the engine if it makes 5 strokes per second.

Here $P = 10^5$ newtons/meter$^2$

$L = 1$ metre

$A = 0.15$ sq. metre

$N = 5$ strokes/second

$$\therefore \quad \text{Power } P = \frac{2 \times PLAN}{1000}$$

$$= \frac{2 \times 10^5 \times 1 \times 0.15 \times 5}{1000}$$

$= 150$ kilowatts.

## 4.31 Internal Combustion Engine (Otto Engine)

In the otto engine (internal combustion engine), air is the working substance and petrol vapour acts as the fuel. A set of ideal conditions are assumed while discussing an otto cycle as under:

(*i*) The working substance is air all the time and it behaves as a perfect gas.

(*ii*) There is no friction.

On the basis of these assumptions, Otto cycle can be divided into following operations:

Otto Engine

**1.** $E$ to $A$ (fig. 4.20) represents the *charging stroke*. The mixture of air and petrol vapour is allowed to enter at atmospheric pressure. The final volume is $V_2$ at temperature $T_1$.

**2.** $A$ to $B$ represents the *adiabatic compression*. There is no friction and no flow of heat through the walls of the cylinder. The volume changes from $V_2$ to $V_1$ and the temperature changes from $T_1$ to $T_2$. There is a change in pressure also. This represent the *compression stroke*. The compression ratio is nearly 8 *i.e.*, $\dfrac{V_2}{V_1} - 8$ nearly. The pressure changes from one atmosphere to about 18 atmosphere.

**3.** $B$ to $C$ represents the *ignition stage*. A spark is produced and the mixture of air and petrol vapours is ignited. The pressure increases from 18 to about 80 atmospheres and the temperature changes from $T_2$ to $T_3$ (from 500°C to 2000°C).

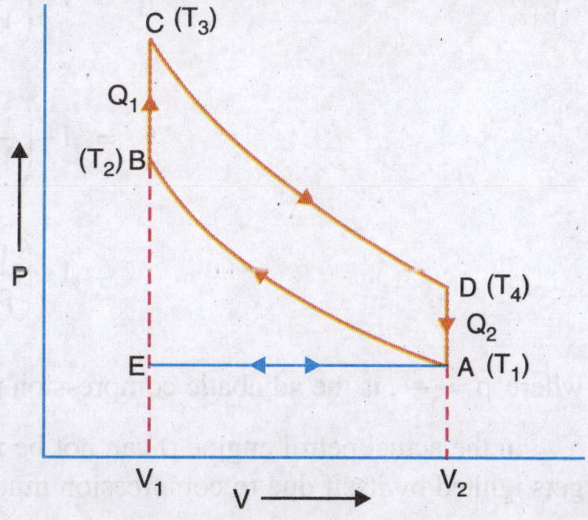


Fig. 4.20

**4.** *C* to *D* represents the *working stroke*. The gas expands adiabatically and the engine works. The volume changes from $V_1$ to $V_2$. The pressure and temperature decreases.

**5.** As the point *D* is reached, the exhaust valve opens. *D* to *A* represents the change of pressure to the atmospheric pressure and the temperature changes from $T_4$ to $T_1$.

**6.** *A* to *E* represents the *exhaust stroke*. The exhaust gases are completely discharged from the cylinder. Thus, the initial condition of the engine is restored.

## Efficiency

Consider that 1 gram of the working substance is used in the process. A quantity of heat $Q_1$ at a higher temperature is absorbed from *B* to *C*,

$$Q_1 = 1 \times C_v (T_3 - T_2)$$

A quantity of heat $Q_2$ is rejected at a lower temperature from *D* to *A*

$$Q_2 = 1 \times C_v (T_4 - T_1)$$

Therefore, dividing we get

$$\frac{Q_2}{Q_1} = \frac{T_4 - T_1}{T_3 - T_2} \qquad ...(i)$$

$\therefore$ Efficiency,

$$\eta = 1 - \frac{Q_2}{Q_1}$$

$$= 1 - \frac{T_4 - T_1}{T_3 - T_2} \qquad ...(ii)$$

The points *D* and *C* lie on the same adiabatic,

$$\therefore \quad T_4 V_2^{\gamma-1} = T_3 V_1^{\gamma-1} \qquad ...(iii)$$

The points *A* and *B* lie on the same adiabatic,

$$T_1 V_2^{\gamma-1} = T_2 V_1^{\gamma-1} \qquad ...(iv)$$

Subtracting (*iv*) from (*iii*),

$$(T_4 - T_1) V_2^{\gamma-1} = (T_3 - T_2) V_1^{\gamma-1}$$

or

$$\frac{T_4 - T_1}{T_3 - T_2} = \left(\frac{V_1}{V_2}\right)^{\gamma-1} \qquad ...(v)$$

Substituting this value in equation (*ii*),

$$\eta = 1 - \left(\frac{V_1}{V_2}\right)^{\gamma-1}$$

$$= 1 - \left(\frac{1}{V_2/V_1}\right)^{\gamma-1}$$

$$= 1 - \left(\frac{1}{\rho}\right)^{\gamma-1} \qquad ...(4.28)$$

where $\rho = \frac{V_2}{V_1}$, is the adiabatic compression ratio.

In the actual petrol engine ρ can not be made greater than 10. If ρ is more than 10, the mixture gets ignited by itself due to compression much before the sparking takes place.

$$\eta = 1 - \left(\frac{1}{\rho}\right)^{\gamma-1}$$

(*i*) If $\rho = 6$ and $\gamma = 1.4$

$$\eta = 1 - \left(\frac{1}{6}\right)^{0.4}$$

$$= 0.5116 = 51.16\%$$

(*ii*) If $\rho = 9$ and $\gamma = 1.4$

$$\eta = 1 - \left(\frac{1}{9}\right)^{0.4}$$

$$= 0.5847 = 58.47\%$$

## 4.32 Internal Combustion Engine (Petrol Engine)

It consists of a cylinder with an air-tight piston. It has two valves. I is the inlet valve and *O* is the outlet valve. The valves are opened and closed at the proper times with a suitable mechanism. The petrol vapours and air from the *carburettor* are admitted into the cylinder through the valve *I*. Here *S* acts as a sparking plug. It is also known as *four-stroke engine*.

**1. Charging Stroke.** The mixture of petrol vapours and air enters the cylinder. The piston moves down [Fig. 4.21 (*i*)].

**2. Compression Stroke.** The inlet valve closes and the piston moves up. The mixture is compressed to a high pressure and the temperature rises to about 500°C [Fig. 4.21 (*ii*)].

**3. Working Stroke.** A spark is produced at the sparking plug and the compressed mixture of petrol and air ignites. The temperature rises to about 2000°C and a high pressure of about 15 atmospheres is developed inside the cylinder. The piston is pushed down and the shaft rotates [Fig. 4.21 (*iii*)]. Only in this stroke, the work is done by the piston.

**4. Exhaust Stroke.** The outlet valve opens and the burnt and unburnt gases are exhausted out to the atmosphere [Fig. 4.21 (*iv*)].

The cycle repeats and the engine works.

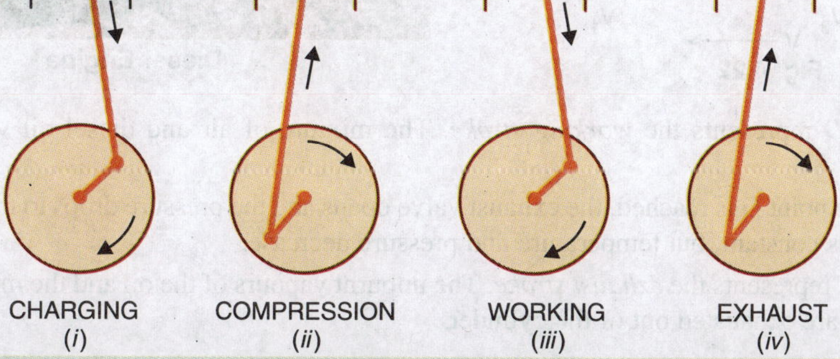


**Fig. 4.21**

**Engine Cooling.** The internal combustion engine requires cooling, although the efficiency of the engine will be higher without cooling. Temperatures of the order of 2000°C are reached inside the cylinder, although this temperature may last only for a very short interval of time before expansion. If the engine is not cooled the following drawbacks will be present:

1. Burning of the piston and exhaust valves.
2. Irregular burning of the fuel.
3. Seizure of the piston due to expansion.
4. Burning of the lubricant, and
5. Melting of the bearing material.

## 4.33 Diesel Engine

In the case of a diesel engine, air is admitted into the cylinder in the beginning. This air is compressed adiabatically so that the temperature is high enough to ignite the oil sprayed into the cylinder. The ideal conditions are assumed while discussing the complete cycle *viz*. (*i*) the working substance is air all the time and it behaves as a perfect gas and (*ii*) there is no friction. The cycle is called the ***diesel cycle***.

**1.** $E$ to $A$ (Fig. 4.22) represents the intake of air and its volume changes to $V_1$. This is the *charging stroke*.

**2.** $A$ to $B$ represents the *compression stroke*. Air is compressed adiabatically. The temperature changes from $T_1$ to $T_2$ and the volume changes from $V_1$ to $V_2$. The temperature rises to about 1000°C and the pressure to about 40 atmospheres.

**3.** $B$ to $C$ represents the stage when the oil is sprayed into the cylinder. The oil burns immediately. The pressure is maintained constant. The temperature changes from $T_2$ to $T_3$ (to 2000°C) and the volume changes from $V_2$ to $V_3$.

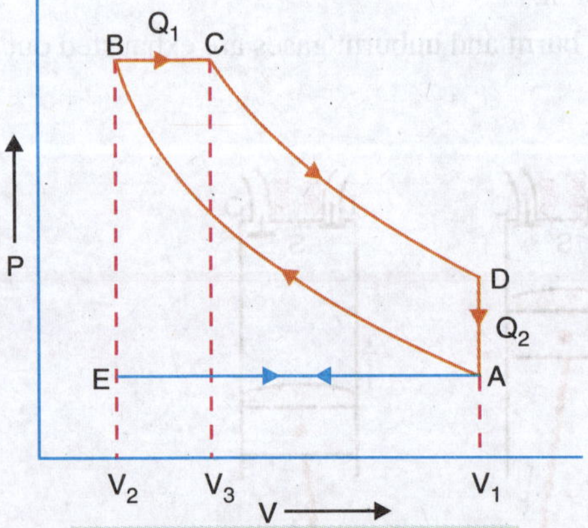

Fig. 4.22

Diesel Engine

**4.** $C$ to $D$ represents the *working stroke*. The mixture of air and diesel oil vapour expand adiabatically.

**5.** As the point $D$ is reached, the exhaust valve opens and the pressure drops to the point $A$. The volume remains constant, but temperature and pressure decrease.

**6.** $A$ to $E$ represents the *exhaust stroke*. The unburnt vapours of the oil and the mixture of gases in the cylinder are exhausted out of the cylinder.

## Efficiency

From $B$ to $C$, the pressure remains constant. Considering 1 gram of the working substance, the quantity of heat absorbed,

$$Q_1 = 1 \times C_p\,(T_3 - T_2)$$

From $D$ to $A$ the volume remains constant.

The quantity of heat rejected,

$$Q_2 = 1 \times C_v\,(T_4 - T_1)$$

Dividing, we get
$$\frac{Q_2}{Q_1} = \frac{C_v}{C_p} \cdot \frac{(T_4 - T_1)}{(T_3 - T_2)}$$

$$= \frac{1}{\gamma} \cdot \frac{(T_4 - T_1)}{(T_3 - T_2)}$$

Efficiency,
$$\eta = 1 - \frac{Q_2}{Q_1}$$

$$= 1 - \frac{1}{\gamma}\left(\frac{T_4 - T_1}{T_3 - T_2}\right) \qquad \text{...(4.29)}$$

To evaluate $\left(\frac{T_4 - T_1}{T_3 - T_2}\right)$, all the temperatures are to be expressed in terms of $T_2$.

Let ρ be the adiabatic expansion ratio. Then $\rho = \frac{V_1}{V_2}$ and $e$ the combustion expansion ratio or fuel cut off ratio $= \frac{V_3}{V_2}$.

**1.** The points $A$ and $B$ are on the same adiabatic,

$$\therefore \quad T_1 V_1^{\gamma-1} = T_2 V_2^{\gamma-1}$$

$$T_1 = T_2\left[\frac{V_2}{V_1}\right]^{\gamma-1}$$

$$= T_2\left[\frac{1}{\rho}\right]^{\gamma-1} \qquad \text{...(4.30)}$$

**2.** The points $B$ and $C$ are at the same pressure

$$\therefore \quad \frac{T_3}{V_3} = \frac{T_2}{V_2}$$

or
$$T_3 = T_2\left[\frac{V_3}{V_2}\right] = T_2[e] \qquad \text{...(4.31)}$$

**3.** The points $C$ and $D$ are on the same adiabatic.

$$\therefore \quad T_4 V_4^{\gamma-1} = T_3 V_3^{\gamma-1}$$

$$[\text{But } V_4 = V_1]$$

$$T_4 = T_3\left[\frac{V_3}{V_1}\right]^{\gamma-1}$$

$$= T_3\left[\frac{V_3}{V_2} \times \frac{V_2}{V_1}\right]^{\gamma-1}$$

$$T_4 = T_3\left[\frac{V_3}{V_2}\right]^{\gamma-1}\left[\frac{V_2}{V_1}\right]^{\gamma-1}$$

$$T_4 = T_3\,[e]^{\gamma-1}\left[\frac{1}{\rho}\right]^{\gamma-1}$$

$$T_4 = T_2\,[e][e]^{\gamma-1}\left[\frac{1}{\rho}\right]^{\gamma-1}$$

$$T_4 = T_2\,[e]^{\gamma}\left[\frac{1}{\rho}\right]^{\gamma-1} \qquad ...(4.32)$$

$$\eta = 1-\left[\frac{T_4-T_1}{T_3-T_2}\right]$$

$$= 1-\frac{1}{\gamma}\left[\frac{T_2[e]^{\gamma}\left[\frac{1}{\rho}\right]^{\gamma-1}-T_3\left[\frac{1}{\rho}\right]^{\gamma-1}}{T_2[e]-T_2}\right]$$

$$\eta = 1-\frac{1}{\gamma}\left(\frac{1}{\rho}\right)^{\gamma-1}\left[\frac{e^{\gamma}-1}{e-1}\right] \qquad ...(4.33)$$

For the same compression ratio, the efficiency of an Otto engine is more than a diesel engine. In practice the compression ratio for an Otto engine is from 7 to 9 and for a diesel engine it is from 15 to 20. Due to the higher compression ratio, an actual diesel engine has higher efficiency than the Otto (Petrol) engine. The cylinder must be strong enough to withstand very high pressure.

## 4.34 Multicylinder Engine

With an engine having one cylinder, the engine works only during the working stroke. The piston moves during the rest of the three strokes due to the momentum of the shaft. In a multicylinder engine (say 4-cylinder engine) the four cylinders are coupled. The working of each cylinder is given below:

| | *First* | *Second* | *Third* | *Fourth* |
|---|---|---|---|---|
| First quarter | Working | Exhaust | Compression | Charging |
| Second quarter | Exhaust | Charging | Working | Compression |
| Third quarter | Charging | Compression | Exhaust | Working |
| Fourth quarter | Compression | Working | Charging | Exhaust |

In this way, the power of the engine increases and the shaft gets momentum during each quarter cycle.

## NUMERICAL EXAMPLES

### EXAMPLE 4.1

*A motor car tyre has a pressure of 2 atmospheres at the room temperature of 27°C. If the tyre suddenly bursts, find the resulting temperature.* **(Purvanchal 2006, Garhwal 2001)**

**SOLUTION.** Here

$$P_1 = 2 \text{ atmospheres}$$
$$T_1 = 273 + 27 = 300 \text{ K}$$
$$P_2 = 1 \text{ atmosphere}$$
$$T_2 = ?$$
$$\gamma = 1.4$$

$$\frac{P_1^{\gamma-1}}{T_1^{\gamma}} = \frac{P_2^{\gamma-1}}{T_2^{\gamma}} \quad \text{or} \quad \left(\frac{P_2}{P_1}\right)^{\gamma-1} = \left(\frac{T_2}{T_1}\right)^{\gamma}$$

$$\left(\frac{1}{2}\right)^{0.4} = \left(\frac{T_2}{300}\right)^{1.4}$$

$$2.304 \times 0.4 \log_{10}(0.5) = 1.4 [\log_{10} T_2 - \log_{10} 300] \times 2.304$$

$$-0.1204 = 1.4 \log_{10} T_2 - 3.4680$$

$$1.4 \log_{10} T_2 = 3.4680 - 0.1204 = 3.3476$$

$$\log_{10} T_2 = \frac{3.3476}{1.4} = 2.3911$$

$$T_2 = \text{Antilog}(2.3911) = 246.1 \text{ K}$$

$$\mathbf{T_2 = -26.9°C}$$

### EXAMPLE 4.2

*A quantity of air at 27°C and atmospheric pressure is suddenly compressed to half its original volume Find the final (i) pressure and (i) temperature.* (*Given* $\gamma = 1.4$, $2^{1.4} = 2.64$)

**(Rohilkhand 2000)**

**SOLUTION.**

(*i*) $P_1 = 1$ atmosphere; $P_2 = ?$, $\gamma = 1.4$

$$V_1 = V \qquad V_2 = \frac{V}{2}$$

During sudden compression, the process is adiabatic.

$$P_1 V_1^{\gamma} = P_2 V_2^{\gamma}$$

$$P_2 = P_1 \left[\frac{V_1}{V_2}\right]^{\gamma}$$

$$= 1 [2]^{1.4} = \text{Antilog}[1.4 \log 2] = \text{Antilog}[1.4 \times 0.30]$$

$$= \text{Antilog}(0.42)$$

$$P_2 = 2.636 \text{ atmospheres}$$

(*ii*) $V_1 = V$; $V_2 = \frac{V}{2}$, $T_1 = 300$ K ; $T_2 = ?$, $\gamma = 1.4$

$$T_1 (V_1)^{\gamma-1} = T_2 (V_2)^{\gamma-1}$$

$$T_2 = T_1 [2]^{1.4-1}$$

$$= 300 [2]^{0.4}$$

$$= 395.9 \text{ K}$$

$$\mathbf{T_2 = 122.9°C}$$

### EXAMPLE 4.3

*Air at N.T.P. is compressed adiabatically to half of its volume. Calculate the change in its temperature.*

**SOLUTION.** Let the initial temperature be $T_1$ *K* and the final temperature $T_2$ *K*.

$$\text{Initial volume} = V_1$$

$$\text{Final volume} = V_2 = \frac{V_1}{2}$$

During an adiabatic process

$$T_1 V_1^{\gamma-1} = T_2 V_2^{\gamma-1}$$

$$T_2 = T_1\left[\frac{V_1}{V_2}\right]^{\gamma-1}$$

$$T_2 = T_1 [\,2\,]^{\gamma-1}$$

$$\text{But } \gamma \text{ for air} = 1.4$$

$$T_2 = \mathrm{T}_1 [\,2\,]^{0.4}$$

$$= 1.319\, T_1$$

$$\therefore \quad \text{Change in temperature} = T_2 - T_1$$

$$= 1.319\, T_1 - T_1$$

$$= \mathbf{0.319\, T_1\ K}$$

### EXAMPLE 4.4

*1 gram molecule of a monoatomic* $\left(\gamma = \frac{5}{3}\right)$ *prefect gas at 27°C is adiabatically compressed in a reversible process from an initial pressure of 1 atmospher to a final pressure of 50 atmospheres. Calculate the resulting difference in temperature.* **(Mysore 2000)**

**SOLUTION.** In a reversible adiabatic process

$$\frac{P_1^{\gamma-1}}{T_1^{\gamma}} = \frac{P_2^{\gamma-1}}{T_2^{\gamma}}$$

or

$$\left(\frac{P_2}{P_1}\right)^{\gamma-1} = \left(\frac{T_2}{T_1}\right)^{\gamma}$$

Here

$$P_2 = 50$$

$$P_1 = 1$$

$$T_1 = 273 + 27 = 300\ \mathrm{K}$$

$$T_2 = ?$$

$$\gamma = \frac{5}{3}$$

$$\therefore \quad \left(\frac{50}{1}\right)^{\frac{2}{3}} = \left(\frac{T_2}{300}\right)^{\frac{5}{3}}$$

$$\frac{2}{3}\log 50 = \frac{5}{3}\,[\log T_2 - \log 300]$$

$$T_2 = \mathbf{1434\ K}$$

$$= \mathbf{1161°C}$$

### EXAMPLE 4.5

*A quantity of dry air at 27°C is compressed (i) slowly and (ii) suddenly to $\frac{1}{3}$ of its volume. Find the change in temperature in each case, assuming $\gamma$ to be 1.4 for dry air.*

**(Rohilkhand 2002)**

**SOLUTION.**

1. When the process is slow, the temperature of the system remains constant. Therefore, there is no change in temperature.
2. When the compression is sudden, the process is adiabatic.

Here $V_1 = V$, $V_2 = \frac{V}{3}$, $T_1 = 300$ K, $T_2 = ?$, $\gamma = 1.4$

$$T_2 (V_2)^{\gamma-1} = T_1 (V_1)^{\gamma-1}$$

$$T_2 = T_1 \left(\frac{V_1}{V_2}\right)^{\gamma-1}$$

$$= 300 \left[\frac{3V}{V}\right]^{\gamma-1} = 300 [\,3\,]^{1.4-1} = 300 [\,3\,]^{0.4}$$

$$T_2 = 465.5 \text{ K}$$

$$= \mathbf{192.5°C}$$

The temperature of air increases by

$$= 192.5 - 27$$

$$= \mathbf{165.5°C = 165.5\ K}$$

### EXAMPLE 4.6

*Find the efficiency of the Carnot's engine working between the steam point and the ice point.*

**(Bangalore 2001)**

**SOLUTION.** Here

$$T_1 = 273 + 100 = 373 \text{ K}$$

$$T_2 = 273 + 0 = 273 \text{ K}$$

$$\eta = 1 - \frac{T_2}{T_1}$$

$$= 1 - \frac{273}{373} = \frac{100}{373}$$

$$\% \text{ efficiency} = \frac{100}{373} \times 100$$

$$= \mathbf{26.81\%}$$

### Example 4.7

*Find the efficiency of a Carnot's engine working between 127°C and 27°C. It absorbs 80 cals of heat. How much heat is rejected.* **(Nagpur 2008, 2001, Meerut 2005, 04, 03, Purvanchal 2007)**

**SOLUTION.**

$$T_1 = 273 + 127 = 400 \text{ K}$$

$$T_2 = 273 + 27 = 300 \text{ K}$$

$$\eta = 1 - \frac{T_2}{T_1}$$

$$= 1 - \frac{300}{400} = 0.25$$

$$\% \eta = 25\%$$

$$W = n\,Q_1 = 0.25 \times 80 = 20 \text{ cals.}$$

$$\therefore \quad \text{Heat rejected } Q_2 = Q_1 - W = 80 - 20$$

$$= \mathbf{60\ cals.}$$

**EXAMPLE 4.8**

*A Carnot's engine whose temperature of the source is 400 K takes 200 calories of heat at this temperature and rejects 150 calories of heat to the sink. What is the temperature of the sink? Also calculate the efficiency of the engine.* **(Nagpur 2005/w, Panjab 2005)**

**SOLUTION.** $Q_1 = 200$ Cal ; $Q_2 = 150$ Cal

$T_1 = 400$ K ; $T_2 = ?$

$$\frac{Q_1}{T_1} = \frac{Q_2}{T_2} \qquad \text{Efficiency, } \eta = 1 - \frac{T_2}{T_1}$$

$$T_2 = \frac{Q_2}{Q_1} \times T_1 \qquad = 1 - \frac{300}{400} = 0.25$$

$$= \frac{150}{200} \times 400$$

$$\mathbf{T_2 = 300\ K} \quad ; \quad \mathbf{\%Efficiency = 25\%}$$

**EXAMPLE 4.9**

*A Carnot's engine working as a refrigerator between 260 K and 300 K receives 500 calories of heat from the reservoir at the lower temperature. Calculate the amount of heat rejected to the reservoir at the higher temperature. Calculate also the amount of work done in each cycle to operate the refrigerator. (1 calorie = 4.2 Joules).* **(Kerala 2001)**

**SOLUTION.** $Q_1 = ?$ $\qquad Q_2 = 500$ Cal

$T_1 = 300$ K $\qquad T_2 = 260$ K

$$\frac{Q_1}{Q_2} = \frac{T_1}{T_2}$$

$$Q_1 = Q_2 \frac{T_1}{T_2} = \frac{500 \times 300}{260} = \mathbf{576.92\ Cal}$$

Work done, $W = Q_1 - Q_2 = 76.92$ Cal

$$= 76.92 \times 4.2 \text{ joules}$$

$$W = \mathbf{323.08\ joules.}$$

**EXAMPLE 4.10**

*A Carnot's engine whose lower temperature heat-sink is at 27°C has its efficiency 40%. What is the temperature of the heat sources. By how much should the temperature of the source be raised if the efficiency is to be raised to 70%.* **(Kanpur 2001, Nagpur 2004/w)**

**Solution.** $T_2 = 273 + 27 = 300$ K, $\eta_1 = \dfrac{40}{100} = 0.4$, $\eta_2 = \dfrac{70}{100} = 0.7$

**(*i*) Temperature of the source:**

$$\eta_1 = 1 - \frac{T_2}{T_1} \quad \text{or} \quad \frac{T_2}{T_1} = 1 - \eta_1$$

$$\therefore \quad T_1 = \frac{T_2}{1 - \eta_1} = \frac{300}{1 - 0.4} = \frac{300}{0.6} = \mathbf{500\ K = 227°C}$$

(*ii*) **Required rise in source temperature:**

$$T_1' = \frac{T_2}{1-\eta_2} = \frac{300}{1-0.7} = \frac{300}{0.3} = 1000 \text{ K} = 727°\text{C}$$

∴ Required rise in the source temperature

$$= T_1' - T_1 = 1000 - 500 = \mathbf{500\ K = 500°C}$$

## Example 4.11

*A Carnot's refrigerator takes heat from water at 0°C and discards it to a room temperature at 27°C. 1 kg of water at 0°C is to be changed into ice at 0°C. How many calories of heat are discarded to the room? What is the work done by the refrigerator in this process? What is the coefficient of performance of the machine? (1 cal = 4.2 Joule).* **(Patna 2000)**

**SOLUTION.** Here

$$Q_1 = ?\ Q_2 = 1000 \times 80 = 80{,}00 \text{ cal}$$

$$T_1 = 300 \text{ K},$$

$$T_2 = 273 \text{ K}$$

(*i*) **Heat rejected to the room:**

$$\frac{Q_1}{Q_2} = \frac{T_1}{T_2}$$

∴

$$Q_1 = \frac{T_1}{T_2} Q_2 = \frac{300}{273} \times 80{,}000$$

$$Q_1 = 87{,}900 \text{ cal}$$

(*ii*) **Work done by the Refrigerator:**

$$W = \text{J}\,(Q_1 - Q_2)$$

$$= 4.2\,(87{,}900 - 80{,}000)$$

$$= 4.2 \times 7900$$

$$\mathbf{W = 3.183 \times 10^4 \text{ Joules}}$$

(*iii*) **Coefficient of performance:**

$$P = \frac{Q_2}{Q_1 - Q_2}$$

$$= \frac{80{,}000}{87{,}900 - 80{,}000} = \frac{80{,}000}{7900}$$

$$\mathbf{P = 10.13}$$

## Example 4.12

*A Carnot engine whose low temperature reservoir is at 7°C has an efficiency of 50%. It is desired to increase the efficiency to 70%. By how many degrees should the temperature of the high temperature reservoir be increased?*

**SOLUTION.** In the first case

$$\eta = 50\% = 0.5,\ T_2 = 273 + 7 = 280 \text{ K}.$$

$$T_1 = ?$$

$$\eta = 1 - \frac{T_2}{T_1}$$

or

$$0.5 = 1 - \frac{280}{T_1}$$

or $$T_1 = 560 \text{ K}$$

In the second case

$$\eta' = 70\% = 0.7,$$
$$T_2 = 280 \text{ K},$$
$$T_1' = ?$$
$$\eta' = 1 - \frac{T_2}{T_1'}$$
$$0.7 = 1 - \frac{280}{T_1'}$$

or $$T_1' = 840 \text{ K}$$

Increase in temperature= 840 – 560 = **280 K**

**EXAMPLE 4.13**

*An inventor claims to have developed an engine working between 600 K and 300 K capable of having an efficiency of 52%. Comment on his claim.* **(Nagpur 2009, Lucknow 2003, Purvanchal 2002)**

**SOLUTION.** The efficiency of a Carnot engine working between 600 K and 300 K

$$\eta = 1 - \frac{T_1}{T_2}$$
$$= 1 - \frac{300}{600} = 0.5$$
$$= 50\%$$

The efficiency claimed is 52%. It means that the efficiency of the engine developed is more than the efficiency of Carnot's engine working between the same two temperature limits. But no engine can have an efficiency more than a Carnot's engine. Hence his claim is invalid.

**EXAMPLE 4.14**

*When a system is taken from the state A to the state B, along the path ACB, 80 joules of heat flows into the system, and the system does 30* joules *of work (Fig. 4.23).*

(*a*) *How much heat flows into the system along the path ADB, if the work done is 10 joules.*

(*b*) *The system is returned from the state B to the state A along the curved path. The work done on the system is 20* joules. *Does the system absorb or liberate heat and how much?*

(*c*) *If $U_A = 0$, $U_D = 40$ joules, find the heat absorbed in the process AD and DB.*

**(Nagpur 2000)**

**SOLUTION.**

Along the path *ACB*,

$$Q_{ACB} = U_B - U_A + W$$

Here $$Q = +80 \text{ joules}$$
$$W = +30 \text{ joules}$$
$$\therefore \quad +80 = U_B - U_A + 30$$
$$U_B - U_A = 80 - 30 = \mathbf{50\ joules}$$

(*a*) Along the path *ADB*,

$$W = +10 \text{ joules}$$
$$Q_{ADB} = U_B - U_A + W$$
$$Q = 50 + 10 = \mathbf{60\ joules}$$

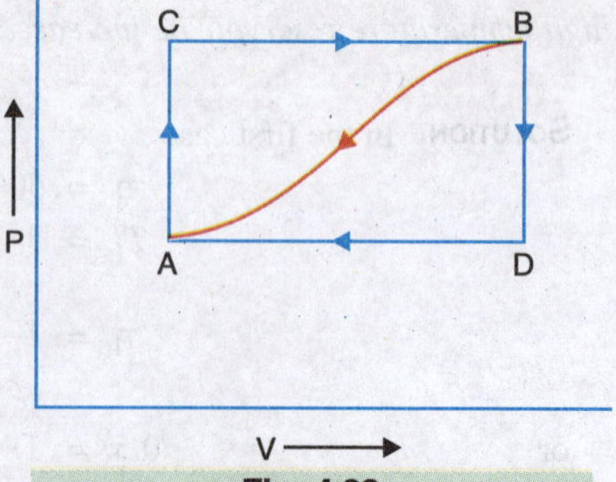

Fig. 4.23

(*b*) For the curved path from *B* to *A*,

$$W = -20 \text{ joules}$$
$$Q = (U_A - U_B) + W$$
$$= -50 - 20 = \mathbf{-70 \text{ joules}}$$

(– *ve* sign shows that heat is liberated by the system)

(*c*) $U_A = 0, \quad U_D = 40$ joules

$$U_B - U_A = 50$$

$\therefore$ $U_B = 50$ joules

In the process *ADB*, 10 joules of work is done. Work done from *A* to *D* is + 10 joules and from D to *B* is zero.

For *AD*,

$$Q_{AD} = (U_D - U_A) + W$$

For *DB*

$$= 40 + 10 = 50 \text{ joules}$$
$$Q_{DB} = U_B - U_D + W$$
$$= 50 - 40 + 0 = 10 \text{ joules}$$

### EXAMPLE 4.15

*A certain mass of gas at NTP is expanded to three times its volume under adiabatic conditions. Calculate the resulting temperature and pressure. γ for the gas is 1.40.* **(Osmania 2001)**

**SOLUTION.**

(*a*) Here, $V_1 = V, \quad V_2 = 3V$

$$T_1 = 273 \text{ K}, \quad T_2 = ?$$
$$T_1 V_1^{\gamma-1} = T_2 V_2^{\gamma-1}$$

or

$$T_2 = T_1\left[\frac{V_1}{V_2}\right]^{\gamma-1}$$
$$T_2 = 273\left[\frac{1}{3}\right]^{1.4-1}$$
$$T_2 = \mathbf{176 \text{ K} = -97°\text{C}}$$

(*b*) Here, $V_1 = V, \quad V_2 = 3V$

$$P_1 = 1 \text{ atmosphere}, \quad P_2 = ?$$
$$P_1 V_1^{\gamma} = P_2 V_2^{\gamma}$$
$$P_2 = P_1\left[\frac{V_1}{V_2}\right]^{\gamma}$$

or

$$P_2 = 1\left(\frac{1}{3}\right)^{1.4}$$
$$P_2 = \mathbf{0.2148 \text{ atmosphere}}$$

### EXAMPLE 4.16

*A gas occupying 1 litre at 80 cm of Hg pressure is expanded adiabatically to 1190 c.c. If the pressure falls to 60 cm of Hg in the process, deduce the value of γ.*

**SOLUTION.** Here $P_1V_1^{\gamma} = P_2V_2^{\gamma}$

$$\left(\frac{P_1}{P_2}\right) = \left(\frac{V_2}{V_1}\right)^{\gamma}$$

Taking logs,

$$\log\left(\frac{P_1}{P_2}\right) = \gamma\left(\log\frac{V_2}{V_1}\right)$$

or

$$\gamma = \frac{\log\left(\frac{P_1}{P_2}\right)}{\log\left(\frac{V_2}{V_1}\right)}$$

$$\gamma = \frac{\log\left(\frac{80}{60}\right)}{\log\left(\frac{1190}{1000}\right)}$$

$$\gamma = \frac{0.1249}{0.0755}$$

$$\gamma = \mathbf{1.787}$$

**EXAMPLE 4.17**

*The initial temperature of a gas is 27°C. Calculate the rise in temperature when the gas is compressed suddenly to 8 times its original pressure ($\gamma = 1.5$)*

**(Osamania 2004, Nagpur 2004/s, Mysore 2004)**

**SOLUTION.** For an adiabatic process,

$$\frac{P_1^{\gamma-1}}{T_1^{\gamma}} = \frac{P_2^{\gamma-1}}{T_2^{\gamma}}$$

$$\left(\frac{P_2}{P_1}\right)^{\gamma-1} = \left(\frac{T_2}{T_1}\right)^{2}$$

Here $\frac{P_2}{P_1} = 8$ and $T_1 = 273 + 27 = 300$ K

$$\gamma = 1.5$$

$\therefore$

$$(8)^{\gamma-1} = \left(\frac{T_2}{300}\right)^{\gamma}$$

$$(8)^{0.5} = \left(\frac{T_2}{300}\right)^{1.5}$$

$$0.5 \log 8 = 1.5\,[\log T_2 - \log 300]$$

$$0.5\,(0.9031) = 1.5 \log T_2 - 1.5 \times 2.4771$$

$$1.5 \log T_2 = 0.5 \times 0.9031 + 1.5 \times 2.4771$$

$$1.5 \log T_2 = 0.4516 + 3.7157$$

$$1.5 \log T_2 = 4.1673$$

$$\log T_2 = \frac{4.1673}{1.5}$$

$$\log T_2 = 2.7782$$

$$T_2 = \mathbf{600\ K}$$

$$T_2 = \mathbf{327°C}$$

### EXAMPLE 4.18

*Calculate the work done when a gram molecule of an ideal gas expands isothermally at 27°C to double its original volume. [R = 8.3 J/deg mole]*

**(Rohilkhand 2001, Meerut 2003)**

**SOLUTION.**

Work done $= \int_{V_1}^{V_2} P\,dV$

But $PV = RT$

$$P = \frac{RT}{V}$$

$$W = RT\int_{V_1}^{V_2}\left(\frac{dV}{V}\right)$$

or $W = RT \log_e\left(\frac{V_2}{V_1}\right)$

Here $\frac{V_2}{V_1} = 2$

$$R = 8.3 \text{ J/deg mole}$$

$$T = 273 + 27 = \mathbf{303\ K}$$

$$W = 8.3 \times 300 \log_e 2$$

$$W = \mathbf{1725.9\ J}$$

### EXAMPLE 4.19

*A Carnot engine has an efficiency of 30% when the temperature of the sink is 27°C. What must be the change in temperature of the source to make its efficiency 50%.*

**(Purvanchal 2005, Calcutta 2001)**

**SOLUTION.** In the first case,

$$\eta = 30\% = 0.3,\ T_2 = 300\text{ K},\ T_1 = ?$$

$$\eta = 1 - \frac{T_2}{T_1}$$

$$0.3 = 1 - \frac{300}{T_1}$$

$$T_1 = \mathbf{428.57\ K}$$

In the second case,

$$\eta' = 50\% = 0.5,\ T_2 = 300\text{ K},\ T_1' = ?$$

$$\eta' = 1 - \frac{T_2}{T_1'}$$

$$0.5 = 1 - \frac{300}{T_1'}$$

$$T_1' = 600\text{ K}$$

Increase in temperature

$$= \mathbf{171.43\ K}$$

**Example 4.20**

*One of the most efficient engines ever developed by a manufacturer has an efficiency of 40% while operating between 2100 K and 700 K. What percentage is this efficiency if the maximum possible efficiency of an engine working between these very temperatures?*

**(Rajasthan 2000)**

**Solution.**

$$T_1 = 2100 \text{ K}$$

$$T_2 = 700 \text{ K}$$

$$\eta = 1 - \left(\frac{T_2}{T_1}\right)$$

or

$$\eta = 1 - \left(\frac{700}{2100}\right) = \frac{2}{3}$$

Actual efficiency,

$$\eta' = 40\% = \frac{2}{5}$$

$$\frac{\eta'}{\eta} = \frac{2/5}{2/3} = \frac{3}{5} = 60\%$$

**Example 4.21**

*An inventor claims to have developed an engine working between 600 K and 300 K capable of having an efficiency of 52%. Comment on his claim.* **(Nagpur 2009, Lucknow 2001)**

**Solution.**

$$T_1 = 600 \text{ K}, \quad T_2 = 300 \text{ K}$$

$$\eta = 1 - \left(\frac{T_2}{T_1}\right)$$

$$\eta = 1 - \left(\frac{300}{600}\right) = \frac{1}{2}$$

*i.e.* $\eta = 50\%$.

The efficiency claimed = 52%

It means that the efficiency of the engine is more than the efficiency of a Carnot's engine working between the same two temperature limits. But no engine can have an efficiency more than a Carnot's engine, so his **clam is invalid**.

**Example 4.22**

*A Carnot's engine absorbs $10^4$ calories of heat from a reservoir at 627°C and rejects heat to a sink at 27°C. What is its efficiency? How much work does it perform? [J = 4.2 J/Calorie]*

**(Osmania 2004)**

**Solution.**

Here

$$T_1 = 627 + 273 = 900 \text{ K}$$

$$T_2 = 27 + 273 = 300 \text{ K}$$

Efficiency,

$$\eta = 1 - \frac{T_2}{T_1}$$

$$\eta = 1 - \frac{300}{900}$$

$$\eta = \frac{2}{3} \text{ or } 66.67\%$$

Also

$$\frac{Q_1}{Q_2} = \frac{T_1}{T_2}$$

Here

$$Q_1 = 10^4 \text{ Calories}$$

$\therefore \qquad Q_2 = \dfrac{Q_1 \times T_2}{T_1}$

$$Q_2 = \frac{10^4 \times 300}{900} = \frac{10^4}{3} \text{ Calories}$$

Work done $\qquad W = Q_1 - Q_2$

$$W = 10^4 - \left(\frac{10^4}{3}\right)$$

$$W = \frac{2 \times 10^4}{3} \text{ Calories}$$

$$W = \frac{2 \times 10^4}{3} \times 4.2$$

$$\mathbf{W = 2.8 \times 10^4\ J}$$

**EXAMPLE 4.23**

*A 100 k-W engine is operating between 217°C and 17°C. Calculate*

*(i) the amount of heat absorbed*

*(ii) the amount of heat rejected and*

*(iii) the efficiency of the engine.* **(Osmania 2005)**

**SOLUTION. Given:** $\qquad W = 100 \text{ kW} = 10^5 \text{ watts} = 10^5 \text{ J/s}$

$$T_1 = 217°\text{C} = 400 \text{ K}$$

$$T_2 = 17°\text{C} = 300 \text{ K}$$

Efficiency, $\qquad \eta = 1 - \left(\dfrac{T_2}{T_1}\right)$

$$\eta = 1 - \frac{300}{400}$$

$$\eta = 0.25 = 25\%$$

Also $\qquad \dfrac{Q_1}{Q_2} = \dfrac{T_1}{T_2} = \dfrac{400}{300} = \dfrac{4}{3}$

or $\qquad \left(\dfrac{Q_1}{Q_2}\right) - 1 = \left(\dfrac{4}{3}\right) - 1$

$$\frac{Q_1 - Q_2}{Q_2} = \frac{1}{3}$$

But $\qquad W = Q_1 - Q_2 = 10^5 \text{ J/s}$

$\therefore \qquad \dfrac{10^5}{Q_2} = \dfrac{1}{3}$

$$Q_2 = 3 \times 10^5 \text{ J/s}$$

But $\qquad \dfrac{Q_1}{Q_2} = \dfrac{4}{3}$

$$Q_1 = \left(\frac{4}{3}\right) Q_2 = \left(\frac{4}{3}\right) \times 3 \times 10^5$$

or $\qquad Q_1 = 4 \times 10^5 \text{ J/s}$

Hence (*i*) Amount of heat absorbed per second

$$= 4 \times 10^5 \text{ J/s}$$

(*ii*) Amount of heat rejected per second

$$= 3 \times 10^5 \text{ J/s}$$

(*iii*) Efficiency of the engine

$$= 0.25 = 25\%$$

### EXAMPLE 4.24

*The temperature of 5 kg of air is raised by 1 °C at constant volume. Caluclate the increase in its internal energy. Give: $C_P = 993$ J kg$^{-1}$ K$^{-1}$, $\gamma = 5/3$.*

**(Madras 2004/w, Meerut 2004)**

**SOLUTION.** From first law of thermodynamics

$$\delta Q = dV + \delta W$$

where all quantities have same unit, *i.e.*, say kilocal.

Here air is heated at constant volume, and hence no external work is done *i.e.*,

$$\delta W = P.dV = 0$$

Let $m$ be the mass, $C_V$ the sp. heat at constant volume and $\delta T$ is the rise in temperature, the heat taken by air is

$$\delta Q = m \times C_v \times \delta T \text{ in kilocals.}$$

Then from first law of thermodynamics,

$$dU = \delta Q - \delta W = \delta Q$$

$$= m\, C_V\, dT$$

$$= m \times \frac{C_P}{\gamma} \times dT = 5 \times \frac{993 \times 3}{5} \times 1 \qquad \left(\because \gamma = \frac{C_P}{C_V}\right)$$

$$= \textbf{2979 Kilocals.}$$

### EXAMPLE 4.25

*A carnot engine has its source at 100°C and its sink is maintained at a constant temperature by means of ice at 0°C. If it is working at the rate of 100 watts, how much ice will melt in one minute?*

**(Rohilkhand 2002)**

**SOLUTION.** $P = 100$ watts $= 100$ J/s

Work done in one minute $= 100 \times 60 = 6000$ J

$$T_1 = 100°\text{C} = 373 \text{ K}$$

$$T_2 = 0°\text{C} = 273 \text{ K}$$

$$\frac{Q_1 - Q_2}{Q_1} = \frac{T_1 - T_2}{T_1}$$

$$\frac{W}{Q_1} = \frac{T_1 - T_2}{T_1} \qquad ...(i)$$

$$\frac{Q_1}{Q_2} = \frac{T_1}{T_2} \qquad ...(ii)$$

From (*i*) and (*ii*)

$$\frac{W}{Q_2} = \frac{T_1 - T_2}{T_2}$$

$$Q_2 = \left[\frac{WT_2}{T_1 - T_2}\right] = \frac{6000 \times 272}{100}$$

$$Q_2 = 16380 \text{ J}$$

Suppose, $m$ kg of ice melts

$$Q_2 = mL$$

$$m = \frac{Q_2}{L} = \frac{16380}{80\times10^3\times4.2} = 0.04875 \text{ kg}$$

## EXAMPLE 4.26

*Air in the cylinder of a combustion engine at 20°C is compressed from an initial pressure of 1 atmosphere and volume $8 \times 10^{-4}$ m$^3$ to a volume $6 \times 10^{-5}$ m$^3$. Assume that air behaves an ideal gas ($\gamma = 1.40$) and the process is adiabatic, calculate.* **(Agra 2000)**

*(a) final pressure* *(b) final temperature of the gas*

**SOLUTION.** Process is adiabatic

(*a*)
$$P_1V_1^\gamma = P_2V_2^\gamma$$

$$P_2 = P_1\left[\frac{V_1}{V_2}\right]^\gamma$$

Here
$$P_1 = 1 \text{ atmosphere}$$
$$V_1 = 8 \times 10^{-4} \text{ m}^3$$
$$V_2 = 6 \times 10^{-5} \text{ m}^3$$
$$T_1 = 273 + 20 = 293 \text{ K}$$

$$P_2 = 1\left[\frac{8\times10^{-4}}{6\times10^{-5}}\right]^{1.4}$$

$$P_2 = \mathbf{37.6 \text{ atmospheres}}$$

(*b*) Since $\frac{PV}{T}$ is always constant during the process as no gas has escaped from the cylinder

$$\frac{PV}{T} = nR = \text{constant}$$

$$\frac{P_1V_1}{T_1} = \frac{P_2V_2}{T_2}$$

$$T_2 = \left(\frac{P_2}{P_1}\right)\left(\frac{V_2}{V_1}\right)T_1$$

$$T_2 = \left(\frac{37.6}{1}\right)\left(\frac{6\times10^{-5}}{8\times10^{-4}}\right)293$$

$$\mathbf{T_2 = 826.2 \text{ K}}$$

$$\mathbf{T_2 = 553.2°C}$$

## EXAMPLE 4.27

*Calculate the rise in temperature of a gas initially at 27°C, if its pressure is suddenly doubled, $\gamma = 1.4$* **(Kerala 2001)**

**SOLUTION.**

Here,
$$\frac{P_1^{\gamma-1}}{T_1^\gamma} = \frac{P_2^{\gamma-1}}{T_2^\gamma}$$

$$\left(\frac{P_2}{P_1}\right)^{\gamma-1} = \left(\frac{T_2}{T_1}\right)^\gamma$$

Here,
$$P_2 = 2P_1$$
$$T_1 = 27°C = 27 + 273\ K$$
$$= 300\ K$$
$$T_2 = ?$$
$$\gamma = 1.4$$

$$\therefore \quad (2)^{1.4-1} = \left(\frac{T_2}{300}\right)^{1.4}$$
$$0.4 \log 2 = 1.4 [\log T_2 - \log 300]$$
$$0.4 \times 0.3010 = 1.4 [\log T_2 - 2.4771]$$
$$\log T_2 = 0.086 + 2.4771$$
$$= 2.5631$$
$$T_2 = 365.7\ K$$
$$= 365.7 - 273 = 92.7°C$$
$$\text{Rise in temperature} = (92.7 - 27)\ °C$$
$$= \mathbf{65.7°C}$$

### EXAMPLE 4.28

*Calculate the work done when one litre of a mono atomic perfect gas at N.T.P. is compressed adiabatically to half its volume $\gamma = 1.67$.* **(Calicut 2000)**

**SOLUTION.** Here
$$V_1 = 1 \text{ litre} = 10^3\ cm^3 = 10^{-3}\ m^3$$
$$V_2 = 0.5 \text{ litre} = 500\ cm^3 = 5 \times 10^{-4}\ m^3$$
$$\gamma = 1.67$$

Work done
$$W = \left(\frac{1}{1-\gamma}\right)\left[\frac{1}{V_2^{\gamma-1}} - \frac{1}{V_1^{\gamma-1}}\right]$$
$$W = \left(\frac{1}{1-1.67}\right)\left[\left(\frac{1}{5\times10^{-4}}\right)^{0.67} - \left(\frac{1}{10^{-3}}\right)^{0.67}\right]$$
$$= -\left(\frac{1}{0.67}\right)[(2\times10^3)^{0.67} - (10^3)^{0.67}]$$
$$= -\left(\frac{1}{0.067}\right)[162.8 - 102.3]$$
$$= -90.3\ J$$

Negative sign shows, work is done on the gas.

Hence work done on the gas = **90.3 J**

### EXAMPLE 4.29

*Two Carnot engines X and Y are operating in series. The first one X receives heat at 1200 K and rejects to a reservoir at temperature T K. The second engine Y receives the heat rejected by X and inturn rejects to a heat reservoir at 300 K. Calculate the temperature T for the situation when.*

(*i*) The work outputs of two engines are equal.

(*ii*) The efficiency of two engines are equal. **(Nagpur 2001)**

**SOLUTION.**

(*i*) When work outputs are equal.

For the first engine,

$$W_1 = Q_1 - Q_2$$

For the second engine,

$$W_2 = Q_2 - Q_3$$

Here $W_1 = W_2$

$$Q_1 - Q_2 = Q_2 - Q_3$$

$$\left(\frac{Q_1}{Q_2}\right) - 1 = 1 - \left(\frac{Q_3}{Q_2}\right)$$

Also $\dfrac{Q_1}{Q_2} = \dfrac{1200}{T}$ and $\dfrac{Q_2}{Q_3} = \dfrac{T}{300}$ or $\dfrac{Q_3}{Q_2} = \dfrac{300}{T}$

$\therefore$
$$\left(\frac{1200}{T}\right) - 1 = 1 - \frac{300}{T}$$

or $T = \mathbf{750\ K}$

(*ii*) When efficiencies are equal.

$$\eta_1 = 1 - \left(\frac{Q_2}{Q_1}\right)$$

$$\eta_2 = 1 - \left(\frac{Q_3}{Q_2}\right)$$

$$\eta_1 = \eta_2$$

$$1 - \left(\frac{Q_2}{Q_1}\right) = 1 - \left(\frac{Q_3}{Q_2}\right)$$

$$1 - \left(\frac{T}{1200}\right) = 1 - \left(\frac{300}{T}\right)$$

$$\frac{T}{1200} = \frac{300}{T}$$

$$T = \mathbf{600\ K}$$

**EXAMPLE 4.30**

*A Carnot engine working between a source at temperature 27°C and a sink at – 73°C delivers 300 calories of heat to the latter in one cycle. Calculate the work performed in joules by the engine per cycle.* **(Mumbai 2001)**

**SOLUTION.** Here,

$$T_1 = 27 + 273 = 300\text{ K}$$

$$T_2 = -73 + 273 = 200\text{ K}$$

$$Q_2 = 300\text{ calories.}$$

$$Q_1 = ?$$

$$\frac{Q_1}{Q_2} = \frac{T_1}{T_2}$$

$$\frac{Q_1}{300} = \frac{300}{200}$$

$$Q_1 = 450\text{ calories}$$

Work performed per cycle,

$$W = Q_1 - Q_2$$
$$= 450 - 300$$
$$= 150 \text{ calories}$$
$$= 150 \times 4.2 \text{ J}$$
$$= \mathbf{630 \text{ J}}$$

**EXAMPLE 4.31** *A Carnot's engine is operated between two reservoirs at temperatures of 450 K and 350 K. If the engine receives 1000 calories of heat from the source in each cycle, calculate the amount of heat rejected to the sink in each cycle. Calculate the efficiency of the engine and the work done by the engine in each cycle. (1 calorie = 4.2 joules)* **(Madurai Kamraj 2001, Nagpur 2007)**

**SOLUTION.**

$$T_1 = 450 \text{ K} \; ; \qquad T_2 = 350 \text{ K}$$
$$Q_1 = 1000 \text{ cal} \; ; \qquad Q_2 = ?$$
$$\frac{Q_2}{Q_1} = \frac{T_2}{T_1}$$
$$Q_2 = Q_1 \times \frac{T_2}{T_1}$$
$$= \frac{1000 \times 350}{450} = 777.77 \text{ cals}$$
$$\eta = 1 - \frac{T_2}{T_1} = 1 - \frac{350}{450} = \frac{100}{450}$$
$$= 0.2222$$
$$\% \text{ efficiency} = \mathbf{22.22\%}$$

Work done in each cycle

$$= Q_1 - Q_2$$
$$= 1000 - 777.77$$
$$= 222.23 \text{ cal}$$
$$= 222.23 \times 4.2 \text{ joules}$$
$$= \mathbf{933.33 \text{ joules}}$$

**EXAMPLE 4.32** *Efficiency of a Carnot's cycle changes from $\frac{1}{6}$ to $\frac{1}{3}$ when source temperature is raised by 100 K. Calculate the temperature of the sink.* **(Purvanchal 2004)**

**SOLUTION.**

$$\eta = 1 - \frac{T_2}{T_1} = \frac{1}{6}$$

or

$$\frac{T_2}{T_1} = 1 - \frac{1}{6} = \frac{5}{6}$$

or

$$T_1 = \frac{6}{5} T_2 \qquad \text{...}(i)$$

or

$$\eta' = 1 - \frac{T_2}{T_1 + 100} = \frac{1}{3}$$

or

$$\frac{T_2}{T_1 + 100} = \frac{2}{3}$$

or $$T_2 = \frac{2}{3}(T_1 + 100) = \frac{2}{3}T_1 + \frac{2}{3}\times 100$$

or $$T_2 = \frac{2}{3}T_1 + \frac{200}{3}$$

$$= \left(\frac{2}{3}\times\frac{6}{5}T_2\right) + \frac{200}{3}$$

$$= \frac{4}{5}T_2 + \frac{200}{3}$$

or $$\frac{T_2}{5} = \frac{200}{3}$$

or $$T_2 = \frac{1000}{3}$$

$$= \mathbf{333.3\ K}$$

**EXAMPLE 4.33**

*A heat engine working between two temperatures could theoretically convert one-eighth of the heat supplied into work. If the lower temperature is reduced by 95 °C, the theoretical efficiency would be doubled. Find the initial temperatures.* **(Madras 2005, 2004/w)**

**SOLUTION.** (*i*) As the engine converts $\frac{1}{8}$ th of the heat input in to work, its efficiency $\eta = \frac{1}{8}$.

Let $T_1$ and $T_2$ be the temperatures of the source and sink respectively, then efficiency

$$\eta = 1 - \frac{T_2}{T_1}$$

or $$\frac{1}{8} = 1 - \frac{T_2}{T_1}$$

or $$\frac{T_2}{T_1} = 1 - \frac{1}{8} = \frac{7}{8}$$

(*ii*) When the temperature of the sink is reduced by 95 °C *i.e.*, 95 K, then

Efficiency $$\eta' = 1 - \frac{T_2 - 95}{T_1}$$

As the efficiency is doubled, $n' = 2\times\frac{1}{8} = \frac{1}{4}$

∴ $$\frac{1}{4} = 1 - \frac{T_2}{T_1} + \frac{95}{T_1}$$

$$= 1 - \frac{7}{8} + \frac{95}{T_1}$$

or $$\frac{95}{T_1} = \frac{1}{4} - 1 + \frac{7}{8} = \frac{1}{8}$$

or $$T_1 = 95 \times 8$$

$$= \mathbf{760\ K}$$

and $$T_2 = \frac{7}{8}\times T_1 = \frac{7}{8}\times 760$$

$$= \mathbf{665\ K}$$

## EXERCISE–IV

**4.1** Explain the term 'Heat' and 'Temperature' **(Nagpur 2009, Madras 2005)**

**4.2** Explain thermodynamic system and thermodynamic variables. **(Nagpur 2009)**

**4.3** Explain the terms:
(*i*) open system
(*ii*) closed system
(*iii*) isolated system

**4.4** State and explain zeroth law of thermodynamics. What is its importance. On the basis of this law introduce the concept of temperature. **(Nagpur 2008, 2007, Madras 2004/w, Rohilkhand 2001, Ranchi 2004, Calcutta 2003, 2005, Lucknow 2003, Purvanchal 2005, 2004, 2003, Osmania 2004, M.S. Uni. Tamil Nadu 2007)**

**4.5** What is meant by thermodynamic equilibrium. What are the basic requirements to be strictly followed? **(Purvanchal 2007, Rohilkhand 2000, G.N.D.U. 2000)**

**4.6** What do you mean by quasi-static process? Can it be achieved in practice? State its importance? **(Punjab 2000, Meerut 2003, Calcutta 2003, Purvanchal 2004)**

**4.7** Show that heat and work both are path dependent functions with the help of indicator diagram. **(Purvanchal 2007, 2003)**

**4.8** What is internal energy of a system? "Internal energy is state function and not a path function", Explain. **(Agra 2005, Nagpur 2009, Kanpur 2002, Lucknow 2002, G.N.D.U. 2004)**

**4.9** Explain the terms:
(*i*) Isochoric process **(Rohilkhand 2003, Lucknow 2000)**
(*ii*) Isobaric process **(Rohilkhand 2003, Lucknow 2004)**
(*iii*) Cyclic process **(Kerala 2001)**
(*iv*) Isothermal process **(Madras 2004/w, Rohilkhand 2003, Mysore 2004, Purvanchal 2004, Lucknow 2004)**
(*v*) Adiabatic process **(Madras 2004/w, Rohilkhand 2003, Mysore 2004, Purvanchal 2004, Lucknow 2004)**

**4.10** Show that change in the internal energy of a non-isolated system is always zero. **(Nagpur 2000)**

**4.11** What is indicator diagram? State its importance. **(Kuvempa 2005, Garhwal 2005)**

**4.12** Explain the concepts of reversible and irreversible processes. **(Nagpur 2008, Agra 2006, 2005, 2003, 2000, Madras 2004/w, Kanpur 2001, Garhwal 2000)**

**4.13** Define a thermodynamic system. Distinguish between open and closed system.

**4.14** Write notes on "Thermal equilibrium and zeroth law of thermodynamics." **(Rohilkhand 2003)**

**4.15** What is equation of state? Give its limitations. **(Garhwal 2001)**

**4.16** What do you understand by variables of state of a system and equation of state?

**4.17** The conduction of heat from a hot body to a cold body is reversible or irreversible process? Explain. **(Kanpur 2000)**

**4.18** Explain different forms of internal energy and show that the internal energy of a system is a function of state of the system. **(Nagpur 2004 – Winter)**

**4.19** State first law of thermodynamics. Give its physical significance. What are the limitations of first law? **(Rohilkhand 2003, Nagpur 2009, 2008, 2007, 2004/s, 2005/w, 2001, Kerala 2001, Meerut 2001, 2000, 2002, Madras 2003, Bangalore 2005, Purvanchal 2003)**

**4.20** Give applications of the first law to (*a*) isobaric (*b*) adiabatic processes and (*c*) isothermal process. **(Nagpur 2008, 2004/s, Purvanchal 2003)**

**4.21** Explain how first law of thermodynamics leads to the concept of internal energy.

**4.22** What is the first law of thermodynamics? Show that this law is a particular form of general law of conservation of energy. **(Purvanchal 2004, 2007, Agra 2006, Rohilkhand 2000)**

**4.23** What do you mean by isobaric process and isochoric process? Calculate the work done by an ideal gas in these processes. **(Kanpur 2002, Utkal 2003, G.N.D.U. 2004)**

**4.24** Obtain an expression for the external work done, when one mole of an ideal gas is suddenly expanded. **(Agra 2005)**

**4.25** From the first law of thermodynamics, prove that $C_P - C_V = R$. **(Agra 2004, Garhwal 2004)**

**4.26** Calculate the work done by an ideal gas during an isothermal expansion. **(Rohilkhand 2002, Nagpur 2008, 2005/w, 2005/s, Kuvempa 2005, Mysore 2004, Bangalore 2005)**

**4.27** Derive an expression for work done during an adiabatic process. **(Kerala 2001, Nagpur 2001, G.N.D.U. 2004, Garhwal 2005, Purvanchal 2004, Alageppa 2005, M.S. Uni. Tamil Nadu 2007, Bangalore 2005, Kuvempa 2005)**

**4.28** Show that the slope of an adiabatic is $\gamma$ times the slope of the isothermal. **(Nagpur 2001, 2003, Kerala 2001)**

**4.29** Prove that $PV^\gamma$ = constant in an adiabatic transformation. **(Madras 2005/Summer, Rohilkhand 2001, Nagpur 2003)**

**4.30** Derive the relation between volume and pressure of a gas undergoing adiabatic changes. **(Bangalore 2000, G.N.D.U. 2000, 2001, Calcutta 2003)**

**4.31** Why does the temperature of a gas drops when it is subjeted to adiabatic expansion? Explain.

**4.32** Show that the work done during an adiabatic process depends only upon the initial and final temperratures. **(Rohilkhand 2001 Imp.)**

**4.33** Explain why the gases have two specific heats? Explain why the specific heat at constant pressure $C_p$ is greater than that $C_V$; the specific heat at constant volume. Prove $C_P - C_V = R$. **(Madurai 2003, Garhwal 2005, Rohilkhand 2000, 2001, 2002, 2003, Lucknow 2001, Kerala 2001)**

**4.34** State the first law of thermodynamics. Express it mathematically and explain its physical significance. **(Meerut 2001, Nagpur 2009, 2008, 2003, 2001 Kerala 2001, G.N.D.U. 2001, Alagappa 2005)**

**4.35** Distinguish between isothermal and adiabatic changes. Show that for an adiabatic change in a perfect gas

$$PV^\gamma = \text{constant}.$$ **(Delhi 2000)**

**4.36** For a quasi-static adiabatic process of an ideal gas, prove $\dfrac{T}{(P)^{(\gamma-1)/\gamma}}$ = constant.

**(Bangalore 2000, Utkal 2003, Purvanchal 2003, Madras 2004/w)**

**4.37** Show that the equation of an adiabatic change is $V^{\gamma-1}\,T$ = constant. **(Madras 2004/Summer, Bangalore 2000)**

**4.38** Show that the work done in adiabatic expansion of an ideal gas from a state $(P_1, V_1)$ a state $(P_2, V_2)$ is given by

$$W = \frac{1}{\gamma-1}[P_1V_1 - P_2V_2]$$ **(Meerut 2004)**

**4.39** What is meant by adiabatic demagnetisation? **(Madras 2004/w)**

**4.40** Define specific heat at constant volume and at constant pressure. Why a gas has two specific heats? Explain. **(Alagappa 2003, 2004, Madras 2005, Madurai Kamraj 2003)**

**4.41** Distinguish between isothermal and adiabatic process. **(Mysore 2004/s, Bangalore 2004)**

**4.42** Draw $P - V$ diagrams representing isothermal and adiabatic processes. **(Mysore 2004/w, Bangalore 2004)**

**4.43** Obtain an expression for the ratio of heat capacities $C_P/C_V$. **(Gujrat 2004, 2005, Alagappa 2004)**

**4.44** Derive an expression for cooling produced by adiabatic demagnetisation of a paramagnetic salt. **(Punjab 2005, Meerut 2003)**

**4.45** Why does a rubber sting heats up on stretching? **(Punjab 2005)**

**4.46** A metallic wire is stretched reversibly and adiabatically. Calculate the temperature change in the process.

Show that when a rubber band is quickly strestched, it warms up. **(Agra 2000)**

**4.47** Prove that adiabatic elasticity of a gas is $\gamma$ times the isothermal elasticity. or $\frac{E_S}{E_\gamma} = \gamma$.
**(Calcutta 2003, G.N.D.U. 2001, Madurai Kamraj 2003)**

**4.48** Derive thermadynamic relation for stretching of a wire treating it as a thermodynamic system. Interpret the relation. **(Mumbai 2005)**

**4.49** Define efficiency of a heat engine. Obtain an expression for the efficiency of a reversible Carnot's engine with a perfect gas as the working substance. Mention the effective ways to increase its efficiency.
**(Garhwal 2001, Rohilkhand 2001 Imp., Meerut 2005, 2003, Bangalore 2005, Nagpur 2004–Winter 2005, 2009)**

**4.50** What is meant by a reversible process? Describe Carnot's cycle and show that all reversible engines working between the same two temperatures have the same efficiency.
**(Kanpur 1997, Nagpur 2009, 2007, 2001)**

**4.51** Calculate the work done in a Carnot's cycle of operations. Deduce the efficiency of a Carnot's engine in terms of the temperatures between which it works.**(Agra 2004, Nagpur 2007, 2006, H.P.U. 2001)**

**4.52** Derive an expression for the efficiency of an ideal heat engine, working between the temperatures $T_1$ and $T_2$ K. Prove that the efficiency of a reversible engine is maximum for the given source and sink.
**(Meerut 2004, 2002, 2001, Guwahati 2000, Pbi 2000, Kerala 2001, Punjab 2001)**

**4.53** State the Second Law of thermodynamics. Describe Carnot's cycle and deduce the efficiency of an ideal heat engine.
**(Kanpur 2001, Allahabad 1999, Meerut 1996, Purvanchal 2005, Bangalore 2004, Alagappa 2004, Garhwal 2005, 2004)**

**4.54** Describe Carnot's cycle and obtain an expression for the efficiency of an ideal heat engine working between two temperatures $T_1$ and $T_2$. **(Rohilkhand 2002, 2003, Meerut 2000, Nagpur 2007)**

**4.55** Show that no engine between two given temperatures can be more efficient than a reversible Carnot's engine working between the same two temperatures. **(Purvanchal 2003, Calcutta 2004)**

**4.56** What is the principle used in the working of a refrigerator. Define coefficient of performance. Is it greater than 1? Explain. **(Garhwal 2001)**

**4.57** Give two versions of second law of thermodynamics.
**(Nagpur 2005/s, Meerut 2001, Mysore 2004/s, 2004/w)**

**4.58** State and prove Cannot's theorem. Show that the temperature below zero Kelvin is not possible.
**(Kanpur 2001, Agra 2007, Rohilkhand 2001 Imp., 2003. Nagpur 2007, 2005/w. Alagappa 2004, Meerut 2001, Bangalore 2005, Purvanchal 2007, 2005, 2002, Agra 2006)**

**4.59** Explain the concepts of reversible and irreversible processes. Obtain an expression for the efficiency of a reversible Carnot engine with a perfect gas as the working substance. Show that for the same pair of working temperatures, the efficiency of all reversible engines is the same and is the maximum value attainable.
**(Garhwal 2001)**

**4.60** State Carnot's Theorem and show that it is a necessary consequence of Second Law of Thermodynamics.

Prove that the efficiency of a Carnot engine using an ideal gas as a working substance is $\eta = \frac{T_1 - T_2}{T_1}$.
**(Mysore 2004, Bangalore 2004, Kuvempa 2005, Agra 2003)**

**4.61** What are Kelvin–Planck and Clausius statement of second law of thermodynamics? Prove that they are correct in terms of principle of increase of entropy
**(Utkal 2003, Calcutta 2003, G.N.D.U. 2001, Calicut 2004)**

**4.62** State the second law of thermodynamics in atleast two distinctly different ways. Are these statements equivatent to each other? Justify your answer. **(Agra 2008, Kanpur 2002, Alagappa U. 2004)**

**4.63** Prove the equivalence of the Clausius and Kelvin–Planck's statement of the second law of thermodynamics. **(Meerut 2005, Calcutta 2005, 2003)**

**4.64** Describe with necessary theory, the construction and working of a Diesel engine. Explain its merits over otto engine. **(Madras 2004/w, 2005, Calcutta 2004)**

**4.65** Explain Clement and Desorme's method of determing the ratio of two specific heat capacities of a gas. **(Alagappa 2004)**

**4.66** Discuss with necessary theory the construction and working of a Otto engine. **(Madras 2004)**

**4.67** Write short notes on:

(*i*) Cornot cycle **(Meerut 2001, 2000, Rohilkhand 2000)**

(ii) Carnot's theorem **(Patna 2005)**

(*iii*) Efficiency of Carnot's Heat engine. **(Alagappa 2005, Patna 2005)**

(*iv*) Second law of thermodynamics. **(Agra 2000, Rohilkhand 2001 Imp., Kanpur 1995)**

(*v*) Zeroth law of thermodynamics. **(Kanpur 2002, Rohilkhand 2000)**

(*vi*) First law of thermodynamics. **(Meerut 2001, Patna 2004)**

(*vii*) Internal energy.

(*viii*) Quasi-static process. **(Rohilkhand 2000)**

(*ix*) External and Internal work.

(*x*) Thermodynamic equilibrium.

(*xi*) Reversible and Irreversible process. **(Punjab 2005, Kuvempa 2005, Ranchi 2005, Calcutta 2003, 2004, Madras 2005, Autonomus)**

**4.68** A monoatomic ideal gas of volume 1 litre and pressure 8 atmosphere undergoes adiabatic expansion unitill the pressure drops to 1 atmosphere. What is the final volume and how much work is done? Given : 1 atms. = $10^5$ N/m$^2$. **[Ans.** $3.47 \times 10^{-3}$ m$^3$, 686 Joules]

**4.69** A motor car tyre has a pressure of 3 atmospheres at the room temperature of 27°C. If the tyre suddenly bursts what is the resulting temperature? **[Ans.** 218.6 K = – 54.4°C] **(Ranchi 2000)**

**4.70** A quantity of air ($\gamma = 1.4$) at 27°C is compressed suddenly to $\frac{1}{4}$ of its original volume. Find the final temperature. **[Ans.** 522.3 K = 249.3°C] **(Rohilkhand 2001, Alagappa 2004, Nagpur 2009)**

**4.71** A quantity of air at 27°C and atmospheric pressure is suddenly compressed to $\frac{1}{5}$ of its original volume. Find (*i*) the final pressure and (*ii*) the final temperature.

**[Ans.** (*i*) 8.29 atmospheres (*ii*) 571.1 K = 298.1°C] **(Agra 2001)**

**4.72** A certain mass of an ideal gas at 27°C temperature and 8 atmospheric pressure, is expanded suddenly to 4 times of its volume. Find final pressure and final temperature. (Given: $\gamma = 1.5$) **(Nagpur 2009, 2004/w, 2003)**

**4.73** A carnot engine has an efficiency of 30% when the temperature of the sink is 27°C. Find the temperature of the source. **(Madras 2005 – Summer)**

**4.74** The temperature of 5 kg of air is raised by 1°C at constant volume. Calculate the increase in its internal energy. Given $C_p = 999$ J kg$^{-1}$ K$^{-1}$, $\gamma = \frac{5}{3}$. **(Madras 2004 – Winter)**

**4.75** If the temperature of the sink is 20°C and that of the source 250°C, calculate the efficiency of the carnot's engine. **[Ans.** 43.98%] **(Madras 2004 – Summer)**

**4.76** Initial pressure of an ideal gas at 30 °C is 'P'. Calculate the rise in temperature at which the pressure suddenly increases to 10 times its original pressure. (Given: $\gamma = 1.4$). **(Nagpur 2005/w)**

**[Hint :** $T_1 = 303$ K, $P_2 = 10\,P_1$, $\gamma = 1.4$,

$$T_2 = T_1\left(\frac{P_2}{P_1}\right)^{\gamma-1/\gamma} = 303\times(10)^{\frac{0.4}{1.4}} = 303\times 10^{\frac{2}{7}} = 303\times 10^{0.2875} = 585\ K$$

$\therefore$ Rise in temperature = 585 – 303 = 282 K].

**4.77** Dry air at N.T.P. (normal temperature and pressure) is compressed adiabatically to one-third of its original volume. Calculate the resulting pressure. Given that the ratio $C_p/C_v$ of air is 1.40. **(Bangalore 2001)**

**4.78** A Carnot's engine whose temperature of the source is 400 K takes 500 calories of heat at this temperature and rejects 400 calories of heat to the sink. What is the temperature of the sink? Calculate the efficiency of the engine. [**Ans.** (*i*) 320 K, (*ii*) 20%] **(Mysore 2001)**

**4.79** A Carnot's engine is operated between two reservoirs of 500 K and 400 K. If the engine receives 2000 calories of heat from the source in each cycle, calculate (*a*) the amount of heat rejected to the sink in each cycle, (*b*) the efficiency of the engine and (*c*) the work done by the engine in each cycle in (*i*) Joule (*ii*) Kilo-Watt hours

[**Ans.** (*a*) 1600 calories, (*b*) 20%, (*c*) (*i*) 1780 Joules (*ii*) $4.94 \times 10^4$ kWh] **(Nagpur 2005/w)**

**4.80** A Carnot's engine takes in 1000 kilocalories of heat a reservoir at 627°C and exhausts it to a sink at 27°C. (*a*) What is its efficiency, (*b*) How much work does it perform? Express it in (*i*) ergs (*ii*) kilo-watt hours (*iii*) electron volts.

[**Ans.** (*a*) 66.67%, (*b*) (*i*) $14 \times 10^{12}$ ergs (*ii*) 0.39 kWh (*iii*) $8.75 \times 10^{24}$ eV] **(Rajasthan 2001)**

**4.81** A Carnot's engine working as a refrigerator between 250 K and 300 K receives 1000 calories of heat from the reservoir at the lower temperature. (*i*) Calculate the amount of heat rejected to the reservoir at the higher temperature. (*ii*) Calculate also the amount of work done in each cycle to operate the refrigerator. [**Ans.** (*i*) 1200 cal, (*ii*) 840 joules] **(Calcutta 2000)**

**4.82** A reversible engine works between two temperatures whose difference is 100 °C. If it absorbs 746 Joule of heat from the source and gives 546 Joule to sink, calculate the temperatures of the source and the sink. [**Ans.** 100°C, 0°C] **(Meerut 1999)**

**4.83** A Caront engine has the same efficiency between 1000 K and 500 K and between *x* K and 1000 K (This being the temperature of the sink in this case). Calculate *x*.

[**Ans.** 2000 K] **(Bangalore 1995)**

**4.84** The efficiency of a Carnot's cycle is 1/6. If on reducing the temperature of the sink by 65 °C, the efficiency becomes 1/3. Find the initial and final tmepratures between which the cycle is working.

[**Ans.** 117°C, 52°C]

**4.85** The efficiency of an ideal engine is 0.2. If the temperaure of the sink is lowered by 20°C, the efficiency becomes 0.25. Find the temperature of th source and sink. **(Madras 2004)**

**4.86** A Carnot engine working between a source at 400 K and a sink at $T_2$ K has an efficiency of 50%. If the temperature of both the source and the sink are increased by 100 K, what is the efficiency of the engine?

[**Hint :** $\eta = 1 - \frac{T_2}{400} = \frac{1}{2}$ and $\eta' = 1 - \frac{T_2 + 100}{500}$] **(Mysore 2004/w, Bangalore 2005)**

**4.87** A carnot's engine has an efficiency 0.5 when the temperature of the sink is 300 K. What is the temperature of the source? By how much should it be raised so that the efficieney becomes 0.75 ? **(Bangalore 2002)**

**4.88** Two samples of air having same composition and initially at the same temperature and pressure are compressed from a volume *V* to a volume *V*/2, one isothermally and the other adiabatically. In which sample, is the final pressure greater? **(Lucknow 2003)**

**4.89** A thermos bottle containing coffee is vigorously shaken and thereby the temperature of the coffee rises. Regard the coffee as system: (*i*) Has heat been transferred to coffee? (*ii*) Has work been done in coffee? **(Lucknow 2003)**

**4.90** When a refrigerator is switched off, the ice stored in a cold storage melts at the rate of 36 kg/hr when the external temperature is 30°C. Find the minimum output power of the motor of the refrigerator required to prevent the ice from melting. (Latent heat of fusion of ice = 80 cal/gm, *J* = 4.2 Joules/cal.)

**(Calcutta 2003)**

**4.91** Find the work done to stretch a copper wire 1 m long and 2 mm in diameter through 1 cm. (Given $q = 12.5 \times 10^{10}$ N/m$^2$) **(Mysore 2004/s, Bangalore 2004)**

## OBJECTIVE QUESTIONS

**Select the correct answer from the given options**

**4.1** Out of the following, the physical quantity that relates with first law of thermodynamics is:
(*a*) Temperature (*b*) Pressure
(*c*) Energy (*d*) Number of moles **(Agra 2003)**

**4.2** In producing cooling by adiabatic demagnetisation, we use:
(*a*) magnetic substance (*d*) paramagnetic substance
(*c*) diamagnetic substance (*d*) ferromagnetic substance **(Agra 2003)**

**4.3** Zeroth law of thermodynamics is:
(*a*) kinetic energy of molecules of a gas is zero.
(*b*) ideal gas does not contain molecules
(*c*) if two systems are seperated in thermal equilibrium with a third system then they themselves are in thermal equilibrium with each other.
(*d*) absolute zero temperature cannot be attained. **(Garhwal 2005)**

**4.4** Transfer of heat from a body at low temperature to a body at high temperature:
(*a*) is impossible.
(*b*) is possible by keeping both the bodies in contact.
(*c*) is possible by doing some external work.
(*d*) none of these. **(Garhwal 2005)**

**4.5** In Carnot cycle, the first step is:
(*a*) Isothermal expansion. (*b*) Isothermal compression.
(*c*) Adiabatic expansion. (*d*) Adiabatic compression. **(Garhwal 2005)**

**4.6** An engine works between the temperatures 30 K and 300 K. What is its efficiency?
(*a*) 50% (*b*) 47%
(*c*) 90% (*d*) 10% **(Garhwal 2002)**

**4.7** The efficienay of Carnot engine working between steam point and ice point is:
(*a*) 1 (*b*) 0
(*c*) 26.81% (*d*) 16.81% **(Garhwal 2000, Rohilkhand 2003)**

**4.8** An ideal mono atomic gas initially at 17°C is suddenty compressed to 1/8th of its initial volume. The temperature of gas after compression is:
(*a*) 17 °C (*b*) 136 °C
(*c*) 887 °C (*d*) 1010 K **(Garhwal 2002)**

**4.9** The ratio of adiabatic and isothermal elasticities of a gas is:
(*a*) $\gamma$ (*b*) $\gamma^2$
(*c*) $\gamma^{-1}$ (*d*) $\left(d.\sqrt{\pi}\sqrt{\gamma}\right)$ **(Garhwal 2002)**

**4.10** The efficiency of Carnot's engine working between 127 °C and 27 °C is:
(*a*) 25% (*b*) 50%
(*c*) 75% (*d*) 100% **(Agra 2004, Kanpur 2001)**

**4.11** The efficiency of a reversible Carnot's engine working between temperature $T_1$ and $T_2$ $(T_1 > T_2)$ is.
(*a*) $\frac{T_2}{T_1}$ (*b*) $\frac{T_1}{T_2}$
(*c*) $\left(1-\frac{T_2}{T_1}\right)$ (*d*) $\left(\frac{T_1}{T_2}-1\right)$ **(Ranchi 2005)**

**4.12** Internal energy of a real gas depends upon:
(*a*) only on the temperature of the gas.
(*b*) only on the volume of the gas.
(*c*) only on the pressure of the gas.
(*d*) size of the molecule **(Rohilkhand 2002)**

**4.13** An adiabatic process occurs at constant
(*a*) temperature (*b*) pressure
(*c*) heat (*d*) none of these **(Rohilkhand 2002)**

**4.14** A diatomic gas at 18 °C is compressed adiabatically to $\frac{1}{32}$ times of its original volume. The temperature after compression will be:
(*a*) 18 °C (*b*) 887 °C
(*c*) 891 °C (*c*) 144 °C

**4.15** The gas law $\frac{PV}{T}$ = constant is true for
(*a*) isothermal change only (*b*) adiabatic change only
(*c*) both isothermal and adiabatic changes (*d*) none of these **(Rohilkhand 2002)**

**4.16** The work done during an adiabatic expansion of an ideal gas is:
(*a*) $W = \frac{1}{1-\gamma}(P_1V_1 - P_2V_2)$ (*b*) $W = \frac{1}{\gamma-1}(P_1V_1 - P_2V_2)$
(*c*) $W = R\,(T_2 - T_1)$ (*d*) $W = RT_1 \log_e \frac{V_2}{V_1}$ **(Rohilkhand 2002)**

**4.17** A Carnot's engine is operating between 100 ° and 50 °C. It's efficiency will be:
(*a*) 13.4% (*b*) 15.2%
(*c*) 50% (*d*) 100% **(Rohilkhand 2002)**

**4.18** A reversible heat engine can be 100% efficient, if the temperature of the sink is:
(*a*) less than that of source (*b*) equal to that of source
(*c*) 0 °C (*d*) 0 K

**4.19** The efficiency of a Carnot engine is 0.4. If the temperaute of the sink is 27 °C, the temperature of source is:
(*a*) 127 °C (*b*) 500 °C
(*c*) 500 K (*d*) 400 K **(Kanpur 2002)**

**4.20** The physics underlying the working of a refrigerator closely resembles the plysics underlying:
(*a*) Ice formation (*b*) Heat engine
(*c*) Vapour compression machine (*d*) Vaporization of water

**4.21** The door of running refrigerator inside a room is left open. Mark the correct statement:
(*a*) the room will be cooled slightly
(*b*) the room will be warmed up gradually
(*c*) the room will be cooled to the temperature inside the refrigerator
(*d*) the temperature of the room will remain unaffected

## ANSWERS

| | | | |
|---|---|---|---|
| **4.1** (*c*) | **4.2** (*b*) | **4.3** (*c*) | **4.4** (*c*) |
| **4.5** (*a*) | **4.6** (*c*) | **4.7** (*c*) | **4.8** (*c*) |
| **4.9** (*a*) | **4.10** (*a*) | **4.11** (*c*) | **4.12** (*a*) |
| **4.13** (*c*) | **4.14** (*c*) | **4.15** (*c*) | **4.16** (*b*) |
| **4.17** (*a*) | **4.18** (*d*) | **4.19** (*c*) | **4.20** (*b*) |
| **4.21** (*b*) | | | |

# 5

# Entropy

## AT A GLANCE

• Introduction • Concept of Entropy • Change in Entropy • Change in Entropy in Adiabatic Process • Change in Entropy in Reversible Cycle • Principle of Increase of Entropy • Change in Entropy in Irreversible Cycle • The T–S Diagram • Physical Significance of Entropy • Entropy of a perfect gas • Entropy of a steam • Kelvin's Thermodynamic Scale of Temperature • The Size of a Degree • Zero of Absolute or Thermodynamic Scale • Identity of Perfect Gas Scale and Absolute Scale • Third Law of thermodynamics: Nernst's heat theorem • Zero Point energy • Negative Temperature • Heat Death of Universe

## INTRODUCTION

The concept of entropy was first introduced by Clausius in 1854 during his work on the formulation and application of the second law of thermodynamics. It is a very important thermodynamic quantity and has proved to be very useful in the study of behaviour of heat engines.

## 5.1 Concept of Entropy

In order to describe the condition of a working substance completely, Clausius felt the need of another variable of state, in addition to volume, pressure, temperature, internal energy etc. This new variable is entropy. The quantity entropy found to remain constant in adiabatic process just as temperature remains constant in an isothermal process. Thus, the entropy can be defined as the thermal property of a working substance which remains constant during an adiabatic process.

## 5.2 Change in Entropy

Let us consider reversible Carnot's cycles bounded by the same two adiabatics $L$ and $M$ and isothermals $T_1$, $T_2$ and $T_3$ as shown in an indicator diagram (Fig. 5.1) for an ideal gas. Then all along the adiabatics $L$ and $M$, there is a change in volume and temperature with change in pressure. Let $ABCD$ and $DCEF$ represent the Carnot's reversible cycles. During Carnot's cycle $ABCD$, an amount of heat $Q_1$ is absorbed in going from $A$ to $B$ at constant temperature $T_1$ and an amount of heat $Q_2$ is rejected at constant temperature $T_2$. Then efficiency of Carnot's engine is given by

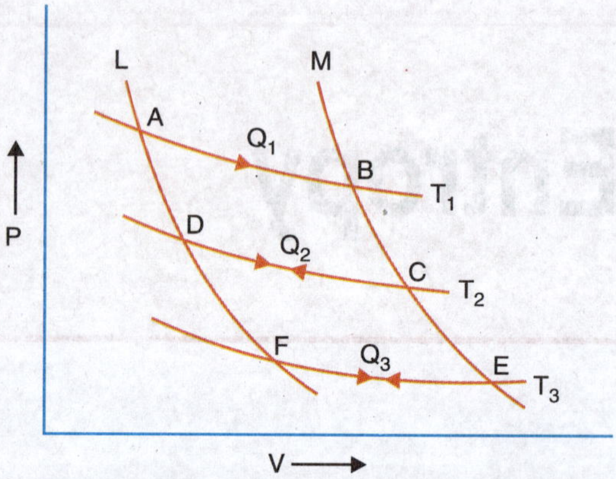


Fig. 5.1

$$\eta = \frac{Q_1 - Q_2}{Q_1} = \frac{T_1 - T_2}{T_1}$$

$$1 - \frac{Q_2}{Q_1} = 1 - \frac{T_2}{T_1}$$

$$\frac{Q_2}{Q_1} = \frac{T_2}{T_1}$$

$$\therefore \quad \frac{Q_1}{T_1} = \frac{Q_2}{T_2} \qquad ...(5.1)$$

Similarly considering the Carnot's cycle $DCEF$ in which an amount of heat $Q_2$ is absorbed at constant temperature $T_2$ and heat $Q_3$ is rejected at constant temperature $T_3$.

$$\therefore \quad \frac{Q_2}{T_2} = \frac{Q_3}{T_3} \qquad ...(5.2)$$

From equations (5.1) and (5.2) we have

$$\frac{Q_1}{T_1} = \frac{Q_2}{T_2} = \frac{Q_3}{Q_3} = \text{......... constant.}$$

In general, if $Q$ is the amount of heat absorbed or rejected at a temperature $T$ in going from one adiabatic to the other, then

$$\frac{Q}{T} = \text{constant}$$

If the two adiabatics are very close to each other and if $\delta Q$ is the small quantity of heat absorbed at constant temperature $T$ in going from one adiabatic to another, then

$$\frac{\delta Q}{T} = \text{constant}$$

This constant ratio is called the 'change in entropy' between the states represented by the two adiabatics. It is denoted by $\delta S$

$$\therefore \quad \delta S = \frac{\delta Q}{T}$$

The change in entropy for a finite reversible change in the state of working substance from A to B is given by

$$\int_{S_A}^{S_B} ds = \int_A^B \frac{dQ}{T}$$

$$\therefore \qquad S_B - S_A = \int_A^B \frac{dQ}{T} \qquad ...(5.3)$$

The expression $\int_A^B \frac{dQ}{T} = \int_{S_A}^{S_B} ds$ is a function of the thermodynamic coordinates of a system. This function is represented by symbol $S$ and is called entropy. Hence "entropy of a system is a function of the thermodynamical coordinates defining the state of the system *viz.*, the pressure, volume, temperature or internal energy and its change between two states is equal to $\int \frac{dQ}{T}$ between the states along any reversible path joining them." $dS$ is an exact differential. Entropy is an extensive property since it depends on the mass of the working substance.

## Physical Concept of Entropy

In fact, it is very difficult to form a *physical concept of entropy* as there is nothing physical to represent it. Moreover it can not be felt like temperature, pressure etc. But according to definition.

$$\text{Change of entropy} = \frac{\text{Heat added or subtracted}}{\text{Absolute temperature}} = \frac{dQ}{T}$$

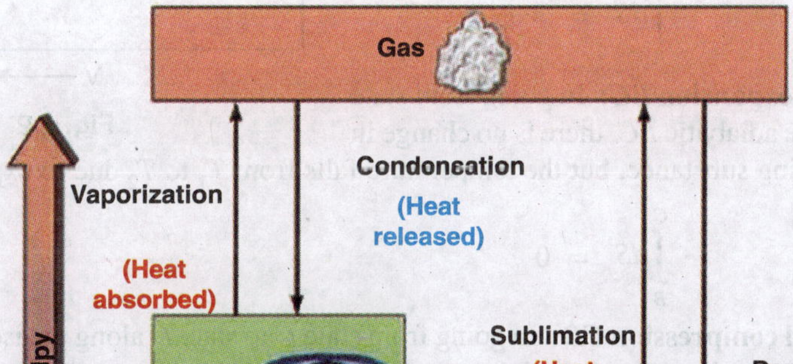


One can say that heat has the same dimensions as the product of entropy and absolute temperature. Since the gravitational potential energy of a body is proportional to the product of its mass and height above some zero level, likewise we may take temperature analogous to height and entropy as analogous to mass or inertia. Thus, we may take *entropy as **thermal inertia** which bears to heat motion, a relation similar to that which mass bears to linear motion or moment of inertia bears to rotational motion*.

## Unit of Entropy

Unit of entropy depends on the unit of heat used and the absolute temperature. It is measured in calories (or ergs or Joules) per degree Kelvin *i.e.*, cal./K. S.I. unit is Joule/K or Kcal/K.

## 5.3 Change in Entropy in Adiabatic Process

In an adiabatic process, no heat is allowed to enter or leave the system. Hence $\delta Q = 0$

$$\therefore \qquad \delta S = \frac{\delta Q}{T} = 0 \qquad \text{...(5.4)}$$

Thus, there is no change of entropy during an adiabatic process, or the entropy remains constant during an adiabatic reversible process. Hence the adiabatic curves on the $P - V$ diagram are called as *isoentropics* or curves of constant entropy.

## 5.4 Change of Entropy in Reversible Cycle

Consider a complete reversible Carnot's cycle *ABCD* as shown in Fig. 5.2 for an ideal gas formed by two isothermals *i.e.* *AB* at a temperature $T_1$ and *CD* at temperature $T_2$ and the two adiabatics *BC* and *DA*.

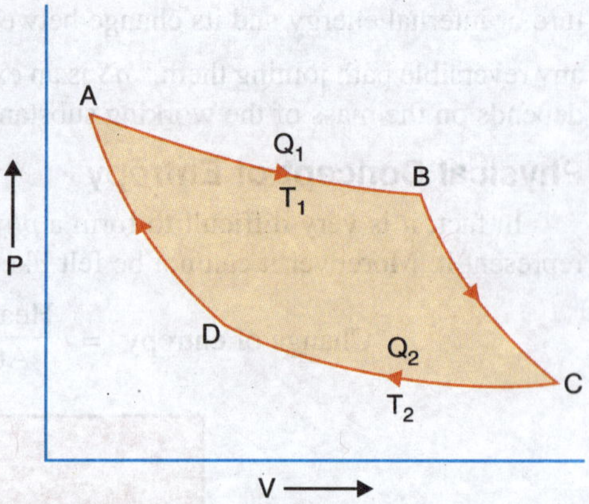


Fig. 5.2

***(i)* Isothermal expansion AB:** Let $Q_1$ be the quantity of heat absorbed by the working substance in going from state *A* to state *B* during isothermal expansion *AB* at a constant temperature $T_1$. The increase in entropy of the working substance is given by

$$\int_A^B dS = +\frac{Q_1}{T_1}$$

***(ii)* Adiabatic expansion *BC*:** In going from state *B* to state *C* along the adiabatic *BC*, there is no change in entropy of the working substance, but the temperature falls from $T_1$ to $T_2$ due to expansion.

$$\int_B^C dS = 0$$

***(iii)* Isothermal compression *CD*:** In going from state *C* to state *D* along the isothermal *CD*, the working substance rejects heat $Q_2$ to the sink at temperature $T_2$. The entropy of the working substance decreases and the change in entropy is given by

$$\int_C^D dS = -\frac{Q_2}{T_2}$$

***(iv)* Adiabatic compression *DA*:** In going from *D* to *A* along the adiabatic *DA*, there is no change in entropy but temperature rises from $T_2$ to $T_1$.

$$\therefore \qquad \int_D^A dS = 0$$

Thus the net gain in entropy of the working substance in the whole cycle ABCDA

$$= \int_A^B dS + \int_B^C dS + \int_C^D dS + \int_D^A dS$$

$$\oint dS = \frac{Q_1}{T_1} - \frac{Q_2}{T_2}$$

But for a reversible Carnot's cycle

$$\frac{Q_1}{T_1} = \frac{Q_2}{T_2} \quad \text{or} \quad \frac{Q_1}{T_1} - \frac{Q_2}{T_2} = 0$$

Substituting, we get

$$\oint dS = \frac{Q_1}{T_1} - \frac{Q_2}{T_2} = 0 \qquad ...(5.5)$$

where the integral sign with a circle refers to a complete cycle.

Thus in a cycle of reversible process, the entropy of the system remains unchanged or remains constant. In other words, the total change in entropy is always zero.

## 5.5 Principle of Increase of Entropy

Let us consider an engine performing irreversible cycle of changes in which the working substance absorbs heat $Q_1$ at temperature $T_1$ from the source and rejects heat $Q_2$ to the sink at temperature $T_2$. Then the efficiency of this cycle is given by

$$\eta' = \frac{Q_1 - Q_2}{Q_1} = 1 - \frac{Q_2}{Q_1}$$

But according to Carnot's theorem, this efficiency is less than that of a reversible engine working between the same two temperatures $T_1$ and $T_2$ for which efficiency is given by

$$\eta = 1 - \frac{T_2}{T_1}$$

Thus,

$$\eta' < \eta$$

$$1 - \frac{Q_2}{Q_1} < 1 - \frac{T_2}{T_1}$$

$$\frac{Q_2}{Q_1} > \frac{T_2}{T_1} \quad \text{or} \quad \frac{Q_2}{T_2} > \frac{Q_1}{T_1}$$

or

$$\frac{Q_2}{T_2} - \frac{Q_1}{T_1} > 0 \qquad ...(5.6)$$

Considering the whole system, the source loses (– *ve* sign) the entropy by an amount $\frac{Q_1}{T_1}$ and the sink gains (+ *ve* sign) an entropy $\frac{Q_2}{T_2}$. Therefore, the net change in entropy for the whole system is $\left(\frac{Q_2}{T_2} - \frac{Q_1}{T_1}\right)$, which is clearly greater than zero *i.e.*, positive ($\because$ equation 5.6). Thus, there is an increase in entropy of the system during an irreversible process.

To make it more clear, let us consider the natural process of conduction of heat from a body $A$ at temperature $T_1$ to another body $B$ at a lower temperature $T_2$. This process is irreversible. Since heat always flows from higher to a lower temperature the quantity of heat thus transmitted be $Q$, then

$$\text{Decrease in entropy of hotter body} = \frac{Q}{T_1}$$

Increase in entropy of colder body = $\frac{Q}{T_2}$

∴ Net increase in entropy of the system

$$dS = \frac{Q}{T_2} - \frac{Q}{T_1}$$

$$= + \text{ve quantity as } T_1 > T_2.$$

Thus, we conclude that the entropy of a system increases in all irreversible processes. This is known as the law or principle of increase of entropy. All natural processes taking place in the universe are irreversible. It means the entropy of the universe increases.

$$dS \text{ [universe]} \geqslant 0$$

## 5.6 Change of entropy in Irreversible Process

The thermodynamical state of a system can be defined with the help of the thermodynamical coordinates of the system. The state of a system can be changed by altering the thermodynamical coordinates. Changing from one state to the other by changing the thermodynamical coordinates is called a *process*.

Consider two states of a system *i.e.*, state $A$ and state $B$. Change of state from $A$ to $B$ or *vice versa* is a process and the direction of the process will depend upon a new thermodynamical coordinate called **entropy**. All processes are not possible in the universe.

### Consider the following processes:

**1.** Let two blocks $A$ and $B$ at different temperatures $T_1$ and $T_2$ ($T_1 > T_2$) be kept in contact but the system as a whole is insulated from the surroundings. Conduction of heat takes place between the blocks, the temperature of $A$ falls and the temperature of $B$ rises and thermodynamical equilibrium will be reached.

**2.** Consider a flywheel rotating with an angular velocity $\omega$. Its initial kinetic energy is $\frac{1}{2} I\omega^2$. After some time the wheel comes to rest and kinetic energy is utilised in overcoming friction at the bearings. The temperature of the wheel and the bearings rises and the increase in their internal energy is equal to the original kinetic energy of the fly wheel.

**3.** Consider two flasks $A$ and $B$ connected by a glass tube provided with a stop cock. Let $A$ contain air at high pressure and $B$ is evacuated. The system is isolated from the surroundings. If the stop cock is opened, air rushes from $A$ to $B$, the pressure in $A$ decreases and the volume of air increases.

All the above three examples though different, are thermodynamical processes involving change in thermodynamical coordinates. Also, in accordance with the first law of thermodynamics, the principle of conservation of energy is not violated because the total energy of the system is conserved. It is also clear that, with the initial conditions described above, the three processes will take place.

Let us consider the possibility of the above three processes taking place in the reverse direction. In the first case, if the reverse process is possible, the block $B$ should transfer heat to $A$ and initial conditions should be restored. In the second case, if the reverse process is possible, the heat energy must again change to kinetic energy and the fly wheel should start rotating with the initial angular velocity $\omega$. In the third case, if the reverse process is possible the air in $B$ must flow back to $A$ the initial condition should be obtained.

But, it is a matter of common experience, that none of the above conditions for the reverse processes are reached. It means that the direction of the process cannot be determined by knowing the thermodynamical coordinates in the two end states. To determine the direction of the process a new thermodynamical coordinate has been devised by Clausius and this is called the entropy of the system. Similar to internal energy, entropy is also a function of the state of a system. For any possible process, the entropy of an isolated system should increase or remain constant. The process in which there is a possibility of decrease in entropy cannot take place.

If the entropy of an isolated system is maximum, any change of state will mean decrease in entropy and hence that change of state will not take place.

To conclude, *processes in which the entropy of an isolated system decreases do not take place or for all processes taking place in an isolated system the entropy of the system should increase or remain constant.* It means a process is irreversible if the entropy decreases when the direction of the process is reversed. A process is said to be irreversible if it cannot be retraced back exactly in the opposite direction. During an irreversible process, heat energy is always used to overcome friction. Energy is also dissipated in the form of conduction and radiation. This loss of energy always takes place whether the engine works in one direction or the reverse direction. Such energy cannot be regained. In actual practice all the engines are irreversible. If electric current is passed through a wire, heat is produced. If the direction of the current is reversed, heat is again produced. This is also an example of an irreversible process. All chemical reactions are irreversible. In general, all natural processes are irreversible.

## 5.7 The T–S Diagram

The temperature-entropy diagram is used in engineering and meteorology. Consider the Carnot's cycle *ABCDA* (Fig. 5.3).

Thermodynamic changes in the state of a substance can be represented by plotting entropy ($S$) along horizotal axis and temperature ($T$) along perpendicular axis. Such a diagram is called temperature entropy diagram (*i.e.* $T$–$S$ diagram) for the substance.

A Carnot's reversible cycle is shown on $P$–$V$ diagram (Fig. 5.3). It is composed of two isothermals $AB$ and $CD$ at constant temperatures $T_1$ and $T_2$ respectively.

On the $T$–$S$ diagram (Fig 5.4), the isothermal curves are shown by two straight lines $AB$ and $CD$ parallel to S-axis and the adiabatic curves are shown by the straight lines $BC$ and $DA$ parallel to T-axis. The resulting $T$–$S$ diagram is a rectangle $ABCD$.

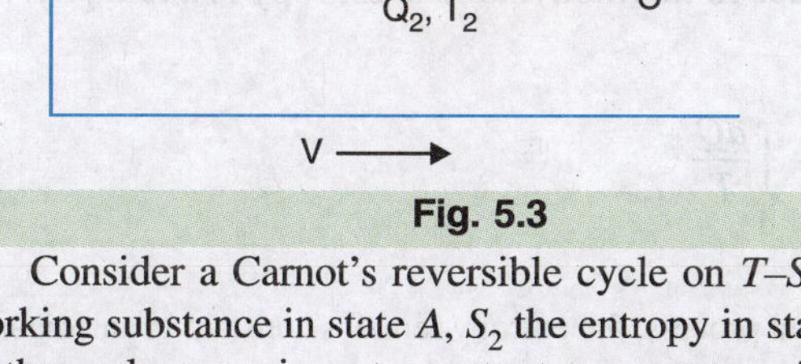


Fig. 5.3

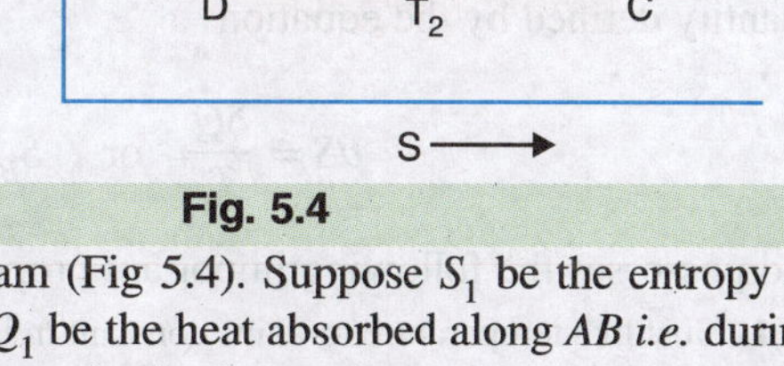


Fig. 5.4

Consider a Carnot's reversible cycle on $T$–$S$ diagram (Fig 5.4). Suppose $S_1$ be the entropy of working substance in state $A$, $S_2$ the entropy in state $B$, $Q_1$ be the heat absorbed along $AB$ *i.e.* during isothermal expansion at constant temperature $T_1$ and $Q_2$ is the heat rejected in the isothermal compression at constant temperature $T_2$ along $CD$.

In going from $A$ to $B$, along isothermal expansion $AB$, the gain in the entropy of the working substance is given by

$$S_2 - S_1 = \frac{Q_1}{T_1} \qquad \text{...(5.7)}$$

In going from $B$ to $C$, along adiabatic expansion $BC$, there is no change in the entropy.

In going from $C$ to $D$, along isothermal compression $CD$, the loss in the entropy of the working substance is given by

$$S_2 - S_1 = \frac{Q_2}{T_2} \qquad ...(5.8)$$

Along adiabatic compression $DA$, there is no change in the entropy. Therefore, from equations (5.7) and (5.8), we get

$$Q_1 = T_1 (S_2 - S_1)$$

and $$Q_2 = T_2 (S_2 - S_1)$$

$$\therefore \quad Q_1 - Q_2 = (T_1 - T_2)(S_2 - S_1) \qquad ...(5.9)$$

The quantity $(Q_1 - Q_2)$ represents the external work done in the cycle and $(T_1 - T_2)(S_2 - S_1)$ is the area of the rectangle on the $T$–$S$ diagram.

Thus, the area of the rectangle on the $T$–$S$ diagram represents the external work done in a reversible Carnot's cycle.

### Efficiency of Carnot's engine

Using $T$–$S$ diagram, we can write

$$\text{Efficiency of the engine} = \frac{\text{external work done in one cycle}}{\text{heat absorbed from the source}}$$

$$\eta = \frac{Q_1 - Q_2}{Q_1}$$

$$= \frac{(T_1 - T_2)(S_2 - S_1)}{T_1 (S_2 - S_1)}$$

$$= \frac{T_1 - T_2}{T_1}$$

$$\eta = 1 - \frac{T_2}{T_1} \qquad ...(5.10)$$

## 5.8 Physical Significance of Entropy

Although it is very difficult to conceive the idea of entropy as there is no physical method to demonstrate it i.e, it cannot be felt like temperature, pressure, volume and does not produce any effect which can be demonstrated; it has a great significance in thermodynamics. Entropy is a real physical quantity defined by the equation

$$dS = \frac{\delta Q}{T} \quad \text{or} \quad S_B - S_A = \int_A^B \frac{dQ}{T}$$

and possesses the following important properties:

1. Just like temperature remains constant in isothermal process, the entropy remains constant in an adiabatic process.
2. It is a definite single valued function of the thermodynamic variables describing the state of a working substance.
3. In every natural process (*i.e.* irreversible change) there is always an increase in entropy.
4. The second law of thermodynamics can be stated in terms of entropy of a system.
5. Due to increase in entropy, unavailable anergy increases.

6. According to Freeman Dyson, the entropy is a measure of disorderness. This disorderness can be evaluated by using the relation.

$$S = k \log W$$

where $k$ is the Boltzmann's constant and $W$ is the disorder parameter.

## 5.9 Entropy of a Perfect Gas

Consider 1 gm. of a perfect gas, occupying a volume $V$ at pressure $P$ and temperature $T$. Let $dQ$ be the heat supplied to the gas, then according to first law of thermodynamics we have

$$dQ = dU + dW \qquad ...(i)$$

Let $C_V$ be the specific heat at constant volume, $dT$ be the rise of temperature and $dV$ be the change in volume then

$$dU = C_V\, dT \quad \text{and} \quad dW = P.dV$$

Substituting in eqn. (*i*), we get

$$dQ = C_V\, dT + PdV \qquad ...(ii)$$

If S is the entropy per unit mass of the gas, then

$$S = \int \frac{dQ}{T} = \int \frac{1}{T}(C_V dT + PdV)$$

$$= \int C_V \frac{dT}{T} + \int \frac{P}{T} dV \qquad ...(iii)$$

We can calculate $S$ in terms of different thermodynamic variables. These are:

***(i) S* in terms of temperature and volume**

for a perfect gas, $\quad PV = rT$

$$\therefore \quad \frac{P}{T} = \frac{r}{V}$$

Substituting in equation (*iii*), we get

$$S = \int C_V \frac{dT}{T} + \int r \frac{dV}{V}$$

$$= C_V \log_e T + r \log_e V + \text{constant}$$

But $\quad r = C_P - C_V$, as per Mayer's relation.

$$\therefore \quad S = C_V \log_e T + (C_P - C_V) \log_e V + \text{constant} \qquad ...(iv)$$

where $C_P$ is specific heat at constan pressure.

***(ii) S* in terms of temperature and pressure**

For a perfect gas, $PV = rT$

$$\therefore \quad V = \frac{rT}{P}$$

and $\quad PdV + VdP = rdT$

or $\quad PdV = rdT - VdP$

Substituting in eqn. (*iii*), we get

$$S = \int \left[ C_V \frac{dT}{T} + \frac{rdT - VdP}{T} \right]$$

$$= \int \left[ C_V \frac{dT}{T} + \frac{rdT}{T} - \frac{rdP}{T} \right] \qquad \left(\because \frac{1}{T} = \frac{r}{PV} \text{ from } PV = rT\right)$$

$$= \int \left[ C_V \frac{dT}{T} + r\left(\frac{dT}{T} - \frac{dP}{P}\right)\right]$$

$$= \int \left[ C_V \frac{dT}{T} + (C_P - C_V)\left(\frac{dT}{T} - \frac{dP}{P}\right) \right] \qquad (\because C_P - C_V = r)$$

$$= \int \left[ C_V \frac{dT}{T} + C_P \frac{dT}{T} - C_V \frac{dT}{T} - (C_P - C_V)\frac{dP}{P} \right]$$

$$= \int \left[ C_P \frac{dT}{T} - (C_P - C_V)\frac{dP}{P} \right]$$

$$= C_P \log_e T - (C_P - C_V) \log_e P + \text{constant} \qquad ...(v)$$

**(*iii*) *S* in terms of pressure and volume**

For ideal gas, $PV = rT$

$$\therefore \qquad T = \frac{PV}{r} \quad \text{and} \quad dT = \frac{PdV + VdP}{r}$$

Substituting in equation (*iii*), we get

$$S = \int \left[ C_V \frac{PdV + VdP}{rT} + \frac{PdV}{PV} \right]$$

$$= \int \left[ C_V \frac{PdV + VdP}{PV} + \frac{PdV \cdot r}{PV} \right] \qquad \left(\because PV = rT, \frac{L}{T} = \frac{r}{PV}\right)$$

$$= \int \left[ C_V \left(\frac{dV}{V} + \frac{dP}{P}\right) + (C_P - C_V)\frac{dV}{V} \right]$$

$$= \int \left( C_V \frac{dP}{P} + C_P \frac{dV}{V} \right)$$

$$= C_V \log_e P + C_P \log_e V + \text{constant} \qquad ...(vi)$$

while deriving the relations for $S$ given by eqns. (*iii*), (*iv*), (*v*) and (*vi*), we have assumed the entropy of the gas to be zero at absolute zero of temperature.

However, in actual practice, we are intersted in change in entropy of a gas when it goes from its state of pressure $P_1$, volume $V_1$ and temperature $T_1$ to another state having the values $P_2$, $V_2$ and $T_2$ respectively. This is obtained by integrating eqn. (*iii*) between the limits $T_1$ and $T_2$. The change of entropy $\Delta S = S_2 - S_1$ for each of the above three cases and can be calculated as under:

In terms of temperature and volume, the change of entropy

$$\Delta S = S_2 - S_1 = C_V \log_e \frac{T_2}{T_1} + (C_P - C_V) \log_e \frac{V_2}{V_1}. \qquad ...(vii)$$

In terms of temperature and pressure, $\Delta S$ will be

$$\Delta S = S_2 - S_1 = C_P \log_e \frac{T_2}{T_1} - (C_P - C_V) \log_e \frac{P_2}{P_1} \qquad ...(viii)$$

and in terms of pressure and volume, $\Delta S$ will be

$$\Delta S = S_2 - S_1 = C_V \log_e \frac{P_2}{P_1} + C_P \log_e \frac{V_2}{V_1} \qquad ...(ix)$$

The above equations are evaluated for a gas of unit mass. For a gas of *m* gm., the change of entropy can be obtained by multiplying these relations by mass *m* of the gas.

## 5.10 Entropy of a Steam

Consider 1 gm. of ice at temperature $T_1$. Let us calculate the total change in entropy when this ice changes into steam at absolute temperature $T_2$.

If $dQ$ is the amount heat supplied to a substance at temperature $T$, the change in entropy is given by

$$dS = \frac{dQ}{T}$$

In order to convert 1 gm. of ice at $T_1$ K into water at $T_1$ K, an amount of heat absorbed is equal to latent heat of ice, $L_i$, say.

$\therefore$ Change in entropy during this process = $\frac{L_i}{T_1}$ ...(*i*)

Now 1 gm. at water of temperature $T_1$ is heated to $T_2$ K. The change in entropy during process

$$= \int_{T_1}^{T_2} \frac{dQ}{T} = \int_{T_1}^{T_2} C\frac{dT}{T} = C\log_e \frac{T_2}{T_1} \quad ...(ii)$$

where $C$ is the specific heat of water, a constant quantity.

Further, 1 gm of water at $T_2$ K is converted into steam at the same temperature $T_2$ K. It will absorb heat $L_s$, equal to the latent heat of steam. Thus,

Change of entropy during this process = $\frac{L_S}{T_2}$ ...(*iii*)

$\therefore$ Adding (*i*), (*ii*) and (*iii*), we get total change in entropy during the complete process,

$$\Delta S = \frac{L_i}{T_1} + C\log_e \frac{T_2}{T_1} + \frac{L_S}{T_2} \quad ...(iv)$$

This equation gives the total energy of 1 gm. of steam of temperature $T_2$ K. For $m$ gm. of substance, we will multiply eqn. (*iv*) by $m$, to give the total change in entropy during the process. *i.e.*, for converting the ice into steam.

## 5.11 Kelvin's Thermodynamic Scale of Temperature

The efficiency of reversible Carnot's engine depends only upon the two temperatures between which it works and is independent of the properties (nature) of working substance. Using this property of Carnot's reversible engine which solely or absolutely depends on temperature and nothing else, Lord Kelvin in 1848 suggested a new scale of temperature, known as 'absolute scale of temperature'. Lord Kelvin worked out the theory of such an absolute scale, called the **Kelvin's work** or **thermodynamic scale** and showed that it agrees with the ideal gas scale.

### Theory :

Suppose a reversible engine absorbs heat $Q_1$ at temperature $\theta_1$ and rejects heat $Q_2$ at temperature $\theta_2$ (measured on any arbitrary scale), then since the efficiency of the engine is a function of these two temperatures

$$\eta = \frac{Q_1 - Q_2}{Q_1} = f(\theta_1, \theta_2)$$

$$1 - \frac{Q_2}{Q_1} = f(\theta_1, \theta_2)$$

$$\therefore \quad \frac{Q_1}{Q_2} = \frac{1}{1-f(\theta_1,\theta_2)} = F(\theta_1,\theta_2) \qquad ...(i)$$

where $F$ is some other function of $\theta_1$ and $\theta_2$.

Similarly, if the reversible engine works between a pair of temperature $\theta_2$ and $\theta_3$ ($\theta_2 > \theta_3$), absorbing a heat $Q_2$ and rejecting $Q_3$, we can write

$$\frac{Q_2}{Q_3} = F(\theta_2, \theta_3) \qquad ...(ii)$$

Also if it works between $\theta_1$ and $\theta_3$ ($\theta_1 > \theta_3$), then

$$\frac{Q_1}{Q_3} = F(\theta_1, \theta_3) \qquad ...(iii)$$

Multiplying equation (*i*) and (*ii*), we have

$$\frac{Q_1}{Q_2} \times \frac{Q_2}{Q_3} = \frac{Q_1}{Q_3} = F(\theta_1, \theta_2) \times F(\theta_2, \theta_3)$$

Now comparing with equation (*iii*), we have

$$F(\theta_1, \theta_3) = F(\theta_1, \theta_2) \times F(\theta_2, \theta_3) \qquad ...(iv)$$

This is called the **functional equation**. It does not contain $\theta_2$ on the left hand side, therefore, function $F$ should be so chosen that $\theta_2$ disappears from the right hand side also. This is possible if

$$F(\theta_1, \theta_2) = \frac{\phi(\theta_1)}{\phi(\theta_2)} \quad \text{and} \quad F(\theta_2, \theta_3) = \frac{\phi(\theta_2)}{\phi(\theta_3)}$$

where $\phi$ is another unknown function of temperature.

From equation (*iv*),

$$F(\theta_1, \theta_3) = \frac{\phi(\theta_1)}{\phi(\theta_2)} \times \frac{\phi(\theta_2)}{\phi(\theta_3)} = \frac{\phi(\theta_1)}{\phi(\theta_3)}$$

Equation (*i*) can now be written as

$$\frac{Q_1}{Q_2} = F(\theta_1, \theta_2) = \frac{\phi(\theta_1)}{\phi(\theta_2)} \qquad ...(v)$$

Since $\theta_1 > \theta_2$ and $Q_1 > Q_2$, the function $\phi(\theta_1) > \phi(\theta_2)$. Thus, function $\phi(\theta)$ is a linear function of $\theta$ and can be used to measure temperature.

Thus, Lord Kelvin suggested $\phi(\theta)$ should be taken proportional to $\theta$. *i.e.* $\phi(\theta_1) \propto \theta_1$ and $\phi(\theta_2) \propto \theta_2$, we have

$$\frac{Q_1}{Q_2} = \frac{\theta_1}{\theta_2}$$

$$\text{or} \quad \frac{\theta_1}{\theta_2} = \frac{Q_1}{Q_2} \qquad ...(5.11)$$

This equation shows that the ratio of the two temperatures on this scale is equal to the ratio of the heat absorbed to the heat rejected. This temperature scale is called Kelvin's thermodynamic (or absolute or work) scale of temperature.

### Alternative Method (using Carnot's cycle)

Consider a Carnot's reversible cycle *ABCD* (Fig. 5.5). The two isothermals are *AB* and *CD*. The temperature on the work scale is $\theta_1$ and the ideal scale is $T_1$ for the isothermal *AB* and the temperature on

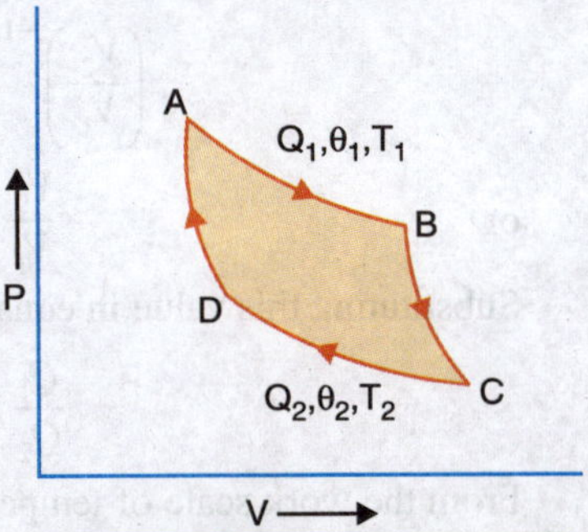


Fig. 5.5

the work scale is $\theta_2$ and the ideal gas scale is $T_2$ for the isothermal *CD*. Let the pressure and volume for the points *A, B, C* and *D* be $P_1$, $V_1$; $P_2$, $V_2$ ; $P_3$, $V_3$ and $P_4$, $V_4$ respectively. Consider that 1 gram of an ideal gas is used as the working substance.

From *A* to *B*, the process is isothermal.

$$\text{Work done} = \int_{V_1}^{V_2} PdV$$

$$= rT_1 \int_{V_1}^{V_2} \frac{dV}{V}$$

$$\therefore \quad Q_1 = rT_1 \log \frac{V_2}{V_1} \quad ...(i)$$

Here *r* is expressed in heat units.

From *C* to *D*, the process is isothermal and a quantity of heat $Q_2$ is rejected by the gas.

$$Q_2 = -\int_{V_3}^{V_4} PdV$$

$$= -rT_2 \int_{V_3}^{V_4} \frac{dV}{V}$$

$$Q_2 = rT_2 \log \frac{V_2}{V_4} \quad ...(ii)$$

Dividing (*i*) by (*ii*)

$$\frac{Q_1}{Q_2} = \frac{rT_1 \log \frac{V_2}{V_1}}{rT_2 \log \frac{V_3}{V_4}} \quad ...(iii)$$

Also, *A* and *B* lie on the same isothermal

$$P_1 V_1 = P_2 V_2$$

$$\text{or} \quad \frac{P_2}{P_1} = \frac{V_1}{V_2} \quad ...(iv)$$

*C* and *D* lie on the same isothermal

$$P_3 V_3 = P_4 V_4$$

$$\text{or} \quad \frac{P_3}{P_4} = \frac{V_4}{V_3} \quad ...(v)$$

*B* and *C* lie on the same adiabatic

$$P_2 V_2^{\gamma} = P_3 V_3^{\gamma} \quad ...(vi)$$

*A* and *D* lie on the same adiabatic

$$P_1 V_1^{\gamma} = P_4 V_4^{\gamma}$$

Dividing (*vi*) by (*vii*)

$$\left(\frac{P_2}{P_1}\right)\left(\frac{V_2}{V_1}\right)^{\gamma} = \left(\frac{P_3}{P_4}\right)\left(\frac{V_3}{V_4}\right)^{\gamma}$$

$$\text{or} \quad \left(\frac{V_1}{V_2}\right)\left(\frac{V_2}{V_1}\right)^{\gamma} = \left(\frac{V_4}{V_3}\right)\left(\frac{V_3}{V_4}\right)^{\gamma}$$

$$\left(\frac{V_2}{V_1}\right)^{\gamma-1} = \left(\frac{V_3}{V_4}\right)^{\gamma-1}$$

or $$\frac{V_2}{V_1} = \frac{V_3}{V_4} \qquad ...(viii)$$

Substituting this value in equation (*iii*)

$$\frac{Q_1}{Q_2} = \frac{T_1}{T_2} \qquad ...(ix)$$

From the work scale of temperature,

$$\frac{Q_1}{Q_2} = \frac{\theta_1}{\theta_2} \qquad ...(x)$$

$$\therefore \quad \frac{\theta_1}{\theta_2} = \frac{T_1}{T_2} \qquad ...(xi)$$

## Special considerations

(*i*) If $\theta_1$ is zero, $T_1$ is also zero. It means the two scales are identical at absolute zero temperature.

(*ii*) Consider that the Carnot's engine works between the steam point and the ice point and there are 100 degrees between the ice point and the steam point on the two scales.

$$\frac{\theta_{steam}}{\theta_{ice}} = \frac{T_{steam}}{T_{ice}}$$

$$\frac{\theta_{ice}+100}{\theta_{ice}} = \frac{T_{ice}+100}{T_{ice}}$$

or $$\theta_{ice} = T_{ice}$$

Therefore, the ice point is the same on the two scales.

Similarly $$\theta_{steam} = T_{steam}$$

Hence the two scales are completely identical.

**Efficiency.** The efficiency of a Carnot's engine,

$$\eta = 1-\frac{Q_2}{Q_1}$$

$$= 1-\frac{\theta_2}{\theta_1} = 1-\frac{T_2}{T_1}$$

Let the adiabatic expansion ratio be $\rho$

$$\rho = \frac{V_3}{V_2}$$

The points *B* and *C* lie on the same adiabatic

$$\therefore \quad T_1 V_2^{\gamma-1} = T_2 V_3^{\gamma-1}$$

$$\frac{T_2}{T_1} = \left(\frac{V_2}{V_3}\right)^{\gamma-1}$$

$$= \left(\frac{1}{\rho}\right)^{\gamma-1}$$

$$\eta = 1-\frac{T_2}{T_1} = 1-\left(\frac{1}{\rho}\right)^{\gamma-1} \qquad ...(5.12)$$

## 5.12 The size of a degree

The efficiency of a reversible engine can be written as

$$\eta = 1-\frac{Q_2}{Q_1} = 1-\frac{\theta_2}{\theta_1} \quad ...(i)$$

Let us now consider a reversible Carnot's engine operating between the steam point and ice point. Let *ABCD* represents the indicator diagram for the Carnot's cycle of its operation, where *AB* and *CD* are the isothermals for steam point and ice-point, respectively (Fig 5.6), then

$$\eta = 1-\frac{\theta_{ice}}{\theta_{steam}} = \frac{\theta_{steam}-\theta_{ice}}{\theta_{steam}} \quad ...(ii)$$

The work done by the engine is numerically given by the area *ABCD*. Let this area be divided into 100 equal parts by drawing isothermals parallel to *CD* or *AB*. Then any isothermal will be at a temperature of 1° lower than isothermal just above it and the area of each part will correspond to one degree on the absolute scale.

Hence, a thermodynamic degree is defined as the difference in temperature between which a Carnot's engine should be worked to get energy equal to 1/100th of the energy obtained from Carnot's engine working between the temperatures of steam point (boiling point of water) and ice point (melting of ice).

The difference in temperature between the steam point and the ice point is 100 on this scale also. In this way, the work scale (Kelvin) of temperature is fixed.

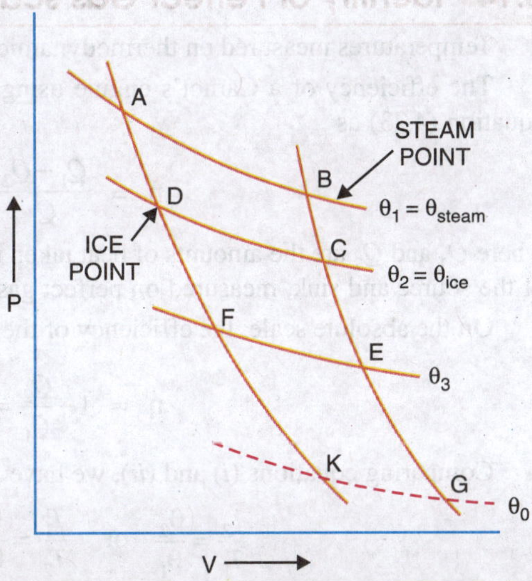


**Fig. 5.6**

## 5.13 Zero of Absolute or Thermodynamic scale

Isothermals below the ice point ($\theta_{ice}$) can be drawn and an isothermal representing absolute zero can be drawn (Fig. 5.6). The isothermal GK represents absolute zero.

Consider an engine working between the steam point ($\theta_{steam}$) and absolute zero ($\theta_o = 0$), then

$$\eta = 1-\frac{\theta_o}{\theta_{steam}} = 1$$

*i.e.* the engine has efficiency equal to 1.

Similarly an engine working between ice point ($\theta_{ice}$) and absolute zero ($\theta_o = 0$), will have an efficiency

$$\eta' = 1-\frac{\theta_o}{\theta_{ice}} = 1$$

But efficiency $\eta = \frac{Q_1-Q_2}{Q_1} = 1-\frac{Q_2}{Q_1}$

If $\eta = 1$ *i.e.* $1-\frac{Q_2}{Q_1} = 1$

$$\frac{Q_2}{Q_1} = 0$$

or $$Q_2 = 0 \qquad ...(5.13)$$

It means that if the sink is at the temperature of zero degree absolute, no heat is rejected and whole of the available energy has been used up in doing useful work and in this case, the engine will have 100% efficiency.

Thus, the absolute zero temperature on the thermodynamic (or work) scale is defined as that temperature at which a system undergoes a reversible isothermal process without any transfer of heat.

Absolute zero on thermodynamics (or work) scale can also be defined as that temperature of the sink at which the reversible heat engine undergoing Carnot's cycle rejects no heat to it.

According to Kelvin-Planck statement of second law of thermodynamics, no engine can have 100 percent efficiency. Thus, the absolute zero temperature is not attainable.

## 5.14 Identity of Perfect Gas scale and Absolute scale

Temperatures measured on thermodynamic scale agrees with those measured on ideal gas scale.

The efficiency of a Carnot's engine using a perfect gas as the working substance is given by equation (4.23) as

$$\eta = \frac{Q_1 - Q_2}{Q_1} = \frac{T_1 - T_2}{T_1} = 1 - \frac{T_2}{T_1} \qquad ...(i)$$

where $Q_1$ and $Q_2$ are the amounts of heat taken in and rejected respectively at temperatures $T_1$ and $T_2$ of the source and sink, measured on perfect gas scale.

On the absolute scale, the efficiency of the engine is given by

$$\eta = 1 - \frac{Q_2}{Q_1} = 1 - \frac{\theta_2}{\theta_1} \qquad ...(ii)$$

Comparing equations (*i*) and (*ii*), we have

$$\frac{T_2}{T_1} = \frac{\theta_2}{\theta_1} \quad \text{or} \quad \frac{T_1}{T_2} = \frac{\theta_1}{\theta_2} \qquad ...(iii)$$

Thus, the ratio of any two temperatures on the perfect gas scale is the same as on the thermodynamic scale.

If in equation (*iii*), $T_2 = 0$, we will have $\theta_2 = 0$. It means the zero of the absolute scale is identical with the zero of the perfect gas scale.

If $T_{steam}$ and $T_{ice}$ are the steam and ice points measured on the perfect gas scale

$$T_{steam} - T_{ice} = 100 \qquad ...(iv)$$

On the thermodynamic scale, for the same two fixed points, as shown above, we have

$$\theta_{steam} - \theta_{ice} = 100 \qquad ...(v)$$

On the perfect gas scale, the efficiency is given by

$$\eta = \frac{T_{steam} - T_{ice}}{T_{steam}} = \frac{100}{T_{steam}}$$

and on using a thermodynamic scale,

$$\eta = \frac{\theta_{steam} - \theta_{ice}}{\theta_{steam}} = \frac{100}{\theta_{steam}}$$

Since, the efficiency must be same in either case

$$\frac{100}{T_{steam}} = \frac{100}{\theta_{steam}}$$

$$\therefore \qquad T_{steam} = \theta_{steam} \qquad \text{...(vi)}$$

*i.e.* the steam point on a perfect gas scale has the same numerical value as on the absolute scale of temperature.

Comparing equation (*iv*) and (*v*) and using (*vi*), we have

$$T_{ice} = \theta_{ice}$$

*i.e.* the ice point is the same on both the scales.

Considering equation (*iii*) and that the zero's of two scales are identical, we conclude

$$T = \theta$$

*i.e.* the temperature on the perfect gas scale represented by $T$ agrees with the corresponding temperature $\theta$ on the thermodynamic scale.

Thus, the Kelvin's thermodynamical scale is completely identical with the perfect gas scale.

## 5.15 Third Law of Thermodynamics : Nernst's Heat Theorem

In 1906, Nernst proposed a general principle supported by a series of experimental evidences on the problem of atomic heat at low temperature. Actually, it was proposed as '*the new heat theorem*' but its importance is so great that today it is recognized as 'the *third law of thermodynamics*'.

### Statement :

According to Nernst, "*The heat capacities of all solids tend to zero as the absolute zero of temperature is approached and that the internal energies and entropies of all substances become equal there, approaching their common value asymptotically tending to zero*". This simple statement of Nernst follows neither from the first law (law of conservation of energy) nor from the second law (law of transmutability of energy) and is thus of the nature of a new law, usually called as *third law of thermodynamics*.

This theorem is useful in explaining the nature of bodies in the neighbour hood of absolute zero temperature. Its importance lies in the fact that it permits the calculations of absolute values of entropy, Helmholtz and Gibb's free energies, etc. In terms of entropy, the theroem may also be stated as:

It states that "at absolute zero temperature, the entropy tends to zero and the molecules of a substance or a system are in perfect order (well arranged)".

In all heat engines, there is always loss of heat in the form of conduction, radiation and friction.

Therefore, in actual heat engines $\frac{Q_1}{T_1}$ is not equal to $\frac{Q_2}{T_2}$

$\therefore \quad \left(\frac{Q_1}{T_1} - \frac{Q_2}{T_2}\right)$ is not zero but it is a positive quantity. When cycle after cycle is repeated, the entropy of the system increases and attains a maximum value. When the system has attained the maximum value, a stage of stagnancy is reached and no work can be done by the engine at this stage. In the universe; the entropy is increasing and ultimately the universe will also reach a maximum value of entropy when no work will be possible. With the increase in entropy, the disorder of the molecules of a substance increases. The entropy is also a measure of the disorder of the system. With decrease in entropy, the disorder deceases. At absolute zero temperature, the entropy tend to zero and the molecules of a substance or a system are in perfect order (well arranged). This is known as the third law of thermodynamics.

### Example :

The molecules are more free to move in the gaseous state than in the liquid state. The entropy in the gaseous state is more than in the liquid state. The entropy in liquid state is more than in the solid state. Thus, when a substance is converted from solid to a liquid and then from the liquid to the gaseous state, the entropy increases and vice-versa. When ice is converted into water and then into steam, the entropy and disorder of the molecules increase. When steam is converted into water and then into ice, the entropy and disorder of the molecules decrease. Hence, entropy is a measure of the disorder of the molecules of the system.

By any ideal procedure, it is impossible to bring any system to absolute zero temperature performing a finite number of operations. This is called the principle of unattainability of absolute zero. Thus, according to Fowler and Guggenheim, the unattainability principle is called the third law of thermodynamics.

## 5.16 Zero Point Energy

According to Kinetic theory, the energy of a system at absolute zero should be zero. It means the molecules of the system do not possess any motion. But according to the modern concept, even at absolute zero, the molecules are not completely deprived of their motion and hence possess energy. The energy of the molecules at absolute zero temperature is called **zero point energy**.

## 5.17 Negative Temperatures (Not Possible)

From the validity point of view of the laws of thermodynamics, there cannot be any temperature less than zero Kelvin *i.e.*, a negative temperature is not possible on Kelvin's scale. This can be logically explained as under:

If at all negative temperature exists and is attainable, then let us assume the sink of a heat engine be at some negative temperature $T_2 = -T_2$ say. Then the efficiency would be

$$\eta = 1-\left(\frac{-T_2}{T_1}\right)=1+\frac{T_2}{T_1}$$

$$\therefore \quad \eta > 1 \qquad ...(i)$$

*i.e.*, the efficiency of engine is greater than unity which is against the second law thermodynamics and hence impossible. In other words, the above conclusion obtained is in contradictionwith the laws of thermodynamics, as it violates them. Hence the assumption made that the negative temperature exists is not true. Thus, there cannot be a negative temperature.

The laws of thermodynamics prohibits a heat engine to have 100% efficiency. The necessary condition for 100% efficiency is that the sink should be at the absolute zero and since nature does not provide us with a sink at absolute zero (which is unattainable) an engine with 100% efficiency is not possible.

The specific heat of a substance decreases with increase in temperature. However, the specific heat does not tend to be zero as the temperature tends to infinity. This shows that the temperature† has a + ve sign only.

† However, experiments by Ramsey (1956) have shown that a part of a system *i.e.*, the nucleus of a solid, can have a negative temperature. This sub-system is considered isolated from the main system (*i.e.*, solid lattice). The specific heat of the sub-system tends to zero at high temperature. A small amount of heat energy tends to raise the temperature of the system to infinity. It is possible to add still more energy to the sub-system at infinity and it forces the sub-system into the negative temperature region. It has been shown by microscopic statistical analysis that there is no distinction between the temperature of $+\infty$ and $-\infty$. In thermodynamics, the parameter $1/T$ is more significant than $T$. The negative temperatures are hotter than the positive temperatures.

The negative temperature is not possible with the system as a whole and is only an exception to the rule that only positive temperatures exist. The negative temperatures are possible only for isolated sub-systems. For all normal purposes the temperatures are always positive.

## 5.18 Heat Death of Universe

In the universe, the natural processes that are taking place such as flow of heat, conduction, radiation, mixing, diffusion and so on are all irreversible and in these processes, the system tends to have maximum value of entropy by attaining a uniform state of temperature, pressure, composition etc. Therefore we conclude that the entire universe would attain a state of absolute uniformity at some distant future. In such a state all physical, chemical and biological processes would stop, although there would be no change in the total energy of the universe. As there will be no temperature difference, it will not be possible to convert any amount of heat into even a small amount of useful work. This means that we are heading towards the **Heat death** of the universe. This state being inevitably approached; although it may take millions of billions of years.

## NUMERICAL EXAMPLES

### EXAMPLE 5.1

*Calculate the change in entropy when 5 kg of water at 100°C is converted into steam at the same temperature. (Given; Latent heat of steam = 540 cal/gram)* **(Alagappa 2004)**

**SOLUTION.** Heat absorbed by 5 kg of water at 100°C when it is converted into steam at 100°C

$$\delta Q = \text{mass} \times \text{latent heat of vaporisation of steam}$$

$$= 5000 \times 540$$

$$\therefore \quad \delta Q = 2700000 \text{ cal}$$

The gain in entropy,

$$dS = \frac{\delta Q}{T}$$

$$= \frac{2700000}{373}$$

$$dS = \mathbf{7240\ cal/K}$$

### EXAMPLE 5.2

*Calculate the change in entropy when 10 grams of ice at 0°C is converted into water at the same temperature. (Given: Latent heat of ice. = 80 cal/gram).*

**(Rohilkhand 2001 Imp. Nagpur 2008, 2005/s, Meerut 2004, 2001, Purvanchal 2002)**

**SOLUTION.** Heat absorbed by 10 gram of ice at 0°C to convert it into water at 0°C is $10 \times 80 = 800$ cal.

$$\therefore \quad \delta Q = 800 \text{ cal}$$

$$T = 0°C = 273 \text{ K}$$

The gain in entropy,

$$dS = \frac{\delta Q}{T}$$

$$= \frac{800}{273}$$

$$dS = \mathbf{2.93\ cal/K}$$

### Example 5.3

*Calculate the increase in entropy when 1 gram of ice at – 10°C is converted into steam at 100°C. (Given: specific heat of ice – 0.5, Latent heat of ice – 80 cal/gram, Latent heat of steam = 540 cal/gram).* **(Rohilkhand 2001, Lucknow 2004)**

**SOLUTION.**

1. Increase in entropy when the temperature of 1 gram of ice increases from – 10°C to 0°C

$$dS = \int_{T_1}^{T_2} \frac{\delta Q}{T}$$

$$= ms \int_{T_1}^{T_2} \frac{dT}{T}$$

$$= ms \log_e \frac{T_2}{T_1}$$

$$= ms \times 2.303 \log_{10} \frac{T_2}{T_1}$$

$$= 1 \times 0.5 \times 2.303 \log_{10} \frac{273}{263}$$

$$= \mathbf{0.01865\ cal/K}$$

**2.** Increase in entropy when 1 gram of ice at 0°C is converted into water at 0°C

$$dS = \frac{\delta Q}{T}$$

$$= \frac{1 \times 80}{273} = \mathbf{0.293\ cal/K}$$

**3.** Increase in entropy when 1 gram of water is raised from 0°C to 100°C

$$dS = \int_{T_1}^{T_2} \frac{\delta Q}{T}$$

$$= ms \times 2.303 \log_{10} \frac{T_2}{T_1}$$

$$= 1 \times 1 \times 2.303 \times \log_{10} \frac{373}{273}$$

$$= \mathbf{0.312\ cal/K}$$

**4.** Increase in entropy when 1 gram of water at 100°C is converted into steam at 100°C

$$dS = \frac{\delta Q}{T}$$

$$= \frac{1 \times 540}{373} = 1.447 \text{ cal/K}$$

∴ Total increase in entropy

$$= 0.01865 + 0.293 + 0.312 + 1.447$$

$$= \mathbf{2.07065\ cal/K}$$

## Example 5.4

*One gram molecule of a gas expands isothermally to four times of its volume. Calculate the change in its entropy in terms of the gas constant.* **(Agra 2000)**

**Solution.** Work done $= \int_{V_1}^{V_2} PdV$

But $PV = RT$

or $P = \frac{RT}{V}$

∴ $W = RT \int_{V_1}^{V_2} \frac{dV}{V} = RT \log_e \frac{V_2}{V_1}$

Here, $\frac{V_2}{V_1} = 4$

$\therefore \quad W = RT \times 2.303 \times \log_{10}(4)$

Here $W$ and $R$ are in units of work

$$\therefore \quad \text{Gain in entropy} = \frac{\delta Q}{T}$$

$$= \frac{W}{JT} = \frac{2.303\, RT}{JT} \log_{10}(4)$$

$$dS = \mathbf{1.387}\ \frac{R}{J}\ \textbf{cal/K}$$

### Example 5.5

*50 grams of water at 0°C is mixed with an equal mass of water at 80°C. Calculate the resultant increase in entropy.* **(Calcutta 2001)**

**SOLUTION.**

(*i*) $m_1 = 50$ gram ; $T_1 = 273$ K

$m_2 = 50$ gram, $T_2 = 273 + 80 = 353$ K

Let the final temperature of the mixture be T K

$$m_1 s \times (T - T_1) = m_2 s (T_2 - T)$$

$$50 \times 1 \times (T - 273) = 50 \times 1 \times (353 - T)$$

$$\therefore \quad T = 313 \text{ K}$$

(*ii*) Change in entropy of 50 gm of water when its temperature rises from 273 K to 313 K will be

$$dS = \frac{\delta Q}{T}$$

$$= ms \int_{T_1}^{T} \frac{dT}{T}$$

$$= 50 \times 1 \log_e \frac{313}{273}$$

$$= 50 \times 2.303 \log_{10} \frac{313}{273}$$

$$= \mathbf{+\ 6.829\ cal/K}$$

Here, the + ve sign indicates gain in entropy.

(*iii*) Change in entropy when temperature of water falls from 353 K to 313 K

$$dS = \frac{\delta Q}{T} = ms \int_{T_2}^{T} \frac{dT}{T}$$

$$= 50 \times 1 \times \log_e \frac{313}{353}$$

$$= 50 \times 2.3026 \times \log_{10} \frac{313}{353}$$

$$= \mathbf{-\ 6.023\ cal/K}$$

Here, – ve sign indicates loss in entropy.

Therefore, the total gain in entropy of the system

$$= 6.829 - 6.023$$

$$= \mathbf{0.806\ cal/K}$$

### EXAMPLE 5.6

*1 gram of water at 20°C is converted into ice at – 10°C at constant pressure. Heat capacity at 1 gm of water is 4.2 J/g.K and that of ice is 2.1 J/g. K. Heat of fusion of ice at 0°C = 335 J/g. Calculate the total change in the entropy of the system.* **(Bangalore 2000)**

**SOLUTION.**

(*i*) Change in entropy when the temperature of 1 gram of water at (273 + 20) = 293 K falls to 273 K.

$$dS = \frac{\delta Q}{T} = ms\int_{T_1}^{T_2}\frac{dT}{T}$$

$$= 1 \times 4.2 \int_{293}^{273}\frac{dT}{T}$$

$$= 4.2 \times 2.3026 \log_{10}\left(\frac{273}{293}\right)$$

$$= -\,0.2969 \text{ J/K}$$

(*ii*) Change in entropy when 1 gm of water at 273 K is converted into ice at 0°C i.e. 273 K

$$dS = \frac{\delta Q}{T} = \frac{-1\times 335}{273} = -\,1.227 \text{ J/K}$$

(*iii*) Change of entropy when the temperature of 1 gram of ice at 273 K falls to 263 K

$$dS = \frac{\delta Q}{T} = ms\int_{T_1}^{T_2}\frac{dT}{T}$$

$$= 1 \times 2.1 \times 2.3026 \times \log_{10}\left(\frac{263}{273}\right)$$

$$= -\,0.07834 \text{ J/K}$$

∴ Total change in entropy of the system

$$= -\,0.2969 - 0.07834 - 1.227$$

$$= \mathbf{-\,1.60224 \text{ J/K}}$$

Negative sign indicates that there is a decrease in entropy of the system.

### Example 5.7

*1 kg of water at 273 K is brought in contact with a heat reservoir at 373 K. Then (1) what is the change in entropy of water when its temperature reaches 373 K? (2) what is the change in entropy of (i) the reservoir and (ii) the universe.* **(Purvanchal 2000)**

**Solution.**

**1.** Increase in entropy when the temperature of 1000 gram of water is raised from 273 K to 373 K

$$dS = \int_{T_1}^{T_2}\frac{\delta Q}{T}$$

$$= ms \times 2.3026 \log_{10}\frac{T_2}{T_1}$$

$$= 1000 \times 1 \times 2.3026 \times \log_{10}\frac{373}{273}$$

$$= \mathbf{+\,312 \text{ cal/K}}$$

**2.** (*i*) Change in entropy of the reservoir

$$dS = -\frac{\delta Q}{T}$$

$$= -\frac{1000 \times 1 \times 100}{373} = \mathbf{-268.1\ cal/K}$$

Negative sign shows decrease in entropy.

(*ii*) Change in entropy of the universe

$$= 312 - 268.1 = \mathbf{43.9\ cal/K}$$

Therefore, the net increase in entropy of the universe

$$= \mathbf{43.9\ cal/K}$$

## EXAMPLE 5.8

*Calculate the change in entropy when 10 g of water at 60°C is mixed with 30 g of water at 20°C.* **(Purvanchal 2005, Calcutta 2004)**

**SOLUTION.**

(*i*) $m_1 = 10\text{g} \quad T_1 = 333\ \text{K}$

$m_2 = 30\text{g} \quad T_2 = 293\ \text{K}$

Let the final temperature be $T$ K

$$m_1 s\,(T_3 - T) = m_2 s\,(T - T_2)$$

$$10 \times 1 \times (333 - T) = 30 \times 1\,(T - 293)$$

$$T = 303\ \text{K}$$

(*ii*) Change in entropy when the temperature of 10 g of water changes from 333 K to 303 K

$$\frac{\delta Q}{T} = ms \int_{T_1}^{T} \frac{dT}{T}$$

$$- 10 \times 1 \times \log_e \frac{303}{333}$$

$$= 10 \times 2.3026 \times \log_{10} \left(\frac{303}{333}\right)$$

$$= \mathbf{-0.9443\ cal/K°}$$

Negative sign indicates decrease in entropy.

(*iii*) Change in entropy when the temperature of 30 g of water change from 293 K° to 303 K°,

$$= \frac{\delta Q}{T} = ms \int_{T_2}^{T} \frac{dT}{T}$$

$$= 30 \times 1 \log_e \frac{303}{293}$$

$$= 30 \times 2.3026 \log_{10} \frac{303}{293}$$

$$= \mathbf{+1.002\ cal/K°}$$

The total gain in entropy of the system

$$= +1.0020 - 0.9443$$

$$= \mathbf{0.0577\ cal/K°}$$

### EXAMPLE 5.9

*Calculate the increase in entropy of 10 kg of water at 100°C when it changes to vapour.*

**(Lucknow 2001)**

**SOLUTION.**

Increase in entropy when 10 kg of water at 100 °C is changed into vapour

$$dS = \frac{\delta Q}{T} = \frac{10 \times 10^3 \times 1 \times 540}{373}$$

$$dS = \mathbf{14{,}477\ cal/K}$$

### EXAMPLE 5.10

*When 50 g of water is heated from 10°C to 90°C, by how much does its entropy change?*

**(Rajasthan 2002)**

**SOLUTION.** Here $T_1 = 273 + 90 = 363$ K

$T_2 = 273 + 10 = 283$ K

Increase in entropy when 50 g of water is heated from 10°C to 90°C

$$dS = \int_{T_1}^{T_2} \frac{\delta Q}{T}$$

$$dS = ms \times 2.3026 \log_{10}\left(\frac{T_2}{T_1}\right)$$

$$dS = 50 \times 1 \times 2.3026 \log_{10}\left(\frac{363}{283}\right)$$

$$dS = \mathbf{12.448\ cal/K}$$

### EXAMPLE 5.11

*Calculate the change in entropy when 10 grams of ice at 0°C is converted into steam at 100°C.*

**SOLUTION.**

1. Increase in entropy when 10 grams of ice at 0°C is converted into water at 0°C.

$$dS = \frac{\delta Q}{T}$$

$$= \frac{10 \times 80}{273} = 2.93\ \text{cal/K}$$

2. Increase in entropy when the temperature of 10 g of water is raised from 0°C to 100°C

$$dS = \int_{T_1}^{T_2} \frac{\delta Q}{T}$$

$$= ms \times 1 \times 2.3026 \log_{10}\left(\frac{T_2}{T_1}\right)$$

$$= 10 \times 1 \times 2.3026 \log_{10}\left(\frac{373}{273}\right)$$

$$= \mathbf{3.12\ cal/K}$$

3. Increase in entropy when 10 g of water at 100°C is converted into steam at 100°C

$$dS = \frac{\delta Q}{T}$$

$$= \frac{10 \times 540}{373} = 14.47 \text{ cal/K}$$

Total increase in entropy = 2.93 + 3.12 + 14.47

$$= \mathbf{20.52 \text{ cal/K}}$$

**EXAMPLE 5.12**

*Calculate the change in entropy of 1 gram of nitrogen when its temperature rises from 50°C to 100°C while its volume is kept constant. Molar specific heat, $C_v = 0.18$ and molecular weight of nitrogen is 28.* **(Madras 2001)**

**SOLUTION.** Specific heat of nitrogen = $\frac{C_v}{28} = \left(\frac{0.18}{28}\right)$

Change in entropy, $$dS = \int_{T_1}^{T_2} \frac{\delta Q}{T}$$

$$dS = \frac{1 \times 0.18}{28} \int_{T_1}^{T_2} \frac{dT}{T}$$

$$dS = \left(\frac{0.18}{28}\right) \int_{323}^{373} \frac{dT}{T}$$

$$dS = \left(\frac{0.18}{28}\right) \times 2.3026 \log_{10} \left(\frac{373}{323}\right)$$

$$dS = \frac{0.18 \times 2.3026 \times 0.0625}{28}$$

$$dS = \mathbf{9.253 \times 10^{-4} \text{ cal/K}}$$

**EXAMPLE 5.13**

*1 kg of water at 273 K is brought in contact with a heat reservoir at 373 K. What is the change in entropy of water as its temperature reaches 373 K?* **(Kanpur 2001)**

**SOLUTION.** Increase in entropy when the temperature of 1 kg of water is raised from 273 K to 373 K

$$dS = \int_{T_1}^{T_2} \frac{\delta Q}{T}$$

$$m = 1 \text{ kg}, \; s = 10^3 \text{ cal/kg-K}$$

$$dS = ms \times 2.3026 \log_{10} \left(\frac{T_2}{T_1}\right)$$

$$dS = 1 \times 10^3 \times 2.3026 \log_{10} \left(\frac{373}{273}\right)$$

$$dS = \mathbf{312 \text{ cal/K}}.$$

**EXAMPLE 5.14**

*Calculate the change in entropy when 0.0273 kg of ice of 0°C is converted into water at the same temperature* **(Garhwal 2003)**

**SOLUTION.** Here $m = 0.0273$ kg

$$L = 80 \text{ Cal/g} = 80 \times 10^3 \text{ Cal/kg}$$

$$T = 273 \text{ K}$$

$$dS = \frac{\delta Q}{T}, \; dS = \frac{mL}{T}$$

$$dS = \frac{0.0273 \times 80 \times 10^3}{273}$$

$$dS = 8 \text{ cal/K} = 8 \times 4.2 \text{ J/K}$$

$$= \mathbf{33.6 \ J/K}$$

## Example 5.15

*Calculate the increase in entropy of 1 kg of ice when it is converted into steam. Given the specific heat of water is 1 k cal kg °C, latent heat of ice is 80 kcal/kg and the latent heat of steam is 540 k cal/kg.*

**Solution.**

1. Increase in entropy when 1 kg of ice at 0°C is converted into water at 0°C

$$dS = \frac{\delta Q}{T} = \frac{80}{273} = 0.293k \text{ cal/K}$$

2. Increase in entropy when the temperature of 1 kg of water is raised from 0°C to 100°C.

$$dS = \int_{T_1}^{T_2} \frac{\delta Q}{T} = ms \times 2.3026 \log_{10}\left(\frac{T_2}{T_1}\right)$$

$$= 1 \times 1 \times 2.3026 \log_{10}\left(\frac{373}{273}\right) = 0.312 \text{ k cal/K}$$

3. Increase in entropy when 1 kg of water at 100°C is converted into steam at 100°C.

$$dS = \frac{\delta Q}{T} = \frac{540}{373} = 1.447 \text{ k cal/K}$$

Total increase in entropy,

$$dS = 0.293 + 0.312 + 1.447$$

$$= \mathbf{2.052 \ k \ cal/K.}$$

## Example 5.16

*Calculate the change of entropy on converting a mole of a gas occupying 20 litres at a pressure of $2 \times 10^5$ N/m² to occupying 50 litres at a pressure of $5 \times 10^5$ N/m². Given*

$R = 8.4 \text{ J mol}^{-1} \text{ K}^{-1}$

$C_v = 21 \text{ J mol}^{-1} \text{ K}^{-1}$ **(Kerala, 2001)**

**Solution.** Here,

$$P_1 = 2 \times 10^5 \text{ N/m}^2$$
$$V_1 = 20 \text{ litres}$$
$$P_2 = 5 \times 10^5 \text{ N/m}^2$$
$$V_2 = 50 \text{ litres}$$

For a perfect gas,

$$\frac{P_1 V_1}{T_1} = \frac{P_2 V_2}{T_2}$$

$$\therefore \quad \frac{T_2}{T_1} = \frac{P_2 V_2}{P_1 V_1}$$

$$= \frac{5 \times 10^5 \times 50}{2 \times 10^5 \times 20}$$

$$= 6.25$$

Also, $\frac{V_2}{V_1} = \frac{50}{25} = 2.5$

$C_v = 21$ J/mol – K

$R = 8.4$ J/mol – K

The change in entropy,

$$dS = C_v \times 2.3026 \log_{10}\left(\frac{T_2}{T_1}\right) + R \times 2.3026 \log_{10}\left(\frac{V_2}{V_1}\right)$$

$$= 21 \times 2.3026 \log_{10}(6.25) + 8.4 \times 2.3026 \log_{10}(2.5)$$

$$= 21 \times 2.3026 \times 0.7959 + 8.4 \times 2.3026 \times 0.3979$$

$$= 38.4852 + 7.6961$$

$$= 46.1813$$

$$= \mathbf{46.18\ J/K}$$

Therefore, increase in entropy = **46.18 J/K**

### EXAMPLE 5.17

*12.5 gm of ice at – 24°C is converted into steam at 100°C. Calculate the change in entropy. It is given that, latent heat of steam = 5.36 cal/gm, latent heat of ice = 80 cal/gm. specific heat of ice = 0.5 cal/gm–K.*

**SOLUTION.** [In *SI* system],

$m$ = 12.5 grams = $12.5 \times 10^{-3}$ kg

Latent heat of ice = 80 cal/gram = $80 \times 10^3 \times 4.2$ J/kg

= $3.36 \times 10^5$ J/kg

Latent heat of steam = 536 cal/gram = $536 \times 10^3 \times 4.2$ J/kg

= $2.25 \times 10^6$ J/kg

Specific heat of ice = 0.5 cal/gram – K

= $0.5 \times 10^3 \times 4.2$ J/kg – K

= $2.1 \times 10^3$ J/kg – K

Specific heat of water = 1 cal/gram – K

= $1 \times 10^3 \times 4.2$ J/kg – K

= $4.2 \times 10^3$ J/kg – K

**1.** Increase in entropy when temperature of $12.5 \times 10^{-3}$ kg of ice is increased from – 24°C to 0°C (249 K to 273 K)

$$dS = \int_{T_2}^{T_1} \frac{\delta Q}{T}$$

$$= ms \int_{T_1}^{T_2} \frac{dT}{T}$$

$$= ms \log_e\left(\frac{T_2}{T_1}\right)$$

$$= 12.5 \times 10^{-3} \times 2.1 \times 10^{3} \log_e \left(\frac{273}{249}\right)$$

$$= 12.5 \times 2.1 \times 2.3026 \log_{10} \left(\frac{273}{249}\right)$$

$$= \mathbf{2.42\ J/K}$$

**2.** Increase in entropy when $12.5 \times 10^{-3}$ kg of ice at 0°C is converted into water at 0°C

$$dS = \frac{\delta Q}{T} = \frac{mL}{T}$$

$$= \frac{12.5 \times 10^{-3} \times 3.36 \times 10^{5}}{273}$$

$$= \mathbf{15.38\ J/K}$$

**3.** Increase in entropy when the temperature of $12.5 \times 10^{-3}$ kg of water is raised from 0°C to 100°C (273 K to 373 K)

$$dS = \int_{T_1}^{T_2} \frac{\delta Q}{T}$$

$$= ms \times 2.3026 \log_{10} \left(\frac{T_2}{T_1}\right)$$

$$= 12.5 \times 10^{-3} \times 4.2 \times 10^{3} \times 2.3026 \log_{10} \left(\frac{373}{273}\right)$$

$$= \mathbf{16.38\ J/K}$$

**4.** Increase in entropy when $12.5 \times 10^{-3}$ kg of water at 100°C is converted into steam at 100°C (373 K)

$$dS = \frac{\delta Q}{T} = \frac{mL}{T}$$

$$= \frac{12.5 \times 10^{-3} \times 2.25 \times 10^{6}}{373}$$

$$= \mathbf{75.4\ J/K}$$

Total increase in entropy

$$= 2.42 + 15.38 + 16.38 + 75.4$$

$$= \mathbf{109.58\ J/K}$$

### EXAMPLE 5.18

*How much time will it take for a layer of ice of thickness 20 cm to increase by 10 cm, on the surface of a pond, when the temperature of surroundings is – 15°C?*

*$K = 0.005$ C.G.S. units; $L = 80$ Cal/g, $\rho = 0.90$ g/cm$^3$* **(Purvanchal 2003)**

**SOLUTION.** [In *CGS* system]

$$\int dt = \frac{\rho L}{K\theta} \int x\, dx$$

$$t = \frac{\rho L}{2K\theta}\left[x_2^2 - x_1^2\right]$$

Here,

$$\rho = 0.9 \text{ g/cm}^3$$

$$L = 80 \text{ cal/cm}^3$$

$$K = 0.005 \text{ C G S units}$$

$$\theta = 15°\text{C}$$

$$x_1 = 20 \text{ cm}$$
$$x_2 = 30 \text{ cm}$$
$$t = \left[\frac{0.9\times 80}{2\times 0.005\times 15}\right]\left[(30)^2-(20)^2\right]$$
$$= 2.4\times 10^5\ s$$
$$= \textbf{66 hrs – 40 min}$$

## Example 5.19

*Assuming the specific heat capacity of water, find the change in entropy when 0.1 kg. of water at 15°C is mixed with 0.16 kg of water at 40°C* **(Bhagalpur 2001)**

**SOLUTION.** [In *SI* system],

Specific heat of water,

$$s = 1 \text{ k cal/kg – °C}$$
$$= 4.2\times 10^3 \text{ J/kg – °C}$$

Here,
$$m_1 = 0.1 \text{ kg}$$
$$T_1 = 15 + 273 = 288 \text{ K}$$
$$m_2 = 0.16 \text{ kg}$$
$$T_2 = 40 + 273 = 313 \text{ K}$$

Let the final temperature be $T$

$$m_1\times s\times[T-T_1] = m_2\times s\times[T_2-T]$$
$$m_1\,[T-T_1] = m_2\,[T_2-T]$$
$$0.1\,[T-288] = 0.16\,[313-T]$$
$$T = \mathbf{303.4\ K}$$

(*i*) Change in entropy when the temperature of 0.1 kg of water rises from 288 K to 303.4 K

$$\frac{\delta Q}{T} = ms\int_{T_1}^{T}\left(\frac{dT}{T}\right)$$
$$= 0.1\times 4.2\times 10^3\times 2.3026\log_{10}\left(\frac{303.4}{288}\right)$$
$$= \mathbf{21.88\ J/K}$$

(*ii*) Change in entropy when the temperature of 0.16 kg of water decreases from 313 K to 303.4 K

$$\frac{\delta Q}{T} = ms\int_{T_2}^{T}\left(\frac{dT}{T}\right)$$
$$= 0.16\times 4.2\times 10^3\times 2.3026\log_{10}\left(\frac{303.4}{313}\right)$$
$$= -20.93 \text{ J/K}$$

Therefore, the net change in entropy of the system = + 21.88 – 20.93

$$= \mathbf{0.95\ J/K}$$

Hence, net increase in entropy,

$$= \mathbf{0.95\ J/K}$$

### EXAMPLE 5.20

*Prove that in a T-S diagram the slop of isochoric curve is* $\frac{T}{C_V}$ *and that of isobaric curve is* $\frac{T}{C_P}$.

**(Calcutta 2002)**

**SOLUTION.** Suppose the heat supplied is $dQ$ at temperature $T$. The increase in entropy is

$$dS = \frac{dQ}{T}$$

$$\therefore \quad dQ = TdS$$

**1.** In case of isochoric process, volume is constant,

$$\therefore \quad dQ = C_V\,dT$$

$$\text{or} \quad TdS = C_V\,dT$$

$$\therefore \text{ change in entropy,} \quad dS = \frac{C_V\,dT}{T}$$

and, slope of isochoric curve, $\frac{dT}{dS} = \frac{T}{C_V}$ ...(*i*)

**2.** In case of isobaric process, pressure is constant,

$$\therefore \quad dQ = C_P\,dT$$

$$\text{or} \quad T.dS = C_P\,dT$$

$$\therefore \text{ change in entropy,} \quad dS = \frac{C_P\,dT}{T}$$

and, slope of isobaric curve, $\frac{dT}{dS} = \frac{T}{C_P}$ ...(*ii*)

### EXAMPLE 5.21

*A mass m of a liquid at temperature* $T_1$ *is mixed with an equal mass of same liquid at temperature* $T_2$. *The system is thermally insulated. Show that the entropy change of the universe is* $2mC_P \log_e \frac{(T_1+T_2)}{2\sqrt{T_1 T_2}}$.

**(Purvanchal 2004)**

**SOLUTION.** Since the liquid of mass $m$ at temperature $T_1$ is mixed with the same liquid of same $m$ at temperature $T_2$, then the temperature of mixture attained will be $\frac{T_1+T_2}{2}$. Let us calculate the change in entropy of each liquid.

(*i*) The change of entropy of one liquid of mass $m$ in changing temperature from $T_1$ to $\frac{T_1+T_2}{2}$ will be

$$\Delta S_1 = \int_{T_1}^{\frac{T_1+T_2}{2}} \frac{dQ}{T} = \int_{T_1}^{\frac{(T_1+T_2)}{2}} \frac{mC_P\,dT}{T}$$

$$= mC_P\,[\log_e T]_{T_1}^{(T_1+T_2)/2}$$

$$= mC_P\left[\log_e\left(\frac{T_1+T_2}{2}\right) - \log_e(T_1)\right]$$

$$= mC_P\left[\log_e \frac{T_1+T_2}{2T_1}\right] \quad \text{...}(i)$$

(*ii*) The change of entropy of another liquid of mass $m$ in changing its temperature from $T_2$ to $\frac{(T_1+T_2)}{2}$, will be

$$\Delta S_2 = \int_{T_2}^{\frac{(T_1+T_2)}{2}} \frac{dQ}{T} = \int_{T_2}^{\frac{(T_1+T_2)}{2}} \frac{mC_p\, dT}{T}$$

$$= mC_p \left[\log_e T\right]_{T_2}^{\frac{(T_1+T_2)}{2}} = mC_P \log_e \frac{T_1 \quad T_2}{2} - \log_e T_2$$

$$= mC_p \left[\log_e \frac{T_1+T_2}{2T_2}\right] \quad ...(ii)$$

Hence, the entropy change of the universe is obtained by adding the two *i.e.*, eqn. (*i*) and (*ii*), we have

$$\Delta S = \Delta S_1 + \Delta S_2 = mC_p \left[\log_e \frac{T_1+T_2}{2T_2}\right] + mC_P \left[\log_e \frac{T_1+T_2}{2T_2}\right]$$

$$= mC_P \log_e \frac{T_1+T_2}{2T_1} \times \frac{T_1+T_2}{2T_2} = mC_P \log_e \left[\frac{T_1+T_2}{2\sqrt{T_1 T_2}}\right]^2$$

$$= 2\, mC_p \log_e \frac{(T_1+T_2)}{2\sqrt{T_1 T_2}}$$

## EXERCISE–V

**5.1** Define entropy. Discuss the physical meaning of entropy.
**(Garhwal 2005, 2001, Madras 2005, Rohilkhand 2003, Agra 2006, Nagpur 2007, 2005/s, Lucknow 2004, Purvanchal 2007, 2005, 2003, M.S. Uni. Tamil Nadu 2007)**

**5.2** Show that during a reversible adiabatic process the entropy of the system remains constant.
**(Nagpur 2004/s, 2005/s, Meerut 2005, Madras 2005, Madurai Kamraj 2003)**

**5.3** Prove that for a complete reversible cycle of change in the state of substance

$\oint dS = 0$ **(Calcutta 2001)**

**5.4** What happens to change in entropy of a system which undergoes

(a) a reversible process (b) an irreversible process

(c) an adiabatic process **(Purvanchal 2005, M.S. Uni. Tamil Nadu 2007)**

**5.5** State and prove the principle of increase of entropy.

**5.6** Show that entropy of an irreversible process, entropy of the universe increases.
**(Nagpur 2009, 2007, 2004/w, Meerut 2001, Lucknow 2004, Patna 2005, Bangalore 2004, Mysore 2004)**

**5.7** Explain the term entropy and isoentropics. Show that the entropy of a system tends to be a maximum. Represent the Carnot's cycle on a temperature -entropy diagram and prove that its area represents available energy. **(Nagpur 2005/s)**

**5.8** Establish the relation for efficiency of a Carnot's engine using T–S diagram in

$$\eta = \frac{T_1 - T_2}{T_1}$$

**5.9** Explain the concept of entropy and disorder. **(Kanpur 2000, Alagappa 2004, Bangalore 2005)**

**5.10** Give the statistical definition of entropy. State its unit **(Nagpur 2005/w, Ranchi 2005, Punjab 2005)**

**5.11** What do you mean by entropy? Show that entropy remains constant in reversible process but increases in irreversible process. **(Meerut 2005, Kanpur 2001, Rohilkhand 2000; Patna 2004, Bangalore 2004, Mysore 2004, Nagpur 2005/w, Purvanchal 2007, Agra 2006)**

**5.12** What is T-S diagram? What is the importance of T-S diagram? Find the expression for efficiency of a reversible carnot's engine with the help of T-S diagram.
**(Meerut 2003, Ranchi 2005, Punjab 2005, Nagpur 2005/w, Osmania 2004)**

**5.13** Define entropy. Derive an expression for entropy of an ideal gas in terms of its pressure, volume and specific heat. **(Meerut 2005, Garhwal 2000, Agra 2003)**

**5.14** Derive an expression for the change of entropy of a perfect gas. **(Kanpur 2000)**

**5.15** Prove that $Tds = C_V dT + T\left(\frac{\partial P}{\partial T}\right)_V dV$. **(Meerut 2003, Agra 2004)**

**5.16** Obtain an expression for the change in entropy when ice changes in to steam.
**(M.S. Uni. Tamil Nadu 2007, Meerut 2003)**

**5.17** Mention whether the change in entropy of a mole of an ideal gas is positive, zero or negative, when the gas undergoes free expansion. **(Meerut 2003)**

**5.18** Describe Carnot's reversible heat engine. Deduce an expression for its efficiency. Is Carnot's cycle a practical possiblility? Explain **(Agra 2005, Garhwal 2005)**

**5.19** Explain how concept of entropy leads to $2^{nd}$ law of thermodynamics. **(Punjab 2005)**

**5.20** What is entropy? Give the expression for entropy? What is the unit of entropy? **(Madras 2005)**

**5.21** Show that the entropy of *n* moles of an ideal gas of molar heat capacity $C_V$ at constant volume at a temperature *T* and volume *V* is given by

$$S = nC_V \log T + nR \log V + S_0$$

assuming that $C_V$ is independent of temperature. **(Calcutta 2003)**

**5.22** (*a*) Explain Kelvin's thermodynamic scale of temperature.
**(Nagpur 2009, 2005/s, Madras 2005, 2004, Purvanchal 2005, 2002)**

(*b*) Show that the ratio of two temperatures of Kelvin's scale in the ratio of heats absorbed and rejected by a Carnot's engine operating between these temperatures. **(Punjab 2001)**

**5.23** Discuss absolute zero on the basis of Kelvin's thermodynamic scale of temperature. How size of 1 degree represents on a P-V indicator diagram. Hence Define a thermodynamic degree.
**(Agra 2001, Madras 2000)**

**5.24** Explain thermodynamic scale of temperature and show that the thermodynamic and the ideal gas scale are identical. Explain why? **(Rohilkhand 2001, Garhwal 2000, Agra 2008, Meerut 2000, Purvanchal 2005, 2002, Nagpur 2004/s, 2005/w)**

**5.25** Why the zero on absolute scale is considered to be the lowest possible temperature? Is negative temperature possible on absolute scale of temperature.
**(Purvanchal 2002, Rohilkhand 2001, Meerut 2003, 2004, Agra 2004, 2003)**

**5.26** How do you define zero on absolute scale of temperature?

**5.27** What do you mean by thermodynamical scale of temperature? How did Kelvin introduce this scale? How is this scale realised in practice? **(Patna 2004)**

**5.28** Assuming temperature to be a thermodynamic coordinate of a system, show how Kelvin derived a scale of temperature which is independent of the properties of the measuring system. **(Calcutta 2003)**

**5.29** State and explain with example the third law of thermodynamics. Prove that it is impossible to obtain absolute zero temperature.
**(Nagpur 2005/s, 2005/w, Alagappa 2004, Madras 2005, Bangalore 2005, Agra 2006)**

**5.30** What is Nernst's heat theorem? State and explain. What do you infer from that ?
**(Garhwal 2005, Madras 2004)**

**5.31** Explain the concept of 'Heat Death of Universe'. **(Mumbai 2000)**

**5.32** Write notes on the following:

(*i*) Kelvin's absolute scale of temperature. **(Meerut 2001, Bangalore 2005)**
(*ii*) Entropy and disorder. **(Osmania 2005)**
(*iii*) T-S diagram. **(Rohilkhand 2001 Imp.)**
(*iv*) Absolute scale of temperature. **(Rohilkhand 2000)**
(*v*) Principle of increase of entropy. **(Osmania 2000)**
(*vi*) Entropy of a gas. **(Agra 2000)**
(*vii*) Third law of thermodynamics. **(Meerut 2000)**
(*viii*) Heat death of universe.

**5.33** Calculate the change in entropy when 10 kg of water at 100°C is converted into steam at the same temperature. **[Ans.** 14480 Cal/K] **(Calcutta 2001)**

**5.34** Two gram molecules of a gas expands isothermally to four times its initial volume. Calculate the change in its entropy in terms of gas constant. **[Ans.** 2.774 $\frac{R}{J}$ Cal/K] **(Madurai 2002)**

**5.35** Calculate the change in entropy when 100 g of water at 15°C is mixed with 160 g of water at 40°C. Specific heat of water may be assumed to be equal to 1. **[Ans.** – 0.130 Cal/K] **(Kanpur 2000, Agra 2006)**

**5.36** Calculate the change in entropy when one gram of ice is mixed with 2 millilitres of water at 55°C. Assume that specific heat and specific gravity of water remains constant (*i.e.* equal to 1 in both cases) between 0°C and 55°C. latent heat of ice = 80 Cal/g. **[Ans.** + 0.03372 Cal/K] **(Madras 2001)**

**5.37** 3 gm. of nitrogen doubles in volume at constant temperature. Calculate the chahge in entropy. Given for nitrogen, molecular weight = 28. **[Ans.** 0.621 Joule/K]

**5.38** 10 gm. of water is heated from 40 °C to 80 °C. Calaulate the change in entropy. **[Ans.** 12.1 Cal/K] **(Punjab 2005)**

**5.39** Find the change in entropy when 1 gm. of ice at 0°C changes to water at 10°C. **(Garhwal 2005)**

**5.40** The T-S diagram of a reversible engine is given below. Find its efficiency. **(Calcutta 2004)**

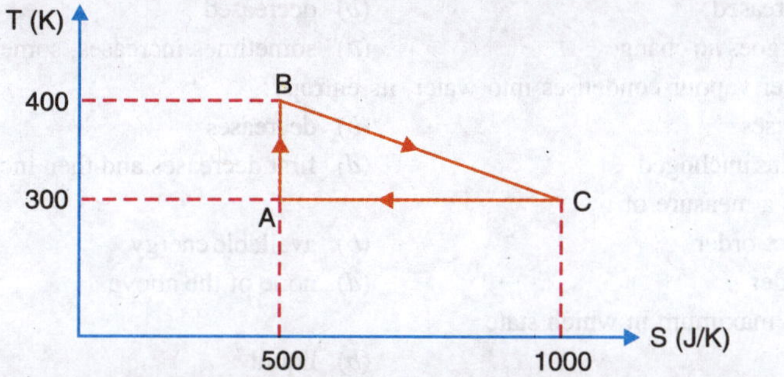


**5.41** 10 gm of steam at 100 °C is blown into 90 gms. of water at 0 °C contained in a calorimeter of water equivalent 10 gms. The whole of the steam is condensed. Calculate the increase in the entropy of the system. **(Purvanchal 2003)**

**5.42** Calculate the change in entropy when temperature of 1 kg of water is raised from 0 °C to 100°C. Specific heat capacity of water is 1 cal/gm. K. **[Ans.** 1305.4 J/K] **(Nagpur 2005/w)**

## OBJECTIVE QUESTIONS

**5.1** In a reversible adiabatic process, entropy:
(*a*) increases (*b*) remains unchanged
(*c*) decreases (*d*) none of these **(Garhwal 2005)**

**5.2** Which of these is intensive variable?
(*a*) area (*b*) entropy
(*c*) temperature (*d*) volume. **(Panjab 2000)**

**5.3** The change in entropy of a mole of an ideal gas, when the gas undergoes free expansion is
(*a*) positive (*b*) zero
(*c*) negative **(Meerut 2003)**

**5.4** Which relation represents Clausius theorem?
(*a*) $\oint_R \frac{dQ}{T} = 0$ (*b*) $\oint_R \frac{dQ}{T} \neq 0$
(*c*) $\oint_R \frac{dQ}{T} > 0$ (*d*) $\oint_R \frac{dQ}{T} < 0$ **(Rohilkhand 2003)**

**5.5** Entropy remains constant in:
(*a*) adiabatic process (*b*) isothermal process
(*c*) isochoric process (*d*) isolated process **(Rohilkhand 2003)**

**5.6** In a reversible process, the entropy of the system:
(*a*) increases (*b*) decreases
(*c*) remains zero (*d*) remains constant **(Kanpur 2002)**

**5.7** The entropy of a system in an irreversible process:
(*a*) increases (*b*) decreases
(*c*) remains constant (*d*) none of the above **(Rohilkhand 2002, 2003)**

**5.8** The unit of entropy is
(*a*) Joule/Kelvin (*b*) Cal/Kelvin
(*c*) Both (*a*) and (*b*) (*d*) None of the above **(Rohilkhand 2002)**

**5.9** Choose the correct statement:
(*a*) All quasi-static processes are reversible.
(*b*) All revessible processes are quasistatic.
(*c*) Adiabatic process is quasi-static.

**5.10** A piece of ice is added to water in a cup. The entropy
(*a*) is increased (*b*) decreased
(*c*) undergoes no change (*d*) sometimes increases, sometimes not

**5.11** When water vapour condenses into water, its entropy:
(*a*) increases (*b*) decreases
(*c*) remains unchnged (*d*) first decreases and then increases

**5.12** Entropy is a measure of
(*a*) prefect order (*b*) available energy
(*c*) disorder (*d*) none of the above

**5.13** Entropy is maximum in which state
(*a*) Solid (*b*) liquid
(*c*) gas (*d*) can be any

**5.14** In a natural process entropy
(*a*) increases (*b*) decreases
(*c*) remains the same

**5.15** Net entropy change of a system in Carnot's cycle
(*a*) zero (*b*) positive
(*c*) negative (*d*) more than 1

**5.16** Which of the following represents a reveresible process?
(*a*) $dS < 0$ (*b*) $dS = 0$
(*c*) $dS > 0$ (*d*) none of these

**5.17** In the two gases at the same temperature:
(*a*) the average kinctic energy per molecule is equal.
(*b*) the internal energy is equal
(*c*) the entropy is equal.

## ANSWERS

| | | | |
|---|---|---|---|
| **5.1** (*b*) | **5.2** (*c*) | **5.3** (*a*) | **5.4** (*a*) |
| **5.5** (*b*) | **5.6** (*d*) | **5.7** (*a*) | **5.8** (*c*) |
| **5.9** (*b*) | **5.10** (*a*) | **5.11** (*b*) | **5.12** (*c*) |
| **5.13** (*c*) | **5.14** (*a*) | **5.15** (*a*) | **5.16** (*b*) |
| **5.17** (*b*) | | | |

# Thermo-dynamical Relationship

# 6

## AT A GLANCE

• Thermodynamic Variables • Extensive and Intensive Variables • Maxwell's Thermodynamical Relations • Applications of Maxwell's Thermodynamic Relations • Thermodynamic Potentials • Significance of Thermodynamic Potentials • Relation of Thermodynamical Potentials with Their Variables • Relations Between $C_p$, $C_v$ and $\mu$ • The T-dS Equations • Clapeyron's Latent Heat Equation(Using Thermodynamical Relations) • Clapeyron's Latent Heat Equation (Using Carnot's Cycle) • Adiabatic Stretching of a Wire • Internal Energy of Ideal and Real Gases • Clausius Inequality • Entropy and the Second Law of Thermodynamics • Joule – Kelvin Coefficient • Equilibrium Between Liquid and its Vapour • First Order Phase Transitions • Second Order Phase Transitions • E.M.F. of a Reversible Cell

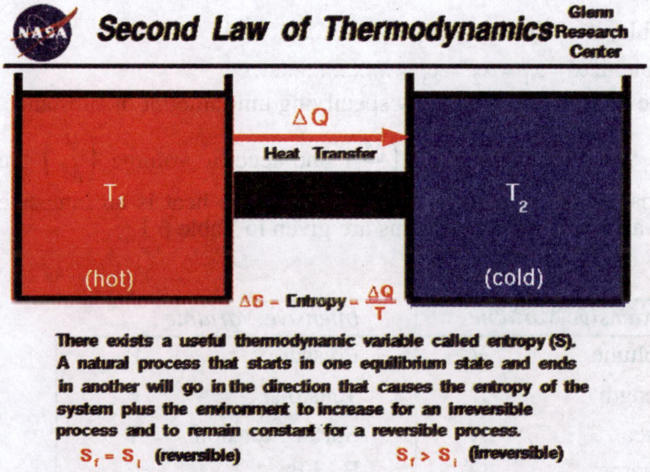


## 6.1 Thermodynamic Variables

The thermodynamic state of a substance is specified by some of its properties like pressure, volume, temperature, internal energy and entropy. These properties undergo a change when the system passes from one state to another. These variables are known as thermodynamic variables or co-ordinates. These are called macroscopic co-ordinates. They require the specification of few measurable properties of the system and do not require the knowledge of microscopic structure of matter composing the system.

## 6.2 Extensive and Intensive Variables

An *extensive variable* of a system is a macroscopic parameter which describes a system in equilibrium and which has a value equal to the sum of its values in each part of the system. The extensive variable depends upon the mass or the size of the substance present in the system.

### Examples

Mass, volume, internal energy, entropy, length, area, heat-capacity, electric charge, magnetisation etc.

An *intensive variable* of a substance is a macroscopic parameter which describes the system in equilibrium and has the same value in any part of the system. It is independent of mass or size of the system. It is a characteristic of the substance present in the system.

### Examples

Pressure, temperature, viscosity, refractive index, density, specific volume, magnetic induction, surface tension, electromotive force etc.

### Distinction between Extensive and Intensive variables

Consider a homogeneous system in equilibrium. Suppose the system is divided into many parts. If a macroscopic variable $x$ of the system has the values $x_1, x_2, x_3,$ .......... in each of these parts of the system respectively, then

(*i*) $x$ is said to be extensive variable, if $x = x_1 + x_2 + x_3 +$ ................, and

(*ii*) $x$ is said to be intensive variable, if $x = x_1 = x_2 = x_3 =$ ......................

An extensive variable may become an intense variable by specifying unit amount of substance. Thus, mass and volume are extensive variables but density $\left(\frac{M}{V}\right)$ and specific volume $\left(\frac{V}{M}\right)$ are intensive variables. Similarly heat capacity is extensive variable but specific heat is an intensive variable. The extensive and intensive variables of some systems are given in Table 6.1.

**TABLE 6.1**

| *System* | *Extensive Variable* | | *Intensive Variable* | |
|---|---|---|---|---|
| Chemical system | Volume | V | Pressure | P |
| Stretched wire | Length | L | Tension | F |
| Surface film | Area | A | Surface tension | T |
| Electric cell | Charge | Q | E.M.F. | E |
| Paramagnetic solid | Magnetisation | I | Magnetic intensity | H |

## 6.3 Maxwell's Thermodynamical Relations (General Relationship)

From the first and second law of thermodynamics, Maxwell was able to derive six fundamental thermodynamical relations. The state of a system can be specified by any pair of quantities, *viz.* pressure ($P$), volume ($V$), temperature ($T$), and entropy ($S$). In solving any thermodynamical problem, the most suitable pair is chosen and the quantities constituting the pair are taken as independent variables.

From the first law of thermodynamics,

$$\delta Q = dU + \delta W$$

$$\delta Q = dU + PdV$$

or

$$dU = \delta Q - PdV$$

For the second law of thermodynamics,

$$dS = \frac{\delta Q}{T}$$

or $$\delta Q = TdS$$

Substituting this value of $\delta Q$, we get

$$\delta U = TdS - PdV \qquad ...(6.1)$$

Considering $U$, $S$ and $V$ to be function of two independent variables $x$ and $y$ [here, in general, $x$ and $y$ can be any two variables out of $P$, $V$, $T$ and $S$],

$$dU = \left(\frac{\partial U}{\partial x}\right)_y dx + \left(\frac{\partial U}{\partial y}\right)_x dy$$

$$dS = \left(\frac{\partial S}{\partial x}\right)_y dx + \left(\frac{\partial S}{\partial y}\right)_x dy$$

and $$dV = \left(\frac{\partial V}{\partial x}\right)_y dx + \left(\frac{\partial V}{\partial y}\right)_x dy$$

Substituting these values in equation (6.1), we get

$$\left(\frac{\partial U}{\partial x}\right)_y dx + \left(\frac{\partial U}{\partial y}\right)_x dy = T\left[\left(\frac{\partial S}{\partial x}\right)_y dx + \left(\frac{\partial S}{\partial y}\right)_y dy\right] - P\left[\left(\frac{\partial V}{\partial x}\right)_y dx + \left(\frac{\partial V}{\partial y}\right) dy\right]$$

$$= \left[T\left(\frac{\partial S}{\partial x}\right)_y - P\left(\frac{\partial V}{\partial x}\right)_x\right] dy + \left[T\left(\frac{\partial S}{\partial y}\right)_x - P\left(\frac{\partial V}{\partial y}\right)_x\right] dy$$

Comparing the coefficients of $dx$ and $dy$, we get

$$\left(\frac{\partial U}{\partial x}\right)_y = T\left(\frac{\partial S}{\partial x}\right)_y - P\left(\frac{\partial V}{\partial x}\right)_y \qquad ...(6.2)$$

$$\left(\frac{\partial U}{\partial y}\right)_x = T\left(\frac{\partial S}{\partial y}\right)_x - P\left(\frac{\partial V}{\partial y}\right)_x \qquad ...(6.3)$$

Differentiating equation (6.2) with respect to $y$ and equation (6.3) with respect to $x$,

$$\frac{\partial^2 U}{\partial y.\partial x} = \left(\frac{\partial T}{\partial y}\right)_x \left(\frac{\partial S}{\partial x}\right)_y + T\frac{\partial^2 S}{\partial y.\partial x} - \left(\frac{\partial P}{\partial y}\right)_x \left(\frac{\partial V}{\partial x}\right)_y - P\frac{\partial^2 V}{\partial V.\partial x}$$

and $$\frac{\partial^2 U}{\partial x.\partial y} = \left(\frac{\partial T}{\partial x}\right)_y \left(\frac{\partial S}{\partial y}\right)_x + T\frac{\partial^2 S}{\partial x.\partial y} - \left(\frac{\partial P}{\partial x}\right)_y \left(\frac{\partial V}{\partial y}\right)_x - P\frac{\partial^2 V}{\partial x.\partial y}$$

The change in internal energy brought about by changing $V$ and $T$, whether $V$ is changed by $dV$ first and $T$ by $dT$ later or vice versa is the same.

It means $dU$ is a perfect differential.

$$\therefore \qquad \frac{\partial^2 U}{\partial y.\partial x} = \frac{\partial^2 U}{\partial x.\partial y} \text{ and}$$

$$\left(\frac{\partial T}{\partial y}\right)_x \left(\frac{\partial S}{\partial x}\right)_y + T\frac{\partial^2 x}{\partial y.\partial x} - \left(\frac{\partial P}{\partial y}\right)_x \left(\frac{\partial V}{\partial x}\right)_y - P\frac{\partial^2 V}{\partial y.\partial x}$$

$$= \left(\frac{\partial T}{\partial x}\right)_y \left(\frac{\partial S}{\partial y}\right)_x + T\frac{\partial^2 S}{\partial x.\partial y} - \left(\frac{\partial P}{\partial x}\right)_y \left(\frac{\partial V}{\partial y}\right)_x - P\frac{\partial^2 V}{\partial y.\partial x} \qquad ...(6.4)$$

Since $dS$ and $dV$ are also perfect differentials, we have

$$\frac{\partial^2 S}{\partial x.\partial y}=\frac{\partial^2 S}{\partial y.\partial x} \quad \text{and} \quad \frac{\partial V}{\partial x.\partial y}=\frac{\partial V}{\partial y.\partial x}$$

Equation (6.4), therefore reduces to

$$\left(\frac{\partial T}{\partial y}\right)_x\left(\frac{\partial S}{\partial x}\right)_y-\left(\frac{\partial P}{\partial y}\right)_x\left(\frac{\partial V}{\partial x}\right)_y = \left(\frac{\partial T}{\partial x}\right)_y\left(\frac{\partial S}{\partial y}\right)_x-\left(\frac{\partial P}{\partial x}\right)_y\left(\frac{\partial V}{\partial y}\right)_x \quad ...(6.5)$$

This is the general expression for Maxwell's thermodynamical relations. In place of the independent variables $x$ and $y$, any two of the four variables $S, T, P$ and $V$ can be substituted so that there may be one mechanical variable ($P$ or $V$) and one thermal variable ($S$ or $T$). Thus, there may be four sets of possible substitutions $(S, V)$, $(T, V)$, $(S, P)$ and $(T, P)$, providing the four Maxwell's thermodynamical relations.

### First Relation:

Put $x = S$ and $y = V$ in equation (6.5), so that

$$\frac{\partial S}{\partial x}=1 \quad , \quad \frac{\partial V}{\partial y}=1$$

and
$$\frac{\partial S}{\partial y}=0 \quad , \quad \frac{\partial V}{\partial x}=0$$

Substituting in equation (6.5), we get

$$\left(\frac{\partial T}{\partial y}\right)_x = -\left(\frac{\partial P}{\partial x}\right)_y$$

But $\partial y = \partial V$ (as $y = V$) and $\partial s = \partial S$ (as $x = S$). Hence

$$\left(\frac{\partial T}{\partial V}\right)_S = -\left(\frac{\partial P}{\partial S}\right)_V \quad ...(6.6)$$

This is Maxwell' first thermodynamical relation.

### Second Relation :

Put $x = T$ and $y = V$ in equation (6.5),

then
$$\frac{\partial T}{\partial x}=1 \quad , \quad \frac{\partial V}{\partial y}=1$$

and
$$\frac{\partial T}{\partial y}=0 \quad , \quad \frac{\partial V}{\partial x}=0$$

Substituting in equation (6.5), we get

$$\left(\frac{\partial S}{\partial V}\right)_T = \left(\frac{\partial P}{\partial T}\right)_V \quad ...(6.7)$$

This is Maxwell's second thermodynamical relation.

### Third Relation:

Put $x = S$ and $y = P$, in equation (6.5) then

$$\frac{\partial S}{\partial x}=1, \frac{\partial P}{\partial y}=1, \frac{\partial S}{\partial y}=0, \frac{\partial P}{\partial x}=0$$

Substituting these in equation (6.5), we get

$$\left(\frac{\partial T}{\partial P}\right)_S = \left(\frac{\partial V}{\partial S}\right)_P \quad ...(6.8)$$

This is Maxwell's third thermodynamical relation

### Fourth Relation:

Put $x = T$ and $y = P$, then equation (6.5) gives

$$\frac{\partial T}{\partial x} = 1, \frac{\partial P}{\partial y} = 1, \frac{\partial T}{\partial y} = 0 \text{ and } \frac{\partial P}{\partial x} = 0$$

Substituting these values in equation (6.5), we get

$$\left(\frac{\partial S}{\partial P}\right)_T = -\left(\frac{\partial V}{\partial T}\right)_P \qquad ...(6.9)$$

This is Maxwell's fourth thermodynamical relation. These are the four Maxwell's fundamental thermodynamic relations.

Further there are two more relations within the mechanical variables ($P$, $V$) and thermal variables ($T$, $S$).

### Fifth Relation:

Put $x = P$ and $y = V$

$$\frac{\partial P}{\partial x} = 1, \frac{\partial V}{\partial y} = 1, \frac{\partial P}{\partial y} = 0 \text{ and } \frac{\partial V}{\partial x} = 0$$

Substituting these values in equation (6.5), we get

$$\left(\frac{\partial T}{\partial P}\right)_V \left(\frac{\partial S}{\partial V}\right)_P - \left(\frac{\partial T}{\partial V}\right)_P \left(\frac{\partial S}{\partial P}\right)_V = 1 \qquad ...(6.10)$$

### Sixth Relation:

Put $x = T$ and $y = S$

$$\frac{\partial T}{\partial x} = 1, \frac{\partial S}{\partial y} = 1, \frac{\partial T}{\partial y} = 0 \text{ and } \frac{\partial S}{\partial x} = 0$$

Substituting in equation (6.5), we get

$$\left(\frac{\partial P}{\partial T}\right)_S \left(\frac{\partial V}{\partial S}\right)_T - \left(\frac{\partial P}{\partial S}\right)_T \left(\frac{\partial V}{\partial T}\right)_S = 1 \qquad ...(6.11)$$

Out of these six thermodynamic relations, the one suited for a particular problem is used and the problem is solved. Let us see, some of the important applications of these Maxwell's thermodynamic relations.

## 6.4 Applications of Maxwell's Thermodynamic Relations

### 6.4.1 Specific Heat Equation

The specific heat at constant pressure is given

as $$C_P = \left(\frac{\partial Q}{\partial T}\right)_P = T\left(\frac{\partial S}{\partial T}\right)_P \qquad (\because \partial Q = T.\partial S)$$

and the specific heat at constant volume is

$$C_V = \left(\frac{\partial Q}{\partial T}\right)_V = T\left(\frac{\partial S}{\partial T}\right)_V$$

Now, if the entropy $S$ is regarded as a function of $T$ and $V$ and since $dS$ is a perfect differential,

$$dS = \left(\frac{\partial S}{\partial T}\right)_V dT + \left(\frac{\partial S}{\partial V}\right)_T dV$$

$$\therefore \quad \left(\frac{\partial S}{\partial T}\right)_P = \left(\frac{\partial S}{\partial T}\right)_V \left(\frac{\partial T}{\partial T}\right)_P + \left(\frac{\partial S}{\partial V}\right)_T \left(\frac{\partial V}{\partial T}\right)_P$$

or
$$T\left(\frac{\partial S}{\partial T}\right)_P = T\left(\frac{\partial S}{\partial T}\right)_V + T\left(\frac{\partial S}{\partial V}\right)_T \left(\frac{\partial V}{\partial T}\right)_P$$

But
$$\left(\frac{\partial S}{\partial V}\right)_T = \left(\frac{\partial P}{\partial T}\right)_V$$
, from Maxwell's second relation (Eqn. 6.7)

$$\therefore \quad T\left(\frac{\partial S}{\partial T}\right)_P - T\left(\frac{\partial S}{\partial T}\right)_V = T\left(\frac{\partial P}{\partial T}\right)_V \left(\frac{\partial V}{\partial T}\right)_P$$

$$C_P - C_V = T\left(\frac{\partial P}{\partial T}\right)_V \left(\frac{\partial V}{\partial T}\right)_P \qquad \text{...(6.12)}$$

**(a) For a perfect gas:** The equation of state is

$$PV = RT$$

$$\therefore \quad \left(\frac{\partial P}{\partial T}\right)_V = \frac{R}{V} \quad \text{and} \quad \left(\frac{\partial V}{\partial T}\right)_P = \frac{R}{P}$$

Hence, equation (6.12) becomes

$$C_P - C_V = T\left(\frac{R}{V}\right)\left(\frac{R}{P}\right) = \frac{R^2T}{PV} = \frac{R^2T}{RT} \qquad (\because PV = RT)$$

$$\therefore \quad C_P - C_V = R \qquad \text{...(6.13)}$$

**(*b*) Van der Waals :** The equation of state is

$$\left(P + \frac{a}{V^2}\right)(V - b) = RT$$

where $a$ and $b$ are constants.

$$\left(P + \frac{a}{V^2}\right) = \frac{RT}{(V - b)}$$

Differentiating with respect to $T$ at constant volume, we get

$$\left(\frac{\partial P}{\partial T}\right)_V = \frac{R}{(V - b)}$$

and differentiating with respect to $T$ at constant pressure we get

$$0 - \frac{2a}{V^3}\left(\frac{\partial V}{\partial T}\right)_P = -\frac{RT}{(V - b)^2}\left(\frac{\partial V}{\partial T}\right)_P + \frac{R}{(V - b)}$$

or
$$\left(\frac{\partial V}{\partial T}\right)_P \left[\frac{RT}{(V - b)^2} - \frac{2a}{V^3}\right] = \frac{R}{(V - b)}$$

or
$$\left(\frac{\partial V}{\partial T}\right)_P = \frac{\left[\dfrac{R}{(V - b)}\right]}{\left[\dfrac{RT}{(v - b)^2} - \dfrac{2a}{V^3}\right]}$$

Substituting these values of $\left(\frac{\partial P}{\partial T}\right)_V$ and $\left(\frac{\partial V}{\partial T}\right)_P$ in equation (6.12), we get

$$C_P - C_V = -\frac{T\left(\frac{R}{V-b}\right)\left(\frac{R}{V-b}\right)}{\left[\frac{RT}{(V-b)^2} - \frac{2a}{V^3}\right]} = \frac{R}{1 - \frac{2a}{V^3} \cdot \frac{(V-b)^2}{RT}}$$

Neglecting $b$ in comparison to $V$,

$$C_P - C_V = \frac{R}{1 - \frac{2a}{V^3} \cdot \frac{V^2}{RT}} = \frac{R}{1 - \frac{2a}{VRT}} = R\left(1 - \frac{2a}{VRT}\right)^{-1}$$

Expanding binomially and neglecting $a$ in higher powers, as $a$ is also small as compared to $V$, we have

$$C_P - C_V = R\left(1 + \frac{2a}{VRT}\right) \qquad ...(6.14)$$

[Here $R$ is in heat units]

### 6.4.2 Joule-Thomson Cooling

As we know, in Joule-Thomson porous plug experiment, if a gas at constant high pressure is forced through a porous plug to a region of constant low pressure, the temperature of escaping gas changes. This is called Joule-Thomson effect.

Due to throttling, the gas suffers expansion. Although there is a pressure difference on two sides of the porous plug, the enthalpy $H$ of the gas remains constant. This is the necessary condition i.e.

$$H = U + PV = \text{a constant}$$

$\therefore$ $$dH = dU + PdV + VdP = 0$$

But $$dQ = dU + PdV \qquad (\because \text{ First law of thermodynamics})$$

Also $$dQ = TdS \qquad (\because \text{ Second law of thermodynamics})$$

$\therefore$ $$dQ = TdS + VdP = 0 \qquad ...(6.15)$$

Now $dS$ being a perfect differential and $S$ is a function of $P$ and $T$ i.e. $S = f(P, T)$, we have

$$dS = \left(\frac{\partial S}{\partial T}\right)_P dT + \left(\frac{\partial S}{\partial P}\right)_T dP$$

Substituting in equation (6.15),

$$T\left(\frac{\partial S}{\partial T}\right)_P dT + \left[T\left(\frac{\partial S}{\partial P}\right)_T + V\right] dP = 0$$

Now $$C_P = \left(\frac{\partial Q}{\partial T}\right)_P = T\left(\frac{\partial S}{\partial T}\right)_P$$

where $C_P$ is specific heat at constant pressure.

$\therefore$ $$C_P\, dT = -\left[T\left(\frac{\partial S}{\partial P}\right)_T + V\right] dP \qquad ...(6.16)$$

According to Maxwell's fourth thermodynamical relation

$$\left(\frac{\partial S}{\partial P}\right)_T = -\left(\frac{\partial V}{\partial T}\right)_P$$

$\therefore$ $$C_P\, dT = \left[T\left(\frac{\partial V}{\partial T}\right)_P - V\right] dP$$

or $$dT = \frac{1}{C_P}\left[T\left(\frac{\partial V}{\partial T}\right)_P - V\right] dP \qquad ...(6.17)$$

### 6.4.3 Joule-Thomson Coefficient (μ) :

The Joule-Thomson coefficient is given by

$$\mu = \left(\frac{\partial T}{\partial P}\right)_H = \frac{1}{C_P}\left[T\left(\frac{\partial V}{\partial T}\right)_P - V\right]$$

Now $\left(\frac{\partial V}{\partial T}\right)_P = \alpha V$, where α is the coefficient of increase of volume at constant pressure, because $\alpha = \frac{1}{V}\left(\frac{\partial V}{\partial T}\right)_P$.

∴ From equation (6.17), we have

$$dT = \frac{V}{C_P}[\alpha T - 1]\, dP \qquad ...(6.18)$$

Equation (6.18) gives the change in temperature due to Joule-Thomson effect.

Here $dP$ represents a fall in pressure in porous plug experiment and it is always a negative quantity. All other quantities in equation (6.18) are + ve. Hence

(*i*) There is cooling effect *i.e.* $dT$ is negative if $(\alpha T - 1)$ is + ve or $\alpha T > 1$.

(*ii*) There is neither a cooling nor a heating effect *i.e.* $dT = 0$ when $(\alpha T - 1) = 0$ or $\alpha T = 1$

(*iii*) There is a heating effect *i.e.* $dT$ is positive if $(\alpha T - 1)$ is negative or $\alpha T < 1$.

Let us apply the thermodynamical treatment of heating and cooling effect to a perfect gas and Van der Walls gas.

**(*a*) For a perfect gas :**

According to the gas equation.

$$PV = RT \qquad \text{(For perfect gas)}$$

At constant pressure,

$$P\partial V = R\partial T$$

∴ $$\left(\frac{\partial V}{\partial T}\right)_P = \frac{R}{P}$$

or $$T\left(\frac{\partial V}{\partial T}\right)_P = \frac{RT}{P} = V$$

Hence $$\left[T\left(\frac{\partial V}{\partial T}\right)_P - V\right] = 0$$

Using equation (6.17), we have

For a perfect gas, $dT = 0$

In other words, there is no Joule-Thomson effect for a perfect gas. Hence, the porous plug experiment provides us a method to decide whether the given gas is perfect or not. For a perfect gas, the Joule-Thomson coefficient (μ) is zero.

∴ $$\mu = \left(\frac{\partial T}{\partial P}\right)_H = \frac{1}{C_P}\left[T\left(\frac{\partial V}{\partial T}\right)_P - V\right] = 0$$

**(b) For Van der Waals gas**

Van der Waals equation for one mole of a real gas is given by

$$\left(P+\frac{a}{V^2}\right)(V-b) = RT \qquad \text{...(i)}$$

or

$$PV - Pb + \frac{a}{V} - \frac{ab}{V^2} = RT$$

Since both the Van der Waals' constants $a$ and $b$ are both very small quantities, their product $ab$ is very small as compared to $V^2$ and hence $\frac{ab}{V^2}$ can be neglected.

$$\therefore \quad PV - Pb + \frac{a}{V} = RT$$

Differentiating both sides with respect to $T$, keeping $p$ constant, we get.

$$P\left(\frac{dV}{dT}\right)_P - \frac{a}{V^2}\left(\frac{dV}{dT}\right)_P = R$$

or

$$\left(\frac{dV}{dT}\right)_P\left(P-\frac{a}{V^2}\right) = R$$

or

$$\left(\frac{dV}{dT}\right)_P = \frac{R}{\left(P-\frac{a}{V^2}\right)} \qquad \text{...(ii)}$$

From equation (*i*)

$$P = \frac{RT}{V-b} - \frac{a}{V^2}$$

Substituting for $P$ in equation (*ii*), we have

$$\left(\frac{dV}{dT}\right)_P = \frac{R}{\left(\frac{RT}{V-b} - \frac{2a}{V^2}\right)} = \frac{R\,(V-b)}{\left[RT - \frac{2a\,(V-b)}{V^2}\right]}$$

$$= \frac{R\,(V-b)}{RT - \frac{2a}{V}} \qquad [\because \text{As } b \text{ is very small, } V - b = V]$$

$$\therefore \quad T\left(\frac{dV}{dT}\right)_P = \frac{RT\,(V-b)}{RT - \frac{2a}{V}} = \frac{V-b}{1-\frac{2a}{VRT}}$$

$$= (V-b)\left(1-\frac{2a}{VRT}\right)^{-1}$$

$$= (V-b)\left(1+\frac{2a}{VRT}\right)$$

Neglecting squares and higher powers of $\frac{2a}{VRT}$ as $\frac{2a}{VRT} << 1$.

$$\therefore \quad T\left(\frac{dV}{dT}\right)_P = (V-b)\left(1+\frac{2a}{VRT}\right)$$

$$= V - b - \frac{2a(V-b)}{VRT}$$

$$= V - b + \frac{2a}{RT} \qquad \text{[As } b \text{ is very small } V - b = V\text{]}$$

$$\therefore \qquad T\left(\frac{dV}{dT}\right)_P - V = \frac{2a}{RT} - b \qquad \text{...(6.19)}$$

According to Joule-Thomson effect, the change in temperature

$$dT = \frac{1}{C_P}\left[T\left(\frac{\partial V}{\partial T}\right)_P - V\right]dP$$

where $C_P$ is molar specific heat at constant pressure.

Substituting the value of $\left[T\left(\frac{\partial V}{\partial T}\right)_P - V\right]$ from equation (6.19), we get

$$dT = \frac{1}{C_P}\left[\frac{2a}{RT} - b\right]dP$$

$$dT = \frac{dP}{C_P}\left[\frac{2a}{RT} - b\right] \qquad \text{...(6.20)}$$

Equation (6.20) gives the change in temperature for Van der Waals gas *i.e.* real gas due to Joule-Thomson effect.

Thus, we conclude that

(*i*) If $\frac{2a}{RT} > b$ *i.e.* $T$ is less than $\frac{2a}{Rb}$, then $\delta T$ is positive. Since $\delta P$ is – ve, $dT$ will be – ve. Hence there will be a cooling effect.

(*ii*) If $\frac{2a}{RT} < b$ i.e. $T$ is greater than $\frac{2a}{Rb}$, then $\delta T$ is negative, since $\delta P$ is – ve, $dT$ will be + ve. Hence there will be a heating effect.

(*iii*) If $\frac{2a}{RT} = b$ or $T = \frac{2a}{Rb}$, then $dT = 0$, there will be no change of temperature.

### 6.4.4 Temperature of Inversion

The temperature at which Joule-Thomson effect is zero and changes sign is known as *temperature of inversion* and is denoted by $T_i$

$$\therefore \qquad T_i = \frac{2a}{Rb} \qquad \text{...(6.21)}$$

When the room temperature is above the temperature of inversion ($T > T_i$) or $\frac{2a}{RT} < b$, there will be a heating effect.

Hence, during the expansion of the gas through a porous plug (1) cooling will take place if temperature of the gas is less than its temperature of inversion $\left[T < \frac{2a}{Rb}\right]$ and (2) heating will take place if temperature of the gas is more than its temperature of inversion $\left[T > \frac{2a}{Rb}\right]$.

### 6.4.5 Heating Effect for Hydrogen and Helium at Room Temperature

For most of the gases, the ordinary working temperatures are below the temperature of inversion and hence they show cooling effect.

The temperature of inversion for hydrogen and helium is much below the ordinary room temperature. Hence, at ordinary room temperatures, these gases show a heating effect. For hydrogen, at about 100 atmospheres pressure, the temperature of inversion ($T_i$) is – 80°C. In hydrogen, a cooling effect is, therefore, observed if it is cooled below – 80°C and then passed through the porous plug.

For helium, the temperature of inversion ($T_i$) is 35 K (*i.e.* – 238°C). Hence at ordinary temperature, it shows a heating effect.

### 6.4.6 Temperature Change in Adiabatic Process

From first thermodynamical relation

$$\left(\frac{\partial T}{\partial V}\right)_S = -\left(\frac{\partial P}{\partial S}\right)_V$$

Multiplying and dividing the right hand side by $T$, we get

$$\left(\frac{\partial T}{\partial V}\right)_S = -T\left(\frac{\partial P}{T\partial S}\right)_V = -T\left(\frac{\partial P}{\partial Q}\right)_V \qquad \text{...(6.22)}$$

An increase in the amount of heat ($\delta Q$) at constant volume always results in an increase in pressure. Therefore, $\left(\frac{\partial P}{\partial Q}\right)_V$ is positive. Hence the term $\left(\frac{\partial T}{\partial V}\right)_S$ should be negative. In other words, the temperature should decrease with an increase in volume at constant entropy (*i.e.* adiabatic process). Thus, *an adiabatic expansion must result in a fall of temperature.*

Recalling third thermodynamical relation

$$\left(\frac{\partial T}{\partial P}\right)_S = \left(\frac{\partial V}{\partial S}\right)_P$$

or

$$\left(\frac{\partial T}{\partial P}\right)_S = T\left(\frac{\partial V}{T\partial S}\right)_P = T\left(\frac{\partial V}{\partial Q}\right)_P \qquad \text{...(6.23)}$$

Since, an increase in the quantity heat ($\partial Q$) at constant pressure, always results in an increase in its volume the term $\left(\frac{\partial V}{\partial Q}\right)_P$ is positive, hence the term $\left(\frac{\partial T}{\partial P}\right)_S$ must also be positive i.e. the temperature must increase with increase in pressure at constant entropy (*i.e.* adiabatic process). Thus, *an adiabatic compression must result in an increase in temperature.*

### 6.4.7 Clausius-Clapeyron's Equation (First Latent Heat Equation)

The equation $\frac{dP}{dT} = \frac{L}{T(V_2 - V_1)}$ was first derived by Clapeyron using Carnot's reversible cycle.

Therefore, it is some times called Clapeyron's equation.

Whenever there is a change of state, either from a solid state to liquid state or from liquid state to vapour state, the temperature remains constant; as far as the change takes place. This temperature (either melting point or boiling point) depends upon the pressure and is a characteristic of each substance. The above relation shows how melting or boiling points vary with pressure and was derived by Clapeyron by applying the second law of thermodynamics. Later Clausius obtained the same relation by using the Maxwell's thermodynamic relations, as under :

### Derivation :

From second thermodynamical relation

$$\left(\frac{\partial S}{\partial V}\right)_T = \left(\frac{\partial P}{\partial T}\right)_V$$

Multiplying both sides by $T$, we have

$$T\left(\frac{\partial S}{\partial V}\right)_T = T\left(\frac{\partial P}{\partial T}\right)_V$$

But $$T\partial S = \partial Q$$

$$\therefore \quad \left(\frac{\partial Q}{\partial V}\right)_T = T\left(\frac{\partial P}{\partial T}\right)_V \qquad \text{...(6.24)}$$

The quantity $\left(\frac{\partial Q}{\partial V}\right)_T$ represents the quantity of heat absorbed or liberated per unit change in volume at constant temperature. As there is a change in volume due to the heat absorbed at constant temperature, the heat represents the latent heat used when a substance changes from solid to liquid (melting) or liquid to vapour (boiling) state when the temperature remains constant, during the change of state.

If $L$ is the quantity of heat required to change the state of a unit mass of the substance, $V_2$ and $V_1$ the corresponding specific volumes (Volume per unit mass) then

$$\partial Q = L \text{ and } \partial V = V_2 - V_1.$$

Hence $$\left(\frac{\partial Q}{\partial V}\right)_T = \frac{L}{V_2 - V_1}$$

Therefore, $$\frac{L}{V_2 - V_1} = T\left(\frac{\partial P}{\partial T}\right)_V$$

or $$\frac{dP}{dT} = \frac{L}{T(V_2 - V_1)} \qquad \text{...(6.25)}$$

Where $dT$ is the change in melting point or boiling point due to a change in pressure $dP$. This is Clausius-Clapeyron latent heat equation.

### Effect of Pressure on Boiling Point of Liquid

When a liquid converts into vapour state, there is an increase in volume. Hence $V_2 > V_1$; $T$ and $L$ being positive quantities, $\left(\frac{\partial P}{\partial T}\right)_V$ is + ve. This shows that the boiling point of the liquid is raised by increasing the pressure. This is known as elevation of boiling point.

### Effect of Pressure on Melting Point of Solid

When a solid melts, there may be increase in volume, as in case of wax or there may be decrease in volume, as in case of ice. In melting of wax

$$(V_2 - V_1) = +\text{ve}$$

$$\therefore \quad \frac{dP}{dT} = +\text{ve}$$

Hence the melting point is raised by increase of pressure.

In case of ice, $(V_2 - V_1) = -$ ve, hence melting point is lowered by increase of pressure. This is known as depression in melting point. In general, substances which expands on melting, have their melting points raised by increase of pressure and the substance which contract on melting have their melting points lowered by increase of pressure.

## 6.5 Thermodynamic Potentials

The state of a system can be completely described by any two of the five state variables *P, V, T, S* and *U*. Out of these, *U* the internal energy state variable is determined by using the remaining four, as proved below:

According to first law of thermodynamics,

$$\delta Q = \delta U + PdV$$

and from the second law of thermodynamics,

$$\delta Q = TdS$$

$$\therefore \quad dU + PdV = TdS$$

or

$$dU = TdS - PdV$$

Thus, *U* can be eliminated and we are left with four state variables *P, V, T* and *S* only. This is the reason, why only four state variables are defined. Taking two of the four state variables *P, V, T* and *S* at a time, there are six possible pairs, *i.e.* (*P, V*), (*P, T*), (*P, S*), (*V, T*), (*V, S*) and (*T, S*). The pair (*P, V*) is connected with composite and in exact differential quantity $\delta W$ as $\delta W = PdV$ and pair (*T, S*) with $\delta Q$ as $\delta Q = TdS$. Hence these two pairs can be eliminated. Thus, we are left with four pairs of thermodynamical variables (*P, T*), (*P, S*), (*V, T*) and (*V, S*). Corresponding to each pair, we can write a thermodynamical relation. These four thermodynamical relations are known as Maxwell's thermodynamic relations.

There are four thermodynamic potentials:

(*i*) Internal energy, *U*

(*ii*) Helmholtz free energy, $F = U - TS$

(*iii*) Enthalpy $H = U + PV$

(*iv*) Gibbs function $G = U + PV - TS$

Each of Maxwell's four thermodynamical relations can be derived from one of these thermodynamic potentials *U, F, H* and *G*. Hence, only four thermodynamic potentials are defined.

**1. Internal Energy *U*.** The internal energy or the intrinsic energy is the total energy of a system. According to the first law of thermodynamics,

$$dQ = dU + dW = dU + PdV$$

$$\therefore \quad dU = dQ - PdV$$

and from second law of thermodynamics

$$dQ = TdS$$

Substituting for *dQ*, we get

$$dU = TdS - PdV \qquad ...(6.26)$$

This equation gives the change in internal energy of the system in terms of four thermodynamical variables *P, V, T* and *S*. Internal energy (*U*) is called *first thermodynamical potential.*

**(*a*) For an adiabatic process :**

$$dQ = 0$$

$$\therefore \quad dU = -PdV$$

*i.e.* the work done by the system in an adiabatic process is at the expense of its internal energy.

**(*b*) For an isochoric adiabatic process :**

$$dV = 0 \quad \text{and} \quad dQ = 0$$

$$\therefore \quad dU = 0 \quad \text{or} \quad U = \text{a constant.}$$

*i.e.* the internal energy of system remains constant in an isochoric adiabatic process.

**2. Helmholtz Free Energy *F*.** Helmholtz free energy is defined as

$$F = U - TS$$

As *U, T* and *S* are state variables, *F* is also a state variable. *F* has dimensions of energy. According to first and second law of thermodynamics,

$$dU = TdS - dW$$

If the system is maintained at a constant temperature by exchanging heat continuously with the surroundings,

then, $$TdS = d(TS)$$

$\therefore$ $$dU = d(TS) - dW$$

or $$d(U - TS) = -dW$$

or $$dF = -dW \qquad \text{...(6.27)}$$

where $F = (U - TS)$ is known as *Helmholtz free energy or Helmholtz work function.*

Now $$dF = d(U - TS) = dU - TdS - SdT$$

But $$dU = TdS - PdV$$

$\therefore$ $$dF = -PdV - SdT \qquad \text{...(6.28)}$$

This equation gives the change in Helmholtz free energy during an infinitesimal reversible process.

**(*a*) For reversible isothermal process:**

$$dT = 0$$

$\therefore$ $$dF = -PdV \quad \text{or} \quad PdV = -dF$$

Thus, the work done in a reversible isothermal process is equal to the decrease in Helmholtz free energy.

**(*b*) For isothermal isochoric process:**

$$dT = 0 \quad \text{and} \quad dV = 0$$

$\therefore$ $$dF = 0 \quad \text{or} \quad F = \text{a constant.}$$

*i.e.* the Helmholtz free energy remains constant during isothermal isochoric process.

**3. Enthalpy *H*.** This is known as the total heat and is given by

$$H = U + PV$$

As *U, P* and *V* are state variables, *H* is also a state variable. *H* has dimensions of energy. If the system undergoes an infinitesimal reversible process, then

Change in enthalpy, $$dH = dU + PdV + VdP$$

But $$dU = TdS - PdV$$

$\therefore$ $$dH = TdS + VdP \qquad \text{...(6.29)}$$

**(*a*) For reversible isobaric process :**

$$dP = 0$$

$\therefore$ $$dH = TdS = dQ$$

*i.e.* for an isobaric process, the change in enthalpy is equal to the heat absorbed.

**(*b*) For an isobaric adiabatic process :**

$$dP = 0 \; ; \quad dQ = 0$$

$\therefore$ $$dH = 0 \quad \text{or} \quad H = \text{a constant}$$

*i.e.* enthalpy remains constant in a reversible isobaric adiabatic process.

**4. Gibbs Function G or Gibbs Free Energy.** Gibbs function (*G*) was called by Duhem the thermodynamical potential at constant pressure. This is defined by the equation

$$G = U - TS + PV \qquad \text{...(6.30)}$$

As $F = U - TS$, therefore, we can write

$$G = F + PV \quad \text{...(6.31)}$$

This is the relation between Gibbs function and Helmholtz function. An enthalpy $H = U + PV$, therefore, equation (6.30) becomes

$$G = H - TS$$

or

$$H = G + TS \quad \text{...(6.32)}$$

∴ Enthalpy = Gibbs free energy + latent heat

**(*a*) For an isothermal process:** $TdS = d(TS)$

**(*b*) For an isobaric process:** $dP = 0$

Hence, if the process is isothermal and isobaric then

$$dH = d(TS)$$

$$d(H - TS) = 0$$

$$dG = 0 \quad \text{or} \quad G = \text{a constant}$$

where $G = H - TS$ is known as thermodynamic potential at constant pressure; or Gibbs function or Gibbs free energy.

Thus, Gibbs function ($G$) or Gibbs free energy remains constant in an isothermal-isobaric process.

## 6.6 Significance of Thermodynamic Potentials

A mechanical system is said to be in stable equilibrium when the potential energy of the system is minimum. It means that the system must proceed in such a direction so as to acquire minimum potential energy. This what we observe in nature, viz. water flows from a higher level to lower level, electric current flows from higher to lower potential, heat flows from higher to lower temperature, a body falls from higher to lower potential due to gravitational field and so on.

In thermodynamics, the behavior of internal energy ($U$), Helmholtz free energy ($F$), enthalpy ($H$), and Gibbs free energy ($G$) is similar to potential energy in *mechanics*. As we have seen, the direction of isothermal-isochoric process is to make Helmholtz free energy ($F$) minimum. In isothermal-isobaric process, Gibbs free energy ($G$) tends to be minimum. In an isobaric-adiabatic process, the enthalpy ($H$) tends to be minimum.

Since the four functions $U$, $F$, $H$ and $G$ play in thermodynamics the same role as played by potential energy in mechanics, they are called thermodynamic potentials.

## 6.7 Relation of Thermodynamical Potentials with their Variables

The four quantities $U(S, V)$, $F(T, V)$, $H(S, P)$ and $G(P, T)$ are called thermodynamic potentials because the thermodynamic variables $S$, $T$, $P$ and $V$ can be derived from them by their differentiations with respect to the independent variables associated with them. Let us derive them:

### 1. Thermodynamic Potential $U(S, V)$

According to first and second law of thermodynamics,

Change in internal energy, $dU = \delta Q - PdV$

$$= TdS - PdV \quad (\because \delta Q = TdS)$$

Taking partial differentials of internal energy $U$ with respect to variable $S$ (entropy), we have

$$\left(\frac{\partial U}{\partial S}\right)_V = T \quad \text{and} \quad \left(\frac{\partial U}{\partial V}\right)_S = -P \quad \text{...(6.33)}$$

These are the relations connecting the internal energy $U$ with thermodynamical variables $S$, $V$, $T$ and $P$.

Since $dU$ is a perfect differential (*i.e.* $U$ is a single valued), we have

$$\left[\frac{\partial}{\partial V}\left(\frac{\partial U}{\partial S}\right)_V\right]_S = \left[\frac{\partial}{\partial S}\left(\frac{\partial U}{\partial V}\right)_S\right]_V$$

or
$$\left(\frac{\partial T}{\partial V}\right)_S = -\left(\frac{\partial P}{\partial S}\right)_V \qquad ...(6.34)$$

This is Maxwell's *first* thermodynamical relation.

## 2. Thermodynamic Potential *F* (*T*, *V*)

The Helmholtz function or Helmholtz free energy is given by

$$F = U - TS$$

$\therefore$ $$dF = d(U - TS) = dU - TdS - SdT$$

But $$dU = TdS - PdV$$

$\therefore$ $$dF = -PdV - SdT$$

Taking partial differential of Helmholtz function $F$ with respect to $T$ and $V$, we get

$$\left(\frac{\partial F}{\partial T}\right)_V = -S \quad \text{and} \quad \left(\frac{\partial F}{\partial V}\right)_T = -P \qquad ...(6.35)$$

Since $dF$ is a perfect differential,

$$\left[\frac{\partial}{\partial V}\left(\frac{\partial F}{\partial T}\right)_V\right]_T = \left[\frac{\partial}{\partial T}\left(\frac{\partial F}{\partial V}\right)_T\right]_V$$

or
$$\left(\frac{\partial S}{\partial V}\right)_T = \left(\frac{\partial P}{\partial T}\right)_V \qquad ...(6.36)$$

This is Maxwell's *second* thermodynamical relation.

## 3. Thermodynamic Potential *H* (*S*, *P*)

The enthalpy ($H$) is given by

$$H = U + PV$$

Differentiating, $$dH = dU + PdV + VdP$$

But $$dU = TdS - PdV$$

$\therefore$ $$dH = TdS + VdP$$

Taking partial differential of enthalpy $H$ with respect to $S$ and $P$, we get

$$\left(\frac{\partial H}{\partial S}\right)_P = T \quad \text{and} \quad \left(\frac{\partial H}{\partial P}\right)_S = V \qquad ...(6.37)$$

Since $dH$ is perfect differential,

$$\left[\frac{\partial}{\partial P}\left(\frac{\partial H}{\partial S}\right)_P\right]_S = \left[\frac{\partial}{\partial S}\left(\frac{\partial H}{\partial P}\right)_S\right]_P$$

or
$$\left(\frac{\partial T}{\partial P}\right)_S = \left(\frac{\partial V}{\partial S}\right)_P \qquad ...(6.38)$$

This is Maxwell's *third* thermodynamical relation.

## 4. Thermodynamic Potential *G* (*P*, *T*)

The Gibbs function ($G$) is given by

$$G = H - TS$$

Differentiating, we get

$$dG = dH - TdS - SdT$$

But $$dH = TdS + VdP$$

$\therefore$ $$dG = VdP - SdT$$

Taking partial differential of Gibbs function $G$ with respect to $P$ and $T$, we get

$$\left(\frac{\partial G}{\partial P}\right)_T = V \quad \text{and} \quad \left(\frac{\partial G}{\partial T}\right)_P = -S \qquad \text{...(6.39)}$$

As $dG$ is perfect differential,

$$\left[\frac{\partial}{\partial T}\left(\frac{\partial G}{\partial P}\right)_T\right]_P = \left[\frac{\partial}{\partial P}\left(\frac{\partial G}{\partial T}\right)_P\right]_T$$

or $$\left(\frac{\partial V}{\partial T}\right)_P = -\left(\frac{\partial S}{\partial P}\right)_T \qquad \text{...(6.40)}$$

This is Maxwell's *fourth* thermodynamical relation.

Thus, the thermodynamical variables $S$, $T$, $P$ and $V$ can be written by using equation (6.33), (6.35), (6.37) and (6.39) as

$$S = -\left(\frac{\partial G}{\partial T}\right)_P = -\left(\frac{\partial F}{\partial T}\right)_V$$

$$T = \left(\frac{\partial U}{\partial S}\right)_V = \left(\frac{\partial H}{\partial S}\right)_P$$

$$P = -\left(\frac{\partial U}{\partial V}\right)_S = -\left(\frac{\partial F}{\partial V}\right)_T$$

$$V = \left(\frac{\partial H}{\partial P}\right)_S = \left(\frac{\partial G}{\partial P}\right)_T \qquad \text{...(6.41)}$$

These equations give the values of thermodynamical variables in terms of thermodynamical potentials.

Thus, Maxwell's thermodynamical relations are tabulated as under :

**TABLE 6.2 Maxwell's thermodynamical relations**

$$\left(\frac{\partial S}{\partial V}\right)_T = \left(\frac{\partial P}{\partial T}\right)_V \qquad \text{...}(i)$$

$$\left(\frac{\partial S}{\partial P}\right)_T = -\left(\frac{\partial V}{\partial T}\right)_P \qquad \text{...}(ii)$$

$$\left(\frac{\partial T}{\partial V}\right)_S = -\left(\frac{\partial P}{\partial S}\right)_V \qquad \text{...}(iii)$$

$$\left(\frac{\partial T}{\partial P}\right)_S = \left(\frac{\partial V}{\partial S}\right)_P \qquad \text{...}(iv)$$

### How to Remember:

***Note.*** The first four Maxwell's thermodynamical relations need not be memorised. These can be easily written by remembering the word SP or $T_i$ $V_e$ (sportive). In this, *SPTV* represent entropy, pressure, temperature and volume. In deriving the first two relations, $\partial S$ is written in the numerator on the left hand side of the equation and the rest of the quantities $\partial P$, $\partial T$ and $\partial V$ are written (*i*) in the clockwise order and (*ii*) in the anticlockwise order. In the clockwise direction, the right hand side expression is with a + sign and for the anticlockwise direction, it is – *ve*. These equations are

$$\left(\frac{\partial S}{\partial V}\right)_T = \left(\frac{\partial P}{\partial T}\right)_V \quad ...(i)$$

$$\left(\frac{\partial S}{\partial P}\right)_T = -\left(\frac{\partial V}{\partial T}\right)_P \quad ...(ii)$$

Similarly to write the other equations, $\partial S$ is written in the denominator of the right hand side of the equation and the other quantities are written (*i*) in the anticlockwise direction and (*ii*) in the clockwise direction. These equations are

$$\left(\frac{\partial T}{\partial V}\right)_S = -\left(\frac{\partial P}{\partial S}\right)_V \quad ...(iii)$$

$$\left(\frac{\partial T}{\partial P}\right)_S = \left(\frac{\partial V}{\partial S}\right)_P \quad ...(iv)$$

## 6.8 Relation between $C_P$, $C_V$ and μ

The specific heat at constant pressure $C_P$, the specific heat at constant volume $C_V$ and the Joule-Kelvin coefficient μ are defined as follows:

$$C_P = \left(\frac{\partial H}{\partial T}\right)_P \quad ...(i)$$

$$C_V = \left(\frac{\partial U}{\partial T}\right)_V \quad ...(ii)$$

$$\mu = \left(\frac{\partial T}{\partial P}\right)_h \quad ...(iii)$$

These three quantities are defined in terms of the thermodynamic properties *viz*. pressure, volume, temperature, internal energy and enthalpy. Hence $C_P$, $C_V$ and μ are also thermodynamic properties of a substance.

## 6.9 The *T. dS* Equations

**1. *First T.dS Equation.*** The entropy $S$ of a pure substance can be taken as a function of temperature and volume.

$$\therefore \quad dS = \left(\frac{\partial S}{\partial T}\right)_V dT + \left(\frac{\partial S}{\partial V}\right)_T dV$$

Multiplying both sides by $T$

$$T.dS = T\left(\frac{\partial S}{\partial \mathrm{T}}\right)_V dV + T\left(\frac{\partial S}{\partial V}\right)_T dV$$

∴ But

$$C_V = T\left(\frac{\partial S}{\partial T}\right)_V$$

and from Maxwell's relations

$$\left(\frac{\partial S}{\partial V}\right)_T = \left(\frac{\partial P}{\partial T}\right)_V$$

Substituting these values in equation (*ii*)

$$T\,dS = C_V dT + T\left(\frac{\partial P}{\partial T}\right)_V dV \quad ...(6.42)$$

Equation (6.42) is called the first *T. dS* equation.

2. ***Second T.dS Equation.*** The entropy $S$ of a pure substance can also be regarded as a function of temperature and pressure

$$\therefore \qquad dS = \left(\frac{\partial S}{\partial T}\right)_P dT + \left(\frac{\partial S}{\partial P}\right)_T dP$$

Multiplying both sides by $T$

$$T.dS = T\left(\frac{\partial S}{\partial T}\right)_P dT + T\left(\frac{\partial S}{\partial P}\right)_T dP$$

But

$$C_P = T\left(\frac{\partial S}{\partial T}\right)_P$$

and from Maxwell's relations

$$\left(\frac{\partial S}{\partial P}\right)_T = -\left(\frac{\partial V}{\partial T}\right)_P$$

Substituting these values in equation ($v$)

$$T.dS = C_P\, dT - T\left(\frac{\partial V}{\partial T}\right)_P dP \qquad \text{...(6.43)}$$

Equation (6.43) is called the second $T\,dS$ equation.

## 6.10 Clapeyron's Latent Heat Equation using Maxwell's Thermodynamical Relations

From Maxwell's second thermodynamical relation

$$\left(\frac{\partial S}{\partial V}\right)_T = \left(\frac{\partial P}{\partial T}\right)_V \qquad (\because \text{ eqn. 6.36})$$

Multiplying by $T$, we get

$$T\left(\frac{\partial S}{\partial V}\right)_T = T\left(\frac{\partial P}{\partial T}\right)_V$$

$$\left(\frac{\partial Q}{\partial V}\right)_T = T\left(\frac{\partial P}{\partial T}\right)_V \qquad (\because T\,\partial S = \partial Q)$$

Here $\left(\frac{\partial Q}{\partial V}\right)_T$ represents the quantity of heat absorbed per unit increase in volume at constant temperature. This quantity of heat asborbed at constant temperature is the latent heat ($L$).

Thus, $\partial Q = L$ and $\partial V = V_2 - V_1$, for unit mass of a substance.

Substituting,

$$\left(\frac{L}{V_2 - V_1}\right)_T = T\left(\frac{\partial P}{\partial T}\right)_V$$

or

$$\frac{L}{V_2 - V_1} = T\frac{dP}{dT}$$

or

$$\frac{dP}{dT} = \frac{L}{T(V_2 - V_1)} \qquad \text{...(6.44)}$$

This is **Clapeyron's latent heat equation.**

## 6.11 Clapeyron Latent Heat Equation using Carnot's Cycle

Consider the isothermals FBAE at temperature $T + dT$ and $GCDH$ at temperature $T$. Here $EA$ and $HD$ show the liquid state of the substance. At $A$ and $D$, the substance is purely in the liquid state (Fig. 6.1). From $A$ to $B$ or $D$ to $C$, the substance is in transition from the liquid to the gaseous state and *vice versa*. At $B$ and $C$, the substance is purely in the gaseous state. From $B$ to $F$ or $C$ to $G$, the substance is in the gaseous state. Join $A$ to $D$ and $B$ to $C$ by dotted lines.

The cycle $ABCD$ represents a complete cycle and Carnot's theorem can be applied. Suppose the volume at the point $A$ is $V_1$ and temperature is $T + dT$. The pressure is just below its saturation pressure and the liquid begins to evaporate and at the point $B$ the volume is $V_2$. The substance is in the vapour state. Suppose the mass of the liquid at $B$ is one gram the amount of heat absorbed is $Q_1$, here $Q_1 = L + dL$, where $L + dL$ is the latent heat of the liquid at temperature $(T + dT)$.

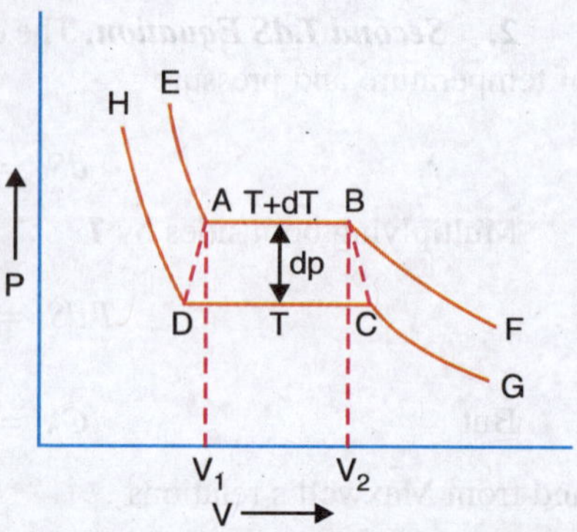

Fig. 6.1

At the point $B$, the pressure is decreased by $dP$. The vapour will expand and its temperature falls. The temperature at $C$ is $T$. At this pressure and temperature $T$, the gas begins to condense and is converted into the liquid state. At the point $D$, the substance is in the liquid state. From $C$ to $D$, the amount of heat rejected (given out) is $Q_2$. Here $Q_2 = L$ where $L$ is latent heat at temperature $T$. By increasing the pressure a little, the original point $A$ is restored. The cycle $ABCDA$ is completely reversible. Applying the principle of the Carnot's reversible cycle

$$\frac{Q_1}{T_1} = \frac{Q_2}{T_2}$$

or

$$\frac{Q_1}{Q_2} = \frac{T_1}{T_2}$$

$$\frac{Q_1 - Q_2}{Q_2} = \frac{T_1 - T_2}{T_1}$$

Here,

$$Q_1 = L + dL, Q_2 = L,$$

$$T_1 = T + dT, T_2 = T$$

$$Q_1 - Q_2 = L + dL - L = dL$$

$$T_1 - T_2 = T + dT - T = dT$$

$$\therefore \quad \frac{dL}{L} = \frac{dT}{T}$$

The area of the figure

$$ABCD = Q_1 - Q_2 = dL$$

$$= dP\,(V_2 - V_1)$$

$$\therefore \quad \frac{dP(V_2 - V_1)}{L} = \frac{dT}{T}$$

$$\frac{dP}{dT} = \frac{L}{T(V_2 - V_1)} \qquad ...(6.45)$$

This is called the **Clapeyron's latent heat equation.**

## Applications

1. ***Effect of change of pressure on the melting point.***

When a solid is converted into a liquid, there is change in volume.

(*i*) If $V_2$ is greater than $V_1$.

$\frac{dP}{dT}$ is a positive quantity. It means that the rate of change of pressure with respect to temperature is positive. In such cases, the melting point of the substance will increase with increase in pressure and *vice versa*.

(*ii*) If $V_2$ is less than $V_1$.

$\frac{dP}{dT}$ is a negative quantity. It means that the rate of change of pressure with respect to temperature is negative. In such cases, the melting point of the substance will decrease with increase in pressure and *vice versa*. In the case of melting ice, the volume of water formed is less than the volume of ice taken. Hence $V_2 < V_1$.

Therefore, the melting point of ice decreases with increase in pressure. Hence ice will melt at a temperature lower than zero degree centigrade at a pressure higher than the normal pressure.

Ice melts at 0°C only at a pressure of 76 cm of Hg.

**2. *Effect of change of pressure on the boiling point.***

When a liquid is converted into a gaseous state, the volume $V_2$ of the gas is always greater than the corresponding volume $V_1$ of the liquid *i.e.*, $V_2 > V_1$.

Therefore, $\frac{dP}{dT}$ is a + *ve* quantity.

With increase in pressure, the boiling point of a substance increases and *vice versa*. The liquid will boil at a lower temperature under reduced pressure. In the case of water, the boiling point increases with increase in pressure and *vice versa*. Water boils at 100°C only at 76 cm of Hg pressure. In the laboratories, while preparing steam, the boiling point is less than 100°C because the atmospheric pressure is less than 76 cm of Hg. In pressure cookers, the liquid boils at a higher temperature because the pressure inside is more than the atmospheric pressure.

## 6.12 Adiabatic Stretching of a Wire

Let us derive again (already discussed in 4.16.2), the relation using Maxwell's thermodynamic relations as under:

From first law of thermodynamics,

$$dQ = dU + PdV$$

and from second law of thermodynamics, we have

$$dQ = TdS$$

Combining the two, we have

$$TdS = dU + PdV$$

In case of stretching a wire, the work done, $dW = -F.dl$ instead of $P.dV$. Here $F$ is the tension applied for the change in length $dl$.

Therefore,

$$T.dS = dU - F.dl$$

or

$$dU = T.dS + F.dl \quad ...(i)$$

Now writting $-F$ and $l$ in place of $P$ and $V$ respectively, in Maxwell's first thermodynamic relation

$$\left(\frac{\partial T}{\partial V}\right)_S = -\left(\frac{\partial P}{\partial S}\right)_V$$

we get

$$\left(\frac{\partial T}{\partial l}\right)_S = \left(\frac{\partial F}{\partial S}\right)_l$$

$$= T\left(\frac{\partial F}{T\partial S}\right)_l = T\left(\frac{\partial F}{\partial Q}\right)_l$$

or $$\left(\frac{\partial T}{\partial l}\right)_S = T\left(\frac{\partial F}{\partial T}\right)_l\left(\frac{\partial T}{\partial Q}\right)_l \qquad ...(ii)$$

Now taking length $l$ as a function of $F$ and $T$, we put $dl$ as sum of partial differentials *w.r.* to $F$ and $T$.

Thus, $$dl = \left(\frac{\partial l}{\partial F}\right)_T dF + \left(\frac{\partial l}{\partial T}\right)_F dT$$

under condition of no elongation *i.e.*, $dl = 0$, we get

$$0 = \left(\frac{\partial l}{\partial F}\right)_T dF + \left(\frac{\partial l}{\partial T}\right)_F dT$$

or $$0 = \left(\frac{\partial l}{\partial F}\right)_T\left(\frac{\partial F}{\partial T}\right)_l + \left(\frac{\partial l}{\partial T}\right)_F$$

or $$\left(\frac{\partial F}{\partial T}\right)_l = -\frac{\left(\frac{\partial l}{\partial T}\right)_F}{\left(\frac{\partial l}{\partial F}\right)_T} \qquad ...(iii)$$

Substituting in equ. (*ii*), we get

$$\left(\frac{\partial T}{\partial l}\right)_S = -T\frac{\left(\frac{\partial l}{\partial T}\right)_F}{\left(\frac{\partial l}{\partial F}\right)_T}\times\left(\frac{\partial T}{\partial Q}\right)_l \qquad ...(iv)$$

But, the coefficient of linear expansion $\beta = \frac{1}{l}\left(\frac{\partial l}{\partial T}\right)_F$ and isothermal Young's modulus $Y_T = \frac{l}{A}\left(\frac{\partial F}{\partial l}\right)_T$,

so that $\left(\frac{\partial l}{\partial F}\right)_T = \frac{l}{Y_T . A}$, where $A$ is the area of cross-section of the wire. Moreover, $\left(\frac{\partial Q}{\partial T}\right)_l = C_l$, the specific heat at constant length in *erg*. $\text{deg}^{-1}$ $\text{gm}^{-1}$. Putting these values in equation (*iv*), we get

$$\left(\frac{\partial T}{\partial l}\right)_S = -T\frac{\beta l Y_T A}{l}\cdot\frac{1}{C_l} = -\frac{T\beta Y_T A}{C_l}$$

$$= -\frac{T\beta Y_T A}{mC} \qquad ...(v)$$

where $m$ is the mass per unit length of the wire and $C$ is the specific heat of the material of wire in ergs $\text{deg}^{-1}$ $\text{gm}^{-1}$.

Equation (*v*) may be in the form $\left(\frac{\partial T}{\partial F}\right)_S$.

Therefore, $$\left(\frac{\partial T}{\partial F}\right)_S = \left(\frac{\partial T}{\partial l}\right)_S\cdot\left(\frac{\partial l}{\partial F}\right)_S$$

$$= -\frac{T\beta l Y_T A}{mC}\left(\frac{1}{Y_S A}\right)$$

This is because $$Y_S = \frac{l}{A}\left(\frac{\partial F}{\partial l}\right)_S$$

$\therefore$ $$\left(\frac{\partial T}{\partial F}\right)_S = -\frac{T\beta}{mC} \qquad ...(vi)$$

where we have put $Y_T = Y_S$, which is closely true for all solids.

Therefore,
$$dT = \frac{TP}{mC} dF \qquad \text{... (6.46)}$$

If $\beta$ is $+ve$, then an increase in tension would cool the wire. Thus, the wires of substances which expand on heating should show a cooling effect when stretched adiabatically.

In case of *a rubber string*, $\alpha$ is negative *i.e.*, rubber contracts on heating. Therefore, for a positive value of $dF$ *i.e.*, for an increase in stretching force, the temperature rises and there is a heating effect. This is why, a rubber string gets heated up on stretching.

## 6.13 Internal Energy of Ideal and Real Gases

The internal energy of a perfect gas is only dependent upon temperature whereas the internal energy of a real gas (or Van der Waals gas) is a function of both temperature and volume.

For a perfect gas, the internal energy is only kinetic energy of its molecules. It is proportional to $kT$ and hence only depends upon temperature, $k$ being Boltzmann's constant. The internal energy of a perfect gas is independent of its volume.

For a real gas, say a gas obeying Van der Waals equation, the internal energy is the sum of the kinetic and potential energies of molecules due to their mutual forces of attraction. The force of attraction between the molecules depends upon the intermolecular distance and is thus a function of volume. This is why the internal energy of a real gas or a gas obeying Van der Waals equation is a function of both volume and temperature. As the volume of an actual gas increases (at constant temperature) the internal energy increases. This is due to the fact that when the volume increases, the molecules move farther apart against intermolecular attraction. Hence work is done thereby causing an increase in potential energy. (for details refer 6.5.1)

## 6.14 Clausius Inequality

Before considering entropy, it is necessary to establish the Clausius inequality which states that
$$\oint \frac{\delta Q}{T} \leq 0$$

This relation is between the temperatures of an arbitrary number of heat reservoirs and the amounts of heat transferred (given up or lost).

Consider three heat reservoirs at temperature $T_1$, $T_2$ and $T_0$ (Fig. 6.2)

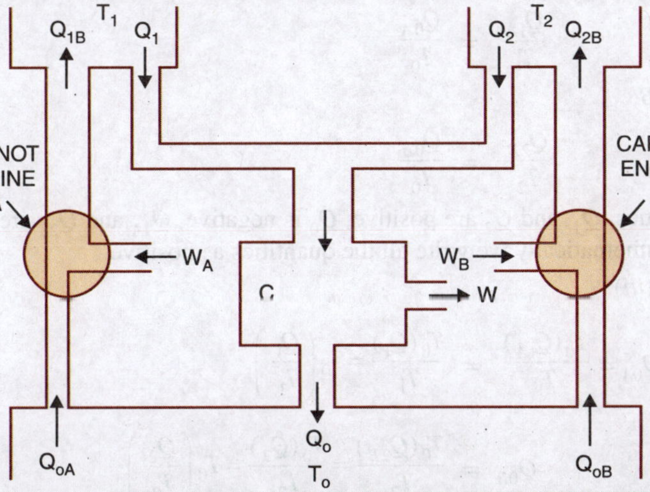


Fig. 6.2

The engine $C$ works as a heat engine. It absorbs that $Q_1$ from reservoir at temperature $T_1$ and $Q_2$ from reservoir at temperature $T_2$. The system as a whole does an amount of work $W$ and rejects a quantity of heat $Q_0$ to the reservoir at temperature $T_0$.

Carnot's engines $A$ and $B$ work as refrigerators. The engine $A$ absorbs heat $Q_{0A}$ from the reservoir at temperature $T_0$, an amount of work $W_A$ is supplied to it and it rejects a quantity of heat $Q_{1A}$ to the reservoir at temperature $T_1$. The engine $B$ absorbs heat $Q_{0B}$ from the reservoir at temperature $T_0$, an amount of work $W_B$ is supplied to it and it rejects a quantity of heat $Q_{2B}$ to the reservoir at temperature $T_2$.

In the general discussion given below, no restriction on the direction of interchanges of the energy is made except the interchanges must be consistent with the first and the second laws of thermodynamics.

The engine $A$ working as a refrigerator rejects heat $Q_{1A}$ to the reservoir at temperature $T$.

Here, $$Q_{1A} + Q_1 = 0 \quad ...(i)$$

Similarly the engine $B$ working as a refrigerator rejects heat $Q_{2B}$ to the reservoir at temperature $T_2$.

Here, $$Q_{2B} + Q_2 = 0 \quad ...(ii)$$

It means that when the three engines work simultaneously, there is no outstanding change in the temperature of reservoirs at temperature $T_1$ and $T_2$.

Thus, there has been no change except for the reservoir at temperature $T_0$ and the mechanical system that supplied or absorbed work. If $Q_0$ is not equal to the sum of $Q_{0A}$ and $Q_{0B}$, the reservoir at temperature $T_0$ must have gained or lost heat. Similarly, if $W$ is not equal to the sum of $W_A$ and $W_B$, the mechanical system must have gained or lost work. According to the first law, the net flow of heat from reservoir must necessarily be equal to the amount of work done on or by the mechanical system. As far as the first law is concerned it is a matter of indifference whether the reservoir gains or loses heat as long as the mechanical system loses or gains the same quantity of work. The consideration is different with reference to the second law.

If the heat reservoir loses heat and the mechanical system gains an equal amount of work, it means that the whole of the heat absorbed from the reservoir is converted into work but this violates the Kelvin-Planck statement of the second law of thermodynamics.

Therefore, $Q_0$ must always be greater than the sum of $Q_{0A}$ and $Q_{0B}$ except in the limiting case when they are equal.

For the engine $A$,

$$\frac{Q_{1A}}{T_1} = \frac{Q_{0A}}{T_0} \quad ...(iii)$$

For the engine $B$,

$$\frac{Q_{2B}}{T_2} = \frac{Q_{0B}}{T_0} \quad ...(iv)$$

Also the quantities, $Q_1$, and $Q_2$ are positive, $Q_0$ is negative, $Q_{1A}$ and $Q_{2B}$ are negative, $Q_{0A}$ and $Q_{0B}$ are positive. Mathematically we write all the quantities as positive.

From equation (*iii*)

$$Q_{0A} = \frac{T_0(Q_{1A})}{T_1} = \frac{T_0(Q_1)}{T_1} = T_0\left(\frac{Q_1}{T_1}\right) \quad ...(v)$$

Similarly, $$Q_{0B} = \frac{T_0(Q_{2B})}{T_2} = \frac{T_0(Q_2)}{T_2} = T_0\left(\frac{Q_2}{T_2}\right) \quad ...(vi)$$

The net quantity of heat drawn from the reservoir at temperature $T_0$ is

$$Q_0 + Q_{0A} + Q_{0B}$$

According to the second law, this quantity should be less than or equal to zero.

*i.e.* $$Q_0 + Q_{0A} + Q_{0B} \leqslant 0$$

Substituting the values of $Q_{0A}$ and $Q_{0B}$

$$Q_0 + T_0\left(\frac{Q_1}{T_1}\right) + T_0\left(\frac{Q_2}{T_2}\right) \leqslant 0$$

Dividing by $T_0$

$$\frac{Q_0}{T_0} + \frac{Q_1}{T_1} + \frac{Q_2}{T_2} \leqslant 0 \qquad ...(vii)$$

In general, for a system using a large number of reservoirs, it can be written as,

$$\sum \frac{Q}{T} \leqslant 0 \qquad ...(6.47)$$

When the number of reservoir is infinite and the heat exchanges take place by infinitesimal amounts, it can be written as

$$\int \frac{\delta Q}{T} \leqslant 0 \qquad ...(6.48)$$

Equations (6.47) and (6.48) represent the **Clausius inequality**.

The relation holds good for any cyclic process, reversible or irreversible.

## 6.15 Entropy and the Second Law of Thermodynamics

Consider a closed system undergoing a reversible process from state 1 to state 2 along the path *A* and from state 2 to state 1 along the path *B* (Fig. 6.3). As this is a reversible cyclic process

$$\oint \frac{\delta Q}{T} = 0$$

$$\therefore \qquad \int_{1A}^{2A} \frac{\delta Q}{T} + \int_{2B}^{1B} \frac{\delta Q}{T} = 0 \qquad ...(i)$$

Now consider the reversible cycle from state 1 to state 2 along the path *A* and from state 2 to state 1 along the path *C*.

For this reversible cyclic process

$$\int_{1A}^{2A} \frac{\delta Q}{T} + \int_{2C}^{1C} \frac{\delta Q}{T} = 0 \qquad ...(ii)$$

From equations (*i*) and (*ii*)

$$\int_{2B}^{1B} \frac{\delta Q}{T} = \int_{2C}^{1C} \frac{\delta Q}{T} \qquad ...(iii)$$

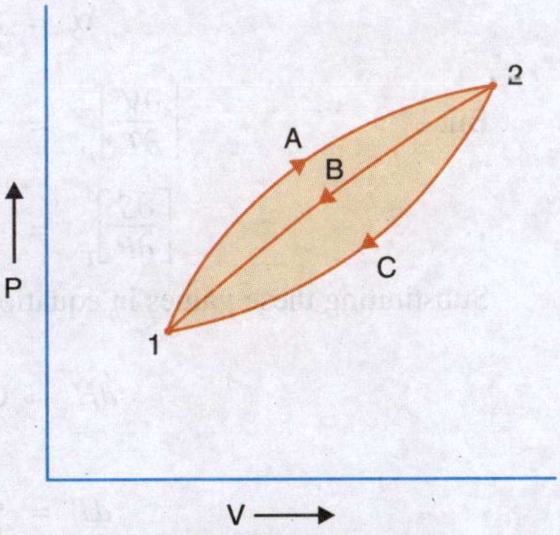

Fig. 6.3

This shows that $\int \frac{\delta Q}{T}$ has the same value for all the reversible paths from state 2 to state 1. The quantity $\int \frac{\delta Q}{T}$ is independent of the path and is a function of the end states only, therefore it is a property.

This property is called entropy. Entropy is a thermodynamical property and is defined by the relation

$$dS = \frac{\delta Q}{T}$$

or $$S_2 - S_1 = \int_1^2 \frac{\delta Q}{T} \qquad ...(6.49)$$

The quantity $S_2 - S_1$ represents the change in entropy of the system when it is changed from state 1 to state 2.

## 6.16 Joule-Kelvin Coefficient

The Joule-Kelvin coefficient is defined as the slope of the is-enthalpic curve at any point on the temperature–pressure curve

$$\mu = \left(\frac{\partial T}{\partial P}\right)_H \qquad ...(i)$$

From the definition of enthalpy

$$H = U + PV \qquad ...(ii)$$

or $$dH = dU + PdV + VdP \qquad ...(iii)$$

Also $$dU = TdS - PdV$$

∴ $$dU + PdV = TdS$$

∴ $$dH = TdS + VdP \qquad ...(iv)$$

Also $$dS = \left(\frac{\partial S}{\partial T}\right)_P dT + \left(\frac{\partial S}{\partial P}\right)_T dP$$

$$dH = T\left(\frac{\partial S}{\partial T}\right)_P dT + \left[T\left(\frac{\partial S}{\partial P}\right)_T + V\right] dP \qquad ...(v)$$

But $$T\left(\frac{\partial S}{\partial T}\right)_P = C_P \qquad ...(vi)$$

The volume coefficient $\alpha$ is given by

$$\alpha = \frac{1}{V}\left[\frac{\partial V}{\partial T}\right]_P$$

But $$\left[\frac{\partial V}{\partial T}\right]_P = -\left[\frac{\partial S}{\partial P}\right]_T = \alpha V$$

∴ $$\left[\frac{\partial S}{\partial P}\right]_T = -\alpha V \qquad ...(vii)$$

Substituting these values in equation (*v*),

$$dH = C_P dT + \left[V - T\left(\frac{\partial V}{\partial T}\right)_P\right] dP \qquad ...(viii)$$

$$dT = \frac{1}{C_P} dH + \frac{1}{C_P}\left[T\left(\frac{\partial V}{\partial T}\right)_P - V\right] dP \qquad ...(ix)$$

But $$dT = \left(\frac{\partial T}{\partial H}\right)_P dH + \left(\frac{\partial T}{\partial P}\right)_H dP \qquad ...(x)$$

Comparing the coefficients of equations (*ix*) and (*x*),

$$\left(\frac{\partial T}{\partial P}\right)_H = \frac{1}{C_P}\left[T\left(\frac{\partial V}{\partial T}\right)_P - V\right]$$

But $$\mu = \left(\frac{\partial T}{\partial P}\right)_H$$

$$\therefore \qquad \mu = \frac{1}{C_P}\left[T\left(\frac{\partial V}{\partial T}\right)_P - V\right]$$

$$\mu C_P = \left[T\left(\frac{\partial V}{\partial T}\right)_P - V\right] \qquad ...(xi)$$

$$\mu C_P = T^2 \frac{\partial}{\partial T}\left(\frac{V}{T}\right)_P \qquad ...(xii)$$

From equation (*viii*),

$$\left(\frac{\partial H}{\partial P}\right)_T = V - T\left(\frac{\partial V}{\partial T}\right)_P \qquad ...(xiii)$$

From equations (*xi*) and (*xiii*)

$$\mu\, C_P = -\left(\frac{\partial H}{\partial P}\right)_T \qquad ...(xiv)$$

The Joule-Kelvin coefficient,

$$\mu = \frac{1}{C_P}\left[T\left(\frac{\partial V}{\partial T}\right)_P - V\right] \qquad ...(xv)$$

$$\mu = \frac{V}{C_P}\left[T\left\{\frac{1}{V}\left(\frac{\partial V}{\partial T}\right)_P\right\} - 1\right] \qquad (xvi)$$

or

$$\mu = \frac{V}{C_P}[\alpha T - 1] \qquad ...(xvii)$$

For a Van der Waals gas

$$\left(P + \frac{a}{V^2}\right)(V - b) = RT$$

$$PV - Pb + \frac{a}{V} - \frac{ab}{V^2} = RT$$

Differentiating, keeping $P$ constant,

$$PdV - \frac{adV}{V^2} + \frac{2ab\,dV}{V^3} = RdT$$

or

$$\frac{dV}{dT} = \frac{R}{P - \frac{a}{V^2} + \frac{2ab}{V^3}}$$

$$\left(\frac{\partial V}{\partial T}\right)_P = \frac{R}{\frac{RT}{V-b} - \frac{a}{V^2} - \frac{a}{V^2} + \frac{2ab}{V^3}}$$

$$\left(\frac{\partial V}{\partial T}\right)_P = \frac{RV^3(V-b)}{RTV^3 - 2a(V-b)^2} \qquad ...(xviii)$$

Substituting this value in equation (*xvi*)

$$\mu = \frac{V}{C_P}\left[\frac{RTV^2(V-b)}{RTV^3 - 2a(V-b)^2} - 1\right]$$

$$= \frac{V}{C_P}\left[\frac{RTV^2(V-b)-RTV^3+2a(V-b)^2}{RTV^3-2a(V-b)^2}\right]$$

$$= \frac{V}{C_P}\left[\frac{2a(V-b)^2-RTV^2b}{RTV^3-2a(V-b)^2}\right] \quad ...(xix)$$

From equation (*xvii*), as $V$ and $C_P$ are always positive, the value of μ and hence the temperature of a Van der Waals gas passing through a porous plug will depend on the value of $\alpha T - 1$.

1. If $\alpha T > 1$, the temperature of the gas will increase.
2. If $\alpha T < 1$, the temperature of the gas will decrease.
3. If $\alpha T = 1$, the temperature will not change.

This temperature at which the inversion takes place is called the temperature of inversion $T_i$. In this case, from equation (*xvii*)

$$\mu = 0 \text{ and from equation } (xix)$$

$$2a(V-b)^2 = RT_iV^2b$$

$$\therefore \quad T_i = \frac{2a(V-b)^2}{RV^2b} \quad ...(xx)$$

If $V >> b$

$$T_i = \frac{2a}{Rb} \quad ...(xxi)$$

**Ideal Gas.** From equation (*xii*)

$$\mu C_P = T^2\left[\frac{\partial}{\partial T}\left(\frac{V}{T}\right)_P\right]$$

$$\therefore \quad \mu = \frac{T^2}{C_P}\left[\frac{\partial}{\partial T}\left(\frac{V}{T}\right)_P\right]$$

For an ideal gas

$$\frac{V}{T} = \frac{R}{P}$$

$$\therefore \quad \mu = \frac{T^2}{C_P}\cdot\frac{\partial}{\partial T}\left[\frac{R}{P}\right]_P = 0 \quad ...(6.50)$$

Thus the Joule-Kelvin coefficient for an ideal gas is zero and the ideal gas passing through the porous gas does not show any change in temperature.

## 6.17 Equilibrium Between Liquid and its Vapour

Applying Gibbs potential ($G$), it is possible to investigate the equilibrium between a liquid and its vapour or between any two phases of a substance. Consider a closed system containing a liquid in equilibrium with its saturated vapour. The temperature and pressure are equal in both the phases. As each of the phases is in equilibrium, the temperature and pressure must remain constant throughout the phase and hence the thermodynamical coordinates $V$, $S$, $U$ and $G$ will be equal to the product of the specific value and the mass of the substance in that phase.

Suppose $m_1$ and $m_2$ are the masses in the liquid and vapour phases and $g_1$ and $g_2$ are the specific values of the Gibbs potential in the two phases. Then, for the whole system

$$G = m_1g_1 + m_2g_2 \quad ...(i)$$

If a small quantity of liquid changes into vapour, differentiating equation (*i*)

$$\delta G = \delta m_1 g_1 + \delta m_2 g_2 \quad ...(ii)$$

As the change takes place at constant temperature and pressure, the process is isothermal and isobaric

$$\therefore \quad \delta G = 0$$

From equation (*ii*)

$$\delta m_1 \, g_1 + \delta m_2 \, g_2 = 0$$

But

$$\delta m_1 = -\,\delta m_2$$

$$\therefore \quad g_1 = g_2 \qquad \text{...}(iii)$$

This shows that the thermodynamical potential per unit mass will be equal in the two phases. Equation (*iii*) is applicable to the processes of evaporation, fusion and sublimation.

## 6.18 First Order Phase Transitions

The changes of phase which take place at constant temperature and pressure and in which heat is either absorbed or evolved during change of phase are called first order phase transitions. In first order phase transitions, the entropy and density (or volume) change. The Gibbs function $G$ remains constant in both phases; while its derivative with respect to temperature and pressure is discontinuous at transition point.

Consider an enclosure containing a liquid and its saturated vapour in equilibrium. If this system undergoes an isothermal, isobaric change, then

$$g_1 = g_2 \qquad \text{...}(i)$$

Let the temperature of the system be increased from $T$ to $T + dT$. For equilibrium

$$g_1 + dg_1 = g_2 + dg_2 \qquad \text{...}(ii)$$

or

$$dg_1 = dg_2 \qquad \text{...}(iii)$$

If the condition of saturation is satisfied,

$$\left(\frac{dg_1}{dT}\right)_{sat.} = \left(\frac{dg_2}{dT}\right)_{sat.} \qquad \text{...}(iv)$$

The pressure also changes from $P$ to $P + dP$,

$$\therefore \quad dg_1 = \left(\frac{\partial g_1}{\partial T}\right)_P dT + \left(\frac{\partial d_1}{\partial P}\right)_T dP \qquad \text{...}(v)$$

or

$$\left(\frac{\partial g_1}{\partial T}\right) = \left(\frac{\partial g_1}{\partial T}\right)_P + \left(\frac{\partial g_1}{\partial P}\right)_T \left(\frac{\partial P}{\partial T}\right) \qquad \text{...}(vi)$$

But, for a unit mass,

$$dg = VdP - SdT \qquad \text{...}(vii)$$

$$\therefore \quad \left(\frac{\partial g}{\partial P}\right)_T = V$$

and

$$\left(\frac{\partial g}{\partial T}\right)_P = -S$$

Substituting these values in equation (*vi*)

$$\left(\frac{\partial g_1}{\partial T}\right)_{sat.} = -S_1 + V_1 \left(\frac{\partial P}{\partial T}\right)_{sat.}$$

Similarly,

$$\left(\frac{\partial g_2}{\partial \mathrm{T}}\right)_{sat.} = -S_2 + V_2 \left(\frac{\partial P}{\partial T}\right)_{sat.}$$

Substituting these values in equation (*iv*)

$$-S_1 + V_1\left(\frac{dP}{dT}\right) = -S_2 + V_2\left(\frac{dP}{dT}\right)$$

or
$$\left(\frac{dP}{dT}\right)_{sat.} = \left(\frac{S_2 - S_1}{V_2 - V_1}\right) \quad ...(viii)$$

But
$$S_2 - S_1 = \frac{\delta Q}{T} = \frac{L}{T}$$

Here $L$ is latent heat of vaporization

$$\therefore \quad \left(\frac{dP}{dT}\right) = \frac{L}{T(V_2 - V_1)} \quad ...(ix)$$

This is **Clausius-Clapeyron latent heat equation.**

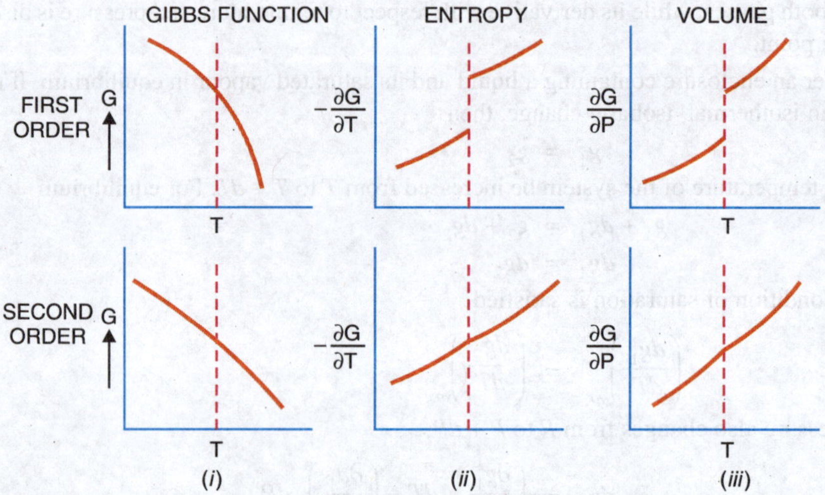


**Fig. 6.4**

This equation holds good in the first order phase transitions. In these processes, there is transference of heat and hence there is change in entropy and volume. Therefore a first order phase transition can be defined as that one in which the Gibbs function with respect to pressure and temperature change discontinuously at the transition point. However, the value of the Gibbs function is the same in both the phases at equilibrium. These changes are represented graphically in (Fig. 6.4).

**Examples:** When heat is given to water at 100°C and 1 atmospheric pressure, it changes from liquid to vapour state. The density of water is 1000 kg/m$^3$, while that of vapour is 0.6 kg/m$^3$. Therefore, *the transformation of water into vapour at constant temperature and volume is first order phase transitions. Similarly the transformation of ice into water at 0°C and 1 atmospheric pressure is an example of first order phase tronsition.*

## 6.19 Second Order Phase Transitions : Ehrenfest's Equations

Some recent investigations have revealed that during phase change there is no transference of heat and there is no change of volume. It has been found in the case of transition from liquid helium I to liquid helium II, that there is no transfer of heat and no change in volume. Such transitions are called second order phase transitions. Second order phase transitions can be defined as the phenomenon that takes place with no change in entropy and volume at constant temperature and pressure.

**Example :** The examples of second order phase transitions are:

1. Transition of liquid helium I to liquid helium II, at λ point (2.19 K).
2. Transition of a ferromagnetic material to a paramagnetic material at the Curie point.
3. Transition of a superconducting metal into an ordinary conductor in the absence of a magnetic field.
4. Order-disorder transitions in chemical compounds and alloys.

In the case of second-order phase transitions there is no discontinuity of $\left(\frac{\partial G}{\partial T}\right)_P$ and $\left(\frac{\partial G}{\partial P}\right)_T$ [Fig. 6.4]. However, the second order derivatives change dicontinuously.

For a phase transition,

$$g_1 = g_2$$

or $$g_2 - g_1 = 0 \quad \text{...}(i)$$

$$-\left(\frac{\partial g_2}{\partial T}\right)_P + \left(\frac{\partial g_1}{\partial T}\right)_P = S_2 - S_1 = 0 \quad \text{...}(ii)$$

$$\left(\frac{\partial g_2}{\partial P}\right)_T - \left(\frac{\partial g_1}{\partial P}\right)_T = V_2 - V_1 = 0 \quad \text{...}(iii)$$

Also

$$C_P = T\left(\frac{\partial S}{\partial T}\right)_P$$

$$\left(\frac{C_P}{T}\right) = \left(\frac{\partial S}{\partial T}\right)_P = \frac{\partial}{\partial T}\left[-\left(\frac{\partial g}{\partial T}\right)_P\right]_P = -\left(\frac{\partial^2 g}{\partial T^2}\right)_P \quad \text{...}(iv)$$

$$K = -\frac{1}{V}\left(\frac{\partial V}{\partial P}\right)_T$$

$$KV = -\left(\frac{\partial V}{\partial P}\right)_T = -\frac{\partial}{\partial P}\left[\left(\frac{\partial g}{\partial P}\right)_T\right]_T$$

or $$KV = -\left(\frac{\partial^2 g}{\partial P^2}\right)_T \quad \text{...}(v)$$

$$\alpha = \frac{1}{V}\left(\frac{\partial V}{\partial T}\right)_P$$

$$\alpha V = \left(\frac{\partial V}{\partial T}\right)_P = \frac{\partial}{\partial T}\left[\left(\frac{\partial g}{\partial P}\right)_T\right]_P$$

or $$\alpha V = \left(\frac{\partial^2 g}{\partial T \partial P}\right) \quad \text{...}(vi)$$

Here $K$ is the isothermal compressibility and α is the volume coefficient of expansion.

From equation (*iv*)

$$\frac{C_{P1}}{T} = -\left(\frac{\partial^2 g_1}{\partial T^2}\right)_P \quad \text{...}(vii)$$

and $$\frac{C_{P2}}{T} = -\left(\frac{\partial^2 g_2}{\partial T^2}\right)_P \qquad ...(viii)$$

Subtracting (*viii*) from (*vii*)

$$\left(\frac{\partial^2 g_2}{\partial T^2}\right)_P - \left(\frac{\partial^2 g_1}{\partial T^2}\right)_P = \frac{C_{P1} - C_{P2}}{T} \qquad ...(ix)$$

Similarly from equations (*v*) and (*vi*)

$$\left(\frac{\partial^2 g_2}{\partial P^2}\right)_T - \left(\frac{\partial^2 g_1}{\partial P^2}\right)_T = V(K_1 - K_2) \qquad ...(x)$$

and $$\left(\frac{\partial^2 g_2}{\partial T.\partial P}\right) - \left(\frac{\partial^2 g_2}{\partial T.\partial P}\right) = V(\alpha_2 - \alpha_1) \qquad ...(xi)$$

For second order phase transitions, there is no change in entropy and volume.

$$\therefore \quad S_1 = S_2 \qquad \text{(at constant temperatures } T \text{ and pressure } P\text{)}$$

$$\therefore \quad S_1 + dS_1 = S_2 + dS_2 \qquad \text{(at temp. } T + dT \text{ and pressure } P + dP\text{)}$$

$$\therefore \quad dS_1 = dS_2 \qquad ...(xii)$$

But $$dS = \left(\frac{\partial S}{\partial T}\right)_P dT + \left(\frac{\partial S}{\partial P}\right)_T dP$$

or $$dS = \left(\frac{C_P}{T}\right) dT - \left(\frac{dV}{dT}\right)_P dP$$

$$dS = \left(\frac{C_P}{T}\right) dT - V\alpha_1\, dP$$

$$\therefore \quad dS_1 = \left(\frac{C_{P1}}{T}\right) dT - V\alpha_1\, dP$$

and $$dS_2 = \left(\frac{C_{P2}}{T}\right) dT - V\alpha_2\, dP$$

$$\therefore \quad \left(\frac{C_{P1}}{T}\right) dT - V\alpha_1\, dP = \left(\frac{C_{P2}}{T}\right) dT - V\alpha_2\, dP$$

$$\therefore \quad \frac{dP}{dT} = \frac{C_{P2} - C_{P1}}{TV(\alpha_2 - \alpha_1)} \qquad ...(6.51)$$

Similarly by assuming $V_1 = V_2$ for second order phase transition,

$$V_1 = V_2 \qquad \text{(at temp. } T \text{ and pressure } P\text{)}$$

$$V_1 + dV_1 = V_2 + dV_2 \qquad \text{(at temp. } T + dT \text{ and pressure } P + dP\text{)}$$

$$\therefore \quad dV_1 = dV_2$$

But $$dV = \left(\frac{\partial V}{\partial T}\right)_P dT + \left(\frac{\partial V}{\partial P}\right)_T dP$$

$$dV = V\alpha dT - VKdP$$

$$\therefore \quad dV_1 = V\alpha_1 dT - VK_1 dP$$

and $$dV_2 = V\alpha_2 dT - VK_2 dP$$

$$\therefore \quad V\alpha_1 dT - VK_1 dP = V\alpha_2 dT - VK_2 dP$$

$$\therefore \quad \frac{dP}{dT} = \left(\frac{\alpha_2 - \alpha_1}{K_2 - K_1}\right) \quad ...(6.52)$$

Equations (6.51) and (6.52) are called **Ehrenfest's Equations.**

These equations represent the condition of equilibrium between the two phases.

## 6.20 EMF of a Reversible Cell

The first *TdS* equation is given by

$$TdS = C_V dT + T\left(\frac{\partial P}{\partial T}\right)_V dV \quad ...(i)$$

In the case of a reversible cell composed of solids and liquids in the above equation replace charge $q$ for $V$ and $EMF$, $-E$ for $P$.

$$\therefore \quad TdS = C_q\, dT - T\left(\frac{\partial E}{\partial T}\right)_q dq \quad ...(ii)$$

For a reversible cell, in which the E.M.F. depend only on temperature,

$$TdS = C_q\, dT - T\left(\frac{\partial E}{\partial T}\right)_q dq \quad ...(iii)$$

For a reversible cell, when a charge $(q_f - q_i)$ is transferred isothermally, taking $TdS = \delta Q$, we get

$$\delta Q = -T\left(\frac{\partial E}{\partial T}\right)(q_f - q_i) \quad ...(iv)$$

Here $(q_f - q_i)$ is negative when positive charge flows from positive to the negative electrode externally. The work done by the cell,

$$\delta W = -E\,(q_f - q_i) = -E\,\Delta q \quad ...(v)$$

Here $\Delta q = (q_f - q_i)$

$$\therefore \quad \delta Q = -T\left(\frac{\partial E}{\partial T}\right)\Delta q \quad ...(vi)$$

From the first law of thermodynamics

$$\delta Q = dU + \delta W$$

$$\therefore \quad dU = \delta Q - \delta W$$

$$dU = \Delta q\left[E - T\left(\frac{\partial E}{\partial T}\right)\right] \quad ...(vii)$$

For a thermodynamic process taking place at constant pressure and negligible change in volume, the change in enthalpy is obtained as follows

$$H = U + PV$$

$$dH = dU + PdV + VdP$$

But $dP = 0$ and $dV$ is negligible

$$\therefore \quad dH = dU \quad ...(viii)$$

From equation (*vii*) and (*viii*)

$$dH = \Delta q\left[E - T\left(\frac{\partial E}{\partial T}\right)\right] \quad ...(ix)$$

$dH$ represents the net transfer of heat between the final and the initial states of the reaction.

From equation (*ix*)

$$\frac{dH}{\Delta q} = E - T\left(\frac{\partial E}{\partial T}\right)$$

or
$$E = \left(\frac{dH}{\Delta q}\right) + T\left(\frac{\partial E}{\partial T}\right)$$

Let
$$\left(\frac{dH}{\Delta q}\right) = Q$$

Here $Q$ is the heat transferred per unit charge

$$\therefore \quad E = Q + T\left(\frac{\partial E}{\partial T}\right) \quad \text{...(6.53)}$$

Equation 6.53 is called **Gibbs Helmholtz equation.**

## NUMERICAL EXAMPLES

### EXAMPLE 6.1

*Show that $C_P - C_V = TE\alpha^2 V$ where $T$ is the absolute temperature, $E$ the modulus of isothermal elasticity, $\alpha$ the coefficient of volume expansion and $V$ the specific volume.*

**(Meerut 2004; Agra 2006, 2000, 2002; Garhwal 2001; Rohilkhand 2000; Kerala 2002; Utkal 2003; Lucknow 2002; Madras 2004, 2005; Purvanchal 2003; Mumbai 2005; Calcutta 2005)**

**SOLUTION.** From Maxwell's second law of thermodynamical relation

$$\left(\frac{\partial S}{\partial V}\right)_T = \left(\frac{\partial P}{\partial T}\right)_V$$

and from the definitions of $C_P$ and $C_V$ as

$$C_P = \left(\frac{\partial Q}{\partial T}\right)_P = T\left(\frac{\partial S}{\partial T}\right)_P$$

or
$$C_V = \left(\frac{\partial Q}{\partial T}\right)_V = T\left(\frac{\partial S}{\partial T}\right)_V$$

Considering $S = f(T, V)$, we write

$$dS = \left(\frac{\partial S}{\partial T}\right)_V dT + \left(\frac{\partial S}{\partial V}\right)_T dV$$

$$\therefore \quad \left(\frac{\partial S}{\partial T}\right)_P = \left(\frac{\partial S}{\partial T}\right)_V \left(\frac{\partial T}{\partial T}\right) + \left(\frac{\partial S}{\partial V}\right)_T \cdot \left(\frac{\partial V}{\partial T}\right)_P$$

But from the equation

$$\left(\frac{\partial S}{\partial V}\right)_T = \left(\frac{\partial P}{\partial T}\right)_V$$

$$\therefore \quad T\left(\frac{\partial S}{\partial T}\right)_P = T\left(\frac{\partial S}{\partial T}\right)_V + T\left(\frac{\partial P}{\partial T}\right)_V \cdot \left(\frac{\partial V}{\partial T}\right)_P$$

$$\therefore \quad C_P = C_V + T\left(\frac{\partial P}{\partial T}\right)_V \left(\frac{\partial V}{\partial T}\right)_P \quad \text{...}(i)$$

Again taking the general equation of state for a gas as $P = f(V, T)$

$$\therefore \quad dP = \left(\frac{\partial P}{\partial T}\right)_V dT + \left(\frac{\partial P}{\partial V}\right)_T dV$$

$$\frac{dP}{dT} = \left(\frac{\partial P}{\partial T}\right)_V + \left(\frac{\partial P}{\partial V}\right)_T \left(\frac{\partial V}{\partial T}\right)_P$$

At constant pressure

$$dP = 0$$

$$\therefore \quad \left(\frac{\partial P}{\partial T}\right)_V = -\left(\frac{\partial P}{\partial V}\right)_T \left(\frac{\partial V}{\partial T}\right)_P$$

Substituting this value in equation (*i*)

$$C_P - C_V = -T\left(\frac{\partial P}{\partial V}\right)_T \left(\frac{\partial V}{\partial T}\right)_P^2 \qquad ...(ii)$$

$$C_P - C_V = -T\left(\frac{\partial V}{\partial T}\right)_P^2 \left(\frac{\partial P}{\partial V}\right)_T \qquad ...(iii)$$

But, the bulk modulus of elasticity is given by

$$E = -V\left(\frac{\partial P}{\partial V}\right)_T$$

and the coefficient of volume expansion is

$$\alpha = \frac{1}{V}\left(\frac{\partial V}{\partial T}\right)_P$$

Substituting these values in equation (*iii*), we get

$$C_P - C_V = -T E \alpha^2 V \qquad ..(iv)$$

**EXAMPLE 6.2**

*Deduce the second latent heat equation of Clausius*

$$C_2 - C_1 = \frac{dL}{dT} - \frac{L}{T}$$

*were $C_1$ and $C_2$ represent the specific heat of a liquid and its saturated vapour and L is the latent heat of the vapour.*

**(Meerut 2000, M.D.U. 2001, Kerala 2000, Nagpur 2003, Purvanchal 2003, 2004)**

**SOLUTION.** For a change of state from liquid to vapour

$$S_2 - S_1 = \frac{L}{T} \qquad ...(i)$$

Here $S_1$ and $S_2$ are the entropies in the liquid and vapour states respectively.

Differentiating equation (*i*) with respect to $T$

$$\frac{dS_2}{dT} - \frac{dS_1}{dT} = -L\frac{1}{T^2} + \frac{1}{T}\cdot\frac{dL}{dT}$$

$$T\left(\frac{dS_2}{dT}\right) - T\left(\frac{dS_1}{dT}\right) = -\frac{L}{T} + \frac{dL}{dT}$$

$$\therefore \quad C_2 - C_1 = \frac{dL}{dT} - \frac{L}{T} \qquad ...(ii)$$

This equation is known as Clausius latent heat equation.

**EXAMPLE 6.3**

*Calculate under what pressure water would boil at 150°C if the change in specific volume when 1 gram of water is converted into steam is 1676 CC. Given latent heat of vaporization of steam = 540 cal per gram; $J = 4.2 \times 10^7$ erg/cal and one atmosphere pressure = $10^6$ dynes/cm$^2$.*

**SOLUTION.** Here

$$\delta Q = 540 \text{ cal} = 540 \times 4.2 \times 10^7 \text{ erg}$$
$$\partial V = 1676 \text{ CC}$$
$$T = 373 \text{ K and } 150°\text{C} = 423 \text{ K}$$
$$\partial T = 423 - 373 = 50 \text{ K}$$
$$\partial P = 1$$

using these values in Maxwell's second thermodynamical relation

$$\left(\frac{\partial S}{\partial V}\right)_T = \left(\frac{\partial P}{\partial T}\right)_V$$

$$T\left(\frac{\partial S}{\partial V}\right)_T = T\left(\frac{\partial P}{\partial T}\right)_V$$

$$\left(\frac{\partial Q}{\partial V}\right)_T = T\left(\frac{\partial P}{\partial T}\right)_V$$

$$\frac{540 \times 4.2 \times 10^7}{1676} = 373 \times \left(\frac{\partial P}{50}\right)$$

$$\delta P = \frac{540 \times 4.2 \times 10^7 \times 50}{373 \times 1676}$$

$$= \mathbf{1.814 \times 10^6 \text{ dynes/cm}^2}$$

$$= \mathbf{1.814 \text{ atmospheres}}$$

Therefore, the pressure at which water would boil at 150°C

$$= 1.8814 + 1.000$$

$$= \mathbf{2.814 \text{ atmospheres}}$$

### EXAMPLE 6.4

*Calculate the specific heat of saturated steam given that the specific heat of water at 100°C = 1.01 and latent heat of vaporization decreases with size in temperature at the rate of 0.64 cal/K. Latent heat of vaporization of steam is 540 cal.* **(Kanpur 2000)**

**SOLUTION.** Here

$$C_1 = 1.01, \quad C_2 = ? \quad T = 373 \text{ K}$$

$$\frac{dL}{dT} = -0.64 \text{ cal/K}$$

$$C_2 - C_1 = \frac{dL}{dT} - \frac{L}{T}$$

$$C_2 = C_1 + \frac{dL}{dT} - \frac{L}{T}$$

$$= 1.01 + (-0.64) - \frac{540}{373}$$

$$= \mathbf{-1.077 \text{ cal/g}}$$

### EXAMPLE 6.5

*Calculate the specific heat of saturated steam at 100°C from the following data:*

$$L \text{ at } 90°\text{C} = 545.25 \text{ cal}$$
$$L \text{ at } 100°\text{C} = 539.30 \text{ cal}$$
$$L \text{ at } 110°\text{C} = 533.17 \text{ cal}$$

*specific heat of water at 100°C = 1.013 cal/g* **(Kerala 2001, 2002)**

**SOLUTION.** Here

$$\frac{dL}{dT} = \frac{533.17 - 545.25}{20}$$

$$= \frac{-12.08}{20} = -0.604 \text{ cal/K}$$

$$L = 539.3 \text{ cal}$$

$$T = 373 \text{ K}$$

$$C_1 = 1.013 \;,\; C_2 = ?$$

$$C_2 - C_1 = \frac{dL}{dT} - \frac{L}{T}$$

$$\therefore \quad C_2 = C_1 + \frac{dL}{dT} - \frac{L}{T}$$

$$= 1.013 - 0.604 - \frac{539.3}{373}$$

$$\mathbf{C_2 = -1.036 \text{ cal/g}}$$

## EXAMPLE 6.6

*Calculate under what pressure ice freezes at 272 K if the change in specific volume when 1 kg of water freezes is $91 \times 10^{-6}$ m$^3$. Given latent heat of ice = $3.36 \times 10^5$ Jkg$^{-1}$.*

**(Calcutta 2000)**

**SOLUTION.** Latent heat of ice $L = 3.36 \times 10^5$ J kg$^{-1}$

Freezing point of ice under normal pressure $T = 273$ K, change in freezing point, $dT = (273 - 272) = 1$ K

Change in specific volume = $(V_2 - V_1) = 91 \times 10^{-6}$ m$^3$

According to Clapeyron's latent heat equation

$$\frac{dP}{dT} = \frac{L}{T(V_2 - V_1)}$$

$$\therefore \quad dP = \frac{dT \times L}{T(V_2 - V_1)} = \frac{1 \times 3.36 \times 10^5}{273 \times 91 \times 10^{-6}} = 1.352 \times 10^7 \text{ N/m}^2$$

Taking 1 Atm = $10^5$ N, we have

$$dP = \frac{1.352 \times 10^7}{10^5} = 135.2 \text{ Atm.}$$

∴ Pressure under which ice would freeze at 272 K.

$$= 1 + 135.2$$

$$= \mathbf{136.2 \text{ Atm.}}$$

## EXAMPLE 6.7

*Calculate the change in boiling point of water when the pressure is increased by 1 atmosphere. Boiling point of water is 373 K. Specific volume of steam = 1.671 m$^3$ kg$^{-1}$ and latent heat of steam $2.268 \times 10^6$ J kg$^{-1}$.* **(Agra 2002, Kanpur 2001)**

**SOLUTION.** Specific volume of steam $V_2 = 1.671$ m$^3$ kg$^{-1}$, specific volume of water $V_1 = 1 \times 10^{-3}$ m$^3$ kg$^{-1}$ = 0.001 m$^3$ kg$^{-1}$

$$\therefore \quad V_2 - V_1 = (1.671 - 0.001) = 1.670 \text{ m}^3 \text{ kg}^{-1}$$

Boiling point of water $T = 373$ K

Latent heat of steam = $2.268 \times 10^6$ J kg$^{-1}$

$$\text{Now} \quad \frac{dP}{dT} = \frac{L}{T(V_2 - V_1)} = \frac{2.268 \times 10^6}{373 \times 1.670} = 3.64 \times 10^3 \text{ N/m}^2\text{K}$$

Increase in pressure $dP$ = 1 Atm = $10^5$ N/m$^2$

$$\therefore \quad dT = \frac{10^5}{3.64\times10^{-3}}$$

$$dT = \mathbf{27.47\ K}$$

**Example 6.8**

*Calculate the depression of melting point of ice produced by one atmosphere increase of pressure. Given that latent heat of ice = 80 cal/gm and specific volume of ice and water at 0°C are 1.091 cm³ and 1.0 cm³ respectively.*

**(Nagpur 2009, 2007, 2001, Lucknow 2001, Kanpur 1995, Purvanchal 2004, 02, Kuvempa 2005)**

**Solution.** Increase of pressure = $dP$ = 1 × 76 × 13.6 × 981 dynes/cm$^2$

Latent heat $L$ = 80 cal = 80 × 4.2 × $10^7$ ergs

Change in volume when 1 gm of ice melts = $V_2 - V_1$

$$= (1.091 - 1.0) = 0.091 \text{ cm}^3$$

Melting point of ice $T$ = 0°C = 273 K

Let $dT$ be the depression of melting point, then

$$\frac{dP}{dT} = \frac{L}{T(V_2 - V_1)}$$

or
$$\frac{76\times13.6\times981}{dT} = \frac{80\times4.2\times10^7}{273\times0.091}$$

$$\therefore \quad dT = \frac{76\times13.6\times981\times273\times0.091}{80\times4.2\times10^7}$$

$$= \mathbf{0.007497\ K}$$

$$= \mathbf{0.007497\ °C}$$

**Example 6.9**

*Find the increase in the boiling point of water at 100°C when the pressure is increased by one atmosphere. Latent heat of vaporisation of steam is 540 cal/gram and 1 gram of steam occupies a volume of 1677 cm³.* **(Madras 2001)**

**Solution.**

$$dP = 76 \times 13.6 \times 980 \text{ dynes/cm}^2$$

$$T = 100 + 273$$

$$= 373 \text{ K}$$

$$L = 540 \times 4.2 \times 10^7 \text{ ergs}$$

$$V_1 = 1.00 \text{ cm}^3$$

$$V_2 = 1677 \text{ cm}^3$$

$$\frac{dP}{dT} = \frac{L}{T(V_2 - V_1)}$$

$$dT = \frac{dP\times T(V_2 - V_1)}{L}$$

$$= \frac{76\times13.6\times980\times373\times1676}{540\times4.2\times10^7}$$

$$= 27.92°\text{C}$$

Therefore, the increase in the boiling point of water with an increase in pressure of one atmosphere

$$= \mathbf{27.92°C}$$

$$= \mathbf{27.92\ K}$$

### EXAMPLE 6.10

*Prove the Clausius inequality* $\sum \frac{Q}{T} \leqslant 0$ *with examples.* **(Rajasthan 2000)**

**SOLUTION. 1.** Consider two blocks at temperature 300 K and 200 K that are in contact. The block at higher temperature conducts heat to the block at the lower temperature. Suppose, 200 joules of heat is transferred.

$$Q_1 = 200 \text{ joules}; \qquad T_1 = 300 \text{ K}$$

$$Q_2 = -200 \text{ joules}, \qquad T_2 = 200 \text{ K}$$

$$\therefore \quad \sum \frac{Q}{T} = \frac{Q_1}{T_1} + \frac{Q_2}{T_2}$$

$$= \frac{200}{300} + \frac{(-200)}{200} = -\frac{1}{3}$$

$$= \mathbf{-0.33} \text{ joule/degree}$$

**2.** Consider an actual heat engine working between the temperatures 500 K and 300 K.

Suppose its efficiency is 20% and it takes 1000 joules of heat from the high temperature reservoir. As the efficiency is 20%.

$$\eta = 1 - \frac{Q_2}{Q_1}$$

$$\frac{20}{100} = 1 - \frac{Q_2}{100}$$

or

$$Q_2 = 800 \text{ joules}$$

$$\therefore \quad \sum \frac{Q}{T} = \frac{Q_1}{T_1} + \frac{Q_2}{T_2}$$

Here

$$Q_1 = +1000 \text{ joules}$$

$$Q_2 = -800 \text{ joules (since heat is rejected)}$$

$$T_1 = 500 \text{ K}$$

$$T_2 = 300 \text{ K}$$

$$\sum \frac{Q}{T} = \frac{1000}{500} + \frac{-800}{300}$$

$$= -\frac{2}{3} \text{ joule/degree}$$

**3.** Consider a Carnot's reversible engine working between the temperatures 500 K and 300 K. Suppose 1000 joules of heat energy is drawn from the high temperature reservoir.

Here

$$\frac{Q_1}{T_1} = \frac{Q_2}{T_2}$$

$$\frac{1000}{500} = \frac{Q_2}{300}$$

or

$$Q_2 = 600 \text{ joules}$$

$$\sum \frac{Q}{T} = \frac{Q_1}{T_1} + \frac{Q_2}{T_2}$$

Here

$$Q_i = +1000 \text{ joules}$$

$$Q_2 = -600 \text{ joules}$$

$$T_1 = 500 \text{ K}$$

$$T_2 = 300 \text{ K}$$

$$\therefore \quad \sum \frac{Q}{T} = \frac{1000}{500} + \frac{(-600)}{300}$$

or

$$\sum \frac{Q}{T} = 0$$

This example shows $\sum \frac{Q}{T} = 0$, only in the limiting case and in no case $\sum \frac{Q}{T}$ is greater than zero.

### EXAMPLE 6.11

*Calculate the change in temperature of boiling water when the pressure is increased by 27.12 mm of Hg. The normal boiling point of water at atmospheric pressure is 100°C.*

*Latent heat of steam* = *537* cal/g

*and specific volume of steam* = *1674* cm³ **(Bangalore 2002)**

**SOLUTION.**

$$dP = 2.712 \times 13.6 \times 980 \text{ dynes/cm}^2$$

$$T = 100 + 273 = 373 \text{ K}$$

$$L = 537 \times 4.2 \times 10^7 \text{ ergs}$$

$$V_1 = 1.000 \text{ cm}^3$$

$$V_2 = 1674 \text{ cm}^3$$

$$\frac{dP}{dT} = \frac{L}{T(V_2 - V_1)}$$

$$dT = \frac{dP \times T\,(V_2 - V_1)}{L}$$

$$= \frac{2.712 \times 13.6 \times 980 \times 373 \times 1673}{537 \times 4.2 \times 10^7}$$

$$= \mathbf{1°C \text{ or } 1 \text{ K}}$$

### EXAMPLE 6.12

*Calculate the change in the melting point of naphthalene for one atmosphere rise in pressure given that its melting point is 80°C. Latent heat of fusion is 4563 cal/mol and increase in volume on fusion is 18.7 cm³ per* mol *and 1* calorie = *4.2* × *10⁷ ergs.*

**(Meerut 2001, Kanpur 2000, Punjab 2002)**

**SOLUTION.**

$$\frac{dP}{dT} = \frac{L}{T\,(V_2 - V_1)}$$

$$dP = 1 \text{ atmosphere}$$

$$= 76 \times 13.6 \times 980 \text{ dynes/cm}^2$$

$$dT = ?$$

$$L = 4563 \text{ cal/mol}$$

$$= 4563 \times 4.2 \times 10^7 \text{ ergs/mol}$$

$$V_2 - V_1 = 18.7 \text{ cm}^3\text{/mol}$$

$$T = 80 + 273$$

$$= 353 \text{ K}$$

$$dT = \frac{dP.T.(V_2 - V_1)}{L}$$

$$= \frac{76 \times 13.6 \times 980 \times 353 \times 18.7}{4563 \times 4.2 \times 10^7}$$

$$= \mathbf{0.03488 \text{ K}}$$

Therefore, the increase in the melting point of naphthalene with an increase in pressure of one atmosphere

$$= \mathbf{0.03488\ K}$$
$$= \mathbf{0.03488°C}$$

**EXAMPLE 6.13**

*Calculate the boiling point of benzene under a pressure of 80 cm of mercury. The normal boiling point is 80°C. Latent heat of vaporization is 380 joules/g, density of vapour at the boiling point is 4 g/litre and that of the liquid 0.9 g/cm³.* **(Punjab 2000)**

**SOLUTION.** Here

$$dP = 80 - 76$$
$$= 4 \text{ cm of Hg}$$
$$= 4 \times 13.6 \times 980 \text{ dynes/cm}^2$$
$$T = 80°C = 80 + 273$$
$$= 353 \text{ K}$$
$$L = 980 \text{ joules/g}$$
$$= 380 \times 10^7 \text{ ergs/g}$$
$$V_1 = \frac{1}{0.9} = 1.11 \text{ cm}^3$$
$$V_2 = \frac{1{,}000}{4} = 250 \text{ cm}^3$$
$$\frac{dP}{dT} = \frac{L}{T[V_2 - V_2]}$$
$$dT = \frac{dP \times T[V_2 - V_1]}{L}$$
$$dT = \frac{4 \times 13.6 \times 980 \times 353[250 - 1.11]}{380 \times 10^7}$$
$$= \mathbf{1.233\ K}$$
$$= \mathbf{1.233°C}$$

∴ The boiling point of benzene at a pressure of 80 cm of Hg

$$= 80 + 1.233$$
$$= \mathbf{81.233°C}$$

**EXAMPLE 6.14**

*Calculate the change in the boiling point of water when the pressure of steam on its surface is increased from 1 atmosphere to 1.10 atmospheres.* **(Madurai 2001)**

*Latent heat of water at 100°C*

$$= 537 \text{ cal/g}$$

*Volume of one gram of steam at 100°C*

$$= 1{,}676 \text{ cm}^3$$

**SOLUTION.**

$$dP = 1.10 - 1$$
$$= 0.10 \text{ atmosphere}$$
$$dP = 0.10 \times 76 \times 13.6 \times 980 \text{ dynes/cm}^2$$
$$T = 100 + 273$$
$$= 373 \text{ K}$$

$$L = 537 \text{ cals/g}$$
$$= 537 \times 4.2 \times 10^7 \text{ ergs/g}$$
$$V_1 = 1.00 \text{ cm}^3$$
$$V_2 = 1{,}676 \text{ cm}^3$$
$$\frac{dP}{dT} = \frac{L}{T[V_2 - V_1]}$$
$$dT = \frac{dP \times T[V_2 - V_1]}{L}$$
$$dT = \frac{0.1 \times 76 \times 13.6 \times 980 \times 373 \times (1{,}675)}{537 \times 4.2 \times 10^7}$$
$$= \mathbf{2.792 \text{ K}}$$
$$= \mathbf{2.792°C}$$

Therefore, the increase in the boiling point of water with an increase of 0.1 atmosphere pressure

$$= \mathbf{2.792 \text{ K}}$$
$$= \mathbf{2.792°C}$$

**EXAMPLE 6.15**

*Calculate the change in the melting point of ice when it is subjected to a pressure of 100 atmospheres.* **(Bundelkhand 2001)**

*Density of ice* $= 0.917 \text{ g/cm}^3$ *and*

*Latent heat of ice* $= 336 \text{ J/g}$

**SOLUTION.**

$$\frac{dP}{dT} = \frac{L}{T(V_2 - V_1)}$$
$$dP = 100 - 1$$
$$= 99 \text{ atmospheres}$$
$$dP = 99 \times 76 \times 13.6 \times 980 \text{ dynes/cm}^2$$
$$L = 336 \text{ J/g}$$
$$= 336 \times 10^7 \text{ ergs/g}$$
$$T = 273 \text{ K}$$
$$(V_2 - V_1) = 1 - \frac{1}{0.917}$$
$$= -\frac{0.083}{0.917}$$
$$= -0.091 \text{ cm}^3$$
$$\therefore \quad dT = \frac{T\,dP(V_2 - V_1)}{L}$$
$$dT = \frac{273 \times 99 \times 76 \times 13.6 \times 980 \times (-0.091)}{336 \times 10^7}$$
$$dT = -0.7326 \text{ K}$$
$$= \mathbf{-0.7326°C}$$

The decrease in the melting point of ice with a pressure of 100 atmospheres

$$= \mathbf{0.7326°C}$$

**EXAMPLE 6.16**

*Water boils at a temperature of 101°C at a pressure of 787 mm of* Hg. *1 gram of water occupies 1,601 cm³ on evaporation. Calculate the latent heat of steam, J = 4.2 × 10⁷* ergs/cal.

**(Nagpur 2001)**

**SOLUTION.**

$$\frac{dP}{dT} = \frac{L}{T(V_2 - V_1)}$$

$$dP = 787 - 760$$
$$= 27 \text{ mm of Hg}$$
$$= 2.7 \text{ cm of Hg}$$
$$= 2.7 \times 13.6 \times 980 \text{ dynes/cm}^2$$
$$dT = 1°\text{C} = 1 \text{ K}$$
$$T = 373 \text{ K}$$
$$V_2 - V_1 = 1{,}601 - 1 = 1{,}600 \text{ cm}^3$$
$$L = ?$$
$$L = \frac{T\, dP(V_2 - V_1)}{dT}$$
$$L = \frac{373 \times 2.7 \times 13.6 \times 980 \times 1{,}600}{1} \text{ ergs/g}$$
$$L = \frac{373 \times 2.7 \times 13.6 \times 980 \times 1{,}600}{4.2 \times 10^7} \text{ cal/g}$$
$$L = \mathbf{511.3 \text{ cal/g}}$$

**EXAMPLE 6.17**

*When lead is melted at atmospheric pressure, (the melting point is 600 K) the density decreases from 11.01 to 10.65 g/cm³ and the latent heat of fusion is 24.5 J/g. What is the melting point at a pressure of 100 atmospheres?*

**(Bangalore 2001)**

**SOLUTION.**

$$\frac{dP}{dT} = \frac{L}{T(V_2 - V_1)}$$

$$dP = 99 \text{ atmospheres}$$
$$dP = 99 \times 76 \times 13.6 \times 980 \text{ dynes/cm}^2$$
$$L = 24.5 \text{ J/g}$$
$$= 24.5 \times 10^7 \text{ ergs/g}$$
$$V_1 = \frac{1}{11.01} \text{ cm}^3$$
$$V_2 = \frac{1}{10.65} \text{ cm}^3$$
$$V_2 - V_1 = \frac{1}{10.65} - \frac{1}{11.01}$$
$$= \left(\frac{0.36}{10.65 \times 11.01}\right) \text{cm}^3$$
$$dT = \frac{T\,dP\,(V_2 - V_1)}{L}$$
$$dT = \frac{600 \times 99 \times 76 \times 13.6 \times 980 \times 0.36}{10.65 \times 1.01 \times 24.5 \times 10^7}$$

$$dT = \mathbf{0.7539\ K}$$
$$= \mathbf{0.7539°C}$$

∴ Melting point of lead at 100 atmospheres pressure

$$= 600 + 0.7539$$
$$= \mathbf{600.7539\ K}$$

## Example 6.18

*Calculate under what pressure water will boil at 120°C if the change in specific volume when 1 gram of water is converted into steam is 1,676 cm$^3$.* **(Nagpur 2009, Delhi 2001)**

*Latent heat of steam* $= 540$ cal/g

$J = 4.2 \times 10^7$ *ergs/cal*

*1 atmosphere pressure* $= 10^6$ *dynes/cm$^2$*

**Solution.** Here

$$\frac{dP}{dT} = \frac{L}{T(V_2 - V_1)}$$
$$dT = 120 - 100$$
$$= 20\text{ K}$$
$$T = 373\text{ K}$$
$$(V_2 - V_1) = 1{,}676\text{ cm}^3$$
$$L = 540\text{ cal/g}$$
$$= 540 \times 4.2 \times 10^7\text{ ergs/g}$$
$$dP = ?$$
$$dP = \frac{L.dT}{T(V_2 - V_1)}$$
$$dP = \frac{540 \times 4.2 \times 10^7 \times 20}{373 \times 1{,}676}\text{ dynes/cm}^2$$
$$dP = \frac{540 \times 4.2 \times 10^7 \times 20}{373 \times 1{,}676 \times 10^6}\text{ atmospheres}$$
$$dP = 0.7254\text{ atmosphere}$$

Pressure required $= 1 + 0.7254$

$$= \mathbf{1.7254\ atmospheres}$$

## Example 6.19

*Calculate the pressure required to lower the melting point of ice by 1°C. Given L = 79.6 cal/g. Specific volumes of water at 0°C = 1 cm$^3$.*
*Specific volume of ice, at 0°C = 1.091 cm$^2$. 1 atmospheric pressure = 1.013 × 10$^6$ dyne/cm$^2$.* **(Mumbai 2001)**

**Solution.**

$$\frac{dP}{dT} = \frac{L}{T(V_2 - V_1)}$$
$$dT = -1\text{ K}$$
$$T = 273\text{ K}$$
$$V_2 - V_1 = -0.091\text{ cm}^3$$

$$L = 79.6 \text{ cal/g}$$
$$= 79.6 \times 4.18 \times 10^7 \text{ ergs/g}$$
$$dP = \frac{L\,dT}{T(V_2 - V_1)} = \frac{79.6 \times 4.18 \times 10^7 \times 1}{273 \times 0.091} \text{ dynes/cm}^2$$
$$dP = 1.33932 \times 10^8 \text{ dynes/cm}^2$$
$$dP = \frac{1.33932 \times 10^8}{1.013 \times 10^6} \text{ atmospheres}$$
$$dP = \mathbf{132.21} \text{ atmospheres}$$

∴ Pressure required $= 132.21 + 1$

$= \mathbf{133.21}$ atmospheres

## EXAMPLE 6.20

*Calculate the temperature at which ice will freeze if the pressure is increased by 135.2 atmospheres. Given that the increase in the specific volume when 1 gram of water freezes into ice is 0.091 cm³. Latent heat of fusion of ice is 80 cal/gram.* **(Agra 2001)**

**SOLUTION.**
$$\frac{dP}{dT} = \frac{L}{T(V_2 - V_1)}$$

Here
$$dP = 135.2 \text{ atmospheres}$$
$$dP = 135.2 \times 1.013 \times 10^6 \text{ dynes/cm}^2$$
$$V_2 - V_1 = -0.091 \text{ cm}^3$$
$$L = 80 \text{ cal/g}$$
$$L = 80 \times 4.18 \times 10^7 \text{ ergs/g}$$
$$T = 273 \text{ K}$$

∴
$$dT = \frac{dP \, . \, T(V_2 - V_1)}{L}$$
$$dT = \frac{135.2 \times 1.013 \times 10^6 \times 273 \times (-0.091)}{80 \times 4.18 \times 10^7}$$
$$dT = -1 \text{ K}$$

or
$$dT = -1°\text{C}$$

Therefore, the temperature at which ice will freeze is

$= \mathbf{-1°C \text{ or } 272 K.}$

## EXAMPLE 6.21

*If the initial temperature is 27°C, calculate the drop in temperature produced by adiabatic throttling in the case of oxygen when the pressure is reduced by 50 atmospheres. Given*

$$a = 13.2 \times 10^{-2} \text{ Nm}^4 \text{ (mole)}^{-2}$$
$$b = 31.2 \times 10^{-6} \text{ m}^3 \text{ (mole)}^{-1}$$
$$R = 8.3 \text{ JK}^{-1} \text{ (mole)}^{-1}$$
$$C_P = 3.5\,R$$

**(Kanpur 2002)**

**SOLUTION.**
$$T = \left(\frac{\delta P}{JMC_P}\right)\left(\frac{2a}{RT} - b\right)$$

Here
$$\delta P = 50 \text{ atmospheres}$$
$$= 50 \times 0.76 \times 13.6 \times 10^3 \times 9.8 \text{ N/m}^2$$

$$J = 4.2 \text{ J/cal}$$
$$M = 32$$
$$C_P = 3.5\,R = 3.5 \times 8.3 \text{ JK}^{-1} \text{ (mole)}^{-1}$$
$$T = 300 \text{ K}°$$
$$\delta T = \left(\frac{50\times0.76\times13.6\times10^3\times9.8}{4.2\times32\times3.5\times8.3}\right) \times \left(\frac{2\times13.2\times10^{-2}}{8.3\times300} - 31.2\times10^{-6}\right)$$
$$\mathbf{\delta T = 0.0969 \text{ K}°}$$

**EXAMPLE 6.22**

*Water boils at a temperature of 101°C at a pressure of 787 mm of Hg. 1 gram of water occupies 1601 cm³ on evaporation. Calculate the lalent heat of steam. (J = 4.42 × 10⁷ ergs/cal)*

**(Kerala 2000)**

**SOLUTION.**
$$\frac{dP}{dT} = \frac{L}{T(V_2 - V_1)}$$
$$dP = 787 - 760 = 27 \text{ mm of Hg}$$
$$= 2.7 \text{ cm of Hg}$$
$$= 2.7 \times 13.6 \times 980 \text{ dynes/cm}^2$$
$$dT = 1°\text{C} = 1 \text{ K}$$
$$T = 373 \text{ K}$$
$$V_2 - V_1 = 1601 - 1 = 1660 \text{ cm}^3$$
$$L = ?$$
$$L = \frac{TdP(V_2 - V_1)}{dT}$$
$$= \frac{373\times2.7\times13.6\times980\times1600}{1} \text{ ergs/g}$$
$$= \mathbf{2147 \times 10^7} \text{ ergs/g}$$
$$L = \frac{2147.6\times10^7}{4.2\times10^7}$$
$$= \mathbf{511.33 \text{ cal/g}}$$

**EXAMPLE 6.23**

*For a metallic copper disc at 300 K, the following values are known*

**(Madras 2001)**

$$C_P = 24.5 \text{ J/mol K}$$
$$\alpha = 50.4 \times 10^{-6} \text{ K}^{-1}$$
*Isothermal compressibility,* $K = 7.78 \times 10^{-12} \text{ N/m}^2$
$$V = 7.06 \text{ cm}^3\text{/mol}$$

*Determine* $C_V$.

**SOLUTION.**
$$C_P - C_V = TE\alpha^2 V$$

and
$$E = \frac{1}{K}$$

$$C_P - C_V = \frac{TV\alpha^2}{K}$$

$$V = 7.06 \text{ cm}^3/\text{mol} = 7.06 \times 10^{-6} \text{ m}^3/\text{mol}$$

$$= \frac{300 \times 7.06 \times 10^{-6} \times (50.4 \times 10^{-6})^2}{7.78 \times 10^{-12}}$$

$$C_P - C_V = 0.6915 \text{ J/mol-K}$$

$$C_V = C_P - 0.6915$$

$$C_V = 24.5 - 0\ 0.6915$$

$$C_V = \mathbf{23.8085} \text{ J/mol-K}$$

**EXAMPLE 6.24**

*Calculate the drop in temperature produced by adiabatic throttling process in the case of oxygen when the pressure is reduced by 50 atmospheres. Initial temperature of the gas is 27°C. Given that the gas obeys Van der Waals' equation and*

$$a = 1.32 \times 10^{12} \text{ cm}^4 \text{ dynes/mole}^2$$

$$b = 31.2 \text{ cm}^3\text{/mole}$$

*and* $C_P = 7$ *cal/mole-K* **(Punjab 2002)**

**SOLUTION.** Here

$$\frac{\delta T}{\delta P} = \frac{1}{C_P}\left[\frac{2a}{RT} - b\right]$$

$$\delta T = \frac{\delta P}{C_P}\left[\frac{2a}{RT} - b\right]$$

$$\delta P = 50 \text{ atmospheres}$$

$$= 50 \times 76 \times 13.6 \times 980 \text{ dynes/cm}^2$$

$$C_P = 7 \text{ cal/mole-K}$$

$$= 7 \times 4.2 \times 10^7 \text{ ergs/mole-K}$$

$$T = 27°\text{C} = 300 \text{ K}$$

$$R = 8.31 \times 10^7 \text{ ergs/mole-K}$$

$$\therefore \quad \delta T = \left(\frac{50 \times 76 \times 13.6 \times 980}{7 \times 4.2 \times 10^7}\right)\left(\frac{2 \times 1.32 \times 10^{12}}{8.31 \times 10^7 \times 300} - 31.2\right)$$

$$\delta T = \left(\frac{0.5064}{7 \times 4.2}\right)(317.7 - 31.2)$$

$$\delta T = \mathbf{4.935 \text{ K}}$$

$$= \mathbf{4.935°C}$$

**EXAMPLE 6.25**

*In a cold storage ice melts at the rate of 3 kg per hour when the external temperature is 28°C. Find the minimum power output of the motor used to run the refrigerator just to prevent melting of ice.*

*Specific latent heat of ice fusion = 3.3 × 10⁵ J kg⁻¹* **(Calcutta 2002)**

**SOLUTION.**

$$m = 3 \text{ kg}$$

$$L = 3.3 \times 10^5 \text{ J kg}^{-1}$$

$$s = 4.2 \times 10^3 \text{ J kg}^{-1} - °\text{C}$$

Heat taken by ice to melt into water and reach a temperature of 28°C in one hour.

$$Q = mL + ms\,\theta$$
$$= 3 \times 3.33 \times 10^5 + 3 \times 4.2 \times 10^3 \times 28$$
$$= 1.343 \times 10^6 \text{ J}$$

To prevent melting of ice, the refrigerator should have the power out put,

$$P = 1.343 \times 10^6 \text{ J/hour}$$
$$= \frac{1.343 \times 10^6}{3600} \text{ J/s}$$
$$= \mathbf{373 \text{ Watts.}}$$

**EXAMPLE 6.26**

*Calculate the depression of melting point of ice by one atmosphere increase of pressure, given, latent heat of ice = 3.35 × $10^5$ J/kg and the specific volumes of 1 kg of ice and water at 0°C are 1.090 × $10^{-3}$ $m^3$ and $10^{-3}$ $m^3$ respectively.* **(Nagpur 2007, Kerala 2000)**

**SOLUTION.** Here,

$$L = 3.35 \times 10^5 \text{ J/kg}$$
$$T = 273 \text{ K}$$
$$V_1 = 1.090 \times 10^{-3} \text{ m}^3$$
$$V_2 = 10^{-3} \text{ m}^3$$
$$dP = 1 \text{ atmosphere}$$
$$= 0.76 \times 13.6 \times 10^3 \times 9.8$$
$$= 1.01 \times 10^5 \text{ N/m}^2$$

$$\frac{dP}{dT} = \frac{L}{T(V_2 - V_2)}$$

$$dT = \frac{dP \times T(V_2 - V_1)}{L}$$

$$= \frac{1.01 \times 10^5 \times 273\,(10^{-3} - 1.09 \times 10^{-3})}{3.35 \times 10^5}$$

$$= \mathbf{-7.4 \times 10^{-3} \text{ K}}$$
$$= \mathbf{-7.4 \times 10^{-3}\ °C}$$

Therefore, decrease in melting point of ice with an increase in pressure of one atmosphere.

**EXAMPLE 6.27**

*Considering mercury at 0°C and 1 atmospheric pressure, determine $\gamma\,(C_P/C_V)$. Given $C_P$ = 28.0 J/mol K, specific volume V = 1.47 × $10^{-5}$ $m^3$/mol, volume expansivity β = 1.81 × $10^6$ $K^{-1}$ and compressibility is 3.94 × $10^{-6}$ $atm^{-1}$ (3.89 × $10^{-11}$ $Pa^{-1}$).* **(Alagappa 2001)**

**SOLUTION.**

$$C_P - C_V = \frac{TV\beta^2}{k}$$

Here,

$$T = 273 \text{ K}, V = 1.47 \times 10^{-5} \text{ m}^3\text{/mol}$$
$$\beta = 1.81 \times 10^6 \text{ K}^{-1}$$
$$k = 3.89 \times 10^{-11} \text{ Pa}^{-1}$$

$$\therefore \quad \frac{TV\beta^2}{k} = \frac{273 \times 1.47 \times 10^{-5} \times (1.81 \times 10^{-6})}{3.89 \times 10^{-11}}$$

$$= 338 \times 10^{1-6}$$

$$\therefore \quad C_V = C_P - 338 \times 10^{-6} = 28 - 338 \times 10^{-6}$$
$$= 27.999662 \text{ J mol K}$$
$$\gamma = \frac{C_P}{C_V} = \frac{28}{27.999662} = 1 \text{ (approx.)}$$

**EXAMPLE 6.28**

*Find the change in boiling point of water for 100 cm change in mercury pressure if the specific volume of steam is 1601 cm³/gram at 100°C and 76 cm mercury pressure. Take latent heat as 5366 cal/gram.* **(Rajasthan 2001)**

**SOLUTION.** [In SI system]

Here,
$$dP = 100 \text{ cm Hg} = 1 \text{ m Hg}$$
$$= 1 \times 13.6 \times 10^3 \times 9.8 \text{ N/m}^2$$
$$= 1.33 \times 10^5 \text{ N/m}^2$$
$$V_1 = 1 \text{ cm}^3/\text{gram} = 10^{-3} \text{ m}^3/\text{kg}$$
$$V_2 = 1601 \text{ cm}^3/\text{gram} = 1601 \times 10^{-3} \text{ m}^3/\text{kg}$$
$$L = 536 \text{ cal/gram}$$
$$= 536 \times 4.2 \times 10^3 \text{ J/kg}$$
$$= 2.25 \times 10^6 \text{ J/kg}$$
$$T = 100°\text{C} = 373 \text{ K}$$
$$dT = ?$$
$$\frac{dP}{dT} = \frac{L}{T(V_2 - V_1)}$$
$$dT = \frac{dP \times T(V_2 - V_1)}{L}$$
$$= \frac{1.33 \times 10^5 \times 373\,(1601 \times 10^{-3} - 10^{-3})}{2.25 \times 10^6}$$
$$= \mathbf{35.28 \text{ K}}$$
$$= \mathbf{35.28°C}$$

Therefore, increase in boiling point of water with an increase of 100 cm in mercury pressure

$$= \mathbf{35.28 \text{ K} = 35.28°C}$$

**EXAMPLE 6.29**

*The ratio of the densities of ice and water at 0°C is $\frac{10}{11}$. Calculate the decrease in the melting point of ice, given that latent heat of fusion of ice = 80 cal/gm and change in specific volume – 0.1 c.c./gm. The increase in pressure is one atmosphere.* **(Osmania, 2000)**

**SOLUTION.** [In SI system],

$$dP = 0.76 \times 13.6 \times 10^3 \times 9.8 = 1.01 \times 10^5 \text{ N/m}^2$$
$$L = 80 \text{ cal/gram}$$
$$= 80 \times 4.2 \times 10^3 \text{ J/kg}$$
$$= 3.36 \times 10^5 \text{ J/kg}$$
$$T = 273 \text{ K}$$
$$V_2 - V_1 = -0.1 \text{ cm}^3/\text{gram}$$
$$= -0.1 \times 10^{-3} \text{ m}^3/\text{kg}$$

$$dT = ?$$

$$\frac{dP}{dT} = \frac{L}{T(V_2 - V_1)}$$

$$dT = \frac{dP \times T(T_2 - V_1)}{L}$$

$$= \frac{1.01 \times 10^5 \times 273(-0.1 \times 10^{-3})}{3.36 \times 10^5}$$

$$= \mathbf{-8.2 \times 10^{-3}\ K}$$

$$= \mathbf{-8.2 \times 10^{-3}\ °C}$$

Therefore, the decrease in the melting point of ice with an increase in pressure of one atmosphere

$$= \mathbf{8.2 \times 10^{-3}\ K = 8.2 \times 10^{-3}\ °C}$$

### EXAMPLE 6.30

*Using Maxwell's thermodynamical relations, prove that for any substance, the ratio of the adiabatic and isothermal elasticities is equal to the ratio of the two specific heats.*

**(Garhwal 2005, 2001, Lucknow 2002)**

**SOLUTION.** From definition,

Isothermal elasticity $E_T = -V\left(\frac{\partial P}{\partial V}\right)_T$

Adiabatic elasticity $E_S = -V\left(\frac{\partial P}{\partial V}\right)_S$

$$\therefore \quad \frac{E_S}{E_T} = \frac{-V\left(\frac{\partial P}{\partial V}\right)_S}{-V\left(\frac{\partial P}{\partial V}\right)_T}$$

$$= \frac{\left(\frac{\partial P}{\partial V}\right)_S}{\left(\frac{\partial P}{\partial V}\right)_T}$$

$$= \frac{\left(\frac{\partial P}{\partial T}\right)_S \left(\frac{\partial T}{\partial V}\right)_S}{\left(\frac{\partial P}{\partial S}\right)_T \left(\frac{\partial S}{\partial V}\right)_T}$$

$$\frac{E_S}{E_T} = \frac{\left(\frac{\partial S}{\partial P}\right)_T \left(\frac{\partial T}{\partial V}\right)_S}{\left(\frac{\partial T}{\partial P}\right)_S \left(\frac{\partial S}{\partial V}\right)_T} \quad \text{...}(i)$$

But from the first four Maxwell's equations

$$\left(\frac{\partial S}{\partial P}\right)_T = -\left(\frac{\partial V}{\partial T}\right)_P$$

$$\left(\frac{\partial T}{\partial V}\right)_S = -\left(\frac{\partial P}{\partial S}\right)_V$$

$$\left(\frac{\partial T}{\partial P}\right)_S = \left(\frac{\partial V}{\partial S}\right)_P$$

$$\left(\frac{\partial S}{\partial V}\right)_T = \left(\frac{\partial P}{\partial T}\right)_V$$

∴ Substituting these values in equation (*i*)

$$\frac{E_S}{E_T} = \frac{\left(\frac{\partial V}{\partial T}\right)_P \left(\frac{\partial P}{\partial S}\right)_V}{\left(\frac{\partial V}{\partial S}\right)_P \left(\frac{\partial P}{\partial T}\right)_V}$$

$$\frac{E_S}{E_T} = \frac{\left(\frac{\partial S}{\partial V}\right)_P \left(\frac{\partial V}{\partial T}\right)_P}{\left(\frac{\partial S}{\partial P}\right)_V \left(\frac{\partial P}{\partial T}\right)_V}$$

$$= \frac{\left(\frac{\partial S}{\partial T}\right)_P}{\left(\frac{\partial S}{\partial T}\right)_V}$$

$$= \frac{\left(\frac{\partial Q}{\partial T}\right)_P}{\left(\frac{\partial Q}{\partial T}\right)_V}$$

But $\left(\frac{\partial Q}{\partial T}\right)_P = C_P$

and $\left(\frac{\partial Q}{\partial T}\right)_V = C_V$

$$\therefore \quad \frac{E_S}{E_I} = \frac{C_P}{C_V} = \gamma \qquad \text{...(ii)}$$

### EXAMPLE 6.31

*Using Maxwell's thermodynamical relations, prove that the ratio of the adiabatic to the isobaric coefficient of expansion is* **(Nagpur 2001)**

$$\frac{1}{(1-\gamma)}$$

**SOLUTION.** Adiabatic volume coefficient of expansion,

$$\alpha_S = \frac{1}{V}\left(\frac{\partial V}{\partial T}\right)_S \qquad \text{...(i)}$$

Isobaric volume coefficient of expansion,

$$\alpha_P = \frac{1}{V}\left(\frac{\partial V}{\partial T}\right)_P \qquad \text{...(ii)}$$

$$\therefore \quad \frac{\alpha_S}{\alpha_P} = \frac{\frac{1}{V}\left(\frac{\partial V}{\partial T}\right)_S}{\frac{1}{V}\left(\frac{\partial V}{\partial T}\right)_P} \qquad \text{...(iii)}$$

$$= \frac{\left(\frac{\partial V}{\partial T}\right)_S}{\left(\frac{\partial V}{\partial T}\right)_P} = \frac{1}{\left(\frac{\partial T}{\partial V}\right)_S \left(\frac{\partial V}{\partial T}\right)_P}$$

But $$\left(\frac{\partial T}{\partial V}\right)_S = -\left(\frac{\partial P}{\partial S}\right)_V$$

$$\therefore \quad \frac{\alpha_S}{\alpha_P} = -\frac{1}{\left(\frac{\partial P}{\partial S}\right)_V \left(\frac{\partial V}{\partial T}\right)_P}$$

$$= \frac{1}{-\left[\left(\frac{\partial P}{\partial T}\right)_V \left(\frac{\partial T}{\partial S}\right)_V \left(\frac{\partial V}{\partial T}\right)_P\right]}$$

$$= \frac{\left(\frac{\partial S}{\partial T}\right)_V}{-\left(\frac{\partial P}{\partial T}\right)_V \left(\frac{\partial V}{\partial T}\right)_P}$$

$$= \frac{T\left(\frac{\partial S}{\partial T}\right)_V}{-T\left(\frac{\partial P}{\partial T}\right)_V \left(\frac{\partial V}{\partial T}\right)_P}$$

But $$C_P - C_V = T\left(\frac{\partial P}{\partial T}\right)_V \left(\frac{\partial V}{\partial T}\right)_P$$

and $$C_V = T\left(\frac{\partial S}{\partial T}\right)_V$$

$$\therefore \quad \frac{\alpha_S}{\alpha_P} = \frac{C_V}{-(C_P - C_V)}$$

or $$\frac{\alpha_S}{\alpha_P} = \frac{1}{1-\left(\frac{C_P}{C_V}\right)}$$

or $$\frac{\alpha_S}{\alpha_P} = \frac{1}{1-\gamma} \quad \text{...}(iv)$$

**EXAMPLE 6.32**

*Using Maxwell's thermodynamical relations, prove that the ratio of the adiabatic to the isochoric pressure coefficient of expansion is equal to*

$$\frac{\gamma}{(\gamma-1)}$$

**(Calicut 2003, Calcutta 2003)**

**SOLUTION.** Adiabatic pressure coefficient of expansion,

$$\beta_S = \frac{1}{P}\left(\frac{\partial P}{\partial T}\right)_S \quad \text{...}(i)$$

Isochoric pressure coefficient of expansion,

$$\beta_V = \frac{1}{P}\left(\frac{\partial P}{\partial T}\right)_V \quad \text{...}(ii)$$

$$\therefore \qquad \frac{\beta_S}{\beta_V} = \frac{\frac{1}{P}\left(\frac{\partial P}{\partial T}\right)_S}{\frac{1}{P}\left(\frac{\partial P}{\partial T}\right)_V} \qquad ...(iii)$$

$$= \frac{\left(\frac{\partial P}{\partial T}\right)_S}{\left(\frac{\partial P}{\partial T}\right)_V}$$

$$= \frac{1}{\left(\frac{\partial T}{\partial P}\right)_S \left(\frac{\partial P}{\partial T}\right)_V}$$

But

$$\left(\frac{\partial T}{\partial P}\right)_S = \left(\frac{\partial V}{\partial S}\right)_P \qquad ...(iv)$$

$$\frac{\beta_S}{\beta_V} = \frac{1}{\left(\frac{\partial V}{\partial S}\right)_P \left(\frac{\partial P}{\partial T}\right)_V} \qquad ...(v)$$

$$= \frac{1}{\left(\frac{\partial V}{\partial T}\right)_P \left(\frac{\partial T}{\partial S}\right)_P \left(\frac{\partial P}{\partial T}\right)_V}$$

$$= \frac{\left(\frac{\partial S}{\partial T}\right)_P}{\left(\frac{\partial V}{\partial T}\right)_P \left(\frac{\partial P}{\partial T}\right)_V}$$

$$= \frac{T\left(\frac{\partial S}{\partial T}\right)_P}{T\left(\frac{\partial V}{\partial T}\right)_P \left(\frac{\partial P}{\partial T}\right)_V}$$

$$= \frac{C_P}{C_P - C_V}$$

or

$$\frac{\beta_S}{\beta_V} = \frac{\left(\frac{C_P}{C_V}\right)}{\left(\frac{C_P}{C_V} - 1\right)} = \frac{\gamma}{\gamma - 1} \qquad ...(vi)$$

**EXAMPLE 6.33**

*Using Maxwell's thermodynamical relations, show that*

$$\left(\frac{\partial C_V}{\partial V}\right) = T\left(\frac{\partial^2 S}{\partial V . \partial T}\right) = T\left(\frac{\partial^2 P}{\partial T^2}\right)_V$$ **(Agra 2002, M.S. Uni. Tamil Nadu 2007)**

**SOLUTION.** From Maxwell's relation,

$$\left(\frac{\partial S}{\partial V}\right)_T = \left(\frac{\partial P}{\partial T}\right)_V \qquad ...(i)$$

Differentiating equation (*i*) with respect to temperature,

$$\left(\frac{\partial^2 S}{\partial T.\partial V}\right) = \left(\frac{\partial^2 P}{\partial T^2}\right)_V \quad ...(ii)$$

But

$$C_V = T\left(\frac{\partial S}{\partial T}\right)_V \quad ...(iii)$$

Differentiating equation (*iii*) with respect to volume,

$$\left(\frac{\partial C_V}{\partial V}\right) = T\left(\frac{\partial^2 S}{\partial T.\partial V}\right) \quad ...(iv)$$

From equation (*ii*) and (*iv*)

$$\left(\frac{\partial C_V}{\partial V}\right) = T\left(\frac{\partial^2 S}{\partial T.\partial V}\right) = T\left(\frac{\partial^2 P}{\partial T^2}\right)_V \quad ...(v)$$

**EXAMPLE 6.34**

*Using Maxwell's thermodynamical relations, show that*

$$\left(\frac{\partial C_P}{\partial P}\right) = T\left(\frac{\partial^2 S}{\partial P \partial T}\right) = -T\left(\frac{\partial^2 V}{\partial T^2}\right)_P$$

**(Agra 2000, M.S. Uni. Tamil Nadu 2007)**

**SOLUTION.** From Maxwell's relations,

$$\left(\frac{\partial S}{\partial P}\right)_T = -\left(\frac{\partial V}{\partial T}\right)_P \quad (i)$$

Differentiating equation (*i*) with respect to temperature

$$\left(\frac{\partial^2 S}{\partial T.\partial P}\right) = -\left(\frac{\partial^2 V}{\partial T^2}\right)_P \quad ...(ii)$$

But

$$C_P = T\left(\frac{\partial S}{\partial T}\right)_P \quad ...(iii)$$

Differentiating equation (*iii*) with respect to pressure,

$$\left(\frac{\partial C_P}{\partial P}\right) = T\left(\frac{\partial^2 S}{\partial T \partial P}\right) \quad (iv)$$

From equations (*ii*) and (*iv*)

$$\left(\frac{\partial C_P}{\partial P}\right) = T\left(\frac{\partial^2 S}{\partial T \partial P}\right) = -T\left(\frac{\partial^2 V}{\partial T^2}\right)_P \quad ...(v)$$

**EXAMPLE 6.35**

*Show that for a perfect gas*

$$\left(\frac{\partial U}{\partial V}\right)_T = 0$$

**SOLUTION.** From the Maxwell's first equation,

$$\left(\frac{\partial S}{\partial V}\right)_T = \left(\frac{\partial P}{\partial T}\right)_V$$

$$\frac{1}{T}\left(\frac{\partial Q}{\partial V}\right)_T = \left(\frac{\partial P}{\partial T}\right)_V$$

or

$$\left(\frac{\partial Q}{\partial V}\right)_T = T\left(\frac{\partial P}{\partial T}\right)_V$$

But $$dQ = dU + PdV$$

$\therefore$ $$\left(\frac{dU + PdV}{dV}\right) = T\left(\frac{\partial P}{\partial T}\right)_V$$

$$\frac{dU}{dV} + P = T\left(\frac{\partial P}{\partial T}\right)_V$$

or $$\left(\frac{\partial U}{\partial V}\right)_T = T\left(\frac{\partial P}{\partial T}\right)_V - P$$

But from perfect gas equation,

$$PV = RT$$

$\therefore$ $$\left(\frac{\partial P}{\partial T}\right)_V = \frac{R}{V}$$

$$\left(\frac{\partial U}{\partial V}\right)_T = T\left(\frac{R}{V}\right) - P$$

or $$\left(\frac{\partial U}{\partial V}\right)_T = P - P = 0$$

$\therefore$ $$\left(\frac{\partial U}{\partial V}\right)_T = 0$$

**EXAMPLE 6.36**

*If L = 800 – 0.705 T, show that the specific heat of steam is negative.*

**(Lucknow 2002; Kanpur 2002; Meerut 2000)**

**SOLUTION.** Given : $$L = 800 - 0.705\ T$$

$\therefore$ $$\frac{dL}{dT} = -0.705$$

At normal boiling point of water *i.e.*, at 100 °C = 373 K, the value of $L$ would be

$$L = 800 - 0.705 \times 373 = 537$$

The specific heat of water, in liquid state, $C_1 = 1$.

Using second latent heat equation

$$C_2 - C_1 = \frac{dL}{dT} - \frac{L}{T}$$

$\therefore$ $$C_2 = C_1 + \frac{dL}{dT} - \frac{L}{T}$$

$$= 1 + (-0.705) - \frac{537}{373}$$

$$= \mathbf{-1.14467}$$

Thus, the specific heat of saturated vapour pressure at 100 °C is **negative.**

**EXAMPLE 6.37**

*Prove the thermodynamical relation*

$$\left(\frac{\partial Q}{\partial P}\right)_T = -T\left(\frac{\partial V}{\partial T}\right)_P = -TV\alpha$$

*and show that heat is generated when a substance which expands on heating is compressed and for substance which contract a cooling takes place.* **(Rohilkhand 2000, 2001, 2002, G.N.D.U. 2001, Punjab 2002, Meerut 2004, Garhwal 2003, Agra 2003)**

**SOLUTION.** From Maxwell's fourth thermodynamical relation,

$$\left(\frac{\partial S}{\partial P}\right)_T = -\left(\frac{\partial V}{\partial T}\right)_P$$

Multiplying both sides by $T$, we get

$$\left(\frac{T.\partial S}{\partial P}\right)_T = -T\left(\frac{\partial V}{\partial T}\right)_P$$

But $\quad T.\partial S = \partial Q$, substituting, we get

$$\left(\frac{\partial Q}{\partial P}\right)_T = -T\left(\frac{\partial V}{\partial T}\right)_P \qquad ...(i)$$

which is the required relation.

Now, $\frac{1}{V}\left(\frac{\partial V}{\partial T}\right)_P$ represents the increase in volume per unit rise of temperature at constant pressure. This is nothing but the coefficient of volume expansion α. Hence

$$\frac{1}{V}\left(\frac{\partial V}{\partial T}\right)_P = \alpha$$

or

$$\left(\frac{\partial V}{\partial T}\right)_P = V\alpha \qquad ...(ii)$$

Substituting in eqn (*i*), we get

$$\therefore \quad \left(\frac{\partial Q}{\partial P}\right)_T = -TV\alpha \qquad ...(iii)$$

This is another form of required relation.

From this equation, it is very clear that if α is positive *i.e.*, if the substance expands on heating, then $\left(\frac{\partial Q}{\partial P}\right)_T$ is negative. It means heat must be withdrawn from the substance in order to keep its temperature constant when the pressure is increased *i.e., the increase in pressure heats a body that expands on rise of temperature.*

On the other hand, if α is negative i.e., the substance contracts on heating, the $\left(\frac{\partial Q}{\partial P}\right)_T$ is positive, which means heat must be added to keep its temperature constant, when the pressure is increased *i.e., a cooling is produced when a substance, which contracts on heating, is suddenly compressed.*

## EXERCISE—VI

**6.1** Explain thermodynamic system. What are thermodynamic variables? Define intensive and extensive variables. **(Kerala 2002, Bangalore 2000, Nagpur 2004, 2005, 2008, Agra 2003)**

**6.2** Define extensive and intensive variables. 'An extensive variable may become intensive variable, explain with example. **(Nagpur 2007, 2005, Calcutta 2004)**

**6.3** Derive Maxwell's thermodynamical general equations connecting the thermodynamic quantities. **(Meerut 2003; Garhwal 2001, 2005; Bangalore 2000; Kerala 2000, 2002; Utkal 2003; Nagpur 2009, 2007, Lucknow 2001; M.D.U. 2002; Madras 2005)**

**6.4** Show that for a homogeneous fluid.

$$C_P - C_V = T\left(\frac{\partial P}{\partial T}\right)_V\left(\frac{\partial V}{\partial T}\right)_P$$

where the symbols have their usual meaning.

**(Garhwal 2000, Madras 2004, Calcutta 2004, M.S. Uni. Tamil Nadu 2007, Purvanchal 2003, 2004)**

**6.5** Use Maxwell's relations to obtain $C_P - C_V = R$ for an ideal gas where $C_P$ and $C_V$ are molar specific heats at constant pressure and constant volume respectively.

**(Garhwal 2005; Lucknow 2004; Garhwal 2000, 2004)**

**6.6** Show that for a Van der Waals' gas

$$C_P - C_V = R\left(1+\frac{2a}{VRT}\right)$$

**(Kerala 2000, 2001, 2002; Purvanchal 2004, 2002)**

**6.7** Write down the Maxwell's four thermodynamic relations and prove the relation

$$\frac{E_S}{E_T} = \frac{C_P}{C_V}$$ **(Rohilkhand 2000, Agra 2008, 2007)**

**6.8** Discuss the effect of change of pressure on melting point of a solid.

**(Nagpur 2007, Rohilkhand 2001, Ranchi 2002)**

**6.9** Using Carnot's cycle derive Clausius-Clopeyron's latent heat equation

$$\frac{dP}{dT} = \frac{L}{T(V_2 - V_1)}$$

where $T$ is the absolute temperature at which a body changes its state, $L$ the latent heat and $V_1$ and $V_2$ the specific volumes in the first and second state respectively.

**(Kerala 2000, 2001; Guwahati 2000, Meerut 2003, 2004)**

**6.10** Deduce and state first latent heat equation,

$$\frac{dP}{dT} = \frac{L}{T(V_2 - V_1)}$$ **(Purvanchal 2005, Garhwal 2005)**

**6.11** Explain Joule-Thomson effect using the Maxwell's thermodynamical relations.

**(Nagpur 2005/w, Purvanchal 2003)**

**6.12** 'There is no Joule-Thomson effect for a perfect gas' Explain. **(Agra 2008, Madras 2002)**

**6.13** Explain Joule-Thomson effect for Van der Waal's gas. Obtain the relation for temperature of inversion.

**(Kanpur 2001, Purvanchal 2003)**

**6.14** Establish the Clausius-Clapeyron's equation

$$\frac{dP}{dT} = \frac{L}{T(V_2 - V_1)}$$

from Maxwell's thermodynamical relations and explain the effect of pressure on

(*a*) boiling point of a liquid, and

(*b*) melting point of a solid. **(Agra 2006, 2004, Purvanchal 2007, 2003; Nagpur 2009, 2008, 2007, Patna 2005; Kanpur 1998; 2001; Ranchi 2004; Osmania 2004; Kerala 2000, 2001, 2002; Punjab 2001, 2002, 2005; Utkal 2003; Lucknow 2001, 2002; Meerut 2001, 2002, 2005; Bangalore 2002, 2004; Garhwal 2004)**

**6.15** Prove that

$$\left(\frac{\partial T}{\partial P}\right)_S = \frac{TV\alpha}{C_P}$$ **(Purvanchal 2000)**

**6.16** Prove the thermodynamic relation

$$\left(\frac{\partial S}{\partial V}\right)_T = \left(\frac{\partial P}{\partial T}\right)_V$$

**(Kanpur 2001, Purvanchal 2007, Meerut 2005, Osmania 2002)**

and $$\left(\frac{\partial S}{\partial P}\right)_T = \left(\frac{\partial P}{\partial T}\right)_P$$ **(Purvanchal 2007)**

**6.17** Prove the thermodynamic relations

(*a*) $T.dS = C_V dT + T\left(\frac{\partial P}{\partial T}\right)_V dV$

**(Calicut 2004, Garhwal 2004, Agra 2004, Purvanchal 2002, Gujarat 2004, 2005)**

(*b*) $T.dS = C_P dT - T\left(\frac{\partial V}{\partial T}\right)_P dP$ **(Garhwal 2000, 2001, Ranchi 2005, Gujarat 2004, 2005)**

**6.18** What is the most important application of Joule-Kelvin effect? **(Kanpur 2000; Utkal 2002)**

**6.19** The internal energy of a perfect gas is only dependent on temperature whereas internal energy of Van der Waals' gas is a function of both temperature and volume. Why? **(G.N.D.U. 2003)**

**6.20** Prove Maxwell's second thermodynamic relation

$$\left(\frac{\partial S}{\partial V}\right)_T = \left(\frac{\partial P}{\partial T}\right)_V$$

from it, establish Clausius-Clapeyron relation

$$\left(\frac{\partial P}{\partial T}\right) = \frac{L}{T(V_2 - V_1)}$$

and find out the effect of pressure on boiling point of a liquid and melting point of a solid.
**(Rohilkhand 2001, Nagpur 2007, Meerut 2005)**

**6.21** Derive Maxwell's four thermodynamical relations. Use one of these to obtain Clausius-Clapeyron's latent heat equation. **(Purvanchal 2002, Meerut 2001)**

**6.22** How internal energy of ideal gas differ from a real gas? **(Punjab 2005, Calicut 2002)**

**6.23** Give two statements of second law of thermodynamics. Prove the Clausius inequality $\sum_i \frac{Q_i}{T_i} \leq 0$.
Hence develop the concept of entropy. **(Patna 2004, Mumbai 2001)**

**6.24** Prove the thermodynamic relation

$$\left(\frac{\partial T}{\partial V}\right)_S = -\left(\frac{\partial P}{\partial S}\right)_V$$ **(Calcutta 2003, Agra 2005, 2001, Meerut 2002)**

**6.25** Derive an expression for the work done on a paramagnetic material filled in a current carrying solenoid and placed in an external field. **(Mumbai 2005)**

**6.26** Prove that $H = \left[\frac{\partial (G/T)}{\partial (1/T)}\right]_P$, where $H$ is enthalpy and $G$ is free energy. **(Mumbai 2005)**

**6.27** Derive Maxwell's four thermodynamical equations and hence find two *T.dS* equations.
**(Purvanchal 2004)**

**6.28** What do you mean by first order phase transition? Prove that $\left(\frac{dP}{dT}\right) = \frac{S_2 - S_1}{V_2 - V_1}$, where the terms have their usual meaning. **(Magadh 2002, Purvanchal 2004, Nagpur 2001, 2003)**

**6.29** Deduce Maxwell's four thermodynamical relations. Discuss the usefulness and importance of these relations. **(Lucknow 2003, Ranchi 2002, Patna 2002)**

**6.30** Discuss about the first order phase transition and hence Clausius-Clapeyron's latent heat equation. Can latent heat become zero? If yes, under what condition ?
**(M.S. Uni. Tamil Nadu 2007, Lucknow 2003)**

**6.31** Discuss about second order phase transition and hence derive Ehrenfest's theorem.
**(M.S. Uni. Tamil Nadu 2007)**

**6.32** Prove the following thermodynamical relation

$$\left(\frac{\partial U}{\partial V}\right)_T = T\left(\frac{\partial P}{\partial T}\right)_V - P$$

Find also the value of $\left(\frac{\partial U}{\partial V}\right)_T$ for an ideal gas.

Also prove

$$\frac{\partial C_V}{\partial V}_T = T\left(\frac{\partial^2 P}{\partial T^2}\right)$$

**(Calcutta 2005)**

**6.33** Define thermodynamic potentials $U$, $F$, $H$ and $G$. Derive their relationship with state variables.
**(Rohilkhand 2000, 2001; Kanpur 2000, Lucknow 2004)**

**6.34** Obtain the values of thermodynamical variables $P$, $V$ in terms of thermodynamic potentials. **(Kerala 2003)**

**6.35** Define the Helmholtz function and for an isochoric process establish the relation

$$U = F - T\left(\frac{\partial F}{\partial T}\right)_V$$ **(Agra 2008, Garhwal 2000, Sagar 2002)**

**6.36** Define the Gibbs potential function and show that for simultaneous isobaric and isothermal process, Gibbs free energy remains constant. **(Agra 2000)**

**6.37** Deduce an expression for the change in temperature of wire thermodynamically when it is stretched adiabatically. **(Agra 2002, Punjab 2002, 2003, 2005; G.N.D.U. 2001; 2003)**

**6.38** What do you mean by thermodynamic potentials? Give the physical significance. **(Garhwal 2001, Agra 2000)**

**6.39** What do you mean by enthalpy $H$ and Gibbs function G. Show that

$$\left(\frac{\partial H}{\partial T}\right)_P = C_P$$

[**Hint :** $dH = dU + P.dV + V.dP = T.dS - P.dV + P.dV + VdP = T.dS + VdP$

$$\therefore \quad \left(\frac{\partial H}{\partial T}\right)_P = T\left(\frac{\partial S}{\partial T}\right)_P = \left(\frac{\partial Q}{\partial T}\right)_P = C_P \,]$$ **(Rajasthan 2001, Kerala 2000)**

**6.40** Define the thermodynamic potential enthalpy and obtain Maxwell thermodynamic relation

$$\left(\frac{\partial T}{\partial P}\right)_S = \left(\frac{\partial V}{\partial S}\right)_P$$ **(Kanpur 2002, Agra 2001)**

**6.41** What do you mean by order of phase. Distinguish between first and second order phase transitions. **(Purvanchal 2003, 2006)**

**6.42** Starting from four thermodynamical potentials, derive Maxwell's thermodynamic relations. **(M.S. Uni. Tamil Nadu 2007, Pb. 2001; Kerala 2001, Garhwal 2000, Kanpur 2001)**

**6.43** What do you understand by the thermodynamic potentials. What is the importance of these potentials? Deduce from thermodynamic potentials the corresponding Maxwell's thermodynamic relations. **(Purvanchal 2006, Rohilkhand 2002, Garhwal 2004, 2005; Ranchi 2005, Osmania 2005)**

**6.44** Define Helmholtz function and Gibbs function and derive related Maxwell's thermodynamic relations. **(Garhwal 2000)**

**6.45** Establish the following relation

$$C_P - C_V = -T\left(\frac{\partial V}{\partial T}\right)_P^2\left(\frac{\partial P}{\partial V}\right)_T$$ **(Calcutta 2004)**

**6.46** A measure of the result of a Joule-Kelvin expansion (throttling process ) is provided by the Joule-Kelvin coefficient $\mu = \left(\frac{\partial T}{\partial P}\right)_H$. Derive the expression

$$\mu = \frac{1}{C_P}\left[T\left(\frac{\partial V}{\partial T}\right)_P - V\right]$$

and show that for an ideal gas $\mu = 0$. Explain the existence of inversion temperature. **(Kanpur 2000, Agra 2001)**

**6.47** Deduce the expression for the Joule-Thomson coefficient

$$\mu = \left(\frac{\partial T}{\partial P}\right)_H = \frac{1}{C_P}\left[T\left(\frac{\partial V}{\partial T}\right)_P - V\right]$$

by using Maxwell's relations and show that for an ideal gas $\mu = 0$ and for a Van der Waals' gas

$$\mu = \frac{1}{C_P}\left(\frac{2a}{RT} - b\right)$$ **(Nagpur 2007, Purvanchal 2003, G.N.D.U. 2001)**

**6.48** Explain enthalpy, Helmholtz and Gibbs functions in thermodynamics. Derive Gibbs Helmholtz equation. **(Purvanchal 2002, Garhwal 2005, Patna 2005)**

**6.49** Heat is produced by adiabatic compression of a substance. Explain it by using Maxwell relations. **(Punjab 2005, Kerala 2002)**

**6.50** Why does a rubber string heat up on stretching. **(Punjab 2005, Garhwal 2000, 2001)**

**6.51** Show that for a system undergoing processes at constant pressure and constant temperature, the Gibbs free energy will be minimum at equilibrium. **(Calcutta 2004, Agra 2001, 2003, 2008)**

**6.52** Establish the following relation between enthalpy (*H*), Helmholtz free energy (*F*) and Gibbs free energy (*G*),

$$G = H + T\left(\frac{dG}{dT}\right)_P$$ **(Purvanchal 2007, 2006, 2003, Madras 2001)**

**6.53** Prove that enthalpy remains constant (isoenthalpic) in Joule-Kelvin expansion, *i.e.*, in a throttling process. **(Nagur 2009, 2007, 2006, Purvanchal 2003, Lucknow 2003)**

**6.54** Give the difference between enthalpy and internal energy. **(Gujarat 2005, 2004, G.N.D.U. 2001)**

**6.55** Establish that for a system of constant volume and in contact with a heat reservoir, the equilibrium corresponds to the mimimum of Helmholtz free energy. **(Calcutta 2005)**

**6.56** State Gibbs phase rule. Explain the rule with example. What is meant by a higher order phase transition? **(Calcutta 2005)**

**6.57** Derive the Gibbs-Helmholtz equation for e.m.f. of a reversible cell

$$E = H + T\left(\frac{dE}{dT}\right)$$

using *T–dS* equations. Examine the cases when,

(*i*) $\frac{\partial E}{\partial T} = 0$ (*ii*) $\frac{\partial E}{\partial T} > 0$ and (*iii*) $\frac{\partial E}{\partial T} < 0$

**6.58** Write short notes on

(*a*) Thermodynamic potentials. **(Agra 2002)**

(*b*) Joule-Kelvin effect. **(Rohilkhand 2001, Meerut 2003, Madras 2005, Purvanchal 2004)**

(*c*) Maxwell's thermodynamic relations. **(Agra 2000)**

(*d*) Clausius-Clapeyron equation. **(Rohilkhand 1995)**

(*e*) Specific heat at saturated vapour. **(Meerut 2001)**

(*f*) First order phase transformation. **(Magadh 2001)**

**6.59** Calculate the pressure required to make ice freeze at –2°C change of specific volume when 1 gram of water freezes into ice = 0.091 cc. **[Ans.** *271.4 atmospheres*]

**6.60** Calculate the change in temperature of boiling water when pressure is increased by 27.12 mm of Hg. The normal boiling point of water at atmospheric pressure is 100°C.

[**Ans.** = 1°C, 1K] **(Kanpur 2000)**

**6.61** Calculate the change in melting point of ice when it subjected to a pressure of 100 atmosphere. Density of ice = 0.917 g/cm$^3$, Latent heat of ice = 336 J/g.

[**Ans.** $dT$ = – 0.7326°C = – 0.7326 K] **(Osmania 2001)**

**6.62** One mole of an ideal gas at temperature *T* undergoes free expansion which doubles its volume. Show that change of entropy $\Delta S = Nk \log_e (2)$.

**6.63** Calculate the temperature of inversion of *He*. Given: $a$ = 0.0341 atm. lit$^2$/mole$^3$, $b$ = 0.0237 litre/mole, R = 8.3 Joule/mloe–K. [**Ans.** – 238 °C] **(Mysore 2000)**

## OBJECTIVE QUESTIONS

**6.1** Four thermodynamic potentials are: **(Garhwal 2005)**

(*a*) Pressure, volume, temperature and internal energy function.

(*b*) Pressure, volume, internal energy and Helmholtz function.

(*c*) Internal energy function, Helmholtz function, enthalpy and Gibbs function.

(*d*) None of these.

**6.2** Specific heat of saturated vapur pressure: **(Garhwal 2005)**

(*a*) zero (*b*) positive

(*c*) negative (*d*) sometimes positive sometimes negative.

**6.3** From Maxwell's thermodynamic relations $\frac{E_S}{E_T}$ = ? **(Agra 2004, Rohilkhand 2002)**

(*a*) 2 (*b*) $\frac{1}{\gamma}$

(*c*) $\gamma$ (*d*) 2

**6.4** The Clausius-Clapeyron equation is:

(*a*) $\frac{dP}{dT} = \frac{L}{T(V_2 - V_1)}$ (*b*) $\frac{dP}{dT} = \frac{L}{L(V_2 - V_1)}$

(*c*) $\frac{dP}{dT} = TL(V_2 - V_1)$ (*d*) None of these **(Rohilkhand 2002)**

**6.5** Gibbs potential is defined as:

(*a*) $G = U - PV + TS$ (*b*) $G = U + PV + TS$

(*c*) $G = U - PV - TS$ (*d*) $G = U + PV - TS$ **(Kanpur 2001)**

**6.6** The value of $C_P - C_V$ for an ideal gas is:

(*a*) $TE \alpha V^2$ (*b*) $T^2 E \alpha V$

(*c*) $TE \alpha^2 V$ (*d*) $TE^2 \alpha^2 V^2$ **(Rohilkhand 2002)**

**6.7** Helmholtz free energy function is defined by:

(*a*) $F = U + TS$ (*b*) $F = U - TS$

(*c*) $F = U + PV$ (*d*) $F = U + PV - TS$ **(Kanpur 2002)**

**6.8** Hydrogen and helium at normal temperature show:

(*a*) Heating effect (*b*) Cooling effect

(*c*) Sometimes cooling and some times heating

(*d*) None of these

**6.9** Maxwell's thermodynamic relation is:

(*a*) $\left(\frac{\partial S}{\partial T}\right)_T = \left(\frac{\partial P}{\partial V}\right)_T$ (*b*) $\left(\frac{\partial T}{\partial V}\right)_S = -\left(\frac{\partial P}{\partial S}\right)_V$

(*c*) $\left(\frac{\partial T}{\partial P}\right)_P = \left(\frac{\partial V}{\partial P}\right)_P$ (*d*) None of the above **(Rohilkhand 2003)**

**6.10** Rice takes longest time to cook

(*a*) in submarine 100 m below the surface of the sea.

(*b*) at the sea level.

(*c*) at Simla.

(*d*) at Mount–Everest.

**6.11.** Paraffin wax contracts on solidification. The melting point of wax will

(*a*) increase with pressure (*b*) decrease with pressure

(*c*) no change with pressure (*d*) decrease linearly with pressure.

## ANSWERS

| | | | |
|---|---|---|---|
| **6.1** (*c*) | **6.2** (*c*) | **6.3** (*c*) | **6.4** (*a*) |
| **6.5** (*d*) | **6.6** (*c*) | **6.7** (*b*) | **6.8** (*a*) |
| **6.9** (*b*) | **6.10** (*d*) | **6.11** (*a*) | |

# Liquefaction of Gases

# 7

## AT A GLANCE

• Introduction • Different Methods of Liquefaction of Gases • Method of Freezing Mixture • Cooling by Evaporation Under Reduced Pressure • Cooling by Adiabatic Expansion • Joule-Thomson Expansion • Liquefaction of Gases • Principle of Regenerative Cooling • Liquefaction of Air (Linde's Process) • Principle of Cascaded Cooling: Liquefaction of Oxygen • Liquefaction of Hydrogen • Liquefaction of Helium (Onnes Method) • Helium I and Helium II • Some Peculiar Properties of Helium II • Helium 'A Unique Liquid' • Production of Low Temperatures • Adiabatic Demagnetisation • Conversion of Magnetic Temperature into Kelvin Temperature • Measurement of Very Low Temperatures • Superconductivity • Meissner Effect • Refrigeration Cycle: Electrolux Refrigerator

## Introduction

It was believed since ancient timethat nothing could be colder than ice and hence temperature of ice was considered as the zero (or starting point) of the scale used in temperature measurements in all the earlier systems. More over, it was also thought that air remains in the gaseous state permanently at all temperatures. Later on, Andrew's experiments on $CO_2$ gas in 1862 showed that below the critical temperature, a gas can be liquefied by mere application of pressure. The study of the laws of perfect gases categorically stated that the lowest conceivable temperature on centrigrades scale is – 273 °C, much below the temperature of ice (0°C) and now – 273 °C is universally taken as the zero of the absolute temperature scale, known as *absolute zero temperature*. Different methods are employed to reach the absolute zero temperature.

In 1877, Pictet liquefied oxygen by cascade process. Lande, was able to liquefy air in 1895 by Joule-Thomson effect. Using the principle of Joule-Thomson effect hydrogen and helium were also liquefied. The critical temperatures and boiling points of some important gases are given in Table 7.1. Moreover, a thermometer showing the significant temperatures of few gases are plotted by using the logarithmic thermodynamic scale (Fig. 7.1).

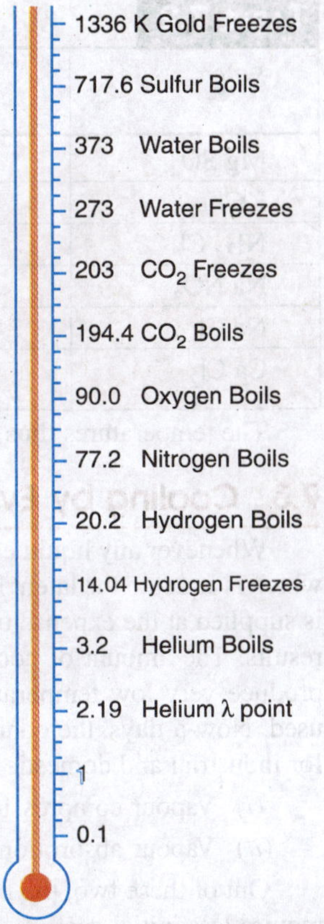


**Fig. 7.1** Thermometer showing temperatures on Kelvin scale.

**TABLE 7.1**

| *Gas* | *Critical Temperature* | *Boiling Point* |
|---|---|---|
| $CO_2$ | – 31.1 °C | – 78.6 °C |
| Oxygen | – 118.8 °C | – 183 °C |
| Nitrogen | – 146 °C | – 195.8 °C |
| Hydrogen | – 240 °C | – 252.8 °C |
| Helium | – 267.8 °C | – 269.8 °C |

## 7.1 Different Methods of Liquefaction of Gases

The production of very low temperatures is essentially linked with the liquefaction of gases. The usual method is to first liquefy the suitable gas by purely mechanical process and then evaporating it. The latent heat of evaporation is obtained at the expense of the heat content of the liquid itself and an intense cooling results. The different methods employed to produce very low temperatures are:

(*i*) Method of freezing mixture.

(*ii*) Evaporation of liquid under reduced pressure.

(*iii*) Adiabatic expansion of a gas.

(*iv*) Joule-Thomson expansion.

(*v*) Regenerative cooling.

(*vi*) Adiabatic demagnetisation of a paramagnetic salt.

The last method is useful in the region of liquid helium temperatures and is used in producing the lowest temperatures. Some of them are discussed below.

## 7.2 Method of Freezing Mixture

The easiest method for production of temperatures much lower than that of ice is adding a salt to ice. When a salt (NaCl say) is added to ice, the freezing mixture is formed. The salt is dissolved by the water adhering to ice and more ice melts. The heat necessary for this process *viz.* heat of solution and the latent heat necessary for melting ice is extracted from the mixture itself. As a result, the temperature of the mixture falls down. This is the principle involved in *freezing mixtures*. The process cannot go indefinitely. The lowest temperature that can be attained by the method of freezing mixture is definite and depends upon the salt used and is called the **eutectic temperature**. For different salts, the eutectic temperatures are as given in Table 7.2.

**TABLE 7.2**

| *Salt* | *Amount of salt in 100 gm. of mixture* | *Eutectic temperature in °C* |
|---|---|---|
| $Mg\ SO_4$ | 19.0 gm. | – 3.9 |
| KCl | 19.7 gm. | – 11.1 |
| $NH_4\ Cl$ | 18.6 gm. | – 15.8 |
| $Na\ NO_3$ | 37.0 gm. | – 18.5 |
| Na Cl | 22.0 gm. | – 21.2 |
| $Ca\ Cl_2$ | 29.8 gm. | – 55.0 |

The temperatures thus reached by the method of freezing mixtures are not low enough.

## 7.3 Cooling by Evaporation Under Reduced Pressure

Whenever any liquid evaporates, some heat is required to convert it from liquid to gaseous state, which is equal to the latent heat of vaporisation. If the liquid is isolated, the latent heat of vaporisation is supplied at the expense of the total heat content of the liquid itself and hence an intensive cooling results. The amount of cooling depends upon the nature of liquid and the rate of evaporation. To produce very low temperatures generally liquefied gases, *e.g.*, ammonia, sulphur-dioxide etc. are used. Now-a-days, the contrivance utilising the process is employed in refrigerating machines used for industrial and domestic purpose. The machines used are :

(*i*) Vapour compression machine

(*ii*) Vapour absorption machine

Out of these two, the machines of first type are much popular as they have more efficiency and require less initial cost.

**Vapour Compression Machine.** Fig. 7.2 shows the schematic arrangement of vapour compression machine. It consists of the following main parts:

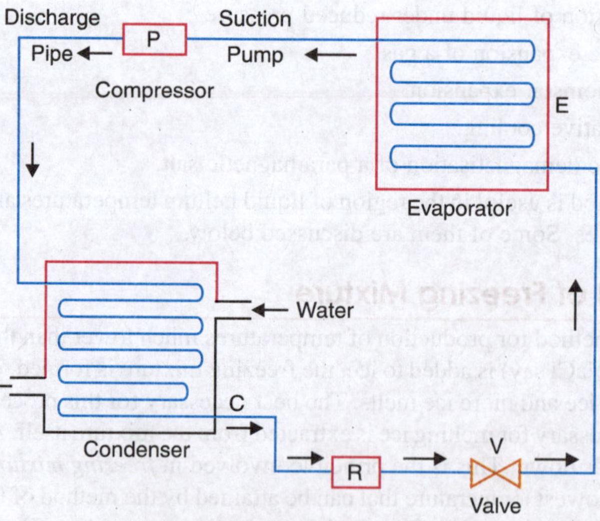


Fig. 7.2

1. The compressor $P$.
2. The condenser $C$.
3. The evaporator $E$.

**1. The compressor $P$:** The electrically driven compressor $P$ consists of a piston and two valves, one for the suction of low pressure vapour from the evaporator and the other for the discharge of the compressed vapour to the condenser.

**2. The condenser $C$ :** The condenser is generally a copper coil embedded in a vessel through which a coolant like cold water or brine water is passed.

### Working

When the piston of the compressor P moves upwards, the pressure in the cylinder of compressor falls much below the pressure in the evaporator. As a result, the low pressure vapour is sucked up through the valve and the suction pipe. When the piston moves downwards, the vapour is compressed and the compressed vapour is then delivered to the condenser $C$ through the discharge valve and discharge pipe. The vapour at high pressure enters the condenser $C$, maintained at low temperature liquefies the vapour in condenser $C$. The liquid passes through the expansion or regulator valve $V$, which reduces the pressure of the liquid and delivers it to the evaporator $E$. Due to the low pressure, the liquid evaporates, thereby extracting its latent heat from the cold storage space. As a result, the temperature of cold storage space is cooled further. The low pressure vapour is again sucked in the compressor $P$ and the cycle of operation continues.

The liquid used in the cycle of operation as refrigerant should have large latent heat of evaporation, smaller specific volume and sufficient vapour pressure at the temperature of the evaporator.

## 7.4 Cooling by Adiabatic Expansion

In adiabatic process, the working gas is perfectly insulated from the surroundings so that no heat is allowed to enter or leave the system. Under this condition, when the gas expands against some external pressure, the gas has to do external work at the expense of its internal energy. As the internal energy (U) of the gas is a function of its temperature, during adiabatic expansion of the gas, its internal energy and hence its temperature falls. If a highly compressed gas is suddenly allowed to expand adiabatically, it has to perform large amount of external work and an intense cooling results.

### Theory

If $P$ is the pressure and $T$ the temperature of a perfect gas, then during adiabatic process

$$PV^\gamma = \text{constant} \qquad ...(i)$$

Here $\gamma$ is the ratio of two specific heats ($\gamma = C_P/C_V$).

Using perfect gas equation, $PV = RT$, Equation ($i$) becomes

$$TP^{(1-\gamma)/\gamma} = \text{constant} \qquad ...(ii)$$

If $P_1$ and $T_1$ are the initial pressure and temperature of the given mass of gas, while $P_2$ and $T_2$ are its final pressure and temperature, we have

$$T_1 P_1^{(1-\gamma)/\gamma} = T_2 P_2^{(1-\gamma)/\gamma}$$

$$\therefore \qquad T_2 = T_1\left(\frac{P_1}{P_2}\right)^{(1-\gamma)/\gamma} \qquad ...(iii)$$

From equation ($iii$), it is clear that a gas at initial ordinary temperature $T_1$ and compressed at a very high pressure $P_1$ is suddenly allowed to expand, the gas may be liquefied or even solidified. For example: if a $CO_2$ gas at room temperature compressed to a very light pressure (say 150 atmosphere) is suddenly released to atmospheric pressure, the final temperature $T_2$ reached is

$$T_2 = T_1\left(\frac{P_2}{P_1}\right)^{(\gamma-1)/\gamma}$$

$$= 300\left[\frac{1}{150}\right]^{(1.3-1)/1.3}$$

$$= 96 \text{ K}$$

The final temperature reached is so low that the expanding $CO_2$ gas gets even solidified, usually known as **dry ice**.

## Gas Expansion Machine

A machine which uses above principle for cooling purpose is known as gas expansion machine and is schematically shown by Fig. 7.3. The machine consists of the following major parts:

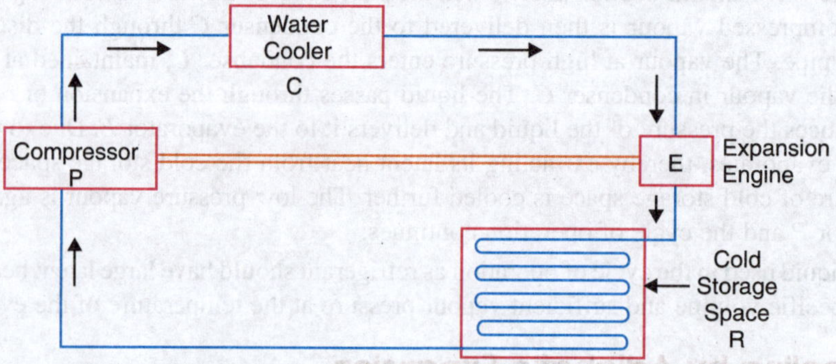


**Fig. 7.3**

1. Compressor $P$ driven by an electric motor.
2. Water cooler $C$.
3. An expansion engine $E$.
4. A reservoir or cold storage space.

### Working

The working substance (*e.g.* gas or air) is first compressed in a compressor $P$ adiabatically to a desired pressure. This compressed gas (or air) is passed through a water cooler $C$ maintained at suitable temperature at constant pressure. The cooled gas is then delivered to expansion engine where it suffers adiabatic expansion to the pressure equal to that of the intake of the compressor. The gas gets cooled. This cold gas is further passed to the reservoir or the cold storage space where it absorbs heat at constant pressure and finally returns to the compressor $P$ and a fresh cycle of operation begins.

## 7.5 Joule-Thomson Expansion

In 1852, Joule in collaboration with Lord Kelvin (Thomson) devised a very sensitive technique known as porous plug experiment. The principle involved is as under:

### Principle:

When a gas is allowed to escape adiabatically, through a porous plug, from the region of constant high pressure to the region at constant low pressure, it undergoes a *change* of temperature. This phenomenon is known as *Joule-Thomson effect* and the process is called the *Joule-Thomson* or *adiabatic throttling*. The change of temperature may be positive or negative, depending upon the nature of gas and the initial temperature. This phenomenon is explained with theory in detail in Chapter 2. Let us summarize the results of the experiment :

## Results :

**1.** On suffering Joule-Thomson expansion (*i.e.* throttling), all gases undergo a change in temperature. At ordinary temperatures, all gases except hydrogen and helium show cooling effect. Hydrogen and helium show heating instead of cooling at room temperature.

**2.** At sufficiently low temperatures, all gases show a cooling effect.

**3.** The fall in temperature is directly proportional to the difference in pressure on the two sides of the porous plug.

**4.** The fall in temperature for a given difference of pressure decreases with the rise in initial temperature of the gas. It was found that the cooling effect is decreased with the increase of initial temperature and becomes zero at a certain temperature and at a temperature higher than this temperature, instead of cooling, heating was observed. This particular temperature at which the Joule-Thomson effect changes sign is called the **temperature of inversion**. The following important conclusions can be drawn from the above result:

**Conclusions :** Any gas below its temperature of inversion will cool on suffering Joule-Thomson expansion.

The fall in temperature (Refer chapter 2) $dT$ for a gas obeying Van der Waals' equation of state is given by

$$dT = \frac{1}{C_P}\left(\frac{2a}{RT} - b\right)dP \quad ...(i)$$

where $dP$ is the difference of pressure on the two sides of the porous plug. From equation (i), we infer that

(*a*) $dT$ will be positive, if $\frac{2a}{RT} > b$

or $$T < \frac{2a}{Rb} \quad ...(ii)$$

Hence, cooling will result if the initial temperature of the gas is less than $\frac{2a}{Rb}$.

(*b*) $dT$ will be zero, if $\frac{2a}{Rb} - b = 0$ *i.e.*, $T = \frac{2a}{Rb}$

This is called temperature of inversion and is represented by $T_i$. Thus,

$$\therefore \quad T_i = \frac{2a}{Rb} \quad ...(iii)$$

(*c*) $dT$ will be negative if $\left(\frac{2a}{Rb} - b\right)$ is negative.

or $$T > \frac{2a}{Rb}$$

or $$T > T_i \quad ...(iv)$$

Hence, heating will take place if the temperature of the gas is greater than its temperature of inversion.

**Inversion Curve :** In order to plot first isenthalpic curve and then inversion curve, let us consider a series of Joule-Thomson experiments are performed on the same gas.

Let $P_1$ and $T_1$ be the initial pressure and temperature (arbitrarily chosen) on the high pressure side of the porous plug and $P_2$ be the pressure ($P_2 < P_1$) on the other side of the plug which is taken to be constant. The temperature of escaping gas after throttling is measured, let it be $T_2$. Now pressure $P_2$ is varied (keeping $P_1$, $T_1$ constant) on the other side and every time temperature $T_2$ is noted. If now a graph is plotted between various values of $P_2$ and the corresponding $T_2$, a curve is obtained as shown in Fig. 7.4. Since in Joule-Thomson expansion, the initial and final enthalpies are equal, the enthalpies at all the points on the curve, *i.e.*, *a, b, c, d, e, f* etc. is the same. Such a curve with constant enthalpy is called **isenthalpic curve.**

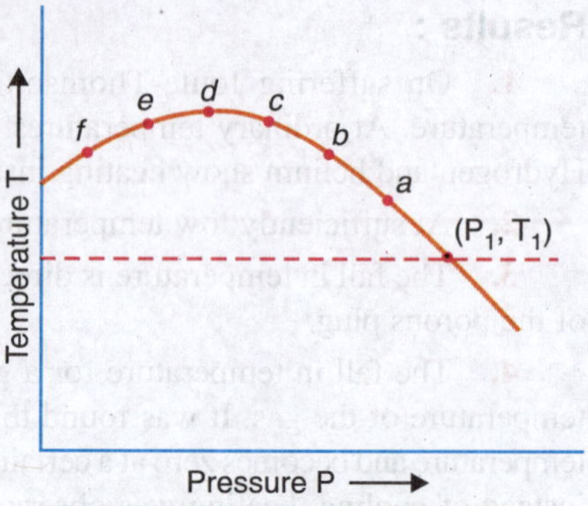


**Fig. 7.4** Isenthalpic curve

Keeping Pressure $P_1$ constant, the temperature $T_1$ on higher pressure side is changed. Again pressure $P_2$ is varied and every time the temperature $T_2$ is noted. The experiment is repeated for various constant values of $T_1$. A family of curves corresponding to different values of enthalpy $H$ are obtained as shown in Fig. 7.5. Every curve possesses a maximum temperature (inversion temperature). The locus of such inversion points is called the **inversion curve.** The dotted curve in Fig. 7.5 represents the inversion curve. The region inside the inversion curve represents the *cooling region* (in which gas shows *cooling effect*) and region out side the inversion curve represents the *region of heating* (in which gas shows *heating effect*).

It can be noted that at any given pressure less than the extreme value there are two inversion temperatures. One is called the 'ordinary' inversion temperature (higher $T_i$) at which the heating effect gives place to cooling and the other (lower $T_i$) inversion temperature at which cooling gives place again to heating. The ratio of these two inversion temperatures for a Van der Waals' gas is approximately 9. Thus, in order to produce cooling effect the gas must be precooled first below the inversion temperature. For most of the gases, ordinary temperature is already below the maximum inversion temperatures, so precooling is not necessary. However, for hydrogen and helium, even the maximum inversion temperature are much below the ordinary temperatures, hence at ordinary temperatures, they show heating effect. These gases must, therefore, be precooled up to – 80 °C and – 243 °C respectively, then they will also show a cooling effect.

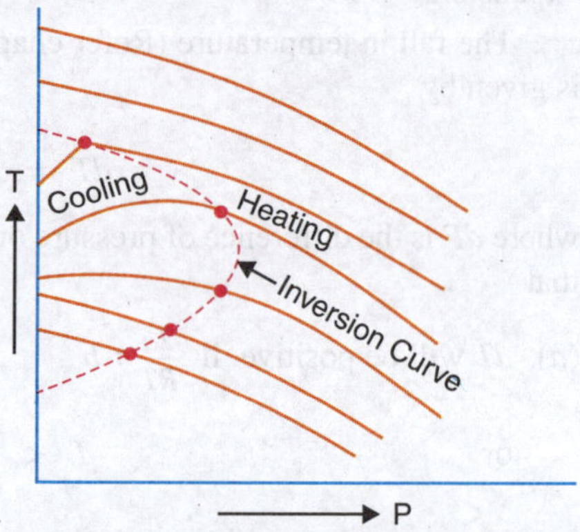


**Fig. 7.5.** Inversion curve

Note that the **inversion temperature is not constant** for a given gas but varies with the initial pressure of the gas.

## 7.6 Liquefaction of Gases

The systematic study of liquefaction of gases started in 1823 when Faraday successfully liquefied chlorine, $CO_2$, hydrogen sulphide etc. only by cooling them and applying pressure. However-gases like oxygen, nitrogen, hydrogen and helium etc. could not be liquefied even by applying very high pressure. Hence, they were termed as **permanent gases**. Andrew's experiment on $CO_2$ in 1862 showed that a gas must be cooled below its critical temperature before being liquefied under pressure. Real success was obtained when Joule-Thomson effect is coupled with regenerative cooling.

## 7.7 Principle of Regenerative Cooling

The cooling produced in Joule-Thomson expansion of a gas depends on the difference of pressure on the sides of the porous plug and the initial temperature for most of the gases, the Joule-Thomson cooling is very small. For example: at a temperature of 20°C when the pressure on the two sides are 50 atmospheres and 1 atmosphere respectively, the air cools only by 11.7°C. In order to increase the cooling, the pressure difference was increased; but the cooling did not increase very much. For example, at a temperature of 20°C when the pressure on two sides were kept 210 atmospheres and 1 atmosphere, the air was cooled only by 42°C. However, the cooling effect can be intensified by employing the process called the regenerative cooling.

**Principle :** In this process a portion of the gas which suffers Joule-Thomson expansion and becomes cooled is employed to cool other portions of the incoming gas before the later reaches the nozzle to suffer Joule-Thomson expansion. After suffering the Joule-Thomson expansion at the nozzle, the incoming gas becomes further cooled. By continuous repetition of this process the temperature of the gas coming out of the nozzle falls continuously till the gas starts liquefying on the low pressure side. The commutative process used to cool the gas continuously is called the regerative cooling.

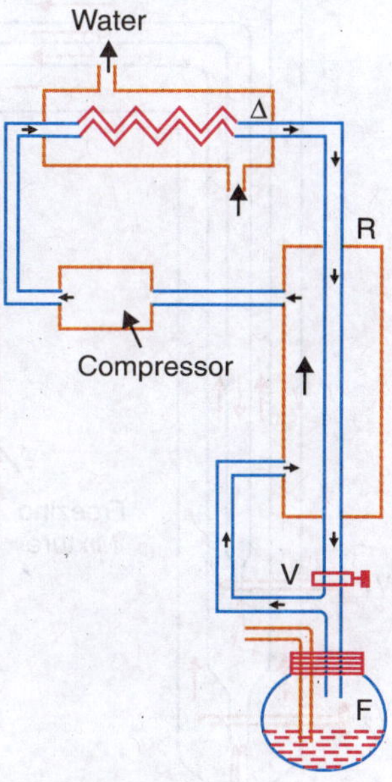

Fig. 7.6

**Experimental Arrangement :** The schematic arrangement of regenerative principle is shown in Fig. 7.6. The gas is compressed to a high pressure in the compressor. The heat of compression evolved is removed by passing the gas through spiral contained in the water cooled jacket *A*. The gas next enters the inner tube of regenerator *R* and suffers Joule-Thomson expansion at the nozzle *V* so that it cools by a small amount. This cooled gas returns by the outer tube and absorbs heat from the incoming high pressure gas before it reaches the compressor at the same temperature as the incoming gas at *R*. The gas is again compressed, cooled by *A* and re-enters at *R*. As time passes, the gas approaching *V* becomes more and more cooled till the Joule-Thomson cooling at *V* is sufficient to liquefy the gas. A portion of the escaping gas then condenses inside the Dewar flask *F*. At this stage, the temperature of the whole apparatus becomes steady and the temperature of the low pressure gas in *R* which is less than the temperature of the adjacent high pressure gas at every point of the inner tube. Thus, the low pressure gas extracts heat from the incoming gas.

## 7.8 Liquefaction of Air-Linde's Process

Linde in 1895 used *Joule-Thomson effect* and the *principle of regenerative cooling* in the liquefaction of air. The principle involved in cooling the incoming gas by the gas which has already undergone cooling due to Joule-Thomson effect. By maintaining a continuous cycle of operation, the initial temperature can be progressively lowered and hence the cooling effect is enormously enhanced.

The apparatus employed by Linde for liquefaction of air is schematically shown in Fig. 7.7. It consists of two-stage compressors $P_1$ and $P_2$. First of all air is compressed to a pressure of about 25 atmospheres by pump $P_1$ and then passed through a tube surrounded by a jacket through which cold water is circulated. This compressed air is passed through KOH solution so that $CO_2$ and water vapour may be absorbed and removed from air. Then the air, coming out of KOH solution, free from $CO_2$ and water vapour, is compressed to a pressure of 200 atmosphere by the pump $P_2$ and is passed through a spiral tube surrounded by a jacket containing a freezing mixture, so that air at high pressure is cooled to – 20 °C. This cooled air at high pressure is allowed to come out of the throttle valve $V_1$ so that Joule-Thomson effect takes place. In the course of expansion, the pressure is suddenly lowered to 20 atmospheres and this causes the temperature to fall further to – 78 °C. The cooled air is then made to flow through the outer tube cooling the air flowing through inner pipe and finally returns back to compressor $P_2$ where it is again compressed to a pressure of 200 atmospheres. It is again circulated through the inner pipe.

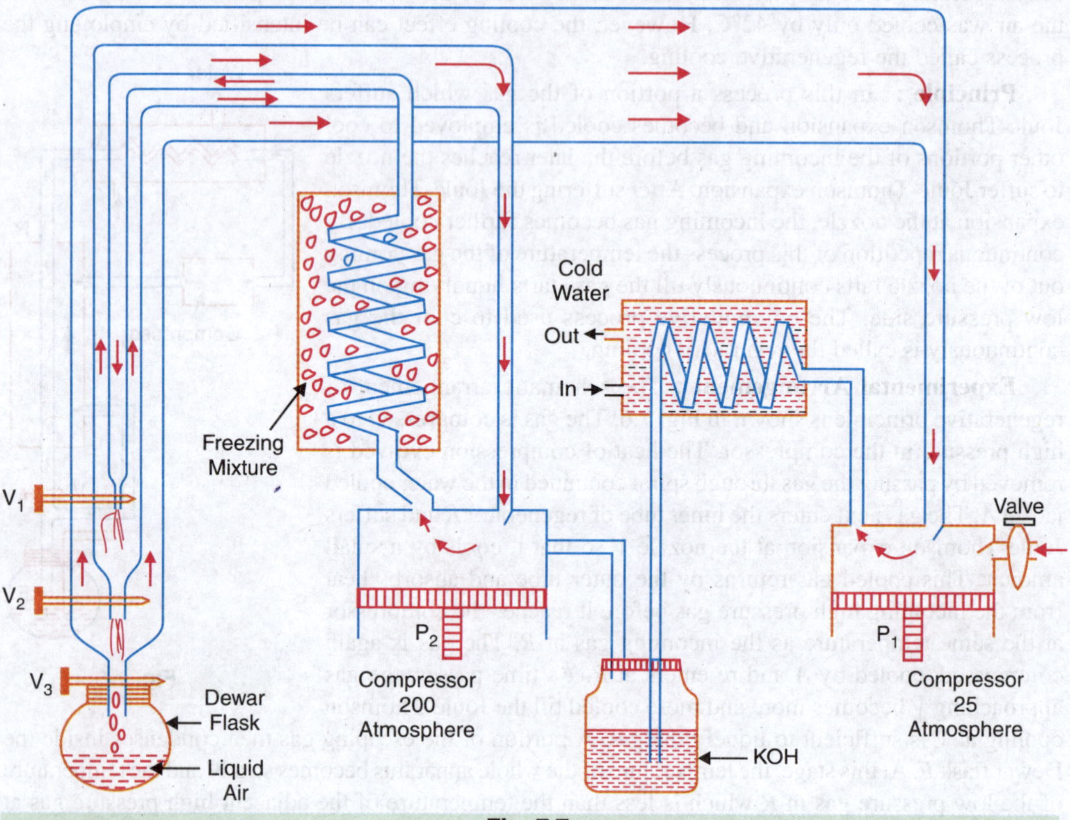


**Fig. 7.7**

After the completion of a few cycles, the temperature of the incoming gas falls sufficiently low and the second throttle valve $V_2$ is opened. The air is allowed to expand to a pressure of one atmosphere when it becomes liquid and is collected in a Dewar flask through the valve $V_3$. The unliquefied gas is again led through the outermost tube back to compressor $P_1$ as shown by the arrow heads. The process is repeated again and again.

The entire arrangement is enclosed in cotton wool to avoid any conduction or radiation.

### Uses of Liquid Air

1. Liquid air is used to cool the substances to a very low temperatures.
2. It is used in developing high vacuum by solidifying the gas less volatile than air.
3. It is used in chemical industries for separation and purification of gases.

Constituents of liquid air (oxygen, nitrogen, argon, neon etc.) can be easily separated and are used commercially as:

(*i*) Oxygen is used in metal cutting such as breaking up of old ships and other machines. It is also used in breathing apparatus in aeroplanes, mines and diving suits.

(*ii*) Nitrogen is used in the manufacture of ammonia and fertilizers.

(*iii*) Argon is used in gas filled electric lamps.

(*iv*) Neon is used in discharge tubes required for illumination and advertising sign boards.

## 7.9 Principle of Cascaded Cooling–Liquefaction of Oxygen

The principle of cascaded cooling can be well understood by considering process employed in liquefaction of oxygen. The apparatus suggested by Pictet (1878) consists of number of compression pumps $P_1$, $P_2$ and $P_3$ arranged in series as shown in Fig 7.8. *A, B* and *C* are the outer jackets containing cold water, liquid methyl chloride and liquid ethylene. The pump $P_1$ compresses methyl chloride gas and it is cooled by cold water circulating in the jacket *A*. As the critical temperatures of methyl chloride is 143°C, it is liquefied. Liquid methyl chloride circulates in the outer jacket *B*.

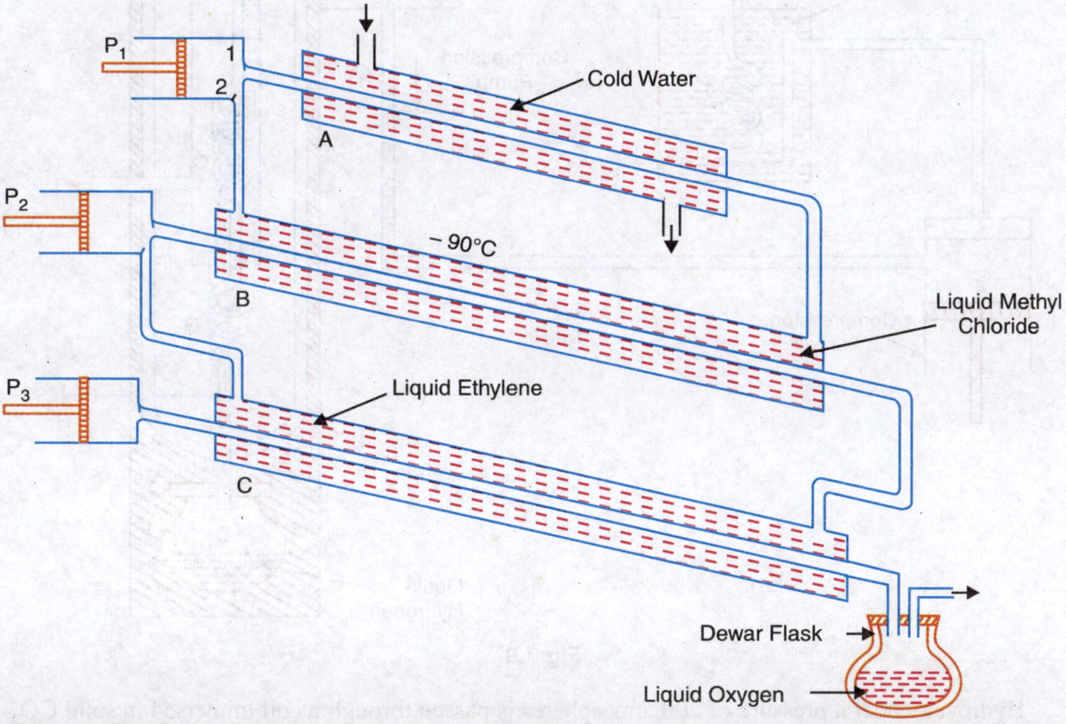


**Fig 7.8**

The pump $P_2$ compresses ethylene gas and it is surrounded by the outer jacket *B* containing liquid methyl chloride at about – 24°C. The liquid methyl chloride is allowed to boil under reduced pressure with the help of pump $P_1$. The temperature of *B* reaches –90°C. The critical temperature of ethylene is 10°C. It is cooled to – 90°C and due to compression, it gets liquefied.

The liquid ethylene in the jacket *C* is allowed to boil under reduced pressure with the help of the pump $P_2$ and finally a temperature of – 160°C is reached. The critical temperature for oxygen is – 118.8°C. Oxygen is compressed to a pressure of about 25 atmospheres and is passed through a spiral tube surrounded by the jacket *C*. Oxygen gets liquefied and is collected in a Dewar flask. Oxygen in the form of gas in the Dewar flask is circulated back to the pump $P_3$ and the process is repeated. This repeated process used is called cascading.

## 7.10 Liquefaction of Hydrogen

Hydrogen cannot be liquefied by cascade process because its critical temperature is –240°C. Linde's ordinary apparatus used for the liquefaction of air cannot be used for hydrogen because the temperature of inversion for hydrogen is –83°C. The gas must initially be cooled to a temperature lower than the temperature of inversion for the cooling to take place due to Joule - Kelvin effect. The original apparatus designed by Dewar (1898) was improved later by Travers, Olszewski, Nernst and others (Fig. 7.9). To have complete insulation, the whole apparatus in enclosed in an outer Dewar flask *L*.

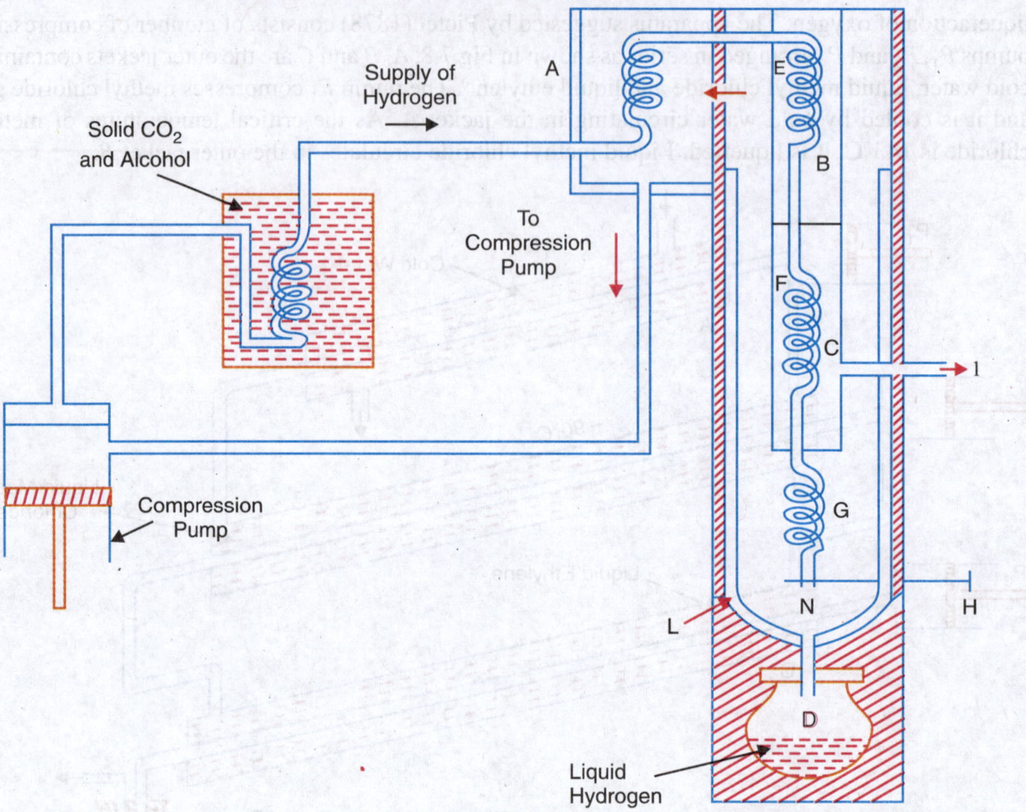


Fig 7.9

Hydrogen under a pressure of 200 atmospheres is passed through a coil immersed in solid $CO_2$ and alcohol. It enters the coil in the chamber *A* where it is further cooled by the outgoing hydrogen.

The chamber *B* contains liquid air and cools hydrogen in the coil *E*. In the chamber *C*, liquid air is allowed to boil under reduced pressure (10 cm of Hg) and hydrogen in the coil *F* is cooled to a temperature of –200°C. This cooled hydrogen passes through the regenerative coil *G* and nozzle *N*. Hydrogen is cooled further due to Joule-Kelivin effect. The cooled hydrogen coming from the nozzle *N* is allowed to circulate back to the pump as shown in Fig 7.9. The process of regenerative cooling continues and after some time hydrogen gets *liquefied* and is collected in the Dewar flask *D*.

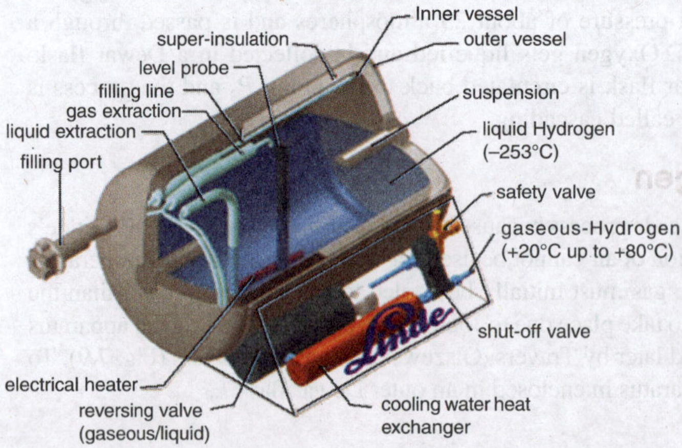


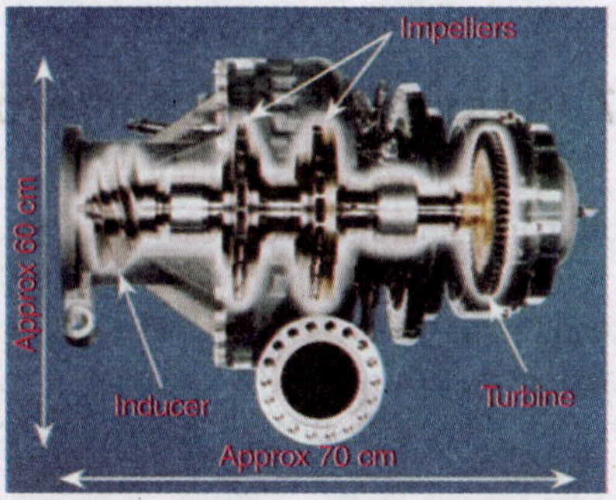

## 7.11 Liquefaction of Helium (K. Onnes Method)

*K.* Onnes liquefied helium in 1908. The temperature of inversion of helium is – 240°C. Kapitza had liquefied helium by precooling it and passing it through a coil surrounded by a bath containing liquid hydrogen boiling under reduced pressure. In this way helium was cooled to – 258°C.

Liquifaction of Helium

The apparatus used for liquefaction of helium is shown in Fig. 7.10. Helium gas at a pressure of 40 atmospheres enters the spiral tube at *A*. It is divided into two portions. Helium passing through the spiral $S_1$ in *E* is cooled because it is surrounded by hydrogen boiling under reduced pressure. The other portion of helium gas passing through the spiral $S_2$ in *F* is cooled due to outgoing cooled helium gas. Similar processes take place in the spirals $S_3$ and $S_4$. The process is repeated and when the temperature of the helium is sufficiently low, it gets liquefied after passing through the nozzle *N*. At *N*, cooling take place due to Joule-Kelvin effect. This outgoing helium is compressed again by *P* and fed back to the spirals $S_1$ and $S_2$. Liquefied helium is collected in the Dewar flask. The whole apparatus is surrounded by Dewar flasks to provide perfect heat insulation.

Simon has been able to liquefy helium by using activated charcoal. When helium is adsorbed by charcoal, heat is evolved and when it is desorbed (removed), the temperature of the gas falls. Helium is adsorbed in large quantities by activated charcoal immersed in liquid hydrogen. The apparatus is enclosed in an evacuating vessel to minimize heat exchanges with the surroundings. On pumping off helium, its temperature falls considerably and it gets liquefied.

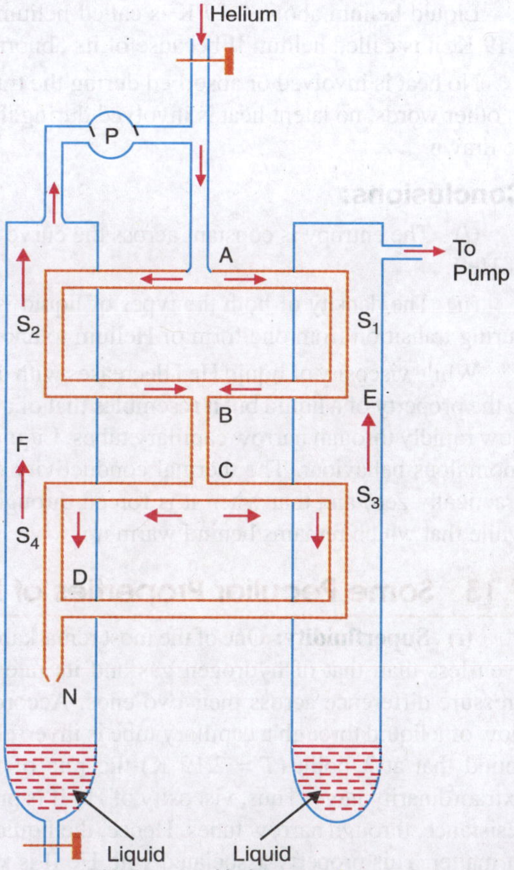


Fig. 7.10

## 7.12 Helium I and Helium II

K. Onnes observed no sign of solidification when liquid helium was cooled at ordinary pressures. It was found that at a temperature of 2.19 K, the liquid which was contracting when cooled, suddenly began to expand (Fig. 7.11).

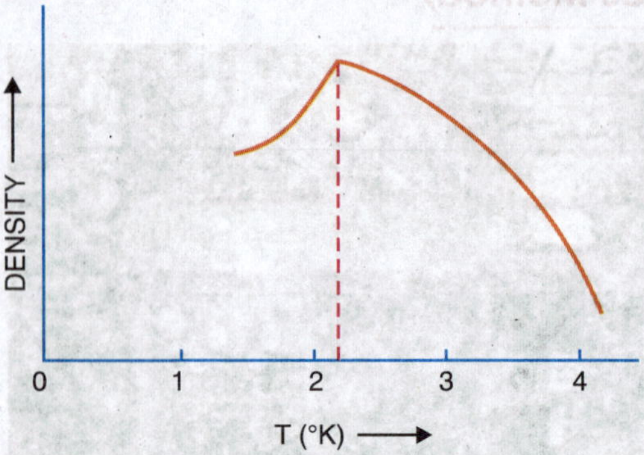


Fig. 7.11

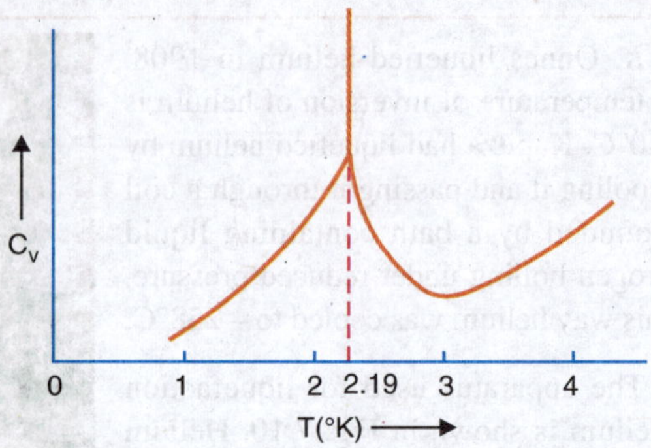


Fig. 7.12

The specific heat of liquid helium increases up to 2.19 K and *at this temperature there is a sudden* and abnormal increase in the specific heat. Beyond 2.19 K, the specific heat first decreases and then increases (fig. 7.12).

The specific heat-temperature graph resembles λ and hence this temperature at which the specific heat ($C_v$) changes abruptly (2.19 K) is called the **λ point**.

Liquid helium above 2.19 K is called helium I because it behaves in a normal way and below 2.19 K, it is called helium II because of its abnormal properties.

No heat is involved or absorbed during the transition from Helium I to Helium II and *vice versa*. In other words, no latent heat is involved during this transition. Hence the following conclusions can be drawn.

### Conclusions:

(*i*) The entropy is constant across the curve *i.e.*, entropy of He II is practically the same as that of He I.

(*ii*) The density of both the types of liquid is about the same *i.e.*, there is no change of density during transition from one form of Helium to another.

While viscosity of liquid He I decreases with decrease in temperature and this property is contrary to the property of a liquid but it resembles that of a gas. The viscosity of He II is almost zero and it can flow rapidly through narrow capillary tubes. Liquid He I is a normal liquid while He II presents a very anomalous behaviour. The thermal conductivity of He II is abnormally high, its internal friction is practically zero and that when it is forced through a capillary tube, the emerging liquid He II cools while that which remains behind warm up.

## 7.13 Some Peculiar Properties of Helium II

(*i*) **Superfluidity:** One of the most remarkable properties of He II is its abnormally low viscosity, even less than that of hydrogen gas and its rate of flow through capillaries is independent of the pressure difference across their two ends. According to Poiseuille's law of liquid flow, the rate of flow of a liquid through a capillary tube is inversely proportional to its viscosity. In case of He II, it is found that at λ point ($T$ = 2.19 K) the rate of flow increases abruptly and below it the flow is extraordinarily large. Thus, viscosity of He II is practically zero and can flow rapidly, almost without resistance, through narrow tubes. Hence, the liquid He II is called **superfluid**, a new frictionless state of matter. This property associated with He II is known as **superfluidity**.

(*ii*) **High heat conductivity:** Liquid He I has a quite normal heat conductivity, of the order of

$10^{-4}$, which is of the same order for gases at ordinary temperatures. But He II is found to have an extraordinarily high coefficient of thermal conductivity. He II is nearly 800 times more conducting than copper and about $13.5 \times 10^6$ times more than liquid He I, the absolute value being 820 cal $cm^{-1}$ $deg^{-1}$ $sec^{-1}$. It is also observed that the heat flow in not proportional to the temperature gradient.

**(*iii*) High electrical conductivity:** In 1947, Daunt and Mendelssohn noticed the striking similarity between the flow of helium atoms in liquid He II and the phenomenon of *superconductivity*. In a superconducting state, no trace of electrical resistance is observed in a material. In other words, the electrons move freely in the conductor, unhindered by any electrical resistance.

**(*iv*) Formation of rolling film:** The behavior of He II is very peculiar. Liquid He II forms a thin film (of the order of several hundred atoms thick) on all solid surfaces. This film is called **rolling film** (also called rolling-sim on film). It is through this film that He II flows from one vessel to another. If a tube containing He II be placed in a He II bath, it is observed that:

(*a*) If the liquid level inside the tube is lower than that out side, the liquid helium II from outside starts creeping in to the tube along its surface in the direction of arrows shown in Fig. 7.13 (*a*). The process of creeping continues till both the inside and outside levels of the tube become equal.

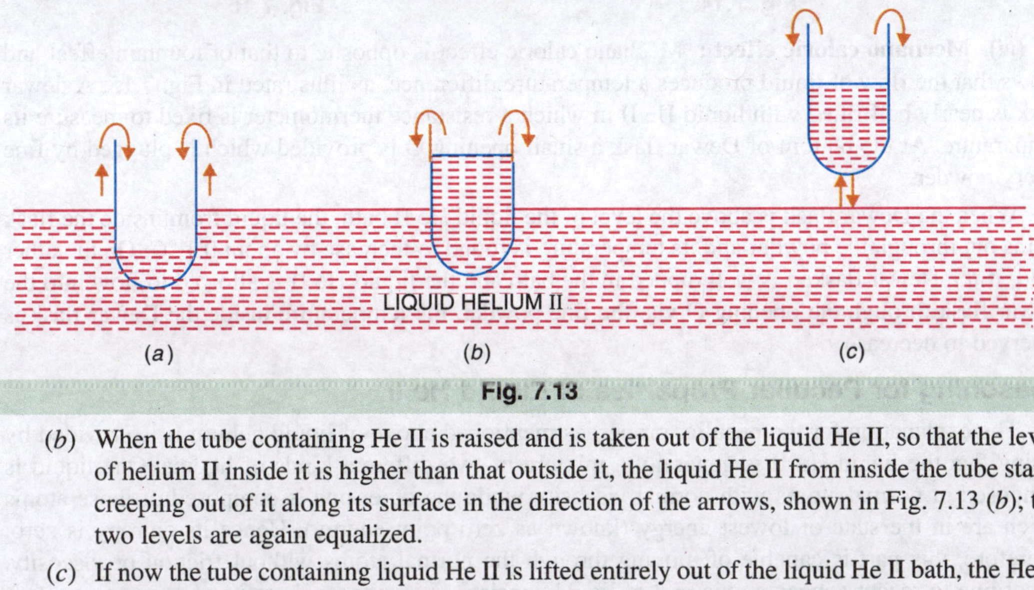


**Fig. 7.13**

(*b*) When the tube containing He II is raised and is taken out of the liquid He II, so that the level of helium II inside it is higher than that outside it, the liquid He II from inside the tube starts creeping out of it along its surface in the direction of the arrows, shown in Fig. 7.13 (*b*); till two levels are again equalized.

(*c*) If now the tube containing liquid He II is lifted entirely out of the liquid He II bath, the He II inside the tube now creeps out and flows down along the surface of the tube, collects at its bottom in the form of drops and falls into the liquid bath. It continues to flow out until the tube is empty Fig. 7.13(*c*).

**(*v*) The 'Fountain' effect:** Liquid He II exhibits a peculiar property in the form of fountain effect as shown in Fig. 7.14. Heat radiation falling locally on a system containing liquid He II produces a pressure difference. The apparatus was first suggested, in 1938, by Allen and Misener. The apparatus consists of liquid He II taken in a Dewar flask. Liquid He II is also taken in a tube *AB* open at both the ends such that part A is in the form of capillary and is kept projecting out of the liquid He II bath. The lower part B has a small orifice O at the bottom of the tube. A fine *emery* powder is packed in part *B* of the tube. The interspacing between emery fine particles provide fine capillaries for the liquid He II to flow in the tube *AB*. When heat radiation from a light torch fall on the powder, liquid He II is found to spurt out of the capillary tube in the form of fountain which is even of the height 30 cm. This phenomenon is due to the absorption of heat energy by the fine emery powder and the capillary action of the fine emery particles.

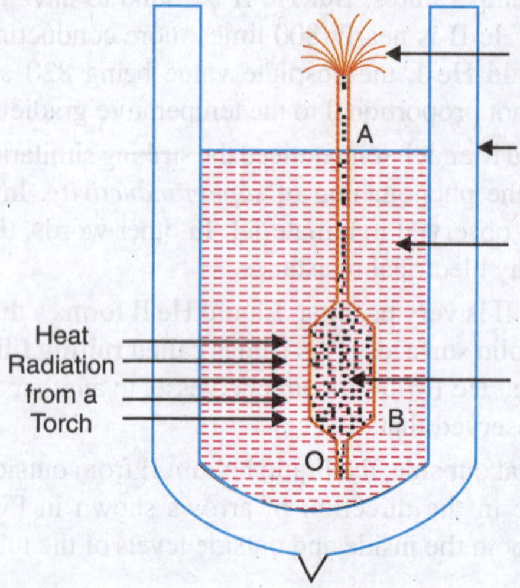

Fig. 7.14

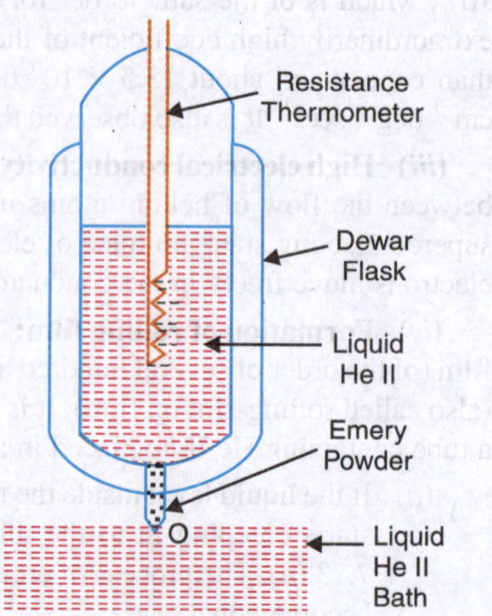

Fig. 7.15

***(vi)* Mechano caloric effect :** Mechano caloric effect is opposite to that of fountain effect and shows that the flow of liquid produces a temperature difference, as illustrated in Fig. 7.15. A dewar flask is nearly half filled with liquid He II in which a resistance thermometer is fixed to measure its temperature. At the bottom of Dewar flask a small opening O is provided which is plugged by fine emery powder.

When the Dewar flask is above the level of the liquid He II bath, the liquid from inside the flask drains out through the powder and its temperature is found to increase by about 0.1 °C. On the other hand, if the Dewar flasked is well dipped in the bath and the liquid He II is allowed to flow into the Dewar flask through the opening O via. the fine powder, the temperature inside the Dewar flask is observed to decrease.

## Reasoning for Peculiar Properties of Liquid He II

The explanation for the peculiar properties mentioned above of liquid helium II is provided by stating that the liquid He II is generally a mixture of two different kinds of liquids. One liquid is composed of normal atoms with normal viscosity while the other one is composed of those atoms which are in the state of lowest energy (known as zero point energy). Hence its entropy is zero. Therefore, this part is capable of moving through the normal atoms without friction or viscosity. According to recent concepts, this part of liquid consists of 'condensed' atoms and appears at its λ point for the first time. This liquid is called a 'superfluid' which can flow through capillary tubes without showing any resistance. Such a fluid property is known as superfluidity. On lowering the temperature, the superfluid grows in quantity at the expense of normal component, untill at absolute zero, the whole liquid becomes superfluid.

## 7.14 Helium 'A Unique Liquid'

Helium was the last gas to be liquefied on account of having the lowest critical temperature (– 267.8° C) of all known gases. It is a colourless, transparent, very

volatile liquid and has the lowest boiling point of 4.2 K at a pressure of 1 atmosphere. It has no normal freezing point and shows some of the peculiar properties (see Fig. 7.13) which single it out as almost an exceptional and unique substance.

An amazing property of helium system is that solid helium cannot be obtained merely by lowering the temperature of helium liquid. Kamerlingh Onnes failed to solidify helium despite the fact that he reached a temperature of 0.84 K. Solid helium was first obtained by Keesom at a pressure of 250 atmospheres at 4.2 K, while only under 23 atmospheres at 1.1 K. Further at 5800 atmospheres, solid helium is obtainable at a temperature as high as 42 K. This is a curious property of helium system that although liquid cannot exist above the critical temperature (5.2 K), solid can exist if sufficiently high pressure is applied. Hence, at high enough pressures, the melting point exceeds the critical temperature and so the solid helium 'melts' to form the gas. Moreover, the three phases solid, liquid and vapour are never found to coexist *i.e.*, helium has *no tripple point.* Hence helium is a unique liquid.

Helium II has a higher heat of vaporization and smaller surface tension. The large specific heat anomaly of liquid helium at 2.19 K is due to rapid decrease in its entropy with decreasing temperatures. This is a complicated phenomenon and statistical mechanics cannot be applied to get the proper solution. However, Bose-Einstein statistics can be successfully applied in this case.

## 7.15 Production of Low Temperatures

Quite long ago, it was noticed that there is a lower limit for a temperature scale. Today, we know that there is no limit for high temperature but there is a limit for low temperature. The lowest temperature corresponds to 0°K (– 273.16°C) called the absolute zero temperature. Attempts have been made to reach absolute zero temperature.

Temperatures below zero degree centrigrade can be obtained with the help of freezing mixtures. Temperatures up to –65°C can be obtained with KOH and ice. With the liquefaction of gases, temperatures lower than –65°C could be achieved. With liquid helium boiling under normal pressure at temperature of –268.9°C can be reached. By boiling liquid helium under reduced pressure, temperature of the order of 1 K could be obtained. With liquid helium isotope ($He^3$) boiling under reduced pressure, a temperature of 0.4 K can be reached. Temperatures below 0.4 K can be reached by 'adiabatic demagnetisation methods' as per Debye (1926) and Giauque (1927).

## 7.16 Adiabatic Demagnetisation

Debye, Giauque and Macdougall were able to produce temperatures below 1 K with the help of gadolinium sulphate. Haas and Kramers used the magnetic balance for a number of paramagnetic substances. They found that potassium and chromium alum give a much lower temperature.

**Method.** The apparatus used is shown in Fig 7.16. The paramagnetic specimen (salt) is suspended in a vessel A, which is surrounded by liquid helium. Liquid helium taken in Dewar flask *D*, is boiled under reduced pressure. It is surrounded by another Dewar flask $D_2$ containing liquid hydrogen. The salt is in contact with the helium gas. A magnetic field of the order of 30,000 Gauss is applied.

When the magnetic field is switched on the specimen (salt) is magnetised. The heat due to magnetisation is removed by first introducing hydrogen gas into A and then pumping it off with high vacuum pump, so that the specimen is thermally isolated. In the mean time, the specimen (salt) picks up the temperature of the liquid helium *i.e.*, the specimen and the liquid helium are at the same temperature. Now the magnetic field is switched off. Adiabatic demagnetisation of the specimen takes place and its temperature falls. The temperature of the specimen is determined by fitting a co-axial solenoid coil round the tube A and measuring the self-inductance and hence susceptibility of the substance at the beginning and at the end of experiment. Then temperature *T* is calculated by Curie law $\chi = \frac{C}{T}$.

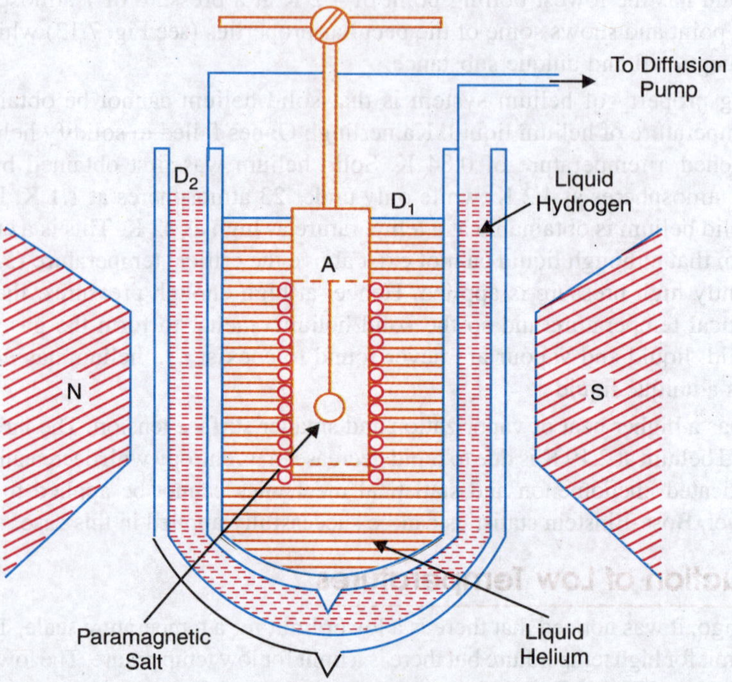


**Fig. 7.16**

Haas, in 1944, was able to produce temperatures upto 0.002 K using a double sulphate of potassium and aluminium. Klerk, Stenland and Gorter used powdered mixed crystals of chromium alum and aluminium alum and went down to a temperature of **0.0014 K**.

**Theory.** When a paramagnetic material is placed in a magnetising field $H$, its elementary magnetic dipoles get aligned parallel to the direction of the field. *The magnetic moment per unit volume* thus produced is called the **intensity of magnetisation** (*I*). According to Curie's law, this intensity of magnetisation is directly proportional to the magnetising field $H$ and inversely proportional to temperature $T$ of the paramagnetic substance. Thus

$$I \propto \frac{H}{T}$$

$$I = C\left(\frac{H}{T}\right) \qquad ...(i)$$

where $C$ is a constant, known as Curie constant.

If $V$ is the volume of 1 mole (*i.e.*, molar volume), then intensity of magnetisation of 1 mole of paramagnetic substance

$$M = IV$$

Substituting for I from eqn (*i*), we get

$$M = CV\left(\frac{H}{T}\right) \qquad ...(ii)$$

In the experiment, let 1 mole of paramagnetic substance is placed in magnetising field $H$. Then its thermodynamic behaviour can be expressed in terms of thermodynamic variables $P$, $V$, $T$ and $S$. In thermodynamic system, an increase in pressure $P$ results in decrease in volume $V$, analogously in our case any increase in $H$ results in an increase in $M$. Hence replacing $P$ by – H and $V$ by $M$ in Maxwell's third thermodynamic relation

$$\left(\frac{\partial T}{\partial P}\right)_S = \left(\frac{\partial V}{\partial S}\right)_P$$

We have
$$\left(\frac{\partial T}{\partial H}\right)_S = -\left(\frac{\partial M}{\partial S}\right)_H$$

$$= \frac{-\left(\frac{\partial M}{\partial T}\right)_H}{\left(\frac{\partial S}{\partial T}\right)_H}$$

Multiplying $N^r$ and $D^r$ by $T$, we get

$$\left(\frac{\partial T}{\partial H}\right)_S = \frac{-T\left(\frac{\partial M}{\partial T}\right)_H}{T\left(\frac{\partial S}{\partial T}\right)_H}$$

$$= \frac{-T\left(\frac{\partial M}{\partial T}\right)_H}{\left(\frac{\partial Q}{\partial T}\right)_H} \qquad [\because T.\partial S = \partial Q]$$

$$= -\frac{T}{C_H}\left(\frac{\partial M}{\partial T}\right)_H \qquad ...(iii)$$

where $\left(\frac{\partial Q}{\partial T}\right)_H = C_H$, the specific heat of the substance at constant magnetic field $H$. The unit is Joule/mole K.

Since the process is carried out adiabatically (in which entropy $S$ remains unchanged, *i.e.*, constant), we may write,

$$dT = -\frac{T}{C_H}\left(\frac{\partial M}{\partial T}\right)_H .dH$$

Therefore, when a field changes from $H_1$ to $H_2$, the change in temperature

$$\Delta T = -\frac{T}{C_H}\int_{H_1}^{H_2}\left(\frac{\partial M}{\partial T}\right)_H .dH \qquad ...(iv)$$

Differentiating eqn. (*ii*) *w.r.* to $T$, we get

$$\left(\frac{\partial M}{\partial T}\right)_H = -\frac{CVH}{T^2} \qquad ...(v)$$

Substituting in equation (*iv*), we have

$$\Delta T = -\frac{T}{C_H}\int_{H_1}^{H_2}\left(-\frac{CVH}{T^2}\right)dH$$

$$= \frac{CV}{C_H.T}\int_{H_1}^{H_2} H.dH$$

or,
$$\Delta T = \frac{CV}{2C_H.T}\left(H_2^2 - H_1^2\right) \qquad ...(vi)$$

If the magnetic field is reduced from $H_1 = H$ to $H_2 = 0$, then the change in temperature will be

$$\Delta T = -\frac{CV}{2C_H.T}.H^2 \quad ...(vii)$$

This is the desired relation. The following conclusions can be drawn from this equation (*vii*).

**Conclusions :**

(*a*) The temperature of the paramagnetic substance decreases on decreasing the magnetising field $H$ (indicated by negative sign of $\Delta T$), and

(*b*) Greater is the initial field H and lower is the initial temperature $T$, greater is the temperature fall $\Delta T$.

It should be noted that the quantity $CV$ is the Curie constant per mole. If 1 gm of paramagnetic substance (salt) is taken, then $CV$ would stand for Curie constant per gm.

### *T–S* Diagram

The cooling produced by adiabatic demagnetisation is explained by using *T–S i.e.*, temperature entropy diagram, as shown in Fig. 7.17.

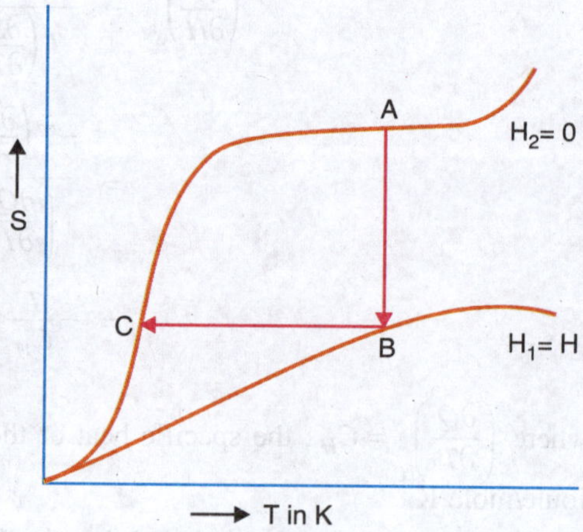


Fig. 7.17

In the diagram, the entropy of the substance is represented schematically as a function of temperature for two different magnetising fields constants, *i.e.*, $H_2 = 0$ and $H_1 = H$. The arrow $AB$ represents the change in entropy during isothermal magnetisation (at constant temperature). This arrow shows the decrease in entropy of the substance during magnetisation (from $H = 0$ to $H$). Now if the field is switched off, the substance demagnetises adiabatically, so that the temperature falls while during this process entropy remains constant, as represented by an arrow $BC$. During this process, *i.e.*, adiabatic demagnetisation the temperature of the order of $10^{-4}$ can be reached. The main hurdle in this method is the measurement of such a low temperature. This problem is resolved by using the Curie law. Assuming the validity of Curie law, and if the susceptibility of the sample is measured, the **Curie temperature** $T_C = \frac{C}{\chi}$ (also known as **magnetic temperature**) can be calculated. Knowing the Curie temperature, the thermodynamical temperature T can be evaluated. The temperature measured by this method is of the order of $10^{-4}$ K. We can still go down up to the order of $10^{-6}$ by making use of *nuclear magnetism*. But it should be rembered that absolute zero can never be attained in actual practice.

## 7.17 Conversion of Magnetic Temperature to Kelvin Temperature

The thermodynamical behaviour of a paramagnetic crystal, placed in a magnetic field H, depends also on the strength of the magnetising field in addition to pressure and volume. Therefore, the usual relation

$$\delta H = dU + dW = dU + PdV$$

is modified in the form

$$\delta H = dU + PdV - BdI \quad ...(i)$$

Here $B$ is the magnetic flux density and $dI$ is the change in the intensity (or strength) of magnetisation per gram molecule. Therefore,

$$I = \chi VB$$

where $\chi$ is the susceptibility per unit volume.

Assuming the change in volume to be negligible *i.e.*, $dV = 0$

$$\delta H = dU - BdI$$

when the magnetic field is kept zero and constant,

$$\text{we have} \qquad \theta = \left(\frac{\partial U}{\partial S}\right)_{B=0}$$

$$\text{or} \qquad \theta = \frac{\left(\frac{\partial U}{\partial T}\right)_{B=0}}{\left(\frac{\partial S}{\partial T}\right)_{B=0}} \qquad ...(ii)$$

Here θ is the temperature expressed on the Kelvin scale and $T$ is the magnetic temperature.

To determine the value of θ, the values of $\left(\frac{\partial U}{\partial T}\right)_{B=0}$ and $\left(\frac{\partial S}{\partial T}\right)_{B=0}$ are to be evaluated. It is done as follows :

The paramagnetic substance is taken at a known initial temperature $\theta_i$. Here $\theta_i$ is found using a helium vapour pressure thermometer (Pl. see article 7.18).

The initial state of the substance, when $B = 0$, is represented by the point $A$ (Fig. 7.18). The entropy at $A$ is say, $S_1$. When the field is increased, keeping temperature constant, the point $B_1$ is reached. Let $S_2$ be the entropy of the substance at $B_1$. The quantity of heat $\delta Q$ produced during magnetisation is directly measured. The change in entropy is $dS$, given by

$$dS = S_2 - S_1 = \frac{\delta Q}{\theta_i}$$

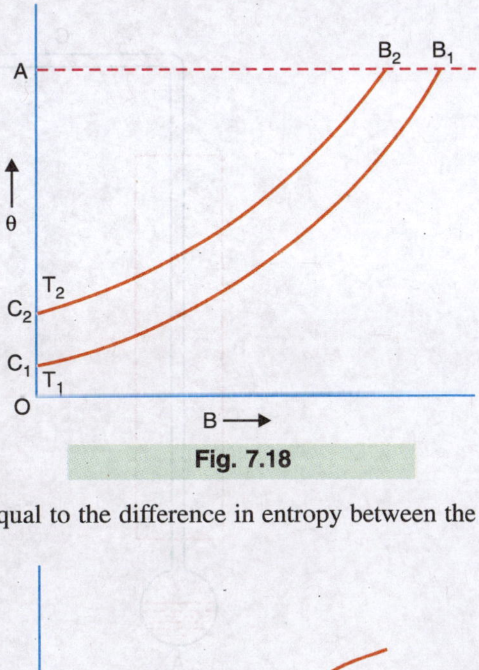

Fig. 7.18

Now the substance is adiabatically demagnetised untill the field becomes zero and the point $C_1$ is reached. During this process, there is no change in entropy. The difference in entropy between the points $C_1$ and $A$ is equal to the difference in entropy between the points $B_1$ and $A$.

Suppose, the temperature at the points $C_1$ is θ degrees Kelvin or $T$ degrees magnetic. The values $T$ and $(S_2 - S_1)$ will depend upon the value of the magnetising field $B$. The experiment is repeated with different initial fields and graph is plotted between $dS$ and $T$ (Fig. 7.19).

From this graph, the value of $\left(\frac{\partial S}{\partial T}\right)_{B=0}$ is found.

To determine the value of $\left(\frac{\partial U}{\partial T}\right)_{B=0}$, the salt is heated from a temperature $T$ to $T'$ and the heat absorbed is estimated directly. This can be done by introducing γ-rays or radiations from a heated filament.

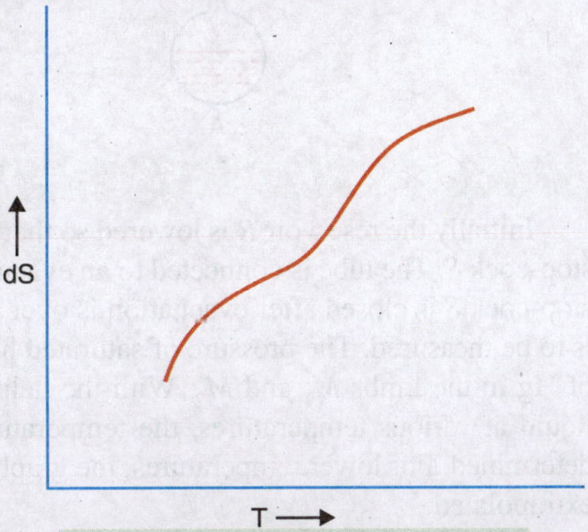

Fig. 7.19

Substituting the values of $\left(\frac{\partial S}{\partial T}\right)_{B=0}$ and $\left(\frac{\partial U}{\partial T}\right)_{B=0}$ in equation (*ii*), the value of θ in degree Kelvin can be calculated.

A lower temperature can be reached by using a mixed salt because magnetic interactions between the paramagnetic ions are weakened when the paramagnetic salt is mixed with a non-paramagnetic salt. A mixed crystal can also be used.

Haas and Wiersma (1935) were able to produce temperatures up to 0.003 K using ferric ammonium alum and potassium chrome alum.

## 7.18 Measurement of Very Low Temperatures: (Helium Vapour Pressure Thermometer)

Helium vapour pressure thermometer is used to measure temperatures up to **0.7 K**. The apparatus consists of a bulb *A* containing liquid helium. This bulb is connected to the manometer limbs $M_1$ and $M_2$ through a connecting tube *C*. The tube *C* is surrounded by a copper tube *B* to ensure uniform temperature of the vapour. *R* is reservoir containing mercury (Fig. 7.20).

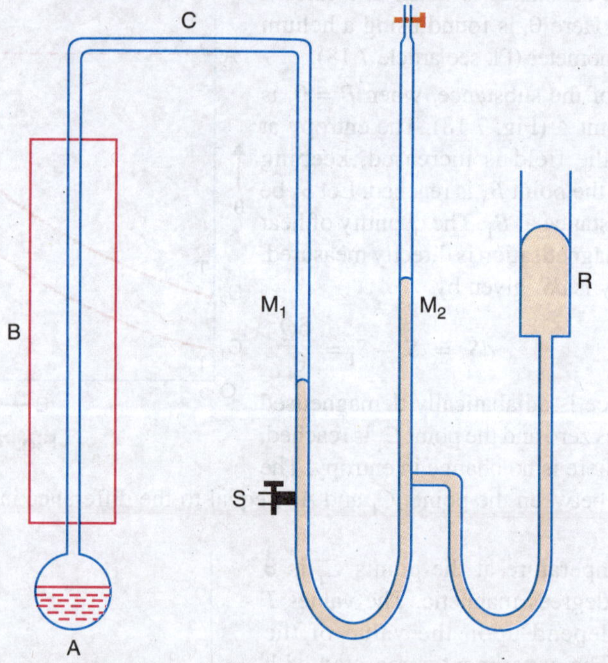


**Fig. 7.20**

Initially the reservoir *R* is lowered so that the mercury in the manometers $M_1$ and $M_2$ is below the stop-cock *S*. The tube is connected to an evacuation pump to remove air in the tube *C* and bulb *A*. The stop-cock *S* is closed after evacuation is over and the bulb *A* is placed in the bath whose temperature is to be measured. The pressure of saturated helium vapour is measured from the difference in levels of Hg in the limbs $M_1$ and $M_2$. With the help of constant tables, giving the vapour pressure of the liquid at various temperatures, the temperature corresponding to any observed vapour pressure is determined. For lower temperatures, the graph between saturated vapour pressure and temperature is extrapolated.

**TABLE 7.3 Vapour pressure of Helium ($He^4$) at different temperatures**

| *Temp* K | *Pressure* mm of Hg |
|---|---|
| 5.00 | 1460 |
| 4.50 | 980 |
| 4.00 | 615 |
| 3.50 | 353 |
| 3.00 | 181 |
| 2.50 | 77 |
| 2.00 | 23.4 |
| 1.50 | 3.6 |
| 1.00 | 0.12 |
| 0.50 | $1.6 \times 10^{-5}$ |
| 0.10 | $3.4 \times 10^{-32}$ |

## 7.19 Superconductivity

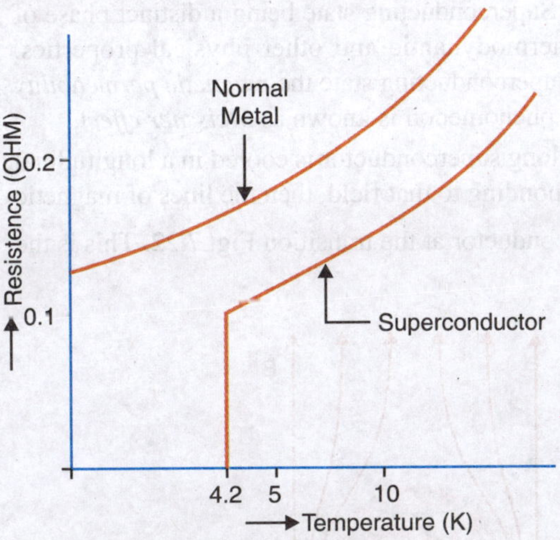


Fig. 7.21

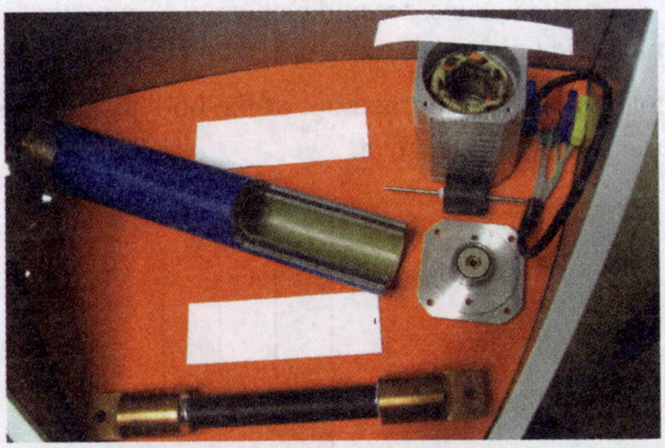

Superconductors

The electrical resistivity of many metals and alloys drops suddenly to zero when cooled to a sufficiently low temperature, particularly in the *liquid helium range*. This phenomenon was first observed by Kamerling Onnes in 1911, while studying the electrical resistance of Mercury at very low temperatures. He observed that the resistance of mercury first decreases regularly like that of a metal but at 4.2 K, the resistivity *suddenly decreases to zero* (actually becomes one millionth of its original value at the melting point) as shown in Fig. 7.21. The temperature at which this phase transition occurs and the resistance suddenly drops to zero, is called the *critical temperature* ($T_c$). Similar results were obtained for other metals like Al, Pb, Sn, Cd, Zn, In etc. *This phenomenon of disappearance of electrical resistance of material below a certain temperature is called superconductivity and the material in this state is called a superconductor.*

The discovery of superconductivity attracted the attention of many scientists because the materials with no electrical resistance means a lot of power saving. This property could be exploited to fabricate powerful and economical devices which consumes negligible (almost zero) amount of electrical energy. For example: if a magnet is moved in and taken out of solenoid (coil) rapidly, then an induced *e.m.f.* $e = -\dfrac{d\phi}{dt}$ is produced, thereby a momentary current through the solenoid is changing. If, however, the temperature of the solenoid is brought down so that it becomes superconducting, then the current will continue to flow for days and even month after the removal of the magnet. Thus, an electromagnet made-up of a superconducting material can work efficiently for years together, even after the removal of power supply. However, the difficulty was to have very low temperatures for practical purposes. This was not feasible to manufacture such devices. Therefore, a lot of research is going on in this branch of physics.

It was discovered in 1977, that a compound $Nb_3$ Ge to have critical temperature ($T_C$) of 23.2 K. In 1986, Bednorz and Miller discovered the superconducting transition temperature for $La_{1.85}$ $Sr_{0.15}$ $CuSO_4$ was about 40 K. In 1987, a ceramic superconductor of the composition $YBa_2$ $Cu_3$ $O_7$ was discovered showing $T_C$ equal to 90 K. Again higher value of $T_C$ was observed for thallium cuprates giving $T_C$ to be 125 K. The highest superconducting transition temperature reported till today is about 150 K. Higher the $T_C$, better suitable the compound will be for, practical utility.

## 7.20 Meissner Effect

When a metal becomes superconductor, its *thermal conductivity decreases* considerably and tends to zero approximately as the cube of the temperature. Superconducting state being a distinct phase of matter having characteristic electrical, magnetic, thermodynamic and other physical properties. Superconductors have *zero electrical resistance*. In a superconducting state the *magnetic permeability* ($\chi$) *is also zero* and shows *perfect diamagnetism*. This phenomenon is known as *Meissner effect*.

In 1933, Meissner and Ochsenfild found that if a long superconductor is cooled in a longitudinal magnetic field to below its critical temperature corresponding to that field, then the lines of magnetic induction $\vec{B}$ are pushed out of the body of the superconductor at the transition Fig. 7.22. This is the basis of Meissner effect.

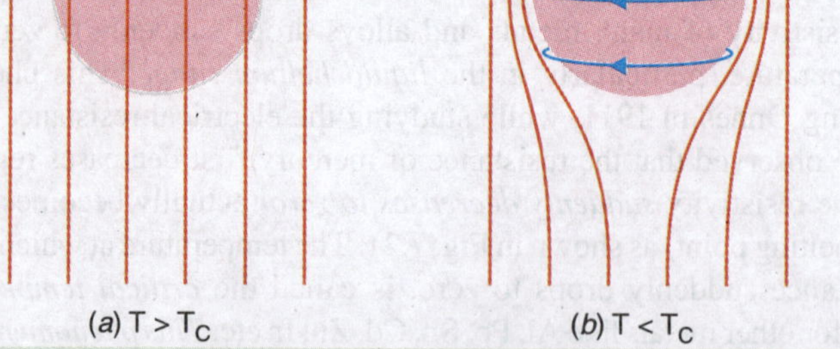


**Fig. 7.22.** The Meissner effect. (*a*) An applied magnetic field can exist inside a superconductor at temperatures above its critical temperature $T_C$. (*b*) When the superconductor is then cooled below $T_{C.\ face}$ currents appear whose effect is to expel the magnetic field from the interior of the superconductor.

## Theory

Using Maxwell's version of Ampere's circuit law, it can be shown that a resistanceless conductor restricts the time dependence of the field $\vec{B}$ through the equation

$$\nabla^2 \frac{d\vec{B}}{dt} = \frac{ne^2\mu}{m}\frac{d\vec{B}}{dt} \qquad ...(i)$$

This equation may be rewritten as

$$\nabla^2 \frac{d\vec{B}}{dt} = \frac{w_s^2}{c^2}\frac{d\vec{B}}{dt} = \frac{1}{\lambda_L^2}\frac{d\vec{B}}{dt} \qquad ...(ii)$$

Here $w_s$ represents the plasma frequency; $c$ is the velocity of light in the conductor and $\lambda_L$ a new parameter called the *penetration depth*. It is seen that the magnetic behaviour is dependent on the shape of the object being magnetised (sphere, cylindrical, ring *etc.*). Let us consider the simplest geometry and assume that our sample, whether it be normal, perfect or superconducting, is in the form of a thin long wire and that the external field is directed parallel to the wire. For this geometry, the wire has zero demagnetising coefficient. We choose the coordinates so that the $z$-axis is perpendicular to the axis of the conductor and $z = 0$ at the surface of the conductor.

Now for particular solution of Equation (*ii*) is

$$\frac{d\vec{B}(z)}{dt} = \frac{d\vec{B}(o)}{dt}e^{-z/\lambda_L} \qquad ...(iii)$$

which says that for a perfect conductor $\frac{d\vec{B}}{dt}$ decays exponentially, with decay constant $\lambda_L$, as we pass from the surface into the body of the conductor. We can readily estimate $\lambda_L$ which comes out to be $10^{-6}$ cm. This means that, apart from a very thin skin, the perfect conductor is characterised by

$$\frac{d\vec{B}(z)}{dt} = 0 \qquad ...(iv)$$

or

$$\vec{B}(z) = \text{constant} \qquad ...(v)$$

This concludes that the flux through the specimen cannot change on cooling through the transition, obviously in contradiction with the Meissner effect. **Thus, perfect diamagnetism and zero resistivity are the two independent essential properties of the superconducting state.**

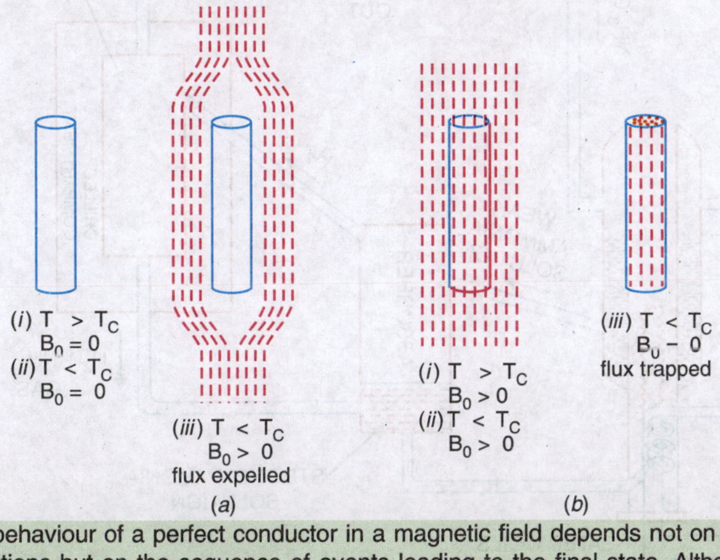


**Fig. 7.23** The behaviour of a perfect conductor in a magnetic field depends not on the final-state conditions but on the sequence of events leading to the final state. Although not shown, the lines of B in (*iii*) (*b*) are continuous as for a bar magnet.

If the field is first switched on after the material is in the perfectly conducting state (Fig. 7.23 *a*), then it cannot penetrate more than a thickness of order $\lambda_L$ of the specimen; on the other hand if the material is transformed to the state of perfect conductivity after the field is applied (Fig. 7.23 *b*), then the field penetrates the specimen and remains frozen there, even if the source of the external field is removed.

Thus, the Meissner effect shows that a superconducting pure metal, in the chosen geometry, *always expels the magnetic flux irrespective of the sequence on which the field is applied or the material becomes superconducting. Thus, a superconductor is not a perfect conductor but a perfect diamagnetic material with zero electrical resistance.*

## 7.21 Refrigeration Cycle: Electrolux Refrigerator

An electrolux refrigerator produces sufficiently low temperature without using a compressor pump. The principle of refrigeration cycle can be explained by the schematic diagram of an electrolux refrigerator shown in Fig. 7.24.

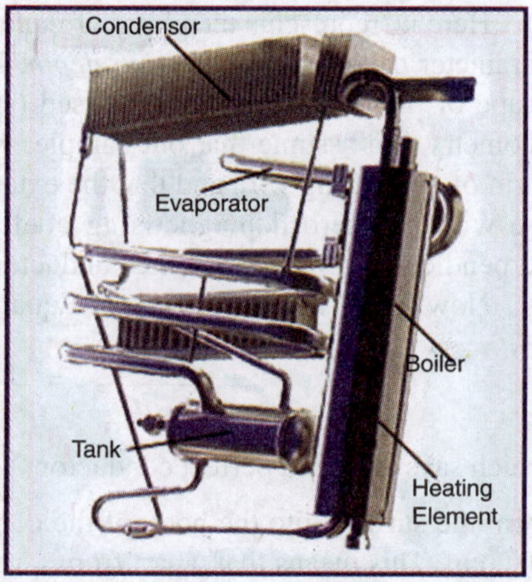


The weak ammonia solution in the boiler is forced into the absorber. The strong ammonia solution goes into the boiler and the gaseous ammonia enters the condenser. It is cooled and is condensed. Liquid ammonia enters the evaporator and is mixed with hydrogen. Hydrogen reduces the partial pressure of ammonia below its saturation point and causes evaporation. This evaporator is surrounded by the chilled compartment. Hydrogen and gaseous ammonia leave the evaporator and enter the absorber. Here they meet with the weak ammonia solution. The ammonia gas is

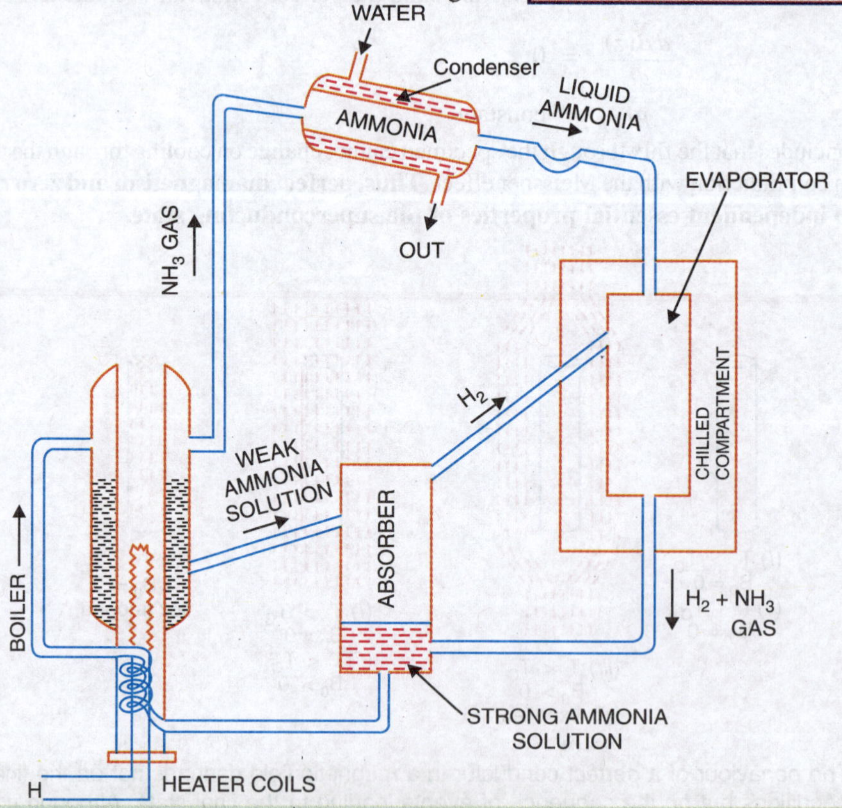


Fig. 7.24

dissolved and hydrogen gas rises through the absorber and enters the evaporator. The strong ammonia solution is forced up into the boiler again. This cycle continues and a sufficiently low temperature is produced in the chilled compartment. The advantage of this apparatus is that no compressor is required and the circulation of the liquid and the gas is automatic.

**EXAMPLE 7.1**

*Calculate the cooling produced by adiabatic demagnetisation of a paramagnetic salt as the field is reduced from 10,000 oersted to zero at initial temperature of 2 K. (Given : Curie constant per gm. mol per c.c. = 0.042 erg degree/gm oersted$^{-2}$ and $C_H$ = 0.42 Joule gm$^{-1}$ deg$^{-1}$.*

**SOLUTION.** The cooling produced is given by

$$\Delta T = \frac{CV}{2C_H T}.H^2$$

Here, $T = 2\,K, C_H = 0.42$ Joule gm$^{-1}$ deg$^{-1}$

$= 0.42 \times 10^7$ erg gm$^{-1}$ deg$^{-1}$

Curie constant, $CV = 0.042$ erg deg gm$^{-1}$ Oersted$^{-2}$.

Substituting, we get

$$\Delta T = -\frac{0.042 \times (10,000)^2}{2 \times (0.42 \times 10^7) \times 2}$$

$$= \mathbf{-0.25\ K}$$

## EXERCISE–VII

**7.1** Explain the principle of regenerative cooling with neat diagram.
**(Nagpur 2009, 2008, 2007, 2005, 04, Garhawal 2004, Bangalore 2004)**

**7.2** Discuss the industrial process for liquefaction of hydrogen. Explain the principle on which it is based.
**(Patna 2004, 05, Nagpur 2005/w, Garhwal 2000)**

**7.3** Explain K. Onnes method for the liquefaction of helium. What is the role of activated charcoal in liquefaction of helium used by Simon? **(Madras 2004)**

**7.4** Discuss the properties of Helium I and Helium II. What do you understand by λ-point? **(Agra 2002)**

**7.5** Explain the peculiar properties of Helium II.

**7.6** What is rolling film. What is its role in Helium II. Which statistics is applicable to explain the behaviour of helium II.

**7.7** Can we produce refrigeration without using a compressor? Explain how.

**7.8** Explain refrigeration cycle. How it is used to obtain sufficiently low temperature in electrolux refrigerator.
**(Osmania 2005)**

**7.9** Describe a process for production of low temperature. **(Patna 2004)**

**7.10** Describe Linde's process for the liquefaction of air.
**(Madras 2004, Madras Autonomous 2005, Mysore 2004, Bangalore 2004)**

**7.11** Discuss K. Onnes method for the liquefaction of helium and explain how the principle of regenerative cooling has been utilised in this method. **(Purvanchal 2002)**

**7.12** Discuss the properties of Helium I and Helium II. What do you understand by λ-point? Give the properties of helium II. **(Kanpur 2001, Meerut 2000, Agra 2003)**

**7.13** Write a short account on the production and measurement of low temperatures.

**7.14** What are the practical uses and applications of low temperatures? **(Madras 2005)**

**7.15** Explain cooling due to adiabatic demagnetisation. **(Nagpur 2004, Mysore 2004, Alagappa 2004)**

**7.16** Describe fully one method of liquefying a gas. State clearly the principle underlying the method and explain how the temperature of liquefaction may be measured.

**7.17** Give an account of liquefaction of gases. Explain regenerative cooling method. **(Kuvempa 2005)**

**7.18** What is Joule-Thomson effect? What is meant by adiabatic demagnetisation? **(Madras 2004)**

**7.19** What do you mean by liquefaction of gases? Give briefly an account of various methods used for liquefaction of gases. Discuss the difficulties encountered in liquefying hydrogen and Helium. **(Nagpur 2004)**

**7.20** When pressure is increased what happens on melting point of ice? Explain adiabatic demagnetisation with necessary theory. **(Madras 2005)**

**7.21** What are the practical uses and applications of low temperatures? **(Madras 2004)**

**7.22** Explain adiabatic demagnetisation with necessary theory. Explain how very low temperature is achieved using it. **(Madras 2004, Madras autonomous 2005, Panjab 2005, Kanpur 1999, Meerut 1999, M.S. Uni. Tamil Nadu 2007)**

**7.23** Explain the principle of cascaded cooling. **(Mysore 2004, Purvanchal 2002, Kanpur 2001)**

**7.24** Explain the principle of adiabatic demagnetisation for reaching temperature near absolute zero. Explain superfluidity and superconductivity with example. **(Bangalore 2004, Garhwal 2000, 2001, 2002, Agra 2002, Purvanchal 2007, Meerut 2000, Agra 2003)**

**7.25** What is the principle of cooling produced by adiabatic expansion of gas? Give a schematic diagram of gas expansion machine. **(Grahwal 2002)**

**7.26** Explain why solidification of Helium is difficult. Explain why the temperature of a gas drops in an adiabatic expansion.

**7.27** Compare Joule-Thomson expansion with adiabatic expansion. **(Madras 2004)**

**7.28** Explain briefly why early attempts to liquefy helium were useful. Give a brief account of method of liquefying this gas. **(Purvanchal 2003)**

**7.29** Explain the theory of Joule-Thomson regenerative cooling. How can you obtain liquid helium by its application? **(Meerut 2001)**

**7.30** What is adiabatic demagnetisation? Deduce thermodynamical expression for cooling produced by adiabatic demagnetisation of a paramagnetic salt. How will you employ the phenomenon to produce and measure very low temperatures? **(Purvanchal 2007, Alagappa 2004, Bangalore 2001, M.S. Uni. Tamil Nadu 2007)**

**7.31** How magnetic temperatures are measured on absolute scale? **(Calcutta 2000, Lucknow 2001)**

**7.32** What is the value of Joule-Thomson coefficient for ideal gas? Show that a real gas will always suffer a cooling effect when undergoes a free Joule expansion. **(Lucknow 2004, Ranchi 2001)**

**7.33** Explain the principle of adiabatic demagnetisation and method of reaching temperature near absolute zero. How are such temperatures measured? **(Garhwal 2000, 2001, Lucknow 2003, Agra 2002)**

**7.34** Explain the principle of producing low temperatures by evaporation of a liquid under reduced pressure. **(Garhwal 1996)**

**7.35** Discuss the process of liquefaction of gases. Explain regenerative cooling method. Give uses of liquid air. **(Kanpur 2000, Garhwal 2001)**

**7.36** What is a refrigerator? Describe vapour compression machine. **(Garhwal 2000, 2001)**

**7.37** What is adiabatic demagnetisation? Deduce expression for change of temperature of paramagnetic salt due to adiabatic demagnetisation. Give its experimental details and then derive the expression for final temperature of the salt in terms of initial temperature and the strength of applied magnetic field. Show your cooling results by drawing curves for change of entropy against temperatures at constant magnetic fields. **(Rohilkhand 2001)**

**7.38** Find the expression of work done when the magnetisation of a magnetic solid is increased by an amount dM quasi-statically. **(Calcutta 2005)**

**7.39** What are liquid Helium I and II. Discuss the important properties of liquid Helium II. **(Purvanchal 2002, Agra 2002)**

**7.40** Describe the necessary theory of the method of production of low temperature by adiabatic demagnetisation of a paramagnetic salt. **(Purvanchal 2007, Meerut 2002)**

**7.41** What is Meissner effect? Prove the equation

$$\nabla^2 \frac{d\vec{B}}{dt} = \frac{w_s^2}{c^2}\frac{d\vec{B}}{dt}$$

Define plasma frequency $w_s$. When a material becomes a superconductor? Explain. Distinguish between Type-I and Type-II superconductors. **(M.S. Uni. Tamil Nadu 2007)**

**7.42** Write short notes on:

(*a*) Adiabatic demagnetisation.
**(Agra 2000, Kanpur 2001, Meerut 2001, Alagappa 2004, Bangalore 2005)**

(*b*) Cooling by evaporation.

(*c*) Joule-Thomson expansion. **(Rohilkhand 2001, Meerut 2003)**

(*d*) Temperature of inversion.

(*e*) Regenerative cooling. **(Agra 2002)**

(*f*) Helium I and Helium II. **(Patna 2004)**

(*g*) Superfluidity.

(*h*) Superconductivity. **(M.S. Uni. Tamil Nadu 2007, Alagappa 2004, Kanpur 2000, 2001)**

(*i*) Meissner effect.

(*j*) Electrolux refrigerator. **(Madras 2005)**

(*k*) Magnetic temperature.

**7.43** Calculate the cooling produced by adiabatic demagnetisation of a paramagnetic salt as field is reduced from 0.5 weber/m$^2$ to zero at initial temperature of 20 K. Given : $C_H = 0.42$ J.gm$^{-1}$ deg$^{-1}$ and Curie constant = 0.06 C.G.S. units. **[Ans.** 0.0009 °C**] (Mysore 2000, Kanpur 1999)**

**7.44** Calculate the fall in temperature produced by adiabatic demagnetisation of a paramagnetic salt at an initial temperature 3 K when magnetic field is reduced from 8400 oersted to zero. Curie constant per cm$^3$, K = 0.6 c.g.s. units, and $C_B = 0.2$ cal gm$^{-1}$ K$^{-1}$. **[Ans.** 1.01 K**] (Purvanchal 2001)**

## OBJECTIVE QUESTIONS

**7.1** Using adiabatic demagnetisation, the minimum temperature produced is **(Garhwal 2001)**

(*a*) 1 K (*b*) $10^{-3}$ K
(*c*) $10^{-4}$ K (*d*) $10^{-5}$ K

**7.2** The temperature of inversion ($T_i$) of a gas is

(*a*) $\frac{a}{Rb}$ (*b*) $\frac{2a}{Rb}$
(*c*) $\frac{8a}{27Rb}$ (*d*) $\frac{27}{8}\frac{Rb}{a}$

**7.3** If $T_i$ is temperature of inversion, $T_B$ is Boyle temperature and $T_C$ is critical temperature, then

(*a*) $T_i < T_B < T_C$ (*b*) $T_c > T_B < T_i$
(*c*) $T_B > T_i > T_c$ (*d*) $T_i = 2T_B = 6.75\, T_C$

**7.4** The critical temperature of a gas is 31 °C. The room temperature is 27 °C. Then the gas in the room

(*a*) can be liquefied (*b*) cannot be liquefied
(*c*) will be solidified (*d*) will break into electrons.

**7.5** The coefficient of performance of a refrigerator working between ice-point and room temperature 30 °C is

(*a*) 2 (*b*) 9
(*c*) 10 (*d*) 6

**7.6** The rate of coefficient of viscosities of He II and He I is

(*a*) $10^{-2}$ (*b*) $10^{-1}$
(*c*) $10^{-3}$ (*d*) $10^{3}$

**7.7** If *A, B, C, D* are the virial coefficients, then

(*a*) $A > B > C > D$ (*b*) $\frac{B}{A} = 10^{-3}$

(*c*) $\frac{C}{A} = 10^{-5}$ (*d*) All the above

**7.8** The Boyle temperature $T_B$ is given by the relation **(Garhwal 2001)**

(*a*) $T_B = \left(\frac{27}{8}\right) T_C$ (*b*) $T_B = \frac{a}{bR}$

(*c*) Both the above (*d*) None of the above

**7.9** The last gas to be liquefied was

(*a*) Oxygen (*b*) Hydrogen

(*c*) Nitrogen (*d*) Helium

**7.10** The decrease in temperature during adiabatic demagnetisation is

(*a*) more when initial magnetic field is high,

(*b*) more when diamagnetic solid is used,

(*c*) more when initial temperature is high,

(*d*) more when initial magnetic field is low

**7.11** Sodium helium

(*a*) can be obtained merely by lowering the temperature of the liquid,

(*b*) melts to form solid,

(*c*) melts to form gas,

(*d*) cannot exist above the critical temperature.

**7.12** Liquid Helium is superconductor at **(Garhwal 2002)**

(*a*) above 2.19 K (*b*) below 4.2 K

(*c*) below 2.19 K (*d*) below ice point

**7.13** The relation between critical temperature $T_C$ and inversion temperature $T_i$ is $T_C = X\, T_i$, then X = ?

(*a*) $\frac{27}{16}$ (*b*) $\frac{4}{27}$

(*c*) $\frac{8}{27}$ (*d*) $\frac{16}{27}$

## ANSWERS

| | | | |
|---|---|---|---|
| **7.1** (*b*) | **7.2** (*b*) | **7.3** (*d*) | **7.4** (*a*) |
| **7.5** (*b*) | **7.6** (*c*) | **7.7** (*d*) | **7.8** (*c*) |
| **7.9** (*d*) | **7.10** (*a*) | **7.11** (*c*) | **7.12** (*b*) |
| **7.13** (*b*) | | | |

# Radiation

# 8

## AT A GLANCE

• Introduction • Thermal Radiation • Some Useful Definitions • Prevost's Theory of Heat Exchanges • Applications of Heat Radiations • Uniform Temperature Encloser • Black Body • Black Body in Practice • Black body Radiation and Temperature Dependence • Emissive Power, Absorptive Power • Kirchhoff's Law • Pressure of Radiation • Stefan-Boltzmann Law • Distribution of Energy in Black Body Spectrum • Wien's Displacement Law • Rayleigh-Jean's Law • The Ultraviolet Catactrophe • Planck's Radiation Law • Planck's Quantum Postulates • Derivation of Planck's Radiation Law • Derivation of Stefan's Law • Newton's Law of Cooling from Stefan's Law • Experimental Verification of Stefan's Law • Determination of Stefan's Constant • Disappearing Filament Optical Pyrometer • Total Radiation Pyrometer • Solar Constant • Temperature of the Sun • Angstrom's Pyrheliometers • Water Flow Pyrheliometers • Water Stir Pyrheliometers • Solar Spectrum • Infrared Spectrum • Ultraviolet Spectrum • Electromagnetic Spectrum • Sources of Solar Energy (Some Everyday Applications) • The Green House Effect

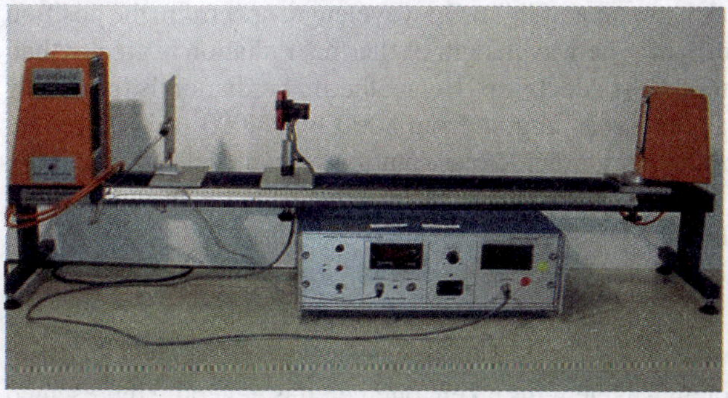

## Introduction

Heat transfers from one point to another by three modes: conduction, convection and radiation. The process of conduction and convection necessarily requires the presence of material medium filling the space between the source and the receiver. However, in radiation, material medium is not necessary. It can travel through vacuum too. The energy radiated by a hot body is termed as radiant energy. It is well established and documented that heat radiations are electromagnetic in nature, like that of light. The position of heat radiation in electromagnetic spectrum is beyond the red end of the spectrum, *i.e.*, infrared region having wavelengths ranging from $8 \times 10^{-5}$ cm to 0.04 cm. Heat is developed when these radiation are absorbed by matter or animal skin. Some substances do absorb radiant heat to a great or less degree and thereby get heated to a more or less extent, e.g. glass, wood, metals etc. absorb radiant heat and become heated, while quartz, sock salt, fluorspar etc. allow radiant heat to pass through themselves, without getting much warmed. The heat radiation obey nearly all the laws of light, like rectilinear propagation, reflection, refraction, interference, diffraction, polarization etc. A concept of 'perfectly black body' and the results obtained from such studies have proved to be of far reaching importance in physics.

## 8.1 Thermal Radiation

According to Maxwell, thermal radiation is defined as the transfer of heat from a hot body to a cooler body without appreciable heating of the intervening medium (or space). Heat or thermal energy is propagated by radiation, just like the light energy propagates, it being an electromagnetic wave. The thermal radiation has the following properties:

1. Thermal radiation has electromagnetic wave nature and hence travels through empty space (no material medium is necessary) with the velocity of light.
2. Like light, thermal radiation travels in straight line.
3. Thermal radiation obeys the law of inverse square.
4. It exhibits reflection, refraction, interference, diffraction and polarisation phenomena.

In short, thermal radiation has the same nature that of light. On wavelength spectrum, the position of thermal radiation is just to red of visible-light. The wavelength of thermal radiation is greater than red colour and is called as **infrared region of E-M spectrum**. Hence thermal radiation is frequently called as **infrared radiation (I.R.)**. Its wavelength is ranging from 8000 Å to 4000 Å. The thermal radiations cannot be detected by the eye or the photographic plate but they can be detected either by **bolometer**, which measures the change in resistance with temperature, or by using **thermopile**. These are discussed in detail at the end of this chapter.

## 8.2 Some Useful Definitions

**(*i*) Total energy density ($u$):** Total energy density of radiations at any point is the total radiant energy per unit volume around that point for all wavelengths taken together. It is generally represented by $u$ and has a unit in S.I. joule/m$^3$ and in C.G.S. erg/cm$^3$.

**(*ii*) Spectral energy density ($u_\lambda$):** Spectral energy density for a particular wavelength is the energy per unit volume per unit range of wavelength. This is denoted by $u_\lambda$.

**(*iii*) Total emissive power or emissivity (E):** The total emissive power of a body is defined as the radiant energy emitted per unit time per unit surface area of the body for all wavelengths taken together. It is denoted by E.

**(*iv*) Spectral emissive power ($e_\lambda$):** The spectral emissive power of a body at a particular wavelength is the radiant energy emitted per unit time per unit surface area of the body within a unit wavelength range. It is denoted by $e_\lambda$. For a perfectly black body, $e_\lambda$ is maximum and is denoted by $E_\lambda$.

**(*v*) Absorptivity or absorptive power ($a_\lambda$):** The absorptive power of a body at a particular temperature and for a particular wavelength is defined as the ratio of the radiant energy absorbed per unit surface area per unit time to the total energy incident on the same area of the body in unit time within a unit wavelength range. It is denoted by $a_\lambda$.

We can easily write the two results from above definitions as :

$$u = \int_0^\infty u_\lambda \, d\lambda$$

and

$$E = \int_0^\infty E_\lambda \, d\lambda$$

**(vi) Relative emittance (e):** The ratio of emittance of a surface to the emittance of a black body is called relative emittance. It is denoted by $e$.

If $E'$ is the emittance of a given surface and $E$ the emittance of a black body, then relative emittance of the surface is given by

$$e = \frac{E'}{E}$$

## 8.3 Prevost's Theory of Heat Exchanges

The ideas regarding *radiant energy*, prior to 1792, were very much confused. People used to talk of *hot radiations* emitted by hot bodies and *cold radiations* emitted by cold bodies like an ice block. Prevost of Geneva was the first to recognise that this was a loose way of talking. Prevost pointed out the conception, that the ice emits cold radiations is incorrect and proved that thermal radiation is essentially an exchange process. The rise or fall of temperature which is observed in a body is due to an exchange of radiant energy with the surroundings.

**Statement:** According to Prevost, every body emits radiations continuously to its surroundings at the rate which depends only on the nature of its surface and temperature and at the same time it receives radiations from its surroundings at the rate depending on the surface and the temperature of surroundings.

Thus, when we stand near a fire, we have the sensation of warmth because our body, *which is also a radiator,* is receiving more energy from the fire than it is losing by its own radiation. When we stand near a block of ice, we feel a sensation of cold because our body being at about 37 °C loses more energy by radiation than it receives from the ice, which is at a much lower temperature, *i.e.*, 0°C. The exchange of process continues till thermal equilibrium is reached.

Thus, all bodies at all temperatures are in a state of dynamical thermal equilibrium when they are at the same temperature. The quantity of heat radiated by a body decreases with the approach towards absolute zero of temperature. When the body is at absolute zero, it will not radiate energy.

## 8.4 Applications of Heat Radiation

**1.** White clothes are preferred in summer and dark coloured clothes in winter.

When heat radiations fall on white clothes, they are reflected back. No heat is absorbed by the clothes and a person does not get heat from outside in summer. Dark clothes in winter will absorb the heat radiations falling on them and keep the body warm.

**2.** Cooking utensils are blackened at the bottom and polished on the upper surface. Black surface will absorb the whole of the heat from the furnace and the upper surface will not allow the heat inside the utensil to flow out.

**3.** Hot water pipes and radiators used in rooms are painted black so that they can radiate maximum amount of heat to the room. The same pipes outside the rooms are painted white so that they do not lose heat to the surroundings.

**4.** The thermocouple junction exposed to heat is blackened to absorb maximum quantity of heat.

**5.** Polished reflectors are used in electric heaters to reflect maximum heat in the room.

## 8.5 Uniform Temperature Encloser

In general, the nature of radiation and the amount of radiation per unit area per second i.e., the quality and quantity of radiation depends on the temperature as well as on the nature of the surface. However, if we consider a non-conducting encloser maintained at a constant temperature, known as *non-conducting isothermal encloser,* the quality and quantity of radiation inside it depends on its temperature and is entirely independent of both the nature of the walls and the nature of the bodies which may be present inside it. Such an encloser is known as uniform temperature (isothermal) encloser.

## 8.6 Black Body

**A perfectly black body is one which absorbs totally all the radiation of any wavelength which fall on it**. As it neither reflects nor transmits any radiation, it appears black; whatever be the colour of incident radiation. The main characteristic of such a body is that when heated to a suitable high temperature, it emits full or total radiation. As it is a perfect absorber, it is also a perfect radiator, its emission being the greatest possible for every wavelength at any given temperature.

## 8.7 Black Body in Practice

In practice, a perfectly black body is not available. Lamp-black and Platinum black are the nearest approach to a black body. However, a body showing close approximation to a perfectly black body can be constructed.

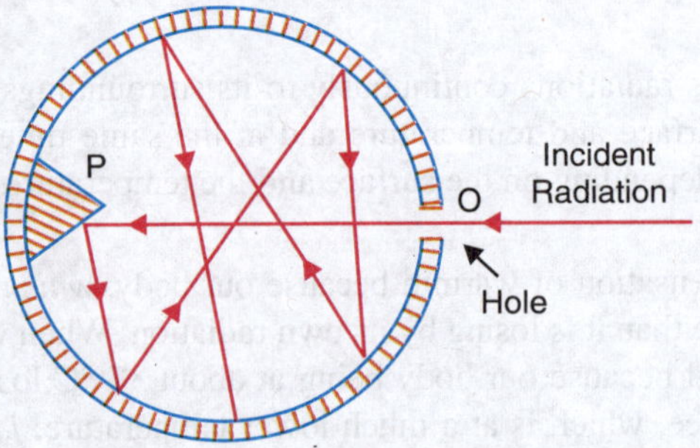


Fig. 8.1: Black body absorber

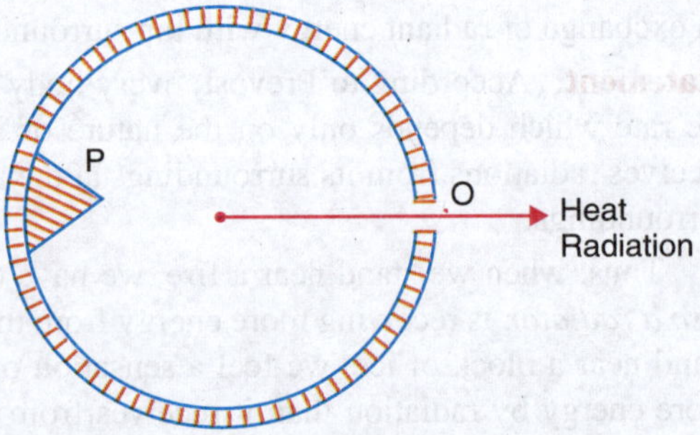


Fig. 8.2: Black body emitter

**8.7.1 Fery's black body :** A double walled hollow copper sphere is taken and coated with lamp black on its inner surface (Fig. 8.1). A fine hole $O$ acts as a very narrow opening. When the radiation enter through the narrow opening they suffer multiple reflections after falling on a pointed projection $P$ which is made just in front of the hole. After suffering multiple reflection, the radiation are completely absorbed. This type of black body shown in Fig. 8.1 was designed by Fery. This body acts as a black body absorber. When this body is placed in a bath at a fixed temperature, the heat radiation come out of the hole (Fig. 8.2). The hole acts a black body radiator. It should be remembered that only the hole and not the walls of the body, acts as the black body radiator. Lamp black absorbs about 96% of visible light, while platinum black absorbs about 98%.

**8.7.2 Wien's black body :** Wien also constructed a black body in the form of cylinder. This black body is commonly used nowadays, as it is more convenient to use.

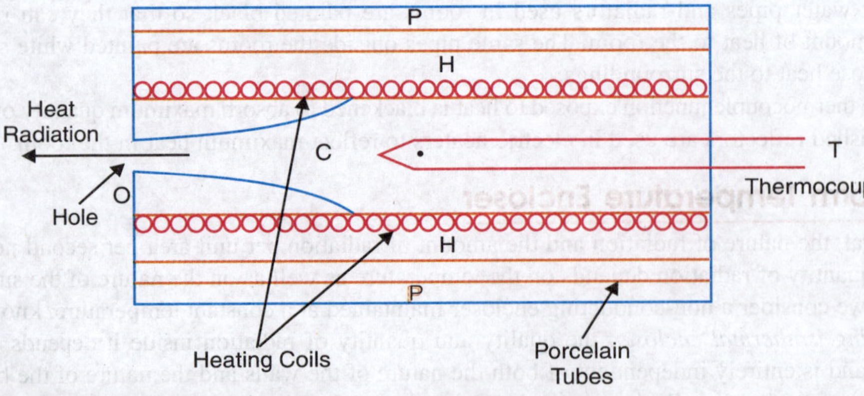


Fig. 8.3

It consists of a hollow cylindrical chamber $C$ of brass or platinum having heating coil of thin platinum wire $HH$ wound over it. The cylinder is protected by co-axial porcelain tubes $PP$. When a suitable current is passed through the coil, the cylinder is heated to a desired temperature. The inner surface of the chamber $C$ is painted black, so that it reaches the equilibrium position in a small time. The radiation from the cylinder is limited by blackened concave diaphragms $D$ and emerges out of the hole $O$. The temperature of the chamber is measured by the thermocouple $T$. The hole $O$ will act as a black body radiator.

## 8.8 Black Body Radiation and its Temperature Dependence

When a black body is placed inside a uniform temperature (isothermal) encloser, it will emit full radiation of the encloser after it is in equilibrium with the encloser. These radiations are independent of the nature of substance, nature of walls of the encloser, and presence of any other body in the encloser but depends only on temperature. Since a perfectly black body is one which absorbs all the radiations that fall on it, of whatever wavelength they may be, the radiation emitted by it will possess a character independent of the property of any particular substance, purely dependent on the temperature to which it is raised. Hence it is solely temperature dependent.

Such radiation in a uniform temperature encloser are known as *black body radiation.*

## 8.9 Emissive Power and Absorptive Power

*Emissive Power:* The emissive power $e_\lambda$ of a body at a particular temperature ($T$) for a wavelength $\lambda$ is defined as the energy emitted per second per unit surface area of the body within a unit wavelength range.

The energy radiated per second per unit surface area lying between $\lambda$ and $\lambda + d\lambda$ is then given by $e\lambda\, d\lambda$. The emissive power of a perfectly black body is maximum and is denoted by $E_\lambda$.

*Absorptive power:* The absorptive power $a_\lambda$ of a body at a particular temperature $T$ and for a particular wavelength $\lambda$ is defined as the ratio of the amount of energy absorbed in a given time by the surface to the amount of energy incident on the surface in the same time.

## 8.10 Kirchhoff's Law

### Statement :

**It states that the ratio of the emissive power to the absorptive power for a given wavelength at a given temperature is the same for all bodies and is equal to the emissive power of a perfectly black body at that temperature.**

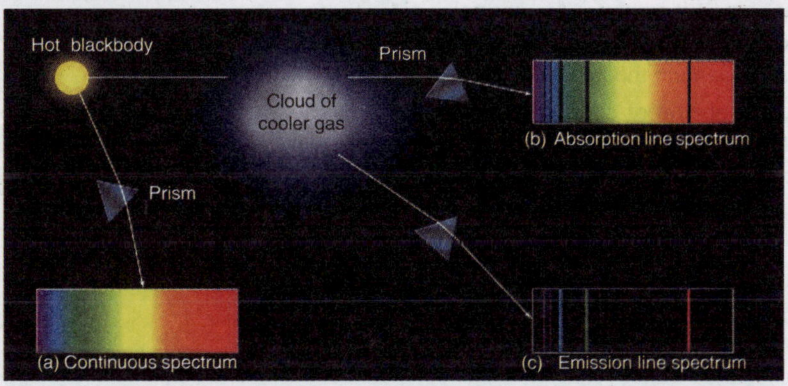


### Proof :

Suppose $a_\lambda$ is the absorptive power of a body. Let $dQ$ be the quantity of heat energy lying within the wavelengths $\lambda$ and $\lambda + d\lambda$, incident on unit area of the surface in one second, then the quantity of heat energy absorbed by the surface is given by $a_\lambda\, dQ$. The remaining energy $(dQ - a_\lambda\, dQ) = dQ\,(1 - a_\lambda)$ will be either reflected or transmitted or both.

If $e_\lambda$ is the emissive power, then the amount of energy emitted per second per unit area between wavelengths $\lambda$ and $\lambda + d\lambda$ by virtue of the temperature, is given by $e_\lambda\, d\lambda$.

Thus, the total energy given out per second per unit area of the surface $= (1 - a_\lambda)\, dQ + e_\lambda\, d\lambda$

In equilibrium state, $$dQ = (1 - a_\lambda)\, dQ + e_\lambda\, d\lambda$$

or $$a_\lambda\, dQ = e_\lambda .\, d\lambda \qquad ...(8.1)$$

For a perfectly black body, $a_\lambda = 1$ and $e_\lambda = E_\lambda$

$$\therefore \quad dQ = E_\lambda\, d\lambda$$

Substituting in equation (8.1), we get

$$a_\lambda\, E_\lambda\, d\lambda = e_\lambda\, d_\lambda$$

or $$E_\lambda = \frac{e_\lambda}{a_\lambda} \qquad ...(8.2)$$

This is known as Kirchhoff's law.

## Applications :

Kirchhoff's law tells us that good absorbers are good emitters. If a body absorbs radiation of a particular wavelength strongly, it also emits the same radiation strongly. A sodium vapour which emits two yellow lines ($D_1$ and $D_2$ of wavelengths 5890 Å and 5896 Å) strongly is also a good absorber of light of these two wavelengths. The origin of Fraunhofer dark lines in sun's spectrum is successfully explained. When the radiation from the central glowing mass which gives continuous spectrum passes through the surrounding atmosphere containing various elements like hydrogen, nitrogen, sodium, caesium, copper etc., in the gaseous state, absorb those wavelengths which they can emit at a higher temperature. As a result, those wavelengths are missing from the sun's spectrum and dark lines are seen in their places. The nature and origin of characteristic spectra of elements is also successfully explained by the Kirchhoff's law.

## 8.11 Pressure of Radiation

**1. Normal Incidence:** The radiation possesses the properties of light. As we know, light exerts a small but definite pressure on the surface on which it is incident. Maxwell proved on the basis of his electromagnetic theory that the pressure is equal to the energy density (*i.e.*, the amount of radiation per unit volume) for normal incidence on a surface. The expression for the pressure of radiation is derived on the basis of quantum theory.

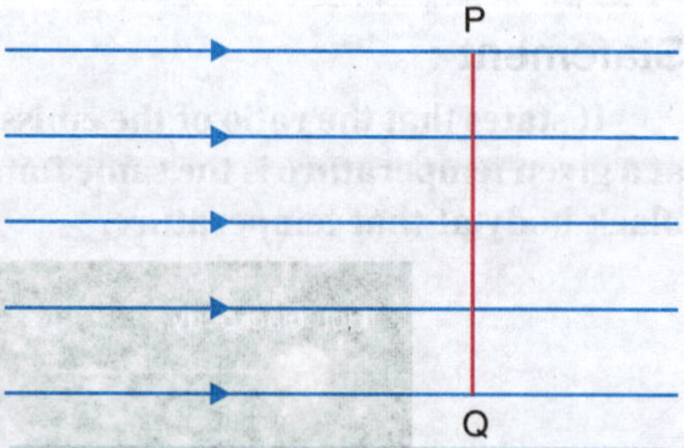


Fig. 8.4

Consider a photon of energy $h\nu$ moves with velocity of light $c$. According to the theory of relativity, this energy of the photon corresponds to a mass $m$ given

$$mc^2 = h\nu$$

$$m = \frac{h\nu}{c^2}$$

Then, the momentum of the photon = mass × velocity

$$= \frac{h\nu}{c^2} \times c = \frac{h\nu}{c}$$

$$= \frac{e}{c}$$

$e$ being energy of a photon.

Now momentum incident on the surface $PQ$ per unit area per second is given by

$$P = \sum \frac{e}{c} = \frac{\sum e}{c} \qquad ...(i)$$

but $\Sigma e$ = total energy incident on the surface per unit area per second = $E$ (say).

Substituting in (*i*), we get

$$P = \frac{E}{c} \qquad ...(ii)$$

If $u$ is the energy density *i.e.*, energy per unit volume, then the total energy passing through area $s$ of the surface normal to the incident radiation per second = $ucs$

∴ The energy flux, *i.e.*, energy radiation per unit area per second,

$$E = \frac{ucs}{s} = uc$$

Substituting in (*ii*), we get

$$P = \frac{uc}{c} = u \qquad ...(iii)$$

If all of this momentum is absorbed, it will give rate of change of momentum per unit area *i.e.*, **the pressure of radiation** $\quad P = u \qquad ...(8.3)$

Thus for normal incidence on the surface, the pressure of radiation is equal to the energy density.

**2. Pressure of Diffuse Radiation:** Suppose the radiation is diffuse, *i.e.*, the photons move in all possible directions. Let the incident beam of radiation make an angle θ with the normal to the surface $LM$.

The energy incident on the surface $KL$ per unit area per second = $uc$.

If $s$ is surface area of plane $KL$, the energy incident on the surface $KL$ in 1 sec

$$= uc \, . \, s$$

But the surface area of the plane $KL$ = surface area of the plane $LM \times \cos\theta$

$$= s' \cos\theta$$

where $s'$ is surface area of plane $LM$.

∴ Energy incident on the surface $KL$ per second

$$= uc \, . \, s' \cos\theta$$

From Fig. 8.5, it is obvious that the total energy crossing the plane $LM$ per second is equal to the energy incident on the plane $LM$ per second.

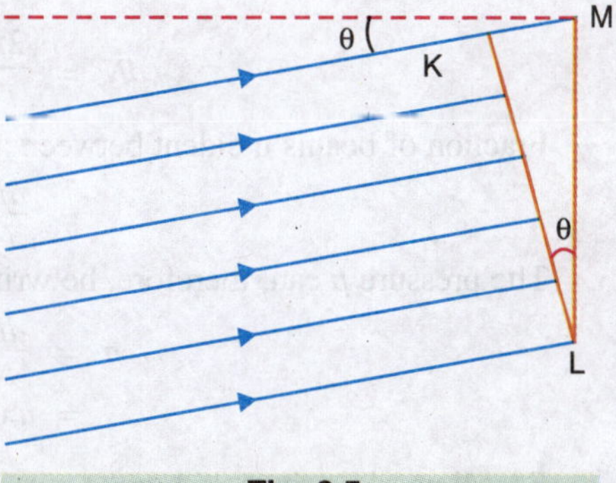


Fig. 8.5

∴ Energy incident on the plane $LM$ per second

$$= uc \, . \, s' \cos\theta$$

∴ Energy incident on per unit area of the plane $LM$ per second

$$= \frac{uc. \, s' \cos\theta}{\text{surface area of plane } LM}$$

$$= \frac{uc \, . \, s' \cos\theta}{s'} = uc \cos\theta \qquad ...(iv)$$

Therefore, the momentum incident the surface $LM$ per unit area per second

$$= \frac{uc \cos\theta}{c}$$

$$= u \cos\theta$$

The component of this momentum in a direction normal to the surface $LM$

$$= (u \cos\theta) . \cos\theta$$

$$= u \cos^2\theta \qquad ...(v)$$

If all this momentum is absorbed, then equation ($v$) gives the total rate of change of momentum per unit area normal to $LM$, *i.e.*, the pressure of radiation on the surface $LM$ is given by

$$= u \cos^2\theta \qquad ...(vi)$$

In the case of diffuse radiations, the radiation is incident from all possible directions with equal probability. Therefore, if there are $N$ beams of radiation meeting the surface, the total radiation pressure on the surface is given by

$$= \Sigma u \cos^2\theta \qquad ...(vii)$$

In order to calculate $\Sigma u \cos^2\theta$, imagine a hemisphere of radius $r$ (Fig. 8.6) round the centre $O$ of the element $LM$.

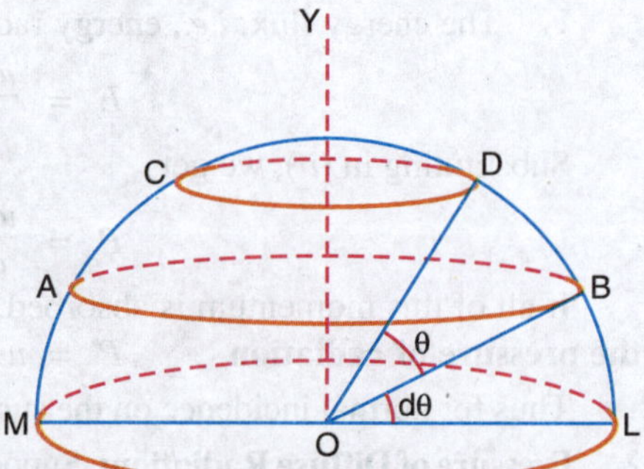

Fig. 8.6

Now from the surface of the hemisphere cut out a ring shaped element ($AB$, $CD$) by means of two cones of semi-angles $\theta$ and $(\theta + d\theta)$ drawn from $O$ as apex and $OY$ (which is perpendicular to $LM$) as axis.

The area of the ring

$$= 2\pi r \sin\theta . r d\theta$$

$$= 2\pi r^2 \sin\theta \, d\theta$$

As $N$ beams are considered to be uniformly distributed over the surface of the hemisphere which is equal to $\left(\frac{1}{2} \times 4\pi r^2\right) = 2\pi r^2$, the number of beam crossing through the area of the ring

$$dN = \frac{2\pi r^2 \sin\theta \, d\theta}{2\pi r^2} N = N \sin\theta \, d\theta \qquad ...(viii)$$

Fraction of beams incident between angle $\theta$ and $(\theta + d\theta)$

$$= \frac{dN}{N} = \sin\theta . d\theta$$

The pressure $p$ can, therefore, be written as

$$p = \frac{dN}{N} . u \cos^2\theta = \sin\theta \, d\theta . u \cos^2\theta$$

$$= u \cos^2\theta \sin\theta \, d\theta \qquad ...(ix)$$

For all possible rings within the hemisphere $\theta$ may vary from $O$ to $\frac{\pi}{2}$, therefore equation ($ix$) in the integral form may be written as

$$P = \int_0^{\pi/2} u \cos^2\theta . \sin\theta \, d\theta$$

$$= u \int_0^{\pi/2} \cos^2\theta . \sin\theta \, d\theta$$

Now put $\quad \cos\theta = x$

$\therefore \quad -\sin\theta \, d\theta = dx$

$$\therefore \quad P = u \int_1^0 x^2 (-dx) = -u \int_0^1 x^2 \, dx$$

$$= u\left[\frac{x^3}{3}\right]_0^1$$

$$P = \frac{u}{3} \qquad ...(8.4)$$

∴ **Pressure of diffuse radiation = $\frac{1}{3}$ total energy density.**

## 8.12 Stefan-Boltzmann Law

***Stefan's Law* : It states that the rate of emission of radiant energy by unit area of a perfectly black body is directly proportional to the fourth power of its absolute temperature.**

Thus, $$E = \sigma T^4 \qquad ...(8.5)$$

where σ is a constant called **Stefan's constant**. The value of σ is $5.67 \times 10^{-8}\ W m^{-2}\ k^{-4}$.

The Stefan's law refers to the emission of heat radiation only by the body itself, and not to the net loss of heat by the body after exchange with the surroundings. It simply deals with the amount of heat emitted by the body by virtue of its temperature irrespective of what it receives from the surroundings. The law can be extended to represent the net loss of heat and may be enunciated as follows:

If a black body at absolute temperature $T$ is surrounded by another black body at absolute temperature $T_o$, the net rate of loss of heat energy per unit area of the surface is given by

$$E \propto (T^4 - T_o^4)$$

or $$E = \sigma (T^4 - T_o^4) \qquad ...(8.6)$$

This is in accordance with the Prevost's theory of heat exchange, *i.e.*, the net loss of heat really the difference in the heat radiated by hot body and that absorbed by it from its surroundings.

This law (given by eqn. 8.6) is known **Stefan Boltzmann's law** as Boltzmann, in 1884, gave a theoretical proof of Stefan's law on the basis of thermodynamics.

If a body has emissitivity or relative emittance is $e$, then total energy radiated by the body per second

$$= e\sigma (T^4 - T_o^4) \qquad ...(8.7)$$

### Thermodynamical proof

The black body radiation exerts pressure similar to a gas. Hence thermodynamical relations are also applicable to heat radiation.

Let $u$ be the energy density of radiation inside a uniform temperature encloser at temperature $T$. Suppose $P$ is the pressure and $V$ is the volume of the encloser.

Applying the first law of thermodynamics $\delta Q = dU + PdV$

Applying the second thermodynamical relation

$$\left(\frac{\partial Q}{\partial V}\right)_T = T\left(\frac{\partial P}{\partial T}\right)_V$$

$$\left(\frac{\partial U + P\partial V}{\partial V}\right)_T = T\left(\frac{\partial P}{\partial T}\right)_V$$

$$\therefore \quad \left(\frac{\partial U}{\partial V}\right)_T = T\left(\frac{\partial P}{\partial T}\right)_V - P \qquad ...(i)$$

Since $V$ is the volume of the encloser, then the total internal energy of radiation inside the encloser is given by

$$U = uV \qquad ...(ii)$$

and, pressure extered $P = \frac{u}{3}$ ($\because$ eqn. 8.4) ...(*iii*)

or differentiating equation (*ii*) with respect to *V*, keeping *T* constant, we get

$$\left(\frac{\partial U}{\partial V}\right)_T = u$$

Here *u* is a function of temperature alone.

Substituting these values in equation (*i*), we get

$$u = \frac{T}{3}\frac{du}{dT} - \frac{u}{3}$$

$$\frac{4u}{3} = \frac{T}{3}\frac{du}{dT}$$

$$\frac{du}{u} = 4\frac{dT}{T}$$

Integrating, we get

$$\log u = 4 \log T + \text{constant} \quad \text{...(iv)}$$

$$= 4 \log T + \log A$$

where log *A* being a constant of integration

$$\therefore \quad \log u = \log T^4 + \log A$$

$$= \log AT^4$$

or $$u = AT^4 \quad \text{...(v)}$$

Also the total rate of emission per unit area of a black body is proportional to the energy density.

$$\therefore \quad E \propto u \propto T^4$$

or $$E = \sigma T^4 \quad \text{...(vi)}$$

where $\sigma$ is Stefan's constant. The value of Stefan's constant is $5.672 \times 10^{-5}$ C.G.S. units and in M.K.S. system it is $5.672 \times 10^{-8}$ M.K.S. units.

## 8.13 Distribution of Energy in Black Body Spectrum

Lummer and Pringesheim investigated the distribution of energy among the radiation emitted by a black body at different temperatures.

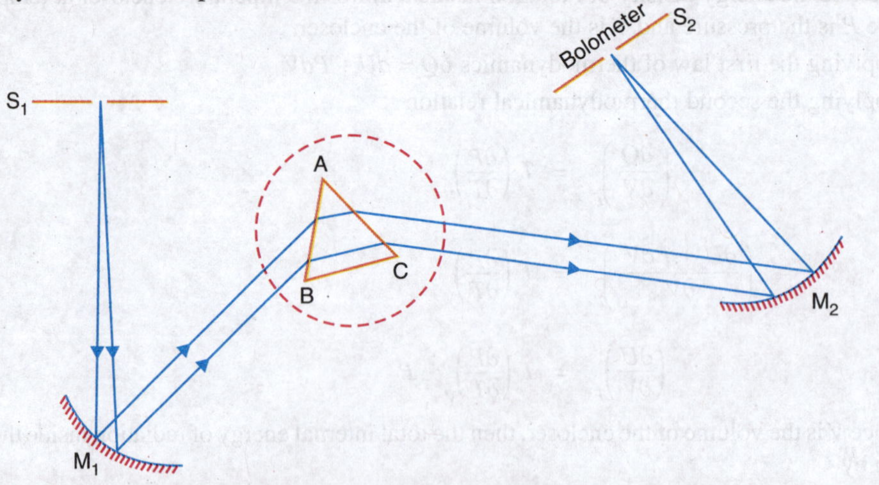


Fig. 8.7

The experimental arrangement of Lummer and Pringesheim is shown in Fig. 8.7. Their experimental black body was a small aperture of an electrical heated chamber whose temperature was measured by a thermocouple.

The radiation from a black body pass through the slit $S_1$ and fall on the reflector $M_1$. After being reflected, the parallel beam of radiation fall on a rock-salt or fluorspar prism *ABC* placed on the rotating (or turn) table of the spectrometer. The emergent light is focused by the reflector (concave mirror) $M_2$ on a line bolometer placed behind the slit $S_2$. The bolometer is connected to a sensitive galvanometer. The turn table is rotated slowly so that the different parts of the radiation spectrum successively fall on the bolometer and the corresponding deflection in galvanometer connected in the bolometer circuit are read. The intensity of each line is proportional to the deflection in the galvanometer. Then a curve (Fig. 8.8) is drawn between intensity and the wavelength.

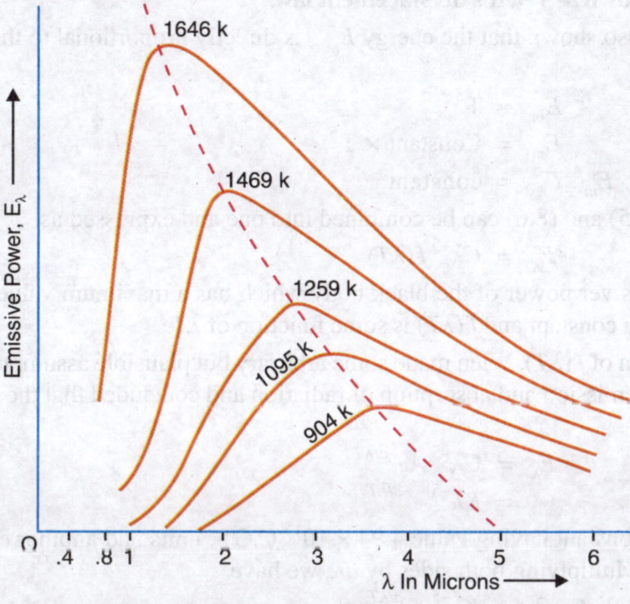


**Fig. 8.8**

Then the body is heated to different temperature and the curves are drawn for various temperatures.

## Results:

(*i*) At a given temperature the energy is not uniformly distributed in the radiation spectrum of a black body.

(*ii*) At a given temperature, the intensity of heat radiation increases with increase in wavelength and at a particular wavelength its value is maximum. With further increase in wavelength, the intensity of heat radiation decreases.

(*iii*) An increase in temperature causes a decrease in $\lambda_m$, where $\lambda_m$ is the wavelength for which the energy emitted is maximum, such that $\lambda_m T =$ constant. This is called Wien's displacement law. The points on the dotted lines represent $\lambda_m$ at various temperature.

(*iv*) An increase in temperature causes an increase in the energy emission, for all wavelengths.

(*v*) The area under each curve represents the total energy emitted by the body at a particular temperature for the range of wavelengths considered. This area increases with increase in temperature of the body. It is found that the area is directly proportional to the fourth power of the temperature of the body, *i.e.*, $E \propto T^4$, this represents Stefan-Boltzmann's law.

## 8.14 Wien's Displacement Law

From thermodynamic considerations, Wien has shown that the product of the wavelength corresponding to maximum energy $\lambda_m$ and the absolute temperature $T$ is constant. *i.e.*,

$$\lambda_m T = \text{Constant} \quad ...(8.8)$$

This constant is called Wien's displacement constant and has a value 0.2892 cm-K.

In laboratory, it is evident that when a blackened platinum wire is electrically heated, it becomes dull red at about 500°C, cherry red at about 900°C, orange red at 1100°C, yellow at 1250°C and finally becomes white at 1600°C. This shows that as the temperature rises, the radiations emitted contain more and more of shorter wavelengths. In other words, as the temperature is raised, the maximum intensity of emission is displaced (shifted) towards the shorter wavelength side.

**This is therefore, known as Wien's displacement law.**

**Deductions:** Wien has also shown that the energy $E_{max}$ is directly proportional to the fifth power of the absolute temperature

$$E_m \propto T^5$$

$$E_m = \text{Constant} \times T^5$$

or

$$E_m \,.\, T^{-5} = \text{constant} \quad ...(8.9)$$

The two relations (8.5) and (8.6) can be combined into one and expressed as

$$E_\lambda = C\lambda^{-5} f(\lambda T) \quad ...(8.10)$$

where $E_\lambda$ is the emissiver power of the black body which has a maximum value at the absolute temperature $T$. Here $C$ is a constant and $f(\lambda T)$ is some function of $\lambda T$.

To determine the form of $f(\lambda T)$, Wien made some arbitrary but plausible assumptions concerning the actual mechanism of emission and absorption of radiation and concluded that the function should have the form $Ae^{-a/\lambda T}$.

$$\therefore \quad E_\lambda = C\lambda^{-5} Ae^{-a/\lambda T}$$

$$= K\lambda^{-5} e^{-a/\lambda T}$$

where $K$ is another constant having value $4.94 \times 10^{15}$ C.G.S. units and another constant and has a value 1.43 cm-degree. Multiplying both sides by $d\lambda$, we have

$$E_\lambda d\lambda = K\lambda^{-5} e^{-a/\lambda T} d\lambda$$

Now $E_\lambda \, d\lambda$ represents the amount of energy $dE$ associated with the spectral region lying in the wavelength range $\lambda$ and $(\lambda + d\lambda)$ emitted by a black body at a temperature $T$.

$$\therefore \quad dE = K\lambda^{-5} e^{-a/\lambda T} d\lambda \quad ...(8.11)$$

This expression is known as **Wien's law of energy distribution.**

This law, however, as found not to accord with the experimental curves of Lummer and Pringsheim. The success of Wien's law is that it holds good only in the region of shorter wavelengths at lower temperatures. It does not hold good at longer wavelengths at higher temperatures.

## 8.15 Rayleigh-Jeans Law

Lord Rayleigh and Sir J. Jeans tried to establish a relation for distribution of energy with wavelengths that could be made to fit with the experimental curves, on the assumption that the electromagnetic radiation spectrum emitted by a black body continuously vary in wavelengths from zero to infinity.

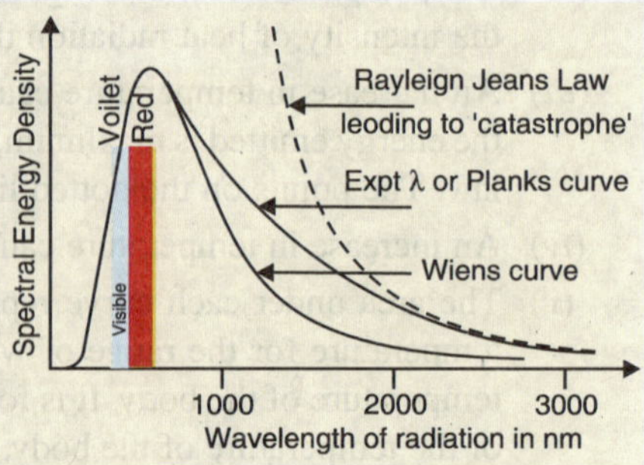

According to Rayleigh-Jeans law the energy distribution is given by the formula

$$E_\lambda \, d\lambda = \frac{8\pi kT}{\lambda^4} d\lambda \quad ...(8.12)$$

where $k$ is Boltzmann's constant.

**Lord Rayleigh, (1842–1919):**

John William Strutt in a wealthy English family was born on Nov, 12, 1842 and died on June 30, 1919. A well-known English physicist, in 1873, succeeded to his father's title and built a research laboratory in his estate. He taught physics at Cambridge University (1879-84) and was secretary of the Royal Society (1884–95). His research included work on electromagnetism, colour, cacoustics and diffraction grating, and his theory explaining the blue colour of the sky evolved into the Rayleigh's scatting law. In 1904, he was awarded the Nobel Prize for Physics for his isolation of ARGON. In 1908, he became Chancellor of Cambridge University. His influential Theory of Sound (1877–78) examines questions of vibrations and resonance of media.

This law, like Wien's was found not to accord with the experimental curves of Lummer and Pringsheim. The Rayleigh-Jeans law holds good in the region of longer wavelengths at higher temperatures but fails for shorter wavelengths. This was shown by Rubens and Kurlbaum. Thus, Wien's law and Rayleigh-Jeans law do not precisely agree with the experimental results. These are the short comings of both these laws.

## 8.16 The Ultraviolet Catastrophe

Rayleigh-Jeans law agrees well with the experimental results at low frequencies but near the maximum in the spectrum and at higher frequencies it is in violent disagreement (Fig. 8.9).

According to this law the energy density will be

$$E_\lambda \, d\lambda = \frac{8\pi \, kT}{\lambda^4} d\lambda$$

or
$$dE = \frac{8\pi \, kT}{c^4} \times \nu^4 \times \frac{c}{\nu^2} d\nu$$

$$\left(\because \lambda = \frac{c}{\nu}, \, d\lambda = \frac{c}{\nu^2} d\nu\right)$$

$$dE = \frac{8\pi \, \nu^2 \, kT \, d\nu}{c^3}$$

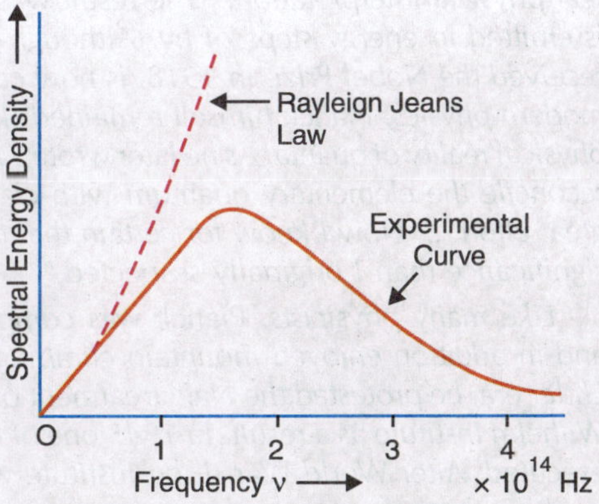


Fig. 8.9

The energy density will continuously increase with increase in frequency $\nu$ and approaches $\infty$ when $\nu \to \infty$.

This is in contradiction to observed results. This fatal objection to the law has been known as ultraviolet catastrophe.

Further, at any temperature $T$, the total energy of radiation,

$$E = \int_0^\infty \frac{8\pi \, \nu^2 \, kT \, d\nu}{c^3}$$

is predicted to be infinite which is against the Stefan's law.

Figure 8.9 shows the comparison of the Rayleigh-Jeans formula for the spectrum of the radiation from a black body at 1500 K with the observed spectrum. This discrepancy known as **ultraviolet catastrophe** (energy density increasing continuously with frequency) was the fatal failure of classical physics. This failure of classical physics led Planck (in 1900) to the discovery that radiation is emitted in quanta whose energy is $h\nu$. In 1900, the German physicist Max Planck used "Luck guesswork" (as he later called it) to come up with a formula for the spectral energy density of black body radiation as

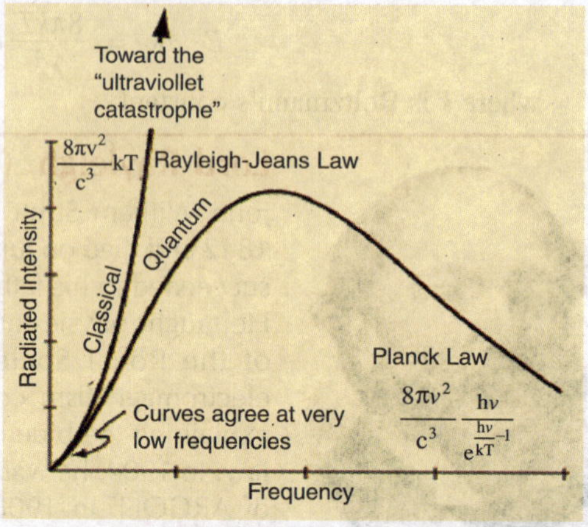


$$E_\nu \, d\nu = \frac{8\pi h}{c^3} \frac{\nu^3 d\nu}{e^{h\nu/kT} - 1}$$

Here $h$ is Planck's constant whose value is $h = 6.626 \times 10^{-34}$ J.S.

**MAX PLANCK** *(1858–1947) was born in Kiel and educated in Munich and Berlin. At the University of Berlin he studied under Kirchhoff and Helmholtz, as Hertz had done earlier. Planck realized that black body radiation was important because it was a fundamental effect independent of atomic structure, which was still a mystery in the late nineteenth century, and worked at understanding it for six years before finding the formula the radiation obeyed. He "strived from the day of its discovery to give it a real physical interpretation." The result was the discovery that radiation is emitted in energy steps of hv. Although this discovery, for which he received the Nobel Prize in 1918, is now considered to mark the start of modern physics, Planck himself remained skeptical for a long time of the physical reality of quanta. As he later wrote, "my vain attempts to somehow reconcile the elementary quantum with classical theory continued for many years and cost me great effort. ... Now I know for certain that the quantum of action has a much more fundamental significance than I originally suspected."*

*Like many physicists, Planck was competent musician (he sometimes played with Einstein) and in addition enjoyed mountain climbing. Although Planck remained in Germany during the Hitler era, he protested the Nazi treatment of Jewish scientists and lost his presidency of the Kaiser Wilhelm Institute as a result. In 1945 one of his sons was implicated in a plot to kill Hitler and was executed. After World War II the Institute was renamed after Planck and he was again its head until his death.*

## 8.17 Planck's Radiation Law

Planck (1901) was able to derive an empirical formula to explain the experimentally observed distribution of energy in the spectrum of a black body, on the basis of his revolutionary hypothesis known as quantum theory heat radiation. According to this theory, the energy distribution is given by

$$E_\lambda \, d\lambda = \frac{8\pi hc}{\lambda^5 \left[ e^{hc/k\lambda T} - 1 \right]} d\lambda \qquad \text{...(8.13)}$$

This relation agrees (and hence completely fit) with the experimental curves.

This formula of distribution of energy with wavelengths, on the basis of quantum theory, was deduced using following assumptions, which may be called as Planck's quantum postulates or Planck's hypothesis.

## 8.18 Planck's Quantum Postulates

**1.** A black body radiation chamber is filled up not only with radiation, but also with simple harmonic oscillators or resonators (energy emitters) of the molecular dimensions, known as Planck's oscillators or Planck's resonators, which can vibrate, with all possible frequencies. The vibration of the resonator entails one degree of freedom only.

**2.** The oscillators (or resonators) cannot radiate or absorb energy continuously, but energy is emitted or absorbed in the form of packets or quanta called photons. Planck assumed that each photon has an energy $h\nu$ where $h$ is the Planck's constant, its value being equal to $6.625 \times 10^{-34}$ Joule-sec, and $\nu$ is the frequency of radiation. This assumption is the most revolutionary in character. In other words, the theory states that the exchange of energy between radiation and matter cannot take place continuously but only in certain multiples of the fundamental frequency of the **resonator** (energy emitter). As the energy of a photon is $h\nu$, the energy emitted (or absorbed) is equal to 0, $h\nu$, $2h\nu$, $3h\nu$, ......... $nh\nu$, *i.e.*, in multiplets of some small unit, called as **quantum**.

## 8.19 Derivation of Planck's Radiation Law

In order to derive Planck's radiation law, let us first derive the number of resonators per unit volume lying in the frequency range $\nu$ and $(\nu + d\nu)$ and then the average energy of Planck's resonator.

**(*a*) Number of resonators per unit volume lying in the frequency range $\nu$ and $(\nu + d\nu)$:** For this, consider the radiation to be enclosed in a hollow cubic encloser, the walls of which are perfectly reflecting. According to electromagnetic theory, the radiation is supposed to consists of a number of waves, travelling in all possible directions in the encloser and suffer multiple reflections from the various walls of the encloser. Inside the encloser, stationary waves are forms with the walls as nodal points, as a result of interference of the reflected waves with the corresponding incident waves. This is just like the vibrations of a stretched string with both end point fixed. With this analogy, it is well known that the nodes are formed at the end points of the string due to stationary vibrations and only certain discrete frequencies of vibrations are allowed. If $l$ is the length of the string, the allowed wavelengths are given by

$$\lambda = \frac{2l}{n} \quad ; \quad n = 1, 2, 3, 4, \ldots\ldots\ldots, \infty. \qquad \ldots(i)$$

The corresponding allowed frequencies are

$$\nu = \frac{c}{\lambda} = \frac{nc}{2l} \quad ; \quad n = 1, 2, 3, 4, \ldots\ldots\ldots, \infty.$$

where $c$ is the speed of the waves.

Every allowed frequency is called a *mode of vibration*. The allowed modes of vibration inside a cube can be calculated just as in the case of string. But in this case, the waves are confined to a three-dimensional space. Let each side of the cube be $a$. If we take the three intersecting edges of the cube as $X$, $Y$ and $Z$-axes of a cartezian coordinate system and $\cos\alpha$, $\cos\beta$ and $\cos\gamma$ as direction cosines of the direction of propagation of a particular wave, then the projections of the edges of the cube on the direction of propagation of the wave are $a\cos\alpha$, $a\cos\beta$ and $a\cos\gamma$ respectively. In this case, only the waves will be allowed for which all the faces of the cube form nodal planes. For this, the allowed wavelengths have to satisfy three conditions of the kind of equation (*i*), in which $l$ is replaced by $a\cos\alpha$, $a\cos\beta$ and $a\cos\gamma$ respectively. Thus, an allowed wavelength $\lambda$ must satisfy

$$\lambda = \frac{2a\cos\alpha}{n_1}, \quad \lambda = \frac{2a\cos\beta}{n_2} \quad \text{and} \quad \lambda = \frac{2a\cos\gamma}{n_3} \qquad \ldots(ii)$$

where $n_1$, $n_2$ and $n_3$ are positive integers.

Using trigonometric condition of direction cosines,

$$\cos^2\alpha + \cos^2\beta + \cos^2\gamma = 1 \qquad ...(iii)$$

Substituting values of cos α, cos β and cos γ from equation (*ii*), we have

$$\frac{n_1^2\lambda^2}{4a^2} + \frac{n_2^2\lambda^2}{4a^2} + \frac{n_3^2\lambda^2}{4a^2} = 1$$

or $$n_1^2 + n_2^2 + n_3^2 = \frac{4a^2}{\lambda^2} = \left(\frac{2a}{\lambda}\right)^2$$

As $\lambda = \frac{c}{\nu}$, we have

$$n_1^2 + n_2^2 + n_3^2 = \left(\frac{2a\nu}{c}\right)^2 \qquad ...(iv)$$

This equation gives the allowed frequencies (or modes of vibration) inside the cube. The total number of modes of vibrations are the total number of possible sets $(n_1, n_2, n_3)$. The number of modes of vibrations within the frequency interval ν and ν + *d*ν can be found with the help of equation (*iv*).

For simplicity, let us first count the modes of vibration in an analogous two-dimensional problem. The two-dimensional equation will be

$$n_1^2 + n_2^2 = \left(\frac{2a\nu}{c}\right)^2 \qquad ...(v)$$

This represents a circle of radius $\left(\frac{2a\nu}{c}\right)$. Therefore, if we plot a graph with $n_1$ along $X$ = axis and $n_2$ along $Y$ = axis, for a given value of ν $\left(i.e. \text{ of } \frac{2a\nu}{c}\right)$, we get the points lying on a circle Fig. 8.10 shows a graph having $n_1$ and $n_2$ all possible integral values upto 10.

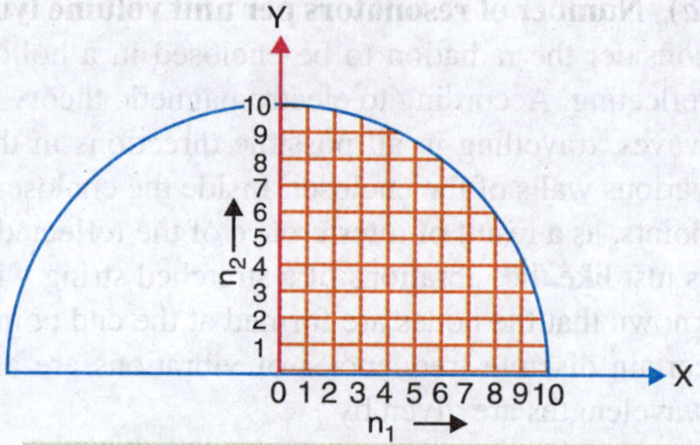

Fig. 8.10

The points on the circle corresponds to the frequency *v*, while those lying inside the circle correspond to frequencies less than *v*. The lines drawn divide the quadrant in a number of unit squares.

Every point of intersection can be associated with one square. Area of each square is unity.

∴ The number of squares = area inside the quadrant of the circle.

Thus, the number modes of vibration within frequency interval *v* and *v* + *dv* is equal to the area in the positive quadrant lying between two circles of radii $\frac{2av}{c}$ and $\frac{2a(v + dv)}{c}$.

The area is $$= \frac{1}{4}\pi\left[\left\{\frac{2a(\nu + d\nu)}{c}\right\}^2 - \left(\frac{2a\nu}{c}\right)^2\right]$$

$$= \frac{\pi}{4} \times \frac{8a^2\nu d\nu}{c^2} = \frac{2\pi a^2\nu}{c^2}\, d\nu$$

Extending the idea to 3-dimensions, the number of modes of vibration within frequency range ν and *v* + *dv* is now equal to the volume of the constant $\left(\frac{1}{8}th\right)$ of the spherical shells with radii $\frac{2a\nu}{c}$ and $\frac{2a(\nu + d\nu)}{c}$.

This volume is $= \frac{1}{8}\left[\frac{4}{3}\pi\left(\frac{2a\nu}{c}\right)^3 - \frac{4}{3}\pi\left\{\frac{2a\,(\nu+d\nu)}{c}\right\}^3\right]$

$$= \frac{1}{8}\times\frac{4}{3}\pi\times\frac{8a^3}{c^3}[\nu^3 - \nu^3 + 3\nu^2 d\nu]$$

$$= \frac{4\pi\, a^3\,\nu^2}{c^3}\,d\nu$$

But $a^3 = V =$ volume of the cube.

Therefore, the modes of vibration inside cubical encloser with frequency range $v$ and $v + dv$ will be

$$= \frac{4\pi\, V\nu^2 d\nu}{c^3}$$

∴ Number of modes of vibration per unit volume with frequency range $v$ and $(v + dv)$ will be

$$= \frac{4\pi\,\nu^2 d\nu}{c^3}$$

Since, the black body radiation travel with velocity of light $c$ and are transverse in character and hence modes of vibrations are double as that for longitudinal waves.

Therefore, for black body radiation or electromagnetic waves, the number of modes of vibration per unit volume within frequency range $v$ and $v + dv$ will be

$$= 2\times\frac{4\pi\,\nu^2 d\nu}{c^3} = \frac{8\pi\,\nu^2 d\nu}{c^3} \qquad ...(vi)$$

**(*b*) Average energy of Planck's oscillator:** If $N$ is the total number of Planck's resonators and $E$ their total energy, then average energy per Planck's oscillator is given by

$$\bar{\varepsilon} = \frac{E}{N} \qquad ...(vii)$$

According to Maxwell's law of molecular motion, if ε is a energies 0, ε, 2ε, 3ε, .........., $r\varepsilon$, ......... are in the ratio $1 : e^{-\varepsilon/kT} : e^{-2\varepsilon/kT} : e^{-3\varepsilon/kT} : .......... : e^{-r\varepsilon/kT}$ .......... etc.

If $N_0$ is the number of resonators having energy zero, then the number of resonators $N_1$ having energy ε will be $N_0\, e^{-\varepsilon/kT}$, the number of resonators $N_2$ having energy 2ε will be $N_0\, e^{-2\varepsilon/kT}$ and in general, the number of resonators $N_r$, having energy $r\varepsilon$ will $N_0\, e^{-r\varepsilon/kT}$ and so on.

$$\therefore \quad N = N_0 + N_1 + N_2 + \dots + N_r + \dots$$

$$= N_0 + N_0 e^{-\varepsilon/kT} + N_0\, e^{-2\varepsilon/kT} + \dots + N_0\, e^{-r\varepsilon/kT} + \dots$$

$$= N_0\,[1 + e^{-\varepsilon/kT} + e^{-2\varepsilon/kT} + \dots + e^{-r\varepsilon/kT} + \dots]$$

Putting $e^{-\varepsilon/kT} = y$, we have

$$N = N_0\,[1 + y + y^2 + \dots + y^r + \dots]$$

or $N = \frac{N_0}{1-y}$

The total energy of Planck's resonator will be

$$E = o\times N_0 + \varepsilon\times N_1 + 2\varepsilon\times N_2 + \dots + r\varepsilon\times N_r + \dots$$

$$= o\times\varepsilon\, N_0\, e^{-\varepsilon/kT} + 2\varepsilon\, N_0\, e^{-2\varepsilon/kT} + \dots + r\varepsilon\, N_0\, e^{-r\varepsilon/kT} + \dots$$

$$= N_0\,\varepsilon\,[e^{-\varepsilon/kT} + \dots + re^{-r\varepsilon/kT} \dots]$$

$$= N_0 \, \varepsilon \frac{y}{(1-y)^2}$$

Therefore, the average energy of a resonator will be

$$\bar{\varepsilon} = \frac{E}{N} = \frac{N_0 \, \varepsilon . \dfrac{y}{(1-y)^2}}{\dfrac{N_0}{1-y}}$$

$$= \frac{\varepsilon \, y}{1-y} = \frac{\varepsilon \, e^{-\varepsilon/kT}}{1-e^{-\varepsilon/kT}} = \frac{\varepsilon}{e^{\varepsilon/kT}-1} \quad ...(viii)$$

According to Planck's hypothesis of quantum theory, $\varepsilon = h\nu$, therefore, the average energy of Planck's oscillator is given by

$$\bar{\varepsilon} = \frac{h\nu}{e^{h\nu/kT}-1} \quad ...(ix)$$

Therefore, the *energy density* belonging to the range $d\nu$ can be obtained by multiplying the average energy of Planck's oscillator by the number of oscillators per unit volume, in this frequency range $\nu$ and $(\nu + d\nu)$.

*i.e.*,

$$E_\nu \, d\nu = \left(\frac{8\pi\nu^2}{c^3} d\nu\right)\left(\frac{h\nu}{e^{h\nu/kT}-1}\right)$$

or

$$= \frac{8\pi \, h\nu^3}{c^3} \cdot \frac{1}{e^{h\nu/kT}-1} d\nu \quad ...(8.14)$$

where $E_\nu \, d\nu$ is energy density (*i.e.*, total energy per unit volume) belonging to the range $d\nu$. Equation (8.14) is called *Planck's radiation law.*

The energy density $E_\lambda \, d\lambda$ belonging to range $d\lambda$ can be obtained by using the relation $\nu = \dfrac{c}{\lambda}$ and hence $|d\nu| = \left|-\dfrac{c}{\lambda^2} d\lambda\right|$ we get

$$E_\lambda \, d\lambda = \frac{8\pi h}{c^3}\left(\frac{c^3}{\lambda^3}\right) \cdot \frac{1}{e^{hc/\lambda kT}-1}\left|-\frac{c}{\lambda^2} d\lambda\right|$$

$$= \frac{8\pi hc}{\lambda^5} \cdot \frac{d\lambda}{e^{hc/\lambda kT}-1} \quad ...(8.15)$$

This is *Planck's radiation law* in terms of wavelength.

## 8.20 Derivation of Stefan's Law

The fact that black body radiations exert pressure similar to a gas, helps in applying thermodynamics to heat radiation.

Let $\psi$ be the energy density of radiation inside a uniform temperature encloser at temperature $T$. $P$ is the pressure and $V$ is the volume. Applying the first law of thermodynamics

$$\delta Q = \delta U + P \, dV \quad ...(i)$$

Applying thermodynamical relation

$$\left(\frac{\partial Q}{\partial V}\right)_T = T\left(\frac{\partial P}{\partial T}\right)_V \quad ...(ii)$$

$$\left(\frac{\partial U + P\partial V}{\partial V}\right)_T = T\left(\frac{\partial P}{\partial T}\right)_v$$

$$\left(\frac{\partial U}{\partial V}\right)_T = T\left(\frac{\partial P}{\partial T}\right)_V - P \qquad ...(iii)$$

Now $U = V\psi$

and $P = \frac{\psi}{3}$

or $\left[\frac{\partial U}{\partial V}\right]_T = \psi$

Here $\psi$ is a function of temperature alone.

Substituting these values in equation (*iii*),

$$\psi = \frac{T}{3}\frac{d\psi}{dT} - \frac{\psi}{3}$$

$$\frac{4\psi}{3} = \frac{T}{3}\frac{d\psi}{dT}$$

or $\frac{d\psi}{\psi} = 4\frac{dT}{T}$

Integrating, $\log \psi = 4 \log T + \text{constant}$

or $\psi = KT^4 \qquad ...(iv)$

Here $K$ is a constant

Also the total rate of emission per unit area of a black body is proportional to the energy density.

$\therefore \quad R \propto \psi \propto T^4$

$\therefore \quad R = \sigma T^4 \qquad ...(v)$

where $\sigma$ is Stefan's constant.

The value of Stefan's constant in C.G.S. system is $5.672 \times 10^{-5}$ C.G.S. units and in M.K.S. system it is $5.672 \times 10^{-8}$ M.K.S. units.

## 8.21 Derivation of Newton's Law of Cooling from Stefan's Law

Stefan's law is applicable for all temperatures of a hot body. But Newton's law is applicable when the difference of temperature between the hot body and the surrounding is small. Consider a hot body at a temperature $T_1$ placed in a uniform temperature enclosure at $T_2$. According to Stefan's law,

$$R = e\sigma (T_1^4 - T_2^4)$$

Here $e$ is the emissivity of the surface of the hot body

$$R = e\sigma (T_1 - T_2)(T_1^3 + T_1^2 T_2 + T_1 T_2^2 + T_2^3)$$

As $(T_1 - T_2)$ is small, $T_1$ can be taken approximately equal to $T_2$.

Then, $R = e\sigma (T_1 - T_2)(T_2^3 + T_2^3 + T_2^3 + T_2^3)$

$$R = 4\, e\sigma T_2^3 (T_1 - T_2)$$

Taking $4\, e\sigma T_2^3 = k$

$$R = k(T_1 - T_2)$$

or $R \propto (T_1 - T_2)$

This equation represents Newton's law of cooling and is true when the difference of temperature is small.

## 8.22 Experimental Verification of Stefan's Law

In 1897, Lummer and Pringsheim experimentally verified Stefan's law over a wide range of temperature (100°C to 1,300°C).

The apparatus consisted of a black body *C*. For temperatures between 200°C and 600°C, a hollow copper sphere coated inside with platinum black was used. The fused nitrates of sodium and potassium having a melting point of 219°C were used as the bath surrounding the black body. For temperature between 900°C and 1,300°C, an iron cylinder coated inside with platinum black was used as a black body and it was enclosed in a double walled gas furnace. A thermocouple *T* was used as a thermometer. A bolometer *B* was used to measure the intensity of the emitted heat radiations. $S_1$, $S_2$ and $S_3$ were the water-cooling shutters (Fig. 8.11).

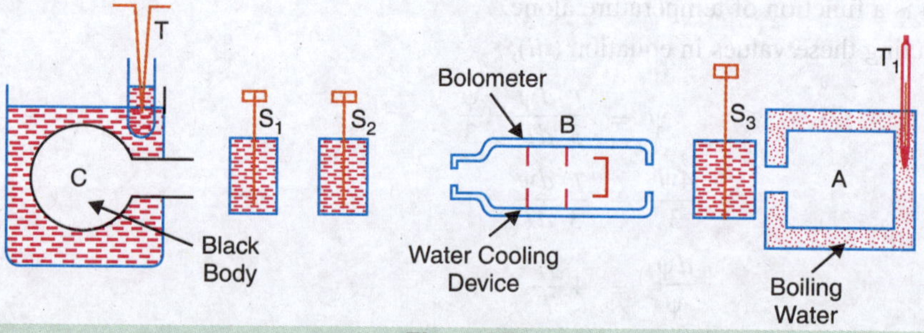


**Fig. 8.11**

Another black body A at 100°C was used to standardize the bolometer. The double walled vessel of the black body *A* contained boiling water at 100°C. The bolometer *B* was allowed to face the opening of the black body *A* and the shutter $S_3$, raised. The deflections in the galvanometer of the bolometer at various distances were noted and it was found that the deflection was inversely proportional to the square of the distance between the bolometer and the opening of the black body *A*. Thus the deflection in the galvanometer was proportional to the intensity of heat radiations.

The shutter $S_3$ was closed and the bolometer *B* was allowed to face the opening of the black body *C*. The shutters of $S_1$ and $S_2$ were raised. The bath surrounding the black body was maintained a constant temperature and the maximum deflection produced in galvanometer of the bolometer was noted. Thus at various constant temperatures of the black body, corresponding to constant deflections (in the galvanometer of the bolometer) were observed. Then the data was reduced to a common arbitrary unit in terms of the total radiations from the black body *A* at 100°C.

Let θ be the deflection in the galvanometer, $T_1$ the temperature of the black body and $T_2$, the temperature at the entrance of the bolometer. It was found that

$$\theta \propto (T_1^4 - T_2^4)$$

But $$\theta \propto R$$

∴ $$\theta \propto (T_1^4 - T_2^4)$$

This verifies Stefan's law.

Recently Coblentz has verified Stefan's law more accurately. He took an electrically heated black body whose temperature was measured by an accurate thermocouple. An absolute bolometer was used to measure the amount of heat radiations emitted by the black body. He was able to show the correctness of Stefan's law experimentally up to 1,600°C.

## 8.23 Determination of Stefan's Constant (Laboratory Method)

The laboratory apparatus used to determine the Stefan's constant is shown in Fig. 8.12. A hollow hemispherical metallic vessel *A* is enclosed in a wooden box *W*. The inner surface of *A* is coated with lamp black and the wooden box $W$ is lined with tin plates. The whole apparatus is placed on a wooden base having a small hole at its centre. The vessel $A$ is heated by passing steam inside the box and $A$ acts as a black body radiator. The thermometers $T\ T$ record the temperature of $A$.

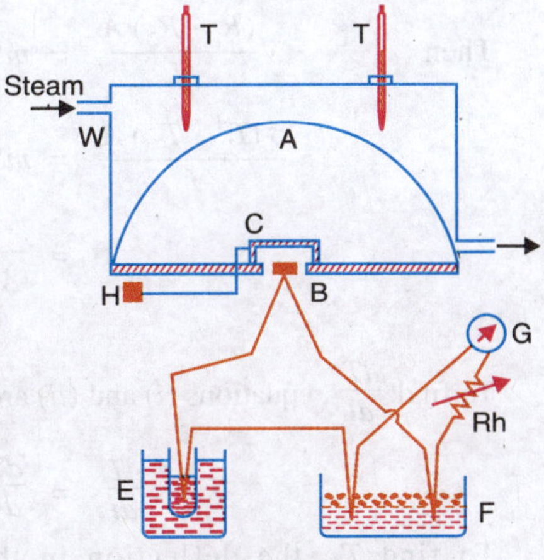

Fig. 8.12

A small silver disc $B$ whose upper surface is coated with lamp black is placed at the central hole. The ebonite covering $C$ is used to cover and uncover the disc $B$ from the radiations of the enclosure. It can be arranged from outside with the help of the handle $H$. The disc $B$ is connected to a thermocouple arrangement. One junction of the thermocouple is immersed in a tube containing oil. The tube is surrounded by a beaker containing water. A sensitive galvanometer $G$ is used in the circuit. The leads connected to the terminals of the galvanometer are immersed in cotton wool in the box $F$ to avoid any distribution effect due to the difference of temperature in the leads. A rheostat Rh can be used in the circuit to obtain the deflection within the range. The actual experiment consists of two parts.

**1.** The thermocouple is first standardized. Before passing steam into the chamber, the disc $B$ is at the room temperature. The water bath $E$ acts as a hot junction. It is heated and at various temperatures of the hot junction, the corresponding deflections in the galvanometer are noted. A graph between the difference of temperature of the hot junction and the room temperature along the $Y$-axis and galvanometer deflection along $X$-axis is plotted (Fig. 8.13). From the graph,

$$\frac{dT}{d\theta} = \tan\alpha = \frac{AB}{BC} \qquad ...(i)$$

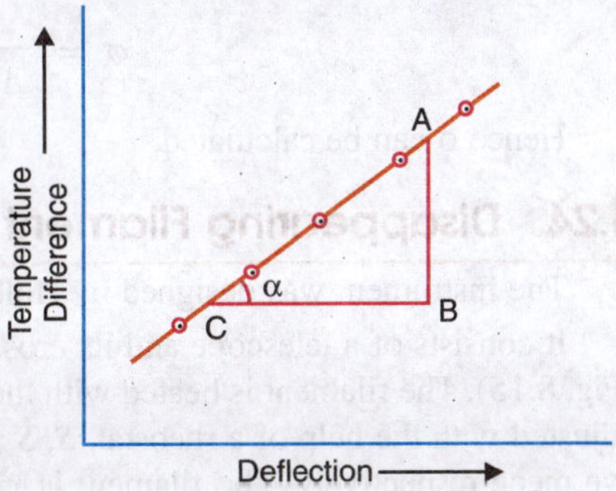

Fig. 8.13

**2.** The disc is completely covered with $C$ and steam is passed into the chamber. After some time, the thermometers $T\ T$ show constant temperature. The bath $E$ is kept at room temperature. With the help of the handle $H$, the cover $C$ is tilted so that the upper surface of the disc $B$ receives the radiations from the enclosure. The deflections in the galvanometer are observed after equal intervals of time (say 10 seconds). A graph is plotted between time and deflection (Fig. 8.14). A tangent is drawn on the curve at a point $D$.

$$\frac{dt}{d\theta} = \tan\beta = \frac{EF}{GF} \qquad ...(ii)$$

**Theory**

Let, at any instant, the temperature of the enclosure and the disc be $T_1$ and $T_2$ (degrees Kelvin) respectively. The disc will absorb more heat from the surroundings and radiate less heat to the surroundings. Its temperature will rise. From Stefan's law,

$$R_1 = \sigma T_1^4 \text{ and } R_2 = \sigma T_2^4$$

$$\therefore \quad (R_1 - R_2) = \sigma (T_1^4 - T_3^4) \qquad ...(iii)$$

Here, $R_1$ is the amount of heat radiation absorbed per unit area per second by the disc and $R_2$ is the amount of heat radiation emitted per unit area per second by the disc.

Let the mass of the disc be $m$, specific heat $S$, rate of rise of temperature $dT/dt$, and area of the upper surface $A$.

Then
$$\frac{(R_1 - R_2)\,A}{J} = mS\,\frac{dT}{dt}$$
$$\frac{\sigma\,(T_1^4 - T_2^4)\,A}{J} = mS\,\frac{dT}{dt}$$
$$\sigma = \frac{JmS}{A\,(T_1^4 - T_2^4)} \times \frac{dT}{dt} \quad ...(iv)$$

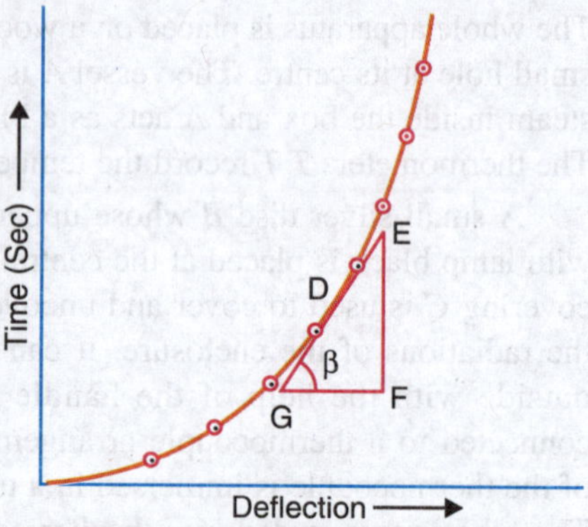

Fig. 8.14

To find $\frac{dT}{dt}$, equations (*i*) and (*ii*) are used.
$$\frac{dT}{dt} = \frac{dT}{d\theta} \times \frac{d\theta}{dt} = \frac{\tan\alpha}{\tan\beta}$$

To find $T_2$, the deflection in the galvanometer corresponding to the point $D$ on graph in Fig. 8.14 is noted and for this deflection, the temperature difference from the graph (Fig. 8.13) is noted. To this reading add the room temperature and find $T_2$ in degrees Kelvin.

Substituting these values in equation (*iv*)
$$\sigma = \frac{JmS\tan\alpha}{A\,(T_1^4 - T_2^4)\tan\beta} \quad ...\ (8.16)$$

Hence σ can be calculated.

## 8.24 Disappearing Filament Optical Pyrometer

The instrument was designed by Holborn, Kurlbaum and Henning.

It consists of a telescope and its cross wires are replaced by the filament *F* of the electric lamp (Fig. 8.15). The filament is heated with the help of a battery and the current passing through it can be adjusted with the help of a rheostat. *S*, *S* are two diaphragms and *O* and *E* are the objective and the eye-piece respectively. The filament is viewed through the eye-piece by placing red glass filter *G* between the eye and the eye-piece.

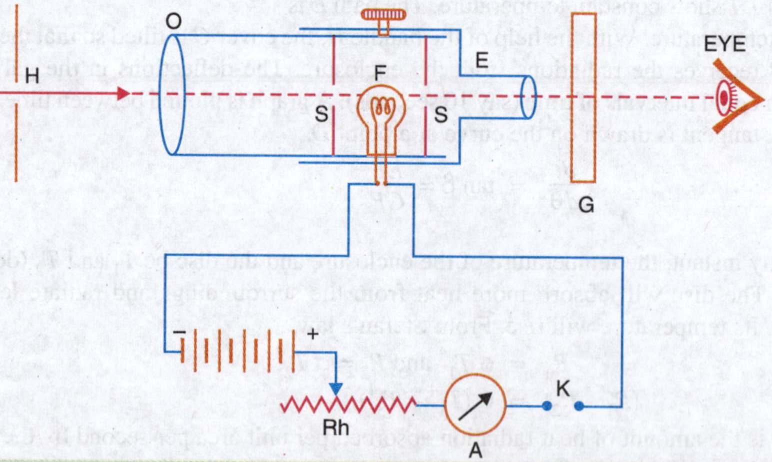

Fig. 8.15

The radiations passing through the hole *H* are allowed to enter the objective of the telescope. By adjusting the position of the telescope, the hole is focused on the filament *F*. The filament is visible against a background. The current through the filament is adjusted such that it begins to glow and the filament *becomes invisible* against the image of a bright hole. The red glass filter $G$ helps in obtaining perfect matching for monochromatic radiation. The current passing through the filament, when it *disappears,* is noted. Since the brightness or heat energy emitted per sq cm is the same in the two cases, the temperature of the source from which radiations are allowed to enter the telescope is the same as that of the filament.

Initially the instrument is calibrated by taking sources at known temperatures and the corresponding currents passing through the filament are noted. A graph is plotted between the current and the temperature. From the graph, for a particular value of current passing through the filament (for disappearing position) the corresponding unknown temperature can be determined. The ammeter can also be calibrated to measure the temperatures directly.

## 8.25 Total Radiation Pyrometer

This pyrometer was first devised by Fery. It is used to measure the total energy of radiation of all wavelengths.

It consists of a concave mirror $M$ made of copper and plated with nickel on the front surface. $E$ is an eye-piece fitted at the pole of the mirror through a hole. $D$ is a diaphragm placed in front of the mirror and at its focus. The position of $M$ can be adjusted with the help of rack and pinion arrangement $P$. Here $S$ is a metal strip painted black on the side facing the mirror and one of the junctions of a thermocouple $T$ is connected to $S$. A sensitive calibrated galvanometer is connected in the thermocouple arrangement. A shield $Q$ is used to surround $S$ so that the radiation is not incident directly on $S$ (Fig. 8.16).

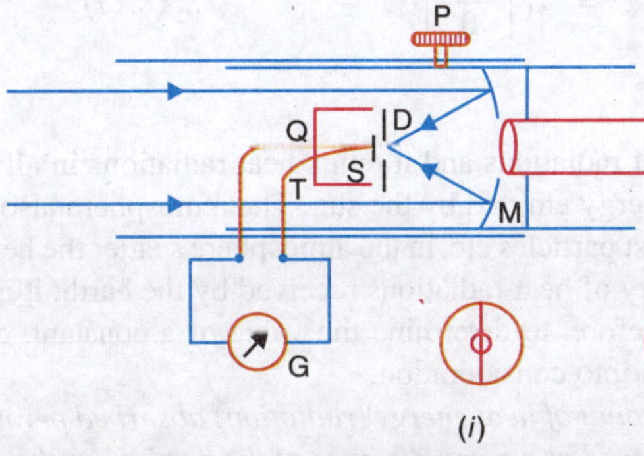


**Fig. 8.16**

When the radiation is incident on the mirror, the focussed image falls on the opening of $D$. Due to the incident radiations falling on $S$, the galvanometer shows a deflection. In order to focus the image accurately, the diaphragm $D$ is made of two semicircular mirrors inclined at a small angle to each other. When the diaphragm $D$ is viewed through the eyepiece it appears to be circular as shown in Fig. 8.16 *(i),* if the focussing is accurate. If focussing is not proper, the two halves appear displaced as shown in Fig. 8.16 *(ii)*.

The image formed on $D$ must completely cover the aperture. The reading of the galvanometer of the thermocouple is independent of the distance of the source provided the aperture of $D$ is completely filled by the image of the source (furnace etc.). The reading of the galvanometer depends only on the intensity of the image. If the distance of the source is doubled, the amount of radiation received by the mirror becomes one-fourth. But the area of the image is also reduced to one-fourth. Thus the intensity of the image remains the same.

### Calibration

If the temperature of the source is $T$ and that of the receiving instruments is $T_o$ the deflections of the galvanometer are not proportional to $(T^4 - T_0^4)$. This is because $T$ is large as compared to $T_o$. The instrument is directly calibrated with source at various temperatures. (The temperature of the source is measured with a standard thermocouple or an optical pyrometer.) A graph is obtained between the logarithm of the deflection of the galvanometer and the logarithm of the absolute temperature $T$.

$$\log \theta = K \log T$$

The graph will be a straight line and this method gives better results. Accurate experiments show that due to the conduction along the wires of the thermocouples and stray e.m.f. generated, Stefan's fourth power law is not strictly applicable. The power of $T$ varies from 3.8 to 4.2.

### Rotating Sector Device

To measure the temperature over a wide range, a rotating sector device is used (Fig. 8.17). It consists of a circular disc with a sector of suitable angle $\theta$ cut from the disc. The disc is rotated about the axis of the pyrometer and it limits the amount of radiation incident on the mirror.

Let $T$ be the temperature of the source and $T_1$ the temperature measured with the instrument. Then

$$\frac{T^4}{T_1^4} = \frac{360}{\theta}$$

$$T^4 = \frac{T_1^4(360)}{\theta}$$

$$T = T_1\left[\frac{360}{\theta}\right]^{\frac{1}{4}} \qquad \text{... (8.17)}$$

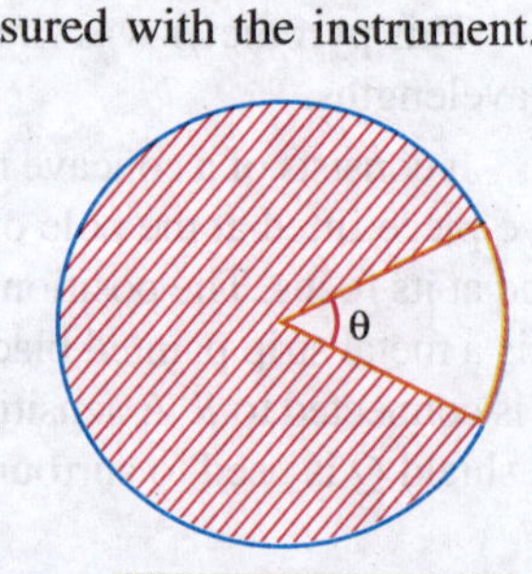


Fig. 8.17

## 8.26 Solar Constant

The sun is the source of heat radiations and it emits heat radiations in all directions. The earth receives only a fraction of the energy emitted by the sun. The atmosphere also absorbs a part of the heat radiations and air, clouds, dust particles etc. in the atmosphere scatter the heat and light radiations falling on them. From the quantity of heat radiations received by the earth, it is possible to estimate the temperature of the sun. Therefore, to determine the value of a constant, called *solar constant,* certain ideal conditions are taken into consideration.

**Solar Constant.** *It is the amount of heat energy (radiation) absorbed per minute by the sq cm of a perfectly black body surface placed at a mean distance of the earth from the sun, in the absence of the atmosphere, the surface being held perpendicular to the sun's rays.*

The instruments used to measure the solar constant are called pyrheliometers. The heat energy absorbed by a known area in a fixed time is found with the help of the pyrheliometer. To eliminate the effects of absorption by the atmosphere, the value of the solar constant is found at various altitudes of the sun on the same day under similar sky conditions. If $S$ is the observed solar constant, $S_0$ the true solar constant and $Z$ the altitude (angular elevation) of the sun, then

$$S = S_0\, a^{\sec Z}$$

or

$$\log S = \log S_0 + \sec Z \log a \qquad \text{... (8.18)}$$

Here $a$ is a constant.

A graph is plotted between $\log S$ along the $y$-axis and $\sec Z$ along the $x$-axis. The graph is a straight line. Produce the graph to meet the $y$-axis. The intercept on the $y$-axis gives $\log S_0$. From the value of $\log S_0$, the value of $S_0$, the solar constant can be calculated. The value obtained varies between 1.90 and 2.60 calories per sq cm per minute.

## 8.27 Temperature of the Sun

The sun consists of a central hot portion surrounded by the photosphere. The central portion has a temperature of the order of $10^7$ K. The photosphere has a temperature of about 6000 K. This temperature is also called the effective temperature of the sun. Considering the sun as a perfect black body radiator, the temperature of the sun can be calculated.

Let the mean distance of the sun from the earth be $R$ and $S$ the solar constant. Then, the total amount of heat energy received by the sphere of radius $R$ in one minute = $4\pi R^2 S$.

If $r$ is the radius of the sun, then the amount of heat energy radiated by 1 sq cm surface of the sun in one minute

$$E = \frac{4\pi R^2 S}{4\pi r^2} = \left(\frac{R}{r}\right)^2 \times S$$

Taking $R = 148.48 \times 10^7$ km

$r = 6.928 \times 10^5$ km

The mean value of

$S = 1.94$ cals per cm$^2$ per minute.

$$E = \left(\frac{148.48\times10^7}{6.928\times10^5}\right)^2 \times \frac{1.94}{60} \text{ cals per second} \quad ...(i)$$

Also $E = \sigma T^4$

But $\sigma = 5.75 \times 10^{-5}$ ergs per cm$^2$ per second

$$= \frac{5.75\times10^{-5}}{4.2\times10^7} \text{ ergs per cm}^2 \text{ per second}$$

$$E = \frac{5.75\times10^{-5}}{4.2\times10^7} . T^4 \quad ...(ii)$$

Equating $(i)$ and $(ii)$

$$\left(\frac{5.75\times10^{-5}}{4.2\times10^7}\right) T^4 = \left(\frac{148.48\times10^7}{6.928\times10^5}\right)^2 \times \frac{1.94}{60} \quad ...(iii)$$

$\therefore$ T = 5730 K

This temperature gives the effective temperature of the sun acting as a black body radiator. The actual temperature of the sun is higher than this value. The temperature of the sun is usually taken as 6000 K.

Temperature of the sun can also be calculated from Wien's displacement law,

$$\lambda_{max} T = 0.2892 \quad ... (8.19)$$

The wavelength of the radiations for which the energy is maximum in the spectrum is $4900 \times 10^{-8}$ cm.

Substituting the value of $\lambda m$, the value of $T$ comes out to be 5902 K. This value is in agreement with the accepted value. Hence, the effective temperature of the sun (photosphere) is about 6000 K.

**Let us calculate the black body temperature of the sun from the following data:**

***(Purvanchal 2003, Bangalore 2004, Mysore 2004)***

Stefan's constant $= 1.37 \times 10^{-12}$ cal/cm$^2$/s

Solar constant $= 2.3$ cal/cm$^2$/minute

Radius of the sun $= 7 \times 10^{10}$ cm

Distance between the sun and the earth

$= 1.5 \times 10^{13}$ cm

$$E = \frac{4\pi R^2 S}{4\pi r^2} = \left(\frac{R}{r}\right)^2 S$$

Here, $R = 1.5 \times 10^{13}$ cm

$r = 7 \times 10^{10}$ cm

$S = 2.3$ cal/cm$^2$/minute

$$\therefore \quad E = \left[\frac{1.5\times10^{13}}{7\times10^{10}}\right]^2 \times \frac{2.3}{60} \text{ cal/s} \qquad ...(iv)$$

$$E = \sigma T^4 \qquad ...(v)$$

Equating (*iv*) and (*v*), we get

$$\sigma T^4 = \left[\frac{1.5\times10^{13}}{7\times10^{10}}\right]^2 \times \frac{2.3}{60}$$

$$1.37 \times 10^{-12} \times T^4 = \left[\frac{1.5\times10^{13}}{7\times10^{10}}\right]^2 \times \frac{2.3}{60}$$

$$T = 5987 \text{ K}$$

## 8.28 Angstrom's Pyrheliometer

Pyrheliometer is an instrument which is used to find the amount of incident heat radiations and the solar constant. Angstrom's pyrheliometer consists of two identical strips *A* and *B* of blackened platinum foil (Fig. 8.18).

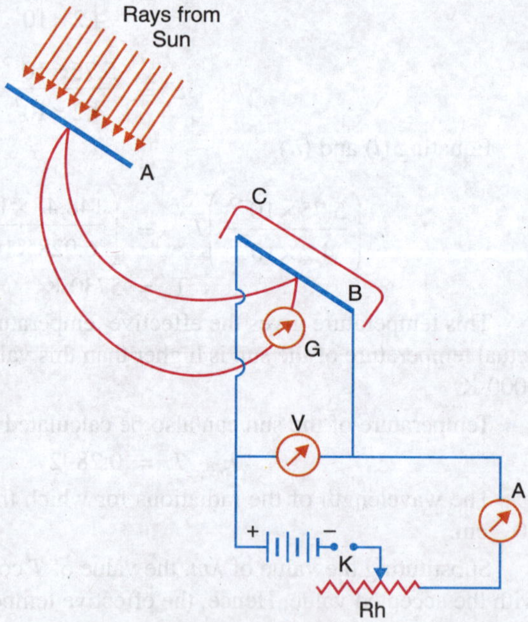

Fig. 8.18

The strip *A* is exposed to the sun and *B* is shielded by a cover C. A thermocouple having a sensitive galvanometer with *A* as one junction and *B* as the other junction is used. The strip *B* can be heated by an electrical arrangement and suitable current passing through *B* can be adjusted with the help of a rheostat.

When both the strips *A* and *B* are shielded from the sun, their junctions are at the same temperature and the galvanometer shows no deflection. The strip *A* is exposed to the sun and *B* is kept shielded. The strip *A* receives heat radiations from the sun and its temperature rises and the galvanometer shows deflection. The strip *B* is gradually heated by passing current through it and the current is adjusted so that there is no deflection in the galvanometer.

It means the strips *A* and *B* are again at the same temperature and they are receiving heat energy at the same rate.

Let *H* calories of heat be incident on one sq cm surface of the strip in one minute. The area of the plate $A = A$ sq cm and absorption coefficient $= a$.

The amount of heat radiations absorbed in one minute by the plate $A = H\,Aa$ calories. ...(*i*)

Heat produced in one minute in the strip $B$

$$= \frac{EI \times 60}{4.2} \text{ cal} \qquad ...(ii)$$

Here $E$ volts is the potential difference across the strip $B$ and $I$ amperes is the current flowing through it.

Equating (*i*) and (*ii*),

$$H\,Aa = \frac{EI \times 60}{4.2} \qquad ... (8.20)$$

Hence $H$ can be calculated.

## 8.29 Water Flow Pyrheliometer

This instrument was designed by Abbot and Fowle. It consists of a cylindrical doubled walled vessel $A$. The lower side is conical in shape and is coated with platinum black on its inner side. Water can be circulated through the walls at a constant rate. $T_1$ and $T_2$ are the platinum resistance thermometers to record the temperatures of incoming and outgoing water (Fig. 8.19).

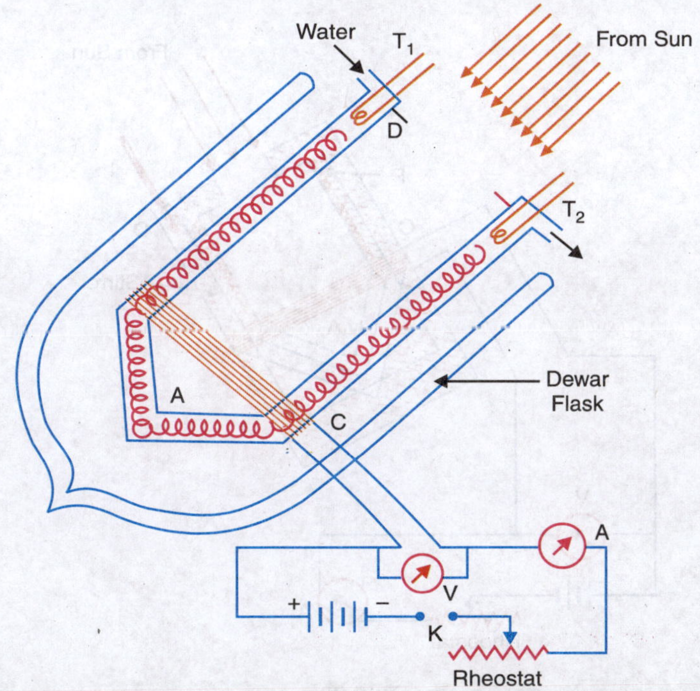


**Fig. 8.19**

$D$ is a diaphragm which allows the heat radiations from the sun through a known area of cross-section. Heat radiations are allowed to enter the vessel $A$ and the ratre of flow of water is adjusted so the thermometers $T_1$ and $T_2$ show constant temperatures (say $\theta_1$ and $\theta_2$). The spiral coil helps water in making good contact with the walls. After the steady state has been reached *i.e.*, $T_1$ and $T_2$ show constant temperatures, the heat radiations from the sun are cut off.

Now a suitable current is passed through the manganin coil *C*, so that the same conditions are restored *i.e.*, the thermometers $T_1$ and $T_2$ show the same constant temperatures $\theta_1$ and $\theta_2$. If *E* volts is the potential difference and *I* amperes is the current flowing through the manganin coil, then the amount of heat produced per second = $\frac{EI}{4.2}$ cal. Therefore, the amount of heat radiations received from the sun in one minute by the exposed surface of the vessel $A = \frac{EI \times 60}{4.2}$ cal. From this value, the solar constant can be calculated.

## 8.30 Water Stir Pyrheliometer

In this type of pyrheliometer, the heat radiations are received from the sun by the conical end of the vessel *A,* which is coated with lamp black. Vessel *A* is surrounded by a calorimeter *D* containing water. The stirrer helps in circulating water vigorously and is driven by an electric motor. The temperature of water is measured with the help of a platinum resistance thermometer *P,* whose wire is wound around *A* (Fig. 8.20). Heat radiations from the sun are allowed to enter the vessel *A* and the rise in temperature of water in one second is calculated. The heat radiations are cut off and a suitable current is passed through the manganin coil *C* for a known time so that the rise in temperature of water in one second is the same.

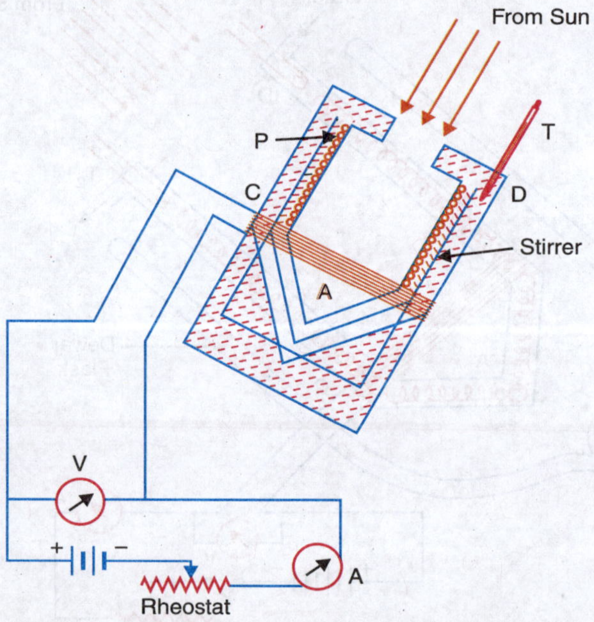


Fig 8.20

If *E* volts is the potential difference and *I* amperes is the current passing through the manganin coil, then the amount of heat produced per second = $\frac{EI}{4.2}$ cal. Therefore, the amount of heat radiations received from the sun in one minute by the exposed surface of the vessel $A = \frac{EI \times 60}{4.2}$ cal. From this value, the solar constant can be calculated.

## 8.31 Solar Spectrum

The radiation received from the sun is similar to that of a perfectly black body. In a solar spectrum, the wavelength corresponding to maximum energy is about 5000 Å for a surface temperature of 5750 K. The solar spectrum consists of a large number of dark lines called **Fraunhofer lines**. These lines were first observed by Wollaston in 1802 and later studied by Fraunhofer in 1814. These lines are observed in the complete range of the spectrum viz. ultraviolet, visible and infrared. Fraunhofer measured the wavelengths of many of these lines accurately and found that they occupied exactly the same positions as the bright lines emitted by different gases and vapours.

According to Kirchhoff's law of radiation, any substance at a lower temperature will absorb the radiations of those wavelengths, which it will emit when excited by electric discharge. Thus the solar spectrum is an absorption spectrum and the Fraunhofer lines are the absorption lines of relatively cooler gases and vapours present in the earth's atmosphere and the sun's outer atmosphere. The central portion of the sun is called the photosphere, which is at a very high temperature. Surrounding the photosphere is the chromosphere, which is at a much lower temperature than the photosphere. The positions of the Fraunhofer lines in the solar spectrum give the information regarding the nature of elements present in the sun's atmosphere. The presence of more than sixty elements in the sun's atmosphere is found this way. In fact, helium was first discovered by the study of the Fraunhofer lines in the solar spectrum before it had been isolated in the laboratory. The most prominent of the Fraunhofer lines are denoted by the letters of the alphabet. Some of the lines, with their wavelengths and the elements responsible for their absorption are given below :

| *Line* | *Element* | *Wavelength in Å* |
|---|---|---|
| A | Atmospheric oxygen | 7594 |
| B | Atmospheric oxygen | 6867 |
| C | Hydrogen | 6563 |
| $D_1$ | Sodium | 5896 |
| $D_2$ | Sodium | 5890 |
| F | Hydrogen | 4861 |
| $G_1$ | Hydrogen | 4341 |
| H | Calcium | 3969 |
| K | Calcium | 3934 |

## 8.32 Infrared Spectrum

Extending on either side of the visible spectrum, there are invisible radiations which do not cause the sensation of sight. The radiations beyond the red end of the visible spectrum are called infrared radiations and their wavelengths extend up to 400,000 Å. Sun is a powerful source of infrared radiation. Beyond the violet end of the visible spectrum, the radiations

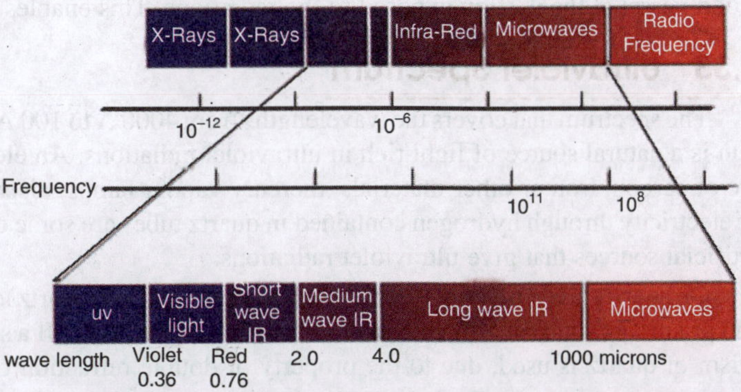


PART OF THE ELECTRO-MAGNETIC SPECTRUM

extending up to a wavelength of 100 Å are called ultraviolet radiations.

The heating effect of the infrared radiations is used in measuring the wavelength of the radiation. An infrared ray spectrometer is shown in Fig. 8.21. Light from a strong source of light such as an electric arc is rendered parallel by reflection from a concave stainless steel mirror $M_1$. This parallel beam is refracted through the rock salt prism $P$ and the emergent dispersed beam after reflection from the mirrors $M_1$ and $M_2$ is incident on a thermopile or a bolometer. $M_1$ is a plane mirror and $M_2$ is a concave mirror. Wien's law of radiation is given by

$$\lambda_m T = 0.2892$$

where $T$ is the absolute temperature and $\lambda_m$ is the wavelength of the radiation. The thermopile readings help in the calculation of temperature and from the temperature, $\lambda$ can be calculated.

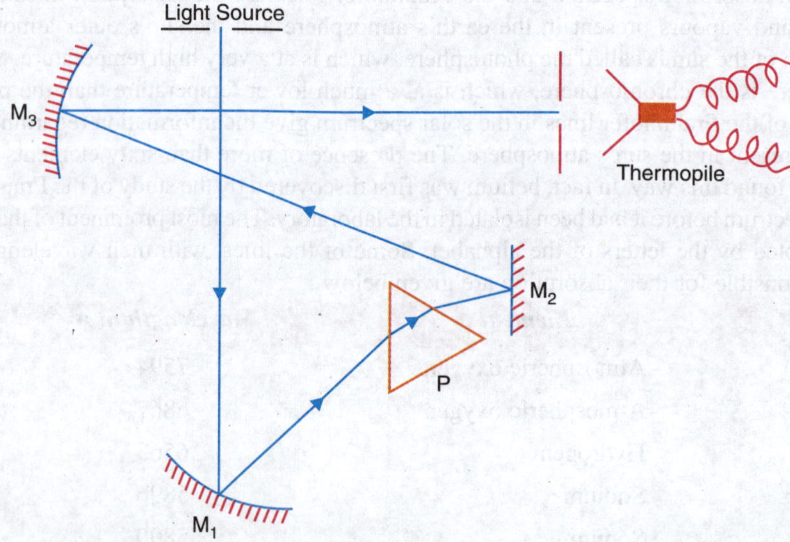


Fig. 8.21

As infrared radiations are not absorbed by air or thick fog, infrared ray photograph can be taken over long distances of fog and mist where visible light cannot penetrate. For this, specially designed photographic plates are used with suitable filters. A solution of iodine in alcohol is a suitable filter, because this transmits the infrared radiations and absorbs the visible light. In the World War II, infrared photosphere played a very useful part in detecting objects in the dark through mist, fog and clouds. Infrared radiations have a wide application in the field of medicine, industry etc. Infrared radiations can penetrate deep into the human body and by their property of heating can dilate the blood vessel at the portion exposed to the radiations. This enables increased flow of blood.

## 8.33 Ultraviolet Spectrum

The spectrum that covers the wavelengths from 4000 Å to 100 Å is called the ultraviolet spectrum. Sun is a natural source of light rich in ultraviolet radiations. An electric arc of carbon, iron or other materials, mercury vapour lamps, discharge of electricity through hydrogen contained in quartz tubes are some of the artificial sources that give ultraviolet radiations.

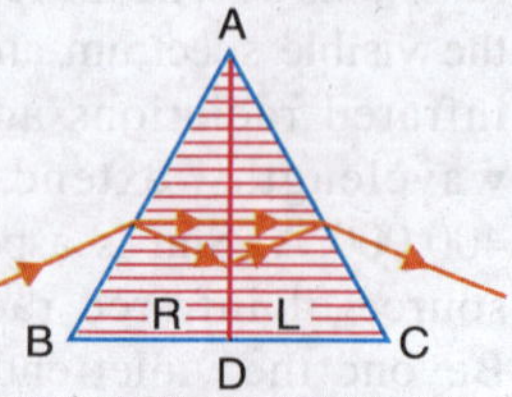


Fig. 8.22

Ordinary glass absorbs the ultraviolet radiations. Hence quartz lenses and prisms are used. But quartz is a doubly refracting material. If a single prism of quartz is used, due to the property of double refraction of the prism, two images of the slit (ordinary and extraordinary) corresponding to a single wavelength are observed. This will reduce the sharpness of each image. To compensate for this, the collimating lens is made from right-handed quartz and the telescope objective is made from left-handed quartz. Similarly the prism used for the dispersion of the incident beam consists of two halves. The two halves are held together by glycerine. One half is made from right-handed quartz and the other half is made from left-handed quartz.

In Fig. 8.22, right angled *ABD* is a prism of right-handed quartz and *ADC* is of left-handed quartz. For recording an ultraviolet spectrum for wavelengths shorter than 1,200Å, a concave reflection grating is used. The diffracted beam is photographed. The grating and the photographic plate are enclosed in a metal chamber which is evacuated.

Ultraviolet radiations have a variety of applications. Sterilisation of rooms in which blood plasma, drugs, vaccines etc. are prepared and sealed in the containers is done by ultraviolet radiations. Drugs, poisons, dyes etc. fluoresce under the action of ultraviolet rays. The resolving power of a microscope is increased when ultraviolet light is used for illumination. Fluorescent tubes depend on the principle of fluorescence effected by ultraviolet radiations.

There are nearly 40 different types of skin diseases. Dermatologists use UV radiations (either artificially produced by UV lamps or natural source, the sun) to treat most of the diseases. In the treatment of skin diseases, two methods are currently used: phototherapy with UVB (311 nm) and photochemotherapy (PUVA). In PUVA therapy, UVA of radiation 365 nm is used along with psoralen as photosensitizer. Both forms have their place in dermatology theory. PUVA therapy has become a standard treatment for severe psoriasis, mycosis tungoides, vitiligo and some other dermatoses.

## 8.34 Electromagnetic Spectrum

Visible spectrum includes those wavelengths which can stimulate the sense of sight. But there is no basic difference between light waves and the electromagnetic waves produced by electrical oscillating circuits. The term electromagnetic spectrum is used for the range of wavelengths from $10^4$ meters to 1Å ($10^{-8}$ cm). There is in fact no limit to the production of electromagnetic waves of very long wavelengths. The frequency of an alternating current generator can be made as low as possible by decreasing the speed of the generator. The wavelength of waves transmitted by a 50 cycle transmission line is $5 \times 10^8$ cm. Waves of shorter wavelength can be produced by electrical oscillators. X-rays and gamma rays represent the waves of very short lengths. The complete electromagnetic spectrum is shown in the following table.

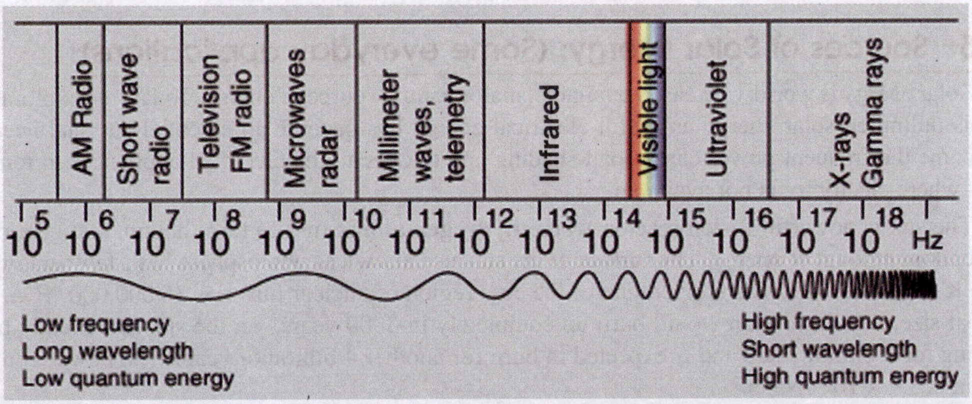


It is interesting to note that the visible range of the spectrum comprises only a small range of the electromagnetic spectrum extending approximately from 4000Å in the extreme violet region to 8000Å in the extreme red. Beyond the violet region of the visible spectrum is the ultraviolet, the X-rays and γ-rays. Beyond the red end of the visible spectrum are the short radio waves and radio broadcast waves.

**ELECTROMAGNETIC SPECTRUM**

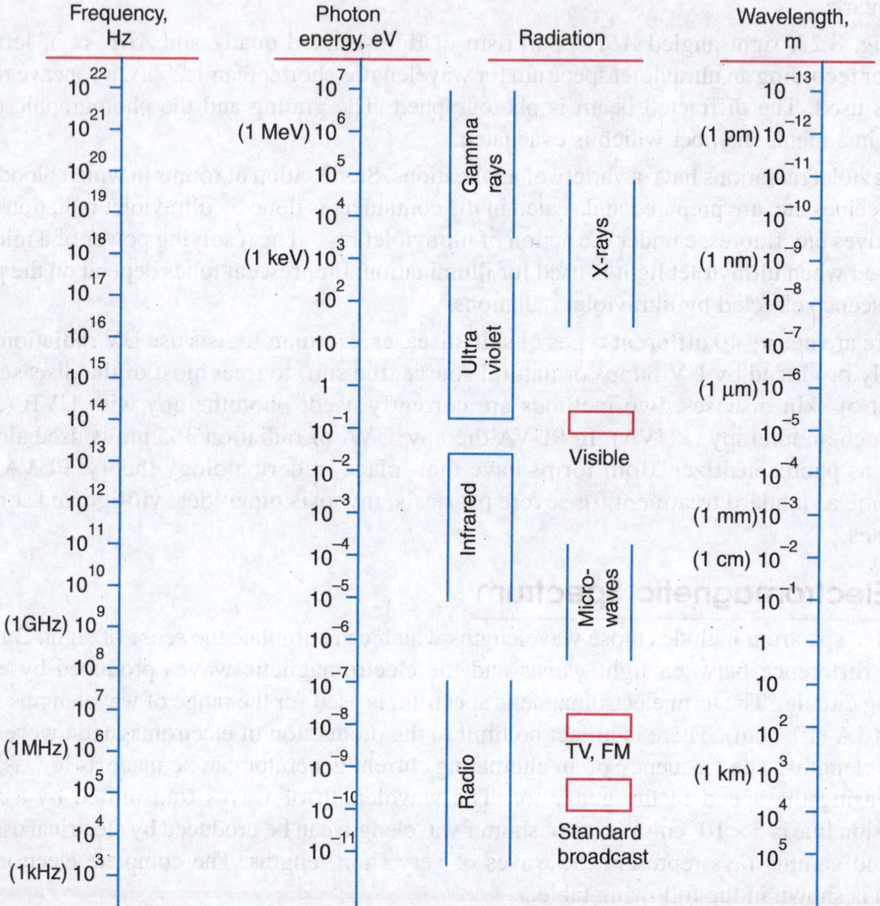


## 8.35 Sources of Solar Energy: (Some everyday applications)

Solar energy is worldwide easily available, major natural source of energy. Today, the exploration of this unlimited solar energy as useful electrical energy has become an essential national need to overcome the frequent power cut or load-shading and to operate the electrical appliances in remote areas where electricity is not available.

The sun is powerful fusion reactor, where **hydrogen atoms** fuse to form helium and give off a tremendous amount of energy. The surface of the sun also known as photosphere has a temperature of 6000 K (or 10,000 °F). The temperature of the core region of nuclear fusion is 36,000,000 °F. A ball of coal size, the size of sun would burn up completely in 3000 years, yet the sun has already been burning for 3 billion years and is expected to burn for another 4 billion of years. The power emitted by sun is $3.9 \times 10^{26}$ Watts.

Solar energy reaches the earth at the rate of about 1.4 kW per square metre of surface perpendicular to the direction of the sun as shown in Figure 8.23. By how much does the mass of the sun decrease per second owing to this energy loss? The mean radius of the earth's orbit is $1.5 \times 10^{11}$ m.

The solar energy reaching the earth's surface is estimated as under. The surface area of a sphere of radius $r$ is $A = 4\pi r^2$. The total power radiated by the sun, which is equal to the power received by a sphere whose radius is that of the earth's orbit, is therefore

$$P = \frac{P}{A}A = \frac{P}{A}(4\pi r^2) = (1.4 \times 10^3 \text{ W/m}^2)(4\pi)(1.5 \times 10^{11} \text{ m})^2 = 4.0 \times 10^{26} \text{ W}$$

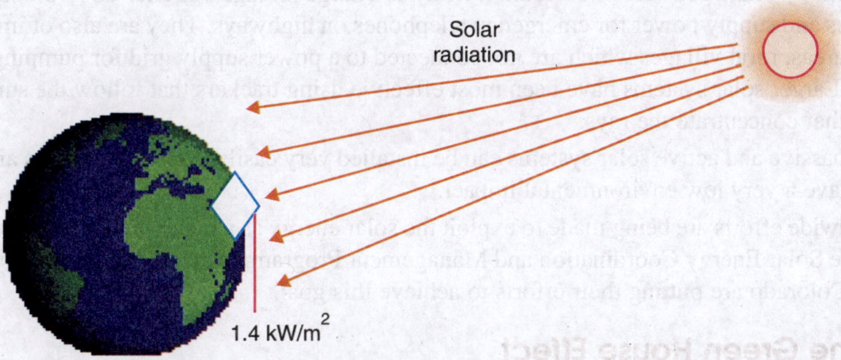


**Fig. 8.23**

Thus the sun loses $E_o = 4.0 \times 10^{26}$ J of rest energy per second, which means that the sun's rest mass decreases by

$$m = \frac{E_o}{c^2} = \frac{4.0 \times 10^{26} \text{ J}}{(3.0 \times 10^8 \text{ m/s})^2} = 4.4 \times 10^9 \text{ kg}$$

per second. Since the sun's mass is $2.0 \times 10^{30}$ kg, it is in no immediate danger of running out of matter. The chief energy-producing process in the sun and most other stars is the conversion of hydrogen to helium in its interior. The formation of each helium nucleus is accompanied by the release of $4.0 \times 10^{-11}$ J of energy, so $10^{37}$ helium nuclei are produced in the sun per second.

Only a small fraction of sun's radiant energy or *insolation*, reaches the earth's atmosphere and only about half of that reaches the surface of the earth. The other half is either reflected back into space by clouds and ice or is absorbed or scattered by molecules within the atmosphere. The sun's energy travels 93,000,000 miles to reach the earth's surface. It takes about 8.5 minutes after leaving the photosphere in various forms of radiant energy with different wavelengths, known as electromagnetic spectrum.

The abudantly available solar energy can be utilised either directly as *heat* or indirectly by converting it into electrical power using **photovoltaic cells**. Accordingly solar energy utility systems are broadly divided into two categories:

(*i*) Active systems, and

(*ii*) Passive systems.

Green-houses and solariams are the most common examples of the direct use of solar energy with glass windows concentrating the visible light from the sun but restricting the heat from escaping. Flat plate collectors are another direct method and mounted on roof tops. They can provide one third of the energy required for space heating. Hence windows and collectors are considered as *passive systems*. Most popular and widely used solar water heaters, solar cookers also belong to this category.

Active solar systems differ from passive systems in that they operate machineries such as pump or electric fans, electronic devices and so many. A huge or small panel of photovoltaic cells, connected in series and framed on a rigid platform. The photovoltaic cells are made of semiconductor materials mainly silicon with small amount of *gallinum arsenide* or *cadmium sulfide* added so that the cell emits electrons, when exposed to sunlight are connected in series, as per power requirements. These cells are capable of absorbing part of the solar flux to produce a direct electric current with about 14% efficiency. A thin film technology is being perfected for the production of these cells and the cost per unit watt will eventually be reduced because less material will be required. These modules based on photovoltaic cells are used economically to charge storage batteries aboard boats, operate light houses and supply power for emergency telephones on highways. They are also of immense use in remote areas, rural villages which are not connected to a power supply grid for pumping water for irrigation. Larger solar systems have been most effective using trackers that follow the sun or mirror reflectors that concentrate the rays.

Both passive and active solar systems can be installed very easily. Neither produces air pollution and both have a very low environmental impact.

Worldwide efforts are being made to exploit the solar energy as a major non-conventional energy source. The Solar Energy Coordination and Management Programme and the Solar Energy Research Institutes Colorado are putting their efforts to achieve this goal.

## 8.36 The Green House Effect

A green house effect is important in heating the earth's atmosphere. Much of the short-wavelengths visible light from the sun that reaches the earth's surface is reradiated as long-wavelength infrared light that is readily absorbed by $CO_2$ and $H_2O$ in the atmosphere. This means that the atmosphere heated mainly from below by the earth rather than from above by the sun. The total energy that the earth and its atmosphere radiate into space on the average equals the total energy that they receive from the sun.

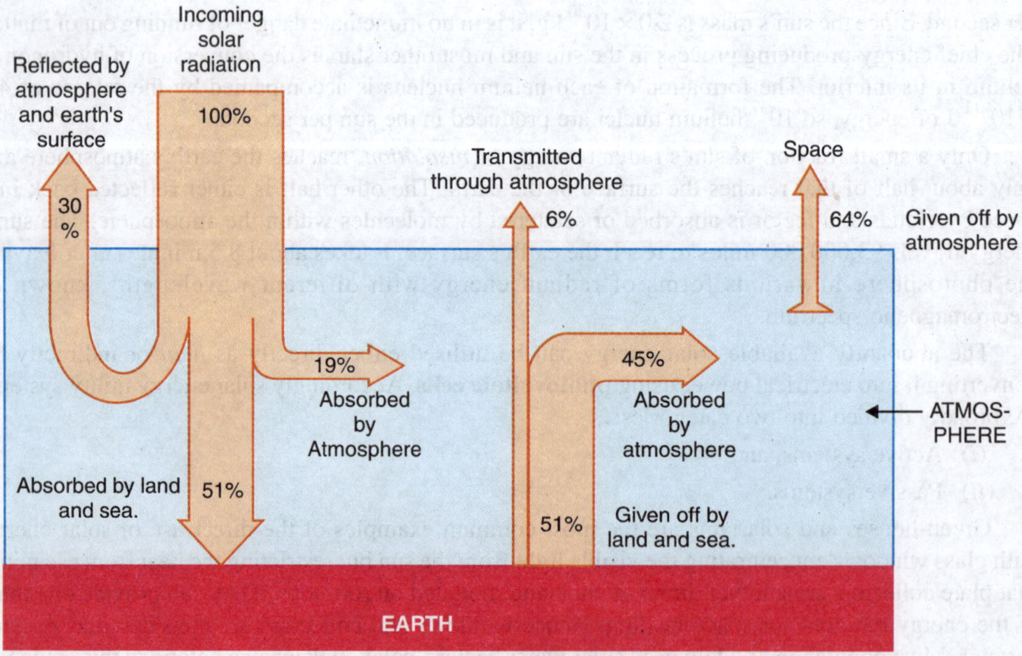


Fig. 8.24

It is worth mentioning that everybody of condensed matter radiates according to Planck's radiation law, regardless of its temperature. An object need not be so hot that it glows conspicuously in the visible region in order to be radiating. The radiation from an object at room temperature is chiefly in the infrared part of the spectrum to which the eye is not sensitive. Therefore, the interior of a green house is warmer than the out side air because sunlight can enter through its windows but the infrared radiation given off by the interior cannot escape through them as shown in Fig. 8.24. (For details, please refer chapter 18).

## NUMERICAL EXAMPLES

### EXAMPLE 8.1

*Show that Planck's law reduces to Wien's law for shorter wavelengths and Rayleigh-Jeans law for longer wavelengths.* **(Avadh 2000)**

**SOLUTION.** Planck's radiation law is

$$E_\lambda \, d\lambda = \frac{8\pi hc}{\lambda^5} \cdot \frac{d\lambda}{e^{hc/\lambda kT} - 1}$$

For *Shorter wavelengths*, $e^{hc/\lambda kT}$ becomes large as compared to unity and hence Planck's law reduces to

$$E_\lambda \, d\lambda = \frac{8\pi hc}{\lambda^5} \cdot \frac{d\lambda}{e^{hc/\lambda kT}}$$

$$= \frac{8\pi hc}{\lambda^5} \cdot e^{-hc/\lambda kT} . d\lambda \qquad ...(8.21)$$

which is *Wien's law*.

For longer wavelengths, $e^{hc/\lambda kT}$ may be approximated to $\left(1 + \frac{hc}{\lambda kT}\right)$ and hence Planck's law reduces to

$$E_\lambda \, d\lambda = \frac{8\pi hc}{\lambda^5} \cdot \frac{d\lambda}{\left(1 + \frac{hc}{\lambda kT} - 1\right)}$$

$$= \frac{8\pi hc}{\lambda^5} \cdot \frac{\lambda kT}{hc} d\lambda = \frac{8\pi kT}{\lambda^4} d\lambda \qquad ...(8.22)$$

which is *Rayleigh-Jeans law*.

Thus, Wien's law holds for shorter wavelengths and Rayleigh-Jeans law for longer wavelengths. We can thus say, the Wien's law and Rayleigh-Jeans law are incorporated in the Planck's law.

### EXAMPLE 8.2

*Derive Wein's displacement law and Stefan's law from Planck's radiation law.* **(Rohilkhand 2002)**

**SOLUTION. (*i*) Wein's displacement law**

Planck's formula is given by

$$E_\lambda \, d\lambda = \frac{8\pi hc}{\lambda^5} \cdot \frac{1}{e^{hc/\lambda kT} - 1} \qquad ...(i)$$

Differentiating eqn (*i*) w.r.t. to $\lambda$ partially, we get the energy for unit range of wavelength, *i.e.*,

$$\frac{\partial E_\lambda}{\partial \lambda} = \frac{1}{e^{hc/\lambda kT} - 1} \times \frac{-5(8\pi hc)}{\lambda^6} + \frac{8\pi hc}{\lambda^5} \times \frac{\frac{hc}{\lambda^2 kT} e^{hc/\lambda kT}}{(e^{hc/\lambda kT} - 1)^2}$$

$$= \frac{-40\pi hc}{\lambda^6} \times \frac{1}{e^{hc/\lambda kT} - 1} + \frac{8\pi hc}{\lambda^5} \times \frac{hc}{\lambda^2 kT} \times \frac{e^{hc/\lambda kT}}{(e^{hc/\lambda kT} - 1)^2}$$

For maximum energy of emission, *i.e.*, for maximum value of $E_\lambda$, $\frac{\partial E_\lambda}{\partial \lambda}$ must be zero. Hence

$$\frac{-40\pi hc}{\lambda^6}\cdot\frac{1}{e^{hc/\lambda kT}-1}+\frac{8\pi hc}{\lambda^5}\times\frac{hc}{\lambda^2 kT}\times\frac{e^{hc/\lambda kT}}{(e^{hc/\lambda kT}-1)^2}=0$$

or $$\frac{8\pi hc}{(e^{hc/\lambda kT}-1)\lambda^6}\left[-5+\frac{hc}{\lambda kT}\times\frac{e^{hc/\lambda kT}}{(e^{hc/\lambda kT}-1)}\right]=0$$

or $$-5+\frac{hc}{\lambda kT}\times\frac{e^{hc/\lambda kT}}{(e^{hc/\lambda kT}-1)} = 0 \qquad ...(ii)$$

Putting $\frac{hc}{\lambda kT} = x$, say in equation (*ii*), we get

$$-5+\frac{xe^x}{e^x-1} = 0$$

or $$\frac{xe^x}{e^x-1} = 5 \qquad ...(iii)$$

It suggests that there must be a root in the neighbourhood of 5. By applying the method of approximation, the exact value of $x$ is found to be equal to 4.965.

∴ $$x=\frac{hc}{\lambda kT} = 4.965 \qquad ...(iv)$$

Here λ represents the wavelength at which energy per unit range of wavelength is maximum. Representing this particular wavelength by λm, equation (iv), may be written as

$$\frac{hc}{\lambda_m kT} = 4.965$$

or $$\lambda_m . T = \frac{hc}{4.965k} = \text{constant} \qquad ...(v)$$

This is **Wein's displacement law.**

**(*ii*) Stefan's law from Planck's formula**

The Planck's radiation formula in terms of frequency is given by eq. (8.11) as

$$E_\nu\, d\nu = \frac{8\pi h\nu^3}{c^3}\cdot\frac{1}{e^{h\nu/kT}-1}d\nu$$

Therefore, the total radiant energy over all frequencies will be

$$E = \int_0^\infty E_\nu d\nu = \frac{8\pi h}{c^3}\int_0^\infty \frac{\nu^3 d\nu}{e^{h\nu/kT}-1} \qquad ...(vi)$$

Putting $\frac{h\nu}{kT} = x$, so that $\nu = \frac{kT}{h}x$ and $d\nu = \frac{kT}{h}dx$

Substituting in equation (*vi*), we get

$$E = \frac{8\pi k^4 T^4}{c^3 h^3}\int_0^\infty \frac{x^3\, dx}{e^x-1}$$

But $$\int_0^\infty \frac{x^3\, dx}{e^x-1} = \frac{\pi^4}{15}$$

∴ $$E = \frac{8\pi^5 k^4 T^4}{15c^3 h^3}$$

or $$E = AT^4 \qquad ...(vii)$$

where $$A = \frac{8\pi^5 k^4}{15 c^3 h^3}$$

But $\frac{Ac}{4} = \sigma$, the Stefan's constant.

$$\therefore \quad E = \sigma T^4 \qquad ...(viii)$$

This is **Stefan's law**, where $\sigma = \frac{2\pi^5 k^4}{15 h^3 c^2}$ gives the Stefan's constant.

### EXAMPLE 8.3

*Two large closely spaced concentric spheres (both are black body radiators) are maintained at temperatures 200 K and 300 K respectively. The space in between the two spheres is evacuated. Calculate the net rate of energy transfer between the two spheres. [$\sigma = 5.672 \times 10^{-8}$ M.K.S. units].*

**SOLUTION.** Here, $T_1 = 300$ K, $T_2 = 200$ K, $\sigma = 5.672 \times 10^{-8}$ M.K.S. units

$$E = \sigma [T_1^4 - T_2^4]$$
$$= 5.672 \times 10^{-8} [(300)^4 - (200)^4]$$
$$E = \mathbf{368.68 \text{ watts/m}^2}$$

### EXAMPLE 8.4

*Calculate the radiant emittance of a black body at a temperature of (i) 400 K (ii) 4000 K. [$\sigma$ is 5.672 × $10^{-8}$ M.K.S. units].* **(Purvanchal 2007, Bhopal 2000)**

**SOLUTION.** According to Stefan's law,

$$E = e\sigma T^4$$
$$= \sigma T^4 \qquad (\because e = 1, \text{ for a black body})$$

(*i*) For $T = 400$ K, $E = 5.672 \times 10^{-8} \times [400]^4$
$$= \mathbf{1452 \text{ watts/m}^2}$$

(*ii*) For $T = 4000$ K, $E = 5.672 \times 10^{-8} \times [4000]^4$
$$= 1452 \times 10^4 \text{ Watts/m}^2$$
$$= \mathbf{14520 \text{ Kilo-watts/m}^2}$$

### EXAMPLE 8.5

*An aluminum foil of relative emittance 0.1 is placed in between two concentric spheres at temperatures 300 K and 200 K respectively. Calculate the temperature of the foil after the steady state is reached. Assume that the spheres are perfect black body radiators. Also calculate the rate of energy transfer between one of the spheres and the foil. [$\sigma = 5.672 \times 10^{-8}$ M.K.S. units].*

**(Agra 2000)**

**SOLUTION.** Here, $T_1 = 300$ K, $T_2 = 200$ K, $e = 0.1$.

(*i*) Let $x$ be the temperature of the foil, after the steady state is reached, then

$$e\sigma (T_1^4 - x^4) = e\sigma (x^4 - T_2^4)$$
$$[(300)^4 - x^4] = [x^4 - (200)^4]$$
$$x^4 = 48.5 \times 10^8$$
$$\therefore \quad x = \mathbf{263.8 \text{ K}}$$

(*ii*) $$E = e\sigma (T_1^4 - x^4)$$
$$= 0.1 \times 5.672 \times 10^{-8} [(300)^4 - (263.8)^4]$$
$$= \mathbf{18.5 \text{ Watts/m}^2}$$

### EXAMPLE 8.6

*Calculate the energy radiated per minute from the filament of an incandescent lamp at 2000 K, if the surface area is 5.0 × $10^{-5}$ sq. metres and its relative emittance is 0.85.* **(Lucknow 2000)**

**SOLUTION.** Here,

$$A = 5 \times 10^{-5}\ \text{m}^2$$
$$e = 0.85$$
$$\sigma = 5.672 \times 10^{-8}\ \text{M.K.S. units}$$
$$t = 60\ \text{sec.}$$
$$T = 2000\ \text{K}$$
$$\therefore \quad E = A e\sigma T^4 \times t$$
$$E = 5 \times 10^{-5} \times 0.85 \times 5.672 \times 10^{-8} \times (2000)^4 \times 60$$
$$E = \mathbf{2315\ Joules}$$

### EXAMPLE 8.7

*An iron furnace radiates 1.53 × $10^5$ calories per hour through an opening of cross-section $10^{-4}$ sq. metre. If the relative emittance of the furnace is 0.80, calculate the temperature of the furnace. Given : $\sigma = 1.36 \times 10^{-8}$ cal/$m^2$–S–$K^4$.* **(Kanpur 2001)**

**SOLUTION.**

$$E = A\, e\sigma t\, (T^4)$$

Here, $A = 10^{-4}\ m^2, \quad e = 0.80, \quad t = 3600$ sec., $\quad T = ?$

Thus, $$T^4 = \frac{E}{Ae\sigma t}$$

$$\therefore \quad T = \left(\frac{E}{Ae\sigma t}\right)^{\frac{1}{4}} = \left[\frac{1.53\times10^5}{10^{-4}\times0.80\times1.36\times10^{-8}\times3600}\right]^{\frac{1}{4}}$$

$$T = \mathbf{2500\ K}$$

### EXAMPLE 8.8

*A black body has its cavity in the shape of a cube. Determine the number of modes of vibration per unit volume in the wavelength region 4990 Å to 5010 Å.* **(Osmania 2004)**

**SOLUTION.** The number of modes per c.c. in the wavelength range $\lambda$ to $\lambda + d\lambda$ in a chamber (or cavity) is

$$N = \frac{8\pi}{\lambda^4} d\lambda$$

Here, $d\lambda$ = (5010 – 4990) Å = 20 Å = 20 × $10^{-8}$ cm. and $\lambda$ = 5000 Å = 5000 × $10^{-8}$ cm

$$\therefore \quad N = \frac{8\times3.14\times20\times10^{-8}}{(5000\times10^{-8})^4}$$
$$= \mathbf{8.038 \times 10^{11}}$$

### EXAMPLE 8.9

*Calculate the surface temperature of the sun and moon, given that $\lambda_m$ = 4753 Å and 14μ respectively, $\lambda_m$ being wavelength of maximum intensity of emission.* **(Meerut 2000)**

**SOLUTION.** We know from Wien's displacement law,

$$\lambda_m T = \text{constant} = 0.2898$$

(*i*) For the sun,

given $\quad \lambda_m$ = 4753 Å = 4753 × $10^{-8}$ cm

∴ the temperature of the sun

$$T_s = \frac{0.2898}{4753 \times 10^{-8}} = \mathbf{6097\ K.}$$

(*ii*) For the moon,

given $\lambda_m = 14\mu = 14 \times 10^{-4}$ cm  [$\because$ 1μ = 1 micron = $10^{-4}$ cm]

$\therefore$ the temperature of the moon,

$$T_m = \frac{0.2898}{14 \times 10^{-4}} = \mathbf{207\ K}$$

**EXAMPLE 8.10**

*A body at 1500 K emits maximum energy of wavelength 2000 nm. If the sun emits maximum energy of wavelength 550 nm, what would be the temperature of the sun?*

**(Nagpur 2007, 2004/s)**

**SOLUTION.** Temperature of the body = 1500 K

Wavelength for maximum energy emission = 2000 × $10^{-9}$ m

Wavelength for maximum energy emission from sun = 550 × $10^{-9}$ m.

If $T$ is the temperature of the sun, then according to Wien's displacement law

$$\lambda_m T = \text{constant}$$

$$550 \times 10^{-9} \times T = 2000 \times 10^{-9} \times 1500$$

$$\therefore \quad T = \frac{2000 \times 10^{-9} \times 1500}{550 \times 10^{-9}} = \mathbf{5454\ K.}$$

**EXAMPLE 8.11**

*What is the wavelength at which human body radiates maximum energy? Temperature of human body is 37°C. Wien's constant is 2.898 × $10^{-3}$ mK.* **(Meerut 2004/s, Nagpur 2009)**

**SOLUTION.** According to Wien's displacement law,

$$\lambda_m T = \text{constant} = 2.898 \times 10^{-3} \text{ m K.}$$

where $\lambda_m$ is the wavelength at which black body radiates maximum energy at temperature $T$.

Temperature of human body = $T$ = 37° C = 273 + 37 = 310 K.

Considering the human body as a black body, the wavelength at which it radiates maximum energy = $\lambda_m$ × 310 = 2.898 × $10^{-3}$ m

$$\therefore \quad \lambda_m = \frac{2.898 \times 10^{-3}}{310} = \mathbf{9.35 \times 10^{-6}\ m.}$$

**EXAMPLE 8.12**

*A black body with temperature 127°C is put in an evacuated enclosure whose walls are blackened and maintained at 27°C. What is net amount of energy lost by body per second per unit area.*

*Stefan's constants* $\sigma$ = *5.672 ×* $10^{-8}$ S.I. units

**SOLUTION.** Here, $T_1$ = 273 + 127 = 400 K

$T_2$ = 273 + 27 = 300 K

$\sigma$ = 5.672 × $10^{-8}$ S.I. units

$$R = \sigma\,[T_1^4 - T_2^4]$$

$$R = 5672 \times 10^{-8}\ [(400)^4 - (300)^4]$$

$$R = \mathbf{992.6\ watts/m^2}$$

**EXAMPLE 8.13**

*If Wien's constant $b = 0.3$ cm-K°, calculate the temperature of the sun whose radiation has maximum energy at wavelength $\lambda = 5500$ Å.* **(Meerut 2004)**

**SOLUTION.** According to Wien's displacement law

$$\lambda_m T = \text{constant} = b$$

$$T = \frac{b}{\lambda_m}$$

Here $b = 0.3$ cm-K

$$\lambda_m = 5500 \text{ Å} = 5500 \times 10^{-8} \text{ cm}$$

$$\therefore \quad T = \frac{0.3}{5500 \times 10^{-8}}$$

$$T = \mathbf{5455\ K}$$

**EXAMPLE 8.14**

*Deduce the temperature at which a black body loses thermal energy at the rate of 1 watt/cm².* **(Bhopal 2000)**

**SOLUTION.**

$$R = 1 \text{ watt/cm}^2 = 10^4 \text{ watts/m}^2$$

$$R = \sigma T^4$$

$$T^4 = \left(\frac{R}{\sigma}\right)$$

$$T = \left(\frac{R}{\sigma}\right)^{\frac{1}{4}}$$

$$T = \left[\frac{10^4}{5.672 \times 10^{-8}}\right]^{\frac{1}{4}}$$

$$T = \mathbf{648\ K.}$$

**EXAMPLE 8.15**

*If a black body at a temperature 6174 K emits 4700 Å with maximum energy; calculate the temperature at which it will emit a wavelength of $1.4 \times 10^{-5}$ m with maximum energy.* **(Mumbai 2005)**

**SOLUTION.** According to Wien's displacement law

$$\lambda_m T = \text{constant}$$

$$\lambda_m T = \lambda_m' T'$$

$$T' = \frac{\lambda_m T}{\lambda_{m'}} = \frac{4700 \times 10^{-10} \times 6174}{1.4 \times 10^{-5}}$$

$$T' = \mathbf{207.27\ K}$$

**EXAMPLE 8.16**

*Assume the sun to be a black body at temperature 5800 K. Use Stefan's law to show that the total radiant energy emitted by sun per second is $3.95 \times 10^{26}$ J. Also show that the rate at which energy is reaching the top of earth's atmosphere is 1.4 k W-m$^{-2}$.* **(Rohilkhand 2000)**

**SOLUTION.** Here radius of the sun, $r = 7 \times 10^8$ m

Surface area of the sun, $A = 4\pi r^2$

$$= 4 \times 3.142 \times (7 \times 10^8)^2 \text{ m}^2$$
$$\sigma = 5.672 \times 10^{-8} \text{ S.I. units}$$
$$T = 5800 \text{ K}$$

Total energy emitted by sun per second

$$U = A \sigma T^4$$
$$U = 4 \times 3.142 \times (7 \times 10^8)^2 \times 5.672 \times 10^{-8} \times (5800)^4$$
$$U = \mathbf{3.95 \times 10^{26} \text{ J}}$$

Distance of the earth's atmosphere from the sun

$$= r_1 = 1.5 \times 10^{11} \text{ m}$$

Energy reaching per unit area per second.

$$R = \frac{U}{4\pi r_1^2}$$
$$R = \frac{3.95 \times 10^{26}}{4 \times 3.142 \times (1.5 \times 10^{11})^2}$$
$$R = 1.3968 \times 10^3 \text{ W/m}^2$$
$$R = 1.3968 \text{ kW/m}^2$$
$$\mathbf{R = 1.4 \text{ kW/m}^2}$$

**EXAMPLE 8.17**

*Calculate the number of modes in a chamber of volume 50 c.c. in the frequency range 4 × 10^14 and 4.01 × 10^14 sec^–1.* **(Indore 2000)**

**SOLUTION.** The no. of modes per c.c. in the frequency range $v$ and $v + dv$ is

$$= \frac{8\pi v^2}{c^3} dv$$
$$= \frac{8 \times 3.14 \times (4 \times 10^{14})^2}{(3 \times 10^{10})^3} \times (4.01 \times 10^{14} - 4 \times 10^{14}) = 1.5 \times 10^{11}$$

∴ The number of modes in the whole chamber of volume $V = 50$ c.c. is

$$= 1.5 \times 10^{11} \times 50$$
$$= \mathbf{7.5 \times 10^{12}}$$

**EXAMPLE 8.18**

*A solid copper sphere (density ρ and specific heat C) of radius r at initial temperature 200 K is suspended inside a chamber whose walls are at almost 0K. The time required for the temperature of the sphere to fall to 100 K is ......* **(Purvanchal 2000)**

**SOLUTION.** Here $\frac{dQ}{dt} = e \sigma A (T^4 - T_0^4)$

Here $e = 1$

and $A = 4\pi r^2$

Also $\frac{dQ}{dt} = mc \left(\frac{dT}{dt}\right)$

Also $m = \left(\frac{4}{3}\pi r^3\right)\rho$

$$\therefore \quad mc\left(\frac{dT}{dt}\right) = \sigma A\,(T^4 - T_0^4)$$

$$T_0 = 0$$

$$dt = \left(\frac{m}{A}\right)\left(\frac{c}{\sigma}\right)\frac{dT}{T^4}$$

Integrating

$$\int dt = \left(\frac{\frac{4}{3}\pi r^3 \rho}{4\pi r^2}\right)\left(\frac{c}{\sigma}\right)\int_{200}^{100}\frac{dT}{T^4}$$

$$t = \left(\frac{r\rho C}{3\sigma}\right)\left(\frac{1}{3}\right)\left[\left(\frac{1}{200}\right)^3 - \left(\frac{1}{100}\right)^3\right]$$

$$t = \left[\frac{7\, r\rho C}{(72\times 10^6)\,\sigma}\right]$$

**EXAMPLE 8.19**

*Compare the radiant emittance of a black body at 200 K and 2000 K. Given σ = 5.672 × 10⁻⁸ M.K.S. units.* **(Delhi 2000)**

**SOLUTION.** Here,

$$T_1 = 200 \text{ K}$$

$$T_2 = 2000 \text{ K}$$

$$R_1 = \sigma T_1^4$$

$$R_2 = \sigma T_2^4$$

$$\left(\frac{R_1}{R_2}\right) = \left(\frac{T_1}{T_2}\right)^4$$

$$= \left(\frac{200}{2000}\right)^4$$

$$= \mathbf{10^{-4}}$$

**EXAMPLE 8.20**

*A black sphere of diameter 8 cm. is heated to 500 K when the surrounding temperature is 300 K. What is the rate at which energy is radiated? Stefan's constant is $6 \times 10^{-8}$ $W\,m^{-2}\,K^{-4}$.* **(Nagpur 2000)**

**SOLUTION.** Here,

$$\sigma = 6\times 10^{-8} \text{ W m}^{-2}\text{ K}^{-4}$$

$$t = 1\ s$$

$$T = 500 \text{ K}$$

$$T_0 = 300 \text{ K}$$

$$d = 8 \text{ cm} = 0.08 \text{ m}$$

$$A = \pi d^2$$

$$= 3.14 \times (0.08)^2$$

$$= 0.02 \text{ m}^2$$

$$e = 1$$

$$R = \sigma A e\,[T^4 - T_0^4]$$

$$= 6\times 10^{-8} \times 0.02 \times 1\,[(500)^4 - (300)^4]$$

$$= 65.28 \text{ J/s}$$

$$= \mathbf{65.28\ watts}$$

**EXAMPLE 8.21**

*An unclothed boy is in a room at 20°C. If the skin temperature of the boy is 37°C, how much heat is lost from his body in 10 minutes. The surface area of the boy is 3 m² and emissivity is 0.90 and*

$$\sigma = 5.67 \times 10^{-8}\ Wm^{-2}\ k^{-4}$$ **(Mumbai 2000)**

**SOLUTION.** Here

$$e = 0.9$$
$$A = 3\ \text{m}^2$$
$$\sigma = 5.67 \times 10^{-8}\ \text{W/m}^2 - \text{k}^4$$
$$\theta_1 = 273 + 37 = 310\ \text{K}$$
$$\theta_2 = 273 + 20 = 293\ \text{K}$$
$$R = eA\,\sigma\,(\theta_1^{\ 4} - \theta_2^{\ 4})$$
$$R = 0.9 \times 3 \times 5.67 \times 10^{-8}\ [(310)^4 - (293)^4]$$
$$R = 286\ \text{J/s}$$
$$t = 10\ \text{minutes} = 600\ \text{s}$$

Total heat lost by the skin in 10 minutes.

$$Q = Rt = 286 \times 600$$
$$= \mathbf{1.716 \times 10^5\ J}$$

**EXAMPLE 8.22**

*The relative emittance of tungsten is approximately 0.35. A tungsten sphere of surface area $10^{-3}$ sq metres is suspended inside a large evacuated enclosure whose walls are at 300 K. What power input is required to maintain the sphere at a temperature of 3000 K? The conduction of heat along the supports can be neglected.* **(Mysore 2000)**

$$[\sigma = 5.672 \times 10^{-8}\ \text{M.K.S. units}]$$
$$R = \sigma A\, e[T^4 - T_0^{\ 4}]$$

**SOLUTION.** Here

$$A = 10^{-3}\ \text{sq metres}$$
$$e = 0.35$$
$$\sigma = 5.672 \times 10^{-8}\ \text{M.K.S. units}$$
$$R = 5.672 \times 10^{-8} \times 10^{-3} \times 0.35\ [(3000)^4 - (300)^4]$$
$$= 1609\ \text{watts}$$

The power input $= \mathbf{1609\ watts.}$

**EXAMPLE 8.23**

*The order of magnitude of the energy received from sun at earth's surface is $10^{-1}$ Joule cm$^{-2}$ sec$^{-1}$. Calculate the order of magnitude of the total force due to solar radiation on the earth (assumed perfectly absorbing). Given: earth's diameter = $10^7$ metre.* **(Lucknow 2001, Purvanchal 2006)**

**SOLUTION.** The energy received from the sun at earth's surface per unit area per second is $E = 10^{-1}$ Joule/cm² sec = $10^{-3}$ Joule m$^{-2}$ sec$^{-1}$.

∴ Radiation pressure

$$P = \frac{E}{C} = \frac{10^{-3}}{3 \times 10^8}\ N/m^2$$
$$= 3.33 \times 10^{-6}\ \text{N/m}^2$$

∴ Total force due to solar radiation on the earth

$$= p \times \text{surface area}$$
$$= 3.33 \times 10^{-6} \times 4\pi r^2$$

$$= 3.33 \times 10^{-6} \times 4 \times 3.14 \times \left(\frac{10^7}{2}\right)^2$$

$$= \mathbf{1.052 \times 10^9\ N}$$

**EXAMPLE 8.24**

*Calculate the average energy $\bar{\varepsilon}$ of an oscillator of frequency $0.60 \times 10^{14}$ $sec^{-1}$ at $T = 1800$ K treating it as (i) classical oscillator, (ii) Planck's oscillator.*

**SOLUTION.** (*i*) The average energy of a classical oscillator

$$\bar{\varepsilon} = kT$$

$$= (1.38 \times 10^{-23} \text{ Joule/K}) \times 1800 \text{ K}$$

$$= \mathbf{2.484 \times 10^{-20}\ Joule}$$

(*ii*) The average energy of a Planck's oscillator

$$\bar{\varepsilon} = \frac{h\nu}{e^{h\nu/kT} - 1} = \frac{h\nu/kT}{e^{h\nu/kT} - 1} \times kT$$

Let us calculate $h\nu/kT$ first.

$$\frac{h\nu}{kT} = \frac{(6.6 \times 10^{-34})(0.60 \times 10^{14})}{1.38 \times 10^{-23} \times 1800} = \frac{3.96 \times 10^{-23}}{2.484 \times 10^{-20}}$$

$$= 1.59$$

$$\therefore \quad \bar{\varepsilon} = \frac{h\nu/kT}{e^{h\nu/kT} - 1} kT = \frac{1.59 \times 2.484 \times 10^{-20}}{e^{1.59} - 1}$$

$$= \frac{1.59}{(2.72)^{1.59} - 1} \times 2.484 \times 10^{-20} \qquad (\because e = 2.72)$$

$$= \frac{1.59 \times 2.484 \times 10^{-20}}{3.909} = \mathbf{1.01 \times 10^{-20}\ Joule}$$

**EXAMPLE 8.25**

*Calculate the maximum amount of heat which may be lost per second by radiation from a sphere of 5 cm. in diameter at a temperature of 600 K when placed in an encloser at a temperature of 300 K. [$\sigma = 5.7 \times 10^{-12}$ Watts × $cm^{-2}$ $(°C)^{-4}$]* **(Garhwal 2005)**

**SOLUTION.** The rate of loss of thermal energy will be maximum, if the sphere radiates as black body. Applying Stefan Boltzmann's law, the net energy lost per unit area of the sphere per second is given by

$$E = \sigma(T^4 - T_o^4)$$

where $T$ and $T_o$ are the absolute temperatures of sphere and the surroundings respectively, and $\sigma$ is Stefan's constant. Total surface area of the sphere = $4\pi r^2$

$\therefore$ The net loss of heat from the total surface of the sphere

$$Q = E \times 4\pi r^2$$

$$= \sigma(T^4 - T_o^4) \times 4\pi r^2$$

$$= \frac{\sigma(T^4 - T_0^4) \times 4\pi r^2}{J} \text{ calories/second}$$

Given:

$$r = 2.5 \text{ cm} = 2.5 \times 10^{-2} \text{ m}$$

$$T = 600 \text{ K}, T_o = 300 \text{ K}.$$

$$\sigma = 5.7 \times 10^{-12} \text{ watts} \times \text{cm}^{-2} \, (°C)^{-4}$$

$= 5.7 \times 10^{-8}$ Joule m$^{-2}$ sec$^{-1}$ K$^{-4}$

$= 5.7 \times 10^{-8}$ Watt m$^{-2}$ (K)$^{-4}$

and $J = 4.2$ Joule/cal

$$\therefore \quad Q = \frac{5.7\times10^{-8}\times[(600)^4-(300)^4]\times4\times3.14\times(2.5\times10^{-2})^2}{4.2}$$

$$= \frac{5.7\times10^{-8}\times1215\times10^8\times4\times3.14\times6.25\times10^{-4}}{4.2}$$

$$= \frac{543651.75\times10^{-4}}{4.2}$$

$$= \mathbf{12.944\ cal/s}$$

## EXERCISE–VIII

**8.1** Define thermal radiation. **(Rohilkhand 2001 Improvement)**

**8.2** Explain the meaning of emissive power and absorptive power. **(Kanpur 2000, Agra 2000, Purvanchal 2002, Calicut 2003)**

**8.3** Explain the characteristics of radiation in a hollow encloser at constant temperature. **(Rohilkhand 2001, Kanpur 2002)**

**8.4** Give a brief account of Prevost's theory of exchanges.

**8.5** What is perfectly black body? Describe how has the idea of a black body been achieved in practice? **(Nagpur 2009, 2008, 2007, Meerut 2001, 2003, Agra 2000)**

**8.6** What is black-body radiation? Explain its temperature dependence. **(Nagpur 2009, 2008, 2007, 2005/s, 2005/w, Ranchi 2005, Madras 2005, Rohilkhand 2001)**

**8.7** Explain the terms emissive power and absorptive power. Prove that at any temperature the ratio of the emissive power to the absorptive power of a substance is constant and is equal to the emissive power of a perfectly black body. **(Purvanchal 2007, Agra 2003, Calcutta 2003)**

**8.8** State and explain Kirchhoff's law. Give its application. **(Lucknow 2004, Meerut 2003, Purvanchal 2005, 2002, 2003, Bundelkhand 2001, Madras 2005, Rohilkhand 2001)**

**8.9** Prove that the pressure due to diffuse radiation on a surface is $u/3$, where $u$ is energy density of radiation. **(Purvanchal 2002, Ranchi 2005)**

**8.10** Show that the diffuse radiation exert pressure on the wall of the encloser and derive a relation between energy and pressure of diffuse radiation. **(Nagpur 2005/w, Calcutta 2004)**

**8.11** Why polished reflectors are used in electric heaters? **(Lucknow 2004)**

**8.12** Describe the main characteristics of radiation emitted by a perfectly black body. Show that the diffused radiation exerts a pressure on the walls of the container equal to one-third of its energy density. **(Garhwal 2005, Purvanchal 2003, 2004, 2005, Rohilkhand 2001)**

**8.13** State Stefan-Boltzmann law of radiation. Deduce this law on thermodynamic considerations. **(Nagpur, 2009, 2007, 2005/w, Meerut 2003, Lucknow 2003, Osmania 2005, Bangalore 2002, Purvanchal 2006)**

**8.14** Explain the distribution of energy of a black body at different temperatures by drawing the graphs. Discuss briefly the different laws which explain the above energy spectrum. **(Nagpur, 2007, 2005/s, 2004/s, Madras 2004, Bangalore 2002, 2005, Meerut 2001, Alagappa 2004, Osmania 2004)**

**8.15** What is a perfectly black body? Draw curves for the distribution of energy in the spectrum of a black body for two different temperatures. Explain the important results obtained from these curves. **(Nagpure 2007, 2009, Lucknow 2004, Meerut 2003)**

**8.16** State and explain the laws relating to radiation and temperature of a radiating body. What do you mean by
(*a*) perfectly black body,
(*b*) coefficient of absorption and
(*c*) coefficient of emission?

**8.17** Deduce Wien's displacement law for the distribution of energy in black body spectrum. Explain its validity in terms of experimental results.
**(Agra 2003, Bangalore 2005, Rohilkhand 2002 Imp. G.N.D.U. 2004, 2003, Gauhati 2000, Kerala 2001, Osmania 2004, Nagpur 2002)**

**8.18** State and explain Rayleigh-Jeans law. Explain its validity in terms of experimental results.
**(Agra 2008, Rohilkhand 2002, Imp.)**

**8.19** State Planck's radiation law. Give the Planck's quantum postulates. **(Nagpur 2005/s, Ranchi 2004)**

**8.20** Derive Planck's radiation formula. Show that Rayleigh-Jeans law and Wien's law are special cases of Planck's law.
**(Nagpur 2005/s, Meerut 2007, 2003, 2005, Ranchi 2004, Mumbai 2005, G.N.D.U. 2001, Punjab 2002, Agra 2006, 2002, Rohilkhand 2002 Imp., M.D.U. 2003, Madurai 2003, Purvanchal 2007, 2005)**

**8.21** Explain the ultraviolet catastrophe according to Rayleigh–Jeans distribution law.
**(Agra 2006, Nagpur 2009, 2007, 2005/s, 2004/s, 2004/w)**

**8.22** Assuming Planck's Law of black body spectrum, obtain Wien's law of distribution for short wavelength.
**(Nagpur 2004/s, Bangalore 2004, Alagappa 2004, Mysore 2004/s, 2004/w)**

**8.23** Derive Wein's displacement law and Stefan's law from Planck's law of black body radiation.
**(Nagpur 2007, Punjab 2005, Alagappa 2004, Madras 2004)**

**8.24** Describe wien's displacement law. Using this law, calculate the temperature of the sun.
**(Purvanchal 2007, Agra 2005)**

**8.25** Give Planck's theory of black body radiation. Show this reduces to Wien's formula for small wavelengths and Rayleigh-Jeans law for large wavelength.
**(Madras 2004, Purvanchal 2003, Rohilkhand 2000, Rajasthan 1996, 1997, 2000, Kumaun 1998, 2003)**

**8.26** Derive Rayleigh-Jeans law from Planck's law.
**(Agra 2004, Rohilkhand 2001 Imp., Bangalore 2004, 2005, Alagappa 2004, Mysore 2004)**

**8.27** Discuss Planck's hypothesis of quantum theory of radiation and obtain an expression for $\overrightarrow{\varepsilon} = \frac{h\nu}{e^{h\nu/kT} - 1}$ mean kinetic energy of Planck's oscillator. **(Purvanchal 2006, Meerut 2004, Sagar 2000)**

**8.28** What is Planck's law of radiation? Derive the Planck's formula for energy distribution of in black body spectrum. **(Nagpur 2007, Meerut 2004, M.S. Uni. Tamil Nadu 2007)**

**8.29** Explain what do you understand by a black body. State Stefan's law of radiation and prove it from thermodynamical considerations. Indicate how it can be verified. How is this law distinguished from Newton's law of cooling ?
**(Madurai Kamraj 2003, Mysore 2004, Purvanchal 2002, Agra 2002, Kanpur 1996)**

**8.30** Show that Planck's law and Rayleigh-Jeans law become identical if the size of the quantum is allowed to vanish or if the temperature becomes very high. **(Mumbai 2005)**

**8.31** State Stefan's law and write its equation. Explain how would you calculate surface temperature of the sun using the law. **(Alagappa 2004, Purvanchal 2003)**

**8.32** What do you understand by a black body? State and explain Stefan's law of heat radiation.
**(Madras 2005)**

**8.33** In what way the Planck's formula overcomes the deficiencies in both Wien's law and Rayleigh-Jeans formula for black body energy distribution? **(Purvanchal 2004)**

**8.34** Prove that the energy density of a black body radiation in a cavity is proportional to the fourth power of absolute temperature. **(Calcutta 2005)**

**8.35** What do you mean by black body? Discuss the distribution of energy in the spectrum of black body on the basis of the spectrum obtained in the experiment performed by Lummer and Pringsheim.
**(M.S. Uni. Tamil Nadu 2007, Agra 2007, 2005, Rohilkhand 2000)**

**8.36** Describe an experiment to verify Stefan's law and the determination of the Stefan's constant. How is the temperature of sun estimated from the data of the solar constant?
**(Madras 2004, 2005, Rohilkhand 2000, Meerut 2000, Kanpur 2002, Gorakhpur 2000)**

**8.37** Define solar constant. How do you find the temperature of the sun from a knowledge of solar constant and Stefan's law. **(Purvanchal 2005, Meerut 2003, Madurai 2003)**

**8.38** Define solar constant. How is it experimentally determined? Comment on the source of energy in the sun. **(Purvanchal 2007, Madras 2004, Osmania 2005, Bangalore 2005, Lucknow 2004)**

**8.39** Derive Planck's law of black body radiation and discuss its experimental verification. **(Patna 2005)**

**8.40** Deduce Kirchhoff's law of radiation and describe an experiment for its verification. **(Patna 2004)**

**8.41** What is the principle of optical pyrometer? With the help of a diagram, explain the construction and working of an optical Pyrometer. **(Osmania 2005, Madras 2005)**

**8.42** Define solar constant. Describe water flow pyrheliometer and explain how solar constant is determined from it. **(Alagappa 2004)**

**8.43** Give an account of the methods for measuring high temperatures by radiation pyrometry. **(Madras 2005, 2004)**

**8.44** Write notes on :

(*i*) Solar spectrum **(Alagappa 2004)**

(*ii*) Green house effect **(Alagappa 2004)**

(*iii*) Wien's displacement law **(Kumaun 2000)**

(*iv*) Solar constant and temperature of sun **(Meerut 2001)**

(*v*) Optical pyrometer **(Osmania 2005)**

(*vi*) Planck law of radiation **(Meerut 2001)**

(*vii*) Kirchhoff's law of black body radiation **(Patna 2005, Rohilkhand 2000)**

(*viii*) Radiation pyrometry **(Madras 2004)**

(*ix*) Sources of solar energy **(Madras 200)**

**8.45** A metallic ball at a temperature of 527 °C is placed in an encloser at 27 °C. If emissivity of the body is 0.4, find the rate of loss of energy per unit area by the body. Given: $\sigma = 5.67 \times 10^{-8}$ J/m$^2$/s/deg$^4$. **(Nagpur 2005/s)**

**8.46** Spectra of three stars A, B and C show maximum brightness for the respective wavelengths of 6500 Å, 4800 Å and 5600 Å. Compare their temperatures $T_A$, $T_B$ and $T_C$. **(Nagpur 2004/s)**

**8.47** In an atomic explosion, the maximum temperature reached was of the order of $10^7$ K. Calculate the wavelength of maximum energy. Given: Wien's constant = 0.2898 cm K. [**Ans.** 2896 Å] **(Osmania 2004)**

**8.48** What is wavelength of maximum intensity radiation radiated from a source at temperature 3000°C? Wien's constant = $2.898 \times 10^{-3}$ m K. [**Ans.** 8854 Å] **(Nagpur 2001)**

**8.49** If the sun emits maximum energy at wavelength 4753 Å; calculate the temperature of its surface. Given: Wien's constant $b$ = 0.288 cm °C. [**Ans.** 6060 K] **(Nagpur 2001, 2004/w, 2005/w, )**

**8.50** Luminosity of rigid star in Orion constellation is 17000 times that of our sun. If the surface temperature of the sun is 6000 K; calculate the temperature of the star. [**Ans.** 68520 K] **(Agra 2008)**

**8.51** Sun light arrives at the earth at the rate of about 1.4 kWm$^{-2}$, when the sun is exactly overhead. The average radius of the earth's orbit is $1.5 \times 10^{11}$ m and the radius of the sun is $7 \times 10^8$ m. Find the surface temperature of the sun, assuming it radiates like a black body. [**Ans.** 5803 K] **(Rohilkhand 2001)**

**8.52** Two large closely spaced concentric spheres (both of which are black body radiators) are maintained at temperatures of 400 K and 600 K respectively. The space in between the two spheres is evacuated. Calculate net rate of energy transfer between the two temperatures.

$$\sigma = 5.672 \times 10^{-8} \text{ M.K.S. units}$$

[**Ans.** 5.90 kilowatts/m$^2$] **(Gorakhpur 2000)**

**8.53** Calculate the radiant emittance of a black body at a temperature of (*i*) 200 K and (*ii*) 2000 K.

$$\sigma = 5.672 \times 10^{-8} \text{ M.K.S. units}$$

[**Ans.** (*i*) 90.75 watts/m$^2$ (*ii*) 907.5 kilowatts/m$^2$] **(Madras 2005, Kanpur 2000)**

**8.54** The relative emittance of tungsten is approximately 0.70. A tungsten sphere of surface area $5 \times 10^4$ sq metres is suspended inside a large evacuated enclosure whose walls are at 300 K. What power input is required to maintain the sphere at a temperature of 3000 K. The conduction of heat along the supports can be neglected.

$$\sigma = 5.672 \times 10^{-8} \text{ M.K.S. units}$$

[**Ans.** 1609 watts] **(Kanpur 1999, 2001, Meerut 2000)**

**8.55** An aluminium foil of relative emittance 0.2 is placed in between two concentric spheres at temperatures 300 K and 200 K respectively. Calculate the temperature of the foil after the steady state is reached. Assume the spheres are perfect black body radiators. Also calculate the rate of energy transfer between one of the sphere,s and the foil.

$$\sigma = 5.672 \times 10^{-8} \text{ M.K.S. units}$$

[**Ans.** (*i*) 263.8 K, (*ii*) 37 watts/m$^2$] **(Nagpur 2000, Indore 2000)**

**8.56** Calculate the energy radiated per minute from the filament of an incandescent lamp at 3000 K, if the surface area is $10^{-4}$ sq metres and its relative emittance is 0.425.

[**Ans.** 11710 joules] **(Allahabad 2001, Kumaun 2000)**

**8.57** What is the wavelength of the maximum intensity radiation, radiated from source having temperature 3000 K, the Wien's constant is 0.29 cm-K. **(Osmania 2005, Bangalore 2004)**

**8.58** If the average energy radiated per unit area of the surface of the sun is $7.452 \times 10^6$ kW. Estimate the surface temperature of the sun assuming it to be a black body. Stefan's constant is $5.67 \times 10^{-8}$ W m$^{-2}$ K$^{-4}$. **(Bangalore 2002, Punjab 2001)**

**8.59** Calculate the number of modes of vibration in a 100 C.C. chamber in the wavelength range (*i*) 5000 to 5002 Å and (*ii*) 8000 to 8005 Å.

[**Ans.** $N_1 = 8.03 \times 10^{12}$
$N_2 = 3.066 \times 10^{12}$]

**8.60** Calculate the energy radiated by unit area of a black body in one second when its temperature is 1000 K. ($\sigma = 5.672 \times 10^{-8}$ J/s/m$^2$) **(Alagappa 2004)**

**8.61** An iron furnace radiates $7.65 \times 10^4$ calories per hour through an opening of cross-section $10^{-4}$ sq metres. If the relative emittance of the furnace is 0.40, calculate the temperature of the furnace.

[**Ans.** 2500 K] **(Indore 2001)**

**8.62** Two large closely spaced concentric spheres (both of which are black body radiators) are maintained at temperatures of 400 K and 600 K respectively. The space in between the two spheres is evacuated. Calculate net rate of energy transfer between the two temperatures.

$$\sigma = 5.672 \times 10^{-8} \text{ M.K.S. units}$$

[**Ans.** 5.90 kilowatts/m$^2$] **(Jiwaji 2000)**

**8.63** Two ideal black bodies A and B at temperatures 227 °C and 327 °C respectively are placed in an evacuated encloser whose walls are blackened and kept at 27 °C. Compare their rates of loss of heat.

[**Ans.** $A : B = 544 : 1215$]

**8.64** A black body with an initial temperature of 300 °C is allowed to cool inside an evacuated encloser surrounded by melting ice at the rate of 0.35 °C per sec. If the mass, specific heat and surface area of the body are 32 g; 0.10 cal/g/°C and 8 cm$^2$ respectively. Calculate the Stefan's constant.

[**Ans.** $5.7 \times 10^{-8}$ Joule m$^{-2}$ sec$^{-1}$ (K)$^{-4}$] **(Agra 2000)**

**8.65** Calculate the number of modes of vibration per unit volume in a black body cavity having frequencies between $5.5 \times 10^{12}$ Hz and $5.55 \times 10^{12}$ Hz. [**Ans.** $1.40718 \times 10^{12}$] **(Mumbai 2005)**

## OBJECTIVE QUESTIONS

**8.1** A perfectly black body is that which:

(*a*) is totally black in colour
(*b*) can radiate all its energy
(*c*) is made of ideal gas names Krishnika
(*d*) absorbs all the radiations incident on it

**(Garhwal 2005)**

**8.2** Average energy of a Planck's oscillator is:

(*a*) $E = h\nu$ (*b*) $E = nh\nu$

(*c*) $E = \dfrac{h\nu}{e^{h\nu/kT} - 1}$ (*d*) $E = mc^2$ **(Garhwal 2005)**

**8.3** Rayleigh–Jeans law of radiation:

(*a*) applies to smaller wavelengths (*b*) applies to longer wavelengths

(*c*) applies to all wavelengths (*d*) does not apply to any wavelength **(Garhwal 2005)**

**8.4** According to Stefan-Boltzmann's law:

(*a*) $E = \sigma T_o^4$ (*b*) $E = \sigma T^2$

(*c*) $E = \sigma (T^2 - T_o^2)$ (*d*) $E = \sigma (T^4 - T_o^4)$ **(Agra 2003)**

**8.5** Wien's displacement law is expressed by the equation:

(*a*) $\lambda_m T$ = constant (*b*) $\lambda_m \sqrt{T}$ = constant

(*c*) $\lambda_m T^2$ = constant (*d*) $\lambda_m T^3$ = constant **(Ranchi 2005)**

**8.6** 'Good absorbers are good emitters' – The statement is called:

(*a*) Prevost's law (*b*) Kirchhoff's law

(*c*) Stefan's law (*d*) Wien's law **(Rohilkhand 2002 Imp.)**

**8.7** Electromagnetic radiation is emitted by

(*a*) all bodies at all temperatures

(*b*) all bodies at 100 °C

(*c*) all bodies at absolute zero

(*d*) only a few bodies at all temperatures. **(Rohilkhand 2003)**

**8.8** For a perfectly black body, the absorptive power is

(*a*) 1 (*b*) 0.5

(*c*) 0 (*d*) ∞ **(Rohilkhand 2002, Imp.)**

**8.9** The absolute temperature of a perfectly black body is increased to twice its value. The rate of emission of energy per unit area will be:

(*a*) 2 times (*b*) 4 times

(*c*) 8 times (*d*) 16 times **(Kanpur 2002)**

**8.10** Plots of intensity verses wavelength for three bodies at temperatures $T_1$, $T_2$ and $T_3$ respectively are shown in Fig. Their temperatures are such that:

(*a*) $T_1 > T_2 > T_3$

(*b*) $T_1 > T_3 > T_2$

(*c*) $T_2 > T_3 > T_1$

(*d*) $T_3 > T_2 > T_1$ **(I.I.T. 2000)**

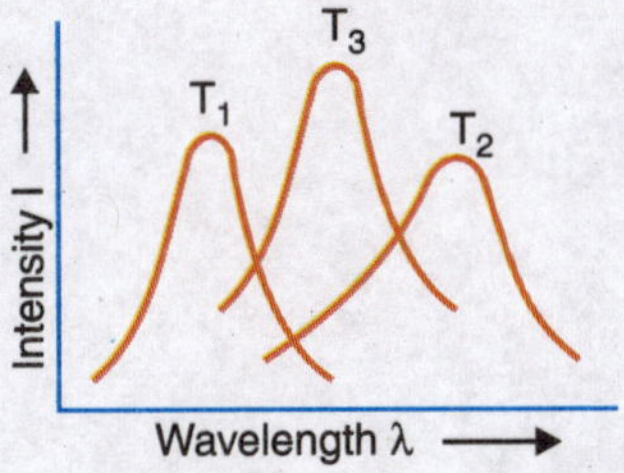


## ANSWERS

| | | | |
|---|---|---|---|
| **8.1** (*d*) | **8.2** (*c*) | **8.3** (*b*) | **8.4** (*d*) |
| **8.5** (*a*) | **8.6** (*b*) | **8.7** (*a*) | **8.8** (*a*) |
| **8.9** (*d*) | **8.10** (*b*) | | |

# SECTION – III

# Statistical Physics

# STATISTICAL PHYSICS

*Statistical physics* aims at studying the macroscopic parameters of a system in equilibrium from a knowledge of the microscopic properties of its constituent particles using the laws of mechanics. This is different from the approach of *thermodynamics* which aims at studying a macroscopic system in equilibrium from a macroscopic phenomenological basis, without considering the microscopic constituents of the system.

The study of statistical physics is mainly classified in two categories. These are:

**1.** Classical statistics or Maxwell-Boltzmann statistics.

**2.** Quantum statistics or B.E. and F.D. statistics.

The classical statistics employs the classical results of Maxwell's laws of distribution of molecular velocities and Boltzmann theorem, relating entropy and probability. Hence classical statistics is known as Maxwell-Boltzmann statistics.

On the other hand, the quantum statistics was developed by Bose, Einstein, Fermi and Dirac scientists. The essential difference between these two B.E. and F.D. statistics is due to the consideration of particles having different spins. B.E. statistics deals the case of indistinguishable, identical particles having zero or integral spin, while F.D. statistics is used in case of indistinguishable identical particles having half-integral spin. Classical statistics is only a limiting case of quantum statistics. Four chapters are devoted for gradual understanding of the subject.

# Statistical Basis of Thermodynamics

# 9

### AT A GLANCE

- Statistical basis • Probability • Principle of Equal a Priori Probability • Probability and Frequency • Some Basic Rules of Probability Theory • Permutations and Combinations • Macrostate and Microstate • Thermodynamic Probability • Fluctuations and their Dependence on *n* • Constraints on a System • Static and Dynamic Systems • Most Probable State • Life Time of a Microstate and Macrostate • Concept of a Cell in a Compartment • Ensemble and Average Properties

## 9.1 Statistical Basis

A study of thermodynamics gives us various macroscopic properties that are related through an equation of state having only two independent parameters. However, the equation of state can not be deduced from the laws of thermodynamics. It has to be obtained experimentally. The ordinary laws of mechanics was the only tool to explain physical phenomena, up to the end of 17$^{th}$ century. In certain cases, particularly where the system contains a large number of particles, ordinary laws of mechanics could not be used, as it is impossible to follow the motion of each particle. For example, a pinch of matter contains a very large number of atoms or molecules.

(Avogadro's number $N_o = 6 \times 10^{26}$ per kg mole). Therefore, it is impossible to apply the ordinary laws of mechanics to a physical system containing large number of particles, particularly that of electrons. Such problems are however, successfully solved by *statistical mechanics*. The larger is the number of particles in the physical system considered, the more nearly correct are the statistical predictions. Smaller the number of particles in a mechanical system, the statistical mechanics goes on becoming meaningless. Before the advent of quantum theory, Maxwell, Boltzmann, Gibbs etc. applied the statistical methods making the use of classical physics. These methods are known as **classical statistics or Maxwell-Boltzmann statistics**. The classical statistics successfully explained the phenomenon like temperature, pressure, energy etc., but failed to explain other observed phenomenon like black body radiation, specific heat at low temperature etc. For this, new approach was introduced by Bose, Einstein, Fermi and Dirac. The Planck's quantum concept of discrete exchange of energy between systems was used. The new statistics was subdivided into two categories (*i*) Bose-Einstein statistics and (*ii*) Fermi-Dirac statistics. Thus, there are three statistics depending upon three different kinds of particles :

1. **Maxwell-Boltzmann statistics:** This is applicable to the identical, distinguishable particles of any spin. The molecules of a gas are the particles of this kind.
2. **Bose-Einstein statistics:** This is applicable to the identical, indistinguishable particles of zero or integral spin. These particles are called *bosons*. The examples of bosons are helium atoms at low temperature and the photons.
3. **Fermi-Dirac statistics:** This is applicable to the identical, indistinguishable particles of half-integral spin. These particles obey Pauli exclusion principle and are called *fermions*. The examples of fermions are electrons, protons, neutrons etc.

It's worth mentioning here that the classical Maxwell Boltzmann statistics is only the limiting case of the other two quantum statistics. Let us consider first some of the basic concepts of classical statistics.

## 9.2 Probability

The probability of an event may be defined as the ratio of the number of cases in which the event occurs to the total number of cases.

$$\text{Thus, The probability of an event} = \frac{\text{Number of cases in which the event occurs}}{\text{Total number of cases}} \qquad ...(9.1)$$

Suppose an event can happen in $a$ ways and fails to happen in $b$ ways, then the probability of happening the event $= \dfrac{a}{a+b}$ and the probability of failing the event $= \dfrac{b}{a+b}$. Here $(a + b)$ represents the total number of equally likely possible ways. It should be noted that the sum of these two probabilities is always 1; since the event must either occur or fail.

To evaluate the probability of random event, we consider a unit of measurement. An event is called a 'sure' event if it occurs in an experiment. Thus, the probability of a 'sure' event is assumed to be equal to 1 and that of an impossible event to be equal to zero. Thus, the probability $P$ of a random event lies between 0 and 1, *i.e.*,

$$0 \leqslant P \leqslant 1 \qquad ...(9.2)$$

This is further explained by the following experiments:

1. **Throwing a coin:** Suppose we toss a coin. Either the 'head' can come upward or the 'tail' *i.e.* an event can occur in a total number of *two* equally likely ways. The number of ways in which the 'head' can come up is only *one*. Therefore, the probability that the 'head' may come up is $\frac{1}{2}$. Similarly, the probability that the 'tail' may come up is also $\frac{1}{2}$.

**2. Throwing a die:** A die is a homogeneous, regular and balanced cube with six faces marked numbers of dots from 1 to 6 engraved on them. It is supposed that the die is symmetrical and cannot fall of its edges. When the die is thrown it falls one of its faces upward *i.e.*, yields one of the six results and no other. All outcomes are equally likely in a single throw of dice. Out of six possible results only one is favourable for the appearance of six spots. In other words, the probability of any one face (say with a number 3) to come up is $\frac{1}{6}$. The set of all possible outcomes can be written as $S$ (1, 2, 3, 4, 5, 6).

The probability of the die coming up with an even number is $\frac{3}{6}$ as there are only three even number on the die 2, 4 and 6.

$$\therefore \qquad P\,(\text{even}) = \frac{3}{6} = \frac{1}{2}$$

Likewise, the probability of the die coming up with odd number is $\frac{3}{6}$ or $\frac{1}{2}$.

The probability of the die coming up with any number less than 6 is given by

$$P\,(\text{a number} < 6) = \frac{5}{6}$$

as there are five numbers 1, 2, 3, 4, 5 which are less than 6.

**Zero Probability.** If we want to know the probability of the die coming up with a face marked with a number 7. The die has only six faces marked serially from 1 to 6. There is no face marked as 7. Therefore, probability of appearing a number 7 is zero, *i.e.*,

$$\therefore \qquad P\,(\text{number } 7) = \frac{0}{6} = 0$$

In other words, impossible event is always zero.

**Probability one.** The probability of appearing any number less than 7 is one. This is because all the six faces of the die are marked from 1 to 6, *i.e.*, the numbers less than 7.

$$P\,(\text{number} < 7) = \frac{6}{6} = 1$$

**Total probability.** If $a$ is the number of cases in which an event occurs and $b$ the number of cases in which an event fails, then

$$\text{Probability of occurrence the event} = \frac{a}{a+b}$$

$$\text{and Probability of failing the event} = \frac{b}{a+b}$$

The sum *i.e.*, the total probability is always one, since the event may either occur or fail.

## 9.3 Principle of Equal A Priori Probability

Suppose we toss a coin it is clear in mind that the coin will fall either with its 'Head' up or 'Tail' up. Similarly if a six faced cubical dice is thrown, it is sure that the dice will fall with one of its six faces upward. In the same way, if we have an open box divided into two equal sized compartments $X$ and $Y$ (Fig. 9.1), and a small particle is thrown from a large distance in such a way that it must fall in either of the two compartments, then the probability of the particle to fall in the compartment marked $X$ is equal

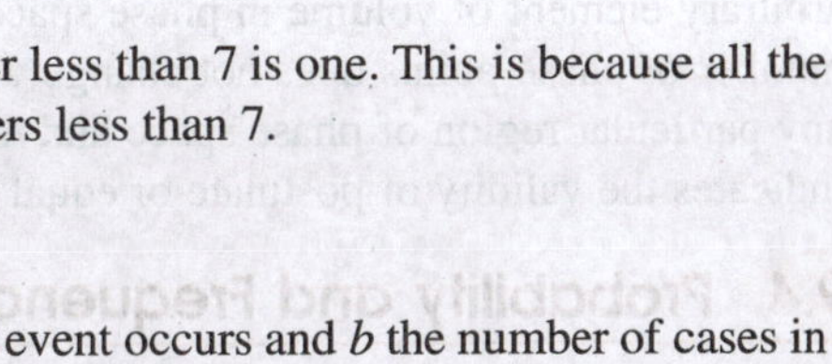


Fig. 9.1

to the probability that it may fall into the compartment marked $Y$. Again there is an equal probability. *This principle of assuming equal probability for events which are equally likely to occur is known as the principle of equal a priori probability.*

A priori means something which exists in our mind prior to and independently of the observations we are going to make.

This principle of equal a priori probability will not hold good, if in the above example, the two compartments are of unequal size.

The postulate of equal a priori probability is the fundamental basis of statistical mechanics. The basis for this postulate is provided by the ergodic hypothesis. According to the principle of conservation of density in phase space, *the density of a group of phase points remains constant along their trajectories in a phase space.* If at any time, the phase points are distributed uniformly in phase space, the phase points move in such a way that *their density is uniform at all times.* So the representative points move in the space in the same fashion *as an incompressible fluid.* For other spaces the density will vary from region to region while there will be no change in phase space. This means, there is no crowding together of phase points in any particular region of phase space. Let at any time $t_0$, the representative points define the line $A_0 B_0$ as shown in Fig. 9.2. At instant $t_0 + \Delta t$, these points define the line $A_1 B_1$ and similarly, let at a time $t_0 + n\Delta t$, they define the line $A_n B_n$. The points which are on line $A_0 B_0$ at time $t_1$ will be on line $A_1 B_1$ at time $t_2$ and so on. Thus, the points behave as an incompressible fluid, the region $A_n B_n B_{n+1}$, 1 $A_{n+1}$ are, therefore all equal in volume. Thus any arbitrary element of volume in phase space bounded by a moving surface and containing a definite number of phase points does not change with time. The property of *no crowding* of phase points in any particular region of phase space and the *constancy of volume* element in phase-space with time indicates the validity of postulate of equal a priori probability.

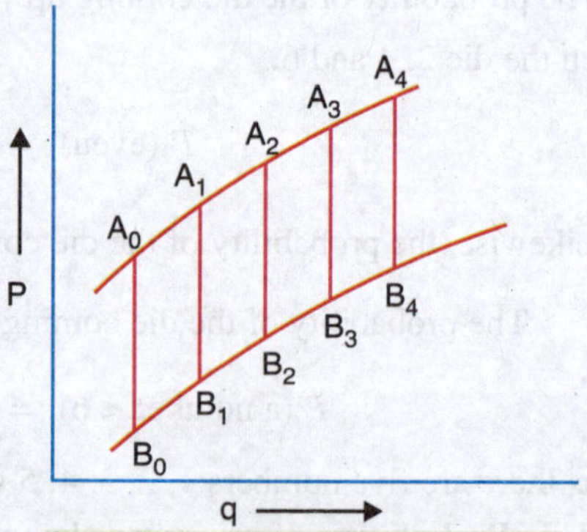


**Fig. 9.2**

## 9.4 Probability and Frequency

All the possible outcomes, in the examples discussed above, are assumed to be known. This is not always possible in practice and hence formula given by equation (9.1) is not a general formula. It can be used only in experiments that possess symmetry. Suppose, the die is asymmetric by adding a little load (by wax) to one of its faces. Now all the outcomes are not equally likely and we cannot say that the probability at appearance of six spots is $\frac{1}{6}$. For all such cases, we adopt a technique as under:

Suppose we toss a coin, say $N$ times and we find that 'Head' appears $M$ times. Here we introduce a term frequency of an event $F$ as

$$F = \frac{\text{No. of trials in which A occurs}}{\text{Total number of trials}} = \frac{\text{H}}{\text{N}} \qquad \text{...(9.3)}$$

Thus, if a coin is tossed 50 time and in 10 of them the coin shows Heads, the frequency of this event is $\frac{10}{50} = 0.2$. From the classical definitions of probability, the probability of occurrence of 'Head' is 0.5 or 50%. Hence, we conclude that frequency is not the same as probability. There must be a relationship between frequency and probability. As the number of trials is increased, the frequency of the event progressively tends to stabilize and gradually approaches a constant value, known as the probability of the event. We define probability in terms of frequency as

$$P = \lim_{N \to \infty} \frac{M}{N} \qquad \text{...(9.4)}$$

Thus, in order to obtain sufficiently accurate result, trials must be conducted until the ratios $\frac{M}{N}$ differ from one another by a very small value.

## 9.5 Some Basic Rules of Probability Theory

### Additive Law of Probability

This is applicable to mutually exclusive events. Two or more events are said to be mutually exclusive if the occurrence of any one of them prevents the occurrence of others. Such events never occurs simultaneously.

For example, consider two small non-overlapping region $\Delta V_1$ and $\Delta V_2$ in a box of volume $V$ as shown in (Fig. 9. 3). A particle in $\Delta V_1$ rules out the possibility of its being present, at the same instant in $\Delta V_2$ and vice-versa. The two events, thus, are mutually exclusive.

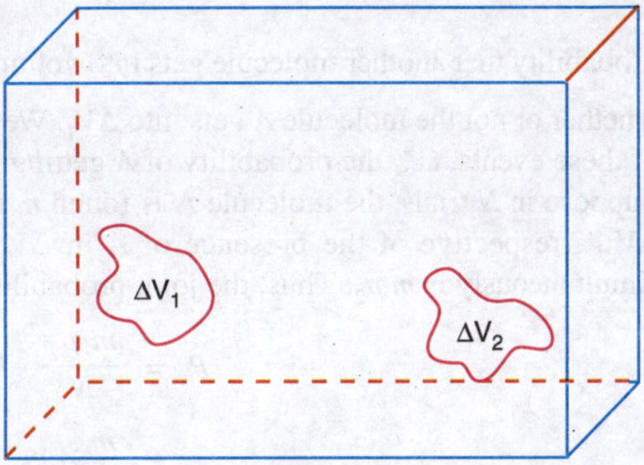

Fig. 9.3

Suppose in $N$ trials the particle is found $m_1$ times in $\Delta V_1$ and $m_2$ times in $\Delta V_2$. The probabilities of finding particles in the two regions are

$$p_1 = \frac{m_1}{N} \quad \text{and} \quad p_2 = \frac{m_2}{N}$$

respectively

The number of times that the particle will be found at least one of the two regions in $N$ trials is $(m_1 + m_2)$. Hence, the probability will be

$$P\,(\Delta V_1 \text{ or } \Delta V_2) = \frac{m_1 + m_2}{N} = \frac{m_1}{N} + \frac{m_2}{N} = p_1 + p_2$$

This law can be generalized to any number of mutually exclusive events, giving probabilities, say, $p_1, p_2, p_3 \text{ ........}, p_n$ then the probability that any one of them occurs is the sum of the probability of these events, *i.e.*,

$$P = p_1 + p_2 + p_3 + \text{ ........ } + p_N = \sum_{i=1}^{N} p_i$$

This is known as additive law of probability.

**EXAMPLE 9.1**

*A card is drawn from a well shuffled pack of 52 cards. Calculate the probability for this card to be either a king or a queen.* **(Kerala 1995)**

**SOLUTION.** We are taking out one card out of 52 cards of pack. So the total number of ways in which the event can occur is, $N = 52$.

But we want to draw a specific card *i.e.* king. There are 4 kings in a pack of cards. Therefore, the number of ways favourable to the first event, $m_1 = 4$

$\therefore$ Probability of drawing a king, $p_1 = \frac{m_1}{N} = \frac{4}{52} = \frac{1}{13}$

The number of ways in which second event *i.e.* drawing a queen may happen, $m_2 = 4$, as there are four queens too.

$\therefore$ Probability of drawing a queen, $p_2 = \frac{m_2}{N} = \frac{4}{52} = \frac{1}{13}$

Both the events are mutually exclusive. Therefore the probability that the card drawn is either a king or a queen is

$$P = p_1 + p_2 = \frac{1}{13} + \frac{1}{13} = \frac{2}{13}$$

### Multiplication Rule: Joint Probability

In the calculations of probabilities, we sometimes come across random events; such that the probability of occurrence of one does not affect the probability of occurrence of the other. For example in fig. 9.3, the probability that a molecule $A$ gets into $\Delta V_1$, at a particular instant is $p_1 = \frac{\Delta V_1}{V}$. The probability that another molecule gets into volume $\Delta V_2$ at the same instant is $p_2 = \frac{\Delta V_2}{V}$, regardless of whether or not the molecule $A$ gets into $\Delta V_1$. We want to calculate the probability of joint occurrence of these events, *i.e.*, the probability of $A$ getting into $\Delta V_1$ and $B$ getting into $\Delta V_2$ *at the same instant.* Suppose in $N$ trials, the molecule A is found $m$ times in $\Delta V_1$. If $p_2$ is the probability that $B$ gets into $\Delta V_2$, irrespective of the presence of $A$ in $\Delta V_1$, the number of times the two events will occur simultaneously is $m\,p_2$. Thus, the joint probability of occurrence of these two events, is

$$P = \frac{m\,p_2}{N}$$

$$= \frac{m}{N} \times p_2$$

$$P = p_1 \times p_2 \qquad \left(\because p_1 = \frac{m}{N}\right)$$

Thus, probability of joint occurence of two independent events is equal to the product of the probabilities of each of these independent events.

**EXAMPLE 9.2**

*We throw a die twice and obtain two numbers. What is the probability that these numbers are 6 and 4 precisely in that order?* **(G.N.D.U. 1996)**

**SOLUTION.** The probability that the first throw gives a 6 is $\frac{1}{6}$. Similarly, the probability that the second throw gives a 4 is also $\frac{1}{6}$. These two events are independent.

$$\therefore \quad \text{Required probability} = \frac{1}{6} \times \frac{1}{6} = \frac{1}{36}$$

### Conditional Probability

The probability for an event $A$ to occur the condition that event $B$ has occurred is called the *conditional probability* and is denoted by $P\left(\frac{A}{B}\right)$.

This can be better understood by the following numerical.

**EXAMPLE 9.3**

*An urn contains 4 black and 3 white balls. What is the probability that on two successive draws, the balls drawn are both black?* **(Allahabad 1997)**

**SOLUTION.** Suppose the first ball drawn is black. Let the event be called as $A$. Its probability is $P(A) = \frac{4}{4+3} = \frac{4}{7}$. This event is followed by second event $B$ in which the second ball drawn is also black (conditional). Since an event $A$ is already realized, the balls remained are 3 black and 3 white, out of which the event $B$ is taking place, *i.e.* second ball out of 6 balls. Hence, the probability of the event $B$ to occur under the condition that $A$ has already realized is $P\left(\frac{B}{A}\right) = \frac{3}{3+3} = \frac{3}{6} = \frac{1}{2}$. The first event thus has an effect on the probability of the second event.

∴ The probability of the composite event $(A + B)$ is

$$P(A+B) = P(A).\ P\left(\frac{B}{A}\right) = \frac{4}{7} \times \frac{1}{2} = \frac{2}{7}$$

## 9.6 Permutations and Combinations

The word permutation means arrangement and combination means formation of groups.

**Permutations:** To understand the meaning of permutation *i.e.* arrangement, let us consider an example of four distinguishable objects marked $a$, $b$, $c$ and $d$. Taking any two objects at a time, the possible arrangements are

*ab, ba, ac, ca, ad, da, bc, cb, bd, db, cd, dc.*

There are total 12 arrangements possible. In arranging these objects the *order* of their placing is also taken into consideration. Thus, *ab* and *ba* are two different permutations. Thus, 4 objects can be arranged in 12 ways by taking 2 objects at a time, *i.e.* the number of permutations is 12.

Symbolically, $\quad {}^4p_2 = 12$

Instead, if 3 objects are taken at a time out of 4 objects $a$, $b$, $c$ and $d$ the various arrangements are

| | | | |
|---|---|---|---|
| *abc* | *abd* | *acd* | *bcd* |
| *acb* | *adb* | *adc* | *bdc* |
| *bca* | *bad* | *cda* | *cbd* |
| *bac* | *bda* | *cad* | *cdb* |
| *cab* | *dab* | *dac* | *dbc* |
| *cba* | *dba* | *dca* | *dcb* |

Thus, 4 objects can be arranged in 24 different ways by taking 3 at a time out of 4 objects. Symbolically,

$${}^4P_3 = 24$$

In general, the number of arrangements of $n$ distinguishable objects by taking $r$ at a time is given by

$${}^nP_r = \frac{n!}{(n-r)!} \qquad \text{...(9.5)}$$

Thus, $\quad {}^4P_2 = \frac{4!}{(4-2)!} = \frac{4!}{2!} = \frac{4\times3\times2\times1}{2\times1} = 12$

and $\quad {}^4P_3 = \frac{4!}{(4-3)!} = \frac{4!}{1!} = \frac{4\times3\times2\times1}{1} = 24$

This includes all types of arrangements, meaningful as well as meaningless.

**Combinations:** Here we consider only combinations *i.e.* groups without considering the order of their placement *i.e.* only meaningful combinations. Thus, the combinations of 4 objects $a$, $b$, $c$ and $d$ taking two at a time are *ab, ac, ad, bc, bd, cd.*

*i.e.* only 6 combinations. Symbolically, ${}^4c_2 = 6$

Similarly, combinations of 4 objects *a, b, c* and *d* taking three at a time are

*abc, abd, acd, bcd*

*i.e.* only 4 combinations. Symbolically, ${}^4c_3 = 4$

In general, the number of groups *i.e.* meaningful combinations of *n* distinguishable objects by taking *r* at a time is given by

$$ {}^nC_r = \frac{n!}{r!(n-r)!} \quad ...(9.6)$$

But

$$ {}^nP_r = \frac{n!}{(n-r)!}$$

$$\therefore \quad {}^nC_r = \frac{{}^nP_r}{r!}$$

or

$$ {}^nP_r = r!\,{}^nC_r \quad ...(9.7)$$

From equation 9.5

$$ {}^nP_n = \frac{n!}{(n-n)!} = \frac{n!}{o!} = n! \qquad (\because 0! = 1)$$

and from equation 9.6

$$ {}^nC_n = \frac{n!}{n!(n-n)!} = \frac{n!}{n!o!} = \frac{n!}{n!} = 1 \qquad (\because 0! = 1)$$

## 9.7 Macrostate and Microstate

**Macrostate:** Consider 4 distinguishable particles and we wish to distribute them into two exactly similar compartments in an open box. Let the particles be *a, b, c* and *d*. When any particle thrown into the box, it must fall into one of the two compartments. Since the compartments are alike, the particles have the same a priori probability of going into either of them and will be $\frac{1}{2}$. The possible ways in which 4 particles can be distributed in two compartments as shown in table 9.1.

**TABLE 9.1**

| Compartment | No. of particles | | | | |
|---|---|---|---|---|---|
| | Distribution 1 | Distribution 2 | Distribution 3 | Distribution 4 | Distribution 5 |
| 1 | 0 | 1 | 2 | 3 | 4 |
| 2 | 4 | 3 | 2 | 1 | 0 |

Thus, there are 5 different distributions (0, 4), (1, 3), (2, 2), (3, 1) and (4, 0). Each compartmentwise distribution of a system of particles is known as a *macrostate*. In general, for a system of *n* particles to be distributed in two similar compartments, the various macrostates are (0, *n*), (1, *n* – 1), (2, *n* – 2), ........ (*n* – 1, 1) and (*n*, 0). Therefore, the total no. of macrostates for *n* particles is (*n* + 1).

**Microstates:** Since the particles are *distinguishable*, the number of different possible arrangements in each compartment is shown below :

**TABLE 9.2**

| Macrostate | Possible arrangements | | No. of microstates |
|---|---|---|---|
| | Compartment 1 | Compartment 2 | $W$ (Thermodynamic probability) |
| 0,4 | 0 | *abcd* | 1 |
| 1, 3 | *a*<br>*b*<br>*c*<br>*d* | *bcd*<br>*cda*<br>*dab*<br>*abc* | 4 |
| 2, 2 | *ab*<br>*ac*<br>*ad*<br>*bc*<br>*bd*<br>*cd* | *cd*<br>*bd*<br>*bc*<br>*ad*<br>*ac*<br>*ab* | 6 |
| 3, 1 | *bcd*<br>*cda*<br>*dab*<br>*abc* | *a*<br>*b*<br>*c*<br>*d* | 4 |
| 4, 0 | *abcd* | 0 | 1 |

Each distinct arrangement is known as the *microstate* of the system. The distribution (0, 4) can have only one arrangement, distribution (1, 3) can have 4 distinct arrangements, distribution (2, 2) can have 6, distribution (3, 1) have 4 and distribution (4, 0) have only one. Thus, a given macrostate may consists of a number of microstates. In the above example of 4 particles, the total number of microstates are $16 = 2^4$. In general, for a system of $n$ particles, the total number of microstates are $2^n$.

Suppose we consider a thermodynamic system of $N$ identical, non-interacting particles occupying a volume $V$. Let $E_1, E_2, E_3, \ldots\ldots\ldots\ldots E_i$ be the energy of particles $n_1, n_2, n_3 \ldots\ldots\ldots\ldots n_i$ respectively, then

$$\text{total energy of the system,} \quad E = n_1 E_1 + n_2 E_2 + \ldots\ldots\ldots + n_i E_i$$

$$= \Sigma n_i E_i \quad \text{and} \quad N = \Sigma n_i$$

According to quantum theory, the possible values of energy of each particle are discrete and hence the total energy $E$ also has a discrete value. But when the volume $V$ is very large, the spacings of different energy levels are so small in comparison with total energy of the system that $E$ may be regarded as almost *continuous*.

The actual values of parameters $N$, $V$ and $E$ defines a particular *macrostate* ($N$, $V$, $E$) of the given system.

There are a large number of different ways in which the individual Ei's can be chosen, so as to make total energy equal to $E$.

Each of the different ways in which the total energy $E$ of the system can be distributed among $N$ particles constituting it specifies a *microstate*. Thus, *a given macrostate corresponds to many microstates* and it is very natural to assume that at any time the system is equally likely to be in any of these *microstates*.

## 9.8 Thermodynamic Probability

*The number of microstates corresponding to any given macrostate is called its thermodynamic probability. In other words, the thermodynamic propability of a particular macrostate is defined as the number of microstates corresponding to that macrostate.* In general, this is a very large number and is represented by $W$ (or $\Omega$).

The number of microstates corresponding to a given macrostate is equal to the number of meaningful arrangements or permutations of various particles in the macrostate excluding those permutations which are meaningless *i.e.* merely interchange the order of particles in a particular compartment (or cell). For the case of $n$ particles and two compartments (or cells), if $r$ is the number of particles in the compartment No. 1 and the remaining $(n-r)$ are in compartment No. 2, then

$$\text{No. of meaningful arrangements} = \frac{n!}{r!(n-r)!} = {}^nC_r$$

Therefore, the number of microstates in a macrostate $(r, n-r)$ or thermodynamic probability

$$W_{(r,n-r)} = \frac{n!}{r!(n-r)!} = {}^nC_r \quad \text{...(9.8)}$$

Applying this to a system of 4 distinguishable particles, for a macrostate (1, 3), $r = 1$, $(n-r) = 3$ and $n = 4$ the number of microstates,

$\Omega(1, 3)$, or

$$W_{(1,3)} = \frac{4!}{1!3!} = 4$$

Similarly, for a macrostate (2, 2), $r = 2$, $(n-r) = 2$ and $n = 4$ the thermodynamic probability *i.e.* no. of microstates will be

$$W_{(2,2)} = \frac{4!}{2!2!} = 6$$

and for macrostate (0, 4) or (4, 0) will be

$$W_{(0,4)} = W_{(4,0)} = \frac{4!}{0!4!} = \frac{4!}{4!} = 1 \quad (\because 0! = 1)$$

This agrees with the table 9.2.

**Probability ($P$).** The probability (or occurrence) of a macrostate is defined as the ratio of the number of microstates (i.e. thermodynamic probability $W$) in it to the total number of possible microstates of the system.

$$\text{Thus,} \quad P_{macro} = \frac{\text{No. of microstates in the macrostate}}{\text{Total no. of microstates of the system}} \quad \text{...(9.9)}$$

The total number of ways of arranging $n$ distinguishable particles in $c$ numbered compartments = $c^n$

If there are only 2 compartments or cells, then the total no. of microstates of the system = $2^n$. For 4 particles system, total no. of microstates will be $2^4 = 16$. From equation (9.9), we get.

$$P_{macro} = \frac{W}{c^n} = \frac{W}{2^n} \quad \text{...(9.10)}$$

Substituting for W from equation (9.8),

$$P_{(r,\,n-r)} = \frac{n!}{r!(n-r)!} \times \frac{1}{2^n} = \frac{1}{2^n}\,{}^nC_r \quad \text{...(9.11)}$$

This gives the probability (of occurrence) of macrostate $(r, n-r)$.

## Combination Possessing Maximum Probability

We know from elementary algebra that if $n$ is even the value of $^nc_r$ is maximum (greatest) when $r = \frac{n}{2}$, refer table 9.2. Hence thé maximum value of probability is given by

$$P_{max} = \frac{1}{2^n}\,{}^nC_{n/2}\,\lambda = \frac{n!}{\frac{n}{2}!\frac{n}{2}!} \times \frac{1}{2^n}$$

$$= \frac{n!}{\left(\frac{n}{2}!\right)^2} \times \frac{1}{2^n}$$

## Combination Possessing Minimum Probability

We know from elementary algebra that $^nC_r$ is minimum, if $r$ is equal to either zero or $n$.
If $r = 0$, the probability is

$$P_{\min} = \frac{n!}{r!(n-r)!} \times \frac{1}{2^n}$$

$$= \frac{n!}{o!(n-o)!} \times \frac{1}{2n} \qquad (\because o! = 1)$$

$$= \frac{1}{2^n} \qquad \text{...}(i)$$

If $r = n$, the probability is

$$P_{min} = \frac{n!}{r!(n-r)!} \times \frac{1}{2^n}$$

$$= \frac{n!}{n!(n-n)!} \times \frac{1}{2^n}$$

$$= \frac{n!}{n!o!} \times \frac{1}{2^n} \qquad (\because o! = 1)$$

$$= \frac{1}{2^n} \qquad \text{...}(ii)$$

Thus, from (*i*) and (*ii*), we conclude that the minimum probability is given by

$$P_{min} = \frac{1}{2^n} \qquad \text{...(9.12)}$$

## Relation between *P* and *W*

The probability of occurrence of a particular microstate is given by

$$p = \frac{1}{2^n}$$

because $2^n$ gives the total no. of microstates of a system of $n$ particles to be arranged in two compartments (or cells) and all the microstates of a system have *a priori* probability. Thus, from equation (9.10), we have

$$P = W \times p$$

= No. of microstates in the macrostate × probability of occurrence of a microstate

or $\qquad P \propto W$

Therefore, the probability (of occurrence) of a macrostate ($P$) is directly proportional to the thermodynamic probability ($W$).

**EXAMPLE 9.4**

*Calculate the probability that in tossing a coin 5 times, we get 3 heads and 2 tails.*

**(Punjab 1998)**

**SOLUTION.** Tossing a coin 5 times is equivalent to tossing 5 similar coins simultaneously. Hence $n = 5$. We wish to get 3 heads and 2 tails i.e.,

$$r = 3 \quad \text{and} \quad (n - r) = 2$$

The probability of distribution ($r$, $n - r$) is given by

$$P_{(r,\, n-r)} = \frac{1}{2^n}\, {}^nC_r = \frac{1}{2^n} \cdot \frac{n!}{r!(n-r)!}$$

$$= \frac{1}{2^5} \times \frac{5!}{3!2!} = \frac{5\times4\times3\times2\times1}{32\times3\times2\times1\times2\times1}$$

$$= \frac{10}{32} = \mathbf{\frac{5}{16}}$$

**EXAMPLE 9.5**

*Calculate the probability that in tossing a coin 10 times, we get (i) all heads (ii) 5 heads & 5 tails and (iii) 3 heads & 7 tails. (iv) 7 heads & 3 tails.* **(Meerut 2004)**

**SOLUTION.** The probability of distribution ($n_1$, $n–n_1$) is given by

$$P_{(r,\, n-r)} = \frac{1}{2^n} \cdot \frac{n!}{r!(r-r)!}$$

(*i*) Tossing a coin 10 times is equivalent to tossing 10 similar coins simultaneously

Here, $\quad n = 10$

We wish to have all heads

$\therefore \quad r = 10 \quad \text{and} \quad (n - r) = 0$

Hence, $\quad P_{(10,\, 0)} = \frac{1}{2^{10}} \cdot {}^{10}C_{10} = \frac{1}{2^{10}} \times \frac{10!}{10!0!} = \frac{10!}{10!} \qquad (\because 0! = 1)$

$\therefore \quad P_{(10,\, 0)} = \frac{1}{2^{10}} = \frac{1}{1024} = \mathbf{0.00098}$

(*ii*) We wish to have 5 heads and 5 tails

$\therefore \quad r = 5 \quad \text{and} \quad (n - r) = 5$

Hence $\quad P_{(5,\, 5)} = \frac{1}{2^{10}}\, {}^{10}C_5 = \frac{1}{2^{10}} \frac{10!}{5!5!}$

$$= \frac{10\times9\times8\times7\times6\times5\times4\times3\times2\times1}{5\times4\times3\times2\times1\times5\times4\times3\times2\times1} \times \frac{1}{1024}$$

$$= \frac{252}{1024} = \mathbf{0.24609}$$

(*iii*) We wish to have 3 heads and 7 tails

$\therefore \quad r = 3 \quad \text{and} \quad (n - r) = 7$

Hence, $\quad P_{(3,\, 7)} = \frac{1}{2^{10}}\, {}^{10}C_3 = \frac{1}{1024} \times \frac{10!}{3!7!}$

$$= \frac{10\times9\times8\times7\times6\times5\times4\times3\times2\times1}{3\times2\times1\times7\times8\times6\times5\times4\times3\times2\times1} \times \frac{1}{1024}$$

$$= \frac{120}{1024} = \mathbf{0.11719}$$

(*iv*) We wish to have 7 heads and 3 tails

Here $\qquad r = 7$ and $(n - r) = 3$

$$\therefore \qquad P_{(7,\,3)} = \frac{1}{2^{10}} \cdot {}^{10}c_7 = \frac{1}{2^{10}} \cdot \frac{10!}{7!3!}$$

$$= \frac{10 \times 9 \times 8 \times 7 \times 6 \times 5 \times 4 \times 3 \times 2 \times 1}{7 \times 6 \times 5 \times 4 \times 3 \times 2 \times 1 \times 3 \times 2 \times 1} \times \frac{1}{1024}$$

$$= \frac{120}{1024} = \mathbf{0.11719}$$

Thus, the probability of distribution (7, 3) is equal to the probability of distribution (3, 7)

## 9.9 Fluctuations and their Dependence on *n*. (Narrowing of Probability Distribution with Increasing *n*)

Let us consider n molecules of an ideal gas enclosed in a container or a box. Let us investigate the positions and distribution of these particles in space. For this purpose, imagine that the box is divided into two equal parts by some not actual partition Fig. 9.4. Let $n_1$ and $n_2$ be the number of molecules in A and B compartments *i.e.*, in equal halves.

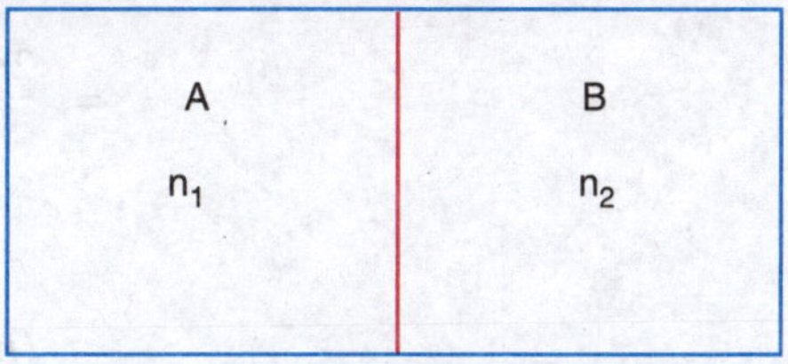


Fig. 9.4

Thus,

$$n_1 + n_2 = n$$

If $n$ is large, it would be ordinarily expected that

$$n_1 \cong n_2.$$

*i.e.* approximately half of the molecules are in each half of the box.

Due to intermolecular interactions, the molecules collide with each other and on the walls of the container; as a result some of the molecules enter left half of the box while some leave it. Thus the number $n_1$ of the molecules actually located in the left half fluctuates constantly with time. For large $n$, these fluctuations are so small that $n$ does not differ too much from $\frac{n}{2}$. The state $\left(\frac{n}{2}, \frac{n}{2}\right)$ is the state of maximum probability and is called **equilibrium state**. We shall show now that the probability distribution narrows rapidly as $n$ increases. For simplicity, again we consider two similar compartments.

From a priori probability, we expect $p = p' = \frac{1}{2}$.

Consider $n = 40$. The most probable distribution is (20, 20). Using equation (9.11), its probability is

$$P_{max} = P_{(20,\,20)} = \frac{1}{2^{40}} \cdot \frac{40!}{20!\,20!} \qquad ...(i)$$

A distribution 25% off from this is with $r = 15$ (or 25). Its probability is

$$P_{(15,\,25)} = \frac{1}{2^{40}} \cdot \frac{40!}{15!\,25!} \qquad ...(ii)$$

The ratio $R$ of the two probabilities is obtained by dividing (*ii*) by (*i*), we get

$$R\ (25\%\ \text{off}) = \frac{20!\,20!}{15!\,25!} = \frac{20 \times 19 \times 18 \times 17 \times 16}{25 \times 24 \times 23 \times 22 \times 21}$$

$$\cong \left(\frac{4}{5}\right)^5 \cong 0.3 \qquad ...(iii)$$

With $n = 400$, a distribution 25% off from the most probable one will have the relative probability

$$R\ (25\%\ \text{off}) = \frac{200!\,200!}{150!\,250!} = \frac{200 \times 199 \times ...... \times 151}{250 \times 249 \times ...... \times 201}$$

$$\cong \left(\frac{4}{5}\right)^{50} \cong 10^{-5} \qquad ...(iv)$$

Thus, as $n$ increases the probability of distributions appreciably far from the most probable distribution becomes negligibly small.

**General Formula:** In order to derive general formula, let us consider $2n$ particles distributed between two compartments with equal probability, the most probable distribution is $(n, n)$. The probability $P_{(n, n)}$ may be compared with the probability $(P_{n+s,\, n-s})$ where $\frac{s}{n} << 1$. From equation (9.11), the ratio of probabilities for these distributions are given by

$$R = \frac{P_{(n+s,\, n-s)}}{P_{(n,n)}} = \frac{{}^{2n}C_{n+s}}{{}^{2n}C_n}$$

$$= \frac{2n \times (2n-1) \times ....... \times (n-s+1)}{1 \times 2 \times 3 ....... \times (n+s)}$$

$$\times \frac{1 \times 2 \times 3 \times ....... \times n}{2n \times (2n-1) \times ....... \times (n+1)}$$

$$= \frac{(n-s+1) \times (n-s+2) \times ....... \times n}{(n+1) \times (n+2) \times ....... (n+s)}$$

$$\simeq \left(1 - \frac{s}{n}\right)^s, \qquad \text{Since } s/n << 1$$

If we write $\frac{s}{n} = f$ to represent the fractional deviation from the most probable distribution, we get (for large $n$ values)

$$\log_e R = s \log_e \left(1 - \frac{s}{n}\right) = s\left[-\frac{s}{n} - \frac{s^2}{n^2} - \frac{s^3}{n^3} - .....\right] = s\left(-\frac{s}{n}\right) \text{ for } s << n = -\frac{s^2}{n^2}\,.\,n$$

$$= -f^2 n$$

$$\therefore \qquad R = e^{-f^2 n} \qquad ...(9.13)$$

Let $P(f)\, df$ represents the probability that the distribution is between fractions $f$ and $(f + df)$ away from the most probable distribution $(n, n)$. Then

$$P(f)\, df = A \,.\, e^{-f^2 n} df$$

where $A$ is a constant determined by the condition

$$\int_{-1}^{+1} P(f)\, df = 1 \qquad \text{(normalization condition)}$$

This gives $\qquad A = \sqrt{\frac{n}{\pi}}$, so that

$$P(f)\,df = \sqrt{\frac{n}{\pi}}\,.\,e^{-f^2 n}\,df \qquad \text{...(9.14)}$$

Thus, as the fractional deviation $f$ from most probable distribution increases, the probability falls exponentially. Further, with larger $n$, the fall will be more rapid (Fig. 9.5). The probability falls to $\frac{1}{e}$ times the maximum, when

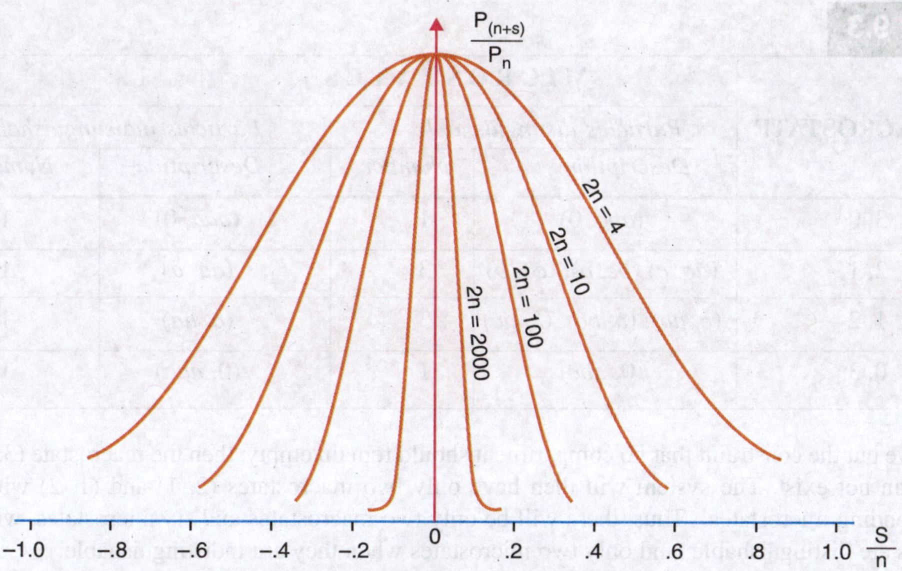


**Fig. 9.5.** The probability distribution narrows rapidly as $n$ increases

$$f^2 n = 1 \quad \text{or} \quad f = \frac{1}{\sqrt{n}} \qquad \text{...(9.15)}$$

A graph is plotted between the relative probability $R$ against the fractional departure $f\left(=\frac{s}{n}\right)$ from equal distribution for different values of $2n$. There is a rapid fall of $R$ and the curve becomes narrower with increasing $2n$ or $n$.

## 9.10 Constraints on a System

*A set of conditions or restrictions that must be obeyed by a system are known as constraints.*

Let us take an example of the distribution of 3 particles in two compartments, the system must obey the constraint that total number of particles in the two compartments must be 3. In general, if there are $N$ particles to be distributed in two compartments and there are $n_1$ particles in compartment No. 1 and $n_2$ particles in compartment No. 2, then we must have

$$n_1 + n_2 = N$$

This relation is known as *equation of constraint* on the system.

A typical set of constraints is

$$\Sigma n_i = \text{constant} = N$$
$$\Sigma E_i n_i = \text{constant} = E$$

where $n_i$ is the number of particles (or molecules) in the $i$th compartment (or $i$th cell), $E_i$ is the energy of each particle (or molecule) in the $i$th compartment (or $i$th cell).

## Accessible states

*Accessible states are the states consistent with the given constraints of the system.*

Suppose we wish to distribute 3 particles in two compartments, then the system will have following macro and microstate (Table 9.3) provided the manner of relative placement of particles is immaterial i.e. only meaningful are considered. The system has four macro and 8 microstates when particles are distinguishable and only four microstates when they are indistinguishable.

**TABLE 9.3**

| MACROSTATE | MICROSTATES | | | |
|---|---|---|---|---|
| | *Particles distinguishable* | | *Particles indistinguishable* | |
| | *Description* | *Number* | *Description* | *Number* |
| 3, 0 | (*abc*, 0) | 1 | (*aaa*, 0) | 1 |
| 2, 1 | (*ab*, *c*) (*bc*, *a*), (*ca*, *b*) | 3 | (*aa*, *a*) | 1 |
| 1, 2 | (*c*, *ab*), (*a*, *bc*), (*b*, *ac*) | 3 | (*a*, *aa*) | 1 |
| 0, 3 | (0, *abc*) | 1 | (0, *aaa*) | 1 |

If we put the constraint that no compartment should remain empty, then the macrostate (3, 0) and (0, 3) can not exist. The system will then have only two macrostates (2, 1) and (1, 2) with their corresponding microstates. Thus, there will be only two macrostates and six microstates, when the particles are distinguishable, and only two microstates when they are indistinguishable.

Suppose again we put the constraint that the particles are indistinguishable, then we shall have four macrostates (3, 0), (2, 1), (1, 2) and (0, 3) but each macrostate will have only one microstate as shown in table 9.3. The total number of micorstates will also be four.

Thus, the constraints decrease the number of macro as well as microstates of the system.

**Accessible macrostates:** *The macrostates which are allowed under a constraint are called accessible macrostates.*

For example, in distributing 3 particles in 2 compartments under the constraint that no compartment will remain empty, the only accessible macrostates are (2, 1) and (1, 2).

**Inaccessible macrostates:** *The macrostates which are not allowed under a constraint are called inaccessible macrostates.*

In the above example, the macrostates (3, 0) and (0, 3) are inaccessible macrostates.

Sometimes we have constraints like this; 'energy of the system in between $E$ and $E + \delta E$'. In this case, all microstates which correspond to energy less than $E$ or greater than $E + \delta E$ are considered inaccessible.

## 9.11 Static and Dynamic Systems

**Static System :** *A system is said to be static if the constituent particles of the system remain at rest in a particular microstate.*

In static system, the particles do not change their relative positions in the compartment in which they are located. They do not move from one compartment to even the neighbouring compartment. In other words, the static systems do not change of their own due to internal forces or the internal conditions. However, the change in possible due to external forces or causes.

Whenever the change in static system takes place due to external causes or forces, the system goes from one macrostate to another macrostate. The system now stays in the new macrostate even after the external cause of force is withdrawn.

**Example:** A system of tossed coin is an example of static system. After every toss, the coins remain permanently latched in the last state.

**Dynamic system:** *A system is said to be dynamic if the constituent particles of the system can move so that the system goes from one microstate to another.*

The particles of the dynamic system change their relative positions within the compartment in which they are. They also go from one compartment to another. Thus, both the micro as well as macrostates of the system change with time. Dynamic systems can change of their own due to internal causes or forces. They can also undergo changes due to external causes or forces affecting.

**Example:** Any gas is an example of dynamic system (except at absolute zero of temperature). The molecules of gas are always in constant, random (haphazard) motion, continuously colliding with one another, following Brownian motion. During this, they change their position, momentum and energy continuously.

If the dynamic system is disturbed by some external causes or forces and then left to itself the system tends to approach the macrostate with maximum thermodynamic probability and remains there or stays very close to the most probable macrostate.

## 9.12 Most Probable State (Equilibrium state)

*The most probable state of a system is that macrostate which has the maximum probability of occurrence.* In the case of $N$ particles distributed in two compartments, the probability of occurrence of the most probable state is given by

$$P_{max} = \frac{{}^{N}C_{N/2}}{2^N}$$

In a **static system**, the particles remain latched *i.e.*, at rest when put in a compartment or placed in a cell.

The thermodynamic probability of the macrostate ($n_1$, $n_2$) in such a system is given by

$$W_{(n_1,\, n_2)} = \frac{N!}{n_1!\,n_2!} = \frac{N!}{n_1!\,(N-n_1)!} = {}^{N}C_{n_1} = {}^{N}C_{n_2}$$

But in **dynamic system**, the particles are in a constant state of motion and are not stationary. At N.T.P, the molecules of 1 $cm^3$ volume of air undergoes $6.6 \times 10^{28}$ collisions per second *i.e.*, one collision in $1.5 \times 10^{-29}$ second, molecules are moving with very high speeds. Suppose, such molecules are enclosed in a vessel having two compartments, using a permeable partition. Due to collisions, the number of molecules in the two compartments go on changing and the system moves from one macrostate to another very very large number of times per second. This number is generally of the order of millions of billions. Some macrostates are more probable than the others depending upon frequency of the microstate. *When the number of particles, in a dynamic system is very large, the system exists practically all the time in the most probable macrostate.* This is because the probabilities of macrostates fall extremely rapidly, even for a very small deviation (Refer Fig. 9.5) from the most probable state. This most probable state is known as the **equilibrium state** of the system. A dynamic system always tends to go from a state of lower probability towards the state of maximum probability *i.e.*, the equilibrium state. Even it is disturbed by some external causes of force, again it quickly comes back to its equilibrium state.

### Importance in Statistical Physics

In statistical physics, particularly in statistical thermodynamics, we handle the systems having a very large number of particles and assume that the system exists in the most probable state *i.e.*, equilibrium state. Since $n$ is very large, no significant error is caused by making this assumption. In other words, the importance of most probable state in determining the behaviour of a macroscopic system in equilibrium lies in the fact that all the properties of the system can be deduced by knowing the most probable or equilibrium state of the system.

## 9.13 Life-time of a Microstate and Macrostate

**Life-time of a Microstate:** The life-time of a microstate is the average time taken by a particle of a system to move from one phase space* compartment to an adjascent phase space compartment.

Let us consider a particle in a phase-plane $(x, p_x)$, moving from compartment 1 to compartment 2, with a change in position coordinate from $(x, p_x)$ to $(x + \Delta x, p_x)$ in time $T_{micro}$, keeping its momentum $p_x$ constant (Fig. 9.6). If m is the mass of the particle, then its velocity

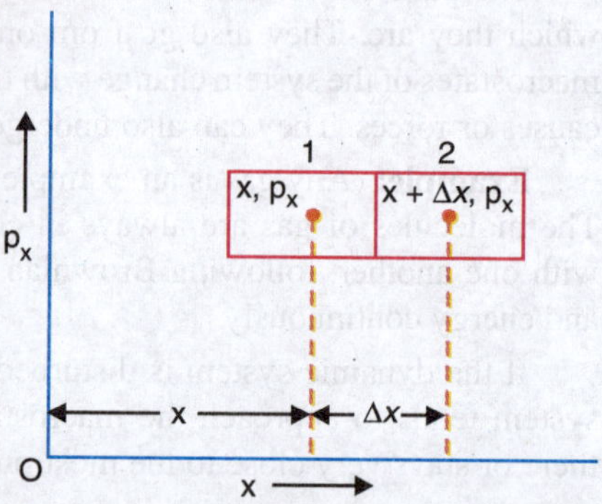

Fig. 9.6

$$v_x = \frac{\Delta x}{T_{micro}} = \frac{p_x}{m}$$

or

$$T_{micro} = \frac{\Delta x}{V_x} \quad \text{...}(i)$$

This $T_{micro}$ is called the life-time of the microstate. According to the law of equipartition of energy, the average K.E. of a gas molecule at temperature T Kelvin is given by

$$\frac{1}{2} m v_x^2 = \frac{1}{2} kT$$

where $k$ is Boltzmann constant ($1.38 \times 10^{-16}$ erg $K^{-1}$).

$$\therefore \quad v_x = \sqrt{\frac{kT}{m}} \quad \text{...}(ii)$$

Substituting in equation (*i*), we get

$$T_{micro} = \frac{(\Delta x)}{\sqrt{\frac{kT}{m}}} = \Delta x . \sqrt{\frac{m}{kT}} \quad \text{...}(iii)$$

**For Hydrogen gas:** Mass of hydrogen molecule m = $3.2 \times 10^{-24}$ gm.

At room temperature, $T = 300$ K, and assuming $\Delta x = 10^{-8}$ cm,

$$T_{micro} = 10^{-8} \sqrt{\frac{3.2 \times 10^{-24}}{1.38 \times 10^{-16} \times 300}}$$

$$= 10^{-13} \text{ second (approx.)}$$

This gives the average life-time of a microstate.

### Life-time of a Macrostate

If we try to examine the state of the system as shown in Fig. 9.6 i.e. the number of particles in each compartment, we will need some **finite time**. This finite time would be of the order of one-millionth ($10^{-6}$) of a second, even if we use a superfast camera. This finite time $T$ would be very large of the order of $10^7$ as compared to the life-time of a microstate, as calculated below :

$$\frac{T}{T_{micro}} = \frac{10^{-6}}{10^{-13}} = 10^7$$

* For phase-space, refer next chapter.

During such a single experimental observation, the system passes through 10 million ($10^7$) microstates.

Further, in a dynamic system, the change from one microstate to another microstate takes place even at very fast rate, the system will spend equal amounts of time in all the microstates. This is because, all the microstates of the system have the *same a priori probability*.

Therefore, the average time spent in each microstate

$$T_{micro} = \frac{T}{\text{Total no. of microstates } (N)}$$

$$= \frac{T}{N} \qquad ...(i)$$

where $T$ is the finite time of observation.

Let $T_{macro} = T_{(n_1, n_2)}$ be the time spent in the macrostate $(n_1, n_2)$ then

$$T_{macro} = \text{Average time spent in each microstate} \times \text{No. of microstate in a given macrostate}$$

$$= \frac{T}{N} \times W_{(n_1, n_2)} \qquad ...(ii)$$

This is because, the no. of microstates in a given macrostate is the thermodynamic probability $W_{(n_1, n_2)}$.

Further, $\frac{1}{N} = P_{micro}$ = probability of microstate.

Hence,
$$T_{macro} = T . P_{micro} \; W_{(n_1, n_2)} \qquad ...(iii)$$

or
$$\frac{T_{macro}}{T} = P_{micro} \; W_{(n_1, n_2)} \qquad ...(iv)$$

or
$$\frac{T_{macro}}{T} \propto W_{(n_1, n_2)} \qquad ...(9.16)$$

Thus, the fraction of the time spent by a system in a macrostate is proportional to the thermodynamic probability of the state.

## 9.14 Concept of a cell in a compartment

In practice, if we wish to distribute $n$ distinguishable particles in $k$ compartments which are of unequal size, we find that the a *priori* probability of a particle going into a particular compartment is not the same. *Large probability in a bigger compartment*. This is because of unequal size of the compartment. This difficulty is overcomed by introducing the concept of a cell.

Each compartment is divided into a very large number of sub-compartments, called as cells, in such a way that each cell is of the same size. Therefore, all the cells have the same a priori probability. The size of the cell is very very small and hence the number of cells in each compartment is exceedingly large. For a given size of the cell, the number of cells in each compartment is fixed.

### Distribution of particles

**(In compartments of unequal size)**

The concept of a cell is well understood by considering the distribution of particles in different compartments marked as 1, 2, 3, ... , *i*..., $k - 1$, $k$ of unequal size (Fig. 9.7). We wish to distribute $n$ distinguishable particles in these k compartments. As the compartments are of unequal size, a priori probability of a particle going into particular compatment is not the same. To overcome this difficulty, we introduce a **concept of a cell**. For this purpose we divide each compartment, from compartment 1 to $k$, into cells of **equal sizes**, so that compartment 1 contains $g_1$ cells, compartment 2 contains $g_2$ cells, compartment 3 contains $g_3$ cells and so on, the $i^{th}$ compartment containing $g_i$ cells and lastly $k^{th}$ compartment has $g_k$ cells.

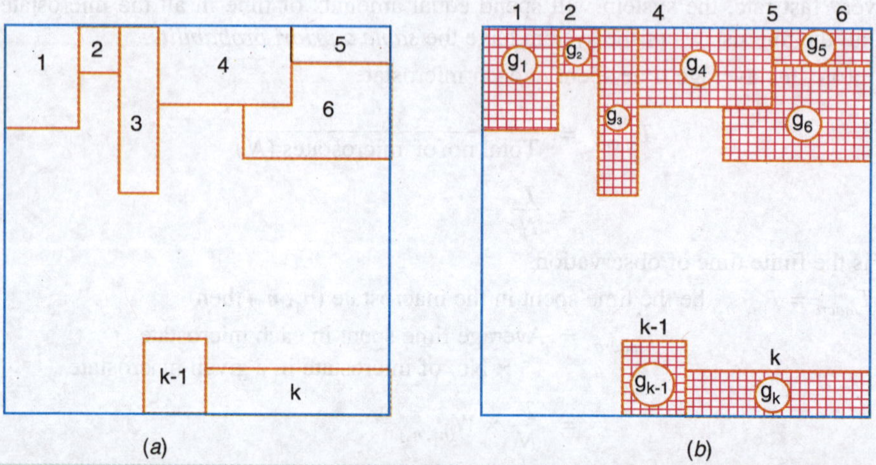


**Fig. 9.7**

As the compartments are of unequal size, the number of cells in each compartment will be different. The total number of cells, each of same size = $g_1 + g_2 + \ldots\ldots\ldots + g_i + \ldots\ldots\ldots g_k = G$

Hence

(*i*) Since all the cells are of same size, there is an equal a priori probability *i.e.* a particle in a given compartment can go into any cell within the compartment with equal probability.

(*ii*) Moreover, since the size of the compartment is big, the number of cells in any compartment are exceedingly large.

Let us find number of ways of placing $n_1$ particles in $g_1$ cells for compartment 1. The first particle can go into any one of the $g_1$ cells of compartment 1. Thus, the number of ways the first particle can be arranged in $g_1$ cells is = $g_1$.

In M.B. statistics*, as there is no restriction on the number of particles that can be placed in a single cell, the 2nd particle can also go into any one of the $g_1$ cells of compartment 1.

∴ Total number of ways in which two particles can be arranged in $g_1$ cells of compartment 1 is given by

$$= g_1 \times g_1 = g_1^2$$

Proceeding in the same manner, the total number of ways in which $n_1$ particles can be distributed in $g_1$ cells of compartment 1 is

$$= g_1 \times g_1 \times g_1 \times \ldots\ldots\ldots \times g_1 \; (n_1 \text{ times}) = g_1^{n_1}$$

The number of ways in which $n_1$ particles can be chosen out of $n$ particles

$$= {}^{n}c_{n_1} = \frac{n!}{n_1!(n-n_1)!}$$

Therefore, the total number of ways in which $n_1$ particles can be chosen out of $n$ particles and placed in $g_1$ cells of compartment 1 will be

$$= \frac{n!}{n_1!(n-n_1)!}(g_1)^{n_1} \qquad \ldots(9.17)$$

+ For this, refer next chapter

Next, let us find the number of ways of placing $n_2$ particles in $g_2$ cells of compartment 2 as under: The number of particles remained after placing $n_1$ particles in compartment 1 are = $(n - n_1)$. Out of these $n_2$ particles are to be placed in $g_2$ cells of compartment 2. The number of ways this can be done is

$$= \frac{(n-n_1)!}{n_2!(n-n_1-n_2)!}(g_2)^{n_2}$$

Thus, the total number of ways in which $n_1$ particles are distributed in compartment 1 and $n_2$ particles distributed in compartment 2 will be

$$= \frac{n!}{n_1!(n-n_1)!}(g_1)^{n_1} \times \frac{(n-n_1)!}{n_2!(n-n_1-n_2)!}(g_2)^{n_2}$$

$$= \frac{n!}{n_1!\,n_2!(n-n_1-n_2)!}(g_1)^{n_1}(g_2)^{n_2} \quad ...(9.18)$$

and so on.

Thus, the total number of ways of placing $n$ particles in all the 1 to $k$ compartments in such a way that $n_1, n_2, \ldots\ldots\ldots n_i, \ldots\ldots\ldots n_k$ particles will go into the compartment numbered as 1, 2, ......... $i$ ......... $k$ respectively i.e. the total number of microstates in a macrostate $(n_1, n_2, n_3 \ldots\ldots\ldots n_k)$ is given by

$$W_{(n_1, n_2, n_3, \ldots\ldots\ldots n_k)} = \frac{n!}{n_1!\,n_2!\,n_3!\ldots\ldots n_k!(n-n_1-n_2-n_3\ldots\ldots-n_k)!} \times (g_1)^{n_1}(g_2)^{n_2}(g_3)^{n_3}\ldots\ldots\ldots(g_k)^{n_k}$$

But $(n - n_1 - n_2 - n_3 \ldots\ldots - n_k) = n - (n_1 + n_2 + n_3 + \ldots\ldots\, n_k) = n - n = 0$

$\therefore \quad (n - n_1 - n_2 - n_3 \ldots\ldots - n_k)! = 0! = 1$

Therefore $\quad W_{(n_1, n_2 \ldots\ldots n_k)} = \dfrac{n!}{n_1!\,n_2!\ldots\ldots n_k!} \times (g_1)^{n_1}(g_2)^{n_2}(g_3)^{n_3}\ldots\ldots(g_k)^{n_k}$

or $$W = n!\prod_{i=1}^{k}\frac{(g_i)^{n_i}}{n_i!} \quad ...(9.19)$$

where $\prod$ denotes multiplication of terms stated above for various values of $i$.

Thus, the concept of a cell plays an important role in overcoming the difficulty in the distribution of particles in different compartments of unequal sizes, keeping the consistency of the principle of a priori probability.

### Distribution of particles in compartments of equal size

The number of different meaningful ways of arranging $n$ distinguishable particles in $k$ compartments of equal size such that $n_1$ are placed in compartment no. 1, $n_2$ in compartment no. 2 and so on is given by

$$W = \frac{n!}{n_1!\,n_2!\,n_3!\ldots\ldots n_k!} = \frac{n!}{\Pi\, n_i!}$$

The probability of the macrostate $(n_1, n_2, n_3, \ldots\, n_k)$ is given by

$$P_{(n_1, n_2, n_3, \ldots\, n_k)} = \frac{n!}{n_1!\,n_2!\,n_3!\ldots\ldots n_k!} \times \frac{1}{k^n}$$

## 9.15 Ensemble and Average Properties

The usual meaning of ensemble[†] is the whole or all the parts taken together. In statistical mechanics an ensemble is a collection of identical systems with all accessible microstates represented in it with

† For this, please refer next chapter.

equal frequency. Consider a system of 4 particles in two similar compartments. The system, under constraint $\Sigma n_i = 4$, has 16 different microstates (table. 9.2). We can draw all the sixteen microstates as 16 identical systems (Fig. 9.8). Consider just one system, which at time $t = 0$ has a particular microstate, say (*ab*, *cd*). Then due to interactions among the particles, the system will pass through all the accessible microstates as time passes. In a short time, we may find that same microstates repeat more frequently than the other. But if we observe for long enough time, then we should expect all the microstates are repeated with equal frequency. In the present case of 16 accessible states, suppose we make 48 observations in all, then we expect that each microstate should appear just 3 times. But in practice, some microstates may occur even 7 times and some microstates may not occur at all in these 48 observations! On the other hand, if we make $48 \times 10^{20}$ observations in all, then the observed frequency of each of the 16 microstates would not be different from $3 \times 10^{20}$ by more than (say) 1 part in $10^{12}$ or so.

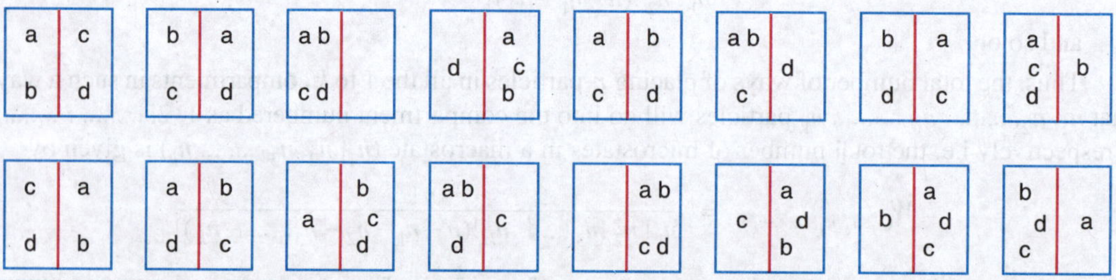


**Fig. 9.8** An Ensemble for a system of 4 particles in two equal compartments

We may either make these $48 \times 10^{20}$ times observations on a single particle system or at a single time on a collection of $48 \times 10^{20}$ identical systems. The conclusion would be the same. Here we take into consideration *the postulate of equal a priori probabilities.* It is a fundamental postulate of statistical mechanics and it says that a system can exist in each of the accessible microstates with equal probability. In other words, no preference is to be given to any one or more microscopic states over others.

The probability of particular macrostate is given by the number of microstates falling under it divided by the total number of accessible microstates. Let a macrostate be designated by suffix *j* and let $W_j$ represent the frequency with which it occurs in the ensemble (*i.e.*, the number of microstates falling under *j*th macrostate. Then, the probability of occurrence of the *j*th macrostate) is given by

$$P_j = \frac{W_j}{\Sigma W_j} \qquad \text{...(9.16)}$$

If any property *x* (may be energy, one component of momentum or any thing) has value $x_j$ in the *j*th macrostate then the average value of this property over the ensemble, denoted by $\bar{x}$ or $< x >$, is given by

$$\bar{x} = < x > = \sum_j x_j p_j = \frac{\Sigma x_j W_j}{\Sigma W_j} \qquad \text{...(9.20)}$$

It may be noted that we have no means of knowing how the system really passes through the various macrostates, much less about the further details of the microstates.

**EXAMPLE 9.6**

*A system has 5 different macrostates, under which there are 6, 20, 42, 12 and 2 microstates. A property x associated with the system has values 4, 4, 2, 6 and 10 respectively for the 5 macrostates. Calculate*

*(i) the probabilities for the different macrostates, and*

*(ii) average values of x and of* $\sqrt{x^2}$

**SOLUTION.** The sum of thermodynamic probabilities or total no. of microstates are

$$\Sigma W_i = 6 + 20 + 42 + 12 + 2 = 82$$

$$\therefore \quad P_j = \frac{6}{82}, \frac{20}{82}, \frac{42}{82}, \frac{12}{82}, \frac{2}{82} \qquad \text{[This gives } \Sigma P_j = 1]$$

(*i*)
$$<x> = \frac{1}{82}(6\times4+20\times4+42\times2+12\times6+2\times10) = 3.41 \text{ and}$$

(*ii*)
$$<x^2>^{\frac{1}{2}} = \left[\frac{1}{82}(6\times16+20\times16+42\times4+12\times36+2\times100)\right]^{\frac{1}{2}}$$

$$= \mathbf{3.85}$$

## EXAMPLE 9.7

*A system consists of 6000 particles distributed in three energy states with equal spacing. The energy of the three states are $E_1 = 0$, $E_2 = x$ and $E_3 = 2x$. All the three states have the same intrinsic probability g. At a certain instant there are 3000 particles in the lower level, 2500 in the middle level and 500 in the upper level. Compare the relative probabilities with the distribution obtained by the transfer of one particle from the middle to the lower level and one particle from the middle to the upper level and the original distribution.*

**SOLUTION.** Let $P_1$ and $P_2$ be the probabilities in the two cases

In the first case,

$$n = 6000, n_1 = 3000, n_2 = 2500, n_3 = 500$$

$$P_1 = \frac{g^n}{n_1!n_2!n_3!}$$

$$P_1 = \frac{g^{6000}}{3000!2500!500!} \qquad ...(i)$$

In the second case

$$n = 6000$$

$$n_1 = 3001, n_2 = 2498, n_3 = 501$$

$$\therefore \quad P_2 = \frac{g^{6000}}{3001!2498!501!} \qquad ...(ii)$$

Dividing (*ii*) by (*i*)

$$\frac{P_2}{P_1} = \frac{3000!2500!500!}{3001!2498!501!}$$

$$\frac{P_2}{P_1} = \frac{2500\times2498}{3001\times501} \qquad ...(iii)$$

$$\frac{P_2}{P_1} = 4.157 \qquad ...(iv)$$

$$\approx \mathbf{4.2}$$

It means the transfer of one particle from the middle to the upper and one particle from the middle to the lower state has changed the probability by a factor 4.2. This shows that both these distributions are not near the equilibrium state.

**EXAMPLE 9.8**

*Eight distinguishable particles are distributed among three compartments of equal size. Find the probability of the macrostate (i) (4, 3, 1) and (ii) (3, 3, 2).* **(Punjab 2001)**

**SOLUTION.** The probability of occurrence of the macrostate $(n_1, n_2, n_3, ... n_k)$ corresponding to the distribution of $n$ distinguishable particles into $k$ compartments of equal size is given by

$$P_{(n_1, n_2, n_3, ... n_k)} = \frac{n!}{n_1!n_2!...n_k!} \times \frac{1}{k^n}$$

Here, total no of particles, $n = 8$
and the total no. of compartments, $k = 3$

(*i*) For the macrostate (4, 3, 1) ; $n_1 = 4$, $n_2 = 3$, $n_1 = 1$

$$\therefore \quad P_{(4, 3, 1)} = \frac{8!}{4!3!1!} \times \frac{1}{3^8}$$

$$= \frac{8 \times 7 \times 6 \times 5}{3 \times 2} \times \frac{1}{3^8} = \frac{280}{6581}$$

$$= \mathbf{0.0427}$$

(*ii*) For the macrostate (3, 3, 2); $n_1 = 3$, $n_2 = 3$, $n_3 = 2$

$$\therefore \quad P_{(3, 3, 2)} = \frac{8!}{3!3!2!} \times \frac{1}{3^8}$$

$$= \frac{8 \times 7 \times 6 \times 5 \times 4}{3 \times 2 \times 2} \times \frac{1}{3^8} = \frac{560}{6561}$$

$$= 0.0854$$

**EXAMPLE 9.9**

*What is the maximum and minimum value of the probability of occurrence of a macrostate for the case of n distinguishable particles distributed in two compartments ? Prove.*

**(Punjab 2003, Kerala 2002)**

**SOLUTION.** The distribution of $n$ distinguishable particles between two halves of a box with equal probability is similar to the tossing of $n$ coins. As in the case of tossing of a coin, either head or tail may be uppermost, in the case of distribution of molecules in two halves of a box, the molecule may either go in first half or in the other half. If $a$ and $b$ denote that a molecule will be in the first and second half of a box respectively, then the possible combinations in the distribution of $n$ molecules will be

$$(a + b)^n = a^n + n_{c_1} a^{n-1} b + n_{c_2} a^{n-2} b^2 + ... + n_{c_r} a^r b^{n-r} + ... + n_{c_n} b^n$$

Total number of terms representing all possible combinations is

$$\Sigma n_{c_r} = 2^n$$

Therefore, the probability of distribution $(r, n - r)$, is given by

$$P_{(r, n-r)} = \frac{n!}{r!\,(n-r)!} \times \frac{1}{2^n} \qquad ..(i)$$

The probability of most probable distribution is

$$P_{max} = \frac{n!}{\left(\frac{n}{2}\right)!\left(\frac{n}{2}\right)!} \times \frac{1}{2^n}$$

$$= \frac{\mathbf{n!}}{\left(\frac{\mathbf{n}}{\mathbf{2}}!\right)^2} \times \frac{\mathbf{1}}{\mathbf{2^n}} \qquad ...(ii)$$

The probability of least or minimum probable distribution is

$$P_{min} = \frac{n!}{n!(n-n)!} \times \frac{1}{2^n} = \frac{1}{2^n} \quad \text{...(iii)}$$

### EXAMPLE 9.10

*A system of n particles has only two allowed states A and B. The probability for state A is p and that for state B is (1 – p). What is the probability for system to be in macrostate defined by distribution (r, n–r).* **(Rohilkhand 2003)**

**SOLUTION.** The probability of state *A* is *p* and that for *B* is (1 – *p*). This means the weight factors *a* and *b* for states *A* and *B* are *p* and ( 1 – *p*) respectively. Therefore, the probability for the distribution of *n* particles in states *A* and *B* in macrostate (*r*, *n–r*), is given by

$$P_{(r,\, n-r)} = \frac{n!}{r!(n-r)!}\, p^r\, (1-p)^{n-r}$$

### EXAMPLE 9.11

*Eight distinguishable particles are distributed in two compartments of unequal sizes. The first compartment is further subdivided into 6 cells and second into 2 cells of equal sizes. Calculate the probability of macrostate (6, 2) and (4, 4).* **(Punjab 2000)**

**SOLUTION.** The probability of occurrence of a macrostate ($n_1$, $n_2$ ... $n_k$) for the distribution of *n* distinguishable particles in *k* compartments in given by

$$P_{(n_1,\, n_2,\, ...\, n_k)} = \frac{n!}{n_1!n_2!\,...n_k!} \times \frac{(g_1)^{n_1}\,(g_2)^{n_2}\,(g_3)^{n_3}\,...(g_k)^{n_k}}{(g_1+g_2+...g_k)^n}$$

where $g_1$, $g_2$ ... $g_k$ are the no. of cells belonging to compartments marked 1, 2, ... *k* respectively.

In the given problem,

Total no. of particles, $n = 8$

Total no. of compartments, $k = 2$

No. of cells in the compartment No. 1, $g_1 = 6$

No. of cells in the compartment No. 2, $g_2 = 2$

For the macrostate (6, 2), $n_1 = 6$ and $n_2 = 2$

$$\therefore \text{ Probability of occurrence, } P_{(6,\,2)} = \frac{8!}{6!2!} \times \frac{(6)^6\,(2)^2}{(6+2)^8} = 28 \times \frac{729}{65536}$$

$$= \mathbf{0.311}$$

For the macrostate (4, 4), $n_1 = 4$ and $n_2 = 4$

$$\therefore \text{ Probability of occurrence, } P_{(4,\,4)} = \frac{8!}{4!4!} \times \frac{(6)^4\,(2)^4}{(6+2)^6}$$

$$= \frac{70 \times 8}{65536}$$

$$= \mathbf{0.0865}$$

### EXERCISE 9.12

*In the above problem, calculate the probability of the most probable macrostate.* **(Punjab 2000)**

**SOLUTION.** Total number of particles, $n = 8$

Total number of cells in the two compartments

$$G = g_1 + g_2 = 6 + 2 = 8$$

For the most probable macrostate, the number of particles in compartment No. 1 = $\frac{n}{G} g_1 = \frac{8}{8} \times 6$ = 6 and the number of particles in compartment No. 2

$$= \frac{n}{G} g_2 = \frac{8}{8} \times 2 = 2$$

∴ Most probable macrostate is (6, 2) for which the probability as calculated above is 0.311.

**EXAMPLE 9.13**

*Calculate the percentage error made in Stirling's formula.*

**(Punjab 2000, 2001, 2005, G.N.D.U. 2004)**

$\log_e (n!) = n \log_e n - n$ when (*i*) $n = 5$ and (*ii*) $n = 10$

**SOLUTION.**

(*i*) $n = 5$ ;

$$\log_e (n!) = \log_e (5 \times 4 \times 3 \times 2 \times 1) = \log_e 120$$
$$= \log_e 1.20 + \log_e 10^2 = 0.1823 + 2 \times 2.3026 = 4.7875$$
$$n \log_e n - n = 5 \log_e (5) - 5 = 5 \times 1.6094 - 5 = 3.047$$

Difference = 4.7875 − 3.047 = 1.7405

$$\therefore \quad \text{Percentage error} = \frac{1.7405}{4.7875} \times 100 = \mathbf{36.38\%}$$

(*ii*) $n = 10$ ;

$$\log_e (10!) = \log_e (10 \times 9 \times 8 \times 7 \times 6 \times 5 \times 4 \times 3 \times 2 \times 1) = \log_e 3628800$$
$$= 15.1044$$
$$n \log_e n - n = 10 \log_e (10) - n = 10 \times 2.3026 - 10$$
$$= 13.026$$
$$\text{Difference} = 15.1044 - 13.026 = 2.0784$$
$$\therefore \quad \text{Percentage error} = \frac{2.0784}{15.1044} \times 100 = \mathbf{13.76\%}$$

**Note :** The values of $\log_e$ may be calculated by using either 'logarithmic table' or calculator as is done in case (*i*) and (*ii*) respectively.

**EXAMPLE 9.14**

*Three similar dice A, B and C each having six equally likely faces marked as 1, 2, 3, 4, 5, 6 are thrown simultaneously and in a toss all the faces have equal probability of appearing up. Calculate the probability of getting the faces of all the dice up marked with 1 number.*

**(Agra 2003)**

**SOLUTION.**

Probability of dice *A* to show face marked 1 up, $p_1 = \frac{1}{6}$

Probability of dice *B* to show face marked 1 up, $p_2 = \frac{1}{6}$

Probability of dice *C* to show face marked 1 up, $p_3 = \frac{1}{6}$

∴ Probability of all the three dice *A*, *B* and *C* to show face marked 1 up,

$$P = p_1 \times p_2 \times p_3 = \frac{1}{6} \times \frac{1}{6} \times \frac{1}{6}$$

$$= \mathbf{\frac{1}{216}}$$

## EXAMPLE 9.15

*What is the probability of drawing two kings in succession from a pack of cards?* **(Punjab 2005)**

**SOLUTION.** The no. of ways in which the first card can be drawn $n_1 = 52$.

As there are 4 kings, the number of ways favourable to the first event, *i.e.*, the first card drawn may be king, $m_1 = 4$

$\therefore$ Probability of getting a king in the first draw,

$$P_1 = \frac{m_1}{n_1} = \frac{4}{52} = \frac{1}{13} \quad ...(i)$$

The number of ways in which the second card can be drawn, $n_2 = 51$, because one king is removed from the pack, the cards left are only 51. The no. of kings left in the pack are only 3. Therefore, the no. of ways favourable to the second event, *i.e.* the second card drawn may also be, king, $m_2 = 3$.

$\therefore$ Probability of getting a king in the second draw,

$$P_2 = \frac{m_2}{n_2} = \frac{3}{51} = \frac{1}{17} \quad ...(ii)$$

Since the events are independent, the total probability of drawing two kings,

$$P = p_1 \times p_2 = \frac{1}{13} \times \frac{1}{17}$$

$$= \frac{1}{221}$$

## EXERCISE 16

*The problem may be extended for 3 kings drawn in succession.*

**SOLUTION.** The probability will be

$$P = p_1 \times p_2 \times p_3$$

Here $$p_3 = \frac{m_3}{n_3} = \frac{2}{50} = \frac{1}{25}$$

$\therefore$ $$P = p_1 \times p_2 \times p_3 = \frac{1}{13} \times \frac{1}{17} \times \frac{1}{25}$$

$$= \frac{1}{5525}$$

## EXERCISE–IX

**9.1** Define probability. When will the probability be (*i*) zero (*ii*) one? What is total probability? **(H.P. 2003, Meerut 2000, Pb. 2002, Purvanchal 2001)**

**9.2** State and explain the principle of equal a priori probability. **(Nagpur U. 2007, Purvanchal 2001)**

**9.3** Prove that the probability of a composite event is equal to the product of the probability of the individual independent events.

**9.4** Discuss additive law and multiplicative law of probability. **(Meerut 2004)**

**9.5** Write a short note on permutations and combinations, giving examples.

**9.6** *n* similar coins are tossed simultaneously for a large number of times. Prove that the probability of getting *r* heads upper most is

$$\frac{n!}{r!\,(n-r)!} \cdot \frac{1}{2^n}$$ **(Purvanchal 2007)**

**9.7** Clearly distinguish between probability and frequency.

**9.8** Define and explain the terms Macrostate and Microstate with the help of an example.
**(Nagpur 2008, 2007, 2006, 2005/winter, 2006/s, Punjab 2001; Pbi 2000, 2001, 2002, H.P. 2001, G.N.D. 2000, Purvanchal 2005, 2002, 2001, Agra 2007)**

**9.9** What is meant by the term thermodynamic probability of macrostate? How it is related to probability of occurrence of that state. How does it differ from mathematical probability.
**(Punjab 2000, 2001, M.D. 2001, 2003, H.P. 2001, Kerala 2000, Nagpur 2009, 2008, 2007, 2004, 2005, Agra 2003)**

**9.10** For $n$ distinguishable particles to be distributed in two compartments, prove that the thermodynamic probability

$$W_{(n_1, n_2)} = \frac{n!}{n_1!\, n_2!} \quad \text{where } n_1 + n_2 = n$$

**9.11** Explain the term constraints on a system. Hence define accessible and inaccessible states.
**(Punjab 2000, 2001, 2003; Purvanchal 2001, Nagpur 2004, Agra 2003)**

**9.12** Derive the general formula to explain that the probability distribution narrows rapidly as $n$ increases.

**9.13** Define an ensemble and average properties. Considering the postulate of equal a priori probability, deduce the formula for the average value over the ensemble. **(Meerut 2005)**

**9.14** Explain in brief static and dynamic systems. **(Kerala 2001, G.N.D. 2004)**

**9.15** Find the expression for the probability of a macrostate corresponding to the distribution of $n$ particles in $k$ compartments of unequal sizes. **(Punjab 2001, 2002, 2005 Pbi. 2000, 2001)**

**9.16** What is the role played by the most probable states in determining the behaviour of a system in equilibrium? What is its importance in statistical physics. **(Kerala 2002)**

**9.17** Show with the help of an equation that on the average the fraction of the time spent by a dynamic system in any particular macrostate is proportional to the thermodynamic probability.
**(Punjab 2001, 2003, 2005)**

**9.18** Distinguish between Accessible and Inaccessible Macrostates. **(Punjab 2005)**

**9.19** Define the terms with examples:

(*i*) Most probable macrostates. **(Rohilkhand 2002 Improvement)**

(*ii*) Thermodynamic probability. **(Rohilkhand 2001, Nagpur 2004)**

(*iii*) Accessible and inaccessible microstates. **(Purvanchal 2001, Meerut 2001, Nagpur 2004)**

(*iv*) Postulate of equal a priori probability. **(Nagpur 2004)**

(*v*) Macrostate and microstate. **(Nagpur 2006, 2004/s, 2004/w, Punjab 2005)**

(*vi*) Probability. **(Nagpur 2004)**

**9.20** If a pair of 6 faced dice with faces marked 1 to 6 is thrown, what the probability that the sum of the numbers which shows up is 8. **[Ans.** $\frac{5}{36}$**] (Punjab 2003)**

**9.21** Two Six faced dice, each marked 1 to 6 are thrown. Calculate the probability that one of the dice shows 6 and the other shows 5? **[Ans.** $\frac{1}{18}$**]**

**9.22** In a system of 14 distinguishable particles distributed in two equally probable halves of a box, determine the probability of distribution (10, 4), (14, 0) and (7, 7).

**Ans.** $\frac{1001}{16384}, \frac{3432}{16384}, \frac{1}{16384}$

**9.23** A six faced dice with faces marked 1 to 6 and in any throw each face is equally likely to land uppermost. It the dice is tossed 5 times, calculate the probability of the following: (*a*) number 1 appearing just once, (*b*) number 2 and 5 appearing just once each. (*c*) Number 4 appearing in all the five tosses.

**9.24** A box is divided into 9 equal sized compartments. A ball thrown in the box is equally likely to go into any one of these compartments. If 9 balls are thrown, calculate the probability for a particular compartment to contain (*i*) one particular ball only (*ii*) all the nine balls (*iii*) none of the balls. State the principle used. **Ans.** $\frac{8^9}{9^9}, \frac{1}{9^9}, \frac{8^9}{9^9}$

**9.25** Calculate the thermodynamic probability of various macrostates corresponding to distribution of 5 particles into two compartments. **(Punjab 2001)**

**9.26** In a system of 8 distinguishable particles distributed in two equal sized compartments, calculate the probability of the macrostate (3, 5), (4, 4), (2, 6). **(Agra 2007, Punjab 2000, Kerala 2002, Pbi 2001)**

**Ans.** $\frac{7}{32}, \frac{35}{128}, \frac{7}{64}$

**9.27** Ten particles are distributed in two equal sized cells. Find the number of macrostates and microstates.
[**Hint :** Number of macrostates = $n + 1 = 10 + 1 = 11$ **(Punjab 2002)**
Number of microstates = $2^n = 2^{10}$]

**9.28** 4 molecules are to be distributed in 2 cells. Find possible no. of macrostates and corresponding no. of microstates. **(Nagpur 2004)**
[**Hint :** No. of macrostates = $n + 1 = 4 + 1 = 5$
No. of microstates = $c^n = 2^4 = 16$
(Here $n = 4$ and $c = 2$)

**9.29** 5 distinguishable particles are to be arranged in 3 compartments of a box. (*i*) calculate total no. of ways in which this can be done; assuming that there is no restriction on the no. of particles that can go in any compartment (*ii*) Also calculate the probability of macrostate (2, 1, 2). **(Punjab 2005)**

## OBJECTIVE QUESTIONS

**9.1** Constraints imposed on a system:
(*a*) increase number of inaccessible microstates.
(*b*) decrecase the number of inaccessible microstates.
(*c*) have no effect.

**9.2** Statistical methods give greater accuracy when the number of observations is:
(*a*) very small (*b*) very large
(*c*) neither very small nor very large.

**9.3** The probability of drawing two kings in succession from a well shuffled pack is:
(*a*) $\frac{1}{52}$ (*b*) $\frac{1}{144}$
(*c*) $\frac{1}{221}$ (*d*) none of these **(Punjab 2005)**

**9.4** The value of probability of an event can not be:
(*a*) zero (*b*) 1
(*c*) $\frac{1}{2}$ (*d*) negative

**9.5** The probability of occurrence of two independent events is equal to their:
(*a*) sum (*b*) difference
(*c*) product (*d*) ratio

**9.6** The number of most probable macrostates for a system having odd number of particles is:
(*a*) 1 (*b*) 2
(*c*) 3 (*d*) 4

**9.7** If 12 particles are distributed randomly between two boxes *A* and *B* with equal probability, what is the probability of the distribution (8, 4)?

(*a*) $4.95 \times 2^{-12}$ (*b*) $495 \times 2^{-12}$

(*c*) $49.5 \times 2^{-12}$ (*d*) $495 \times 2^{-11}$ **(Rohilkhand 2003)**

**9.8** The relative probability between two different energy states having difference $1.1 \times 10^{-20}$ joules at 400 K temperature is:

(*a*) $e^{-1}$ (*b*) $e^{-2}$

(*c*) $e$ (*d*) $e^{2}$ **(Rohilkhand 2003 Improvement)**

## ANSWERS

| | | | |
|---|---|---|---|
| **9.1** (*a*) | **9.2** (*b*) | **9.3** (*c*) | **9.4** (*d*) |
| **9.5** (*c*) | **9.6** (*b*) | **9.7** (*c*) | **9.8** (*d*) |

# Some Universal Laws in Statistical Mechanics

# 10

## AT A GLANCE

• Introduction • Degrees of Freedom • Position Space • Momentum Space • Phase Space • The mu-Space and Gamma Space • Division of Phase Space in to Cells • Applications to : (*i*) One Dimensional Harmonic Oscillator (*ii*) A Free Particle • Fundamental Postulates of Statistical Mechanics • Density of Quantum States of Energy of a Particle • Statistical Ensembles (*i*) Microcanonical Ensemble (*ii*) Canonical Ensemble (*iii*) Grand-canonical Ensemble • Comparison of Ensembles • Equilibrium between Two Systems in Thermal Contact • Bridge with Macroscopic Physics • Theories Based on Statistical Mechanics • Entropy and Probability • Boltzmann's Canonical Distribution Law and it's applications • The Equipartition of Energy • Statistical Interpretation of Second Law of Thermodynamics • Partition Function and Relation with Thermodynamic Quantities • Entropy of an Ideal Gas : Gibbs Paradox.

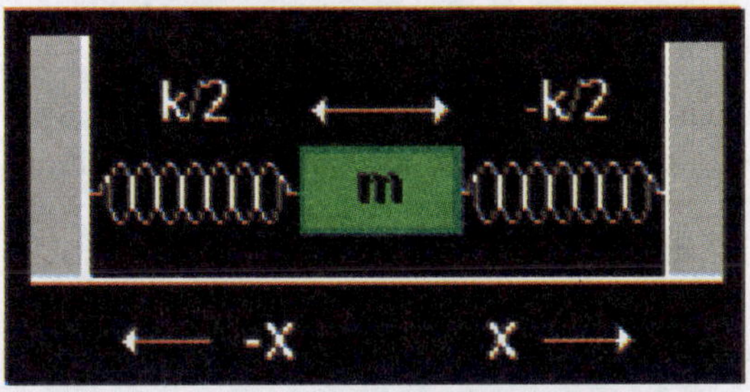


## Introduction

The distribution of particles with a given total energy into a discrete set of energy states as described by the table 9.4 becomes complicated when we deal with very large value of $n$ and possesses a quite complex energy pattern. A special procedure for handling such huge number of molecules has therefore to be developed. This is governed by some universal laws in statistical mechanics. Statistical mechanics is concerned with the properties of matter in equilibrium in the empirical sense used in thermodynamics.

## 10.1 Degrees of Freedom

From the statistical point of view, a monatomic gas constitutes the simplest system. The energy expression for a system involves $f$ co-ordinates of *position* and $f$ co-ordinates of *momentum*. We then say that a system comprising one molecule has $f$ degrees of freedom. If there are of $N$ molecules in the system, the energy expression for the system of $N$ molecules has $Nf$ co-ordinates of position and $Nf$ co-ordinates of momentum. Thus, the system has $Nf$ degrees of freedom.

## 10.2 Position Space

Consider a system consisting of N particles distributed in a given volume V. If the system is **static,** all particles will remain fixed at various points in space. To completely specify the position of any particle in the three dimensional space, we must know the values of its three cartezian coordinates $x$, $y$ and z which are mutually perpendicular to each other. As there are N particles, a knowledge of 3 N co-ordinates gives complete information about such a *static system*.

*The three-dimensional space in which the location of a particle is completely given by the three position coordinates, is known as* **position space**.

A small element in position space is denoted by volume element $dV$ and is expressed as

$$dV = dx\, dy\, dz$$

## 10.3 Momentum Space

If the system is **dynamic**, its particles move about with various velocities and hence possess momenta. A complete specification of such a system can not be described in terms of position co-ordinates only. For a dynamic system, we must specify the three components of momentum* with the help of three velocity components $v_x$, $v_y$ and $v_z$ in addition to the three position coordinates $x$, $y$ and $z$.

If $m$ is the mass of the particle moving with a velocity $v$, its momentum $p = mv$.

The three components of momentum are

$$p_x = mv_x,\ p_y = mv_y\ ,\ p_z = mv_z$$

Just as the position of particle is completely given by the three position co-ordinates $x$, $y$ and $z$ *in three-dimensional position space,* the momentum of the particle is completely specified by the three mutually perpendicular momentum coordinates $p_x$, $p_y$ and $p_z$ in three-dimensional space known as **momentum space** and the small volume element in **momentum space** is given $dp_x\, dp_y\, dp_z$.

## 10.4 Phase Space

*A combination of position space and momentum space is known as phase space*. Thus, phase space has 6 dimensions, *i.e.* three position coordinates and three momentum coordinates, all mutually perpendicular to each other. The position of a particle in phase space is specified by a point with six coordinates $x$, $y$, $z$, $p_x$, $p_y$, $p_z$. Complete information about any particle in a dynamic system can be obtained from a knowledge of these six coordinates which completely determine its position as well as momenta.

A small element in phase space is denoted by $d\tau$ and is given by

$$d\tau = (dx\, dy\, dz)\,(dp_x\, dp_y\, dp_z).$$

* The state of motion of a particle is described in terms of velocity components $\dot{x}$, $\dot{y}$, $\dot{z}$ (dot means derivative with respect to time). Sometimes it is more convenient to use momentum coordinates instead of velocity coordinates.

## 10.5 The mu-space and Gamma Space

Let us limit ourselves to a system comprising of 1 molecule. Moreover, consider it to be mono-atomic as the simplest system. The state of the gas is completely known if the position and momentum of each atom of the gas is specified. The position of an atom can be specified in terms of its rectangular coordinates *x*, *y* and *z*. This is known as *position space*. The momentum of an atom is completely known if its rectangular momentum components $p_x$, $p_y$ and $p_z$ are specified. This is called *momentum space*. These six quantities $x, y, z, p_x, p_y, p_z$ for each atom when known, the state of the gas can be determined. The space covered by the infinite set of these six coordinates is called the *phase-space* for a single monatomic molecule.

Since the energy of a molecule is given by $(p_x^2 + p_y^2 + p_z^2)/2m$ which is equal to $p^2/2m$, the momentums space representation also gives representation of energy of the state.

Such a six-dimensional space for a *single particle* is called *molecular phase space* or *mu-space* or *μ-space*. Since 6-dimensional diagram can not be drawn, phase space is purely a mathematical concept.

Suppose that position or momentum of one of the molecules is changed slightly. The *phase point* of this molecule will undergo a displacement in the phase space; and the microscopic state of the gas will be modified. To explain further, let us consider that the μ-phase space is divided into two-dimensional *energy sheet*, having position coordinate as *x*-axis and momentum coordinate as $p_x$-axis. Let us subdivide the range of variables *x* and $p_x$ into arbitrary small discrete intervals called as *phase cells* of arbitrary size (Fig. 10.1). The minimum size of phase space cell is explained as under:

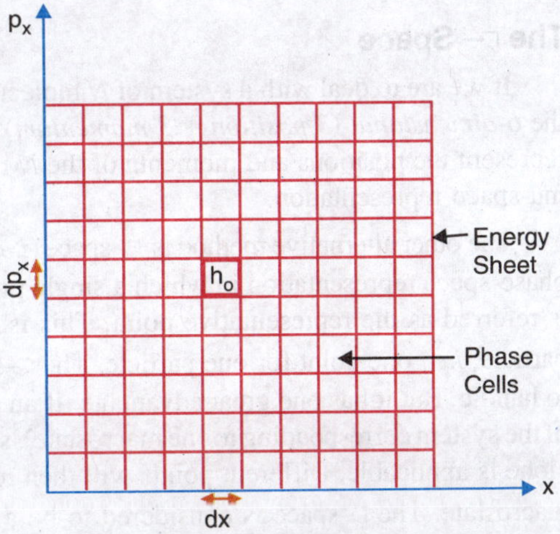

Fig. 10.1

### (i) Minimum size in classical statistics

The meaning of a point in phase space is well understood with the help of **Heisenberg uncertainty principle.**

Let *dx* be the size of the interval on the *x*-axis and $dp_x$, the corresponding length on $p_x$-axis. The phase space is divided into energy sheets of smaller and smaller energy ranges. The energy sheet thus consists of small cells of equal area = $dx \,.\, dp_x = h_o$. The state of the system can then be specified by locating a phase point within a particular cell of phase space. In classical description $h_o$ can be chosen arbitrarily as small as possible and we may set it $10^{-30}$ erg. sec or $10^{-35}$ erg. sec. or even smaller value. It may be noted that $h_o$ has dimensions of *energy* × *time*. The lengths *dx* and $dp_x$ are assumed to be so small that if a phase point within a cell is slightly displaced to any point within the cell, the change in the position and the momentum will be so minute that even the most accurate experiment would be unable to detect the changes occurred. Such an infinisimally small sized phase cell is considered. If the particle lies within the cell, its state is represented by the pair of coordinates $(x, p_x)$. If there is a change in the state of the particle, it will move from this cell to another appropriate cell. The dimension $h_o$ = length × momentum = (metre × Newton. sec) = (Joule. sec).

### (ii) Minimum size in Quantum Statistics

In Boltzmann's time, there was no reason to ascribe any specific value to $h_o$. All the results of classical statistics are independent of the size of $h_o$. Therefore, we permit $h_o$ to go over to zero size ($h_o \to 0$). But the nature itself has provided a finite and constant volume for the unit phase cell with the discovery of the Planck's constant and the advent of quantum theory discussed under quantum statistics. We see that the assumption $dx.\ dp_x$ = h is now consistent with the Heisenberg uncertainty principle given by $(dx.dp_x)_{min} = h = 6.625 \times 10^{-27}$ erg. sec. $= 6.6 \times 10^{-34}$ J-sec. According to this relation,

$$dx\ .\ dp_x \simeq h$$
$$dy\ .\ dp_y \simeq h$$
$$dz\ .\ dp_z = h$$
$$\therefore \quad dx\ .\ dp_x\, dy\ .\ dp_y\ .\ dz\ dp_z = h^3$$
$$\text{or} \quad dx\, dy\, dz\, dp_x\, dp_y\, dp_z = h^3 \qquad ...(10.1)$$

Here $dx\, dy\, dz\, dp_x\, dp_y\, dp_z$ represents the volume of each state $d\tau$.

$$\therefore \quad d\tau = h^3$$

Therefore, according to uncertainty principle, in phase space (mu), the coordinates of a particle can be specified only to the extent that lying within an element of phase space of volume $h^3$. (Planck's constant $h = 6.625 \times 10^{-34}\ J-\text{sec}$). The unit of phase space volume is (Joule – $s$)$^3$.

### The Γ– Space

If we are to deal with a system of $N$ molecules, there are two alternative methods. One is to have the *6-dimensional* (*3 position + 3 momentum*) *phase space* representation, and we take $N$ points to represent the positions and momenta of the $N$ molecules at a given instant. This is called μ-space or mu-space representation.

The other alternative method is Γ–space (*i.e.,* gamma space) representation. It is a 6N–dimensional phase-space representation in which a single point represents a state of entire N–particle system and is referred as the representative point. This is in contrast to the μ-space, which refers to only one particle, *i.e.,* one point for one particle. The Γ–space representation is mathematically more difficult to handle. But it has one great advantage. If an *ensemble* (*i.e.,* assembly of all accessible microstates of the system corresponding to one macrostate) is to be represented, this 6N–dimensional representation alone is applicable. Different points will then represent the corresponding all microstates of another macrostate. The Γ–space is considered to be a conceptual Euclidean space having $2fN$ rectangular axes.

For a general case of $f$ degrees of freedom per molecule, we shall have:

| | | |
|---|---|---|
| (*i*) | position space | $f$ – dimensional |
| (*ii*) | momentum space | $f$ – dimensional |
| (*iii*) | μ – space | $2f$ – dimensional |
| (*iv*) | Γ – space | $2fN$ – dimensional. |

## 10.6 Division of Phase Space into Cells

Let us consider a mu-space representation. It is characterised by $2f$ dimensional phase space configuration. For a $2f$ dimensional phase space (μ–space) defined by position co-ordinates $q_1, q_2, \ldots\ldots\ q_i \ldots\ldots\ q_f$ and momentum coordinates $p_1, p_2, \ldots\ldots\ p_i \ldots\ldots\ p_f$, an element of phase volume may be represented by

$$d\tau = (dq_1\, dq_2 \ldots\ldots\ dq_i \ldots\ldots\ dq_f) \times (dp_1\, dp_2 \ldots\ldots\ dp_i \ldots\ldots\ dp_f). \qquad ...(10.2)$$
$$= \prod_{i=1}^{f} dq_i\, dp_i$$

This quantity has the dimension of (length × momentum)$^f$ or (metre × Newton. sec)$^f$ or (Joule–$s$)$^f$. Now suppose $p_i$ and $q_i$ are allowed to take all possible continuously variable values, then for a finite volume of phase space, there can be infinite number of representation points and hence infinite number of possible microstates. In statistics, however we deal with finite number of microstates, howsoever large. Hence phase space is divided into "Cells", each cell of a certain **fixed** volume $h_0^f$. Therefore, in the element of volume of phase space given by equation (10.2) will contain the number of microstates

$$\frac{d\tau}{h^f} = \frac{(dq_1\, dq_2 \,.......dq_i \,.......dq_f)\,(dp_1\, dp_2 \,.......dp_i \,.......dp_f)}{h_0^f} \qquad ...(10.3)$$

here $h_0$ is a arbitrary constant with the dimension of (erg–sec) or (Joule–sec). This value of $h_0$ is taken to be as small as possible so the number of microstates obtained by using relation (10.3) are large enough to handle with approximations.

For Γ–space representation having $2fN$–dimensional phase space, which is useful to describe a microstate of the system of $N$ particles with a single point, an element of phase volume is given by

$$(dq_1\, dq_2 \,.......\, dq_i \,.......\, dq_{Nf})\,(dp_1\, dp_2 \,.......\, dp_i \,.......\, dp_{Nf}) \qquad ...(10.4)$$

and the number of phase cell in this volume element is given by

$$\frac{d\tau}{h^{Nf}} = \frac{(dq_1\, dq_2 \,.......dq_i \,.......dq_{Nf})\,(dp_1\, dp_2 \,.......dp_i \,.......dp_{Nf})}{h_0^{Nf}} \qquad ...(10.5)$$

## 10.7 Applications

### (i) One–Dimensional Harmonic Oscillator

Consider a linear simple harmonic oscillator of mass $m$ and oscillating freely with frequency ν. Let us calculate the number of phase cells in the energy range from 0 to $E$. The phase diagram contains one position coordinate $x$ and one momentum coordinate $p$. So it is a $p$, $x$ plane. At any instant, the total energy $E$ of the oscillator will comprise of potential energy and kinetic energy.

∴ Total energy,

$$E = P.E + k.E.$$

$$= \frac{1}{2}kx^2 + \frac{1}{2}mv^2$$

$$= \frac{1}{2}kx^2 + \frac{1}{2}\frac{m^2v^2}{m}$$

$$\therefore \quad E = \frac{1}{2}kx^2 + \frac{1}{2}\frac{p^2}{m}$$

$$2E = kx^2 + \frac{p^2}{m}$$

where $k$ is called force constant.

Dividing by $2E$,

$$1 = \frac{x^2}{2E/k} + \frac{p^2}{2mE} \qquad ...(10.6)$$

$$\therefore \quad \frac{x^2}{a^2} + \frac{y^2}{b^2} = 1$$

where $a^2 = \dfrac{2E}{k}$ and $b^2 = 2mE$

Thus, Equation (10.6) represents an ellipse in the $x$, $y$ phase diagram (Fig. 10.2) and the points of constant $E$ lies on this ellipse, having major and minor axes as $\sqrt{\dfrac{2E}{k}}$ and $\sqrt{2mE}$ respectively. For energy range from $O$ to $E$, the phase area is A = π *ab i.e.*, $\pi\left(\frac{2E}{k}\times 2mE\right)^{\frac{1}{2}}$ or $2\pi E\left(\frac{m}{k}\right)^{\frac{1}{2}}$. Each phase cell has an area $q_i \times p_i = h_o$. Hence, the number of phase cells in energy range from $O$ to $E$ will be

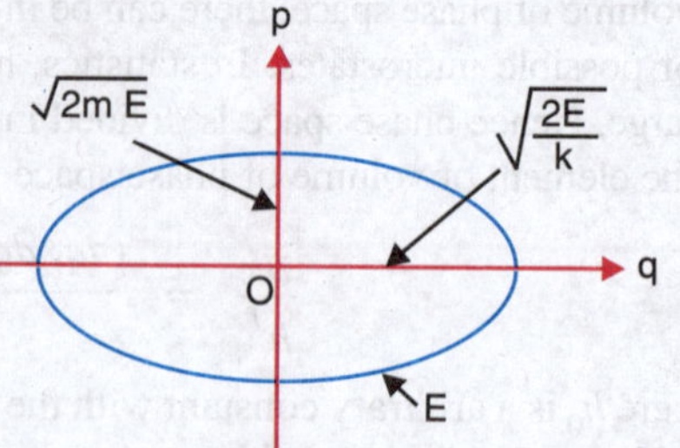


Fig. 10.2

$$\phi(E) = \frac{\pi}{h_o} 2E\left(\frac{m}{k}\right)^{\frac{1}{2}}$$

But the frequency ν is given by

$$\nu = \frac{1}{2\pi}\sqrt{\frac{k}{m}}$$

$$\therefore \quad \phi(E) = \frac{E}{h_o \nu} \qquad ...(10.7)$$

or Energy will be

$$E = h_o \nu \,.\, \phi(E) \qquad ...(10.8)$$

Thus, if the oscillator is so governed that the number of phase cells can take only integral values i.e., $\phi(E) = 0, 1, 2, ....... n$, then we get

$$E = n\, h_o \nu \qquad ...(10.9)$$

This means that energy can occur in discrete states and has values integral multiple of $h_o\nu$ only.

### (*ii*) A Free Particle

For simplicity, consider a phase space representing a single particle of mass $m$ in a volume $V$. Let us calculate the number of phase cells in energy range from 0 to $E$. The volume of each phase cell is say, $h_o^3$. An elemental volume in phase space is

$$dp_x\, dp_y\, dp_z\, dq_x\, dq_y\, dq_z = dp_x\, dq_z\, dp_y\, dq_y\, dp_z\, dq_z = h_0^3$$

Assuming only 3–translational degrees of freedom, the phase space volume is

$$V_{phase} = \iiiint\!\!\iint dp_x\, dp_y\, dp_z\, dq_x\, dq_y\, dq_z$$

Here the integration over all position coordinates $q^{'s}$ gives the total physical volume $V$ and integration over all linear momenta $p^{'s}$ is limited to $(p_x^2 + p_y^2 + p_z^2)_{max} = 2m\,E = (p_{max})^2$. So the phase space volume is

$$V_{phase} = V \times \frac{4\pi}{3}(p_{max})^3$$

$$= \frac{4\pi V}{3}(2m\,E)^{\frac{3}{2}} \qquad [\because (p_{max})^2 = 2m\,E]$$

The number of phase cells will be

$$\phi(E) = \frac{\text{Volume of Phase Space}}{\text{Volume of One Cell}}$$

or

$$\phi(E) = \frac{V_{Phase}}{h_o^3}$$

For a single particle, the number of accessible microstates will be equal to the number of accessible cells in the phase space. Therefore, the number of microstates in the energy range o to E is given by

$$\phi(E) = \frac{4\pi V}{3h_o^3}(2mE)^{\frac{3}{2}} \qquad ...(10.10)$$

**EXAMPLE 10.1**

*For a single particle of mass m enclosed in volume V, show that the number of accessible microstates in the energy range E to E + dE is given by*

$$\Omega(E) = \frac{4\sqrt{2}.\pi V}{h^3} m^{\frac{3}{2}} E^{\frac{1}{2}} \delta E$$

where phase space is divided by the rule $\delta p_i \, \delta q_i = h$ **(Purvanchal 2005, 2001, Mumbai 2005)**

**SOLUTION.** The number of microstates in the phase space having energies between $E$ and $E + dE$ is obtained by differentiating equation (10.10) with respect to $E$. Thus we have

$$\Omega(E) = \frac{\partial \phi(E)}{\partial E}$$

$$= \frac{3}{2} \cdot \frac{4\pi V}{3h^3} (2mE)^{\frac{1}{2}} \, 2m \, \delta E$$

$$= \frac{4\sqrt{2}.\pi V}{h^3} m^{\frac{3}{2}} E^{\frac{1}{2}} \delta E$$

or,
$$\Omega(E) = \frac{2\pi V}{h^3} (2m)^{\frac{3}{2}} E^{\frac{1}{2}} dE$$

## 10.8 Fundamental Postulates of Statistical Mechanics

The application of statistical mechanics to gases, certain fundamental postulates are made. These are:

**1.** *Any gas under consideration may be considered to be composed of large number of molecules which are constantly in motion and behave like very small elastic spheres.*

**2.** *All the cells in phase space are of equal size.*

**3.** *All accessible microstates corresponding to possible macrostates are equally probable.* This is the most fundamental and hence important postulate of statistical mechanics. For example, if we consider as earlier, the distribution of 4 distinguishable particles $a, b, c, d$ in two cells $i$ and $j$, then

Probability of microstate ($a\,b\,c\,d$, o) $= \frac{1}{2^4} = \frac{1}{16}$

Probability of microstate ($a\,b\,c$, $d$) $= \left(\frac{1}{2^3}\right)\left(\frac{1}{2}\right) = \frac{1}{16}$

Probability of microstate ($ab$, $cd$) $= \left(\frac{1}{2^2}\right)\left(\frac{1}{2^2}\right) = \frac{1}{16}$

Probability of microstate ($a$, $bcd$) $= \left(\frac{1}{2}\right)\left(\frac{1}{2^3}\right) = \frac{1}{16}$

Thus, we can conclude that the probability of all accessible microstates of the system is equal. From this postulate it follows that the probability of occurrence of a given macrostate is proportional to the number of microstates, corresponding to the given macrostate. Thus the probability $P(E)$ of system possessing energy $E$ is proportional to the thermodynamic probability $W(E)$. *i.e.*,

$$P(E) \propto W(E)$$

or
$$P(E) = cW(E)$$

where $c$ is a constant of proportionality.

**4.** *The equilibrium state of a gas corresponds to the macrostate of maximum probability.*

**5.** *The total number of molecules is constant.* This is in accordance with the **principle of conservation of matter.** If $n_1$ is the number of molecules in cell 1, $n_2$ in cell 2, $n_3$ in cell 3 etc. If $n$ is the total number of molecules, then

$$n_1 + n_2 + n_3 + ...... + n_i + ...... n_k = \Sigma n_i = n = a \text{ constant.} \quad ...(10.11)$$

**6.** *The total energy of the system is constant.*

This is again in accordance with the **conservation of energy** of the system. If $E_1$ is the energy of each particle in cell 1, $E_2$ the energy of each particle in cell 2 etc. and $E$ the total energy, then

$$E_1 n_1 + E_2 n_2 + E_3 n_3 + ... = \Sigma E_i n_i = E = \text{a constant.} \quad ...(10.12)$$

## 10.9 Density of Quantum States of Energy of a Particle

According to quantum mechanics, the solution of a 3-dimensional time independent Schrodinger's equation

$$\left(\frac{\partial^2}{\partial x^2} + \frac{\partial^2}{\partial y^2} + \frac{\partial^2}{\partial z^2}\right)\psi_k(r) + \frac{2mE}{\hbar^2}\psi_k(r) = 0$$

the allowed energy levels *i.e.* the eigenvalues of energy of a particle of mass $m$ in a 3-dimensional cubic potential box of each side $L$ are given by

$$E_{n_x n_y n_z} = \frac{\pi^2 \hbar^2 n^2}{2mL^2} = \frac{\pi^2 \hbar^2}{2mL^2}(n_x^2 + n_y^2 + n_z^2) \quad ...(i)$$

where,

$$n_x = 1, 2, 3, ......$$
$$n_y = 1, 2, 3, ......$$
$$n_z = 1, 2, 3, ......$$

corresponding to each value of the integer, there is a single quantum state of the particle. The lowest energy level *i.e.* the energy in the ground state, occurs for

$$n_x = n_y = n_z = 1$$

leads to the same value of $n^2 = 6$ for all the three quantum states.

In other words, the particle having the same energy in the first excited state can be in any one of the three quantum states. Such an energy level (with three wave functions $\psi_1$, $\psi_2$, $\psi_3$) is said to be three-fold degenerate. One energy level corresponds to three states. The lowest energy level ($n_x = n_y = n_z = 1$) is however non-degenerate. If the energy levels are to degenerate, there will be a very large number† of

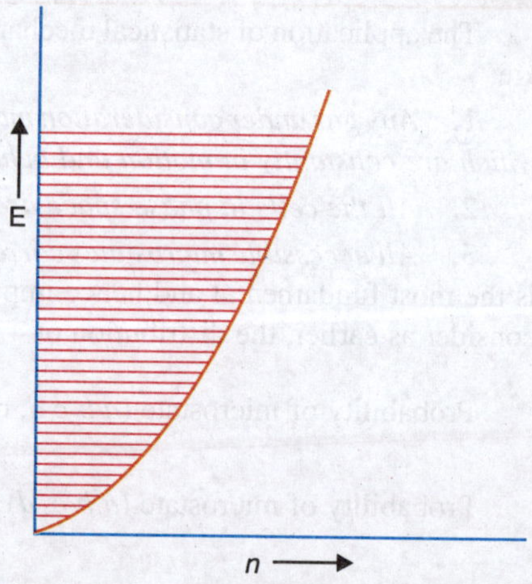


**Fig. 10.3** A parabolic curve energy verses $n$

† The maximum value of the energy difference between consecutive energy levels for a free electron in a cube of metal of side $10^{-2}$ m (1 cm) is less than $10^{-6}$ eV. Therefore, energy levels may be regarded as continuous and we can use calculus method to define the **density of energy states.**

Any small interval $dE$ of energy contains a large number of quantum states. For example, if $dE = 10^{-3}$ eV = 0.001 eV.

Min. number of energy levels contained in $dE$

$$= \frac{dE}{\text{Energy difference between two consecutive energy levels}} = \frac{10^{-3}}{10^{-6}} = 1000.$$

Further, each energy level degenerates (except the ground level) and hence, the number of quantum states are very large *i.e.* quantum states are out numbering the energy levels.

quantum states contained in $dE$. It is represented in terms of density of quantum states, denoted by $g(E)$. These are the available quantum states per unit energy range at the given energy $E$.

**Definition.** *The density of quantum states g (E) is defined as the number of quantum states per unit energy range at the given energy E.*

The quantity $g(E)\,dE$ represents the number of quantum states in the energy range between $E$ and $E + dE$ is given by

$$g(E) = \frac{2\pi}{h^3}(2m)^{\frac{3}{2}} E^{\frac{1}{2}} \qquad ...(ii)$$

Here the spin of the particle is not taken into consideration. For particles of spin $\pm\frac{1}{2}$ like electrons, there will be two allowed quantum states for each representative point corresponding to the two possible spin orientations of the particle. Taking spin in to account, the number of quantum states with energy between $E$ and $E + dE$, is given by

$$g(E)\,dE = 2\times\frac{2\pi}{h^3}(2m)^{\frac{3}{2}} E^{\frac{1}{2}}\, dE$$

or

$$g(E)\,dE = \frac{4\pi}{h^3}(2m)^{\frac{3}{2}} E^{\frac{1}{2}}\, dE \qquad ...(iii)$$

**Distribution Function**

In order to determine average properties of a system of particles, it is necessary to find the number of particles actually occupying quantum states in the energy range between $E$ and $E + dE$. This number is denoted by $n(E)\,dE$ and is expressed as

$$n(E)\,dE = g(E)\,f(E)\,dE \qquad ...(iv)$$

where $g(E)$ is the density of (available) quantum states at energy $E$, $g(E)\,dE$ is the number of quantum states (available) in the energy range between $E$ and $E + dE$ and $f(E)$ is the distribution function of the system, depending upon which statistics (M.B;, B.E. or F.D.) we use corresponding to the spin of the particle under consideration.

From equation (*iv*), the distribution function $f(E)$ is defined as the average number of particles of the system, which occupy a single quantum state of energy $E$. The unit of $n(E)$ is particle/joule.

## 10.10 Statistical Ensembles

In statistical mechanics, Gibbs introduced the concept of ensemble. It is obvious that a very large (in fact infinite) number of states of the gas corresponds to a given macroscopic condition of the gas because there are infinite number of ways to distribute the molecules in space. Through macroscopic measurements, we could not be able to distinguish between two gases existing in different states (2 distinct representative points) but satisfying the same macroscopic conditions. Thus, when we speak of a gas under certain macroscopic conditions, we are in fact referring not to a single state, but to an infinite number of states. In other words, we are referring not a single system but to a collection of systems, identical in composition and macroscopic conditions but existing in different states. According to Gibbs, such a collection of systems is called **an ensemble**, which is geometrically represented by a distribution of representative points in Γ (gamma) space.

Thus an ensemble is defined as a *collection of large number macroscopically identical but essentially* independent systems. By the term, *macroscopically identical*, we mean that each of the systems constituting an ensemble satisfies the same macroscopic conditions, *e.g.*, volume, energy, pressure, temperature, total number of particles etc. By the term independent systems, we mean that the system constituting an ensemble are mutually non-interacting, just like the role of non-interacting molecules do in a gas.

## Kinds of Ensembles

According to Gibbs, there are three standard ensembles to which real experiments could be approximated. These are :

1. The Microcanonical ensemble,
2. The Canonical ensemble, and
3. The Grand-canonical ensemble.

This classification depends on the manner in which their systems interact. It has its own characteristic distribution. Physical systems can remain perfectly insulated from each other or can interact in a variety of ways; in particular, they may exchange energy or matter or both with each other.

## Microcanonical Ensemble

The microcanonical ensemble is a collection of essentially independent systems having the ***same energy E, volume V and number of particles N***. For simplicity, we assume that all the particles are identical. The individual systems of a microcanonical ensemble are separated by rigid, impermeable and well insulated walls such that the values of $E$, $V$ and $N$ for a particular system are not affected at all by the presence of other systems (Fig. 10.4).

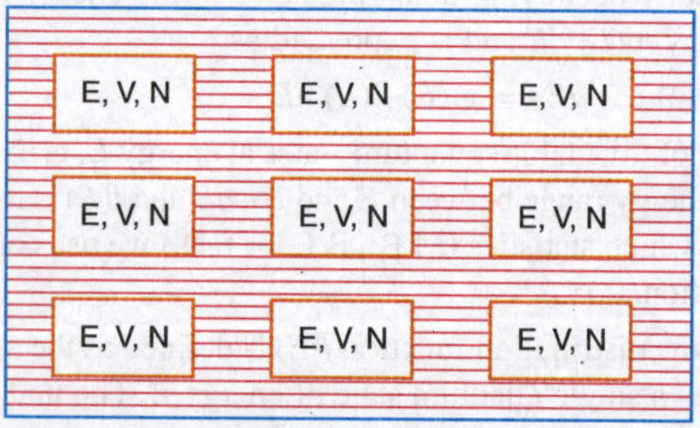


**Fig. 10.4** Microcanonical ensemble

Consider a closed system for which the total energy $H(q, p) = E$ remains constant, according to the equation

$$E(q_1, q_2, \ldots q_f, p_1, p_2, \ldots p_f) = \text{constant.}$$

The locus of all such phase points having equal energies in phase space is called an energy surface or ergodic surface.

A family of such energy surfaces can be drawn in phase space. Consider any two such energy surfaces with energies $E$ and $E + dE$, very close to each other, as shown in Fig. 10.5. Each energy surface divides the phase space into two

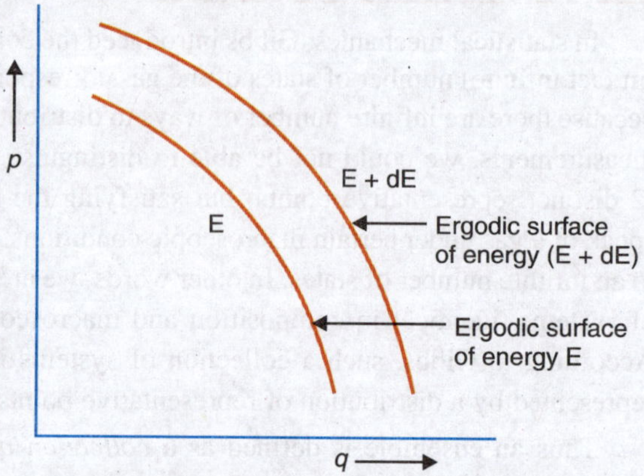


**Fig. 10.5** Energy shell in phase space

parts, one on higher and other of lower energy and hence they will never intersect with each other. As they enclose same phase volume in between them, a certain number of phase points are included. The number of phase points between them will be constant. According to Gibbs, a very useful ensemble can be obtained by taking energy density constant in a narrow selected energy ranges and varnishes outside it *i.e.*,

$$\rho = \text{constant} \qquad \text{[in the range } E \text{ and } (E + dE)]$$

and

$$\rho = 0 \qquad \text{[outside this range]}$$

### 10.10.2 Canonical Ensemble

The canonical ensemble is a collection of essentially independent systems having the **same *temperature T, volume V and number of identical particles N.*** This is achieved by bringing all the systems in thermal contact with a large heat reservoir at constant temperature *T* or by simply bringing all *of the systems in thermal contact with each other*. Fig. 10.6 represents symbolically a canonical ensemble. The individual systems of the canonical ensemble are separated by *rigid, impermeable but diathermic walls i.e., conducting walls*. As the separating walls are conducting, heat can be exchanged between the systems; till they reach a common temperature *T*. Thus, in canonical ensemble, systems can exchange energy but not the matter *i.e.*, particles.

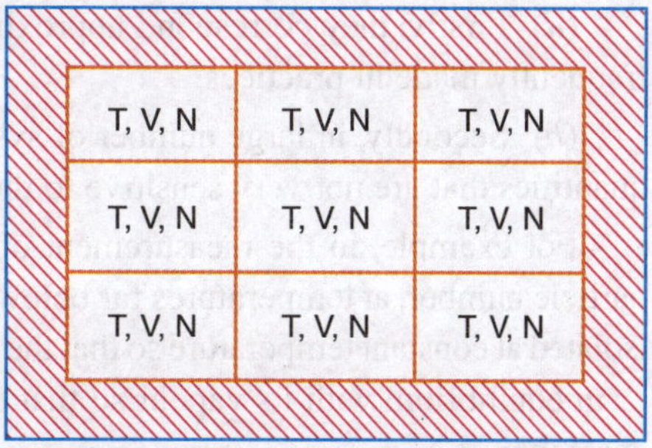


**Fig. 10.6** Canonical ensemble

### 10.10.3 Grand-Canonical Ensemble

In going from microcanonical ensemble (in which independent systems having the same energy *E*, volume *V* and number of particle *N*) to canonical ensemble (in which independent systems have the same temperature *T*, volume *V* and number of particles *N*), the condition of constant energy has been relaxed. This simplifies the calculations in thermodynamics where the exchange of energy takes place. Now the next logical step is to abandon the condition of same number of particles. Actually, in chemical process this number *N* varies in various physical problems, *e.g.*, radio active decay. Thus, the grand-canonical ensemble is a collection of large number of essentially independent system having the ***same temperature T, volume V and chemical potential μ.*** The individual systems of a grand-canonical ensemble are separated by rigid, permeable and conducting walls, as shown in Fig. 10.7. As the separating walls are conducting and permeable, the exchange of heat energy as well as that of particles between the systems take place in such a way that all the systems arrive at common temperature *T* and the chemical potential μ.

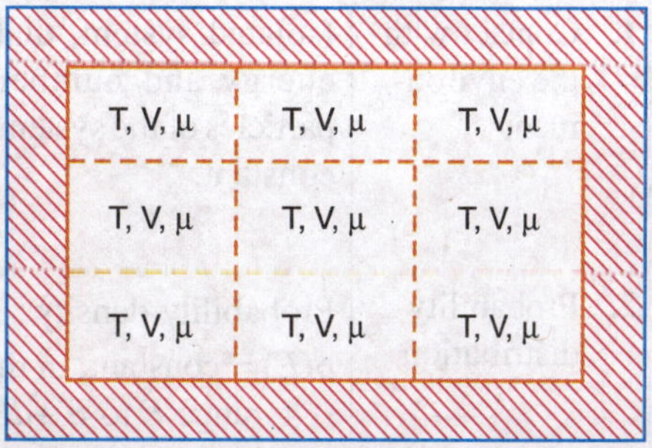


**Fig. 10.7** Grand-canonical ensemble

Thus, an ensemble in which the systems exchange energy but not matter is called a *canonical ensemble*, that in which both energy and matter are exchanged between the systems is called a *grand canonical ensemble*; and that in which neither energy nor matter is exchanged is called a *microcanonical ensemble*. A specific choice of ensembles may be thought of as corresponding to a particular physical situation. The microcanonical ensemble is fundamental and refers to the simple situation in which constituent systems are isolated and not influenced in any way by external disturbances. Therefore, it is best suited for isolated systems. However, there is always some energy exchange with the surroundings. Hence canonical ensembles have been found to be more appropriate for dealing such systems. Besides, canonical approach gives results applicable even when the constituents interact strongly. We will examine critically the Boltzmann's canonical distribution law.

**Uses of Ensemble:** In practice, we consider the three main types of ensembles and each ensemble corresponds to a different experimental situation. In fact, the three ensembles, which we have introduced and discussed above are only examples of the infinite number of ensembles that can be considered. These are particularly useful for two main reasons:

(*a*) Firstly, they correspond approximately to the types of thermodynamic measurements most frequently made in practices.

(*b*) Secondly, in large number of systems, it is useful to find that the values of thermodynamic quantities that are not very sensitive to the method of measurement.

For example, in the measurement of specific heat of a liquid of known mass and thus known particle number, at temperatures far below the boiling point, it matters very little whether the liquid is isolated at constant temperature so that the number of particles is fixed as in a microcanonical ensemble or in equilibrium with its vapour so that the number of particles can fluctuate as in grand-canonical ensemble.

## 10.11 Comparison of Ensembles

| Property | Microcanonical | Canonical | Grand-canonical |
|---|---|---|---|
| 1. Contact with the environment | Isolated system, that is, energy and number of particles of the system are constant. | Systems are in thermal contact with heat reservoir. Hence system can exchange energy but no matter *i.e.* particles | Systems can exchange both energy and matter *i.e.* particles. |
| 2. Probability distribution | Probability density $\rho(E)$ = constant, in the range $E$ to $E + dE$<br>= o, outside this range. | $\rho(E) = Ae^{E/kT}$ | $\rho(E) = e^{[(\Omega + n\mu - E)/kT]}$ |
| 3. Fluctuations | None | Fluctuations in energy only. | Fluctuations in both energy and matter *i.e.*, particles. |
| 4. Partition function z | $Z = \Delta\Gamma$ | $Z = \int e^{-\beta E}\, d\Gamma$ | $Z = \sum_{n=0}^{\infty} e^{\mu n/T} Z_n$ |

## 10.12 Equilibrium Between Two Systems in Thermal Contact

Two systems are said to be in thermal contact when they are capable of exchanging energy of thermal agitation between one another.

Consider two systems $A$ and $B$ with specified volumes $V_A$ and $V_B$ and having number of particles $n_A$ and $n_B$ respectively as shown in Fig. 10.8. Total energy $E$ is also specified. Let us find how this energy $E$ gets shared between two systems in the equilibrium state. This is obtained by deriving the condition of maximum probability under the given constraints.

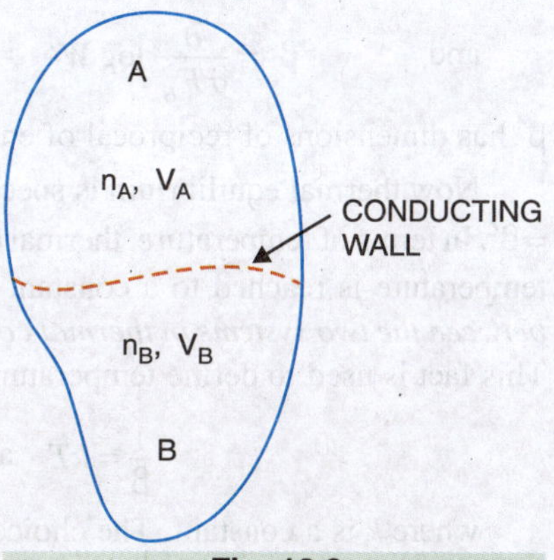

**Fig. 10.8**

Out of the total energy, let $E_A$ be the energy shared by system $A$. Applying the principle of conservation of energy, the energy shared by system $B$ will be

$$E_B = E - E_A. \qquad ...(10.13)$$

The probability that system $A$ will possess energy $E_A$ is given by

$$P(E_A) = C\,W(E_A) \qquad ...(10.14)$$

where $W(E_A)$ gives the number of microstates for a system $A$ in an energy range $\delta E$ at $E$ and $C$ is a arbitrary constant. Similarly for system $B$, we can write

$$P(E_B) = C'\,W'(E_B) \qquad ...(10.15)$$

where $C'$ is another arbitrary constant.

Now applying the multiplication rule for joint probability (Ref. 9.5) of the system, will be

$$P(E_A, E_B) = CC'\,W(E_A)\,W'(E_B) \qquad ...(10.16)$$

The two systems $A$ and $B$ are in perfect thermal contact, and hence they will exchange energy amongst them and an equilibrium state is attained for which the probability $P(E_A, E_B)$ will have a maximum value.

In order to incorporate this fact, take natural logarithms of both sides,

$$\log P(E_A, E_B) = \log CC' + \log W(E_A) + \log W'(E_B) \qquad ...(10.17)$$

Differentiating *w.r.* to $E_A$, and equating to zero,

$$0 = \frac{\partial}{\partial E_A}\log W(E_A) + \frac{\partial}{\partial E_B}\log W'(E_B)\,.\,\frac{\partial E_B}{\partial E_A} \qquad ...(10.18)$$

But $\dfrac{\partial E_B}{\partial E_A} = -1.$ $\qquad$ ($\because$ Eqn. 10.13)

This means, whatever energy is gained by system $A$ is equal and opposite ( –ve sign) to the energy gained by system $B$. This can be seen from equation 10.13.

Hence, Equation 10.18 leads to

$$\frac{\partial}{\partial E_A}\log W(E_A) = \frac{\partial}{\partial E_B}\log W'(E_B) \qquad ...(10.19)$$

Thus, according to statistical mechanics, the two systems in thermal contact will be in equilibrium if their functions $\left[\dfrac{\partial}{\partial E}\log W\right]$ are equal.

This is true for all systems, however complex they may be.

## 10.13 Bridge with Macroscopic Physics

The functions $\dfrac{\partial}{\partial E}\log W$ is generally represented by parameter $\beta$. Using this, we have

$$\beta = \frac{\partial}{\partial E_A}\log W = \frac{1}{W}\frac{\partial W}{\partial E_A} \qquad ...(10.20)$$

and $$\beta' = \frac{\partial}{\partial E_B} \log W' = \frac{1}{W'} \frac{\partial W'}{\partial E_B} \qquad \text{...(10.21)}$$

$\beta'$ has dimensions of reciprocal of energy.

Now thermal equilibrium is specified by the condition given by equation (10.19). This means $\beta = \beta'$. In terms of temperature, thermal equilibrium is specified by $T = T'$ *i.e.* the macroscopic parameter temperature is reached to a constant and maximum value. In other words, *equality of temperature between the two systems in thermal contact means the equality of the β parameters of the two systems.* This fact is used to define temperature, a macroscopic parameter as

$$\frac{1}{\beta} = kT \quad \text{and} \quad \frac{1}{\beta'} = kT' \qquad \text{...(10.22)}$$

where $k$ is a constant. The choice of $k$ will depend upon the scale we select. The dimensions of this constant $k$ are erg, deg$^{-1}$. When this scale matches with the perfect gas scale temperature then this $k$ must be set equal to the Boltzmann's constant which is defined by $R/N_o$ where $R$ is the universal gas constant and $N_o$ is Avogadro's number.

## 10.14 Theories Based on Statistical Mechanics

The entropy and the distribution of molecular speeds, these two theories in physics are based on the statistical concepts. The thermodynamic property entropy ($S$) is explained on the basis of statistical mechanics. According to statistical considerations equilibrium state of an isolated system is the macrostate of maximum thermodynamic probability. Hence in equilibrium state, both entropy and the thermodynamic probability have the maximum value. On the other hand, with the help of Maxwell Boltzmann statistics, the number of molecules per unit volume a gas having speeds between $v$ and $(v + dv)$ has been calculated and hence the value of most probable speed.

## 10.15 Entropy and Probability

We know that when an isolated system (*i.e.* a system which does not interact with its surroundings so that its energy is constant) undergoes an irreversible process, there is a net increase of the entropy of the system. Spontaneous processes, such as spontaneous expansion of a gas into an evacuated space, diffusion of one gas into another, etc., are all irreversible processes. Hence, in all irreversible processes, there is a net increase of entropy, and an increase of randomness.

In general, all spontaneous processes represent changes from a less probable to a more probable state, and since in such processes the entropy increases, there should be relation between the entropy $S$ of the system in a given state and the probability of that state.

Boltzmann, in 1896 derived a relation between entropy (a thermodynamical quantity) and probability (a statistical quantity). Boltzmann started thinking from a simple idea in mind that the equilibrium state of any system is the state of its maximum probability, *i.e.*, the probability of the system in equilibrium state is maximum. But from the thermodynamical point of view, the equilibrium state of a system is the state of its maximum entropy. If the system is not in equilibrium, then changes take place within the system until the equilibrium state is reached and the system attains a maximum entropy. Thus, in the equilibrium state, both the entropy and the thermodynamical probability have their maximum value. This observation inspired Boltzmann to think that there must be corelation between these two quantities. The relation between them is known as Boltzmann's entropy relation.

### Boltzmann's Entropy Relation

According to thermodynamics, the function entropy $S$ of a system is related with temperature $T$ by the relation

$$\frac{1}{T} = \frac{\partial S}{\partial E} \qquad ...(10.23)$$

According to the approach of statistical mechanics, using equations (10.20), (10.21) and (10.22), we write

$$\frac{1}{T} = k\beta = k\frac{\partial}{\partial E}\log W \qquad ...(10.24)$$

Equating equations (10.23) and (10.24), we have

$$\frac{\partial S}{\partial E} = k\frac{\partial}{\partial E}\log W$$

Integrating, we get

$$S = k\log W \qquad ...(10.25)$$

**Aliter.** Boltzmann assumed that entropy $S$ of a physical system in a definite state is a function of the maximum probability $W$ of that state. Thus

$$S = f(W) \qquad ...(i)$$

where $S$ is entropy and $W$ is the thermodynamical probability of the state.

Let us consider two completely independent systems $A$ and $B$ having entropies $S_1$ and $S_2$ respectively. Since entropy is an extensive (*i.e.* additive) quantity, the entropy of the two systems together will be

$$S = S_1 + S_2 \qquad ...(ii)$$

If $W_1$ is the thermodynamic probability of system $A$ and $W_2$ that of system $B$, then the probability $W$ will be the product of their probabilities, *i.e.*,

$$W = W_1 \times W_2 \qquad ...(iii)$$

Thus,

$$S = f(W) = f(W_1 \times W_2) \qquad ...(iv)$$

$$S_1 = f(W_1) \qquad ...(v)$$

$$S_2 = f(W_2) \qquad ...(vi)$$

Substituting (*iv*), (*v*) and (*vi*) in eqn. (*ii*), we get

$$f(W_1 \times W_2) = f(W_1) + f(W_2) \qquad ...(vii)$$

Differentiating partially w.r. to $W_1$ and $W_2$, we get

$$W_2 f'(W_1 \times W_2) = f'(W_1) \qquad ...(viii)$$

$$W_1 f'(W_1 \times W_2) = f'(W_2) \qquad ...(ix)$$

Dividing eqn. (*viii*) by eqn. (*ix*), we get

$$\frac{f'(W_1)}{f'(W_2)} = \frac{W_2}{W_1}$$

or,

$$W_1 f'(W_1) = W_2 f'(W_2) = ...... = k$$

$k$ being any constant.

Thus, $$f'(W_1) = \frac{k}{W_1}, \quad f'(W_2) = \frac{k}{W_2} \text{ and so on.}$$

Integrating, we get

$$f(W_1) = k\log(W_1) + C_1$$

and

$$f(W_2) = k\log(W_2) + C_2$$

In general,

$$f(W) = k\log(W) + C$$

where $C$ is a constant of integration.

Thus, $$S = k \log (W) + C \qquad ...(\because \text{ Eqn. } 1)$$

The constant of integration C is evaluated by putting the condition that at absolute zero ($T = ok$), the entropy $S = O$, and $W = 1$.

$$\therefore \qquad C = O$$

Putting in equation ($x$), we get

$$S = k \log (W) \qquad ...(10.26)$$

This is the required relation between entropy and thermodynamic probability and is called the *Boltzmann's entropy relation*. Here $k$ is universal constant known as Boltzmann's constant. The constant of integration is treated as zero because we are interested in changes in entropy and not in its absolute values. Thus, *the Boltzmann's entropy relation states that the entropy of a system is proportional to the logarithm of thermodynamic probability of that system.*

The law that the entropy of the universe always tends to increase (5.5) now means that $W$ always tends to increase, which in turn means simply that the universe always changes towards statistically more probable states.

## 10.16. Boltzmann's Canonical Distribution Law

In equilibrium state, let $n_1, n_2, n_3, ......\, n_i$ be the number of gas molecules in different cells, say in cell no. 1, 2, 3, ..... $i$ respectively. As the gas molecules are continuously moving, $n_i$'s will change continuously in many different ways, but will always keep value very close to that for the state of equilibrium, *i.e.*, the state of maximum probability. At any instant, the gas molecules obey the following three fundamental postulates of statistical mechanics. Therse are :

**I.** The total number of molecules in constant, *i.e.*,

$$N = n_1 + n_2 + n_3 + .................. + n_i + ................. = \text{constant}$$

or $$\delta N = \delta n_1 + \delta n_2 + \delta n_2 + .................. + \delta n_i + .................. = 0 \qquad ... (10.26\text{-}a)$$

**II.** The total energy of the system is constant, *i.e.*

$$E = E_1\, n_1 + E_2 n_2 + E_3\, n_3 + .................. + E_i n_i + ................ = \text{constant}$$

$$\delta E = E_1\, \delta n_1 + E_2 dn_2 + E_3\, dn_3 + ................. + E_i \delta n_i + ................... = 0 \qquad ... (10.26\text{-}b)$$

**III.** In equilibriunm state, the thermodynamical probability ($W$) is maximum, *i.e.*.,

$$\delta W = 0 \qquad ... (10.26\text{-}c)$$

We know that $\log_e W$ is more convenient than $W$ for mathematical treatment. When $W$ is maximum, $\log W$ is also maximum. Therefore

$$\delta (\log W) = 0 \qquad ...(10.27)$$

The thermodynamical probability $W$ is expressed as

$$W = \frac{N!}{n_1!\, n_2!\, n_3! ........ n_i! ........} \qquad ...(10.28)$$

Taking logarithm of both sides, we get

$$\log W = \log N! - \log n_1! - \log n_2! - \log n_3! \,....\, - \log n_i! \,.... \qquad ...(10.29)$$

As number of molecules ($n_i$'s) are large in number, we can apply Stirling's theorem*. Applying we get

$$\log W = N \log N - N - (n_1 \log n_1 - n_1) - (n_2 \log n_2 - n_2) \,...\, - (n_i \log n_i - n_i) \,...$$

$$= N \log N - N - \sum_i (n_i \log n_i - n_i) \qquad ...(10.30)$$

Differentiating equation (10.30) with respect to $n_1, n_2, ........\, n_i$ ........ and keeping in mind that $N$

* When $x$ large, according to Sterling's theorem $\log_e x! = [x \log_e x] - x$

is constant, we have

$$\delta(\log W) = -\sum_i [n_i(\delta \log n_i) + \log n_i\ \delta n_i - \delta n_i] \quad ...(10.31)$$

or
$$\sum_i \left[ n_i \left(\frac{1}{n_i}\right) \delta n_i + \log n_i\ \delta n_i - \delta n_i \right] = 0 \qquad (\because 10.27) \quad ...(10.32)$$

or
$$\sum_i \log n_i\ \delta n_i = 0$$

or
$$\log n_1\ \delta n_1 + \log n_2\ \delta n_2 + \log n_3\ \delta n_3 + ......... + \log n_i\ \delta n_i + ......... = 0 \quad ...(10.33)$$

It should be noted that the equation (10.33), (10.26-*a*) and (10.26-*b*) are independent of one another and they must be satisfied at any instant. To maintain their independence, let us multiply each by a separate arbitrary constants $\lambda_1$, $\lambda_2$ and $\lambda_3$ respectively and adding them, we get

$$\lambda_1 \{\log n_1\ \delta n_1 + \log n_2\ \delta n_2 + \log n_3\ \delta n_3 + ...... + \log n_i\ \delta n_i + ......\} + \lambda_2 \{\delta n_1 + \delta n_2 + \delta n_3 + ...... + \delta n_i + ......\} + \lambda_3 \{E_1\ \delta n_1 + E_2\ \delta n_2 + E_3\ \delta n_3 + ...... + E_i\ \delta n_i + ......\} = 0$$

Dividing by $\lambda_1$, we get

$$\{\log n_1\ \delta n_1 + \log n_2\ \delta n_2 + ...... + \log n_i\ \delta n_i + ......\} + \frac{\lambda_2}{\lambda_1}\{\delta n_1 + \delta n_2 + ...... + \delta n_i + ......\} + \frac{\lambda_3}{\lambda_1}\{E_1\ \delta n_1 + E_2\ \delta n_2 + ...... + E_i\ \delta n_i + ......\} = 0 \quad ...(10.34)$$

Let $\frac{\lambda_2}{\lambda_1} = \lambda$ and $\frac{\lambda_3}{\lambda_1} = \beta$ where $\lambda$ and $\beta$ are new constants. Substituting in equation (10.34), we have

$$\{\log n_1\ \delta n_1 + \log n_2\ \delta n_2 + ...... + \log n_i\ \delta n_i + ......\} \times \lambda \{\delta n_1 + \delta n_2 + ...... + \delta n_i + ......\} + \beta \{E_1\ \delta n_1 + E_2\ \delta n_2 + ...... + E_i\ \delta n_i + ......\} - 0$$

Rearranging, we get

$$(\log n_1 + \lambda + \beta E_1)\ \delta n_1 + (\log n_2 + \lambda + \beta E_2)\ \delta n_2 + ...... + (\log n_i + \lambda + \beta E_i)\ \delta n_i + ...... = 0 \quad ...(10.35)$$

In this equation, all $\delta n_i$'s are all independent of each other. Therefore, if equation (10.35) is to be satisfied, the coefficient of each term must vanish. Thus

$$\log n_i + \lambda + \beta E_i = 0 \quad \text{for all values of } i. \quad ...(10.36)$$

$\therefore \quad \log n_i = -\lambda - \beta E_i$

or $\quad n_i = e^{-(\lambda + \beta E_i)} = e^{-\lambda} . e^{-\beta E_i}$

or $\quad n_i = A\,e^{-\beta E_i} \quad ...(10.37)$

where $\quad A = e^{-\lambda}$, a new constant.

Equation (10.37) gives the number of molecules in each cell as a function of energy associated with each particle in that cell. This is known as ***Boltzmann's canonical distribution law***.

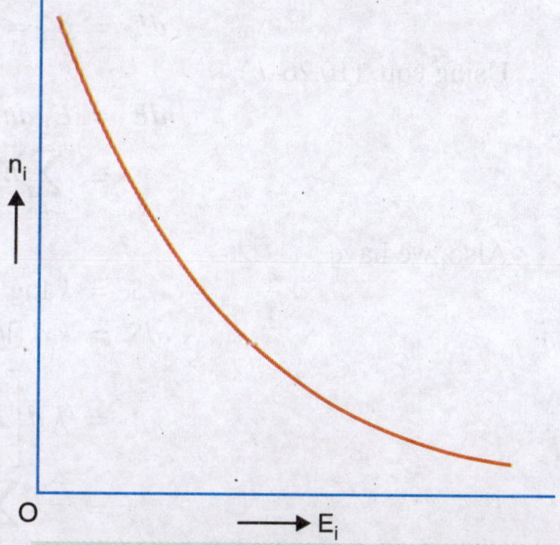

Fig. 10.9

Using equation (10.37), if a graph is plotted between the number of molecules per cell $n_i$ and energy $E_i$ associated with the cell, an exponential decrease nature is observed as shown in (Fig. 10.9). From the graph, we conclude that low energy cells will contain more particles than high energy cells, the number of particles being maximum in zero-energy cell. It does not mean that the number of molecules having zero-energy is greater than any other energy because there are more cells corresponding to high energy.

It is also convenient sometimes to consider a group of adjacent cells in the phase space. Such a group of cells is called a **zone of cells**. As all the cells are of equal size, the size of zone is proportional to the number of cells in it. If $g_i$ is the number of cells in a zone, then the probability of finding a molecule in the zone will be proportional to $g_i$. This $g_i$ is called then a **priori probability** for the zone which solely depends on the size of the zone.

If we consider a zone of cells, all corresponding to about the same energy $E_i$, then the number of molecules in this zone will be

$$n_i = A\, g_i\, e^{-\beta E_i} \qquad ...(10.38)$$

### Evaluation of Constant A

Since the sum of all $n_i$'s must be equal to the total number of molecules, say $N$, then

$$\sum_i n_i = N = A\,\Sigma e^{-\beta E_i} \qquad \text{(From Eqn. 10.37)} \qquad ...(10.39)$$

The summation term $\Sigma e^{-\beta E_i}$ is called the **partition function or sum of states** and is represented by a symbol $Z$, *i.e.*,

Partition function, $$Z = \Sigma e^{-\beta E_i} \qquad ...(10.40)$$

Substituting in equation 10.39, we get

$$N = AZ$$

or $$A = \frac{N}{Z} \qquad ...(10.41)$$

Substituting this value of $A$ in equation 10.37, the number of molecules in the $i^{th}$ cell in its state of maximum thermodynamic probability will be

$$n_i = \frac{N}{Z} e^{-\beta E_i} \qquad ...(10.42)$$

### Evaluation of Constant β

Suppose a small amount of heat $dQ$ is supplied to a gas at constant volume so that some of the molecules move from cell of lower energy to cells of higher energy. Let $dn_1$, $dn_2$, $dn_3$, ... etc. be the change in the number of molecules in cell 1, cell 2, cell 3, ... etc., respectively.

According to thermodynamics, the increase in internal energy at constant volume is given by

$$dE = \delta Q = TdS \qquad ...(10.43)$$

Using equ. (10.26–*b*),

$$dE = E_1\, dn_1 + E_2\, dn_2 + E_3\, dn_3 + ... + E_i\, dn_i + ...$$

$$= \sum_i E_i\, dn_i \qquad ...(10.44)$$

Also, we have

$$S = k \log W$$

or $$dS = k\, d\,(\log W)$$

$$= k\,d\left[N \log N - \sum_i \{n_i \log n_i - n_i\}\right] \qquad \text{[using eqn. 10.31]}$$

$$= -k \sum_i \log n_i\, dn_i \qquad (\because N \text{ is a constant}) \qquad ...(10.45)$$

Since we are assuming that the system is in thermal equilibrium, we may use $n_i = Ae^{-\beta E_i}$, so that we have

$$dS = -k\sum_i [\log (Ae^{-\beta E_i})\, dn_i]$$

$$= -k\sum_i [\log A - \beta E_i)\, dn_i]$$

or

$$dS = -k \log A\, \Sigma dn_i + k\beta\, \Sigma E_i\, dn_i \qquad ...(10.46)$$

But $\sum_i dn_i = 0$, since $\sum_i n_i = N =$ constant.

and

$$dE = \sum_i E_i\, dn_i \qquad (\because \text{Eq. } 10.44)$$

Therefore, eqn. (10.46) becomes

$$dS = k\beta\, dE \qquad ...(10.47)$$

Comparing (10.43) and (10.47), we have

$$k\beta = \frac{1}{T}$$

or

$$\beta = \frac{1}{kT} \qquad ...(10.48)$$

Here $k$ is a Boltzmann constant and $T$ is absolute temperature.

Therefore, the number of molecules in a cell having energy $E_i$ is given by

$$n_i = A\, e^{-E_i/kT}$$

Hence, the probability that a molecule will possess energy $E_i$ will be

$$P_{(E_i)} = \frac{n_i}{N} = \frac{Ae^{-E_i/kT}}{N}$$

where $N$ is the total number of molecules.

$$\therefore \qquad P_{(E_i)} = Ce^{-E_i/kT} \qquad ...(10.49)$$

where $C = \frac{A}{N}$ and can be evaluated by using the fact that the probability of a molecule having energy $E_i$ is definite, i.e.,

$$\sum P_{(E_i)} = 1$$

$\therefore$ Equation (10.41) gives

$$\sum P_{(E_i)} = C\sum e^{-E_i/kT}$$

or

$$1 = C\sum e^{-E_i/kT}$$

or

$$C = \frac{1}{\sum e^{-E_i/kT}}$$

Substituting for $C$ in equation (10.49), the probability $P_{(E_i)}$ is given by

$$P_{(E_i)} = \frac{e^{-E_i/kT}}{\sum e^{-E_i/kT}} \qquad ...(10.50)$$

## 10.17 Application of Boltzmann's Canonical Distribution Law

The most important application of Boltzmann's distribution law expressed in the form given by eqn. (10.50) is to evaluate the probability that a given molecule assembly of $N$ molecules in thermal equilibrium that exist in a state of energy $E_i$. The molecule under consideration is referred as the ***'system'*** and remaining ($N$-1) molecules as the ***'reservoir'*** of temperature $T$. The equation (10.50)

$$P_{(E_i)} = \frac{e^{-E_i/kT}}{\sum e^{-E_i/kT}}$$

is then applied to each one of the N molecules in turn by this procedure.

The Boltzmann's canonical distribution $P_i \alpha \exp(-E_i/kT)$ plays a very crucial role while deriving the theorem of equilibration of energy (article 10.18).

If the system deviates and does not follow the equipartition theorem for particular energy terms which are known to be quadratic type, there is a 'break-down of classical-mechanics.' The assumption of parameters as continuously variable need to be modified. Thus, when the quantum steps (of energy) are much greater than $kT$, the equipatition theorem suffers total break-down. This is the field of 'quantum mechanics'.

## 10.18 The Equipartition of Energy

The theorem of equipartition of energy is usually stated as:

***"For every degree of freedom in a system in thermal equilibrium at temperature T, the average energy of the system is $\frac{1}{2}kT$, where k is Boltzmann's constant."***

For simplicity, let us consider a system of $N$ monotonic molecules. To specify the microstate of the system uniquely, we need 3N position coordinates and 3$N$ momentum coordinates. Thus, in phase space representation, in all we need 6$N$ coordinates. However, the energy $E$ is determined by 3$N$ momentum coordinates only. Thus

$$E = \sum_{i=1}^{3N} \frac{p_i^2}{2m} \qquad ...(10.51)$$

where $p_i$ is the $i^{th}$ momentum coordinate. The number of such coordinate is 3$N$, hence the number of degrees of freedom = 3$N$

$$\therefore \qquad \text{Total energy} = 3N \times \frac{1}{2} kT \qquad ...(10.52)$$

This is kinetic energy due to translational motion.

A system consisting of $N$ harmonic oscillators about mean positions (in case of solid), we need 3$N$ displacement coordinates and 3$N$ momentum coordinates to specify the microstate uniquely. The total energy is comprised of potential energy and kinetic energy. Thus

$$\text{Total Energy} = E = \text{P.E.} + \text{k.E.}$$

$$= \sum_{i=1}^{N} \frac{p_i^2}{2m} + \sum_{i=1}^{N} \frac{1}{2} c_i\, q_i^2 \qquad ...(10.53)$$

where $c_i$ is the $i^{th}$ force constant and $q_i$ is the $i^{th}$ displacement coordinate. Thus, there are (6$N$) squared terms in the energy expression. The number of degrees of freedom is now to be considered as 6$N$, so that

$$\therefore \qquad \text{Total energy} = 6N \times \frac{1}{2} kT\,. \qquad ...(10.54)$$

In case of Van der Waals gas, where interaction energy terms are also present, is much more complex. Here the total energy is given by

$$E = \sum_{i=1}^{N} \frac{p_i^2}{2m} + \sum_a \sum_b \phi(r_{ab}) \qquad \text{...(10.55)}$$

where $r_{ab}$ is the distance between the $a^{th}$ and $b^{th}$ molecule and $\phi(r_{ab})$ is a function of $r_{ab}$, which is not of the quadratic form. The number of these interaction terms is $\frac{1}{2} N(N-1)$. The average energy corresponding to these in **not** given by $\frac{1}{2} kT^{*}$. In this case the degrees of freedom is 3N but the total energy of the system is **not** given by $3N \times \frac{1}{2} kT$. Therefore, the theorem of equipartition of energy stated above is modified and restated as:

***'For every independent quadratic term appearing in the energy expression of a system, the average energy corresponding to temperature T of the system is $\frac{1}{2} kT$.'***

Again let us consider a monatomic gas with $N$ molecules forming a system. We need $f$ position coordinates and $f$ momentum coordinates to specify the system uniquely. The total energy can be expressed as

$$E = E(q_1, q_2, \ldots\ldots\ldots q_f, p_1, p_2, \ldots\ldots\ldots p_f) \qquad \text{...(10.56)}$$

Now imagine that the energy expression is so arranged that there is a term $E_i(p_i)$ which depends on the coordinate $p_i$ alone and this term does not appear in any other energy term. Then we may write

$$E = E_i(p_i) + E'(q_1, q_2, \ldots\ldots q_f, p_1, p_2, \ldots\ldots p_{i-1}, p_{i+1} \ldots\ldots p_f) \qquad \text{...(10.57)}$$

where $E'$ is not a function of $p_i$. The mean value of $E_i$ *i.e.*, $\overline{E_i}$ may now be obtained by using formula (10.50) as

$$\overline{E_i} = \frac{\sum E_i e^{-\beta E}}{\sum e^{-\beta E}} \qquad \text{where } \beta = \frac{1}{kT}$$

and the summation extends to all the microstates which pertains to total energy $E$.

If the $q$ and $p$ coordinates are treated as continuously variable, the summation may be changed into integration as

$$\overline{E_i} = \frac{\iint \ldots\ldots \int E_i - e^{-\beta(E_i + E')} dq_1\, dq_2 \ldots\ldots dq_f . dp_1\, dp_2 \ldots\ldots dp_f}{\iint \ldots\ldots \int e^{-\beta(E_i + E')} dq_1\, dq_2 \ldots\ldots dq_f . dp_1\, dp_2 \ldots\ldots dp_f}$$

since $E_i$ is a function $p_i$ only and $E'$ is not a function of $p_i$, we get

$$\overline{E_i} = \frac{\left[\int E_i e^{-\beta E_i} dp_i\right]}{\left[\int e^{-\beta E_i} dp_i\right]} \times \frac{\left[\iint \ldots\ldots \int e^{-\beta E'} dq_1\, dq_2 \ldots\ldots dq_f\, dp_1 \ldots\ldots dp_{i-1} dp_{i+1} \ldots\ldots dp_f\right]}{\left[\iint \ldots\ldots \int e^{-\beta E'} dq_1\, dq_2 \ldots\ldots dq_f\, dp_1 \ldots\ldots dp_{i-1}\, dp_{i+1} \ldots\ldots dp_f\right]}$$

$$= \frac{\int E_i\, e^{-\beta E_i} dp_i}{\int e^{-\beta E_i} dp_i} \qquad \text{...(10.58)}$$

Thus, all variables other than $p_i$ become irrelevant in deducing $\overline{E_i}$.

Now we introduce the statement that $E_i$ is a quadratic function of $p_i$. If we put

$$E_i = bp_i^2$$

then using standard integrals, we have

$$\overline{E_i} = \frac{\int b\, p_i^2\, e^{-b\beta p_i^2}\, dp_i}{\int e^{-b\beta p_i^2}\, dp_i} = b \cdot \frac{1}{2b\beta}$$

or $$\overline{E_i} = \frac{1}{2\beta} = \frac{1}{2} kT \qquad \left[ Since\ \beta = \frac{1}{kT} \right] \qquad ...(10.59)$$

Here the constant $b$ also does not appear in $E_i$. Equation (10.59) represents the equipartition theorem. Thus, the mean energy corresponding to each quadratic term in the energy expression depends on temperature of the system alone and is equal to $\frac{1}{2} kT$.

The meritorious point to be noted here is that we have considered the position and momentum coordinates as continuously variable. That is the fundamental basis of all classical mechanics. Moreover, we have not laid down any other specification about the system. Therefore, the theorem has a general validity, *viz.* whether the particles have mass large or small, the harmonic oscillator experiences force constant large or small so long as we consider quadratic energy terms and continuously variable coordinates. Again it should be noted that the Boltzmann's canonical distribution $p_i \propto \exp\left(-\frac{E_i}{kT}\right)$ has a crucial role in deriving the above for formula.

## 10.19 Statistical Interpretation of Second Law of Thermodynamics

The second law of thermodynamics is a generalization of experiences and observations and is concerned with the direction in which energy transfer takes place. In practice, there is no engine which can convert heat from the single source in to useful work without rejecting some heat to a heat sink at a lower temperature. In other words, it is impossible to have perpetual motion of the second kind. This is the basis of the second law of thermodynamics.

In statistical mechanics, the basis of the second law is our experience that natural phenomenon must proceed from states of less probability to larger probability. Here, it is a general rule that more ordered states have smaller probability. As a result, we may enunciate that natural processes proceed from more ordered states to less ordered states. This is the statistical interpretation of second law of thermodynamics.

## 10.20 Partition Function and its Relation with Thermodynamic Quantities:

### (*a*) Partition Function (*Z*)

Let us consider an assembly of ideal gas molecules obeying classical statistics i.e., Maxwell–Boltzmann distribution law.

Using this distribution law, let $n_i$ molecules occupy $i^{th}$ state with energy between $E_i$ and $E_i + dE_i$ and degeneracy $g_i$, then

$$n_i = g_i\, e^{-\alpha}\, e^{\frac{-E_i}{kT}}$$
$$= g_i\, A\, e^{-\beta E_i} \qquad ...(i)$$

where $A = e^{-\alpha}$ and $\beta = \dfrac{1}{kT}$

So the total number of molecules in the system.

$$N = \sum_i n_i$$

$$= A \sum_i g_i\, e^{-\beta E_i}$$

$$\therefore \quad \frac{N}{A} = \sum_i g_i\, e^{-\beta E_i} = Z \qquad ...(ii)$$

where $Z$ is called the Boltzmann's partition function or simply the partition function. The term $\sum_i g_i\, e^{-\beta E_i}$ represents the sum of all the $g_i\, e^{-\beta E_i}$ terms for every energy state of the given molecule. Consequently, the *quantity Z indicates that how the gas molecules of a system are distributed* or *partitioned among the various energy levels and hence is called a partition function.*

Thus,

$$\text{Partition function, } Z = \sum_i g_i\, e^{-\beta E_i} \qquad ...(10.60)$$

Taking $\sum_i g_i = 1$, we may write $Z = \sum_i e^{-\beta E_i}$

### (*b*) Correlation with Thermodynamic Quantities

**(*i*) With Entropy, S**

According to Boltzmann's entropy relation between entropy and the maximum thermodynamic probability $W$ is given by $S = k \log W$ ...(*i*)

For a classical system, the total number of molecules are

$$N = \sum_i n_i$$

We have

$$W = N! \prod_i \frac{g_i^{n_i}}{n_i!} \qquad ...(ii)$$

Taking logarithm, we have

$$\log W = \log N! \sum_i (n_i \log g_i - \log n_i!)$$

Using Sterling approximation

$$\log n! = n \log n - n$$

we get

$$\log W = N \log N - N + \sum_i (n_i \log g_i - n_i \log n_i + n_i) \qquad ...(iii)$$

According to Maxwell-Boltzmann distribution law, for the most probable configuration, we have

$$n_i = g_i\, e^{-\alpha} \,.\, e^{-\beta E_i}$$

$$= g_i A\, e^{-\beta E_i}$$

Substituting for $n_i$ in equation (*iii*),

$$\log W = N\log N - N + \sum_i [n_i \log g_i - n_i \log (g_i\, A e^{-\beta E_i}) + n_i].$$

$$= N\log N - N + \sum_i n_i \log g_i - \sum_i n_i \log g_i - \sum_i n_i \log A + \sum_i n_i\, \beta E_i + \sum_i n_i.$$

Putting $\sum_i n_i = N$, the total number of molecules, and

$$\sum_i n_i\, E_i = E, \text{ the total energy of the gas molecules of the system.}$$

we get

$$\log W = N\log N - N - \sum_i n_i \log A + \beta E + N$$

$$= N\log N - \sum_i n_i \log A + \beta A$$

$$= N\log N - N\log A + \beta E$$

$$= N\log \frac{N}{A} + \beta E$$

$$= N\log Z + \beta E \qquad ...(iv)$$

because, $\frac{N}{A} = Z$, the partition function

Substituting eqn (*iv*) in eqn (*i*), we get

$$S = k\log W$$

$$= \text{k}\, N\log \text{Z} + k\beta E$$

$$= k\, N\log \text{Z} + k\,\frac{1}{kT}.E \qquad ...\left(\because \beta = \frac{1}{kT}\right)$$

$$= k\, N\log \text{Z} + \frac{E}{T} \qquad ...(v)$$

But for an ideal gas,

$$E = \frac{3}{2}\, NkT \qquad ...(vi)$$

Substituting (*vi*) in (*v*), we get

$$S = k\, N\log \text{Z} + \frac{3}{2}\, Nk \qquad ...(10.61)$$

This equation gives the entropy of the system of ideal gas molecules.

### (*ii*) With Helmholtz free Energy, *F*

Helmholtz free energy *F* is given by the relation

$$F = E - TS$$

$$= E - T\,(k\, N\log Z + E/T) \qquad (\because \text{Eqn } v)$$

$$= \text{E} - NkT\log Z - E$$

$$= -\,NkT\log Z \qquad ...(10.62)$$

### (*iii*) With total Energy, *E*

Average energy of a system of *N* particles is given by

$$\overline{E} = \frac{E}{N} = \frac{\sum_i n_i E_i}{\sum_i n_i}$$

$$= \frac{\sum_i g_i A e^{-\beta E_i}}{\sum_i g_i A e^{-\beta E_i}} \qquad [\because \text{Eqn. } (a)\ (i)]$$

$$= \frac{\sum_i g_i E_i e^{-\beta E_i}}{Z} \qquad [\because \text{Eqn. } (a)\ (ii)] \qquad ...(10.63)$$

Since,

$$Z = \sum_i g_i e^{-\beta E_i}$$

$$= \sum_i g_i e^{-E_i/kT}$$

For isothermal isochoric transformation

$$\left(\frac{\partial Z}{\partial T}\right)_V = \sum_i g_i \times \frac{E_i}{kT^2} \times e^{-E_i/kT}$$

$$= \frac{1}{kT^2} \times \sum_i g_i e^{-E_i/kT} \times E_i$$

or

$$kT^2 \left(\frac{\partial Z}{\partial T}\right)_V = \sum_i g_i E_i e^{-E_i/kT}$$

$$= Z\overline{E} \qquad (\because \text{Eqn. } 10.63)$$

or

$$\overline{E} = \frac{kT^2}{Z}\left(\frac{\partial Z}{\partial T}\right)_V$$

Multiplying by $N$, we get

$$N\overline{E} = \frac{NkT^2}{Z}\left(\frac{\partial Z}{\partial T}\right)_V$$

Thus, the total energy, $E$ is given by

$$E = N\overline{E} = \frac{NkT^2}{Z}\left(\frac{\partial Z}{\partial T}\right)_V$$

$$\therefore \qquad E = NkT^2\left[\frac{\partial}{\partial T}(\log Z)\right]_V \qquad ...(10.64)$$

### (*iv*) With Enthalpy, *H*

Enthalpy is given by

$$H = E + PV = E + RT \qquad (\text{For ideal gas } PV = RT)$$

$$= NkT^2\left[\frac{\partial}{\partial T}(\log Z)\right]_V + RT \qquad ...(10.65)$$

### (v) With Gibbs Potential, G

Gibbs potential is given by

$$G = H - TS$$

$$= NkT^2\left[\frac{\partial}{\partial T}(\log Z)\right]_V + RT - T\left(kN\log Z + \frac{E}{T}\right)$$

(∵ Using eq. v and 10.65)

$$= NkT^2\left[\frac{\partial}{\partial T}(\log Z)\right]_V + RT - NkT\log Z - E$$

$$= NkT^2\left[\frac{\partial}{\partial T}(\log Z)\right]_V + RT - NkT\log Z - NkT^2\left[\frac{\partial}{\partial T}(\log Z)\right]_V$$

(∵ Eqn. 10.64)

$$= RT - NkT\log Z \qquad \text{...(10.66)}$$

### (vi) With Pressure, P, of the Gas

Pressure is given by $P = -\left(\frac{\partial F}{\partial V}\right)_V$

$$= NkT\left[\frac{\partial}{\partial V}(\log Z)\right]_V \qquad \text{...(10.67)}$$

### (vii) With specific Heat at Constant Volume, $C_v$

$$C_V = \left(\frac{\partial E}{\partial T}\right)_V$$

$$= \frac{\partial}{\partial T}\left[NkT^2\left\{\frac{\partial}{\partial T}(\log Z)\right\}_V\right]$$

$$= Nk\left[2T\frac{\partial}{\partial T}(\log Z) + T^2\frac{\partial^2}{\partial T^2}(\log Z)\right]_V \qquad \text{...(10.68)}$$

## 10.21 Entropy of an Ideal Gas : Gibbs Paradox

According to Boltzmann's relation, the entropy $S$ of an ideal gas is related with the maximum thermodynamic probability $W$ by the relation

$$S = k\log W \qquad \text{...}(i)$$

From the Maxwell - Boltzmann count [Refer eqn 11.11]

$$\log W = \log N! + \sum_i^k (n_i \log g_i - \log n_i!)$$

$$\log W = \log N! + \sum_i n_i \log g_i - n_i \log n_i + n_i$$

$$\log W = \log N! + \sum_i^k n_i(\log g_i - \log n_i + 1)$$

$$= \sum_i^k n_i\left(\log\frac{g_i}{n_i} + 1\right) + \log N! \qquad \text{...}(ii)$$

For maximum value of $W$ or $\log W$, we have

$$n_i = e^{-\alpha} g_i e^{\frac{-E_i}{kT}} \qquad ...(iii)$$

or

$$\frac{g_i}{n_i} = e^{\alpha+\left(\frac{E_i}{kT}\right)}$$

Taking logarithm of both sides, we have

$$\log \frac{g_i}{n_i} = \alpha+\left(\frac{E_i}{kT}\right)$$

$$= \frac{E_i}{kT}+\alpha \qquad ...(iv)$$

Substituting this value in equation (*ii*), we get the maximum value $\log W$ as

$$\log W = \left[\sum_i n_i\left(\frac{E_i}{kT}+\alpha+1\right)\right]+\log N!$$

$$= \frac{1}{kT}\sum_i n_i E_i+\alpha\sum_i n_i+\sum_i n_i+\log N!$$

But $\quad \sum_i n_i E_i = \text{total energy } E = \frac{3}{2} NkT$

and $\quad \sum_i n_i = N$

$$\therefore \quad \log W = \frac{E}{kT}+\alpha N+N+\log N!$$

$$= \frac{1}{kT}\left(\frac{3}{2}NkT\right)+N+\alpha N+\log N!$$

$$= \frac{3}{2}N+N+\alpha N+\log N!$$

$$= \frac{5}{2}N+\alpha N+\log N! \qquad ...(iv)$$

But $\quad \alpha = -\log\left[\frac{N}{V}\left(\frac{h^2}{2\pi mkT}\right)^{\frac{3}{2}}\right] \qquad (\because 10.32)$

Substituting, we get

$$\log W = \frac{5}{2}N - N\log\left[\frac{N}{V}\left(\frac{h^2}{2\pi mkT}\right)^{\frac{3}{2}}\right]+\log N! \qquad ...(v)$$

Now substituting this value in eqn. (*i*), we get

$$S = \frac{5}{2}Nk - Nk\log\left[\frac{N}{V}\left(\frac{h^2}{2\pi mkT}\right)^{\frac{3}{2}}\right]+k\log N!$$

or

$$S = \frac{5}{2}Nk + Nk\log\left[\frac{N}{V}\left(\frac{2\pi mkT}{h^2}\right)^{\frac{3}{2}}\right]+k\log N! \qquad ...(10.69)$$

When two different ideal gases (distinguishable) with number of particles $N_1$ in compartment $A$ of volume $V_1$ and $N_2$ that in comportment $B$ of volume $V_2$. Both the gases are at same pressure and temperature and are separated by a partition (Fig. 10.10). When the partition is removed and gases are allowed to mix in a volume $V = V_1 + V_2$ at the same temperature, then since the process is irreversible, the entropy of the mixture should be greater than the sum of the entropies before mixing.

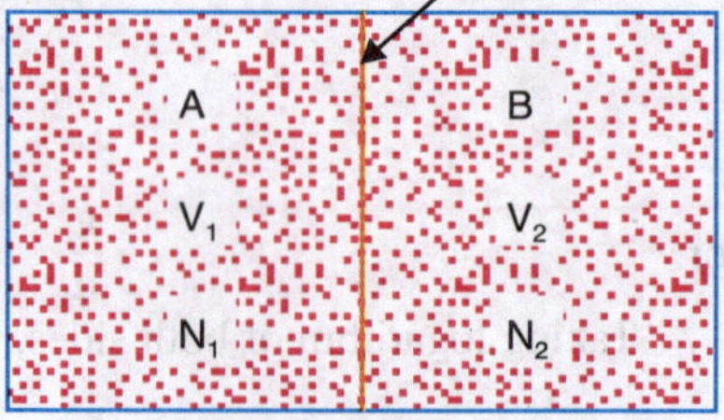


**Fig. 10.10**

If the gases are of the same kind (indistinguishable), the change in entropy on mixing should be zero.

If Eqn. (10.69) is applied to the process, then for mixing of two different gases it gives the correct result. But for mixing of two identical gases (indistinguishable), the equation gives increase in entropy. This is in contradiction. This contradiction or conclusion drawn from this equation is called the **Gibbs paradox.**

**Removal of paradox:** To resolve this paradox, the expression for the Maxwell-Boltzmann count must be divided by N!. In other words, we should subtract the term $k$ log N! from equation (10.69), so that

$$S = \frac{5}{2}Nk + Nk\log\left[\frac{N}{V}\left(\frac{2\pi mkT}{h^2}\right)^{\frac{3}{2}}\right] \qquad ...(10.70)$$

This has been verified experimentally for a gas at high temperatures. It is known as **sackur-Tetrode equation.**

Eqn. (10.70) can be written in the form

$$S = Nk\log\left[\frac{V}{N}\left(\frac{2\pi mkT}{h^2}\right)^{\frac{3}{2}}.e^{\frac{5}{2}}\right] \qquad ...(10.71)$$

where $e = 2.718$ is the base of Napierian logarithm.

## NUMERICAL EXAMPLES

### EXAMPLE 10.2

*In a system in thermal equilibrium at temperature T, two states with energy difference 4.8 × $10^{-14}$ erg occur with relative probability $e^2$ erg $deg^{-1}$. Calculate the temperature. Given k = 1.38 × $10^{-16}$ erg. $deg^{-1}$.*

**SOLUTION.** The probabilities $P_1$ and $P_2$ of the state of energies $E_1$ and $E_2$ are given by

$$P_1 = C\,e^{-E_1/kT}$$

and

$$P_2 = C\,e^{-E_2/kT}$$

$$\therefore \qquad \frac{P_1}{P_2} = \frac{C\,e^{-E_1/kT}}{C\,e^{-E_2/kT}}$$

or

$$\log_e\left(\frac{P_1}{P_2}\right) = \frac{E_2 - E_1}{kT}$$

or
$$T = \frac{E_2 - E_1}{k \log_e \left(\frac{P_1}{P_2}\right)}$$

Here $E_2 - E_1 = 4.8 \times 10^{-14}$ erg and $\frac{P_1}{P_2} = e^2$

$$T = \frac{4.8 \times 10^{-14}}{k \log_e (e^2)} = \frac{4.8 \times 10^{-14}}{k \times 2}$$

$$= \frac{4.8 \times 10^{-14}}{2 \times 1.38 \times 10^{-16}}$$

$$= \mathbf{173.913\ °K}$$

### EXAMPLE 10.3

*The first vibrational energy of a diatomic molecule is 600 cm$^{-1}$ above the ground state. Calculate the relative population of molecules in these two levels at T = 400°K.*

**SOLUTION.** Energy corresponding to 600 cm$^{-1}$ is given by

$$(E_1 - E_0) = E = h\nu = hc/\lambda = hc\bar{\nu}$$

$$= 6.6 \times 10^{-27} \times 3 \times 10^{10} \times 600$$

$$= 1.2 \times 10^{-13} \text{ erg}$$

$$\therefore \quad \frac{P_1}{P_2} = e^{(E_2 - E_1)/kT} = e^{-(E_1 - E_2)/kT}$$

$$= e^{-(1.2 \times 10^{-13})/(1.38 \times 10^{-16} \times 400)}$$

$$\cong \mathbf{0.1}$$

### EXAMPLE 10.4

*Using statistical definition of entropy, show that the difference in entropy between a state of volume $V_f$ and a state of volume $V_i$ (temperature and number of molecules remaining constant) is equal to nR log $\frac{V_f}{V_i}$.*

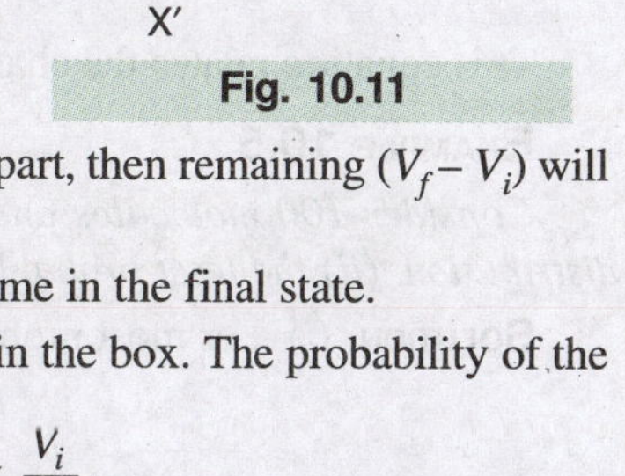


Fig. 10.11

**SOLUTION.** Consider a rectangular-box of volume $V_f$; with a partition wall $XX'$, which may be moved anywhere inside it, and the box gets divided into two compartments. If $V_i$ is volume of one part, then remaining $(V_f - V_i)$ will be the volume of the second part (Fig. 10.11).

Let $V_i$ represents volume in initial state, and $V_f$ represents volume in the final state.

Imagine that there is a single molecule of an ideal gas present in the box. The probability of the molecule to be found in volume $V_i$ at any given instant is obviously $\frac{V_i}{V_f}$.

If the box contains two molecules instead one, then the probability of finding both the molecules in the volume $V_i$ would be $\left(\frac{V_i}{V_f}\right)^2$ and there be p molecules in the box, the probability of finding all

$p$ molecules in volume $V_i$ would be $\left(\frac{V_i}{V_f}\right)^P$.

Instead, suppose the box contains $n$ gram-moles of a perfect gas, then probability of finding all the gas molecules in volume $V_i$ will be

$$W_i = \left(\frac{V_i}{V_f}\right)^{nN}$$

where N is Avogadro's number. If now the partition is removed, the probability of finding all the molecules in volume $V_f$ would be unity, *i.e.*, $W_f = 1$.

Therefore, dividing, we get

$$\frac{W_i}{W_f} = \left(\frac{V_i}{V_f}\right)^{nN}$$

Taking logarithm of both sides, we get

$$\log \frac{W_i}{W_f} = nN \log \frac{V_i}{V_f}$$

Multiplying both sides by Boltzmann constant $k$, we have

$$k \log \frac{W_i}{W_f} = nkN \log \frac{V_i}{V_f}$$

But $kN = R$, universal gas constant,

$$\therefore \quad k \log \frac{W_i}{W_f} = nR \log \frac{V_i}{V_f} \quad ...(i)$$

If $S_i$ and $S_f$ represent the entropy of the gas in volume $V_i$ and $V_f$ respectively, then using the relation $S = k \log W$, we get

$$\Delta S = S_f - S_i = k \log W_f - k \log W_i \quad \text{(By definition of entropy S)}$$

$$= k \log \frac{W_f}{W_i}$$

$$\Delta S = nR \log \frac{V_f}{V_i} \quad [\because \text{ equation } (i)] \; ... (ii)$$

This equation relates the change of entropy for the gas with the change in volume from $V_i$ to $V_f$.

**EXAMPLE 10.5**

*Consider 100 molecules and 10 cells of equal energy. Find log Ω for (i) the most probable distribution, (ii) the lesst probable distribution.*

**SOLUTION.** (*i*) For most probable dissipation,

$$N = 100$$

and $$n_1 = n_2 = .... = n_{10} = \frac{100}{10} = 10$$

$$\therefore \quad \log \Omega = \log \frac{100!}{(10!)^{10}} = \log 100! - 10 \log 10!$$

$$= 100 \log_e 100 - 100 - 10\,[10 \log_e 10 - 10] \quad ...(i)$$

By using Sterling theorem, $\log_e x! = [x \log_e x] - x$ solving (*i*), we get

$$\log \Omega = 100 \times 2.303 \times 2.00 - 100 - 10\,[10 \times 2.303 \times 1.00 - 10]$$
$$= 460 - 100 - 230 + 100$$
$$= \mathbf{230}$$

(*ii*) For least probable distribution,

$$N = 100,\ n_1 = 100 \text{ and } n_2 = n_3 = n_4 = \ldots.\ n_{10} = 0$$

$$\therefore \quad \log \Omega = \log \frac{100!}{100!\,(0!)^9} = \frac{100!}{100!\,(1)^9} \qquad (\because 0! = 1)$$
$$= \log 1$$
$$= \mathbf{0}$$

**EXAMPLE 10.6**

*System can take only three different energy states $E_1 = 0$, $E_2 = 1.38 \times 10^{-14}$ ergs, $E_3 = 2.76 \times 10^{-14}$ ergs. These three states can occur in 2, 5 and 4 different ways respectively. Find the probability that at temperature 100 °K the system may be (i) in one of the microstates of energy $E_3$ (ii) in the ground state $E_1$.*

**SOLUTION.** Let $P_1$, $P_2$ and $P_3$ be the probabilities with which the microstates can occur in three energy values $E_1$, $E_2$ and $E_3$ respectively and are given by

$$P_1 = C e^{-E_1/kT} = C e^{\circ} = C \qquad (\because E_1 = 0)$$

$$P_2 = C e^{-E_2/kT} = C e^{\frac{-1.38\times10^{-14}}{1.38\times10^{-16}\times100}}$$
$$= C e^{-1}$$

and

$$P_3 = C e^{-E_3/kT} = C e^{\frac{-2.76\times10^{-14}}{1.38\times10^{-16}\times100}}$$
$$= C e^{-2}$$

Given that 2, 5, 4 microstates can occur in the three energy states, the probabilities are

$$P_1 = 2\,C,\ P_2 = 5\,C\,e^{-1},\ P_3 = 4\,C\,e^{-2}$$

But

$$\sum P = 1$$

$$\therefore \quad P_1 + P_2 + P_3 = 2\,C + 5\,C\,e^{-1} + 4\,C\,e^{-2}$$
$$= C\,(2 + 5\,e^{-1} + 4\,e^{-2})$$

or

$$C = \frac{1}{2 + 1.84 + 0.54} = \frac{1}{\mathbf{4.38}} \qquad (\text{Since } e = 2.72)$$

(*i*) The probability for the system to be in one of the microstates of energy $E_3$ is

$$P_3 = 4\,C\,e^{-2} = \frac{4C}{e^2} = \frac{4}{4.38 \times (2.72)^2}$$
$$= \mathbf{0.12}$$

(*ii*) The probability for the system to be in ground state is

$$P_1 = 2\,C = \frac{2}{4.38}$$
$$= \mathbf{0.457}$$

## EXERCISE–X

**10.1** Explain the terms

(*i*) mu-space. **(Nagpur 2005)**

(*ii*) Γ-space. **[Nagpur 2004 (S) 2005 (W) Gujarat 2005]**

**10.2** Give two reasons for keeping the size of phase cell $h_o^{Nf}$ small. In classical statistics $h_o$ is an arbitrary choice, but in quantum statistics the corresponding constant $h$ is a fixed natural constant. Why?

**10.3** Calculate the number of phase cells in energy range 0 to $E$ for a linear simple harmonic oscillator of mass $m$ and frequency ν. **(Meerut 2004, Agra 2002)**

**10.4** For the phase space representing a single particle of mass $m$ in a volume $V$. Calculate the number of phase cells in energy range $0$ to $E$, given that volume of each phase cell is $h_o^3$.

**10.5** What is the minimum size of phase space in classical and Quantum Mechanics. **(Purvanchal 2006, Punjab 2005)**

**10.6** Explain the term phase space. **(Purvanchal 2006, Nagpur 2009, Rohilkhand 2003, Agra 2004)**

**10.7** Give the concept of ensemble and phase space. Calculate the number of states per unit volume of the phase space. **(Purvanchal 2001, 2003, Rohilkhand 2000)**

**10.8** Define phase space and find volume of a unit cell. **(Magadh 2001, Purvanchal 2003, Nagpur 2009, 2005 Summer)**

**10.9** Show that the volume of μ-space corresponding to a single quantum state for particles of no spin is $h^3$, where $h$ is Planck's constant. **(Nagpur 2004/w, 2004/s)**

**10.10** Explain the statistical meaning of entropy. **(Nagpur 2004(s), 2005/w)**

**10.11** What do your mean by canonical ensemble? For what type of system is it suitable. **(Lucknow 2000)**

**10.12** Explain concept of ensemble. Differentiate between microcanonical, canonical and grand-canonical ensembles. Where they are used. **(Magadh 2002, Rohilkhand 2000)**

**10.13** What do you understand by a priori probability and thermodynamical probability of distribution. **(Meerut 2000)**

**10.14** What is the significance of partition function (Z) in statistical physics? Obtain the expression for the partition function. How it is related to different thermodynamic functions. **(Magadh 2003, Rohilkhand 2001, Avadh 2002, Purvanchal 2000, 2001, 2007)**

**10.15** What do you understand by thermodynamical probability? How is it related with the entropy of the system? Establish the necessary relation.

**10.16** Deduce Boltzmann's entropy probability relation

$$S = k \log_e \Omega (E)$$

where S is entropy, $\Omega (E)$ is the number of microstates in the energy interval between $E$ and $E + dE$ and $k$ is a Boltzmann's constant. **(Meerut 2004, 2000, Rohilkhand 2000, 2003, Nagpur 2004/s, 2005/s, 2005/w, 2006/s, 2007, 2008, Agra 2004, 2007, Purvanchal 2005, 2007)**

**10.17** Define entropy and probability. Show that the entropy of the system is proportional to the logarithm of probability of that system. **(Rohilkhand 2001 Improvement)**

**10.18** Show that the number of molecules in a cell of energy $E_i$ in the equilibrium state of absolute temperature $T$ is given by

$$n_i = \frac{N}{Z} e^{-E_i / kT}$$

where $Z$ is the partition function. **(Agra 2000)**

**10.19** What are the postulates of statistical mechanics. **(Nagpur 2009, Purvanchal 2006)**

**10.20** Give any two theories in physics based on statistical concepts.

**10.21** Derive the condition of equilibrium between two systems in thermal contact and explain how it plays a role of bridge with macroscopic physics. **(Agra 2004, Nagpur 2004/s)**

**10.22** Derive the condition $\beta_1 = \beta_2$ for equilibrium of two systems in thermal contact. **(Meerut 2002, Agra 2000)**

**10.23** State and prove the Boltzmann's theorem between entropy and probability. **(Rohilkhand 2001, Agra 2001, 2004)**

**10.24** Explain the statistical interpretation of second law of thermodynamics.

**10.25** Show that the number of molecules in a cell of energy $E_i$ in the equilibrium state to temperature $T$ is given by

$$n_i = Ae^{-E_i/kT}$$

where $A$ is any constant.

**10.26** Derive Boltzmann's canonical distribution law. From this calculate the probability for the system. **(Agra 2008, Rohilkhand 2002, Meerut 2002, Gujarat 2005)**

**10.27** Derive the relation for probability

$$P_{(E_i)} = \frac{e^{-E_i/kT}}{\sum e^{-E_i/kT}}$$ **(Agra 2002)**

**10.28** Discuss the applications of Boltzmann's canonical distribution law. **(Meerut 2000, Agra 2000, Nagpur 2005)**

**10.29** State and prove the theorem of equipartition of energy. Give merits of this theorem. Discuss the condition when the theorem of equipartition of energy fails to explain the classical mechanics. **(Purvanchal 2007, Nagpur 2006, Gujarat 2005)**

**10.30** What do you mean by breakdown of equipartition theorem? When it occurs?

**10.31** Show that when two systems are placed in thermal contact

$$\beta(E) = \beta'(E) = \frac{1}{kT} = \frac{\partial\Omega(E)}{\Omega(E).\partial E}.$$ **(Meerut 2002)**

where $\Omega(E)$ represents the number of microstates of energy $E$.

**10.32** Show that in a general interaction between two systems, mechanical equilibrium is reached when the temperatures of the two systems become equal and the pressures of the two becomes equal. **(Mumbai 2005)**

**10.33** Show that the probability of a system in equilibrium with a reservoir at temperature $T$ may exist in a microstate of energy $E_i$ is proportional to $e^{-E_i/kT}$ where $k$ is Boltzmann's constant. What is $e^{-E_i/kT}$ called? Hence show that the partition function $Z = \sum_i e^{-\beta E_i}$. **(Mumbai 2005)**

**10.34** Explain the term Helmholtz free energy. Show that in a natural isothermal change at constant volume Helmholtz free energy decreases. **(Mumbai 2005)**

**10.35** Probability that a system in equilibrium with a reservoir at temperature $T$ may exist in a state of energy $E_i$ is given by $P_i = \frac{1}{Z}e^{-\beta E_i}$, where $Z$ is the partition function. Prove that average energy $\overline{E} = -\frac{\partial(\log Z)}{\partial\beta}$. **(Mumbai 2005)**

**10.36** Show that the partition function $Z_{ij}$ of two independent systems '$i$' and '$j$' is given by $Z_{ij} = Z_i \times Z_j$, where $Z_i$ and $Z_j$ are the partition functions of the systems '$i$' and '$j$' respectively. **(Mumbai 2005)**

**10.37** Using the concept of canonical ensemble obtain the expression of the probability of energy distribution of the any system of the ensemble

$$n_i = Ne^{-\alpha}e^{-\beta E_i}$$

From it, obtain partition function of the system $Z = e^{-\alpha}$. **(Gujarat 2005)**

**10.38** Define thermodynamic probability and state its minimum value. **(Nagpur 2005 summer)**

**10.39** Obtain an expression for the number of microstates accessible to a free particle confined in a box of length L. **(Nagpur 2005 winter)**

**10.40** What is Gibb's paradox. How this paradox is solved? Discuss it by deriving necessary equations. Obtain Suckur Tetrode equation from it. **(Magadh 2003, Gujarat 2004, 2005)**

**10.41** Show that entropy increases when two gases at the same temperature and pressure diffuse into each other. Discuss Gibbs paradox in this connection. **(Calcutta 2003)**

**10.42** Obtain an expression for Maxwell-Boltzmann distribution $n_i = g_i A e^{-\beta E_i}$. Evaluate the values of constants β and A. **(Nagpur 2008)**

**10.43** In M–B statistics, prove that the constant β is equal to $\frac{1}{kT}$ from thermodynamic considerations.

**10.44** Obtain partition function for canonical ensemble ? What is the effect of shifting the uro level of energy on the partition function ? **(Magadh 2001)**

**10.45** What is meant by microcanonical ensemble ? Establish Sackur-Tetrode formula for the entropy of a perfect gas. **(Magade 2001)**

**10.46** What do you understand by microcanonical ensemble and to what kind of systems does it apply ? Use the ensemble derive an expression for the entropy of classical idea gas. **(Purvanchal 2006)**

**10.47** Derive the relation $S = Nk \log_e Z + 3/2\, Nk$, where $N$ = Avogadro's number, $Z$ = Partition function, $k$ = Boltzmann constant, $S$ = entropy. **(Purvanchal 2007)**

**10.48** Write notes on the followings:

(*i*) Density of States. **(Purvanchal 2006, 2000)**

(*ii*) Statistical ensembles. **(Purvanchal 2002, Meerut 2000)**

(*iii*) Canonical Ensemble. **(Rohilkhand 2001)**

(*iv*) Microcanonical Ensemble. **(Purvanchal 2006)**

(*v*) Grand-Canonical Ensemble. **(Magadh 2001)**

(*vi*) Partition function **(Purvanchal 2006, 2002, 2000)**

(*vii*) Phase space. **(Purvanchal 2004, Agra 2000, Rohilkhand 2000, 2002)**

(*viii*) Thermodynamic Probability. **(Agra 2004)**

(*ix*) Entropy and Probability. **(Agra 2000)**

(*x*) Fundamental postulates of statistical mechanics.

(*xi*) Statistics and Entropy.

(*xii*) Boltzmann's factor. **(Agra 2002, Rohilkhand 2001)**

(*xiii*) Gibbs paradox. **(Magadh 2001, Gujarat 2005)**

**10.49** Consider 100 molecules and 10 cells of equal energy. Find $\log_e W$ for (*i*) the most probable distribution and (*ii*) the least probable distribution. **[Ans. (i) 230 (ii) 0]**

**10.50** Given 3 cells and 5 particles. Calculate the thermodynamical probability for

(*i*) $n_1 = 5, n_2 = 0, n_3 = 0$

(*ii*) $n_1 = 4, n_2 = 1, n_3 = 0$

(*iii*) $n_1 = 3, n_2 = 2, n_3 = 0$ **[Ans. (i) = 1, (ii) 5, (iii) 10]**

**10.51** The first vibrational energy level of a particular diatomic molecule is 600 cm$^{-1}$, above the ground state. Calculate the relative population of the molecules in these levels at 127°C. **[Ans. 0.1]**

**10.52** Take a system of 3 distinguishable particles with zero spin. The particle energy levels are non-degenerate and equispaced with energies 0, 1, 2, 3, ... energy units. Calculate the total no. of microstates accessible to the system, if the total energy of the system is 4 units. **(Mumbai 2005)**

## OBJECTIVE QUESTIONS

**10.1** According to Boltzmann canonical distribution law, the number of molecules per cell

(*a*) increases linearly with energy associated with the cell,

(*b*) increases exponentially with energy associated with the cell,

(*c*) decreases linearly with energy associated with the cell,

(*d*) decreases exponentially with energy associated with the cell.

**10.2** In the equilibrium state;
(*a*) probability is maximum,
(*b*) $\beta$ parameters of two systems are equal,
(*c*) both (*a*) and (*b*),
(*d*) number of particles is maximum.

**10.3** Five particles are distributed in two phase cells. Then number of macrostates is;
(*a*) 6 (*b*) 10
(*c*) 32 (*d*) $\frac{5}{2}$

**10.4** According to Boltzmann canonical distribution law :
(*a*) low energy cells contain more particles,
(*b*) high energy cells contain more particles,
(*c*) number of molecules having zero energy is greater than any other energy,
(*d*) zero energy molecules are zero.

**10.5** The thermodynamic probability of a system in equilibrium is
(*a*) maximum (*b*) minimum but not 1
(*c*) 1 (*d*) zero

**10.6** Partition function is
(*a*) $Z = \sum e^{-\beta E_i}$ (*b*) $Z = \sum_i e^{-2\beta E_i}$
(*c*) $Z = \sum e^{-2\beta/E_i}$ (*d*) none of these **(Agra 2004)**

## ANSWERS

| | | | |
|---|---|---|---|
| **10.1** (*d*) | **10.2** (*c*) | **10.3** (*a*) | **10.4** (*a*) |
| **10.5** (*a*) | **10.6** (*a*) | | |

# Maxwell-Boltzmann Statistics

# 11

## AT A GLANCE

- Introduction • Three kinds of Particles • M.B. Statistics applicable to Ideal Gas • Maxwell-Boltzmann Energy Distribution Law • Applications of Maxwell-Boltzmann Distribution Law : (i) Internal Energy and Sp. Heat at const. Volume (ii)Maxwell-Boltzmann Speed Distribution Law • Mean, RMS and Most Probable Speeds • Maxwell's Distribution Law of Velocities • Doppler Broadening of Spectral Lines • Limitations of Maxwell-Boltzmann's Method • Experimental Verification of Maxwellian • Distribution of Molecular Speeds

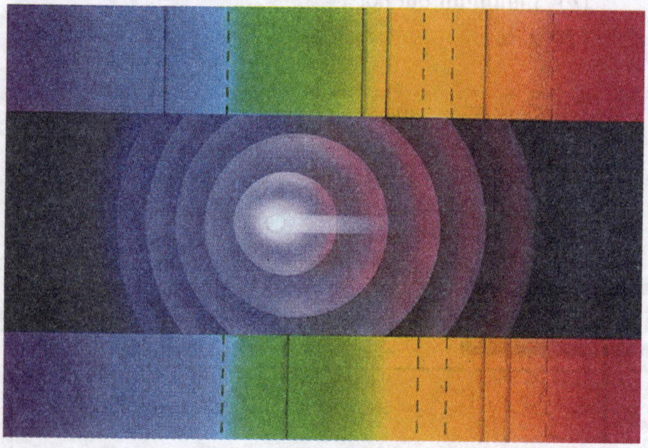

## Introduction

On the basis of assumptions of the kinetic theory of gases and by applying the laws of probability, James Clerk Maxwell in 1859 derived the law of distribution of speeds among the molecules for an ideal gas in thermal equilibrium. Later, in 1872, Boltzmann derived his fundamental non-linear integro-differential relation (known as Boltzmann transport equation) from which he derived the velocity distribution function for the molecules of an ideal gas in thermal equilibrium. From this equation, speed and energy distributions among the molecules of ideal gas can be obtained easily. These distributions are known as Maxwell or Boltzmann or Maxwell-Boltzmann distribution, after the founders of the theory.

In this chapter, we will derive the laws of distribution of energy, speed and velocity among the gas molecules of an ideal gas, followed by the experimental evidences.

## 11.1 Three Kinds of Particles

The assemblies consist of, in general, three kinds of particles :

1. *Identical but distinguishable particles without any spin.* The molecules of a gas are the particles of this kind. In quantum terms, the wave functions of the particles overlap to a negligible extent. The **Maxwell-Boltzmann distribution function** holds for such particles.

2. *Identical and indistinguishable particles of zero or integral spin.* They cannot be distinguished one from another because their wave functions overlap. Such particles are called **bosons** (refer next chapter), and do not obey the Pauli exclusion principle. The **Bose-Einstein distribution function** holds for them. The helium atoms at low temperature and the photons are the particles of this kind and we shall use Bose-Einstein statistics to account for the spectrum of radiation from a black body.

[Supernova explosion resulted in the crab Nebula that was observed in A.D. 1054. The explosion left behind a star believed to consist entirely of neutrons. Statistical Mechanics is needed to understand the properties of neutron stars.]

3. *Identical and indistinguishable particles with odd half-integral spin* $\left(\frac{1}{2}, \frac{3}{2}, \frac{5}{2}, \ldots\right)$. Their wave functions overlap on each other. Such particles are called **fermions** or **fermi particles**, obeying the Pauli exclusion principle. The **Fermi-Dirac distribution function** holds for them. Electrons, protons, neutrons etc. are the examples of this kinds. We shall use F.D. statistics to study the behaviour of the free electrons in a metal that are responsible for its ability to conduct electric current.

## 11.2 M.B. Statistics Applicable to Ideal Gas

A system consisting of molecules of an ideal gas under ordinary conditions of temperature and pressure is governed by the laws of classical *Maxwell-Boltzmann Statictics*. Here the gas molecules are treated as *distinguishable* particles. In the classical statistics, we can choose the size of the cell in the phase space as small as we please. We, therefore, take the size of the cell so small ($h \to 0$) that the total number of cells available are very large as compound to the number of particles, with the result that the probability of a cell containing more than one particle is almost negligible. In other words, *it means that a cell can contain either one or no particle.* In this way, every particle of the system will occupy a different numbered cells. It will, therefore, be possible to *distinguish* one molecule from the other by the number of cell which it occupies. However, there is *no restriction on the number of particles which can occupy a phase space cell.* This is because, *there is a little chance of two molecules occupying the same cell, the molecules of the gas can be treated as distinguishable particles.*

### Statistical Equilibrium

Consider a gas having $N$ molecules at a certain temperature and pressure. Suppose its volume, temperature and pressure are kept constant *i.e.*, a thermodynamically isolated system of $N$ particles. The total energy $U$ of this system remains constant. The energy states available to the particles are $E_1$, $E_2$, $E_3$, ... etc. These energy states may be *quantized* or may be *continuous* and are due to the vibrational and rotational energy of the particles.

Suppose, that at any given instant of time, $n_1$ particles are in state of energy $E_1$, $n_2$ particles in energy state $E_2$ and so on.

The total number of particles in the system is,

$$N = n_1 + n_2 + n_3 + ... = \sum_i n_i \quad ...(i)$$

Here $i$ = 1, 2, 3, ... etc.

The total energy of the system,

$$U = n_1E_1 + n_2E_2 + n_3E_3 + ...$$

or

$$U = \sum_i n_i E_i \quad ...(ii)$$

Equation (*ii*) refers to the total energy of a system in which the particles are *non-interacting*. Here the energy of each particle depends only on the coordinates of the particle in the system. For an **isolated system**, the total energy $U$ is constant. Thus, for an isolated system.

$$U = \sum_i n_i E_i = \text{constant} \quad ...(iii)$$

Since *the system is isolated*, the total energy of the system remains constant. But, the molecules of the gas collide with each other and also with the walls of the container. Consequently, the number of molecules change from one energy state to the other energy state. It means that the values of $n_1$, $n_2$, $n_3$ etc. continuously change. It can be reasonably assumed that for each microscopic state of a system

**JAMES CLERK MAXWELL (1831–1879)** *was born is Scotland shortly before the Faraday's discovery of electromagnetic induction. At the age of 19, he entered Cambridge university as a student of physics and mathematics. During study, he discovered the physics of colour vision. He became known to the scientific world, when he showed that the rings of Saturn could not be solid or liquid but must consists of separate small bodies, at the age of 24. This was the golden era of his life that he became interested in electricity and magnetism and their interrelationship. Maxwell's first step on this regard was in 1856, when he published the paper "On Faraday's Lines of Force."*

*Maxwell left Cambridge in 1856 and joined teaching at a college in Scotland and later at King's College in London. Here he published the well-known theory of 'Four field equations' and showed that "electromagnetic waves" must exists that travel with the velocity of light. It was German scientist Heinrich Hertz, who understood the potential of these equations in 1887, but actual demonstration was made by Marconi and his associates in December 1901 by transmitting letter 'S' across the Atlantic sea, paved the way of radio-communication for today's modern world.*

*Maxwell's contribution to kinetic theory and statistical mechanics is also remarkable. Maxwell along with his wife confirmed in the laboratory that the viscosity of a gas is independent of pressure and directly proportional to the absolute temperature. Maxwell's explanation for this proportionality gave him a way to estimate the size and mass of a molecule. Maxwell shares with Boltzmann credit for the equation that gives the distribution of molecular energies in a gas. In 1871, Maxwell established Cavendish Laboratory in Cambridge. Maxwell died of cancer at the age of 48 only (in 1879), the year in which Albert Einstein was born. (By a similar coincidence, Newton was born in the year of Galileo's death). Maxwell had been the greatest theoretical physicist of the 19th century ; Einstein was to be the greatest theoretical physicist of the 20th century.*

of particles, there is a particular most favoured distribution. When this distribution or partition is reached, the system attains *statistical equilibrium.* For an isolated system, the values of $n_1, n_2, n_3$ etc. vary only near the value corresponding to the most probable distribution. Hence, the basic problem in statistical thermodynamics is to obtain the most probable distribution law for a given composition of the system.

## 11.3. Maxwell-Boltzmann Energy Distribution Law

Suppose the particles are distributed among $k$ energy levels $E_1, E_2, \ldots E_i, \ldots E_k$. Let $n_1, n_2, \ldots n_i, \ldots n_k$ be the number of independent quantum states associated with the energy levels. According to the statistical basic postulates.

$$N = n_1 + n_2 + \ldots + n_i + n_k = \sum_i n_i \qquad \ldots(11.1)$$

and since the force of interaction between particles are negligible, we have

$$U = n_1E_1 + n_2E_2 + \ldots + n_iE_i + \ldots + n_kE_k \qquad \ldots(11.2)$$

Taking differential of equation (11.1) and (11.2), we have

$$dN = \sum_i dn_i = 0 \qquad \ldots(11.3)$$

and

$$dU = \sum_i E_i dn_i = 0 \qquad \ldots(11.4)$$

The set of occupation numbers $n_1, n_2, \ldots$ is a possible mode of distribution of the particles among the energy levels, and it determines a macrostate of the system. According to the fundamental postulate of *equal a priori probabilities,* all the quantum states of energy of the particles in the system in equilibrium corresponding to the constant values of $U$, $V$ and $N$ are equally probable. It means that in the equilibrium state the distribution of the particles among various quantum states has the maximum probability of occurrence. This distribution can be obtained in a maximum number of statistically independent ways. Therefore, let us count (*i.e.*, determine) the total number of independent ways $W$ of distributing the particles among the quantum states corresponding to the set of number $n_1, n_2, \ldots$ and then determine the set of number of which $W$ is maximum.

### Maxwell-Boltzmann Statistical Count

Out of $N$ total number of particles, $n_1$ are in energy level $E_1$, $n_2$ are in energy level $E_2$ and so on as shown in Fig. 11.1. For the first energy level $E_1$, we have to find the number of ways in which $n_1$ particles are chosen from among $N$ distinguishable particles.

The number of ways of choosing the first particle = $N$. When the first particle has been chosen, the number of ways of choosing the second particle = $N - 1$.

Similarly, the number of ways of choosing the third, fourth, ... $n_1$th particles are respectively $(N-2), (N-3), \ldots (N - n_1 + 1)$.

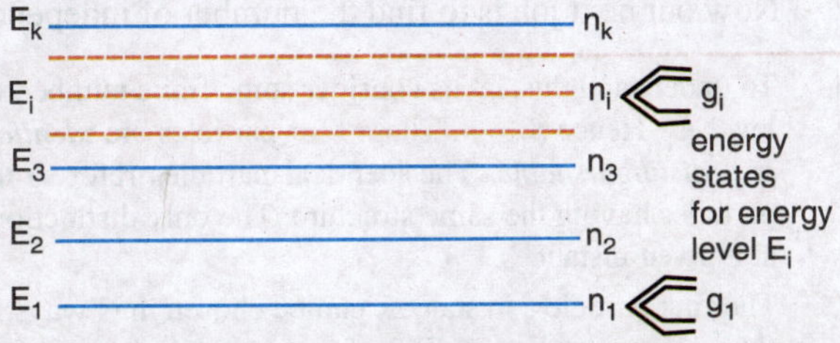


Fig. 11.1

Therefore, the number of ways of choosing $n_1$ particles from $N$ distinguishable particles is

$$= N(N-1)(N-2)\ldots(N-n_1+1)$$

$$= \frac{N(N-1)(N-2)\ldots(N-n_1+1)(N-n_1)!}{(N-n_1)!}$$

$$= \frac{N!}{(N-n_1)!}$$

Let 4 particles out of 100 No. of ways 1st particle out of 100 is = 100

No. of ways 2nd particle out of 99 is = 100 – 1 = 99

No. of ways 3rd particle out of 98 is = 100 – 2 = 98

No. of ways 4th particle out of 97 is = 100 – 3 = 97

$$\therefore \text{ Total no. of ways} = 100 \times 99 \times 98 \times 97 = \frac{100 \times 99 \times 98 \times 97 \times 96}{96!} = \frac{100!}{96!}$$

But this number also includes the $n_1!$ arrangement of $n_1$ particles among themselves†. This is because any sequence of the particles in the first energy level gives the same distribution of $n_1$ particles in the energy level, the number of independence (*i.e.*, meaningful) ways of choosing $n_1$ particles for the first energy level is, thus, given by

$$x_1 = \frac{N!}{n_1!(N-n_1)!} = {}^{N}C_{n1}$$

For the second state $E_2$, only $(N - n_1)$ particles are available and $n_2$ particles are in state $E_2$. The number of independent ways of choosing $n_2$ particles for the second energy level is similarly given by

$$x_2 = \frac{(N-n_1)!}{n_2!(N-n_1-n_2)!} = {}^{N-n_1}C_{n_2}$$

Proceeding in the same way, the number of independent ways of choosing $n_3, n_4, \ldots n_i \ldots n_k$ particles for the third, fourth, ... $i^{\text{th}}$ and $k^{\text{th}}$ energy levels are given by

$$x_3 = \frac{(N-n_1-n_2)!}{n_3!(N-n_1-n_2-n_3)!} = {}^{(N-n_1-n_2)}C_{n_3}$$

$$\ldots \quad \ldots \quad \ldots \quad \ldots$$

$$\ldots \quad \ldots \quad \ldots \quad \ldots$$

$$x_k = \frac{(N-n_1-n_2 \ldots n_{k-1})!}{n_k!(N-n_1-n_2 \ldots n_k)!} = {}^{(N-n_1-n_2 \ldots n_{k-1})}C_{n_k}$$

Since these numbers are independent, the number of distinct ways of obtaining the distribution of the particles among the energy levels is given by

$$x = x_1 x_2 x_3 \ldots x_i \ldots x_k$$

$$= \frac{N!}{n_1!n_2!n_3 \ldots n_k!} \quad \ldots(11.5)$$

Let, Denominator will conatain less term uncancelled as $[N - (n_1 - n_2 + \ldots\ldots n_k]! = 0! = 1$

Now our next job is to find the number of independent ways in which the particles in the various

† To understand this, let us consider some finite number, of particles (say *a*, *b*, *c*, and *d*) of in the first energy level $E_1$. Hence $n_1 = 4$. These four particles are *identical and distinguishable*. The identical particles refer to the particles having the same structure. The only distinction between the particles is due to their energy states at a given instant.

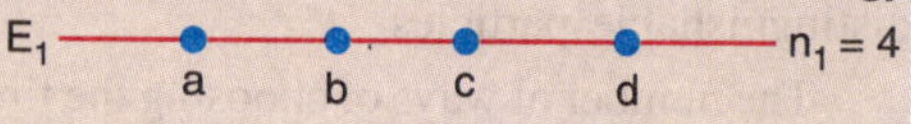


The first particle a in state $E_1$ can be chosen in N ways. The second particle *b* in state $E_1$ can be selected in (N – 1) ways and so on. Therefore, the total number of ways in which the first four particles in state $E_1$ can be selected is given by

$$N(N-1)(N-2)(N-3) = \frac{N!}{(N-4)!}$$

Moreover the four particles in state $E_1$ can be arranged in 4 different orders *i.e.*, in all 24 ways. For example : *abcd*, *bcda*, *cdab* and so on. But it is immaterial for these particles to be arranged in any particular order in state $E_1$, because they are all identical. Thus, the total number of *distinguishable* and meaningful different ways are obtained by dividing the above value by 4!. Hence, we have

$$x_1' = \frac{N!}{4!(N-4)!}$$

energy level can be distributed among the *quantum states* in the respective levels. For example, in the $i^{th}$ energy level, containing $n_i$ particles, there are $g_i$ quantum states. Each of the $n_i$ particles can be in any one of the quantum states. Thus, there are $g_i$ independent choices for the first particle to go in any one of the quantum states. For each of these choices, these are $g_i$ choices for the second particle and so on as there is no restriction of no. of particls to be occupied in any quantum state, unlike to Pauli-exclusion principle in quantum statistics.

Hence there are $g_i^{n_i}$ ways of arranging $n_i$ particles among $g_i$ quantum states available in the $i$th level. Each of these arrangements is a separate independent way of distributing the particles among the quantum states. Thus, all the quantum states belonging to all the energy levels are considered, the total number of ways of arranging all the particles in the given distribution is given by

$$W = \frac{N!}{n_1!n_2!\,.....n_k!} \times g_1^{n_1} \times g_2^{n_2} \times ......g_k^{n_k}$$

$$= N! \prod_{i=1}^{k} \left[\frac{g_i^{n_i}}{n_i!}\right] \qquad ...(11.6)$$

where the symbol $\Pi$ represents the product of the quantities $g_i^{n_i}/n_i!$ for all values of $i$ from 1 to $k$.

## Condition for most Probable Distribution

*The most probable distribution of the particles among the energy states in equilibrium is that for which the probability of occurrence is maximum, i.e. for which* ***W is maximum***. For mathematical convenience, we consider the condition for maximum value of log $W$. The condition for maximum value of log $W$ is

$$d\,(\log W) = 0 \qquad ...(11.7)$$

where $W$ is a function of $n_1, n_2, .....n_k$.

Differentiating equation (11.6), we get

$$d \log (W) = \frac{\partial \log W}{\partial n_1} dn_1 + \frac{\partial \log W}{\partial n_2} dn_2 + ... + \frac{\partial \log W}{\partial n_k} dn_k$$

$$= \sum_{i=1}^{k} \frac{\partial \log W}{\partial n_i} dn_i$$

Applying the condition from eqn. (11.7), we get

$$\sum_{i=1}^{k} \frac{\partial \log W}{\partial n_i} dn_i = 0 \qquad ...(11.8)$$

The solution of this equation is subject to the condition laid down by eqn.(11.3) and (11.4). These conditions are introduced into equation (11.8) by using the method of Langrange's undertmined multipliers.

Let $\alpha$ and $\beta$ be these multipliers independent of $n_i$'s. We multiply eq. (11.3) by $-\alpha$ and eq. (11.4) by $-\beta$ and adding these equations to eq. (11.8), we have

$$\sum_{i=1}^{k} \frac{\partial \log W}{\partial n_i} - \alpha - \beta E_i \;\; dn_i - 0 \qquad ...(11.9)$$

For maximum value of log $W$, eq. (11.9) must be true for the magnitudes of the individual increments $dn_i$. Hence, for all values of $i$, the term in the bracket must be zero.

$$\therefore \quad \frac{\partial \log W}{\partial n_i} - \alpha - \beta E_i = 0 \qquad ...(11.10)$$

## M-B Energy Distribution Law in General Form

Taking logarithms of Eq. (11.6), we have

$$\log W = \log N! + \log\left[\prod_i \left(\frac{g_i^{ni}}{n_i\,!}\right)\right]$$

$$= \log N! + \sum_{i=1}^{k} (n_i \log g_i - \log n_i\,!) \quad ...(11.11)$$

Since the number of particles is very large, $N!$ and $n_i!$ may be approximated by Sterling's formula*, according to which

$$\log N! = N \log N - N \quad ...(11.12\,(a))$$

and

$$\log n_i! = n_i \log n_i - n_i \quad ...(11.12\,(b))$$

$$\therefore \quad \log W = N \log N - N + \sum_{i=1}^{k} (n_i \log g_i - n_i \log n_i + n_i) \quad ...(11.13)$$

Differentiating Eq. (11.13) partially with respect to $n_i$ and remembering that the partial derivatives of the terms, except the $i$th term, are zero. Thus,

$$\frac{d\,(\log W)}{dn_i} = \log g_i - \log n_i + \frac{n_i}{n_i} + 1\ \delta n_i$$

$$= = \sum [\log g_i - \log n_i]\,\delta n_i$$

$$0 = \sum\left[-\log\left(\frac{n_i}{g_i}\right)\right]\delta n_i$$

Substituting this equation into (11.10), we have

$$\sum\left[-\log\left(\frac{n_i}{g_i}\right) - \alpha - \beta E_i\right]^{\delta n_i} = 0 \quad \therefore -\log\left(\frac{n_i}{g_i}\right) - \alpha - \beta E_i$$

or

$$\log\left(\frac{n_i}{g_i}\right) = -\alpha - \beta E_i$$

or

$$n_i = g_i\, e^{(-\alpha - \beta E_i)}$$

or

$$n_i = g_i\, e^{-\alpha}\, e^{-\beta E_i}$$

$$\text{(where } i = 1, 2, 3, ....k) \quad ...(11.15)$$

This equation is known as the **Maxwell – Boltzmann energy distribution law** in the general form.

The value of undetermined multiplier β is evaluated (refer article of chapter 10) as

$$\beta = \frac{1}{kT} \quad ...(11.16)$$

Substituting in eqn. (11.15), we get

$$n_i = g_i\, e^{-\alpha}\, e^{-E_i/kT} \quad ...(11.17)$$

where $k$ is Boltzmann constant and $T$ its absolute temperature. The quantity $e^{-E_i/kT}$ is known as the **Boltzmann factor.**

From the fundamental postulates of statistical mechanics, we write, $N$, the total number of particles as

* Where $x$ is large, according to Sterling's theorem, $\log_e x\,! = [x \log_e x] - x$

$$N = \sum_i n_i$$

$$= \sum_i g_i e^{-\alpha} e^{-E_i/kT}$$

$$= e^{-\alpha} g_i e^{-E_i/kT}$$

$$\therefore \quad e^{-\alpha} = \frac{N}{\sum_i g_i e^{-E_i/kT}} \qquad \text{... (11.18)}$$

Substituting this value in equation (11.17), we get

$$n_i = \frac{N g_i e^{-E_i/kT}}{\sum_i g_i e^{-E_i/kT}} \qquad \text{...(11.19)}$$

The quantity $\sum_i g_i e^{-E_i/kT}$ is the sum over all the states of the system and is called the **partition function** of the system devoted by $Z$. Thus

$$Z = \sum_i g_i e^{-E_i/kT} \qquad \text{...(11.20)}$$

Substituting in equation (11.19), we have

$$n_i = \frac{N}{Z} g_i e^{-E_i/kT} \qquad \text{...(11.21)}$$

## M. B. Energy Distribution Function

The general energy distribution function $f(E_i)$ is the average number of particles per quantum state in the $i^{th}$ energy level $E_i$. It is given by

$$f(E_i) = \frac{n_i}{g_i}$$

Substituting for $\frac{n_i}{g_i}$ from equation (11.17), we have

$$f(E_i) = e^{-\alpha} e^{-E_i/kT} \qquad \text{...(11.22)}$$

Therefore, for energy $E$, the M.B. energy distribution function may be expressed as

$$f(E_i) = e^{-\alpha} e^{-E/kT} \qquad \text{...(11.23)}$$

## M. B. Energy Distribution Function for an Ideal Gas

If the energy levels of the system are very close to each other, then the number of particles $n(E)\,dE$, whose energies lie in the range between $E$ and $E + dE$ is given by

$$n(E)\,dE = f(E)\,g(E)\,dE \qquad \text{...(11.24)}$$

where $g(E)\,dE$ represents the number of quantum states available between the energy $E$ and $(E + dE)$. For an ideal gas, $e^{-\alpha}$ is given by (11.30) as

$$e^{-\alpha} = \frac{N}{V}\left(\frac{h^2}{2\pi mkT}\right)^{3/2} \qquad \text{...(11.25)}$$

Substituting this value in eqn. (11.23), we get M.B. energy distribution function for an ideal gas as

$$f(E) = \frac{N}{V}\left(\frac{h^2}{2\pi mkT}\right)^{3/2} e^{-E/hT} \qquad \text{...(11.26A)}$$

But for an ideal gas consisting of free particles with no spin $g(E)\,dE$ is given by (Example 10.1), as

$$g(E)\,dE = 2\pi V\left(\frac{2m}{h^2}\right)^{3/2} E^{\frac{1}{2}}dE \qquad \text{...(11.26B)}$$

Substituting the value of $f(E)$ in equation (11.24), we get

$$n(E)\,dE = \frac{N}{V}\left(\frac{h^2}{2\pi mkT}\right)^{3/2} e^{-E/kT} \times 2\pi V\left(\frac{2m}{h^2}\right)^{3/2} E^{\frac{1}{2}}\,dE$$

or
$$n(E)\,dE = \frac{2\pi N}{(\pi kT)^{3/2}} E^{\frac{1}{2}} e^{-E/kT}\,dE \qquad \text{...(11.27A)}$$

or
$$n(E)\,dE = \frac{2N}{\sqrt{\pi}}\left(\frac{1}{kT}\right)^{3/2} E^{\frac{1}{2}} e^{-E/kT}dE \qquad \text{...(11.27B)}$$

This equation is the **Maxwell-Boltzmann energy distribution law** for the particles of an ideal gas.

The equation (11.27) gives the number of particles with energies between $E$ and $E + dE$ of an ideal gas containing the total number of particles $N$ (or molecules $N$) at absolute temperature $T$.

**Evaluation of the Multiplier α.**

From eqn. (11.18), we write

$$e^{-\alpha} = \frac{N}{\sum_i g_i e^{-Ei/kT}}$$

Since the variation of energy of free particles of an ideal gas is continuous, we will replace $g_i$ by $g(E)dE$ and $E_i$ by $E$ and the sign of summation is replaced by the sign of integration. Thus, we get

$$e^{-\alpha} = \frac{N}{\int_0^\infty g(E)dE.\, e^{-E/kT}}$$

$$= \frac{N}{\int_0^\infty g(E)\, e^{-E/kT}\,dE} \qquad \text{...(11.28)}$$

Here the limit of integration is taken from 0 to ∞, because energy of the particles of an ideal gas is entirely kinetic and so they can have any kinetic energy. The value of $g(E)dE$ for particles with **no spin** is given by

$$g(E)dE = 2\pi V\left(\frac{2m}{h^2}\right)^{3/2} E^{\frac{1}{2}}\,dE$$

Substituting in eqn. (11.28), we get

$$e^{-\alpha} = \frac{N}{2\pi V\left(\frac{2m}{h^2}\right)^{3/2}\int_0^\infty E^{\frac{1}{2}} e^{\frac{-E}{kT}}\,dE} \qquad \text{...(11.29)}$$

Let us evaluate the definite integral as under :

For this, let $\frac{E}{kT} = x$, so that $dE = kTdx$

$$\therefore \quad \int_0^\infty E^{\frac{1}{2}}\, e^{-E/kT} dE = \int_0^\infty (kTx)^{\frac{1}{2}}\, e^{-x} kT dx$$

$$= (kT)^{3/2} \int_0^\infty x^{\frac{1}{2}}\, e^{-x} dx$$

$$= (kT)^{3/2} \int_0^\infty x^{\frac{3}{2}-1}\, e^{-x} dx$$

The value of the integral on the R.H.S. is the $\Gamma$ (*i.e.* gamma) function,

$$\Gamma\,(3/2) = \frac{\sqrt{\pi}}{2}$$

$$\therefore \quad \int_0^\infty E^{\frac{1}{2}}\, e^{-E/kT} dE = (kT)^{\frac{3}{2}} \frac{\sqrt{\pi}}{2}$$

Substituting in eqn. (11.29), we get

$$e^{-\alpha} = \frac{N}{V\left(\frac{2\pi mkT}{h^2}\right)^{\frac{3}{2}}}$$

$$= \frac{N}{V}\left(\frac{h^2}{2\pi mkT}\right)^{\frac{3}{2}} \qquad \text{...(11.30)}$$

$$\text{or} \qquad e^{\alpha} = \frac{V}{N}\left(\frac{2\pi mkT}{h^2}\right)^{\frac{3}{2}} \qquad \text{...(11.31)}$$

Taking logarithms of both sides, we get

$$\alpha = -\log\left[\frac{N}{V}\left(\frac{h^2}{2\pi mkT}\right)^{\frac{3}{2}}\right]$$

$$\text{or} \qquad \alpha = \log\left[\frac{V}{N}\left(\frac{2\pi mkT}{h^2}\right)^{\frac{3}{2}}\right] \qquad \text{...(11.32)}$$

Here $e^{\alpha}$ is called the **degeneracy parameter**.

**Condition for Application of M.B. Statistics.**

The M.B. Statistics is applicable to a system of particles for which the mean distance between the particles is *greater than* the thermal De Broglie wavelength of the particles.

The volume of per particle $= \left(\frac{V}{N}\right)$

$\therefore$ The mean distance between the particles $= \left(\frac{V}{N}\right)^{\frac{1}{3}}$

The De-Broglie wavelength associated is given by

$$\lambda = \left(\frac{h^2}{2\pi mkT}\right)^{\frac{1}{2}}$$

Thus, the condition for M.B. statistics to became applicable is

$$\left(\frac{V}{N}\right)^{\frac{1}{3}} \geq \left(\frac{h^2}{2\pi mkT}\right)^{\frac{1}{2}}$$

or, $$\left(\frac{V}{N}\right) \geq \left(\frac{h^2}{2\pi mkT}\right)^{\frac{3}{2}}$$

or, $$\frac{V}{N}\left(\frac{2\pi mkT}{h^2}\right)^{\frac{3}{2}} \geq 1$$

or, $$e^{\alpha} \geq 1 \qquad (\because \text{ Eqn. 11.31}) \qquad ...(11.33)$$

**Three cases :**

case (*i*) : When the degeneracy parameter $e^{\alpha}$ satisfies this condition, the gas is said to be **nondegenerate.** *i.e.*

$$e^{\alpha} \geq 1$$

case (*ii*) : When $e^{\alpha} > 1$, but not too large, the gas is said to be **weakly degenerate**.

case (*iii*) : When $e^{\alpha} < 1$, the gas is said to be **strongly degenerate**.

From inequality (11.33), we infer that the M.B. statistics is valid for systems at high temperature. *i.e.*, at low densities $\frac{N}{V}$ has a low value.

**Evaluation of $g_i$**

To evaluate the value of $g_i$, let us consider a *continuous distribution* of molecular energies and within this continuous distribution imagine a very small interval of energy lying between *E* and (*E*+*dE*). As *dE* finally tends to zero, the energy of any molecule lying within this energy range can be taken to be equal to *E*. Now consider a unit interval of energy within mean value *E i.e.*, the interval between (*E* – ½) and (*E* + ½). If *n* (*E*) denotes the number of molecules lying in this energy interval, then eqn. (11.15), the Maxwell-Boltzmann (general) energy distribution law becomes

$$n\,(E) = g\,(E)\,e^{-\alpha}\,e^{-\beta E} \qquad ...(11.34)$$

where *g*(*E*) gives the number of cells in the phase space corresponding to *unit* energy interval with *E* as its mean value.

If *M* is the mass of the molecule and *v* its velocity, then its momentum = *p* = *mv* and its energy $E = \frac{1}{2}mv^2 \text{ or } v = \frac{p^2}{2m}$. Substituting this an equation (11.34), we have

$$n\,(p) = g(p)e^{-\alpha}e^{-\beta\left(\frac{p^2}{2m}\right)} \qquad ...(11.35)$$

where *n*(*p*) gives the number of molecules in *unit momentum interval* between (*p* – 1/2) and (*p* + 1/2) and *g*(*p*) gives the number of cells in the phase space corresponding to this *unit momentum interval.*

The number of molecules in the momentum interval *p* and (*p* + *dp*) is obtained by multiplying (11.35) by *dp*,

$$n(p)dp = g(p)\,e^{-\alpha}\,e^{-\beta\frac{p^2}{2m}}dp \qquad ...(11.36)$$

Here, *g*(*p*)*dp* represents the number of cells in the phase space corresponding to the momentum interval between *p* and (*p* + *dp*). If $h^3$ is the volume of a single cell in the phase space, then

$$g(p)dp = \frac{\text{Volume in phase space corresponding to momentum interval between } p \text{ and } (p+dp)}{h^3}$$

$$= \frac{\iiint dxdydz \iiint dp_x\, dp_y\, dp_z}{h^3}$$

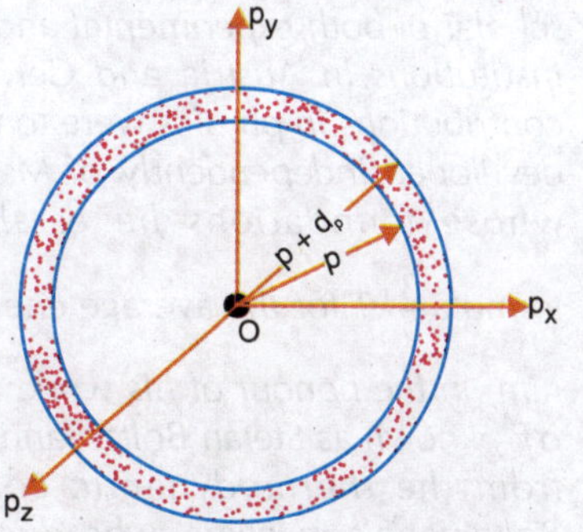

Fig. 11.2. Momentum space

But $\iiint d_x\, d_y\, d_z$ = V, the volume of the system in ordinary *position space i.e.* the volume occupied by gas molecules.

$$\therefore \qquad g(p)dp = \frac{V}{h^3}\iiint dp_x\, dp_y\, dp_z \quad ...(11.37)$$

The quantity $\iiint dp_x\, dp_y\, dp_z$ represents the volume of the system in *momentum space* corresponding to momentum interval between $p$ and $(p+dp)$.

If we draw a sphere with the origin 0 as centre and radius equal to $p$, then all the points lying on this sphere will have a momentum value $p$, given by $p^2 = p_x{}^2 + p_y{}^2 + p_z{}^2$. If we draw another concentric sphere of radius $(p + dp)$, then all the points lying on this sphere will have momentum value $p+dp$ as shown in (Fig. 11.2).

The volume in between these two concentric spheres will be

= (Surface area of the inner sphere of radius $p$) × thickness of the shell.

$$\therefore \qquad \iiint dp_x\, dp_y\, dp_z = 4\pi p^2\, dp \text{ [See details in eqn. (vi) of 11.4]} \quad ...(11.38)$$

Substituting in eqn. (11.37), we have

$$g(p)dp = \frac{V}{h^3}.\, 4\pi p^2\, dp \quad ...(11.39)$$

(Note that here $g(p)$ corresponds to $g_i$).

Substituting eqn. (11.39) in eqn. (11.36), we have

$$n(p)dp = \frac{4\pi V p^2}{h^3}\, e^{\alpha}\, e^{-\beta p^2/2m}\, dp \quad ...(11.40)$$

Substituting the value of $\beta = \frac{1}{kT}$ in eqn. (11.40), we get

$$n(p)dp = \frac{4\pi V p^2}{h^2}.\frac{N}{V}\left(\frac{h^2}{2\pi mkT}\right)^{\frac{3}{2}}.\, e^{-p^2/2mkT}\, dp$$

or

$$n(p)dp = 4\pi N\left(\frac{1}{2\pi mkT}\right)^{\frac{3}{2}}.\, p^2.\, e^{-p^2/2mkT}\, dp \quad ...(11.41)$$

This relation gives the number of molecules having moment between $p$ and $(p + dp)$ and is known as **Maxwell-Boltzmann law of distribution of momentum.**

## 11.4 Applications of M.B. Distribution Law

**1. Total internal energy and specific heat at constant volume of an ideal gas.**

The total internal energy $U$ of an ideal gas is given by

$$U = \int_0^{\infty} E\, n(E)\, dE \quad ...(11.42)$$

**LUDWIG BOLTZMANN (1844–1906)** *was born in Vienna. He was a scholar of both experimental and theoretical research at a number of institutions in Austria and Germany. Boltzmann's most important contributions in physics were to the kinetic theory of gases, which he developed independently of Maxwell, and to statistical mechanics, whose foundations he established. The constant k in the formula $\frac{3}{2}kT$ for the average energy of a gas molecule is named after him, in the honour of his work. Boltzmann derived the relation R = $\sigma T^4$, known as Stefan Boltzmann law for the radiation of a blackbody from the thermodynamic considerations. Stefan was one of Boltzmann's teachers, who discovered this law experimentally 5 years earlier Credit goes to Boltzmann to explain the second law of thermodynamics in terms of order and disorder. The most appreciated work of Boltzmann is the formula s = k log W, he established, which relates the entropy s of the system to its probability W.*

*Although Boltzmann was honoured as champion of the 'atomic theory of matter' still remained controversial in the late 19th century because there was no direct evidence for the existence of atoms and molecules. Battles with non believing scientists deeply upset Boltzmann, and later years asthma and increasingly poor eye sight further depressed his spirit and the great physicist Boltzmann in 1906, committed suicide. Boltzmann Albert Einstein, just after Boltzmann's death, published a paper "Einstein's theory of Brownian Motion" to convince the remaining doubters of the atomic theory of its correctness. Boltzmann was not alone in his despair over doubters of the existence of atoms, Max Planck was driven to an extreme of pessimism too.*

$$= \int_0^\infty E\left[\frac{2\pi N}{(\pi kT)^{\frac{3}{2}}} E^{\frac{1}{2}} e^{-E/kT}\right] dE$$

$$= \frac{2\pi N}{(\pi kT)^{\frac{3}{2}}} \int_0^\infty E^{\frac{3}{2}} e^{-E/kT} dE$$

Suppose $\frac{E}{kT} = x$, so that $E = kTx$ and $dE = kTdx$

$$\therefore \quad U = \frac{2\pi N}{(\pi kT)^{\frac{3}{2}}} \int_0^\infty (kTx)^{\frac{3}{2}} e^{-x}\, kTdx$$

$$= \frac{2\pi N}{(\pi kT)^{\frac{3}{2}}} (kT)^{\frac{3}{2}} KT \int_0^\infty x^{\frac{3}{2}} e^{-x} dx$$

$$= \frac{2\pi NKT}{\pi\sqrt{\pi}} \int_0^\infty x^{\frac{5}{2}-1} e^{-x} dx$$

The integral in the gamma function

$$\Gamma\frac{5}{2} = \frac{3}{2}\Gamma\frac{3}{2} = \frac{3}{2}.\frac{1}{2}\sqrt{\pi} = \frac{3\sqrt{\pi}}{4}$$

Substituting the value of the integral, we have

$$\Gamma\left(\frac{5}{2}\right) = \frac{3}{2}\Gamma\frac{3}{2} = \frac{3}{2}.\frac{1}{2}\Gamma\frac{1}{2} = \frac{3}{2}.\frac{1}{2}\sqrt{\pi} = \frac{3\sqrt{\pi}}{4}$$

Substituting the value of integration, we get

$$U = \frac{2NKT}{\sqrt{\pi}} \cdot \frac{3\sqrt{\pi}}{4}$$

$$U = \frac{3}{2} NKT. \qquad ...(11.43)$$

Therefore, internal energy per molecule *i.e.* average internal energy $\overline{U}$ is given by

$$\overline{U} = \frac{U}{N} = \frac{3}{2} kT \qquad ...(11.44)$$

Thus the average internal energy of an ideal monatomic gas depends only on its temperature, as $k$ is constant.

For 1 mole of an ideal gas, the quantity $N$ in equation (11.43) is nothing but Avogadro's number and hence, the value of $U$ given by equation (11.43) is the total **internal energy of 1 mole of an ideal gas.**

The molar specific heat of the gas at constant volume is given by

$$C_V = \left(\frac{\partial U}{\partial T}\right)_V$$

$$C_V = \frac{\partial}{\partial T}\left(\frac{3}{2} NkT\right) \qquad ... (\because \text{ Eqn. } 11.43)$$

$$= \frac{3}{2} Nk \qquad ...(11.45)$$

or, $$C_V = \frac{3}{2} N\left(\frac{R}{N}\right) \qquad \left(\because k = \frac{R}{N}\right)$$

or, $$C_V = \frac{3}{2} R \qquad ...(11.46)$$

where $R$ is the universal gas constant for 1 mole.

**2. Maxwell - Boltzmann Speed Distribution Law**

The Maxwell-Boltzmann equation for distribution of energy among the molecules of an ideal gas is given by

$$n(E)\, dE = \frac{2\pi N}{(\pi KT)^{\frac{3}{2}}} E^{\frac{1}{2}} e^{-E/kT} dE \qquad ...(i)$$

A classical ideal gas is defined as an assembly of non-interacting molecules, each distinguishable from one another. Therefore, the molecules have no internal degrees of freedom, all the energy of the gas molecules is completely kinetic in nature. Thus,

$$E = \frac{1}{2} mv^2, \text{ Hence } dE = mvdv$$

Substituting these values of $E$ and $dE$ in equation $(i)$, the number of molecules $n(v)dv$ whose speeds lie between $v$ and $(v + dv)$ is given by

$$n(v)dv = \frac{2\pi N}{(\pi kT)^{\frac{3}{2}}} \cdot \left(\frac{1}{2} mv^2\right)^{\frac{1}{2}} e^{-mv^2/2kT} \quad mvdv$$

$$= \frac{2\pi N}{(\pi kT)^{\frac{3}{2}}} \cdot 2\left(\frac{m}{2}\right)^{\frac{3}{2}} v^2 e^{-mv^2/2kT} \quad dv$$

$$n(v)dv = 4\pi N\left(\frac{m}{2\pi kT}\right)^{\frac{3}{2}} v^2 e^{-mv^2/kT} dv \quad ...(11.47)$$

This eqution is known as the **Maxwell or Maxwell-Boltzmann law of distribution of speeds** among the molecules of a gas. In this $n(v)$ is the number of molecules per unit speed range.

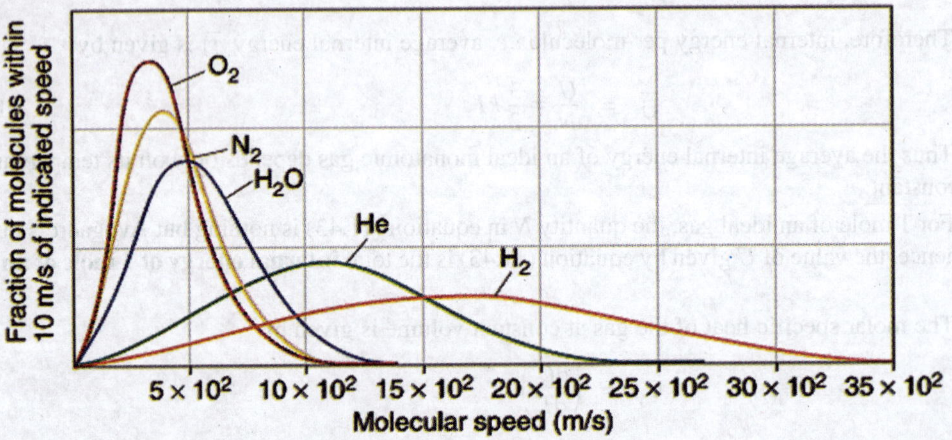


**Fig. 11.2**

**Aliter :** Let us consider an ideal gas in a vessel of volume $V$. In order to derive an expression for the number of molecules of the gas whose speeds lie between $v$ and $v + dv$.

For this, let us consider a zone of cells, *i.e.* a group of cells in the *phase space* corresponding to the range in speed from $v$ to $v + dv$. It $g_i$ represents the number of cells in the zone, then the number of molecules in the $i$th zone, each having energy close to $E_i = \frac{1}{2}mv^2$, is given by Maxwell-Boltzmann law as

$$n_i = Ag_i e^{-\beta E_i} \quad \text{(From eqn. 10.38)} \quad ...(i)$$

where $\beta = \frac{1}{kT}$ where $A = \frac{Z}{N}$

Here $k$ is Boltzmann constant and $T$ the absolute temperature.

As the number of molecules having speed ranging between $v$ and $v + dv$ will be proportional to $dv$, we may write

$$n_i = N(v)dv \quad ...(ii)$$

where $N(v)$ is known as the *distribution function* pertaining to speeds. Substituting for $n_i$ in equation (*i*), we get

$$N(v)dv = Ag_i e^{-E_i/kT} \quad ...(iii)$$

**To evaluate $g_i$**

Let us now express $g_i$ in terms of $v$ and $dv$. We have

$$g_i = \text{Number of cells in the zone under consideration.}$$

$$= \frac{\text{Volume of the zone in the phase space}}{\text{Volume of one cell}}$$

*i.e.*

$$g_i = \frac{\iiint\iiint dx\,dy\,dz\,dp_x\,dp_y\,dp_z}{h^3} \quad ...(iv)$$

If $V$ is the volume occupied by the gas; then it is

$$V = \iiint dx\, dy\, dz \qquad ...(v)$$

Now we have to integrate with respect to $p_x, p_y$, and $p_z$. Here we will consider only the resultant momentum $p$ and do not take care of the momentum components. For this let us imagine that $p_x, p_y$ and $p_z$ are along three mutually perpendicular axes and draw a sphere of radius $p$ about the origin, in the momentum space (refer Fig. 11.3). Then the surface of the sphere will contain all the points for which $p_x^2 + p_y^2 + p_z^2 = p^2$. Now draw another concentric sphere of radius $(p + dp)$. The volume between these two concentric spheres of radius $p$ and $p + dp$ will contain space having momentum between $p$ and $p + dp$.

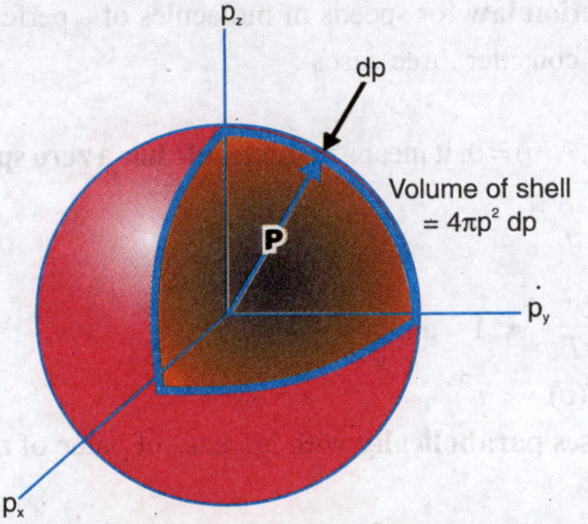


**Fig. 11.3. The coordinates in momentum space are $p_x, p_y, p_z$. The number of momentum states available to a particle with a momentum whose magnitude is between $p$ and $p + dp$ is proportional to the volume of a spherical shell in momentum space of radius $p$ and thickness $dp$.**

The volume between two concentric spheres will be

$$\iiint dp_x\, dp_y\, dp_z = \left(\frac{4}{3}\right)\pi\,(p+dp)^3 - \left(\frac{4}{3}\right)\pi\, p^3$$

$$= \left(\frac{4}{3}\right)\pi\left[(p+dp)^3 - p^3\right]$$

$$= \left(\frac{4}{3}\right)\pi\left[p^3 + 3p^2 dp + 3p(dp)^2 + dp^3 - p^3\right]$$

$$= \left(\frac{4}{3}\right)\pi\left(3p^2 dp\right) \qquad \text{neglecting higher terms } dp^2 \text{ and } dp^3.$$

$$= 4\,\pi\, p^2\, dp \qquad ...(vi)$$

Now substituting eqn. $(v)$ and $(vi)$ in $(iv)$, we get

$$g_i = \frac{V\, 4\pi p^2 dp}{h^3}$$

or,

$$g_i = \frac{4\pi\, V p^2 dp}{h^3} \qquad ...(vii)$$

Substituting for $g_i$ in eqn. $(iii)$, and using $E_i = \frac{1}{2} m v^2$, we get

$$N(v)dv = \frac{4\pi\, AV}{h^3}\, p^2\, e^{-\beta m v^2/2}\, dp$$

putting $p = mv$ and $dp = mdv$

$$N(v)dv = \frac{4\pi AV m^3}{h^3} v^2 e^{-\beta mv^2/2} dv$$

We may put $\frac{4\pi AVm^3}{h^3}$ = B, where B is any constant, we have

$$N(v)dv = Bv^2 e^{-\beta mv^2/2} dv$$

or, $$N(v)dv = Bv^2 e^{-mv^2/2kT} dv \quad \left(\text{Since } \beta = \frac{1}{kT}\right) \quad ...(11.48)$$

This is **Maxwell's distribution law** for speeds of molecules of a perfect gas.

**Discussion.** For this, let us consider three cases :

**Case (*a*) : If $v = 0$**

Then from equation (11.48), $N(v) = 0$. It means **no molecule has a zero speed.** This is an important conclusion.

**Case (*b*) : For small values of $v$**

In this case, $\frac{mv^2}{2kT} \ll 1$ and $e^{\frac{-mv^2}{2kT}} = 1$

$\therefore$ $$N(v) \propto v^2$$

Thus, value of $N(v)$ **increases parabolically** with increase of value of $v$ and attains a maximum value for a particular value of $v$.

**Case (*c*) : For large values of $v$**

In this case, $\frac{mv^2}{2kT} >> 1$ and the exponential term $e^{-mv^2/2kT}$ dominates over $v^2$. Hence

$$N(v) \propto e^{-mv^2/2kT}$$

Thus, value of $N(v)$ **decreases exponentially** with increase of value of $v$.

This behaviour is explained by Fig. 11.4. If a graph is plotted $N(v)$ against $v$, the curve first rises parabolically, reaches a maximum and then falls exponentially to zero as shown in Fig. 11.4.

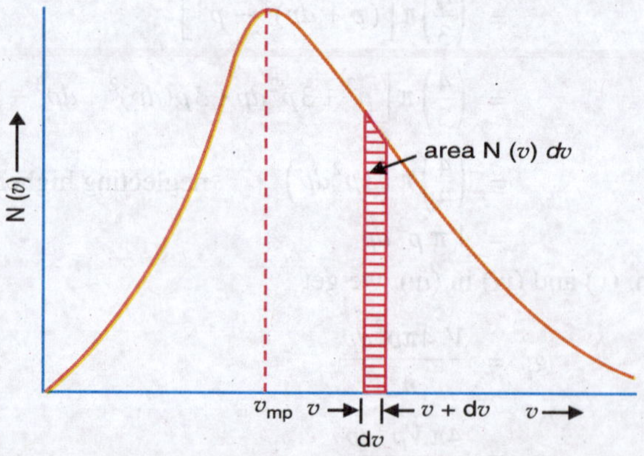


**Fig. 11.4.**

The maximum corresponds to the most probable speed $v_{mp}$. As the area under the curve towards right of maximum is greater than to the left, it means that the mean speed $v$ is greater than $v_{mp}$. The area of the shaded portion gives the number of molecules lying in the speed range $v$ to $v + dv$.

### Evaluation of constant B

The quantity $N(v)\,dv$ represents the number of gas molecules having speed in the range of $v$ to $(v + dv)$ and is given by equation (11.48). In fact, the molecules may have speed ranging from 0 to ∞, hence we may write

$$\int N(v)dv = N$$

where $N$ is total number of molecules in the given state.

Thus,

$$\int Bv^2 e^{-\frac{mv^2}{2kT}}\, dv = N$$

or $$B\int v^2 e^{-\frac{mv^2}{2kT}}\, dv = N$$

or $$B\left[\frac{1}{4}\sqrt{\frac{8\pi k^3T^3}{m^3}}\right] = N$$

or $$B = 4N\sqrt{\frac{m^3}{8\pi k^3T^3}}$$

$$= \frac{4N}{\sqrt{\pi}}\left(\frac{m}{2kT}\right)^{3/2} = 4\pi N\left(\frac{m}{2\pi kT}\right)^{3/2} \qquad ...(11.49)$$

Substituting for $B$ in equation (11.48), the **Maxwell's distribution law of speed** becomes

$$N(v)dv = 4\pi N\left(\frac{m}{2\pi kT}\right)^{3/2} v^2\, e^{-\frac{mv^2}{2kT}}\, dv \qquad ...(11.50)$$

or $$N(v) = 4\pi N\left(\frac{m}{2\pi kT}\right)^{3/2} v^2\, e^{-\frac{mv^2}{2kT}} \qquad ...(11.51)$$

The effect of temperature on the distribution of speed of a given number of molecules can be understood by using this relation. The graphs $N(v)$ against $v$ are plotted using equation (11.51) at different temperatures.

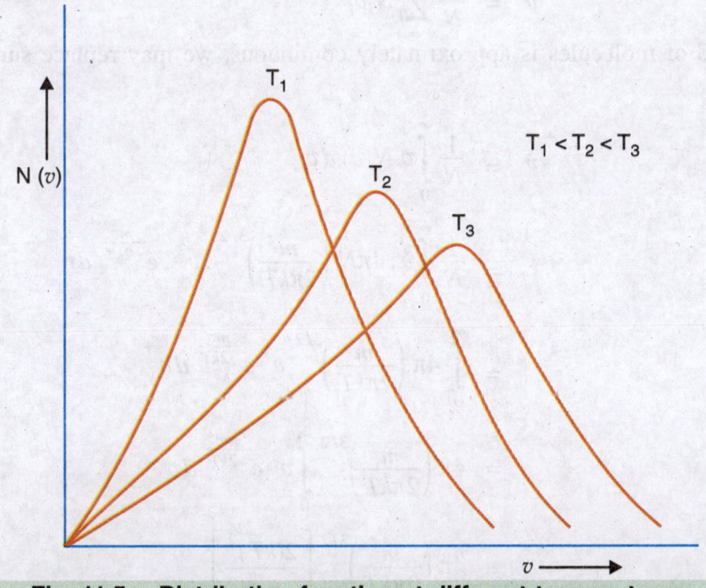


**Fig. 11.5 : Distribution function at different temperatures**

Temperatures $T_1$, $T_2$ and $T_3$ such that $T_1 < T_2 < T_3$ are drawn in Fig. 11.5. From the graphs, we conclude that higher the temperature, higher are the speeds possessed by the gas molecules. However, the area under all the three curves is the same.

The fraction of molecules having speed ranging between $v$ and $v + dv$ is given by

$$\frac{N(v)dv}{N} = 4\pi\left(\frac{m}{2\pi kT}\right)^{3/2} v^2 e^{-\frac{mv^2}{2kT}} dv \qquad ...(11.52)$$

But according to the definition of probability, this ratio represents the probability of molecules to possess the speed between $v$ and $(v + dv)$.

$$P(v)dv = 4\pi\left(\frac{m}{2\pi kT}\right)^{3/2} v^2 e^{-\frac{mv^2}{2kT}} dv$$

The probability function $P(v)$ is given by

$$P(v) = 4\pi\left(\frac{m}{2\pi kT}\right)^{3/2} v^2 e^{-\frac{mv^2}{2kT}} \qquad ...(11.53)$$

## 11.5 Mean, RMS and Most Probable Speeds

### Mean or Average Speed $(\bar{v})$ :

Let there be $N_i$ molecules each having speed $v_i$, then mean or average speed $(\bar{v})$ is defined as

$$(\bar{v}) = \frac{\sum_i N_i v_i}{\sum_i N_i}$$

But $\sum_i N_i = N$, the total number of molecules,

$$\therefore \qquad \bar{v} = \frac{1}{N}\sum N_i v_i$$

As the speed of molecules is approximately continuous, we may replace summation sign by integration, *i.e.*,

$$\bar{v} = \frac{1}{N}\int_0^\infty v\, N(v)\, dv$$

$$= \frac{1}{N}\int_0^\infty v \,.\, 4\pi N\left(\frac{m}{2\pi kT}\right)^{3/2} .\, v^2 \,.\, e^{-\frac{mv^2}{2kT}} dv$$

$$= \int_0^\infty 4\pi\left(\frac{m}{2\pi kT}\right)^{3/2} v^3 e^{-\frac{mv^2}{2kT}} dv$$

$$= 4\pi\left(\frac{m}{2\pi kT}\right)^{3/2} \int_0^\infty v^3 e^{-\frac{mv^2}{2kT}} dv$$

$$= 4\pi\left(\frac{m}{2\pi kT}\right)^{3/2}\left[\frac{2(kT)^2}{m^2}\right]$$

$$\bar{v} = \sqrt{\frac{8kT}{m\pi}} = 1.59\sqrt{\frac{kT}{m}} \qquad ...(11.54)$$

This equation gives the mean or average speed of a gas molecule.

## Root Mean Square Speed ($v_{rms}$)

The mean square speed $\left(\overline{v^2}\right)$ is given by

$$\left(\overline{v^2}\right) = \frac{1}{N}\int_0^\infty v^2 N(v)\,dv$$

$$= \frac{1}{N}\int_0^\infty v^2 \,.\, 4\pi N\left(\frac{m}{2\pi kT}\right)^{3/2} v^2 e^{-\frac{mv^2}{2kT}}\,dv$$

$$= 4\pi\left(\frac{m}{2\pi kT}\right)^{3/2}\int_0^\infty v^4 e^{-\frac{mv^2}{2kT}}\,dv$$

$$= 4\pi\left(\frac{m}{2\pi kT}\right)^{3/2}\left[\frac{3}{8}\sqrt{\left\{\frac{32\pi\,(kT)^5}{m^5}\right\}}\right]$$

$$= \frac{3kT}{m}$$

Taking square root of both sides,

$$v_{rms} = \sqrt{\left(\overline{v^2}\right)} = \sqrt{\frac{3kT}{m}} \qquad ...(11.55)$$

The quantity $\sqrt{\left(\overline{v^2}\right)}$ is called **root mean square speed** of the gas molecules.

## Most Probable Speed ($v_{mp}$)

For the most probable speed the probability $P(v)$ will be maximum. Therefore, to obtain the most probable speed, let us differentiate equation (11.53) with respect to $v$ and equate it to zero. Thus

$$4\pi\left(\frac{m}{2\pi kT}\right)^{3/2}\left[2v\,e^{\left(-\frac{mv^2}{2kT}\right)} - \frac{mv^3}{kT}\,e^{\left(-\frac{mv^2}{2kT}\right)}\right] = 0$$

$$2v\,e^{-\frac{mv^2}{2kT}} = \frac{mv^3}{kT}\,e^{-\frac{mv^2}{2kT}}$$

or $$2v = \frac{mv^3}{kT}$$

or $$v^2 = \frac{2kT}{m}$$

or $$v = \sqrt{\frac{2kT}{m}}$$

This $v$ is the most probable speed, therefore putting it equal to $v_{mp}$, we have

$$v_{mp} = \sqrt{\frac{2kT}{m}} \qquad \text{...(11.56)}$$

Thus, from equations (11.54), (11.55) and (11.56), it is clear that $v_{rms} > \bar{v} > v_{mp}$.

## 11.6 Maxwell's Distribution Law of Velocities

Let us consider an ideal gas contained in a container of volume $V$. Then applying the Maxwell-Boltzmann's canonical distribution law, in equilibrium, the number of molecules in a cell of energy $E_i$ will be

$$n_i = A\,e^{-\beta E_i}$$

The number of molecules having position coordinates in the range $x$ to $x + dx$, $y$ to $y + dy$, $z$ to $z + dz$ and the velocity components in ranges $v_x$ to $v_x + dv_x$, $v_y$ to $v_y + dv_y$, $v_z$ to $v_z + dv_z$ will be proportional to the element of phase of volume $dx\,dy\,dz\,dv_x\,dv_y\,dv_z$.

Therefore, according to Maxwell-Boltzmann's canonical distribution law, the number of molecules having energy $E_i$ and having position coordinates between $x$ and $x + dx$, $y$ and $y + dy$, $z$ and $z + dz$ and velocity components between $v_x$ and $v_x + dv_x$, $v_y$ and $v_y + dv_y$, $v_z$ and $v_z + dv_z$ is given by

$$n_i\,dx\,dy\,dz\,dv_x\,dv_y\,dv_z = A\,e^{-\beta E_i}\,dx\,dy\,dz\,dv_x\,dv_y\,dv_z \qquad \text{...(11.57)}$$

But $E_i$ = energy of the particle $= \frac{1}{2}mv^2 = \frac{1}{2}m\left(v_x^2 + v_y^2 + v_z^2\right)$

Substituting for $E_i$ in equation (11.57), we get

$$n_i\,dx\,dy\,dz\,dv_x\,dv_y\,dv_z = A\,e^{-m\beta\left[\frac{1}{2}\left(v_x^2+v_y^2+v_z^2\right)\right]}\,dx\,dy\,dz\,dv_x\,dv_y\,dv_z \qquad \text{...(11.58)}$$

The constant $A$ is evaluated by integrating equation (11.58) over all available volume and all ranges of velocities, *i.e.*,

$$N = \iiiiii A\,e^{-m\beta\frac{1}{2}(v_x^2+v_y^2+v_z^2)}\,dx\,dy\,dz\,dv_x\,dv_y\,dv_z$$

$$= \iiint dx\,dy\,dz \iiint A\,e^{-m\beta\frac{1}{2}(v_x^2+v_y^2+v_z^2)}\,dv_x\,dv_y\,dv_z$$

But $\iiint dx\,dy\,dz = V$ = Volume of vessel containing gas.

$$\therefore \quad N = AV\iiint e^{-m\beta\left[\frac{1}{2}(v_x^2+v_y^2+v_z^2)\right]}\,dv_x\,dv_y\,dv_z$$

$$= AV\int_{-\infty}^{+\infty} e^{-\frac{\beta m v_x^2}{2}}\,dv_x \int_{-\infty}^{+\infty} e^{-\frac{\beta m v_y^2}{2}}\,dv_y \int_{-\infty}^{+\infty} e^{-\frac{\beta m v_z^2}{2}}\,dv_z \qquad \text{...(11.59)}$$

Using standard definite integrals, we have

$$\int_{-\infty}^{+\infty} e^{-\frac{\beta m v_x^2}{2}}\,dv_x = \sqrt{\frac{2\pi}{m\beta}}$$

$$\int_{-\infty}^{+\infty} e^{-\frac{\beta m v_y^2}{2}}\,dv_y = \sqrt{\frac{2\pi}{m\beta}}$$

and $$\int_{-\infty}^{+\infty} e^{-\frac{\beta m v_z^2}{2}} dv_z = \sqrt{\frac{2\pi}{m\beta}}$$

Substituting these values in equation (11.59), we get

$$N = AV\left(\frac{2\pi}{m\beta}\right)^{3/2}$$

or $$A = \frac{N}{V}\left(\frac{m\beta}{2\pi}\right)^{3/2}$$

or $$A = \frac{N}{V}\left(\frac{m}{2\pi kT}\right)^{3/2} \quad \left(\because \beta = \frac{1}{kT}\right) \qquad ...(11.60)$$

Now substituting values of A and β in equation (11.57), we get

$$n_i\, dx\, dy\, dz\, dv_x\, dv_y\, dv_z = \frac{N}{V}\left(\frac{m}{2\pi kT}\right)^{3/2} e^{-\frac{m(v_x^2+v_y^2+v_z^2)}{2kT}} dx\, dy\, dz\, dv_x\, dv_y\, dv_z \qquad ...(11.61)$$

The number of molecules having velocity coordinates in the range $v_x$ to $v_x + dv_x$, $v_y$ to $v_y + dv_y$, $v_z$ to $v_z + dv_z$ irrespective of the position coordinates, can be found by integrating equation (11.61) with respect to position coordinates, which gives

$$n_i\, dv_x\, dv_y\, dv_z = \frac{N}{V}\left(\frac{m}{2\pi kT}\right)^{3/2} . e^{-\frac{m(v_x^2+v_y^2+v_z^2)}{2kT}} . dv_x\, dv_y\, dv_z . V$$

$$= N\left(\frac{m}{2\pi kT}\right)^{3/2} . e^{-\frac{m(v_x^2+v_y^2+v_z^2)}{2kT}} . dv_x\, dv_y\, dv_z \qquad ...(11.62)$$

Lastly let us find the number of molecules having velocity components in the range $v_x$ to $v_x + dv_x$ irrespective of $v_y$. This will be obtained by integrating eqn. (11.62) with respect to $v_y$ and $v_z$, *i.e.*,

$$n(v_x)\, dv_x = N\left(\frac{m}{2\pi kT}\right)^{3/2} dv_x \iint e^{-m\left(\frac{v_x^2+v_y^2+v_z^2}{2kT}\right)} dv_y\, dv_z$$

$$= N\left(\frac{m}{2\pi kT}\right)^{3/2} . e^{-mv_x^2/2kT}\, dv_x \int_{-\infty}^{+\infty} e^{-mv_y^2/2kT}\, dv_y \int_{-\infty}^{+\infty} e^{-mv_z^2/2kT}\, dv_z$$

$$= N\left(\frac{m}{2\pi kT}\right)^{3/2} e^{-mv_x^2/2kT}\, dv_x \left[\sqrt{\frac{2\pi kT}{m}}\right] \times \left[\sqrt{\frac{2\pi kT}{m}}\right]$$

$$\therefore \quad n(v_x)dv_x = N\left(\frac{m}{2\pi kT}\right)^{1/2} e^{-mv_x^2/2kT}\, dv_x \qquad ...(11.63)$$

This eqation (11.63) gives the number of molecules having $x$-component of velocity in the range between $v_x$ and $v_x + dv_x$. Eq. (11.63) shows that the velocity component $v_x$ is distributed symmetrically about the value $v_x = 0$.

The probability that a molecule will have $x$-component of velocity in the range $v_x$ to $v_x + dv_x$ is given by

$$P(v_x)\,dv_x = \frac{n(v_x)}{N}\,dv_x = \left(\frac{m}{2\pi kT}\right)^{\frac{1}{2}} e^{-\frac{mv_x^2}{2kT}}\,dv_x \quad ...(11.64)$$

Equations (11.63) and (11.64) represent **Maxwell's distribution law of velocities.**

The probability function

$P(v_x)$ is given by $\dfrac{P(v_x)dv_x}{dv_x}$

$$\therefore \quad P(v_x) = \left(\frac{m}{2\pi kT}\right)^{\frac{1}{2}} .\, e^{-mv_x^2/2kT} \quad ...(11.65)$$

Equation (11.65) can also be expressed in terms of momentum component $p_x$. Substituting $mv_x = p_x$ and $mdv_x = dp_x$ on R.H.S. of equation (11.64) and dividing by $dp_x$, we get

$$P(p_x) = \left(\frac{1}{2\pi mkT}\right)^{\frac{1}{2}} .\, e^{-\frac{p_x^2}{2mkT}} \quad ...(11.66)$$

**Discussion :** Consider the following three situations :

The probability that a molecule will have velocity component between $v_x$ and $v_x + dv_x$ is given by

$$P(v_x)dv_x = \left(\frac{m}{2\pi kT}\right)^{\frac{1}{2}} .\, e^{-mv_x^2/2kT}\, dv_x \quad ...(11.67)$$

(*i*) This equation indicates that the function $P(v_x)$ is symmetrically distributed about the value $v_x = 0$, since average value of $v_x$ for Maxwellian distribution is zero, *i.e.*,

$$\bar{v}_x = \int_{-\infty}^{+\infty} v_x\, P(v_x)\, dv_x$$

$$= \int_{-\infty}^{+\infty} \left(\frac{m}{2\pi kT}\right)^{\frac{1}{2}} v_x \,.\, e^{-mv_x^2/2kT}\, dv_x$$

$$= 0 \qquad \left[\text{Since } \int_{-\infty}^{+\infty} v_x \,.\, e^{-mv_x^2/2kT}\, dv_x = 0\right]$$

Also from equation (11.67), it is clear that

$$P(-v_x) = P(v_x) \quad ... (11.68)$$

This again indicates that $P(v_x)$ is symmetrical about $v_x = 0$.

(*ii*) The probability distribution function is maximum for that value of $v_x$ for which

$$\frac{\partial P(v_x)}{\partial v_x} = 0$$

or
$$\frac{\partial}{\partial v_x}\left[\left(\frac{m}{2\pi kT}\right)^{\frac{1}{2}} e^{-mv_x^2/2kT}\right] = 0$$

This gives $v_x = 0$ and the maximum value is

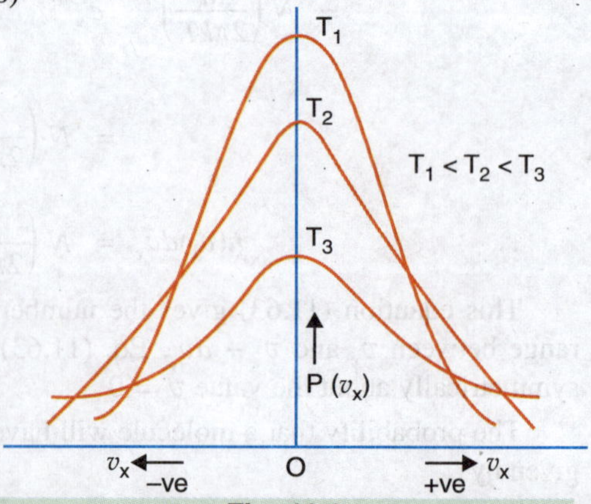


Fig. 11.6

$$P_{max} = \left(\frac{m}{2\pi kT}\right)^{\frac{1}{2}} \qquad ...(11.69)$$

This shows that the $P_{max}$ increases as $m$ increases and $T$ decreases.

(*iii*) $P$ falls to $\frac{1}{e}$ times $P_{max}$ at value of $v_x$, given by

$$P(v_x) = \frac{1}{e}P_{max} = \left(\frac{m}{2\pi kT}\right)^{\frac{1}{2}}$$

or
$$\left(\frac{m}{2\pi kT}\right)^{\frac{1}{2}} e^{-mv_x^2/2kT} = \left(\frac{m}{2\pi kT}\right)^{\frac{1}{2}}$$

or
$$e^{-mv_x^2/2kT} = 1$$

This gives
$$v_x = \sqrt{\frac{2kT}{m}} \qquad ...(11.70)$$

The probability function $P(v_x)$ is plotted against $v_x$ for three different temperatures $T_1$, $T_2$ and $T_3$ such that $T_1 < T_2 < T_3$ as shown in Fig. 11.6. As $T$ increases the peak at $v_x = 0$ lowers down and the distribution spreads out. However, the area under the curve is always *unity*.

## 11.7 Doppler Broadening of Spectral Lines

The Maxwellian distribution law of velocities has been verified quantitatively and proved to be the best from Doppler effect in light. The experimental verification of Maxwell distribution function is obtained by measuring the width of a spectral line emitted by hot gas molecules at low densities. The reason for broadening is Doppler's effect. The broadening (*i.e.* spread) arises from the distribution of velocities of the molecules in a gas.

An atom or molecule at rest emits any particular spectral line with a fairly sharp frequency $\nu$. Let a molecule be moving with velocity $v_x$ towards an observer. Further, let the molecule be emitting a radiation of wavelength $\lambda$ in the direction of motion. The wavelength recorded $\lambda'$ by the observer using spectrometer is related to $\lambda$ by Doppler expression as

$$\lambda' = \lambda\left(1 - \frac{v_x}{C}\right)$$

or,
$$v_x = \frac{C(\lambda - \lambda')}{\lambda} \qquad ...(i)$$

and
$$dv_x = -C\frac{d\lambda'}{\lambda} \qquad ...(ii)$$

The negative sign relates the direction for $v_x$.

The probability that a molecule has component velocity in the range from $v_x$ to $v_x + dv_x$ is given by

$$\left(\frac{m}{2\pi kT}\right)^{\frac{1}{2}} e^{\left\{\frac{-mv_x^2}{2kT}\right\}} dv_x \qquad ...(iii)$$

Substituting the value of $v_x$ and $dv_x$ from Eqn. (*i*) and (*ii*), in Eqn. (*iii*), we get

$$\left(\frac{m}{2\pi kT}\right)^{\frac{1}{2}} e^{\left\{-\frac{mc^2}{2kT}\left(\frac{\lambda-\lambda'}{\lambda}\right)^2\right\}} \cdot \frac{c}{\lambda}d\lambda'$$

or $$A\, e^{\left\{-\frac{mc^2}{2kT}\left(\frac{\lambda-\lambda'}{\lambda}\right)^2\right\}} d\lambda' \qquad ...(iv)$$

where $$A = \left(\frac{m}{2\pi kT}\right)^{\frac{1}{2}} \cdot \frac{c}{\lambda} \qquad ...(v)$$

Expression (*iv*)represents the intensity of radiation $I_{\lambda'} d_{\lambda'}$ received by the observer. Hence

$$I_{\lambda'} d_{\lambda'} = A\, e^{\left\{\frac{-mc^2}{2kT}\left(\frac{\lambda-\lambda'}{\lambda}\right)^2\right\}} d\lambda' \qquad ...(vi)$$

when $$\lambda' = \lambda, \text{ then } A = I_\lambda$$

$\therefore$ $$I_{\lambda'} d_{\lambda'} = I_\lambda \exp\left\{-\frac{mc^2}{2kT}\left(\frac{\lambda-\lambda'}{\lambda}\right)^2\right\} d\lambda' \qquad ...(vii)$$

A graph plotted intensity ($I$) against frequency ($\nu$), we get a intensity distribution curve, which has a Gaussian line shape as shown in Fig, 11.7. The width of the spectral line is usually measured at the half-intensity $k$ points as shown in figure. If eqn. (*vii*) is expressed in terms of frequency, then half-frequency width $\Delta\nu_d$ can be expressed as

$$\frac{\Delta\nu_d}{\nu} = 2\left(\frac{2kT\log_e(2)}{mc^2}\right)^{\frac{3}{2}} \qquad ...(11.71)$$

The theoretical and experimental agreement was found to be excellent.

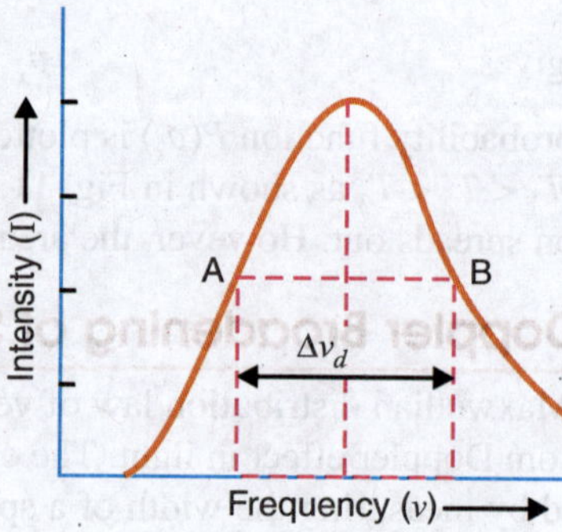


**Fig. 11.7**

It we put $\lambda - \lambda' = s$ = the shift in wavelength, then from equation (*vii*), we write

$$I_{\lambda'} = I_\lambda \exp\left\{-\frac{mc^2 s^2}{2\,kT\,\lambda^2}\right\}$$
$$= I_\lambda \exp\{-bs^2\} \qquad ...(11.72)$$

where $$b = \frac{mc^2}{2kT\lambda^2}$$

Equation (11.72) may be written as,

$$I_{\lambda'} = I_\lambda e^{-bs^2}$$

or, $$\frac{I_{\lambda'}}{I_\lambda} = e^{-bs^2} \qquad ...(11.73)$$

$\therefore$ $$\frac{I_{\lambda'}}{I_\lambda} = e^{bs^2}$$

or $$\log\left(\frac{I_\lambda}{I_{\lambda'}}\right) = bs^2$$

or $$s^2 = \frac{1}{b}\log\left(\frac{I_\lambda}{I_{\lambda'}}\right)$$

or $$s = \left[\frac{1}{b}\log\left(\frac{I_\lambda}{I_{\lambda'}}\right)\right]^{\frac{1}{2}} \qquad ...(11.74)$$

## 11.8 Limitations of Maxwell-Boltzmann's Method

The method has lot of limitations. Some of them are discussed below:

1. It is applicable only to an isolated gas of identical molecules in equilibrium $b$ for which the following conditions are satisfied :

(*i*) The mean potential energy due to mutual interaction between the molecules is very small compared to their mean kinetic energy. (*Under this condition the gas is said to be* ***ideal.***)

(*ii*) The gas is dilute, *i.e.* the number of molecules per unit volume is small, so that the average separation between the molecules is large and hence individual molecules can be distinguished. (*Under this condition the gas is said to be non-degenerate*).

Therefore, the important results, such as expressions for $E$, $C_v$ and $\underline{P}$, obtained by this method are the same as those derived by applying simple kinetic theory.

2. The expression for the Maxwell-Boltzmann count does not give correct expression for the entropy $S$ of an ideal gas, and thus leads to the Gibbs paradox. To resolve the paradox, the expression must be divided by $N$!

3. In the expression for entropy of an ideal gas

$$S = Nk \log\left[\frac{2\pi mkT}{h^2}\right]^{\frac{3}{2}} e^{\frac{5}{2}}$$

if we put, $T = 0$, we get

$$S = Nk \log 0$$
$$= \text{negative quantity}$$

Thus, the expression for $S$ does not satisfy the third law of thermodynamics. According to this law, as the temperature of the system approaches the absolute zero, the entropy aproaches a constant value $S_0$, independent of the system.

4. It cannot be applied to a system of indistinguishable particles. If we apply the Maxwell-Boltzmann distributin law to the thermionic emission, we get the expression for the emission current density

$$J = A_0 T^{\frac{1}{2}} e^{-\phi/kT}$$

which is not correct. The correct expression which has been varified experimentally is

$$J = A_0 e^2 e^{-\phi/kT}.$$

## 11.9 Experimental Verification of Maxwellian distribution of Molecular Speeds

O'Stern tried for the first time to prove experimentally the Maxwell's distribution law of molecular speeds. Eldrige improved this method as the results were not satisfactory. Later a more sophisticated method was suggested by *I.F.* Zartman and C.C. *KO*. Lammert used another method based on Fizeau's toothed wheel model for finding the velocity of light. The basic principle employed is the same in all these methods.

**Zartman and C.C. KO Experiment.** Fig. 11.8 describes the experimental arrangement used by *I.F.* Zartman and C.C. *KO*. Bismuth is taken in the oven $O$ and heated upto 800ºC. Bismuth melts and vaporises. The bismuth molecules in vapour form comes out through a slit $S_1$ and the beam of bismuth molecules is collimated by another slit $S_2$, kept closed to $S_1$. A hollow cylindrical drum $D$ is kept exactly vertically up that of slit $S_2$, capable of rotating with a very high speed, nearly 1000 – 15000 rotations per second about an axis perpendicular to the plane of the paper in anticlockwise direction as shown in fig. 11.8. The drum has a narrow slit $S_3$ in a line that of slit $S_2$. A glass plate $P$ is fixed on the inner side of the drum so that the bismuth molecules can deposit on this glass plate after passing through slits $S_1$, $S_2$ and $S_3$. The whole system is enclosed in a vessel which is highly evacuated to ensure that every bismuth molecule goes from $S_1 - S_2 - S_3$ and deposit ultimately on glass plate $P$ without suffering any scattering due to collision with air molecules on their way.

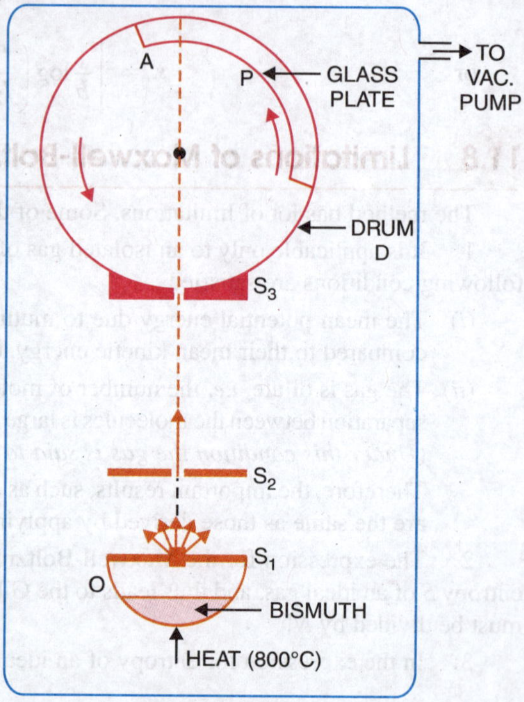


Fig. 11.8

If the drum is stationary, then all the bismuth molecules will strike the glass plate at a single point $A$ just opposite to $S_3$. But when the drum is made to rotate in anticlockwise direction with very high and constant speed, the glass plate $P$ moves towards left and bismuth molecules get deposited on the glass plate, at different points depending upon their speeds. The more rapidly moving bismuth molecules reach the glass plate earlier and condense on it at a point nearer to $A$ than the slowly moving ones which take a longer time to reach the glass plate. After some time, when sufficient amount of bismuth is deposited, the glass plate $P$ is removed and the relative intensity of bismuth molecules at various points is measured with the help of micro-photometer. The relative intensity of bismuth molecules is proportional to the number of bismuth molecules striking at that point and gives the relative number of molecules having different velocities. This gives directly the *velocity spectrum* of bismuth molecules in terms of distance from point $A$. The distance from $A$ is a measure of distribution of molecular speeds. This distribution obtained is found to agree well with the Maxwellian distribution of molecular speeds.

The results obtained from Zartman and C.C. Ko experiment are shown in Fig. 11.9 where the intensity of deposit is plotted on ordinate (Y – axis) and the displacement from the zero-mark or the velocity of the corresponding molecule on the abscissa (X – axis). The points drawn in the diagram are the observed densities of the deposit and the line as calculated on the assumption that the velocities of the molecules are distributed according to Maxwell's law. There is an excellent agreement between the theoretical curve and the observed point. This experiment suggested by Zartman and C.C. Ko proved to be an accurate test of Maxwell's law of distribution of velocities employing classical statistics.

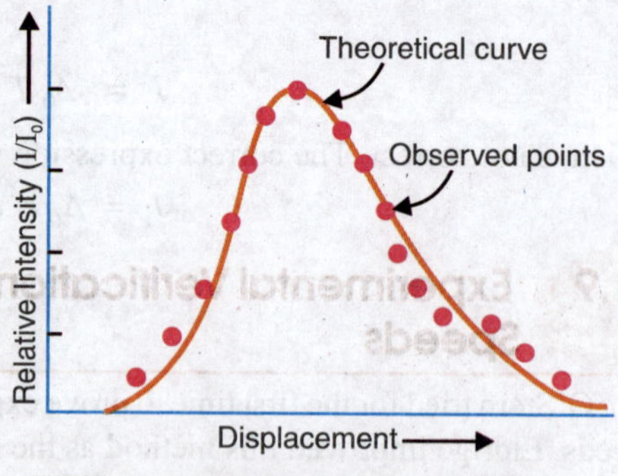


Fig. 11.9

## 11.10 Some Useful Standard Definite Integrals

If $$I_n = \int_0^\infty v^n e^{-xv^2} dv,$$

then the value of $I_n$ for various value of $n$, which may used directly, are given below

$$I_0 = \int_0^\infty e^{-\alpha v^2} dv = \frac{1}{2}\sqrt{\left(\frac{\pi}{\alpha}\right)} \quad \text{...(1)}$$

$$I_1 = \int_0^\infty v e^{-\alpha v^2} dv = \frac{1}{2\alpha} \quad \text{...(2)}$$

$$I_2 = \int_0^\infty v^2 e^{-\alpha v^2} dv = \frac{1}{4}\sqrt{\frac{\pi}{\alpha^3}} \quad \text{...(3)}$$

$$I_3 = \int_0^\infty v^3 e^{-\alpha v^2} dv = \frac{1}{2\alpha^2} \quad \text{...(4)}$$

$$I_4 = \int_0^\infty v^4 e^{-\alpha v^2} dv = \frac{3}{8}\sqrt{\left(\frac{\pi}{\alpha^5}\right)} \quad \text{...(5)}$$

$$I_5 = \int_0^\infty v^5 e^{-\alpha v^2} dv = \frac{1}{\alpha^3} \quad \text{...(6)}$$

If the limits of integration are $-\alpha$ to $+\alpha$ in place of 0 to $\infty$, then the integrals $I_0'$, $I_2'$, $I_4'$ will be double while the integrals $I_1'$, $I_3'$, $I_5'$ will vanish.

### NUMERICAL EXAMPLES

**EXAMPLE 11.1**

*Calculate the value of root mean square speed of a molecule of hydrogen at N.T.P. The Boltzmann's constant is $1.38 \times 10^{-16}$ erg per degree and Avogadro's number is $6 \times 10^{23}$/gm. mol.*

**(Nagpur 2006, Rajasthan 2001)**

**SOLUTION.** The $v_{rms}$ is given by

$$v_{rms} = \sqrt{\frac{3kT}{m}}$$

Here
$$k = 1.38 \times 10^{-16} \text{ erg per degree}$$
$$T = 273 \text{ K}$$
$$m = \text{mass of hydrogen molecule}$$
$$= 2 \text{ a.m.u.} = 2/6 \times 10^{23} \text{ gm}$$

$$\therefore \quad v_{rms} = \sqrt{\frac{3kT}{m}}$$

$$= \sqrt{\frac{3 \times 1.38 \times 10^{-16} \times 273}{(2/6 \times 10^{23})}}$$

$$= \sqrt{\frac{3 \times 1.38 \times 10^{-16} \times 273 \times 6 \times 10^{23}}{2}}$$

$$= \mathbf{1.838 \times 10^5 \text{ cm/sec}}$$

**EXAMPLE 11.2**

*You are given the following group of particles ($n_i$ represents the number of particles with speed $v_i$)*

| $n_i$ | $v_i$ (m/sec) |
|---|---|
| 2 | 1.0 |
| 4 | 2.0 |
| 8 | 3.0 |
| 6 | 4.0 |
| 3 | 5.0 |

(i) *Compute the average speed* ($\overline{v}$),

(ii) *Compute the root mean square speed* ($v_{rms}$)

(iii) *Which one is the most probable speed* ($v_{mp}$) *among the entire group.*

**(Rohilkhand 2000)**

**SOLUTION.** (*i*) The average speed ($\overline{v}$) is given by

$$\overline{v} = \frac{\sum_i N_i v_i}{\sum_i N_i}$$

$$= \frac{2\times1+4\times2+8\times3+6\times4+3\times5}{2+4+8+6+3}$$

$$= \frac{2+8+24+24+15}{23} = \frac{73}{23}$$

$$= \textbf{6.17 m/sec.}$$

(*ii*) The Mean square speed $\left(\overline{v^2}\right)$ is given by

$$\overline{v^2} = \frac{\sum_i N_i v_i^2}{\sum_i N_i}$$

$$= \frac{2\times1^2+4\times2^2+8\times3^2+6\times4^2+3\times5^2}{2+4+8+6+3}$$

$$= \frac{2+16+72+96+75}{23} = \frac{281}{23}$$

∴ The root mean square speed $\left(\sqrt{\overline{v^2}}\right)$ will be

$$v_{rms} = \sqrt{\frac{281}{23}}$$

$$= \textbf{3.36 m/sec.}$$

(*iii*) Out of 23 particles 8 have speed of 3 m/sec., 6 have speed of 4 m/sec., 4 have speed 2 m/sec., while remaining 5 particles have different speeds, therefore, most probable speed $v_{mp}$ is **3 m/sec.**

### EXAMPLE 11.3

*Find the value of $v_x$ for which the probability falls to (i) $\frac{1}{e}$ times (ii) $\frac{1}{10}$ times, the maximum value.* **(Rohilkhand 2002 Imp)**

**SOLUTION.** The probability of a molecule having $x$ component of velocity between $v_x$ and $v_x + dv_x$

$$P(v_x) = \left(\frac{m}{2\pi kT}\right)^{\frac{1}{2}} e^{-mv_x^2/2kT}\, dv_x \qquad ...(i)$$

The probability will be maximum, when $v_x = 0$, *i.e.*,

$$P_{max}(v_x) = \left(\frac{m}{2\pi kT}\right)^{\frac{1}{2}}$$

(*i*) $\frac{1}{e}$ times:

Given $$P(v_x) = \frac{1}{e}(P_{max}) = \frac{1}{e}\left(\frac{m}{2\pi kT}\right)^{\frac{1}{2}}$$

Substituting this value in equation (*i*), we get

$$\frac{1}{e}\left(\frac{m}{2\pi kT}\right)^{\frac{1}{2}} = \left(\frac{m}{2\pi kT}\right)^{\frac{1}{2}} e^{-mv_x^2/2kT}$$

or $$\frac{1}{e} = e^{-mv_x^2/2kT}$$

Taking $\log_e$ of both sides,

$$\log_e e = -\frac{mv_x^2}{2kT}$$

$\therefore$ $$v_x = \sqrt{\frac{2kT}{m}} \qquad (\because \log_e e = 1)$$

(*ii*) $\frac{1}{10}$ times :

Given $$P(v_x) = \frac{1}{10}P_{max} = \frac{1}{10}\left(\frac{m}{2\pi kT}\right)^{\frac{1}{2}}$$

or $$\frac{1}{10}\left(\frac{m}{2\pi kT}\right)^{\frac{1}{2}} = \left(\frac{m}{2\pi kT}\right)^{\frac{1}{2}} e^{-mv_x^2/2kT}$$

or $$\frac{1}{10} = e^{-mv_x^2\backslash 2kT}$$

Taking $\log_e$ of both sides,

$$-\log_e 10 = -\frac{mv_x^2}{2kT}$$

or $$v_x = \sqrt{\frac{2kT}{m}\log_e 10}$$

$$= \sqrt{\frac{2kT}{m} \times 2.303}$$

$$= \sqrt{\frac{4.606\,kT}{m}}$$

**EXAMPLE 11.4**

*Deduce an expression for Doppler half-width of a spectral line.*

**(Rohilkhand 2002, Jiwaji 2000)**

**SOLUTION.** Doppler half-width is that value of shift $s$ at which intensity falls to half the maximum value. Recalling equation (11.73), and putting

$$\frac{I_{\lambda'}}{I_\lambda} = \frac{1}{2} \quad \text{or} \quad \frac{I_\lambda}{I_{\lambda'}} = 2$$

from equ. (11.73), we have $I_{\lambda'} = I_\lambda\, e^{-bs^2}$

$$\therefore \qquad \frac{I_{\lambda'}}{I_\lambda} = e^{-bs^2}$$

Taking $\log_e$, we get, $\log_e\left(\frac{I_{\lambda'}}{I_\lambda}\right) = -bs^2$

$$s = \left[\frac{\log_e \frac{I_\lambda}{I_{\lambda'}}}{b}\right]^{\frac{1}{2}} = \left[\frac{\log_e 2}{b}\right]^{\frac{1}{2}}$$

$$\therefore \qquad (\Delta\lambda)_{half} = s = \left(\frac{0.69 \times 2kT}{m}\right)^{\frac{1}{2}} \cdot \frac{\lambda_o}{c} \qquad \textbf{Ans.}$$

**EXAMPLE 11.5**

*Calculate the Doppler broadening in hydrogen line 4861 Å for $T$ = 400 K. Given $k = 1.380 \times 10^{-16}$ erg per degree.*

**SOLUTION.** Broadening is expressed by half intensity shift, *i.e.*, $S_{half}$ which is experssed as $(\Delta v)_{half}$. This may also be expressed in terms of wavelength shift $(\Delta\lambda)_{half}$.

Thus,

$$\left|\frac{(\Delta\lambda)_{half}}{\lambda_0}\right| = \left(\frac{0.69 \times 2kT}{m}\right)^{\frac{1}{2}} \cdot \frac{1}{c}$$

$$\therefore \qquad (\Delta\lambda)_{half} = \left(\frac{0.69 \times 2kT}{m}\right)^{\frac{1}{2}} \times \frac{\lambda_o}{c}$$

**Given:**

$k$ = Boltzmann's constant = $1.38 \times 10^{-16}$ erg/degree

$T$ = 400 K

$\lambda_0$ = 4861 $A$

$m$ = mass of $H$ atom (and not of $H_2$ molecule)

= mass of proton (rest mass)

= $1.6725 \times 10^{-24}$ gram.

$$\therefore \qquad (\Delta\lambda)_{half} = \left(\frac{0.69 \times 2kT}{m}\right)^{\frac{1}{2}} \cdot \frac{\lambda_0}{c} = \left(\frac{0.69 \times 2 \times 1.38 \times 10^{-16} \times 400}{1.6725 \times 10^{-24}}\right)^{\frac{1}{2}} \times \frac{4861}{3 \times 10^{10}}\ \text{A}^\circ$$

$$= 0.034577\ \text{A}^\circ$$

**EXAMPLE 11.6**

*Calculate the probability that the speed of oxygen molecule lies between 99.5 and 100.5 metre/sec. at 200k.* **(Rohilkhand 2001)**

**SOLUTION.** The probability of a molecule with its speed between $v$ to $v + dv$ is given by

$$P(v)dv = 4\pi\left(\frac{m}{2\pi kT}\right)^{\frac{3}{2}} v^2\, e^{-mv^2/2kT}\, dv$$

**Given :**

$$m = 32 \text{ a.m.u.} = \frac{32}{6\times10^{26}} \text{ kg}$$
$$v = 100 \text{ m/s}$$
$$dv = 100.5 - 99.5 = 1 \text{ metre/sec.}$$
$$k = 1.38\times10^{-23} \text{ Joule/kelvin}$$
$$T = 200 \text{ K}$$

Substituting, we get

$$P(v) = 4\times3.14\left[\frac{\frac{32}{6}\times10^{26}}{2\times3.14\times1.38\times10^{-23}\times200}\right]^{\frac{3}{2}}\times100^2$$
$$\left[\exp-\left\{\frac{32\times100^2/6\times10^{26}}{2\times1.38\times10^{-23}\times200}\right\}\right]\times1$$

$$= \frac{4}{\sqrt{3.14}}\left[\frac{32}{2\times1.38\times10^{-23}\times200\times6\times10^{26}}\right\}^{\frac{3}{2}}\times10^4$$
$$\left[\exp-\left\{\frac{32}{2\times1.38\times10^{-23}\times200\times6\times10^{26}}\right\}^{\frac{3}{2}}\right]$$

$$= \frac{4}{1.77}\left(0.96\times10^{-5}\right)^{\frac{3}{2}}\times10^4\times e^{-9.6\times10^{-2}}$$

$$P(v) = \mathbf{6.11\times10^{-4}}.$$

**EXAMPLE 11.7**

*The first excited state of hydrogen atom is 10.2 eV above its ground state. What temperature is needed to excite hydrogen atoms to the first excited level ?*

**(Gorakhpur 2002)**

**SOLUTION.** The kinetic interpretation of temperature suggests the average kinetic energy per molecule at absolute temperature $T$ is given by

$$K.\,E. = \frac{1}{2}mv^2 = \frac{3}{2}kT$$

as it has 3 degrees of freedom.

Given that, $\frac{1}{2}mv^2 = 10.2\ eV = 10.2\times1.6\times10^{-19}$ Joule

$$\therefore \quad \frac{3}{2}kT = 10.2 \times 1.6 \times 10^{-19} \text{ Joule}$$

or,
$$T = \frac{2}{3} \times \frac{10.2 \times 1.6 \times 10^{-19}}{k}$$

or,
$$T = \frac{2 \times 10.2 \times 1.6 \times 10^{-19}}{3 \times 1.38 \times 10^{-23}} \qquad (\text{Since } k = 1.38 \times 10^{-23} \text{ Joule} / K)$$

$$= \mathbf{7.888 \times 10^4 \text{ Kelvin}}$$

### EXAMPLE 11.8

*Calculate the r.m.s. velocity of $H_2$ at 27°C. Given $k = 1.38 \times 10^{-23}$ J/deg and mass of hydrogen molecule = $3.34 \times 10^{-27}$ kg.*

**(M.D.U. 2003, Agra 2000, Nagpur 2006/s)**

**SOLUTION.** The r.m.s. velocity is given by

$$v_{rms} = 1.73\sqrt{\frac{kT}{m}}$$

$$= 1.73 \times \sqrt{\frac{1.38 \times 10^{-23} \times (273 + 27)}{3.34 \times 10^{-27}}}$$

$$= \mathbf{1.926 \text{ km/s}}$$

### EXAMPLE 11.9

*At what temperature will the average speed of molecules of hydrogen gas be double the average speed of oxygen at 300 K ?*

**SOLUTION.** Average speed $\bar{v} = 1.59\sqrt{\frac{kT}{m}}$

Average speed of oxygen molecule at 300 $K$ is

$$\bar{v}_{oxy} = 1.59\sqrt{\frac{k\,300}{m_{oxy.}}}$$

Let $T$ be the temperature at which speed of hydrogen molecule be double the average speed of oxygen at 300 $K$.

∴ Average speed of hydrogen molecule at $T$ $K$ will be

$$\bar{v}_{Hydro.} = 1.59\sqrt{\frac{kT}{m_{Hydro.}}}$$

Given :
$$\bar{v}_{Hydro.} = 2\,\bar{v}_{oxy.}$$

$$\therefore \quad 1.59\sqrt{\frac{kT}{m_{Hydro}}} = 2 \times 1.59\sqrt{\frac{300k}{m_{oxy}}}$$

or
$$\frac{T}{m_{Hydro}} = 4 \times \frac{300}{m_{oxy.}}$$

or
$$T = 4 \times 300 \times \frac{m_{Hydro}}{m_{oxy.}} = \frac{4 \times 300}{16}$$

$$= \mathbf{75\ K.}$$

**EXAMPLE 11.10**

*The average kinetic energy of a gas molecule at a certain temperature is $6.21 \times 10^{-21}$ joule. Find the temperature. Boltzmann's constant $K = 1.38 \times 10^{-23}$ joule $K^{-1}$.*

**(Guwahati 2000, Nagpur 2003/s, Jiwaji 2000)**

**SOLUTION.** Average kinetic energy $= \frac{3}{2}kT\ 6.21 \times 10^{-21}$ Joule

Given : $k = 1.38 \times 10^{-23}\ JK^{-1}$

$$\therefore \quad T = \frac{2\times 6.21\times 10^{-21}}{3\times 1.38\times 10^{-23}}$$

$$= 300\ K$$

$$= \mathbf{27°C}$$

**EXAMPLE 11.11**

*At what temperature will the mean speed of hydrogen molecules be the same as that of Nitrogen molecules at 35 °C. Molecular weight of $N_2 = 28$ and that of $H_2 = 2$.*

**(Nagpur 2007, 2005, winter 2004/s, Rajasthan 2000, Agra 2003)**

**SOLUTION.** The mean or average speed $\overline{v}$ is given by

$$\overline{v} = \sqrt{\frac{8kT}{\pi m}}$$

For Hydrogen molecules,

$$\overline{v}_{H_2} = \sqrt{\frac{8kT}{\pi m_{H_2}}} = \sqrt{\frac{8kT}{\pi \times 2}} \quad ...(i)$$

For Nitrogen molecules,

$$\overline{v}_{N_2} = \sqrt{\frac{8k\,(273+35)}{\pi\, m_{N_2}}} = \sqrt{\frac{8k\times 308}{\pi\times 28}} \quad ...(ii)$$

But $\quad \overline{v}_{H_2} = \overline{v}_{N_2}$

$$\therefore \quad \sqrt{\frac{8kT}{2\pi}} = \sqrt{\frac{8k\times 308}{28\pi}}$$

or, $\quad \frac{T}{2} = \frac{308}{28}$

or, $\quad T = \frac{308\times 2}{28}$

$$= 22K$$

## EXERCISE–XI

**11.1** Derive Maxwell-Boltzmann distribution law and hence show that

$$n(E)dE = \frac{2\pi N}{(\pi kT)^{\frac{3}{2}}} E^{\frac{1}{2}} \exp(-E/kT)\,dE,$$

when $n(E)dE$ is the number of molecules with energies $E$ and $E + dE$.

**(Agra 2006, 2000, Madras 2001, Rohilkhand 2001, Mumbai 2005, Punjab 2001, Purvanchal 2002, M.D.U. 2003, Magadh 2002, Pbi.U. 2000, 2003, Kerala U. 2000, 2002, Meerut 2000, 2002)**

**11.2** Obtain Maxwell-Boltzmann's distribution of speeds for gas molecules and explain the variation of number of molecules per unit speed range with the speed of molecules.

**(Nagpur 2009, 2005, Kolkata 2004, Agra 2004)**

**11.3** Using M-B distribution obtain relation between the average energy of the particle and its temperature in equilibrium. **(Purvanchal 2007)**

**11.4** Using Maxwell's law of distribution of speeds of molecules in a gas obtain expressions for most probable speed, average speed and root-mean square speed.
Describe a method for the verification of Maxwell's law.
**(Nagpur Uni. 2008, Purvanchal 2003, 2005, 2007, Meerut 2004, 2005, Agra 2002)**

**11.5** Define root mean square speed of the gas molecules. Using Maxwell's distribution law of molecular speeds, show that r.m.s. speed is given by $\sqrt{\frac{3\,kT}{m}}$.
**(Nagpur 2004/w, 2006/s, 2007, G.N.D.U. 2002, Kerala 2001, 2004/W, M.S. Uni. Tamil Nadu 2007, Kanpur 2004/w, 2006/s, Garhwal 2002)**

**11.6** Assuming *M.B.* distribution of molecular speeds, show that the most probable speed is given by $v_{mp} = \sqrt{\frac{2\,kT}{m}}$.
**(Nagpur 2005/w, 2005'/s 20007; Punjab 2000, 2001, 2002; Manonmanian S. Uni. 2007, Garhwal 2000, Kerala 2001)**

**11.7** Assuming M.B. distribution of molecular speeds, show that for the molecules of an ideal gas is given by $\bar{v} = \sqrt{\frac{8\,kT}{\pi\, m}}$.
**(Nagpur 2005, 2009, winter; Punjab 2005, MDU. 2001, 2003, GNDU. 2002, Punjab 2001, Kerala 2001, Calcutta 2001, H.P.U. 2003)**

**11.8** Give the experimental verification of Maxwell-Boltzmann's law of distribution of molecular speeds.
**(Alagappa U. 2004, Kerala 2001, Ranchi 2004, Rohilkhand 2002 Improvement, M.S. Uni. Tamil Nadu 2007, Agra 2006, Lucknow 2002, Punjab 2001)**

**11.9** Show that the probability of a molecule to have its velocity component between $v_x$ and $v_x + dv_x$ is given by

$$P(v_x)\, dv_x = \left(\frac{m}{2\pi kT}\right)^{\frac{1}{2}} e^{-\frac{mv_x^2}{2kT}}\, dv_x$$ **(Nagpur 2008)**

**11.10** Explain the variation of probability function $P(v_x)$ with temperature using Maxwell's distribution law of molecular velocities. **(Rohilkhand 2003)**

**11.11** Explain clearly by plotting graphs, the variation of distribution function $N(v)$ with temperature using Maxwell's distribution law of molecular speeds.
**(GNDU 2001, Punjab 2001, Garhwal 2000, Nagpur 2006/s)**

**11.12** What do you mean by Doppler broadening of spectral lines. Derive the relation for the wavelength shift $s = (\lambda - \lambda')$ given by **(Nagpur 2009, 2007, 2006/s)**

$$I_{\lambda'} = I_{\lambda'}\, e^{-bs^2}$$

where $$b = \frac{mc^2}{2kT\lambda^2}$$

**11.13.** State the assumptions made by M–B for the distribution of velocities of the molecules of a ideal gas and hence derive Maxwell's law of distribution of velocities of the molecules of an ideal gas using Maxwell-Boltzmann's energy distribution formula.
**(Nagpur 2008, 2007, 2006 /s, 2005/w winter, 2004/W, Alagappa U.2004, Patna 2004, Ranchi 2004, Agra 2001, 2003, Kanpur 2000, Kumaun 2001, Rajasthan 2001, Rohilkhand 2002, Imp 2003 M.S. Uni. Tamil Nadu 2007)**

**11.14** It $v_p$, $\overline{v}$ and $v_{rms}$ stand respectively for the most probable, average and root-mean square speeds, show that for the Maxwellian distribution.

$$\frac{v_p}{1} = \frac{\overline{v}}{\sqrt{\frac{4}{\pi}}} = \frac{v_{rms}}{\sqrt{\frac{3}{2}}}$$ **(Nagpur 2005/s)**

**11.15** What are the limitations of Maxwell-Boltzmann method ? **(Nagpur 2005/s)**

**11.16** Applying Maxwell-Boltzmann distribution show that the internal energy of an ideal monatomic gas depends only on its temperature. **(Madras 2005)**

**11.17** A perfect gas of molecules of mass '*m*' each is at absolute temperature *T*. The velocity of a molecule is denoted by $\vec{V}$. Stating the assumptions clearly, find an expression for the average number of molecules per unit volume with x-component of velocity lying between $V_x$ and $V_x + dv_x$ irrespective of the values of the other two components. Find also the expression for the average number of molecules per unit volume having speed between *V* and *V* + *dv*. **(Calcutta 2005)**

**11.18** Draw the graph of the Maxwell-Boltzmann velocity distribution curve and state any two features of the distribution curve. **(Nagpur 2007)**

**11.19** For $O_2$ gas at N.T.P. calculate (*i*) $\overline{v}$ (*ii*) $v_{mp}$ and (*iii*) $v_{rms}$. Given Boltzmann's constant 1.38 × $10^{-16}$ erg/degree, Avogadro's number = 6.02 × $10^{23}$/mole. **(Nagpur 2009, 2006)**
[**Ans.** 4.23 × $10^2$ ms$^{-1}$, 4.61 × $10^2$ ms$^{-1}$, 3.76 × $10^2$ ms$^{-1}$]

**11.20** At what temperature will the average speed of molecules of hydrogen gas is seven times the average speed of nitrogen at 300 K. [**Ans.** 1050 K]

**11.21** Calculate the probability that the speed of Oxygen molecule lie between 100 and 101 ms$^{-1}$ at 200 K. [**Ans.** 6.13 × $10^{-4}$] **(Rohilkhand 2003)**

**11.22** Calculate the fraction of molecules of a gas possessing speeds within 1% of the most probable speed. [**Ans.** 0.016]

**11.23** If the most probable speed of a molecule of Hydrogen gas is 100 ms$^{-1}$ at temperature T *K*; calculate the most probable speed of an Oxygen molecule at temperature 2T *K*. **(Chennai 2001)**

**11.24** Calculate average translational kinetic energy per molecule of a gas at 27°*C* in joules. Given *k* = 1.38 × $10^{-23}$ Joule/mole – *K*. [**Ans.** 6.21 × $10^{-21}$ Joule/mole] **(Nagpur 2005 winter)**

**11.25** The speed of 10 particles in *m/s* is 0, 1, 2, 3, 3, 3, 4, 4, 5 and 6, Calculate the average speed, the root mean square speed and the most probable speed. **Ans.** 3.1, $5\sqrt{5}$, 3

**11.26** Find the Number of possible arrangements of three particles in two cells assuming the particles obey M.B. statistics. **(Nagpur 2009)**

## OBJECTIVE QUESTIONS

**11.1** The formula for the most probable speed of the molecules in a gas is:

(*a*) $C_m = \sqrt{\frac{3kT}{m}}$ (*b*) $C_m = \sqrt{\frac{m}{2kT}}$

(*c*) $C_m = \sqrt{\frac{2kT}{m}}$ (*d*) $C_m = \sqrt{\frac{m}{2kT}}$ **(Agra 2003)**

**11.2** The root mean square speed of molecules of mass '*m*' at temperature *T* is:

(*a*) $\sqrt{\frac{2kT}{m}}$ (*b*) $\sqrt{\frac{3kT}{\pi m}}$

(*c*) $\sqrt{\frac{3kT}{m}}$ (*d*) $\sqrt{\frac{8kT}{m}}$ **(Ranchi 2005)**

**11.3** RMS speed of an ideal gas is inversely proportional to the square-root of its :

(*a*) mass (*b*) temperature

(*c*) none of these (*d*) both of these **(Meerut 2003)**

**11.4** For a gas at *N.T.P.* which shall be maximum?

(*a*) $V_{av}$ (*b*) $V_{r.m.s.}$

(*c*) $V_{m.p.}$ (*d*) none

**11.5** Out of $n$ particles in a gas, the number of particles having exactly the most probable velocity is :
(*a*) zero (*b*) $n$
(*c*) $\frac{n}{2}$ (*d*) 1

**11.6** In Zartman Ko experiment distance from the plate is a measure of
(*a*) electron speed (*b*) molecular speed
(*c*) density of molecules (*d*) electron density

**11.7** The root mean square speed of a gas molecules of mass $m$ at a given temperature $T$ is proportional to
(*a*) $m^{o}$ (*b*) $m$
(*c*) $m^{-1/2}$ (*d*) $m^{1/2}$

**11.8** Four particles have speeds 2, 3, 6, and 5 cm/s respectively. Their average speed is :
(*a*) 4 cm/s (*b*) 2 cm/s
(*c*) 0.25 cm/s (*d*) $\sqrt{2}$ cm/s

**11.9** The *rms* speed of hydrogen gas molecules at *N.T.P.* is *m/s*. The gas is heated at constant volume till the pressure becomes 9 times. The final *rms* speed will be :
(*a*) $3\,v$ (*b*) $9\,v$
(*c*) $18\,v$ (*d*) $v/3$

$$\left[\textbf{Hint : } v_{rms} = \sqrt{\frac{3kT}{m}} \text{ put } PV = RT \text{ i.e. } T = \frac{PV}{R},\ v_{rms} = \sqrt{\frac{3kPV}{Rm}}\right]$$

**11.10** The ratio of most probable speed and average speed of a gas enclosed in a vessel is :
(*a*) $\frac{\sqrt{\pi}}{4}$ (*b*) 1
(*c*) $\frac{\sqrt{\pi}}{2}$ (*d*) $\frac{2}{\sqrt{\pi}}$

**11.11** The speeds of 10 particles in *m/s* are 0, 1.0, 2.0, 3.0, 3.0, 4.0, 4.0, 4.0, 5.0 and 6.0. The most probable speed is :
(*a*) 3 *m/s* (*b*) 4 *m/s*
(*c*) zero (*d*) none of these

**11.12** Four particles have speeds 1, 2, 3 and 4 *cm/s* respectively. Their *rms* speed is :
(*a*) 2.5 (*b*) 10
(*c*) $\frac{\sqrt{30}}{2}$ (*d*) $\sqrt{30}$

**11.13** A gas having *rms* speed at 400 *K* is :
(*a*) twice the value at 100 *K* (*b*) four times the value at 100 *K*
(*c*) half the value at 100 *K* (*d*) same as at 100 *K*

**11.14** The most probable speed of molecules varies with temperature $T$ as $v_{mp} \propto T^n$, the value of $n$ is :
(*a*) 0 (*b*) $\frac{1}{2}$
(*c*) 2 (*d*) $\frac{1}{3}$

**11.15** The particles obeying Maxwell-Boltzmann statistics are :
(*a*) identical (*b*) identical and indistinguishable
(*c*) distinguishable (*d*) photons.

## ANSWERS

| | | | |
|---|---|---|---|
| **11.1** (*c*) | **11.2** (*b*) | **11.3** (*a*) | **11.4** (*b*) |
| **11.5** (*a*) | **11.6** (*b*) | **11.7** (*c*) | **11.8** (*a*) |
| **11.9** (*a*) | **11.10** (*c*) | **11.11** (*b*) | **11.12** (*c*) |
| **11.13** (*a*) | **11.14** (*b*) | **11.15** (*c*) | |

# Quantum Statistics

# 12

## AT A GLANCE

- Need of Quantum Statistics
- Development of Quantum Statistics
- '*h*' as a natural constant
- Indistinguishability of Particles and its Consequences
- Bose-Einstein Distribution Law
- Photon Gas
- Planck's Radiation Law
- Fermi-Dirac Distribution Law
- Free Electrons in Metal : Electron Gas
- Fermi Level and Fermi Energy
- $E_{F_0}$ for Electrons in a Metal
- Fermi-Dirac Velocity and Distribution in Given Direction
- Maxwell-Boltzmann Distribution as the Limiting case of Bose Einstein and Fermi Dirac Distributions
- Application to Liquid Helium
- Comparison of the three Statistics
- Difference between Classic and Quantum Statistics

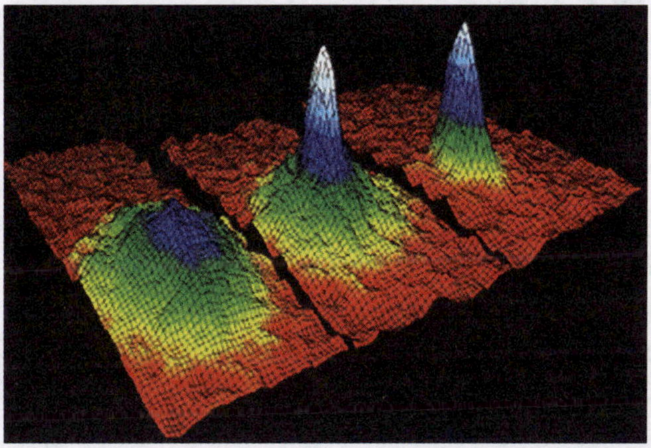

## 12.1 Need of Quantum Statistics

In Classical statistics, Maxwell-Boltzmann explained the energy and velocity distribution of the molecules of an ideal gas to a fairly large degree of accuracy. It has been assumed there that all the energy levels are accessible to all the particles of the systems. However, there are certain energy levels prohibited to a certain group of particles. This is the reason why *M.B.* Statistics failed to account adequately for several other experimentally observed phenomenon such as black body radiation, photoelectric effect, specific heat at low temperatures etc. Altogether a new quantum concept of discrete exchange of energy between systems and along with it a new statistics, known as *quantum statistics* was developed.

In 1924, the quantum statistics was first formulated by Bose in the deduction of Planck's radiation law. Einstein in the same year employed it in kinetic theory of gases as a substitute for the classical Boltzmann statistics, known as Bose-Einstein (*B-E*) statistics. Two years later, in 1926, Fermi and Dirac quite independently, modified *B-E* statistics in certain cases, on the basis of Pauli's exclusion principle, known as Fermi-Dirac (*F-D*) statistics. Thus, the quantum statistics can be put in to two sub-classes as

(*i*) Bose-Einstein statistics

(*ii*) Fermi-Dirac statistics

depending upon symmetric or antisymmetric wavefunctions.

## 12.2 Development of Quantum Statistics

For developing a quantum statistical theory for a system containing a large number of identical and indistinguishable particles, it is necessary to consider the wavefunction for many body state as

$$\Psi\,(\vec{r}_1, \vec{r}_2, \vec{r}_3, \ldots\ldots\ldots\ldots)$$

where $\vec{r}_1, \vec{r}_2, \vec{r}_3, \ldots$ are position vectors of the first, second, third ... particles respectively.

**Symmetric wavefunction :** If on interchanging any pair of identical particles, there is no change of the sign of the wave function, *i.e.*, if

$$\Psi\,(\vec{r}_2, \vec{r}_1, \vec{r}_3, \ldots\ldots\ldots\ldots) = +\Psi\,(\vec{r}_1, \vec{r}_2, \vec{r}_3, \ldots\ldots\ldots\ldots)$$

the wavefunction is said to be **symmetric**.

**Anti-symmetric wavefunction :** If on interchanging any pair of identical particles, the sign of the wavefunction changes, *i.e.*, if

$$\Psi\,(\vec{r}_2, \vec{r}_1, \vec{r}_3, \ldots\ldots\ldots\ldots) = -\Psi\,(\vec{r}_1, \vec{r}_2, \vec{r}_3, \ldots\ldots\ldots\ldots)$$

the wave function is said to be **anti-symmetric.**

For particles having spin angular momenta that are zero or integral multiples of the unit $\hbar \left(=\frac{h}{2\pi}\right)$, the wave-functions are symmetric, and for particles with spin angular momenta that are odd half-integral multiples of $\hbar$, the wave-functions are antisymmetric. Based on this requirement the following two quantum statistics have been developed.

### 1. Bose-Einstein Statistics

This statistic was developed by S.N. Bose for light quanta, called photons and generalised by A. Einstein to describe the energy distribution among particles, whose spin angular momenta are $n\hbar$ where $n = 0, 1, 2, 3, \ldots$ . Such particles are called **bosons**. Wave functions are symmetric on interchanging two bosons. The Pauli exclusion principle which states that in each quantum state there cannot simultaneously be more than one particle, is not applicable to bosons, so that there is no upper limit to the number of bosons in a given quantum state.

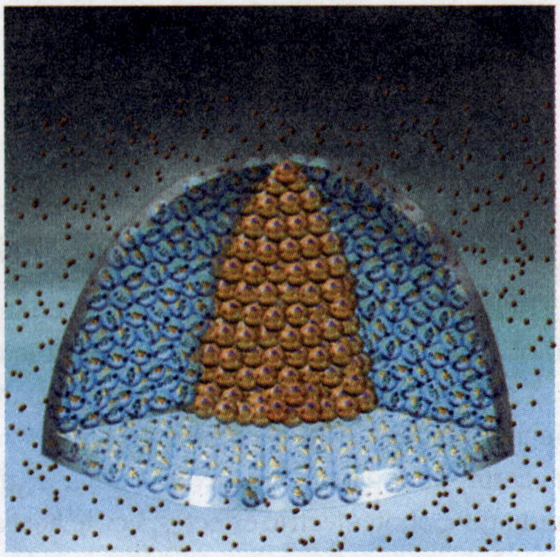

Metallic Phase for Bosons

**Examples of bosons are :** $\alpha$ particles each having spin quantum number $s = 0$, photon for which $s = 1$, deutrons for which $s = 1$, $\pi$ mesons for which $s = 0$, etc.

## 2. Fermi-Dirac Statistics

This statistics was developed by E. Fermi for electrons and its relation to quantum mechanics was established by P.A.M. Dirac. It describes the energy distribution among particles whose spin angular momenta are $\left(n+\frac{1}{2}\right)\hbar$ where $n = 0, 1, 2, 3, \ldots$ . In other words, they are $\frac{1}{2}\hbar, \frac{3}{2}\hbar, \frac{5}{2}\hbar, \ldots$ . Such particles are called **fermions**. Wave functions are antisymmetric on interchanging two fermions. The Pauli exclusion principle is applicable to fermions, so that there can be a maximum of one fermion in a given quantum state.

**Examples of fermions are** electrons for which $s = \frac{1}{2}$, positrons for which $s = \frac{1}{2}$, protons $\left(s=\frac{1}{2}\right)$, neutrons $\left(s=\frac{1}{2}\right)$, μ-mesons $\left(s=\frac{1}{2}\right)$, etc.

## 12.3 'h' as a Natural constant

In classical mechanics $h_o$ is an arbitrary quantity and we may set it as $10^{-30}$ erg. sec or $10^{-35}$ erg. sec or even smaller value. Here we are free to take the cell size *as small as we please*. In other words, the number of cells $g_i$ in a compartment of the phase space can be made as large as we like so that it can be much greater than the number of particles $n_i$ available to the $i$th compartment of the phase space. Therefore, occupation index $\frac{n_i}{g_i} << 1$. In such a case, the probability of a cell containing two or more particles is negligible and particles can, therefore, be treated as distinguishable so that classical mechanics is applicable to them.

However, all natural systems obey Heisenberg's uncertainty principle and we are not free to choose the cell size as small as we like. The size of the phase cell is fixed by nature through this uncertainty principle as

$$(\Delta p \,.\, \Delta x)_{min} = h = 6.6 \times 10^{-27} \text{ erg .sec}$$

The minimum size of a cell in phase space in such a case is $h^3$ as given by

$$dx\, dy\, dz\, dp_x\, dp_y\, dp_z \geq h^3$$

*i.e.* the volume of the phase cell in phase space cannot be less than $h^3$, where h is known as Planck's constant. In contrast to classical mechanics, since the size of the cell is restricted to $h$, the number of cell $g_i$ is limited and hence $\frac{n_i}{g_i}$ cannot made as small as we please. When the occupation index $\frac{n_i}{g_i}$ is of the order of unity, the particles cannot be treated as distinguishable. *i.e.* become indistinguishable. Classical statistics cannot, therefore, be applied. Here lies the necessity of introduction of quantum statistics.

### Two Cases :

**(*i*) Particle in a One Dimensional Box**

Consider a particle of mass $m$ enclosed in a one dimensional box of Length $L$. This case is useful for understanding the role of size of a phase cell which we have taken it to be constant $h_o$ and energy range $\delta E_0$ in statistical mechanics.

In one-dimensional box, the $x$-$p$ plane is the phase space. The 'phase volume' is given by

$$V_{phase} = \iint dx\, dp = \int dx \int dp = L \int dp$$

$$= L\sqrt{2mE} \qquad ...(12.1)$$

Where $L$ is the length within which the particle is constrained in its motion, and the energy range considered is $o$ to $E$. The number of phase cells corresponding to this energy range is

$$g_i = \phi(E) = \frac{L\sqrt{2mE}}{h_o}$$

where unit phase cell of phase volume $h_o$ (and not $h^3_0$ as in 6 dimensional phase space.)

As an example, if we consider a particle of mass $m = 2 \times 10^{-24}$ gm in a 'box' of length $L = 1 \times 10^{-3}$ cm. If $h_o$ is set as $1 \times 10^{-26}$ erg. sec, then in energy range 0 to $10^{-12}$ erg the number of phase cells would be

$$g_i = \phi(E) = 2 \times 10^5$$

This means that the variation of energy within one phase cell would be nearly 1 part in $2 \times 10^5$ of the total energy.

Moreover, the number of microstates in energy range $\delta E_o$ at energy $E$ is, for this case,

$$W(E) = \frac{L\sqrt{2m}}{h_o} \cdot \frac{1}{2} E^{-\frac{1}{2}} \delta E_o \qquad ...(12.2)$$

For the same values of $L$, $m$ and $h_o$ considered above, let us consider an energy sheet of $\delta E_o = 10^{-14}$ erg at $E = 10^{-12}$ erg. This gives

$$W(E) = 10^3$$

In classical mechanics, we are free to set $h_o$ being an arbitrary quantity. We may set it $10^{-30}$ erg. sec or $10^{-34}$ erg. sec or even smaller value. That would make $W(E)$ larger and larger, or it would permit us to divide the phase space into energy sheets of smaller and smaller energy ranges.

But in quantum mechanics, the size of the phase cell is fixed by nature through the uncertainty principle, as we know, to be $h = 6.6 \times 10^{-27}$ erg. sec.

Using this value of $h$ and the values taken for $L$, $m$ and $E$ in the above problem, would make $g_i$ and $W(E)$ some what larger, but small enough to retain the validity of the statistical method. But for the particles of smaller mass, like that of electron ($m = 9.11 \times 10^{-28}$ g), the situation would be worse, but with larger $L$, the validity would improve. Thus, in order to incorporate the particles of smaller mass, particle in a one-dimensional box fully supports to naturally set value of constant $h$ through the uncertainty principle.

**(ii) One-Dimensional Harmonic Oscillator.**

The thermal radiation are electromagnetic waves is well established. In a black body encloser, the equilibrium between the walls and the radiation can be obtained only if the walls contains electrical oscillators of some kind. Therefore, one must conceive electrical oscillators (atoms of the spherical encloser black body or cavity) of corresponding frequencies in the walls of the chamber and lay down that in thermal equilibrium the average energy of these oscillators equals the average energy of the corresponding mode in the waves. For simplicity, let us consider a mechanical linear one-dimensional harmonic oscillator. Let $m$

**Harmonic Oscillator**

be its reduced mass and force constant $c$. Let $p$ be momentum at any instant, then

$$p = mv = m\frac{dx}{dt}$$

potential energy of the oscillator $= \frac{1}{2}cx^2$

and, kinetic energy of the oscillator $= \frac{1}{2}m\left(\frac{dx}{dt}\right)^2 = \frac{p^2}{2m}$

$\therefore$ Total energy $$E = \frac{1}{2}cx^2 + \frac{p^2}{2m} \quad \text{...}(i)$$

The frequency of oscillation is given by

$$\nu = \frac{1}{2\pi}\sqrt{\frac{c}{m}} \quad \text{...}(ii)$$

According to classical theory, radiation of a given frequency would be affected only by oscillators of the *same frequency*. Therefore, in equilibrium the radiation density at a frequency $\nu$ and the average energy of the oscillators of this frequency in the walls must bear a definite ratio. The 'oscillators' are thus these electrical oscillators in the walls and we are concerned with the average energy they possess, when the walls are at temperature $T$.

From the classical theory, which gives the equipartition law, this average energy comes out to be $kT$. Planck realised the failure of Rayleigh–Jeans law, and therefore, *discarded the equipartition principle*. Thus, the classical theory fails to explain black body radiation.

To explain, a new concept of quantum theory was proposed by Planck. To understand this, let us again examine the mechanical case of one-dimensional harmonic oscillator. Consider the $p$, $x$ phase space for such an oscillator as shown in (Fig-12.1). Let us draw ellipse with origin $O$ as centre so as to divide the $x$, $p$ plane into rings of equal area $h$. This ellipse is numbered $o$ and others are numbered as 1, 2, 3, .......... $r$ .......... $n$.

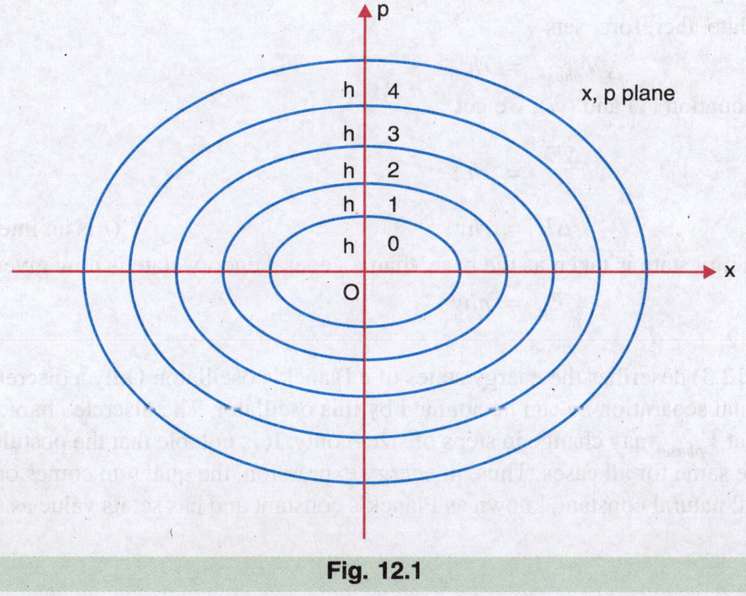


Fig. 12.1

From equation (*i*) it is obvious that for a given energy *E*, the *p*, *x* relation gives a parabola in which

$$P_{max} = (2mE)^{\frac{1}{2}}$$

and $$x_{max} = \left(\frac{2E}{c}\right)^{\frac{1}{2}}$$

where *m* is reduced mass and *c* is a force constant for the oscillator.

The phase volume ($V_{phase}$) (for the 2*f* dimensional phase space, the term phase volume is adopted for $\iint dp\, dx$ for uniformity, here $f = 1$. Hence $V_{phase}$ here means an area in *p*, *x* plane) corresponding to energy range *o* to *E* is therefore, given by area of the ellipse

$$V_{phase} = \pi\, P_{max}\, x_{max}$$

$$= \pi(2mE)^{\frac{1}{2}}\left(\frac{2E}{c}\right)^{\frac{1}{2}}$$

$$= 2\pi\sqrt{\frac{m}{c}}\, E \qquad ...(iii)$$

Eqn. (*ii*) gives $$2\pi\sqrt{\frac{m}{c}} = \frac{1}{\nu}$$

substituting in equation (*iii*), we get

$$V_{phase} = \frac{E}{\nu} \qquad ...(iv)$$

From this equation we predict that if an oscillator of a given frequency ν gains energy Δ*E*, the 'phase volume' change is given by

$$\Delta V_{phase} = \frac{\Delta E}{\nu} \qquad ...(v)$$

Now Planck postulated that an oscillator could absorb or emit energy only in such a way that $V_{phase}$ could change in integral multiples of a certain natural constant *h*.

This postulate, therefore, sets

$$\Delta V_{phase} = rh \qquad ...(vi)$$

Equating equation (*v*) and (*vi*), we get

$$\frac{\Delta E}{\nu} = rh$$

$$\Delta E = rh\nu \qquad (r \text{ is an integer}) \qquad ...(vii)$$

If the lowest energy state is taken as the base, than a general energy state is now given by

$$E_n = nh\nu \qquad ...(12.3)$$

where $n = 0, 1, 2, ...$

Equation (12.3) describes the energy states of a Planck's oscillator. Only a discrete set of energy values, with equal separation *h*ν can be attained by this oscillator. The discrete character arises from the postulate that $V_{phase}$ may change in steps of size *h* only. It is notable that the postulate is universal so far as *h* is the same for all cases. Thus, in energy expression, the quantum comes out as *h*ν. Hence *h* has a universal natural constant, known as Planck's constant and has set its value as $6.6 \times 10^{-27}$ erg. sec.

Albert Einstein

*ALBERT EINSTEIN (1879–1955), bitterly unhappy with the rigid discipline of the schools of his native Germany, went at sixteen to Switzerland to complete his education, and later got a job examining patent applications at the Swiss Patent Office. Then, in 1905, ideas that had been germinating in his mind for years when he should have been paying attention to other matters (one of his math teachers called Einstein a "lazy dog") blossomed into three short papers that were to change decisively the course not only of physics but of modern civilization as well.*

*The first paper, on the photoelectric effect, proposed that light has a dual character with both particles and wave properties. The subject of the second paper was Brownian motion, the irregular zigzag movement of tiny bits of suspended matter, such as pollen grains in water. Einstein showed that Brownian motion results from the bombardment of the particles by randomly moving molecules in the fluid in which they are suspended. This provided the long-awaited definite link with experiment that convinced the remaining doubters of the molecular theory of matter. The third paper introduced the special theory of relativity.*

*After university posts in Switzerland and Czechoslovakia, in 1913 he took up an appointment at the Kaiser Wilhelm Institute in Berlin that left him able to do research free of financial worries and routine duties. Einstein's interest was now mainly in gravitation, and he started where Newton had left off more than two centuries earlier.*

*Einstein's general theory of relativity, published in 1916, related gravity to the structure of space and time. In this theory the force of gravity can be thought of as arising from a warping of spacetime around a body of matter so that a nearby mass tends to move toward it, much as a marble rolls toward the bottom of a saucer-shaped hole. From general relativity came a number of remarkable predictions, such as that light should be subject to gravity, all of which were verified experimentally. The later discovery that the universe is expanding fit neatly into the theory. In 1917 Einstein introduced the idea of stimulated emission of radiation, an idea that bore fruit forty years later in the invention of the laser.*

*The development of quantum mechanics in the 1920s disturbed Einstein, who never accepted its probabilistic rather than deterministic view of events on an atomic scale. "God does not play dice with the world," he said, but for once his physical intuition seemed to be leading him in the wrong direction.*

*Einstein, by now a world celebrity, left Germany in 1933 after Hilter came to power and spent the rest of his life at the Institute for Advanced Study in Princeton, New Jersey, thereby escaping the fate of millions of other European Jews at the hands of the Germans. His last years were spent in an unsuccessful search for a theory that would bring gravitation and electromagnetism together into a single picture, a problem worthy of his gifts but one that remains unsolved to this day.*

## 12.4 Indistinguishability of Particles and its Consequences

The consequences of indistinguishability of particles in quantum statistics may be explained in the following manner :

Let there be two points of collection and only two cells to be occupied. In classical statistics, each particle has a recognizable individuality, hence the particles are distinguishable and each particle is as likely to be in one cell as in the other. This means that in classical statistics, each or both of the particles can occupy any one of the two cells. Thus, in this case, we have four possible arrangements, each of which is counted for assigning a probability to the distribution. This is as shown in Fig. 12.2 (*a*).

(*a*) Classical Statistics

(*b*) Bose-Einstein Statistics

(*c*) Fermi-Dirac Statistics

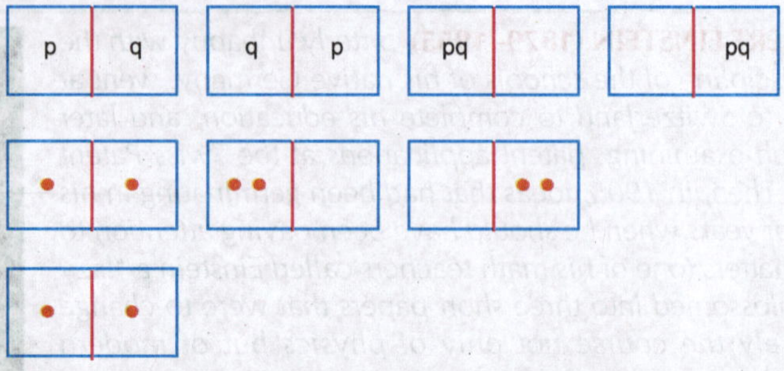


**Fig. 12.2 : Essential difference in three statistics**

In Bose-Einstein statistics, the particles loose their individuality, hence they are indistinguishable. So we must concentrate our attention on the cell rather than the individual particles. Each cell may have any number of particles, so that in this case, we have only three possible arrangements, as shown in Fig. 12.2 (*b*).

In Fermi-Dirac statistics, since the particles are indistinguishable, again we have to concentrate our attention on the cells rather than on individual particles. But in this case, the particles obey Pauli's exclusion principle according to which it is not possible to have more than one representative point in any one cell. Obviously, each cell will contain at the most one particle. Therefore, in this case, only one arrangement is possible, as shown in Fig. 12.2 (*c*).

## 12.5 Bose-Einstein Distribution Law

Bose used Planck's hypothesis, according to which radiation in a temperature encloser are composed of light quantas known as photons each of energy, $E = h\nu$. These photons in the encloser are indistinguishable.

**Basic Postulates :**

In Bose-Einstein statistics, the conditions are:

1. The particles of the system are identical and indistinguishable.
2. Any number of particles can occupy a single cell in the phase space.
3. The size of the cell cannot be less than $h^3$, where $h$ is a Planck's constant ($h = 6.63 \times 10^{-34}$ Joule. sec).
4. The number of phase space cells is comparable with the number of particles, *i.e.*, the occupation index $f(E_i)$ is $\frac{n_i}{g_i} = 1$.
5. Bose-Einstein statistics is applicable to particles with integral spin angular momentum in units of $\frac{h}{2\pi}$. All particles which obey *B.E.* statistics are known as Bosons.

**Satyendranath Bose (1894-1974)**

*Indian mathematician and physicist noted for his collaboration with Albert Einstein in developing a theory regarding the gas-like qualities of* ***Electromagnetic radiation****. After graduating from the University of Kolkata (formerly Calcutta), he taught at the University of Dacca (1921–45) and then at Kolkata (1945–56). His numerous scientific papers which were published from 1918 to 1956, contributed to statistical mechanics, the electromagnetic properties of the IONOSPHERE, the theories of X-ray crystallography and thermoluminescence and unified field theory. Bose's Planck's law and the* ***Hypothesis of light quanta*** *(1924) led Einstein to seek him out for collaboration.*

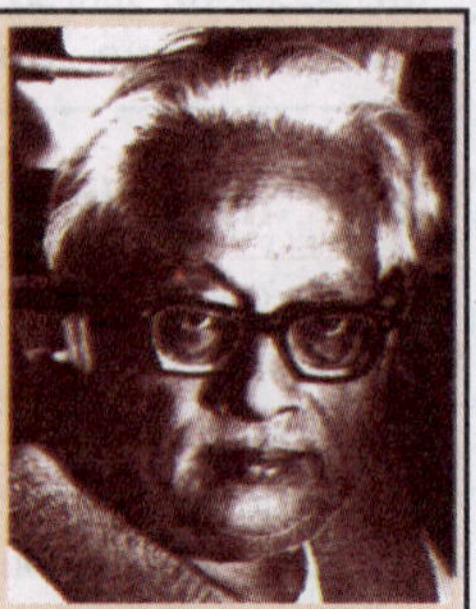

**SN Bose**

Consider a system consisting of $n$ independent and indistinguishable particles. The particles have definite energies (or momenta) and occupy definite positions. Therefore, they can be represented by phase points in the phase space. In order to determine the energy (or momenta) distribution of these particles in the most probable state known as equilibrium state, we divide the available volume in the phase space into large number (say $k$) of compartments (quantum groups or energy levels), each compartment representing a small interval of energy (or momentum) as shown in Fig 12.3 (*a*). Further we divide each compartment into elementary cells, each of size $h^3$ where $h$ is Planck's constants Fig. 12.3(*b*). We suppose that the size of the compartment is very large as compared to the size of the cell so that each compartment contains a very large number of elementary cells.

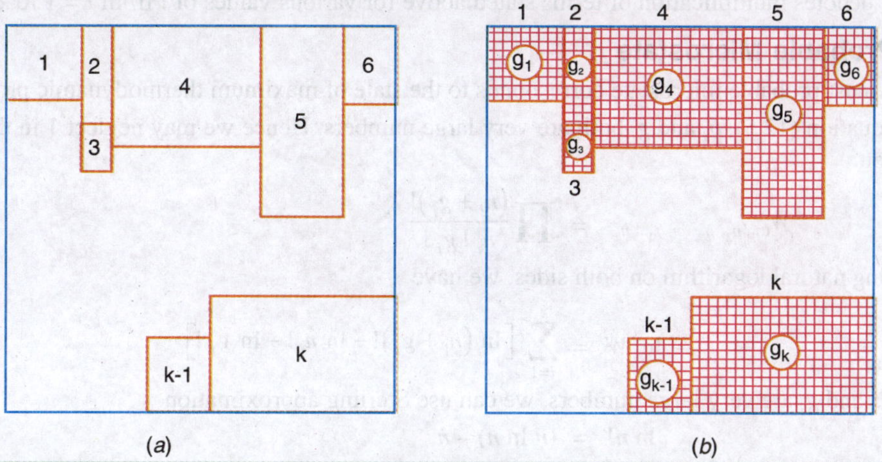


**Fig. 12.3**

Let there be $n_1, n_2, \ldots n_i \ldots n_k$ particles having mean energy values $E_1, E_2, \ldots E_i \ldots E_k$ respectively in compartments numbered as $1, 2, \ldots i \ldots k$ containing $g_1, g_2, \ldots g_i \ldots g_k$ cells respectively in them. The total number of particles in the system is

$$n = n_1 + n_2 + \ldots + n_i + \ldots + n_k$$

Now consider $i$th compartment. It has $n_i$ indistinguishable particles distributed among its $g_i$ cells. Suppose, that $n_i$ particles are arranged in a row and distributed among $g_i$ quantum states with $(g_i - 1)$ partitions in between. The total number of possible arrangements of particles and partitions is equal to the total number of permutations of $(n_i + g_i - 1)$ objects in a row. Therefore, the total possible ways of arranging $n_i$ particles with $g_i - 1$ partitions

$$= (n_i + g_i - 1)!$$

As the particles are indistinguishable, the rearrangement of particles among themselves will not give rise to any new distribution. The number of such meaningless permutations is $n_i$ ! Hence above result would be divided by $n_i$ !. Moreover, this also includes permutations of $(g_i - 1)$ partitions among themselves. These permutations also do not produce different states and hence are meaningless. The above result is further divided by $(g_i - 1)$ !.

Hence, the actual number of ways in which $n_i$ particles are to be distributed in $g_i$ cells in the $i$th compartment is given by

$$W_i = \frac{(n_i + g_i - 1)!}{n_i!(g_i - 1)!} \qquad \ldots(12.4)$$

Similar expressions will be obtained for other compartments. Therefore, the total number of different arrangements for all the $n$ particles of the system gives thermodynamic probability.

$$W_{(n_1, n_2, \ldots\ldots n_i \ldots\ldots n_k)} = \frac{(n_1 + g_1 - 1)!}{n_1!(g_1 - 1)!} \times \frac{(n_2 + g_2 - 1)!}{n_2!(g_2 - 1)!} \times \ldots \times \frac{(n_i + g_i - 1)!}{n_i!(g_i - 1)!} \times \ldots \frac{(n_k + g_k - 1)!}{n_k!(g_k - 1)!}$$

$$= \prod_{i=1}^{k} \frac{(n_i + g_i - 1)!}{n_i!(g_i - 1)!} \qquad \ldots(12.5)$$

where $\prod$ denotes multiplication of terms stated above for various values of $i$ from $i = 1$ to $k$.

## Most Probable Microstate

The most probable microstate corresponds to the state of maximum thermodynamic probability.

In equation (12.5), $n_i$ and $g_i$ both are very large numbers. Hence we may neglect 1 in the above expression

$$\therefore \quad W_{(n_1, n_2, \ldots\ldots n_i \ldots\ldots n_k} = \prod \frac{(n_i + g_i)!}{n_i!\, g_i!} \qquad \ldots(12.6)$$

Taking natural logarithm on both sides, we have

$$\ln W = \sum_{i=1}^{k} \left[ \ln (n_i + g_i)! - \ln n_i! - \ln g_i! \right]$$

As $n_i$ and $g_i$ are very large numbers, we can use Sterling approximation.

$$\ln n! = (n \ln n) - n$$

Applying Sterling approximation, we get

$$\ln W = \sum_{i=1}^{k} \left[ (n_i + g_i) \ln (n_i + g_i) - (n_i + g_i) - n_i \ln n_i + n_i - g_i \ln g_i + g_i \right]$$

$$= \sum_{i=1}^{k} \left[ (n_i + g_i) \ln (n_i + g_i) - n_i \ln n_i - g_i \ln g_i \right] \qquad \ldots(12.7)$$

Here $g_i$ is not subject to variation whereas $n_i$ varies continuously

To get the state of maximum thermodynamic probability, we differentiate equation (12.7) and equate it to zero; *i.e.*,

$$\delta (\ln W) = 0$$

$$\therefore \quad \delta (\ln W) = \sum_{i=1}^{k} \left[ \delta n_i \ln (n_i + g_i) + (n_i + g_i) \frac{1}{(n_i + g_i)} \delta n_i - \delta n_i \ln n_i - n_i \frac{1}{n_i} \delta n_i \right]$$

$$= \sum_{i=1}^{k} \delta n_i \ln (n_i + g_i) - \delta n_i \ln n_i$$

($\because$ $g_i$ is a mere number $\delta g_i = 0$)

$$\text{or} \quad \delta (\ln W) = \sum_{i=1}^{k} \left[ \ln \frac{(n_i + g_i)}{n_i} \right] \delta n_i = -\sum_{i=1}^{k} \left[ \ln \frac{n_i}{(n_i + g_i)} \right] \delta n_i = 0$$

or $$\sum_{i=1}^{k}\left[\ln\frac{n_i}{(n_i+g_i)}\right]\delta n_i = 0 \qquad ...(12.8)$$

In addition, our system must satisfy two subsidiary or auxillary conditions:

(*i*) Conservation of total number of particles, *i.e.*, $n$ = a constant.

$$\therefore \qquad n = \sum_i n_i = \text{constant.}$$

*i.e.* $$\delta n = \sum_i \delta n_i = 0 \qquad ...(12.9)$$

(*ii*) Conservation of total energy of the system, *i.e.*, $E$ = a constant.

$$\therefore \qquad E = \sum_i n_i E_i = \text{ constant.}$$

*i.e.* $$\delta E = \sum_i E_i\,\delta n_i = 0 \qquad ...(12.10)$$

Now we shall apply the Langrangian method of undetermined multipliers. For this let us multiply eqn. (12.9) by $\alpha$ and eqn. (12.10) by $\beta$, and add the resulting expression to eq. (12.8) so that we get

$$= \sum_{i=1}^{k}\left[\ln\left(\frac{n_i}{n_i+g_i}\right)+\alpha+\beta E_i\right]\delta n_i = 0 \qquad ...(12.11)$$

The variations $\delta n_i$ are independent of each other. Hence we get

$$\ln\left(\frac{n_i}{n_i+g_i}\right)+\alpha+\beta E_i = 0$$

or $$\ln\left(\frac{n_i}{n_i+g_i}\right) = -\alpha-\beta E_i$$

or $$\frac{n_i}{n_i+g_i} = e^{-\alpha-\beta E_i} = e^{-(\alpha+\beta E_i)}$$

or $$\frac{n_i+g_i}{n_i} = e^{\alpha+\beta E_i}$$

or $$1+\frac{g_i}{n_i} = e^{\alpha+\beta E_i}$$

or $$\frac{g_i}{n_i} = e^{\alpha+\beta E_i}-1$$

or $$n_i = \frac{g_i}{e^{\alpha+\beta E_i}-1} \qquad ...(12.12)$$

This equation represents the most probable distribution of the particles among various energy levels for a system obeying Bose-Einstein statistics and is therefore, known as **Bose-Einstein's Distribution Law** for an assembly of **bosons**.

Substituting for $\beta=\frac{1}{kT}$, the equation (12.12) becomes

$$n_i = \frac{g_i}{e^{\alpha}.e^{\beta E_i}-1} = \frac{g_i}{e^{\alpha}.e^{E_i/kT}-1} \qquad ...(12.13)$$

**B-E energy distribution function.**

The energy distribution function $f(E_i)$ is the average number of particles per quantum state in the energy level $E_i$. It is given by

$$f(E_i) = \frac{n_i}{g_i}$$

Substituting the value of $n_i/g_i$ from from equation. (12.13), we get

$$f(E_i) = \frac{1}{e^{\alpha} e^{E_i/kT} - 1} \quad \text{...(12.14)}$$

For continuous distribution of energy, the distribution function is written as

$$f(E) = \frac{1}{e^{\alpha} e^{E/kT} - 1} \quad \text{...(12.15)}$$

**B-E energy distribution law for continuous variation of energy.**

When the energy levels of the system are very closely packed, they form a *quasi-continuous* spectrum. In such a case, if $g(E)\,dE$ is the number of energy states between the energy range $E$ and $(E + dE)$, then

$$g_i = g(E)\,dE$$

If $n(E)\,dE$ is the number of particles whose energy lies between $E$ and $(E + dE)$, then $n_i = n(E)\,dE$. Substituting in eqn. (12.13), we get

$$n(E)dE = \frac{g(E)\,dE}{e^{\alpha} . e^{E_i/kT} - 1} \quad \text{...(12.16)}$$

The quantity $g(E)$ denotes the *density of states.*

For the system consisting of free particles with no spin $g(E)dE$ is given by (See Eqn. 11.26B)

$$g(E)dE = 2\pi V \left(\frac{2m}{h^2}\right)^{3/2} E^{\frac{1}{2}} dE \quad \text{...(12.17)}$$

Substituting this value in eqn. (12.14), we get

$$n(E)dE = 2\pi V \left(\frac{2m}{h^2}\right)^{3/2} \frac{E^{\frac{1}{2}} dE}{e^{\alpha} . e^{E/kT} - 1} \quad \text{...(12.18)}$$

This is **Bose-Einstein energy distribution law** for continuous distribution of energy among free particles with no spin.

## 12.6 Photon Gas

S.N. Bose, in 1924, introduced the concept, considering thermal radiation as a photon gas and obtained Planck's formula. Subsequently Einstein developed this idea further into what is now known as Bose-Einstein statistics.

A hollow encloser, say a double walled hollow sphere having a very narrow opening at one point is a perfect absorber as its inner surface is coated with lamp black and a sharp projection opposite to the opening. If the radiation enters through the narrow opening. It is absorbed completely inside by successive reflections on the inner walls, as designed by Fery. If

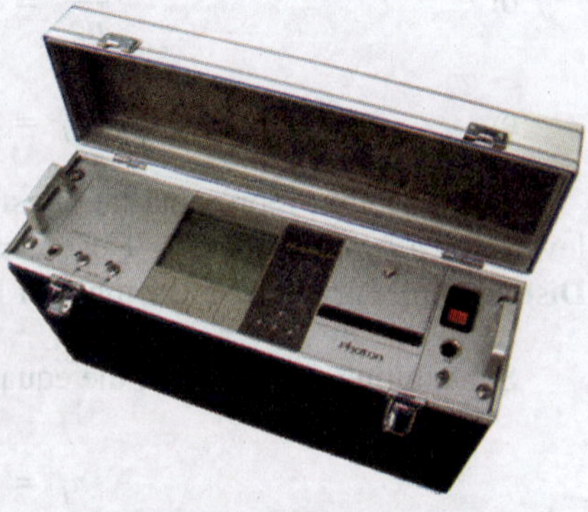

Photon Gas Analyzer

the encloser is maintained at a constant temperature *T*, then atoms of the walls of the encloser emit electromagnetic radiation and at the same time these radiations are absorbed by the atoms in walls. Thus, the atoms of the walls will emit and reabsorb photons continuously. After a certain lapse of time, a situation will be established when emission and absorption of photons attains thermodynamic equilibrium. At this stage, the amount of energy emitted by the atoms per unit time is equal to the amount of energy absorbed by atoms per unit time.

Thus, the interaction of electromagnetic radiation with matter led to the idea that electromagnetic radiation are composed of discrete energy particles called photons. Each photon has an energy $h\nu$ and momentum $\frac{h}{\lambda}$. Hence $\nu$ is the frequency and $\lambda$ is the wavelength of the radiation. The electromagnetic radiation trapped in a cavity and in thermal equilibrium with the walls of the cavity are termed as *black body radiation*. In the equilibrium condition, the black body radiation can be considered as the photon gas.

## 12.7 Planck's Radiation Law

**Statement :** According to the law of radiation of wavelength lying between $\lambda$ and $\lambda + d\lambda$ emitted per unit volume by a perfectly black body at temperature *T* is given by

$$E_\lambda d\lambda = \frac{8\pi hc}{\lambda^5} \frac{d\lambda}{e^{hc/\lambda kT} - 1} \qquad ...(12.19)$$

### Derivation of the law

**Some Important Postulates:** It is assumed that the photons do not interact among themselves. The photons interact only with the atoms of the walls of the cavity. It is further assumed that the photons are indistinguishable and many photons can have the same energy. Photons are considered as bosons and they obey Bose-Einstein statistics.

The most probable distribution of the particles among various energy levels for a system obeying Bose-Einstein's distribution law is

$$n_i = \frac{g_i}{e^{(\alpha+\beta E_i)} - 1} = \frac{g_i}{e^{\alpha}.e^{E_i/kT} - 1} \quad \text{where } \beta = \frac{1}{kT}$$

Most important point to be noted here that photon gas differ from the ideal gas in one respect. Although the total energy of photons inside the hollow encloser at a particular temperature *T* remains constant, the total number of photons may be completely absorbed on striking the wall or the hot wall may emit a new photon of energy $h\nu$, *i.e.*, the photons may be created or destroyed. In other words, for a photon gas system, $\sum_i \delta ni = 0$ is no longer valid. Hence $\delta n = \sum_i \delta ni \neq 0$

Due to this reason, the value of $\alpha$ is equal to zero in the Bose-Einstein distribution law. Thus

$$n_i = \frac{g_i}{e^{E_i/kT} - 1} \qquad (\because \alpha = 0) \qquad ...(12.20)$$

In this case, the cavity is large as compared to the wavelength of radiation, the energy spectrum of photons is taken to be continuous. The energy difference between successive allowed energy values is very small. Thus, replacing $g_i$ by $g(E)\,dE$ and $n_i$ by $n(E)\,dE$, we get

$$n(E)\,dE = \frac{g(E)\,dE}{e^{E/kT} - 1} \qquad ...(12.21)$$

Since $E = h\nu$, the value of $g(E)\,dE$ can be written as $g(\nu)\,d\nu$. The quantity $g(\nu)\,d\nu$ corresponds to the number of oscillatory modes in the frequency range $\nu$ and $(\nu + d\nu)$ is the energy range $E$ and $(E + dE)$.

The number of states in a black body radiation in the frequency range ν and (ν + dν) can be obtained by calculating the spherical volume bounded by two spheres of ratio $\frac{h(\nu+d\nu)}{c}$ and $\frac{h\nu}{c}$.

The volume of the spherical shell

$$= \frac{4}{3}\pi\frac{h^3}{c^3}(\nu+d\nu)^3 - \frac{4}{3}\pi\frac{h^3}{c^3}\nu^3$$

$$= \frac{4}{3}\pi\frac{h^3}{c^3}\left[\nu^3+3\nu^2\,d\nu+.....-\nu^3\right]$$

$$= \frac{4}{3}\pi\frac{h^3}{c^3}\times 3\nu^2\,d\nu \quad \text{(higher terms neglected)}$$

$$= 4\pi\frac{h^3}{c^3}\nu^2\,d\nu \quad ...(12.22)$$

The phase space has volume $V = h^3$ and there are two states of polarization (left hand and right hand) for the radiation. The number of states in the black body radiation in the frequency range ν and (ν + dν) is given by

$$g(\nu)d\nu = 2\times 4\pi\frac{V}{c^3}\nu^2\,d\nu \quad ...(12.23)$$

$$= \frac{8\pi V}{c^3}\nu^2\,d\nu$$

Substituting in equation (12.21), we get the number of photons in the frequency range ν and (ν + dν) given by

$$n(\nu)\,d\nu = \frac{\left(8\pi V\,\nu^2 d\nu\right)}{c^3}\times\frac{1}{e^{h\nu/kT}-1} \quad ...(12.24)$$

Since each photon has energy hν, the energy density E(ν) dν defined as the amount of energy per unit volume lying between the frequency range ν and (ν + dν) is given by

$$E(\nu)\,d\nu = \frac{h\nu.n(\nu)d\nu}{V}$$

$$\therefore \quad E(\nu)\,d\nu = \frac{\left(8\pi h\,\nu^3 d\nu\right)}{c^3}\times\frac{1}{e^{h\nu/kT}-1} \quad ...(12.25)$$

Equation (12.25) represents **Planck's radiation law for black body radiation.**

**In terms of wavelength:** To express the law in terms of wavelength (λ), substituting $\nu = \frac{c}{\lambda}$ and $d(\nu) = -\frac{c}{\lambda^2}$ in equation (12.25) , we get

$$E_\lambda d_\lambda = \frac{8\pi h}{c^3}\cdot\frac{c^3}{\lambda^3}\cdot\frac{c\,d\lambda}{\lambda^2}\cdot\frac{1}{e^{hc/\lambda kT}-1} = \frac{8\pi hc}{\lambda^5}\frac{d\lambda}{e^{hc/\lambda kT}-1}$$

or

$$E_\lambda = \frac{8\pi hc\lambda^{-5}}{e^{hc/\lambda kT}-1} \quad ...(12.26)$$

This equation (12.26) also represents **Planck's law for black body radiation in terms of wavelength.**

**ENRICO FERMI (1901–1954)** *was born in Rome on 29 Sept. 1901. He obtained his doctorate at Pisa. Working with leading scientists in the new quantum mechanics, in 1926 at University of Rome, he investigated the statistical mechanics of those particles that obey Pauli's exclusion principle, such as electrons. Dirac independently arrived at the same conclusion shortly afterword and hence the result is finally named as Fermi-Dirac statistics. In 1933, Fermi introduced the concept of the weak interaction and used it together with Pauli's newly postulated neutrino (as named it by Fermi) to develop a theory of beta decay.*

Enrico Fermi

*Later in 1930s Fermi and a group of collaborators carried out a series of experiments in which radionuclides were produced artificially by bombarding various elements with slow neutrons, which found to be most effective. This giving rise to the formation of transuranic elements. In 1938 Fermi received the Nobel Prize for this work. He then went to the United States. As a part of atom bomb project, Fermi was the leader of designing and construction of first nuclear reactor at the Chicago University. In December 1942, the nuclear reactor became operative. After second World War Fermi shifted to 'high energy particle physics'. He died of cancer in 28 Nov. 1954. An element having atomic number 100 is named as Fermium in his honour, discovered a year after his death.*

**PAUL A.M. DIRAC (1902–1984)** *was born in Bristol on 8 Aug. 1902 in England. Dirac was basically an electrical engineer. He shifted his interest to mathematics and then to physics. He obtained his Ph. D. from Cambridge in 1926. After reading Heisenberg's first paper on quantum machanics in 1925, Dirac soon devised a more general theory and next year, Dirac formulated Pauli's exclusion principle in quantum mechanical terms. He investigated the statistical behaviour of particles like electrons which obey Pauli exclusion principle, which Fermi had investigated independently a little earlier, the result was named as Fermi-Dirac statistics, in the honour of both. In 1928 Dirac applied special relativity to quantum theory. In an attempt to explain why charge is quantized, Dirac in 1931 found it necessary to postulate the existence of magnetic monopoles, isolated N or S magnetic poles. More recent theories show that magnetic monopoles should have been created in profusion just after the Big Bang theory that marked the beginning of the universe. Dirac became Lucasian Professor of mathematics, in 1932, at Cambridge, the post Newton had held two and half centuries earlier. In 1933, Dirac shared the Nobel Prize in physics with Schrödinger. He died on Oct 20, 1984.*

Paul Dirac

## 12.8 Fermi-Dirac Distribution Law

Consider a system consisting of $n$ independent and indistinguishable particles. The particles have definite energies (or momenta) and occupy definite positions. Therefore, they can be represented by phase points in the phase space. In order to determine the energy (or momenta) distribution of these particles in the most probable state known as equilibrium state; we divide the available volume in the phase space into large numbers (say $k$) of compartments (quantum groups or energy levels), each compartment representing a small interval of energy (or momentum). Further each compartment is divided into elementary cells; each of size $h^3$, where $h$ is a Planck's constant.

Further, we suppose that the size of the compartment is very large as compared to the size of the cell so that each compartment contains a very large number of elementary cells. Let the compartments be marked 1, 2, ........ $i$, ........ $k$ and their mean energy values be represented by $E_1$, $E_2$, ........ $E_i$ ........ $E_k$ containing $g_1$, $g_2$, ........ $g_i$ ........ $g_k$ cells respectively in them. The total number of particles in the system is

$$n = n_1 + n_2 + \ldots\ldots + n_i \ldots\ldots + n_k.$$

Our aim is to find the Fermi-Dirac distribution law *i.e.,* the distribution of $n_i$ Fermions out of the total $n$ Fermions in $g_i$ cells of the $i$th compartment. For this, the basic postulates are:

## Basic Postulates

(*i*) The particles of indistinguishable so that there is no distinction between the various ways in which $n_i$ particle are chosen.

(*ii*) The particles obey Pauli's exclusion principle according to which there can be either no particle or only one particle in a given cell. Therefore, the number of cells must be much greater than the number of particles, *i.e.,* $g_i >> n_i$.

Out of $n_i$ number of particles in the $i$th compartment with mean value $E_i$ , the first particle can be placed in any one of the available $g_i$ states, *i.e.,* this particle can be assigned to any of the $g_i$ sets of quantum numbers. Thus, the first particle can be distributed in $g_i$ different ways, in accordance with Pauli's exclusion principle and the remaining $(g_i - 1)$ will remain vacant. The second particle can be arranged in $(g_i - 1)$ different ways and the process continues.

Thus, the total number of different ways of arranging $n_i$ particles among available $g_i$ states with energy level $E_i$ is given by

$$= g_i (g_i - 1)(g_i - 2)(g_i - 3) \ldots\ldots [g_i - (n_i - 1)]$$

$$= \frac{g_i\,!}{(g_i - n_i)!} \qquad \ldots(12.27)$$

Further, since the particles are indistinguishable, it would not be possible to detect any difference when $n_i$ particles are reshuffled into different ways occupied by them in the energy level $E_i$. Thus, out of these $n_i$ permutations of the indistinguishable particles among themselves will be meaningless. Therefore, the total number of different and distinguishable ways is

$$= \frac{g_i\,!}{n_i\,!(g_i - n_i)!} \qquad \ldots(12.28)$$

Therefore, the thermodynamic probability for the microstate $(n_1, n_2, \ldots\ldots n_i, \ldots\ldots n_k)$ of the system is given by

$$W(n_1, n_2, \ldots\ldots i, \ldots\ldots k) = \frac{g_1\,!}{n_1\,!(g_1 - n_1)!} \times \frac{g_2\,!}{n_2\,!(g_2 - n_2)!} \times \ldots\ldots \times \frac{g_i\,!}{n_i\,!(g_i - n_i)!} \ldots\ldots \times \frac{g_k\,!}{n_k\,!(g_k - n_k)!}$$

$$= \prod_{i=1}^{k} \frac{g_i\,!}{n_i\,!(g_i - n_i)!} \qquad \ldots(12.29)$$

where $\prod$ denotes multiplication of terms for various values of $i$ from 1 to $k$.

## Most Probable Microstate

Taking natural logarithm on both sides of eqn. (12.29),

$$\ln W = \sum_{i=1}^{k} [\ln g_i! - \ln n_i! - \ln (g_i - n_i)!]$$

As $n_i$ and $g_i$ are very large numbers, we can use Sterling approximation

$\ln n! = (n \ln n) - n$

Applying Sterling's approximation, we have

$$\ln W = \sum_{i=1}^{k} [g_i \ln g_i - g_i - n_i \ln n_i + n_i$$

$$-(g_i - n_i)\ln(g_i - n_i) + (g_i - n_i)\Big]$$

$$= \sum_{i=1}^{k}\left[g_i \ln g_i - n_i \ln n_i - (g_i - n_i)\ln(g_i - n_i)\right]$$

Here $g_i$ is not subject to variation whereas $n_i$ varies continuously. Differentiating both sides, we have

$$\delta(\ln W) = \sum_{i=1}^{k}\left[-n_i \frac{1}{n_i}\delta n_i - \ln n_i\,\delta n_i + (g_i - n_i)\frac{1}{(g_i - n_i)}\delta n_i + \ln(g_i - n_i)\,\delta n_i\right]$$

$$= \sum_{i=1}^{k}\left[\ln(g_i - n_i) - \ln n_i\right]\delta n_i$$

To get the state of *maximum thermodynamic probability*

$$\delta(\ln W) = 0$$

$$\therefore \quad \sum_{i=1}^{k}\left[\ln(g_i - n_i) - \ln n_i\right]\delta n_i = 0$$

or $$\sum_{i=1}^{k}\left[\ln\frac{g_i - n_i}{n_i}\right]\delta n_i = 0$$

or $$-\sum_{i=1}^{k}\left[\ln\frac{n_i}{g_i - n_i}\right]\delta n_i = 0$$

or $$\sum_{i=1}^{k}\left[\ln\frac{n_i}{g_i - n_i}\right]\delta n_i = 0 \quad \text{...(12.30)}$$

In addition, our system must satisfy the two auxillary conditions:

(*i*) Conservation of total number of particles, *i.e.*, $n = a$ constant

$$\therefore \quad n = \sum_i n_i = \text{constant}$$

*i.e.*, $$\delta E = \sum_i \delta n_i = 0 \quad \text{...(12.31)}$$

(*ii*) Conservation of total energy of the system, *i.e.*, $E$ = a constant.

$$\therefore \quad E = \sum_i n_i E_i \quad \text{constant}$$

*i.e.*, $$\delta E = \sum_i E_i\,\delta n_i = 0 \quad \text{...(12.32)}$$

We shall now apply Langrangian method of undetermined multipliers. For this, let us multiply eqn. (12.31) by $\alpha$ and eqn. (12.32) by $\beta$ and add the resulting expression to eqn. (12.30),

$$\sum_{i=1}\left[\ln\left(\frac{n_i}{g_i - n_i}\right) + \alpha + \beta E_i\right]\delta n_i = 0$$

As the variations $\delta n_i$ are independent of each other, we get

$$\ln\left(\frac{n_i}{g_i - n_i}\right) + \alpha + \beta E_i = 0$$

or $$\frac{n_i}{g_i - n_i} = e^{-(\alpha+\beta E_i)}$$

or $$\frac{g_i - n_i}{n_i} = e^{\alpha+\beta E_i}$$

or $$\frac{g_i}{n_i} - 1 = e^{\alpha+\beta E_i}$$

or $$\frac{g_i}{n_i} = e^{\alpha+\beta E_i} + 1$$

or $$n_i = \frac{g_i}{e^{\alpha+\beta E_i} + 1} \quad ...(12.33)$$

This equation represents the most probable distribution of the particles among various energy levels for a system obeying Fermi-Dirac statistics and is therefore known as **Fermi-Dirac distribution law**, for an assembly of *fermions*.

The parameter $\beta = \frac{1}{kT}$, where $k$ is Boltzmann's constant, has the same role as in case of *M-B* distribution law. Substituting in equation (12.33), we have

$$n_i = \frac{g_i}{e^{\alpha+\beta E_i/kT} + 1} = \frac{g_i}{e^{\alpha}.e^{E_i/kT} + 1} \quad ...(12.34)$$

**Fermi-Dirac energy distribution function.**

Since there can be a *maximum of one particle per quantum state*, the function $f(E_i)$ is the ratio of the number of quantum states of energy $E_i$ occupied by electrons to the total number of quantum states available in the energy level $E_i$. Therefore, the value of $f(E_i)$ for the Fermi-Dirac distribution at a paricular energy $E_i$ is the probability that under equilibrium a quantum state of that energy is occupied by a particle.

From equation (12.34), $f(E_i)$ is given by

$$f(E_i) = \frac{n_i}{g_i}$$

$$= \frac{1}{e^{\alpha} e^{E_i/kT} + 1} \quad ...(12.35)$$

For continuous distribution of energy at energy E, the distribution function is written as

$$f(E) = \frac{1}{e^{\alpha} e^{E/kT} + 1} \quad ...(12.36)$$

**F-D energy distribution law for continuous variation of energy**

If the energy levels are very close together, then the distribution of energy of the particles may be considered continuous. For this distribution the number of particles $n(E)dE$ whose energies lie between $E$ and $E + dE$, then $n_i = n(E)dE$ and is given by

$$n(E)dE = f(E)\ g(E)dE \quad ...(12.37)$$

where $g(E)dE$ is the number of quantum states of energy between $E$ and $E + dE$.

Substituting the value of $f(E)$ in eqn. (12.37), we have

$$n(E)dE = \frac{g(E)dE}{e^{\alpha}.e^{E/kT} + 1} \quad ...(12.38)$$

For particles like electrons of spin angular momentum $\pm\frac{1}{2}\hbar$, there are two possible spin orientations. For a system of such particles $g(E)dE$ is given by (see eqn. 11.26B). Substituting, we get

$$n(E)\, dE = 2\times 2\pi V\left(\frac{2m}{h^2}\right)^{3/2}\frac{E^{\frac{1}{2}}dE}{e^{\alpha}.e^{E/kT}+1} \quad \text{...(12.39)}$$

This is the Fermi-Dirac energy distribution law for continuous variation of energy among free particles with spin $\frac{1}{2}$.

## 12.9 Free Electrons in Metal: Electron Gas

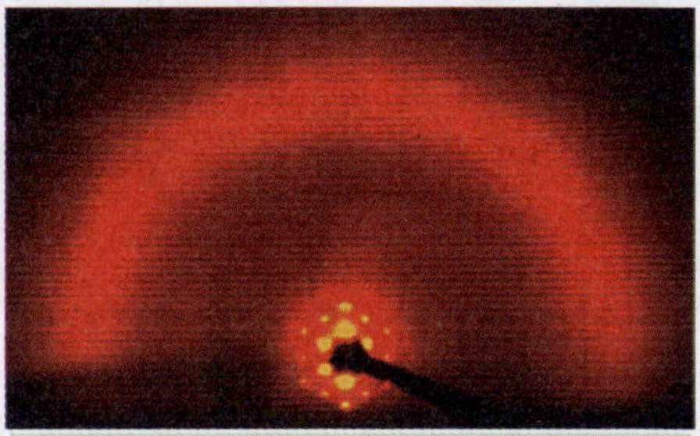

Thermotropic Liquid Crystals.

Metals are very good conductors. The high conductivity of metals is due to the presence of free electrons. These free electrons inside a metallic conductor move freely inside the metal. They continuously collide with the fixed (Localized) atoms and thus behave like an *electron gas*. These free electrons in a metal belong to a most characteristic system of fermions because electrons obey the Pauli's exclusion principle, *i.e.*, the electrons have half integral spin angular momentum in unit of $\frac{h}{2\pi}$ or $\hbar$. They are Fermi particles and obey Fermi-Dirac statistics. The electrons in metal have their energy quantised. Fermi-Dirac derived an expression for the distribution of energies among electrons in the *electron gas* in a metal on the basis of these assumptions.

### Energy Distribution among Electrons in a Metal

Consider an electron gas having *n* free electrons in a metallic conductor whose volume is *V*. Let the total energy of electrons be *E*. Practically at all temperatures, the lower level energy bands are filled with electrons. The upper level energy bands are only partially filled with electrons. The distribution of electrons is to be considered only in the upper bands called the *conduction band*. The zero energy level is taken at the lowest level of conduction band. It is also assumed the electrons have free movement within the conductor, provided the energy associated with the electrons is of the order of upper level energy bands.

As the energy of the electron in the conduction band is continuous, the term $g_i$ replaced by $g(E)$ $dE$. Hence $n(E)\, dE$ electrons have energy in the range $E$ and $E + dE$. According to Fermi-Dirac distribution law.

$$n_i = \frac{g_i}{e^{\alpha}.e^{E_i/kT}+1}$$

Substituting for $g_i = g(E)\, dE$, in above equation and replacing $n_i$ by $n(E)\, dE$ we have

$$n(E)\, dE = \frac{g(E)\, dE}{e^{\alpha}.e^{E/kT}+1}$$

As on electron has spin $\pm\frac{1}{2}$, the total number of states in the sphere is twice. The quantity $\frac{V}{(2\pi)^3}$ refers to the translational states per unit volume in the Fermi space. Fermi sphere of radius $k_F$ has the total number of particles accommodated

$$n(E) = \frac{2\left[\frac{V}{(2\pi)^3}\right]\left(\frac{4}{3}\pi k_F{}^3\right)}{e^{\alpha}.e^{E/kT}+1}$$

Here
$$E = \frac{p^2}{2m} = \frac{\left(\frac{h}{2\pi}\right)^2 k_F^2}{2m}$$

$$\therefore \quad k_F = \frac{p}{h/2\pi} = \frac{2\pi p}{h}$$

substituting, we get

$$n(E) = \frac{2\left[\frac{V}{(2\pi)^3}\right]\left(\frac{4}{3}\pi\left(\frac{2\pi p}{h}\right)^3\right)}{e^{\alpha}.e^{E/kT}+1}$$

But
$$p = (2mE)^{\frac{1}{2}}$$

$$\therefore \quad n(E) = \frac{2\left[\frac{V}{(2\pi)^3}\right]\left(\frac{4}{3}\pi\left(\frac{2\pi}{h}\right)^3(2mE)^{\frac{3}{2}}\right)}{e^{\alpha}.e^{E/kT}+1}$$

$$= \frac{\left[\frac{8\pi V}{(3h)^3}\right](2m)(2m)^{\frac{1}{2}}E^{\frac{3}{2}}}{e^{\alpha}.e^{E/kT}+1}$$

Differentiating,

$$n(E)\,dE = \frac{\left[\frac{8\pi V}{(3h)^3}\right](2m)(2m)^{\frac{1}{2}}\times\frac{3}{2}E^{\frac{1}{2}}\,dE}{e^{\alpha}.e^{E/kT}+1}$$

$$n(E)\,dE = \frac{\left[\frac{8\pi V\left(2m^3\right)^{\frac{1}{2}}}{h^3}E^{\frac{1}{2}}\right]}{e^{\alpha}.e^{E/kT}+1}dE = \frac{8\sqrt{2}\,\pi V}{h^3}\frac{m^{\frac{3}{2}}E^{\frac{1}{2}}\,dE}{e^{\alpha}.e^{E/kT}+1} \quad ...(12.40)$$

Equation 12.40 represents the distribution of energy among free electrons. The value of $\alpha$ is still unknown. To evaluate, let us define first the Fermi energy.

## 12.10 Fermi Level and Fermi Energy

The equation (12.40) reveals that as the value of $T$ decreases, the energy of the electrons goes on decreasing, but at 0 $K$, the energy of electrons is not zero. Moreover according to Pauli's exclusion principle, only one electron can occupy one energy level. Therefore, there will be one electron in the ground state having zero energy. If $n$ is the total number of electrons at $K$, then total number of energy states which will be occupied will also $n$; each state with one electron. Let $E_F$ be the energy of highest state occupied by the electrons at 0 $K$, then all the energy states having energy values greater than $E_F$ will be empty and all the energy states having energy values less than $E_F$ will be fully occupied, each with one electron.

**Definition :** Thus, the energy value up to which all the energy states are completely filled by electrons at absolute zero (*i.e.* 0 *K*) of temperature and above which all the energy states are completely empty is known as **Fermi energy**. It is denoted by $E_F$.

Thus, the occupation index, $f(E) = \dfrac{n(E)}{g(E)} = 0$ for $E > E_F$

and the occupation index, $f(E) = \dfrac{n(E)}{g(E)} = 1$ for $E \leqslant E_F$

From these two conditions, we can prove that $\alpha = -\dfrac{E_F}{kT}$ as under:

For values $E \leqslant E_F$, $T = OK$ and assuming $\alpha = -\dfrac{E_F}{kT}$

$$\text{Occupation Index} = \frac{n(E)}{g(E)} = \frac{1}{\left(e^{-E_F/kT} \cdot e^{E/kT} + 1\right)} = \frac{1}{e^{-\infty} + 1} \cong 1 \qquad \text{...(12.41)}$$

In other words, all energy states having $E \leqslant E_F$, each is occupied by a single electron.

For values $E > E_F$, at absolute zero, $T = 0\ K$ and assuming $\alpha = -\dfrac{E_F}{kT}$.

$$\text{Occupation Index,} \quad \frac{n(E)}{g(E)} = \frac{1}{e^{(E-E_F)/kT} + 1} = \frac{1}{e^{\infty} + 1} \cong 0 \qquad \text{...(12.42)}$$

*i.e.*, This means that at absolute zero all quantum states of energy more than $E_F$ are empty. (At $T = 0\ K$, $E_F$ is expressed as $E_{F_0}$).

This discussion gives the sufficient proof to assign the value of $\alpha = -\dfrac{E_F}{kT}$. Hence substituting for $\alpha$ in equation (12.40), we get

$$n(E)dE = \frac{8\sqrt{2}\,\pi V}{h^3} \cdot \frac{m^{3/2} E^{\frac{1}{2}} dE}{e^{(E-E_F)/kT} + 1} \qquad \text{...(12.42A)}$$

This is the final expression for Fermi-Dirac law of energy distribution of free electrons in a metal.

**Discussion.** The physical interpretation of the Fermi energy at absolute zero is that it represents highest energy level at $T = 0\ K$ in which all the quantum states are occupied,

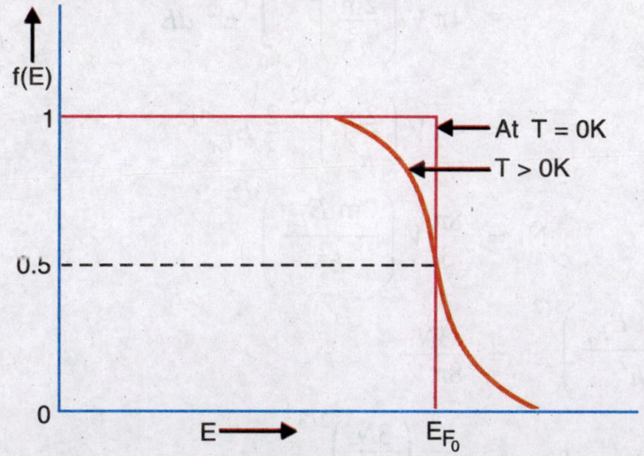


**Fig. 12.4**

Eqn. (12.41) and (12.42) show that at absolute zero, the function $f(E)$ is constant equal to 1 for all values of energies up to $E_{F_O}$ and at $E_{F_O}$, it suddenly falls to zero. Thus, at absolute zero of temperature, $f(E)$ is a *step function,* as shown in Fig. 12.4.

**Fermi energy at $T > OK$**

For $T > 0\,K$ at $E = E_F$, we have

$$f(E) = \frac{1}{e^{O/kT}+1} = \frac{1}{e^0+1} = \frac{1}{1+1} = \frac{1}{2}$$

This gives the physical interpretation of the Fermi energy at a finite temperature ($T > 0\,K$). Thus, the Fermi energy $E_F$ at $T > 0\,K$ is that energy level in which 50% of the quantum states are occupied and 50% are empty.

It can be shown that as $T$ increases, $E_F$ decreases slightly. Therefore, for practical purposes $E_F$ at temperature $T$ may be considered constant and is nearly equal to $E_{F_O}$.

## 12.11 $E_{F_O}$ for Electrons in a Metal

The total number of free electrons in a metal is given by

$$N = \int_0^\infty n(E)dE$$

$$= \int_0^\infty f(E)g(E)dE$$

$$= \int_0^{E_{F_O}} f(E)g(E)dE + \int_{E_{F_O}}^\infty f(E)g(E)dE$$

Now we know that

at $\quad T = 0$, for $E \le E_{F_O}$, $f(E) = 1$

and at $\quad T = 0$, for $E > E_{F_O}$, $f(E) = 0$

Hence the second integral vanishes *i.e.*, becomes zero and the in the first integral we put $f(E) = 1$.

$$\therefore \quad N = \int_0^{E_{F_O}} g(E)dE$$

$$= 4\pi V\left(\frac{2m}{h^2}\right)^{3/2}\int_0^{E_{F_O}} E^{\frac{1}{2}}\,dE$$

$$= 4\pi V\left(\frac{2m}{h^2}\right)^{3/2}\frac{2}{3}E_{F_O}^{3/2}$$

$$N = \frac{8\pi}{3}V\left(\frac{2m\,E_{F_O}}{h^2}\right)^{3/2}$$

or
$$\left(\frac{2m\,E_{F_O}}{h^2}\right)^{3/2} = \frac{3N}{8\pi V}$$

or
$$E_{F_O} = \frac{h^2}{2m}\left(\frac{3N}{8\pi V}\right)^{2/3}$$

or $$E_{F_O} = \frac{h^2}{2m} \times \left(\frac{3n}{8\pi}\right)^{2/3}, \quad ...(12.43)$$

where $n = \frac{N}{V}$, the number of free electrons per unit volume, *i.e.*, electron density.

The values of $E_{F_O}$ calculated from equation (12.43) for a number of metals are of the order of several electron volts. This fact highlights a very important difference between the classical statistics and Fermi-Dirac statistics. According to classical statistics all electrons in a metal at absolute zero would have zero energy.

## 12.12 Fermi-Dirac Velocity Distribution in Given Direction

The Fermi-Dirac energy distribution function for free electrons in a metal at temperature TK is given by

$$f(E) = \frac{1}{e^{(E-E_F)/kT} + 1}$$

for energies $E > E_F$, such that $(E - E_F) \gg kT$. Under this condition, we may omit 1 from the denominator.

Thus, $$f(E) = \frac{1}{e^{(E-E_F)/kT}}$$

$$= e^{E_F/kT} \,.\, e^{-E/kT} \quad ...(i)$$

In a metal if all the electrons are assumed to be free, then the whole energy is kinetic in nature.

$$E = \frac{1}{2} m v^2$$

$$= \frac{1}{2} m \,(v_x^2 + v_y^2 + v_z^2) \quad ...(ii)$$

where the symbols have their meanings.

The function of $E$ in terms of the function of the velocity components $(v_x, v_y, v_z)$ in the form

$$f(v_x, v_y, v_z) = e^{E_F/kT} \,.\, e^{-m(v_x^2 + v_y^2 + v_z^2)/2kT} \quad ...(iii)$$

The number of electrons with velocity components between $v_x$ and $v_x + dv_x$ ; $v_y$ and $v_y + dv_y$; $v_z$ and $z_z + dv_z$ is given by

$$n(v_x, v_y, v_z)\, dv_x\, dv_y\, dv_z = f(v_x, v_y, v_z)\, g(v_x, v_y, v_z)\, dv_x\, dv_y\, dv_z \quad ...(iv)$$

where $g(v_x, v_y, v_z)\, dv_x\, dv_y\, dv_z$ is the number of quantum states within the velocity space element $dv_x dv_y\, dv_z$.

The volume of momentum space corresponding to one quantum state is (taking into account spin $\frac{1}{2}$ of electrons) $\frac{h^3}{2V}$.

Therefore, the number of quantum states in the volume $dp_x\, dp_y\, dp_z$ of momentum space.

$$= \frac{2V}{h^3}\, dp_x\, dp_y dp_z$$

$$= \frac{2m^3 V}{h^3}\, dv_x\, dp_y dp_z$$

Hence, $$g(v_x, v_y, v_z)\, dv_x\, dv_y\, dv_z = \frac{2m^3 V}{h^3}\, dv_x\, dv_y dv_z \quad ...(v)$$

Substituting Eqns. (*iii*) and (*v*) into Eqn. (*iv*), we get

$$n(v_x, v_y, v_z)\, dv_x\, dv_y\, dv_z = \frac{2m^3V}{h^3} e^{E_F/kT} \times e^{-m(v_2^2+v_y^2+v_z^2)/2kT} \times dv_x dv_y dv_z \qquad ...(vi)$$

Now the number $n(v_x)dv_x$ of electrons is obtained by integrating Eqn. (*vi*) over all possible values of $v_y$ and $v_z$. Thus, we get

$$n(v_x)\, dv_x = \int_{-\infty}^{+\infty}\int_{-\infty}^{+\infty} [n(v_x, v_y, v_z)] dv_y\, dv_z$$

$$= \frac{2m^3 V}{h^3} e^{E_F/kT} \times e^{-mv_x^2/2kT} dv_x \times \int_{-\infty}^{\infty} e^{-mv_y^2/2kT} dv_y \times \int_{-\infty}^{\infty} e^{-mv_z^2/2kT} dv_z$$

The integrals are standard integrals, the value of each being $\left(\frac{2\pi kT}{m}\right)^{\frac{1}{2}}$.

Substituting this value in the above equation

$$n(v_x)dv_x = \frac{2m^3 V}{h^3} e^{E_F/kT} \times e^{-mv_x^2/2kT} dv_x \left(\frac{2\pi kT}{m}\right)$$

$$= V\left(\frac{4\pi m^2 kT}{h^3}\right) e^{E_F/kT} \times e^{-mv_x^2/2kT} dv_x \qquad ...(12.44)$$

Eqn. (12.44) gives the Fermi-Dirac velocity distribution law for electrons in a metal, whose *x*-component of velocity lies between $v_x$ and $v_x + dv_x$.

## 12.13 Maxwell-Boltzmann Distribution as a Limiting Case of Bose-Einstein and Fermi-Dirac Distributions

The general form of the Bose-Einstein and Fermi-Dirac distribution laws given by

$$n_i = \frac{g_i}{e^{\alpha} e^{E_i/kT} \pm 1} \qquad ...(i)$$

where the negative sign is for the B-E distribution and the positive sign for the F-D distribution.

If for any value of $E_i$, the condition

$$e^{\alpha} e^{E_i/kT} \gg 1 \qquad ...(ii)$$

is satisfied, then 1 in the denominator of Eq. (*i*) can be neglected. This gives

$$n_i = e^{-\alpha} . g_i\, e^{-E_i/kT} \qquad ...(iii)$$

This is the general form of the Maxwell-Boltzmann distribution law.

Moreover, we can show that the M.B. distribution is applicable to rarefied gases. For this, Equation (*iii*) can be written as

$$\frac{n_i}{g_i} = e^{-\alpha} e^{-E_i/kT}$$

From the inequality (*ii*), we get the inequality as

$$\frac{n_i}{g_i} \ll 1 \qquad ...(iv)$$

The ratio $\frac{n_i}{g_i}$ gives the average number of particles per quantum state. Thus, the inequality (*iv*) shows that the Maxwell-Boltzmann distribution law is true in the limiting case of small number of particles per quantum state, *i.e.*, the distribution law is true for rarefied gases.

## 12.14 Application to Liquid Helium

Helium has proven to be one of the most remarkable elements in nature, particularly from the standpoint of statistical physics. It is the only substance which remains liquid under ordinary pressure at absolute zero; all others freeze to the solid state if sufficiently cooled. Solidification of helium takes place under external pressure exceeding 25 atmospheres. According to classical concepts, at absolute zero, all the atoms of a substance cease to move and occupy fixed positions within the body, causing it to become a solid. The fact that at this low temperature *i.e.* *oK*, helium remains in liquid state, indicates that entirely different concepts of *quantum mechanics* are required for its understanding.

Liquid Helium Cooled Surface

Helium atoms occur in nature in two stable isotopic forms, $^3$He and $^4$He. The former is rarer of these two isotopes. Chemically both these are virtually identical. The only difference between them is a difference of mass and that $^3$He has a nuclear spin $\frac{1}{2}$ and $^4$He has spin zero. Therefore, $^3$He obeys Fermi-Dirac statistics while $^4$He obeys Bose-Einstein statistics.

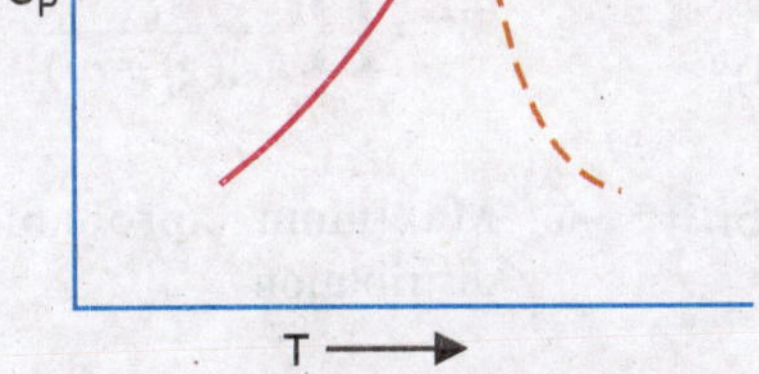


Fig. 12.5

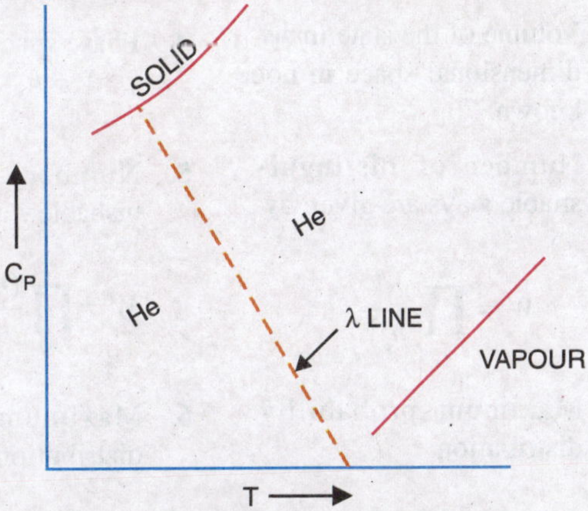


Fig. 12.6

Due to the difference in their statistics, the two systems exhibit very different behaviour upon cooling. $^4$He which is boson liquid exhibits a second order transition to a superfluid state. This was first observed as an abrupt jump in liquid specific heat Fig. 12.5. $^3$He also undergoes a transition to superfluid state, but at much lower temperature nearly $2.7 \times 10^{-3}$ *K*. The mechanisms of the two transitions, however, are different. (Fig. 12.6) shows the co-existence curves for $^4$He. At low temperature, $^4$He has four phases : a solid phase, liquid phases and a vapour phase. As the normal liquid is cooled , a line of λ-point occurs near $2K$, the exact value of temperature depends upon pressure. The λ-line indicates that a continuous phase transition has occurred. Liquid helium above the λ-line is called Helium I (or He I) and that below it Helium II (or He II).

The specific heat-temperature graph resembles λ and hence this temperature at which the specific heat changes abruptly (2.19 $K$) is called the λ-point. Liquid helium above 2.19 $K$ is called He I because it behaves in normal way and below 2.19$K$, it is called He II because of its abnormal properties. Viscosity of liquid He I decreases with decrease in temperature and this property is contrary to the property of normal liquids but it resembles that of a gas. The viscosity of He II is practically almost zero and it can flow rapidly through narrow capillary tubes. It has zero entropy and no viscosity.

The thermal conductivity of He II has an abnormally high value and it is many times more than that of copper and silver.

These properties of He II are described well in quantum statistics by a two-fluid model. It is assumed that He II is a mixture of two fluids: the normal fluid and the superfluid and that there is no viscous interaction between them. While the normal fluid has all the usual properties of a fluid, superfluid has some curious properties *viz.* zero entropy and no viscosity.

## 12.15 Comparison of the Three Statistics

| *Maxwell-Boltzmann* | *Bose-Einstein* | *Fermi-Dirac* |
|---|---|---|
| **1.** Particles are distinguishable. | **1.** Particles are indistinguishable. | **1.** Particles are indistinguishable. |
| **2.** Only particles are taken into consideration. | **2.** Only quantum states are taken into consideration. | **2.** Only quantum states are taken into consideration. |
| **3.** There is no restriction on the number of particles in a given state. | **3.** There is no restriction on the number of parrticles in a given quantum state. | **3.** There is restriction on the number of particles in a given quantum state. |
| **4.** Volume of the state in six dimensional space in not known. | **4.** Phase space is known $V = h^3$ | **4.** Phase space is known $V = h^3$ |
| **5.** Number of distinguishable ways are given by $W = \prod_i^N \frac{g_i^{n_i}}{n_i!}$. | **5.** Number of distinguishable ways are given by $W = \prod_i \frac{(n_i + g_i - 1)!}{n_i!(g_i - 1)!}$ | **5.** Number of distinguishable ways are given by $W = \prod_i \frac{g_i!}{n_i!(g_i - n_i)!}$ |
| **6.** Maximum probability distribution $\propto \frac{1}{e^{(\alpha+\beta E_i)}}$ | **6.** Maximum probability distribution $\propto \frac{1}{[e^{(\alpha+\beta E_i)}]-1}$ | **6.** Maximum probability distribution $\propto \frac{1}{[e^{(\alpha+\beta E_i)}]+1}$ |
| **7.** ________ | **7.** At high temperatures, Bose-Einstein distribution approaches Maxwell-Boltzmann distribution. | **7.** At high temperatures, Fermi-Dirac distribution approaches Maxwell-Boltzmann distribution. |

| *Maxwell-Boltzmann* | *Bose-Einstein* | *Fermi-Dirac* |
|---|---|---|
| **8.** Applicable to ideal gas molecules. | **8.** Applicable to photons and symmetrical particles known as bosons. | **8.** Applicable to electrons and antisymmetric particles known as fermions. |
| **9.** Internal energy of an ideal monatomic gas depends only on its temperature. At absolute zero, energy is zero. | **9.** The energy at absolute zero is taken to be zero. | **9.** Due to Pauli's exclusion principle, all the electrons cannot occupy the lowest energy level. Even at absolute zero temperature, some of the electrons are distributed at higher energy levels. Hence, at absolute zero, the energy is not zero. Therefore, at the complete degenerate state, the energy is in dependent of temperature. |
| **10.** $n_i = \frac{g_i}{e^{(\alpha+\beta E_i)}}$ | **10.** $n_i = \frac{g_i}{[e^{(\alpha+\beta E_i)} - 1]}$ | **10.** $n_i = \frac{g_i}{[e^{(\alpha+\beta E_i)} + 1]}$ |

**11.** All the three distributions can be represented by a single equation

$$\frac{g_i}{n_i} = \left[e^{(\alpha+\beta E_i)} + \delta\right]$$

Here $\delta = 0$ | Here $\delta = -1$ | Here $\delta = +1$

If $\frac{g_i}{n_i} \geqslant 1$ or $n_i \leq g_i$, *i.e.* $\delta$ can be neglected. In the case of diluted systems, the three statistics lead practically to similar results. It means that at high temperatures, the three statistics give practically the same results.

## 12.16. Difference between Classical and Quantum Statistics

The main points of difference between classical and quantum statistics are :

(*i*) Classical statistics is applied to a system of identical particles which obey the laws of classical mechanics, whereas quantum statistics is applied to a system of identical particles which obey the laws of quantum mechanics.

(*ii*) In classical statistics the particles of the system are considered *distinguishable*, whereas in quantum statistics these are considered *indistinguishable*.

(*iii*) In classical statistics the *volume* of the elementary cell in the phase space can be made as small as we please whereas in quantum statistics its value cannot be less than $h^3$, where $h$ is Planck's constant, given by $6.63 \times 10^{-34}$ Joules second. In other words, the number of cells in the phase space can be made as large as we like in classical statistics but this is not so in quantum statistics.

(*iv*) As the number of cells available in phase space in classical statistics is very large as compared to the number of particles, the occupation index $\frac{n_i}{g_i} \ll 1$. In quantum mechanics, the number of cells available is approximately equal to the number of particles. The occupation index $\frac{n_i}{g_i} = 1$.

## NUMERICAL EXAMPLES

### EXAMPLE 12.1

*Calculate the average energy of Planck's Oscillator of frequency ν in thermal equilibrium at temperature T.* **[Kolkata 2000]**

**SOLUTION.** The probability that the oscillator to be in *n*th state is (by Boltzmann's canonical theorem) proportional to $e^{-E_i/kT}$, the phase volume available being equal in all cases. Hence

$\bar{E} = \frac{\sum E_n e^{-E_n/kT}}{\sum e^{-E_n/kT}}$. The summation extends from $n = 0$ to $n = \infty$. Substituting for $E_n = nh\nu$ we get

$$\bar{E} = \frac{\sum nh\nu\, e^{-nh\nu/kT}}{\sum e^{-nh\nu/kT}} = \frac{h\nu}{e^{h\nu/kT} - 1}.$$

This gives the average energy of Planck's oscillator of frequency ν in thermal equilibrium at temp. *T*.

### EXAMPLE 12.2

*Show that if h is treated as arbitrary and allowed to tend to zero, the average energy expressed by* **[Chennai 2001]**

$$\bar{E} = \frac{h\nu}{e^{h\nu/kT} - 1}$$

*would be that given by the equipartition theorem.*

**SOLUTION.** Let $\frac{h\nu}{kT} = x$ then $h\nu = kT.\ x$

substituting, we get

$$\bar{E} = kT \frac{x}{e^x - 1}$$

For $\quad x \to 0,\ e^x - 1 = x + \frac{x^2}{2!} + \ldots = x$

$$\therefore \quad \bar{E} = kT$$

This is the average energy given by equipartition theorem.

### EXAMPLE 12.3

*A system consists of 5 particles arranged in two compartments. The first compartment is divided into 6 cells and the second into 8 cells. The cells are of equal size. Calculate the number of microstates in the macrostate (2, 3), if the particles obey Fermi-Dirac statistics.* **(Purvanchal 2006)**

**SOLUTION.** As the particles obey *F-D* statistics, the number of microstates in the macrostates ($n_1$, $n_2$, ..... $n_k$) or thermodynamic probability of the macrostate is given by

$$W_{(n_1, n_2, \ldots\ldots n_i, \ldots\ldots\ldots n_k)} = \prod_{i=1}^{k} \frac{g_i}{n_i!(g_i - n_i)}$$

Given : $k = 2$

$$\therefore \quad W_{(n_1, n_2)} = \frac{g_1}{n_1!(g_1-n_1)!} \times \frac{g_2}{n_2!(g_2-n_2)!}$$

Now $g_1 = 6$ and $g_2 = 8$

$$\therefore \quad W_{(2,3)} = \frac{6!}{2!(6-2)!} \times \frac{8!}{3!(8-3)!}$$

$$= \frac{6!}{2!4!} \times \frac{8!}{3!5!} = \frac{6\times5}{2\times1} \times \frac{8\times7\times6}{3\times2\times1}$$

$$= 15 \times 56$$

$$= \mathbf{840}$$

**EXAMPLE 12.4**

*Calculate the value of Fermi-energy at absolute zero temperature.* **(Nagpur 2000)**

**SOLUTION.** *At absolute zero temperature, i.e., t = 10 K, the number of electrons is equal to the total number of energy states occupied by the electrons i.e. Fermions ranging from zero to $E_F$, since each energy state can occupy only one electron; i.e.,*

$$n = \int_o^{E_F} g(E)dE = \frac{8\sqrt{2}\pi V}{h^3} m^{\frac{3}{2}} \int_o^{E_F} E^{\frac{1}{2}} dE \left(\because e^{(E-E_F)/kT} = 0 \text{ as } T = 0\right)$$

$$= \frac{8\sqrt{2}\,\pi V}{h^3} m^{\frac{3}{2}} \left[\frac{2}{3} E^{\frac{3}{2}}\right]_0^{E_F}$$

$$= 16\sqrt{2}\,\pi V \frac{m^{\frac{3}{2}}}{3h^3} E_F^{\frac{3}{2}}$$

$$\therefore \quad E_F^{\;3/2} = \frac{3nh^3}{16\sqrt{2}\,\pi V m^{\frac{3}{2}}}$$

$$\text{or} \quad E_F = \frac{h^2}{2m}\left(\frac{3n}{8\pi V}\right)^{\frac{2}{3}}$$

**EXAMPLE 12.5**

*The number of conduction electrons per c.c. is $24.2\times10^{22}$ in beryllium and $0.91\times10^{22}$ in cesium. If the fermi energy of conduction electrons in Be is 14.44 eV. Calculate that is cesium.*

**SOLUTION.**

Electrons per *c.c.* in Be $= \dfrac{n}{V} = 24.2\times10^{22}$

Electrons per *c.c.* in $Cs = \dfrac{n}{V} = 0.91\times10^{22}$

Fermi energy, $E_F = \dfrac{h^2}{2m}\left(\dfrac{3}{8\pi}\right)^{\frac{2}{3}}\left(\dfrac{n}{V}\right)^{\frac{2}{3}} = k\left(\dfrac{n}{V}\right)^{\frac{2}{3}}$

$$\therefore \quad \frac{\text{Fermi energy for } Be}{\text{Fermi energy for } Cs} = \frac{k\times\left(24.2\times10^{22}\right)^{\frac{2}{3}}}{k\times\left(0.91\times10^{22}\right)^{\frac{2}{3}}} = \left(\frac{24.2}{0.91}\right)^{\frac{2}{3}}$$

or Fermi energy for

$$Cs = 14.44 \times \left(\frac{0.91}{24.2}\right)^{\frac{2}{3}}$$

$$= \mathbf{1.587\ eV}$$

### EXAMPLE 12.6

*Fermi energy of conduction electrons in silver is 5.48 eV. Calculate the number of such electrons per cm³ given that h = 6.62×10⁻²⁷ erg sec. and 1 eV= 1.62×10⁻¹² erg.*

**SOLUTION.** Fermi energy is given by

$$E_F = \left(\frac{h^2}{2m}\right)\left(\frac{3n}{8\pi V}\right)^{\frac{2}{3}}$$

$$E_F^{3/2} = \frac{h^3}{(2m)^{\frac{3}{2}}} \cdot \frac{3n}{8\pi V}$$

$\therefore$ Number of electrons per cm³, $\frac{n}{V} = \frac{(2m)^{\frac{3}{2}}\ E_F^{\frac{3}{2}}\ .\ 8\pi}{3h^3}$

$$= \frac{2^{\frac{3}{2}}\times\left(9.1\times10^{-28}\right)^{\frac{3}{2}}\times8\pi\times\left(5.48\times1.62\times10^{-12}\right)^{\frac{3}{2}}}{3\times\left(6.62\times10^{-27}\right)^3}$$

$$= \mathbf{5.9 \times 10^{22}\ cm^{-3}}$$

### EXAMPLE 12.7

*How many photons are present in 1 cm³ of radiation at 727 °C? What is their average energy?*

*Given that* $\int_o^\infty \frac{x^2 dx}{e^x - 1} = 2.405.$

**SOLUTION.** The number of photons per unit volume having frequencies between ν and (ν + *d*ν) is given by

$$n(\nu)d\nu = \frac{8\pi}{c^3}\times\frac{\nu^2\ d\nu}{e^{h\nu/kT}-1} \qquad \text{[Equation 12.24 with } V = 1\text{]}$$

$\therefore$ Total no. of photons in a unit volume having all the frequencies is given by

$$N = \int_o^\infty n(\nu)d\nu = \frac{8\pi}{c^3}\int_o^\infty \frac{\nu^2 d\nu}{e^{h\nu/kT}-1}$$

Substituting $\frac{h\nu}{kT} = x$, we have $\nu = \left(\frac{kT}{h}\right)x$ and $d\nu = \left(\frac{kT}{h}\right)dx$

$$\therefore \quad N = 8\pi\left(\frac{kT}{hc}\right)^3\int_o^\infty \frac{x^2\ dx}{e^x-1} = 2.405\times8\pi\left(\frac{kT}{hc}\right)^3$$

$$= 8 \times 3.14 \times 2.405 \times \left[\frac{1.38\times10^{-16}\times(273+727)}{6.63\times10^{-27}\times3\times10^{10}}\right]$$

$$= 2.027 \times 10^{10} \text{ photons.}$$

Total radiation energy per unit volume is given by

$$E_S = \left(\frac{8\pi^5 k^4}{15\,c^3h^3}\right)T^4$$

$$\text{Average energy per photon} = \frac{E_s}{N}$$

$$= \frac{8\pi^5 k^4 T^4 h^3 c^3}{15\,c^3 h^3\, k^3\, T^3 \times 8\pi\times 2.405}$$

$$= \frac{\pi^4}{15}\cdot\frac{kT}{2.405}$$

$$= \frac{(3.14)^4\times1.38\times10^{-16}\times1000}{15\times2.405}$$

$$= 3.73\times10^{-13} \text{ erg}$$

$$= \frac{3.73\times10^{-13}}{1.6\times10^{-12}}$$

$$= \mathbf{0.233\ eV.}$$

## EXAMPLE 12.8

*A gas has only two particles a and b. Show with the help of diagrams how these two particles can be arranged in three quantum series 1, 2, 3 using (i) Maxwell-Boltzmann (ii) Fermi-Dirac and (iii) Bose-Einstein statistics.* **[Chennai 2004]**

**SOLUTION. (*i*) Maxwell Boltzmann statistics**

In this case, the particles are distinguishable and there is no limit to the number of particles in any one state. The total number of ways is $3^2 = 9$

| *States* | *Possible Distribution in various States* | | | | | | | | |
|---|---|---|---|---|---|---|---|---|---|
| **1.** | *a* | *b* | – | – | *a* | *b* | *ab* | – | – |
| **2.** | *b* | *a* | *a* | *b* | – | – | – | *ab* | – |
| **3.** | – | – | *a* | *a* | *b* | *a* | – | – | *ab* |

**(*ii*) Fermi-Dirac statistics**

In this case, the particles are indistinguishable and not more than one particle can be in any one state. Here $a = 0$. The total number of ways = 3.

| *States* | *Possible Distribution in various States* | | |
|---|---|---|---|
| **1.** | *a* | *a* | – |
| **2.** | *a* | – | *a* |
| **3.** | – | *a* | *a* |

***(iii)* Bose-Einstein statistics**

In this case, the particles are indistinguishable and there is no limit to the number of particles in any one state. Here $a = b$. The total number of ways = 6.

| *States* | *Possible Distribution in various States* | | | | | |
|---|---|---|---|---|---|---|
| 1. | *a* | *a* | – | *aa* | – | – |
| 2. | *a* | – | *a* | – | *aa* | – |
| 3. | – | *a* | *a* | – | – | *aa* |

**EXAMPLE 12.9**

*Which of the statistics will you use for the systems having : (i) electrons, (ii) photons, (iii) mesons, (iv) oxygen molecules, (v) phonons, (vi) holes, (vii) neutrons, (viii) protons, (ix) $He^4$ atom at low temperature, (x) α–particle, (xi) positron.* **[Purvanchal 2007, Agra 2000]**

**SOLUTION.** *(i)* **Electron, Proton, Neutron, Positron and Holes** are spin half particles. They are therefore, *fermions* and hence obey *Fermi-Dirac statistics.*

*(ii)* **Photons, Phonons, Mesons, α-particles and Helium atom at low temperature** are particles of integral spin. Therefore, they are *Bosons* and obey *Bose-Einstein Statistics.*

*(iii)* **Oxygen molecules** are classical particles. Therefore, they obey *Maxwell-Boltzmann statistics.*

**EXAMPLE 12.10**

*Classify the following particles according to B-E and F-D statistics : proton, neutron, electron, photon, α-particle, hydrogen atom, hydrogen molecule, positron, lithium ion ($^6_3Li^+$).* **[Purvanchal 2006, Madras 2005]**

**SOLUTION.** The atoms or molecules having an *even number of fundamental particles have integral spin* value are called *Bosons.*

Similarly, the atoms or molecules having *odd number* of fundamental particles have *half integral spin* value are called *Fermions.* Accordingly, the particles are classified as under :

**Proton, Neutron, Electron and Positron** are fundamental particles having half spin value. Hence, they all are *Fermions.* **Photon** has spin 1 and hence it is a *Boson.*

α-particle ($_2He^4$) = $2p + 2n + 2e$ → Even → *Boson*

Hydrogen atom = $1p + 1n + 1e$ → Odd → *Fermion*

Hydrogen molecule = $2p + 2n + 2e$ → Even → *Boson*

Positron = $e^+$ → odd → *Fermion*

Lithium ion ($^6_3Li^+$) = $3p + 3n + 2e$ → Even → *Boson.*

**EXAMPLE 12.11**

*Three particles are to be distributed in four energy levels a, b, c and d. Calculate all possible ways of this distribution when particles are (i) Fermions, (ii) Bosons, (iii) Classical particles.*

**SOLUTION.** *(i)* **Fermions :** These particles are identical, indistinguishable having half integral spin. They obey F-D statistics, according to which only one particle can occupy one cell. Accordingly, different ways of distribution of $n_i = 3$ particles in $g_i = 4$ cells.

$$= \frac{g_i!}{n_i!(g_i - n_i)!} = \frac{4!}{3!(4-3)!} = \mathbf{4}$$

*(ii)* **Bosons :** These are identical, indistinguishable particles which do not obey Pauli's exclusion

principle. They obey B-E statistics, according to which any number of particles can be in any cell (quantum state) and that all quantun states are equally probable. Thus, the number of ways in which $n_i = 3$ indistinguishable particles can be distributed among $g_i = 4$ cells.

$$= \frac{(n_i + g_i - 1)!}{n_i!(g_i - 1)!}$$

$$= \frac{(3+4-1)!}{3!(4-1)!} = \frac{6!}{3!3!} = \mathbf{20}$$

***(iii)* Classical Particles :** These are identical particles of any spin and are distinguishable from one another. They obey M-B statistics, according to which any number of particles can be in any cell. The number of ways in which $n_i = 3$ distinguishable particles arranged among $g_i = 4$ energy levels.

$$= g_i^{n_i} = (4)^3 = \mathbf{64}$$

## EXERCISE–XII

**12.1** Explain the need for introducing quantum statistics. **(Nagpur 2009, 2004/W)**

**12.2** 'In quantum statistics, '*h*' is selected as a constant by nature', explain by using the case of particle in a one-dimensional box.

**12.3** 'One-dimensional harmonic oscillator employing quantum concept, has set a choice to select *h* as a natural constant, explain. What are its implications? **(Nagpur 2005/W)**

**12.4** What is indistinguishability of particles? What role it plays in quantum statistics? What are its consequences? **(Nagpur 2005/W)**

**12.5** What is Bose-Einstein statistics? What are the basic postulates used? Derive an expression

$n_i = \dfrac{g_i}{e^{\alpha+\beta E_i} - 1}$ for the most probable distribution of the particles of a system obeying *B.E.* statistics.

and hence deduce Planck's black-body radiation formula. **(M.S. Uni. Tamil nadu 2007; Magadh 2001, 2003, Purvanchal 2006, 2000 Nagpur 2006/S)**

**12.6** What are bosons? Which statistics is used to study them? **(Agra 2007, 2002, 2004, Nagpur 2005/W, Madras 2004, Purvanchal 2005, Nagpur 2005/S, Madras 2004/S)**

**12.7** What are Fermions? Write down the postulates of Fermi-Dirac statistics. Derive an expression for the probability distribution of particles governed by Fermi-Dirac statistics. **(Madras 2004/W, 2005/S; Agra 2004, Nagpur 2009, 2004 (W), 2005/W, Purvanchal 2005, 2001, Mumbai 2005, M.S. Uni. Tamil Nadu 2007)**

**12.8** Applying Fermi-Dirac distribution to an electron gas show that the distribution of energy among free electrons of a conductor is

$$n\,(E)\,dE = \frac{8\sqrt{2}\,\pi V}{h^3} \cdot \frac{m^{3/2}\ E^{\frac{1}{2}}\ dE}{e^{(\alpha+\beta E)} + 1}$$

**12.9** What is electron gas? Starting from Fermi-Dirac distribution law, drive the expression for energy distribution of free electrons in a metal.

**12.10** Starting from Bose-Einstein energy distribution law, derive Planck's law of black body radiation. **(Purvanchal 2007, Garhwal 2005, Meerut 2004, Nagpur 2004/W, 2005/S, 2005/W, 2006/S, 2007, 2008, Purvanchal 2003)**

**12.11** What is meant by Fermi-energy of conduction electrons. Calculate the value of Fermi-energy. **(Purvanchal 2002, 2003, Nagpur 2004/S)**

**12.12** What is occupation index. Under what conditions do Bose-Einstein & Fermi-Dirac statistics yield classical statistics. **(Nagpur 2009, Purvanchal 2001)**

**12.13** Explain why behaviour of liquid helium cannot be explained by classical statistics. How it is over - comed by quantum statistics.

**12.14** Describe the peculiar behaviour of liquid helium.

**12.15** What is the difference between classical and quantum statistics.
**(Agra 2005, 2002, Purvanchal 2003, Rohilkhand 2002, Nagpur 2006/S)**

**12.16** Using the expression for Fermi-Dirac distribution of energy among electrons within a metal, prove that at absolute zero, the average *K.E.* is $\overline{E} = \frac{3}{5} E_F$. **(Purvanchal 2007)**

**12.17** Do electrons have zero energy at absolute zero ? If not why? Explain.

**12.18** Define Fermi energy and Fermi temperature. Explain the significance of Fermi energy.
**(Agra 2005, Nagpur 2007)**

**12.19** Discuss the free electron gas model for metals. Obtain the distribution law for the electron gas in a metal at absolute zero. **(Nagpur 2005/W)**

**12.20** Distinguish between Classical Statistics, Fermi-Dirac Statistics and Bose-Einstein Statistics.
**(Nagpur 2009, Purvanchal 2002, 2003, Rohilkhand 2001, 2002, 2002 Imp., Agra 2006, 2000, 2002, 2004, Meerut 2003, Avadh 2000, Madras 2005)**

**12.21** Discuss the thermodynamic probability in B-E statistics. Obtain an expression for Bose-Einstein distribution law and derive Planck's radiation formula from the above law.
**(Purvanchal 2005, 2007, Nagpur 2007, Agra 2000, 2004, Madras 2005)**

**12.22** Is neutron a boson or a fermion? What are bosons? Give example for boson. **(Madras 2005/S)**

**12.23** Discuss the postulates of quantum statistics. **(Purvanchal 2006, Madras 2004/W)**

**12.24** Compare the basic postulates of M-B, B-E and F-D statistics. **(Purvanchal 2001)**

**12.25** Derive Fermi-Dirac distribution formula. How does this distribution tends to classical.
**(Purvanchal 2002, 2003, Agra 2003)**

**12.26** How does F-D statistics differ from B.E. statistics ? **(Magadh 2003, Nagpur 2005/S)**

**12.27** What do you mean by distinguishable and indistinguishable particles. **(Nagpur 2005/S, 2005/W)**

**12.28** Explain the concepts of quantum mechanics which have influenced the transition from classical to quantum mechanics. **(Nagpur 2004/S, 2004/W)**

**12.29** What is Fermi function ? Give its significance. **(Nagpur 2007, 2004/S)**

**12.30** What are symmetric and antisymmetric wave functions ? **(Madras 2004)**

**12.31** In Fermi-Dirac statistics, particles are not obeying Pauli's exclusion principle, is it true ? What is the spin value of Fermions ? **(Madras 2005)**

**12.32** What are classical and quantum particles. **(Madras 2004)**

**12.33** (*i*) State two conditions for applicability of Bose-Einstein statistics.

(*ii*) Using Bose-Einstein statistics, derive the Planck's formula for energy distribution in the black body spectrum.

(*iii*) Draw the energy distribution curve of a black body spectrum. **(Nagpur Uni. 2007)**

**12.34** Derive the relation

$$E_{F_0} = \frac{h^2}{2m}\left(\frac{3n}{8\pi}\right)^{2/3}$$

where letters have their usual meaning. **(Nagpur 2007)**

**12.35** Write short notes on:

(*i*) Statistical equilibrium

(*ii*) Electron gas

(*iii*) Photon gas

(*iv*) Phase gas

(*v*) Quantum statistics
(*vi*) Fermi temperature **(Purvanchal 2007)**
(*vii*) Fermions **(Madras 2004/W)**
(*viii*) Planck's radiation law **(Meerut 2001)**
(*ix*) Statistical thermodynamics
(*x*) Maxwell-Boltzmann distribution law
(*xi*) Fermi-Dirac distribution law **(Magadh 2001)**
(*xii*) Bose-Einstein distribution law

**12.36** Five bosons are distributed in two compartments the first having 3 cells and the second 4. Find the thermodynamic probability for the macrostate (*i*) (5, 0) and (*ii*) (4, 2) **[Ans. 21,150] (Purvanchal 2006)**

**12.37** Calculate the average energy of a Planck's oscillator of frequency $1.5\times10^{14}$ at $T = 1800K$, $h = 6.62\times 10^{-34}$ $Js$ and $k = 1.38\times10^{-23}$ $Jk^{-1}$. **[Ans. $1.86\times10^{-21}$ J]**

**12.38** A system has 7 particles arranged in two compartments. The first compartment has 8 cells and the second has 10 cells. All cells are of equal size. Calculate the number of microstates in the macrostate (3,4) when the particles obey *F.D.* statistics. **[Ans. 11760]**

**12.39** Find the average velocity of electrons at $0°C$ in a metal having $3\times10^{22}$ electrons per $cm^3$. **[Ans. $0.835\times10^8$ cm$^{-1}$]**

**12.40** A system consists of 8000 particles distributed in three energy states with equal spacing. The energy of the three states are $E_1 = 0$, $E_2 = x$, $E_3 = 2x$. All the three states have the same intrinsic probability *g*. At a certain instant, there are 4000 particles in the lower level, 3000 particles in the middle level and 1000 particles in the upper level. Compare the relative probabilities with the distribution obtained by the transfer of one particle from middle to the lower level and one particle from the middle to the upper level and the original distribution. **[Ans. 2.247]**

**12.41** Calculate the number of ways of arranging 8 fermions in 12-phase space cells. **(Punjab 2005)**

## OBJECTIVE QUESTIONS

**12.1** Average energy of a Planck's oscillator is : **(Garhwal 2005)**

(*a*) $E = h\nu$ (*b*) $E = nh\nu$

(*c*) $E = \dfrac{h\nu}{(e^{h\nu/kT} - 1)}$ (*d*) $E = mc^2$

**12.2** From Fermi-Dirac statistics $n_i$ = ? **(Agra 2004)**

(*a*) $\dfrac{g_i}{e^{\alpha+\beta E_i}+1}$ (*b*) $\dfrac{2g_i}{e^{\alpha+\beta E_i}+1}$

(*c*) $\dfrac{2g_i}{e^{\alpha+\beta E_i}-1}$ (*d*) $\dfrac{g_i}{e^{\alpha+\beta E_i}-1}$.

**12.3** Deduction of Planck's law is possible on the basis of **(Agra 2003)**

(*a*) Fermi-Dirac (F.D) statistics (*b*) Classical statistics

(*c*) Maxwell-Boltzmann (M.B) statistics (*d*) Bose-Einstein (B-E) statistics.

**12.4** Boson particles obey Pauli's exclusion principle . **(Agra 2003)**

(*a*) True (*b*) False

(*c*) Can't say (*d*) Sometimes true sometimes false.

**12.5** According to which statistics, the energy at absolute zero cannot be zero.

(*a*) M-B (*b*) B-E

(*c*) F-D (*d*) None of these.

**12.6** Fermions have spin value :

(*a*) $\frac{1}{2}$ (*b*) 1

(*c*) Zero (*d*) Any one.

**12.7** Boson have spin value :

(*a*) 0 (*b*) 1

(*c*) $\frac{1}{2}$ (*d*) 0 or 1.

**12.8** Photons obey the statistics :

(*a*) M-B (*b*) B-E

(*c*) F-D (*d*) None of these.

**12.9** In how many ways two particles can be arranged in three phase cells according to B-E statistics :

(*a*) 6 (*b*) 9

(*c*) 3 (*d*) 27.

**12.10** According to F-D statistics, 3 particles can be distributed in 4 energy states, in number of ways :

(*a*) 16 (*b*) 4

(*c*) 9 (*d*) 12

**12.11** The spin of photon is :

(*a*) Zero (*b*) $\frac{1}{2}\hbar$

(*c*) $\hbar$ (*d*) $\frac{3}{2}\hbar$

**12.12** Pauli's exclusion principle applies to : **[H.P.V 2001]**

(*a*) M.B. Statistics (*b*) B.E. Statistics

(*c*) F.D. Statistics (*d*) None.

**12.13** In which the probability of occupation index can tend to zero :

(*a*) Maxwell-Boltzmann (*b*) Fermi-Dirac

(*c*) Bose-Einstein (*d*) None of these.

**12.14** The number of meaningful ways 4 Fermions can be arranged in 5 compartments :

(*a*) 1 (*b*) 4

(*c*) 5 (*d*) 9.

## ANSWERS

| | | | |
|---|---|---|---|
| **12.1** (*c*) | **12.2** (*a*) | **12.3** (*d*) | **12.4** (*b*) |
| **12.5** (*c*) | **12.6** (*a*) | **12.7** (*d*) | **12.8** (*b*) |
| **12.9** (*a*) | **12.10** (*b*) | **12.11** (*c*) | **12.12** (*c*) |
| **12.13** (*d*) | **12.14** (*c*). | | |

# SECTION – IV

# Heat Flow And Air Conditioning

# Thermometry 13

**At a Glance**

- Concept of Heat and Temperature • Thermometry • Types of Thermometers • Centigrade, Fahrenheit and Rankine Scales • Relation between Celcius, Kelvin, Fahrenheit and Rankine Scales • Liquid Thermometers • Errors and Corrections in Mercury Thermometers • Gas Equation • Advantages of a Gas Thermometer • Callendar's Compensated Constant Pressure Air Thermometer • Jolly's Constant Volume Air Thermometers • Constant Volume Hydrogen Thermometer • Gas Thermometer Corrections (First correction) • Second Correction (Perfect or Ideal Gas Scale) • Platinum Resistance Thermometer • Callendar and Griffith's Bridge • Seebeck Effect • Peltier Effect • Peltier Coefficient (p) •Thomson Effect • Thomson Coefficient (s) • Thermo-electric Power • Thermo-electric Thermometer • Helium Vapour Pressure Thermometer • Standardization and Temperature Scale • Low Temperature Measurement • HighTemperature Measurement

## 13.1 Concept of Heat and Temperature

Heat manifests itself when it flows from a hot body to a cold body due to the difference in temperature. The sensation of heat can be experienced by touching to a hot or a cold body. The hot body is said to possess a higher temperature than the cold one. But the sense of touch is merely qualitative, while scientific precision requires that heat, the physical quantity should be measured and expressed in numerical terms. Further, the measurements must be accurate and easily reproducible. It is our common experience that when two bodies are brought in contact, there is a change in their properties such as volume, pressure etc. due to exchange of heat. Finally an *equilibrium state* is reached, after which there is no further change. The two bodies are then said to be in **thermal equilibrium** with each other. In this state, the two bodies are said to have the *same temperature.* Therefore, heat can be defined as *energy in transit.*

The temperature of a system can be defined *as the property that determines whether or not the body is in thermal equilibrium with the neighbouring systems.* If a number of systems are in thermal equilibrium, this common property of the system can be represented by a single numerical value called the **temperature.** It means that if two systems are not in thermal equilibrium, they are at different temperatures.

Measurement of temperature of a body accurately is one of the most important branches of heat in physics. It also becomes necessary to measure high temperatures and low temperatures. To make this measurement possible, it is essential to construct a suitable scale of temperature. The scale chosen must be precise and consistent and the temperature measured on this scale must be accurate.

## 13.2 Thermometry

The branch of heat pertaining to the measurement of temperature of a body is called *thermometry.* The thermometer is an instrument to measure the temperature of a body.

The essential requisites of a thermometer are :

(1) Construction,
(2) Calibration, and
(3) Sensitiveness.

(1) *Construction.* The physical property of a substance plays an important role in the construction of a thermometer. In a mercury thermometer, the principle of expansion of mercury with rise in temperature is used. The platinum resistance thermometer is based on the principle of the change in resistance with change in temperature. The gas thermometer is based on the principle of change in volume or pressure with change in temperature. Thus, for the construction of a thermometer, the proper choice of a substance, whose physical property varies uniformly with rise in temperature, is essential.

(2) *Calibration.* When a thermometer is constructed, it should be properly calibrated. The standard fixed points are selected for calibrating a thermometer. Melting point of ice, boiling point of water, melting point of silver and melting point of gold are taken as fixed points. The scales are built by dividing the interval between the two fixed points into equal parts. Centigrade scale is built by dividing the interval between the melting point of ice and the boiling point of water (under normal pressure) into 100 equal parts and each part represents 1°C. Similarly, Fahrenheit scale is built by dividing this interval into 180 equal parts.

(3) *Sensitiveness.* The instrument, once constructed and calibrated, should also be sensitive. The thermometer will be sensitive if (*i*) it can detect even small changes in temperature, (*ii*) it shows the temperature of a body in a short time and (*iii*) it does not take large quantity of heat for its own heating from the body whose temperature is being measured.

| *Fixed points* | *Degrees Celsius* |
|---|---|
| Boiling point of oxygen | – 182.97 |
| Ice point | 0.00 |
| Steam point | 100.00 |
| Boiling point of sulphur | 444.60 |
| Melting point of silver | 960.80 |
| Melting point of gold | 1063.00 |
| Melting point of cobalt | 1492.00 |
| Melting point of platinum | 1769.00 |
| Melting point of rhodium | 1960.00 |
| Melting point of tungsten | 3380.00 |

## 13.3 Types of Thermometers

There are different kinds of thermometers :

(1) *Liquid thermometers.* These thermometers are based on the principle of change in volume of a liquid with change in temperature. Mercury and alcohol thermometers are based on this principle.

(2) *Gas thermometer.* These are based on the principle of change in pressure or volume with change in temperature, *e.g.*, Callendar's constant pressure thermometer, constant volume hydrogen thermometer etc.

(3) *Resistance thermometers.* These are based on the principle of change in resistance with change in temperature, *e.g.*, platinum resistance thermometer.

(4) *Thermoelectric Thermometers.* These are based on the principle of thermo-electricity, *i.e.*, production of thermo-E.M.F. in a thermo-couple when the two junctions are at different temperatures.

The various thermocouples commonly used are :

(1) Copper and constantan (2) Iron and constantan

(3) Chromel and constantan (4) Chromel and alumel

(5) Platinum and Rhodium.

(5) *Radiation thermometers.* These are based on the quantity of heat radiations emitted by a body *e.g.*, furnaces. These instruments are known as pyrometer.

(6) *Vapour pressure thermometers.* These are based on the principle of change of vapour pressure with change in temperature. These are used to measure low temperatures, *e.g.*, helium vapour pressure thermometer etc.

(7) *Bimetallic thermometers.* These thermometers are based on the principle of expansion of solids. A bimetallic strip is taken in the form of a spiral. Its one end is fixed and the other end is attached to a long pointer. The pointer moves on a scale, calibrated in degrees. These thermometers are used in meteorology for recording the changes in temperature during the day. They are also used to measure temperatures at high altitudes.

(8) *Magnetic thermometers.* These thermometers are based on the principle of change in the susceptibility of a substance with temperature. These thermometers are useful for measuring low temperatures near the absolute zero temperature.

## 13.4 Centigrade, Fahrenheit and Rankine Scales

The earliest thermometer was constructed by Galileo in 1593. Newton suggested the necessity of the fixed points. The temperature of the melting point of ice is taken as the lower fixed point and the temperature of steam at a pressure of 76 cm of Hg (normal pressure) is taken as the upper fixed point.

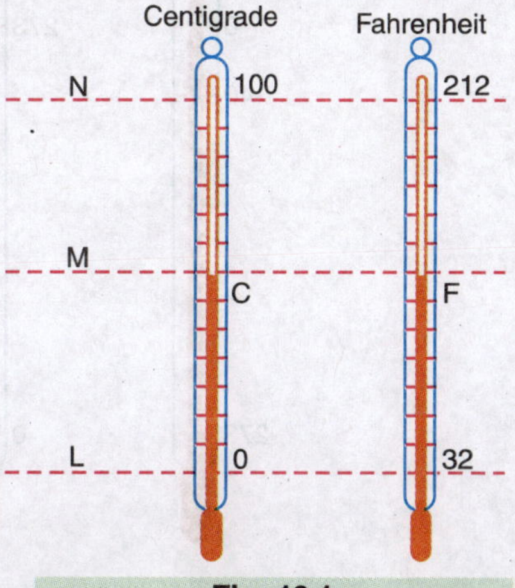

Fig. 13.1

**Centigrade (or Celsius) scale.** Celsius, in 1742, suggested the centigrade system of temperature. He marked zero at the lower fixed point and 100 at the upper fixed point. The interval between the two fixed points is divided into 100 equal parts. Each part of degree represents 1°C or 1° celsius. The scale is also known as Celsius scale.

**Fahrenheit scale.** Fahrenheit, in 1720, suggested this scale by taking zero as the temperature of the freezing mixture. It appears that he took 100 degrees as the temperature of the human body. Later the correct temperature of the human body on this scale was found to be 98.4ºF. The lower fixed point is marked as 32 and the upper fixed point is marked as 212. The interval is divided into 180 equal parts. Each part or degree represents 1ºF.

**Relation.** Consider two identical thermometers marked in Centigrade and Fahrenheit scales. Place the two thermometers in a bath at certain fixed temperature. Mercury in each thermometer stands to the same level *M* (Fig. 13.1).

$$\therefore \quad \frac{ML}{MN} = \frac{C-0}{100-0}$$

$$= \frac{F-32}{212-32}$$

or

$$\frac{C}{100} = \frac{F-32}{180}$$

## 13.5 Relation between Celsius, Kelvin, Fahrenheit and Rankine Scales of Temperature

In Fig. 13.2 the temperatures of the absolute zero, the melting point of ice and the boiling point of water as measured on the Celsius (Centigrade), Kelvin (absolute), Fahrenheit and Rankine scales are shown.

Celsius and Fahrenheit scales show the same reading at – 40º *i.e.,* – 40ºC = – 40ºF. The Kelvin scale and the Rankine scale agree at zero degree only.

The temperatures represented in Fig. 13.2 have been rounded off to the nearest degree.

**Relation.**

$$\frac{C-0}{100} = \frac{F-32}{180} = \frac{K-273}{100} = \frac{R-492}{180}$$

Also

$$\frac{K}{100} = \frac{R}{180}$$

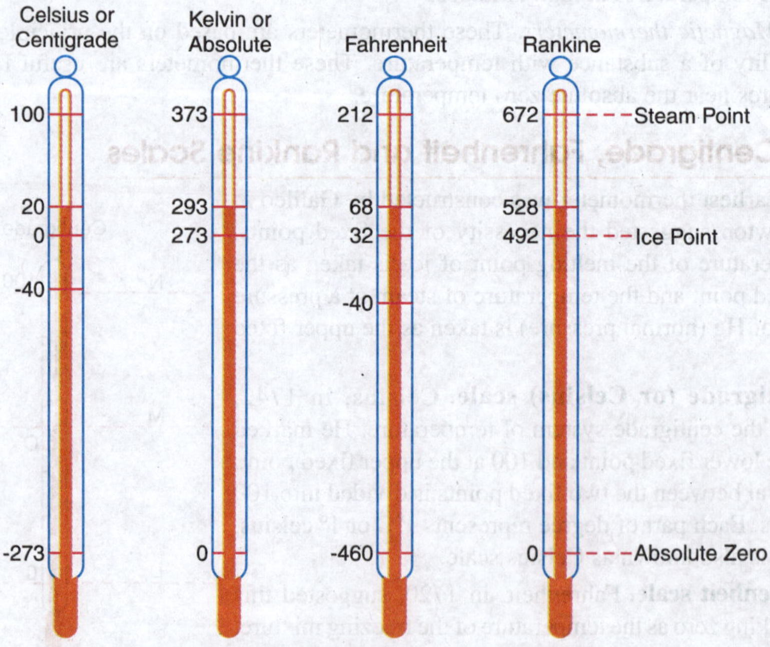


Fig. 13.2

## NUMERICAL EXAMPLES

### EXAMPLE 13.1

*The temperature of the surface of the sun is about 6500°C. What is this temperature (i) on the Rankine scale and (ii) on the Kelvin scale ?*

**SOLUTION.** $$\frac{C-0}{100} = \frac{K-273}{100} = \frac{R-492}{180}$$

Here, $$C = 6500°C$$

$$\frac{6500}{100} = \frac{K-273}{100} = \frac{R-492}{180}$$

$$K = 6500 + 273 = 6773 \text{ K}$$

$$R = 12,192°R$$

The temperature of the surface of the sun corresponding to 6500 °C is (*i*) 6773 K and (*ii*) 12,192 °R.

### EXAMPLE 13.2

*The normal boiling point of liquid oxygen is – 183°C. What is this temperature on (i) Kelvin scale and (ii) Rankine scale ?*

**SOLUTION.** $$\frac{C-0}{100} = \frac{K-273}{100} = \frac{R-492}{180}$$

Here, $$C = -183 °C$$

$$\therefore \quad \frac{-183}{100} = \frac{K-273}{100} = \frac{R-492}{180}$$

$$K = 90° \text{ K}$$

$$R = 162° \text{ R}$$

The boiling point of liquid oxygen corresponding to – 183°C is (*i*) 90 K and (*ii*) 162.6° R.

### EXAMPLE 13.3

*At what temperature do the Kelvin and Fehrenheit scales coincide?*

**SOLUTION.** $$\frac{K-273}{100} = \frac{F-32}{180}$$

$$\frac{x-273}{100} = \frac{x-32}{180}$$

$$x = 574.25°$$

$$574.25 \text{ K} = 574.25 °F$$

### EXAMPLE 13.4

*At what temperature do the Celsius and the Fahrenheit scales coincide?*

**SOLUTION.** $$\frac{C-0}{100} = \frac{F-32}{180}$$

$$\frac{x-0}{100} = \frac{x-32}{180}$$

$$x = -40°$$

$$\therefore \quad -40°C = 40°F$$

### EXAMPLE 13.5

*The boiling point of liquid hydrogen is 20.2 K Convert this temperature into degree Rankine.*

**SOLUTION.**

$$\frac{K-273}{100} = \frac{R-492}{180}$$

Here $K = 20.2° \text{ K}$

$$\therefore \quad \frac{20.2-273}{100} = \frac{R-492}{180}$$

$$R = 36.96° \text{ R}$$

$$\therefore \quad 20.2 \text{ K} = 36.96° \text{ R}$$

## 13.6 Liquid Thermometers

Liquid thermometers are based on the principle of change of volume of the liquid with change in temperature. These thermometers are suitable for a narrow range of measurement. The most commonly used liquids are (*i*) mercury and (*ii*) alcohol. The range of mercury thermometer is – 39°C to 357°C. Alcohol thermometers are used only to measure temperature near ice point.

Liquid Thermometer

### Mercury Thermometer

Mercury is usually selected for use in liquid in glass thermometers for the following reasons :

(1) It has a low specific heat and hence it absorbs little heat from the body whose temperature is being measured.

(2) It is a good conductor of heat and takes the temperature of the body quickly.

(3) It can be easily seen in a fine capillary tube and the thermometer can be made sensitive.

(4) It does not wet the wall of the glass tube. This is an important point for the construction of a thermometer.

(5) It has a uniform coefficient of expansion over a wide range of temperature.

(6) It remains a liquid over a large range. Its freezing point is – 39 °C and the boiling point is 357°C.

## 13.7 Errors and Corrections in a Mercury Thermometer

**(1) Change of zero.** When a thermometer is graduated after sealing, there will be shift in the zero mark. This is due to the fact that when a thermometer is subjected to a high temperature during construction, glass takes a long time for contraction. Consequently when the thermometer is placed in melting ice, the level of mercury will be above the zero mark (say at + 0.5°C). To correct the error, 0.5°C is subtracted from the observed reading. While manufacturing, the sealed thermometer tubes are allowed sufficient time and are graduated after a number of years.

**(2) Recent heating and cooling.** When a thermometer is used to measure a very high temperature, the thermometer bulb expands. If the same thermometer is immediately used to measure low temperature, the level of mercury will be lower than the correct reading. Suppose the thermometer when placed in melting ice, reads – 0.3°C, the correction to be applied is + 0.3°C.

**(3) Due to exposed stem.** While calibrating a thermometer, the thermometer as a whole is kept in steam. While measuring the temperature, only the bulb is immersed in the bath whose temperature is to be measured. Due to this reason the observed reading is lower than the correct reading. To correct this error, maximum possible length of the stem should be immersed in the bath whose temperature is to be measured.

**(4) Inequality of the bore.** The thermometer is calibrated after marking the lower and the upper fixed points. The distance between the two fixed marks is divided into equal parts assuming that the bore of the thermometer is uniform. If the bore is not uniform, at positions where the bore has lesser diameter, the observed reading is higher than the actual reading. Similarly, when the bore has a higher diameter, the observed reading is less. To correct this error the capillary tube is initially calibrated and a graph is provided which gives the correct readings against the observed readings.

**(5) Position and pressure.** While graduating, if the thermometer was held in the horizontal position and while taking the temperature it is kept in the vertical position, the observed readings will be lower than the correct readings. This is due to the pressure of mercury column. To avoid this error, the thermometer is used in the same position in which it was calibrated.

**(6) Surface tension.** Mercury is depressed in a capillary tube because it does not wet the glass tube. The depression is different at different positions of the stem if the bore is not uniform. To correct this error, the reading of the thermometer is taken with a rising mercury column.

**(7) Thermal capacity.** If the thermal capacity of the bulb is large, a part of the heat from the bath is used in heating the bulb. Therefore, the thermometer reads a lower temperature. To correct this error, small thermometer bulbs with low thermal capacity are used.

**(8) Effect of the bulb.** In a mercury thermometer the glass bulb has an insulating effect. Due to this reason the mercury inside the bulb may not attain the temperature of the bath.

## 13.8 Gas Equation

Consider a perfect gas at a pressure $P$, temperature $T$ and volume $V$.

Let the pressure of the gas change from $P$ to $P_1$ at constant temperature $T$ such that the volume changes from $V$ to $v$.

Applying Boyle's law,

$$P_1 v = PV$$

or

$$v = \frac{PV}{P_1} \quad ...(i)$$

Now, let the temperature of the gas change from $T$ to $T_1$ at constant pressure.

| | | |
|---|---|---|
| Initial Pressure = $P_1$ | Volume = $v$, | Temperature = $T$ |
| Final Pressure = $P_1$ | Volume = $V_1$, | Temperature = $T_1$ |

Applying Charles' law,

$$\therefore \quad \frac{v}{T} = \frac{V_1}{T_1} \quad ...(ii)$$

Substituting the value of $v$ in equation (*ii*)

$$\frac{PV}{TP_1} = \frac{V_1}{T_1} \quad \text{or} \quad \frac{PV}{T} = \frac{P_1V_1}{T_1}$$

$$\therefore \quad \frac{PV}{T} = \text{constant} \quad ...(iii)$$

**Universal gas constant.** When one gram molecule of a gas at N.T.P. is taken, the constant in equation (*iii*) is $R$ and is called '*universal gas constant*'. Its value is the same for all gases.

$$\frac{PV}{T} = R \quad \text{or} \quad PV = RT \quad ...(i)$$

One gram molecule of a gas at N.T.P. has a volume = 22, 400 cc. At normal pressure

$$P = 76 \times 13.6 \times 980 \text{ dynes/sq. cm}$$

$$V = 22{,}400 \text{ cc}$$

$$T = 273\text{ K}$$

$$R = \frac{76 \times 13.6 \times 980 \times 22,400}{273}$$

$$= 8.31 \times 10^7 \text{ ergs/g-mol-K}$$

$$R = \frac{8.31 \times 10^7}{4.2 \times 10^7} = 1.986 \text{ cal/g-mol-K}$$

For $n$ gram molecules of a gas, the gas equation is

$$PV = nRT \quad ...(ii)$$

**Ordinary gas constant.** When one gram of a gas at N.T.P. is taken, the gas constant is $r$ and is known as *ordinary gas constant*. Its value is different for different gases.

Here $\frac{PV}{T} = r$ or $PV = rT$

(*i*) *For Oxygen.* One gram of oxygen at N.T.P. has a volume

$$= \frac{22,400}{32} \text{ cc}$$

$$r = \frac{PV}{T} = \frac{76 \times 13.6 \times 980 \times 22,400}{273 \times 32}$$

$$= \left(\frac{8.31}{32}\right) \times 10^7 \text{ / ergs/g-K}$$

(*ii*) *For Hydrogen.* One gram of hydrogen at N.T.P. has a volume

$$= \frac{22,400}{2} \text{ cc}$$

$$r = \frac{PV}{T} = \frac{76 \times 13.6 \times 980 \times 22,400}{273 \times 2}$$

$$= \left(\frac{8.31}{2}\right) \times 10^7 \text{ ergs/g-K}$$

Similarly $r$ for nitrogen

$$= \left(\frac{8.31}{28}\right) \times 10^7 \text{ ergs/g-K}$$

and $r$ for $CO_2$ $= \left(\frac{8.31}{44}\right) \times 10^7$ ergs/g-K

In general,

$$r = \frac{R}{\text{Molecular wt. of the gas in grams}}$$

## 13.9 Advantages of a Gas Thermometer

1. The coefficient of expansion of gases is very large as compared to liquids. Therefore, gas thermometers are sensitive.

2. The coefficient and the rate of expansion of all gases are the same under similar conditions.

3. The coefficient of expansion of the material of the bulb of the thermometer is negligible in comparison to the coefficient of expansion of a gas.

Gas Thermometer

4. The gases expand uniformly and regularly over a wide range of temperature.

5. The thermal capacity of a gas is low as compared to liquids. Hence even small changes of temperature can be recorded accurately.

6. Gases can be obtained in a pure form.

7. Gas thermometers can be used over a wide range of temperature. They are suitable to measure high and low temperatures.

8. The temperatures measured with a gas thermometer agree with the temperatures on the thermodynamic scale.

Gas thermometers are not suitable for routine work. They are large and cumbersome. They can be used only in one position. They are mainly used to standardize and calibrate other thermometers.

## 13.10 Callendar's Compensated Constant Pressure Air Thermometer

It is based on the principle that, pressure remaining constant, the volume of a given mass of gas varies directly as its absolute temperature. It consists of a silica bulb $A$ connected to the reservoir $R$ containing mercury (Fig. 13.3). $EF$ is the connecting tube. The compensating bulb $B$ is exactly similar to the silica bulb $A$. $C_1C_2$ is the compensating tube having a volume equal to the tube $EF$.

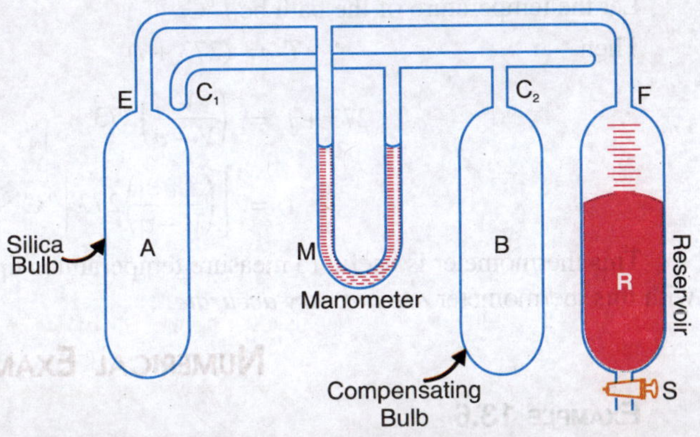


Fig. 13.3

Initially, the reservoir $R$ is filled with mercury up to the zero mark and the stop-cock is closed. The bulbs $A$, $B$ and $R$ are immersed in melting ice. The tubes are sealed when the pressure on the two sides as shown by the manometer is the same. Therefore, pressure of air in $A$ and $B$ is the same.

The manometer $M$ contains sulphuric acid. When the pressure on the two sides of the manometer is the same, the levels of sulphuric acid in the two limbs will be the same. In this way the pressure of the gas or air can be maintained constant at all temperatures.

Suppose, the bulb $A$ and the connecting tube $EF$ contain $n$ gram molecules of air. Also the bulb $B$ and the compensating tube $C_1C_2$ contain $n$ gram molecules of air. The bulb $A$ is immersed in the bath whose temperature is to be measured. The bulb $B$ is kept in melting ice. Tubes $EF$ and $C_1C_2$ are at the room temperature. The air in the bulb $A$ attains the temperature of the bath. Pressure of air in $A$ increases and the manometer shows the difference in level. Mercury is allowed to flow out of the reservoir $R$ through the stop-cock $S$ and the stop-cock is closed just at the moment when the pressure on the two sides of the manometer is the same.

Volume of the bulb $A$ = Volume of the bulb $B = V$

Volume of the connecting tube $EF$ = Volume of the compensating tube $C_1C_2 = x$

Volume of air in the reservoir $R$ at the end $= v$

Pressure of air on each side $= P$

Temperature of the bath $= T$

Room temperature $= T_1$

Temperature of melting ice $= T_0$

For the thermometric part,

$$\frac{PV}{T}+\frac{Px}{T_1}+\frac{Pv}{T_0} = nR \qquad ...(i)$$

For the compensating part

$$\frac{PV}{T_0}+\frac{Px}{T_1} = nR \qquad ...(ii)$$

Equating (*i*) and (*ii*)

$$\frac{PV}{T}+\frac{Px}{T_1}+\frac{Pv}{T_0} = \frac{PV}{T_0}+\frac{Px}{T_1}$$

$$\frac{V}{T} = \frac{V-v}{T_0} \qquad ...(iii)$$

$$T = \left[\frac{1}{V-v}\right]T_o = \left[\frac{V}{V-v}\right]273$$

Here $T$ will be obtained in degrees Kelvin

Let the temperature of the bath be $t$ °C

Then, $T = (273 + t)$

$$\therefore \quad 273 + t = \left(\frac{V}{V-v}\right)273$$

$$t = \left[\left(\frac{v}{V-v}\right)273\right] °\text{C} \qquad ...(iv)$$

This thermometer is useful to measure temperatures upto 600 °C. However, the results obtained with this thermometer *are not very accurate.*

## Numerical Example

**Example 13.6**

*The bulb of the Callendar's compensated constant pressure air thermometer is 1,000 cm*$^3$*. When the bulb is immersed in a bath, 100 cm*$^3$ *of mercury has to be drawn out of the reservoir. Calculate the temperature of the bath on the Celsius scale.*

Here, $V = 1{,}000$ cm$^3$

**Solution.** $v = 100$ cm$^3$

$$t = \left(\frac{v}{V-v}\right)273$$

$$= \left(\frac{100}{1000-100}\right)273$$

$$= 30.33\ °\text{C}$$

## 13.11 Jolly's Constant Volume Air Thermometer

It consists of a glass bulb $B$ connected to a glass tubing. The end of the glass tube is connected to the reservoir of mercury through a rubber tubing. $M$ is a fixed mark on the glass tube. The difference in the levels of mercury in $R$ and $M$ is observed on the scale $S$. The bulb $B$ is filled with 1/7 of its volume with mercury so that the expansion of the bulb $B$ is compensated by the expansion of mercury in the bulb.

Constant Volume Air Thermometer

This keeps the volume of air in the bulb up to the mark $M$ constant (Fig. 13.4).

**Working.** (1) The bulb $B$ is kept in melting ice. The reservoir $R$ is suitably adjusted so that the level of mercury stands at $M$. Suppose the difference in level between $R$ and $M$ is $h_0$. If $P$ is the atmospheric pressure, then

$$P_0 = P + h_0 \qquad ....(i)$$

(2) The bulb $B$ is kept in boiling water. The reservoir $R$ is adjusted so that the level of mercury is at $M$. Let $h_{100}$ be the difference in levels.

$$P_{100} = P + h_{100}. \qquad ...(ii)$$

(3) The bulb is kept in the bath whose temperature is to be measured. The reservoir $R$ is adjusted so that the level of mercury is at $M$. Let $h_t$ be the difference in levels.

$$\therefore \quad P_t = P + h_t. \qquad ...(iii)$$

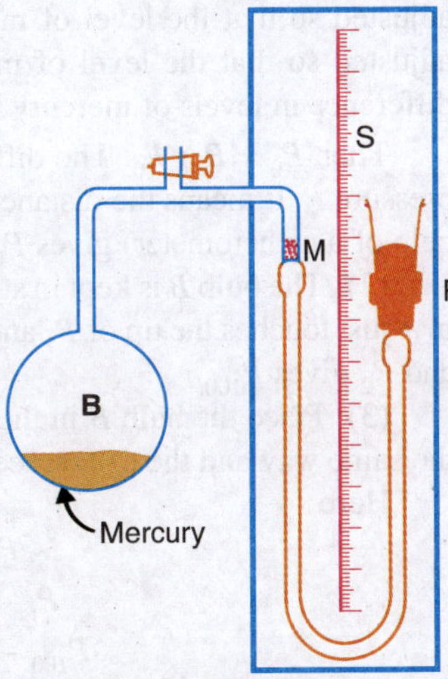


Fig. 13.4

Volume remaining constant, the pressure increases according to the relation

$$P_t = P_0\,(1 + \gamma t)$$

Also $$P_{100} = P_0\,(1 + \gamma\, 100).$$

or $$P_t - P_0 = P_0\,\gamma t. \qquad ...(iv)$$

and $$P_{100} - P_0 = P_0\,\gamma\, 100. \qquad ...(v)$$

Dividing (*iv*) by (*v*)

$$t = \left(\frac{P_t - P_0}{P_{100} - P_0}\right) \times 100 \qquad ...(vi)$$

$$t = \left(\frac{h_t - h_0}{h_{100} - h_0}\right) \times 100 \qquad ...(vii)$$

Hence $t$ can be calculated.

## 13.12 Constant Volume Hydrogen Thermometer

It was first designed by Harker and Chappius and is known as the *International Standard Thermometer*. It is based on the principle that when the volume is kept constant, the pressure of a gas increases with rise in temperature according to the relation

$$P_t = P_0\,(1 + \gamma t)$$

It consists of a platinum-iridium bulb $B$ connected to the tube $A$. The reservoir $K$ containing mercury is connected to the tubes $C$ and $A$ as shown in Fig. 13.5. A movable barometric tube $T$ with a bulb $D$ is clamped to a stand. It can be moved vertically up or down as desired. $P_1$ and $P_2$ are the two ivory tips in the same vertical line. Above the level of mercury in $D$, there is Toricellian vacuum (Fig. 13.5)

**Working.** Initially the bulb is filled with pure dry hydrogen at a pressure higher than the atmospheric pressure.

(1) The bulb $B$ is kept in melting ice. The reservoir $R$ is

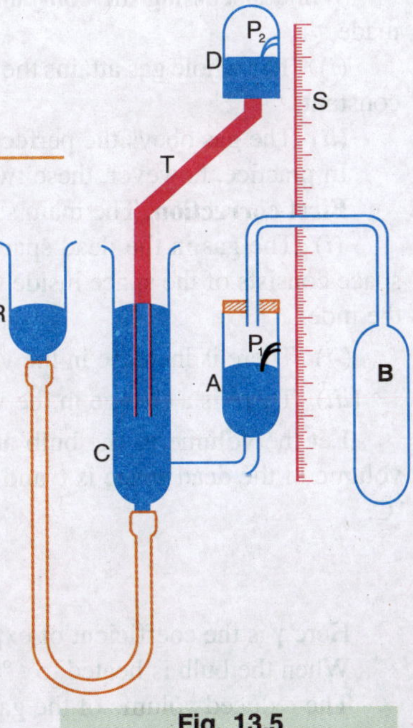


Fig. 13.5

adjusted so that the level of mercury in $A$ just touches the tip of the ivory peg $P_1$. The tube $T$ is adjusted so that the level of mercury in $D$ just touches the tip of the ivory peg $P_2$. Suppose the difference in levels of mercury in $A$ and $C$ is $h_0$ and the atmospheric pressure is $P$.

Then $P_0 = P + h_0$. The difference in levels of mercury in $C$ and $D$ is equal to the atmospheric pressure $P$. It means the distance between the tips of $P_1$ and $P_2$ as measured on the scale (or with the help of a cathetometer) gives $P_0$ directly.

(2) The bulb $B$ is kept in steam or boiling water. $R$ and $T$ are adjusted so that the level of mercury in $A$ just touches the tip of $P_1$ and that in $D$ just touches the tip of $P_2$. The difference in the levels of $P_1$ and $P_2$ gives $P_{100}$.

(3) Place the bulb $B$ in the bath whose temperature is to be measured. $R$ and $T$ are adjusted in the same way and the difference in levels of $P_1$ and $P_2$ gives $P_t$.

Here
$$P_t = P_0 (1 + \gamma t)$$
$$P_{100} = P_0 (1 + \gamma\, 100).$$
$$\therefore \quad P_t - P_0 = P_0 \gamma t. \qquad ...(i)$$
$$P_{100} - P_0 = P_0 \gamma\, 100. \qquad ...(ii)$$

Dividing (*i*) by (*ii*),

$$t = \left(\frac{P_t - P_0}{P_{100} - P_0}\right) \times 100 \qquad ...(iii)$$

The constant volume hydrogen thermometer can be used over a wide range of temperature. With hydrogen and platinum-iridium bulb, the range is – 200 °C to 500 °C.

Beyond 500 °C, hydrogen diffuses through platinum. With a porcelain bulb temperatures up to 1100 °C can be measured. Beyond 1100 °C hydrogen creeps along the wall of the bulb. So it is replaced by nitrogen for measuring temperatures up to 1500 °C. Below – 200°C hydrogen is replaced by helium.

## 13.13 Gas Thermometer Corrections

While discussing the constant volume hydrogen thermometer, the following assumptions were made :

(*i*) The whole gas attains the temperature to be measured and that the volume of the gas remains constant.

(*ii*) The gas obeys the perfect gas laws.

In practice, however, these two assumptions are not true and corrections are to be applied.

**First correction.** The main sources of error in a constant volume gas thermometer are:

(*i*) The gas in the dead space is not at the same temperature as the gas in the bulb. The dead space consists of the space inside the capillary and in the manometer between the mercury level and the index.

(*ii*) There is increase in the volume of the bulb due to rise in temperature.

(*iii*) There is a change in the volume of the bulb due to change in pressure.

Let the volume of the bulb at 0 °C be $V_0$ and $P_0$ the pressure of the gas at 0 °C. Suppose the volume of the dead space is $v$ and the temperature of the dead space throughout the experiment is θ °C.

$$= \left(V_0 + \frac{v}{1+\gamma\theta}\right)\frac{P_0}{76} \qquad ...(i)$$

Here $\gamma$ is the coefficient of expansion of the gas.

When the bulb is heated to $t$ °C, the pressure of the gas becomes $(P_0 + h)$.

The reduced volume of the gas at N.T.P. will be,

$$\left\{\left(\frac{V_0(1+\gamma_0 t)+bh}{1+\gamma t}\right)+\frac{v}{1+\gamma\theta}\right\}\left(\frac{P_0+h}{76}\right) \quad ...(ii)$$

Here, $\gamma_0$ is the volume coefficient of expansion of the bulb and $b$ is the volume coefficient of expansion of the bulb due to internal pressure.

Equations (*i*) and (*ii*) represent the reduced volume of the gas at N.T.P. at 0ºC and $t$ ºC. As the mass of the enclosed gas is constant, these two volumes must be equal.

Equating (*i*) and (*ii*) we get,

$$V_0P_0 = \frac{\{V_0(1+\gamma_0 t)+bh\}(P_0+h)}{1+\gamma t}+\frac{vh}{1+\gamma\theta} \quad ...(iii)$$

or
$$\gamma t = \left(\frac{P_0+h}{P_0}\right)\left(\frac{V_0(1+\gamma_0 t)+bh}{V_0}\right)+\frac{hv}{P_0V_0}\left(\frac{1+\gamma t}{1+\gamma\theta}\right)-1 \quad ...(iv)$$

Here $t$ is the temperature of the gas and it occurs on both sides of the equation (*iv*). In case the above corrections are not taken into account, equation (*iv*) reduces to,

$$\gamma t = \frac{h}{P_0}$$

or
$$t = \frac{h}{\gamma P_0} \quad ...(v)$$

Initially the value of $t$ is calculated from equation (*v*) and this approximate value is substituted on the right hand side of equation (*iv*). After substituting, a new value of $t$ is obtained for the left hand side. This new value is again substituted on the right hand side and the value of $t$ on the left hand side is once again calculated. This method of successive substitution is performed two or three times depending upon the accuracy desired.

## 13.14 Second Correction (Perfect or Ideal Gas Scale)

All practical gases *viz.* hydrogen, oxygen, nitrogen etc, show deviations from Boyle's law at high pressures and low temperatures. Hence the temperatures obtained with a constant volume or a constant pressure gas thermometer have to be corrected so as to compensate for the departure from the ideal gas. On the constant volume scale, the temperature $t_v$ is written as,

$$t_v = \left[\frac{P_t-P_0}{P_{100}-P_0}\right]\times 100 \quad ...(i)$$

Similarly on the constant pressure scale the temperature $t_v$ is written as

$$t_v = \left(\frac{V_t-V_0}{V_{100}-V_0}\right)\times 100 \quad ...(ii)$$

If the mass of the gas enclosed is the same in both the themometers, then at 0 ºC, both will have the same pressure $P_0$ and volume $V_0$. In the case of constant volume gas thermometer, when the temperature is increased to $t$ ºC the pressure becomes $P_t$, the volume is $V_t$. In the constant pressure gas thermometer, at $t$ ºC, the volume is $V_t$. If the gas is ideal, in that case

$$V_0P_t = (PV)_t = V_tP_0 \quad ...(iii)$$

Multiplying and dividing the right hand side of equation (*i*) by $V_0$,

$$t_v = \left[\frac{(P_t-P_0)V_0}{(P_{100}-P_0)V_0}\right]\times 100$$

$$t_v = \left[\frac{P_tV_0-P_0V_0}{P_{100}V_0-P_0V_0}\right]\times 100 \quad ...(iv)$$

But from equation (*iii*).

$$P_t V_0 = (PV)_t$$

and

$$P_0 V_0 = (PV)_0$$

$$P_{100} V_0 = (PV)_{100}$$

$$\therefore \quad t_v = \left[\frac{(PV)_t - (PV)_0}{(PV)_{100} - (PV)_0}\right] \times 100 \qquad ...(v)$$

Similarly it can be shown that

$$t_v = \left[\frac{(PV)_t - (PV)_0}{(PV)_{100} - (PV)_0}\right] \times 100 \qquad ...(vi)$$

Thus for an ideal gas, the temperatures measured on the constant pressure scale and constant volume scale are identical and both can be written as

$$t = \left[\frac{(PV)_t - (PV)_0}{(PV)_{100} - (PV)_0}\right] \times 100 \qquad ...(vii)$$

But in actual practice no gas is ideal.

In the case of real gases, an empirical relation suggested by K-Onnes is

$$PV = A + BP + CP^2 + DP^3 + ....... \qquad ...(viii)$$

Here $A, B, C, D$ etc., are constants.

These constants are functions of temperature but are independent of pressure. The values of these coefficients are determined by performing compression experiments. These values for oxygen at 0 °C are

$$A = 1.00130$$

$$B = -1.30143 \times 10^{-3}$$

$$C = 3.6898 \times 10^{-6}$$

Here $P$ is measured in metres of $Hg$ and when $P = 1$, $V$ is also equal to one.

It can be seen that the constants $A, B, C, D$ are of rapidly decreasing magnitude. In actual practice, for the measurement of temperature, constants $C$ and $D$ are usually neglected. Therefore,

$$PV = A + BP \qquad ...(ix)$$

Using equation (*ix*), it is clear that when $P \to 0$, $PV = A$ or the gas is ideal and obeys Boyle's law.

In order to obtain the temperatures corresponding to an ideal gas, observations are made with a real gas and the values of the temperature, $t_{lim}$ when $P \to 0$ are obtained. Here $t_{lim}$ will correspond to the ideal gas temperature.

**Constant Volume Scale**

On the constant volume scale

$$t_v = \left(\frac{P_t - P_0}{P_{100} - P_0}\right) \times 100 \qquad ...(x)$$

From the equation

$$PV = A + BP, \text{ we have}$$

$$\left.\begin{aligned} P_0 V_0 &= A_0 + B_0 P_0 \\ P_{100} V_0 &= A_{100} + B_{100} P_{100} \\ P_t V_0 &= A_t + B_t P_t \end{aligned}\right\} \qquad ...(xi)$$

Multiplying and dividing the right hand side of equation (*x*) by $V_0$,

$$t_v = \left(\frac{P_t V_0 - P_0 V_0}{P_{100} V_0 - P_0 V_0}\right) \times 100$$

$$\therefore \qquad t_v = \left[\frac{(A_t + B_t P_t) - (A_0 + B_0 P_0)}{(A_{100} + B_{100} P_{100}) - (A_0 + B_0 P_0)}\right] \times 100$$

$$t_v = \left[\frac{(A_t - A_0) + (B_t P_t - B_0 P_0)}{(A_{100} - A_0) + (B_{100} P_{100} - B_0 P_0)}\right] \times 100$$

$$t_v = \left(\frac{A_t - A_0}{A_{100} - A_0}\right) \times 100 \left[\frac{\left(1 + \dfrac{B_t P_t - B_0 P_0}{A_t - A_0}\right)}{\left(1 + \dfrac{B_{100} P_{100} - B_0 P_0}{A_{100} - A_0}\right)}\right]$$

$$t_v = \left(\frac{A_t - A_o}{A_{100} - A_0}\right) \times 100 \left(1 + \frac{B_t P_t - B_0 P_0}{A_t - A_0}\right) \times \left(\frac{1 - B_{100} P_{100} - B_0 P_0}{A_{100} - A_0}\right)$$

$$t_v = \left(\frac{A_t - A_0}{A_{100} - A_0}\right) \times 100 \left(1 + \frac{B_t P_t - B_0 P_0}{A_t - A_0} - \frac{B_{100} P_{100} - B_0 P_0}{A_{100} - A_0}\right) \qquad \text{... (xii)}$$

When $P \to 0$

$$t_{\lim} = \left(\frac{A_t - A_0}{A_{100} - A_0}\right) \times 100$$

$$\therefore \qquad t_v = t_{\lim} \left(1 + \frac{B_t P_t - B_0 P_0}{A_t - A_0} - \frac{B_{100} P_{100} - B_0 P_0}{A_{100} - A_0}\right) \qquad \text{... (xiii)}$$

$$\text{or} \qquad t_{\lim} - t_v = t_{\lim} \left(\frac{B_{100} P_{100} - B_0 P_0}{A_{100} - A_0} - \frac{B_t P_t - B_0 P_0}{A_t - A_0}\right) \qquad \text{...(xiv)}$$

Equation (*xiv*) gives the correction that has to be applied. The value of $t_v$ is obtained with a thermometer and the value of the constants $A_0, B_0, A_t, B_t, A_{100}, B_{100}$ are obtained by actual compression experiments.

Initially the value of $t_v$ is substituted for $t_{lim}$ on the right hand side of equation (*xiv*) and the correction $t_{lim}$ $t_v$ is calculated. This gives a new value of $t_{lim}$. This is again substituted on the right hand side and this process of successive substitution is done two or three times. Finally $t_{lim}$ gives the accurate value of temperature on the **ideal gas scale.**

## Constant Pressure Scale.

On the constant pressure scale

$$t_p = \left(\frac{V_t - V_0}{V_{100} - V_0}\right) \times 100 \qquad \text{...(xv)}$$

From the equation

$$PV = A + BP$$

We have,

$$P_0 V_0 = A_0 + B_0 P_0$$

$$P_0 V_{100} = A_{100} + B_{100} P_0$$

$$P_0 V_t = A_t + B_t P_0$$

Multiplying and dividing the right hand side of equation (*xv*) by $P_0$.

$$t_p = \left(\frac{P_0 V_t - P_0 V_0}{P_0 V_{100} - P_0 V_0}\right) \times 100$$

$$t_p = \left[\frac{(A_t + B_t P_0) - (A_0 + B_0 P_0)}{(A_{100} + B_{100} P_0) - (A_0 + B_0 P_0)}\right] \times 100$$

$$t_p = \left(\frac{A_t - A_o}{A_{100} - A_0}\right)\times 100\left[1 + P_0\left(\frac{B_t - B_0}{A_t - A_0}\right) - P_0\left(\frac{B_{100} - B_0}{A_{100} - A_0}\right)\right]$$

When $P \to 0$

$$t_{\lim} = \left(\frac{A_t - A_0}{A_{100} - A_0}\right)\times 100$$

$$\therefore \quad t_p = t_{\lim}\left[1 + P_0\left(\frac{B_t - B_0}{A_t - A_0}\right) - P_0\left(\frac{B_{100} - B_0}{A_{100} - A_0}\right)\right]$$

$$\therefore \quad t_{\lim} - t_p = t_{\lim}\, P_O\left[\left(\frac{B_{100} - B_0}{A_{100} - A_0}\right) - \left(\frac{B_t - B_0}{A_t - A_0}\right)\right] \qquad \text{...(xvi)}$$

Equation (*xvi*) gives the expression for the correction to be applied. The value of $t_v$ is obtained with a thermometer. By the method of successive substitution, the value of $t_{lim}$ is obtained.

From equations (*xiv*) and (*xvi*) it is clear that in both the cases the value of

$$t_{\lim} = \left(\frac{A_t - A_0}{A_{100} - A_0}\right)\times 100$$

Thus the ideal or perfect gas scale (Celsius scale) can be defined as follows:

$$\underset{\text{(celsius)}}{t} = \left[\frac{\underset{\lim P\to 0}{(PV)_t} - \underset{\lim P\to 0}{(PV)_0}}{\underset{\lim P\to 0}{(PV)_{100}} - \underset{\lim P\to 0}{(PV)_0}}\right]\times 100 \qquad \text{...(xvii)}$$

These corrections obtained from equations (*xiv*) and (*xvi*) are applicable in the range of – 183°C to 400°C. Beyond 400°C, the corrections are negligibly small.

## NUMERICAL EXAMPLES

### EXAMPLE 13.7

*The pressure of air in a constant volume thermometer is 80 cm and 109.3 cm at 0°C and 100°C respectively. When the bulb is placed in hot water, the pressure is 100 cm. Find the temperature of the hot water*

**SOLUTION.**

$$P_t = 100 \text{ cm of Hg}$$
$$P_{100} = 109.3 \text{ cm of Hg}$$
$$P_0 = 80 \text{ cm of Hg}$$
$$t = \left(\frac{P_t - P_0}{P_{100} - P_0}\right)\times 100 = \left(\frac{100 - 80}{109.3 - 80}\right)\times 100$$
$$t = \mathbf{68.28°C}$$

### EXAMPLE 13.8

*A constant volume hydrogen thermometer is used to measure the temperature of a furnace. The excess pressure in the bulb over the atmospheric pressure is found to be equal to 152 cm of Hg. At 0°C the pressure in the bulb is equal to that of the atmosphere. Calculate the temperature of the furnace, assuming that the atmospheric pressure throughout the experiment remains constant.*

**SOLUTION.** Here

$$P_0 = 76 \text{ cm of Hg}$$
$$P_t = 152 + 76 = 228 \text{ cm of Hg}$$
$$T_0 = 273 \text{ K}$$
$$T = ?$$

$$\frac{P_t}{T} = \frac{P_0}{T_0}$$

$$T = \frac{P_t}{P_0} \times T_0$$

$$T = \frac{228 \times 273}{76}$$

$$T = \mathbf{819\ K = 546\ °C}$$

## 13.15 Platinum Resistance Thermometer

It is based on the principle of change of resistance with change in temperature. It was first designed by Siemen in 1871 and later on improved by Callendar and Griffiths.

Plantinum Resistance Thermometer

A platinum resistance thermometer consists of a pure platinum wire wound in a double spiral to avoid inductive effects. The wire is wound on a mica plate. The two ends of the platinum wire are connected to thick copper leads (for lower temperatures) and connected to the binding terminals $B_1$ $B_2$. For higher temperatures the leads are of platinum. $C_1$ and $C_2$ are the compensating leads exactly similar and of the same resistance as the leads used with the platinum wire. The platinum wire and the compensating leads are enclosed in a glazed porcelain tube. The tube is sealed and binding terminals are provided at the top. The leads pass through mica discs which offer the best insulation and also prevent convection currents (Fig. 13.1).

The resistance of a wire at $t°C = R_t$ and at $0°C = R_0$. These resistances are connected by the relation

$$R_t = R_0 (1 + \alpha t + \beta t^2) \qquad ...(i)$$

Here α and β are constants. The values of α and β depend on the nature of the material used. To find the values of α and β, the resistance of the platinum wire is determined at three fixed points (*i*) melting point of ice, (*ii*) boiling point of water, (*iii*) boiling point of sulphur 444.6°C for high temperature measurement and (*iv*) boiling point of oxygen –182.5°C for low temperature measurement.

Using these values of resistance in equation (*i*)

$$R_{100} = R_0 [1 + \alpha\, 100 + \beta\, 100^2] \qquad ...(ii)$$

and

$$R_{444.6} = R_0 [1 + \alpha\, 444.6 + \beta\, (444.6)^2] \qquad ...(iii)$$

The values of α and β can be determined by solving the simultaneous equations (*ii*) and (*iii*).

From (*i*)

$$R_t = R_0 [1 + \alpha t + \beta t^2]$$

Neglecting $\beta t^2$ (because *b* is very small)

$$R_t = R_0 [1 + \alpha t] \qquad ...(iv)$$

and

$$R_{100} = R_0 [1 + \alpha \times 100]$$

$$\therefore \quad R_t - R_0 = R_0\, \alpha t \qquad ...(v)$$

$$R_{100} - R_0 = R_0\, \alpha.\, 100 \qquad ...(vi)$$

Dividing (*v*) and (*vi*)

$$\frac{R_t - R_0}{R_{100} - R_0} = \frac{t}{100}$$

or $$t = \left(\frac{R_t - R_0}{R_{100} - R_0}\right) \times 100 \qquad ...(vii)$$

Knowing the values of $R_0$, $R_{100}$ and $R_t$, $t$ can be calculated.

The resistance of the platinum wire is found accurately using Callendar and Griffith's bridge.

## 13.16 Callendar and Griffith's Bridge

This is a modified form of Wheatstone's bridge, used to measure the change in resistance with temperature of the platinum wire used in a platinum resistance thermometer.

$P$ and $Q$ are two resistances of equal value and $S$ is a standard fractional resistance box. The platinum wire whose resistance is to be calculated is connected in one of the arms of the Wheatstone's bridge as shown in (Fig. 13.6). The compensating leads are connected in series with the resistance $S$. $XY$ is the bridge wire of length $2l$ and of resistance per unit length $\rho$. Let $r$ be the resistance of the compensating leads and $O$ the mid-point of the bridge wire. $R$ is the resistance of the platinum wire at the temperature of the bath. Let $D$ be the balance point, with some suitable value of $S$, $x$ cm to the left of $O$.

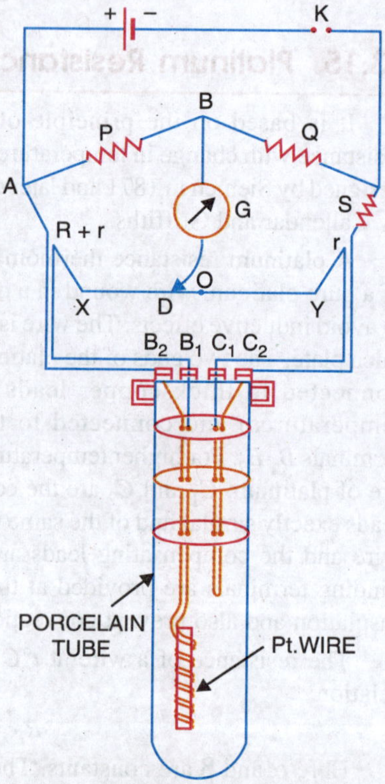


Fig. 13.6

Then length $$XD = l - x$$
$$DY = l + x$$

As $P = Q$, the resistance of the arm $AD$ = resistance of the arm $CD$

$$\therefore \quad R + r + (l - x)\rho = S + r + (l + x)\rho$$
$$R + r + l\rho - x\rho = S + r + l\rho + x\rho$$
$$R = S + 2x\rho$$

(If the balance point is obtained to the right of $O$, it can be shown that $R = S - 2x\rho$.)

Thus, using a Callendar and Griffith's bridge the resistance $R_t$ of the platinum wire can be determined at temperature $t$.

A scale can be fixed along the bridge wire $XY$ to read platinum scale temperatures directly. The range will be different for different values of $S$.

**Correction.** The temperature $t$ measured on the platinum scale according to equation (*vii*) is not accurate because $\beta t^2$ term has been neglected. Therefore, the correct thermodynamic temperature $\theta$ will differ from $t$.

Here $$\theta - t = \delta\left[\left(\frac{\theta}{100}\right)^2 - \left(\frac{\theta}{100}\right)\right] \qquad ...(viii)$$

The value of $\delta$ can be calculated by knowing the values of $\alpha$ and $\beta$.

Equation (*viii*) can be derived as follows:

$$\theta - t = \theta - \left(\frac{R_t - R_0}{R_{100} - R_0}\right) \times 100$$

Take $$R_t = R_0(1 + \alpha\theta + \beta\theta^2)$$

and $$R_{100} = R_0\,(1 + \alpha\, 100 + \beta\,(100)^2]$$

$$\therefore \quad \frac{R_t - R_0}{R_{100} - R_0} = \frac{\alpha\theta + \beta\theta^2}{\alpha 100 + \beta (100)^2}$$

$$\therefore \quad \theta - t = \theta - \left(\frac{\alpha\theta + \beta\theta^2}{\alpha + \beta\, 100}\right)$$

$$\theta - t = \frac{\alpha\theta + 100\beta\theta - \alpha\theta - \beta\theta^2}{\alpha + \beta\, 100}$$

$$\theta - t = \frac{100\beta\theta - \beta\theta^2}{(\alpha + \beta\, 100)}$$

$$\theta - t = \frac{-(100)^2\,\beta}{(\alpha + \beta 100}\left[\frac{\theta^2}{(100)^2} - \frac{\theta}{100}\right]$$

$$\theta - t = \delta\left[\left(\frac{\theta}{100}\right)^2 - \left(\frac{\theta}{100}\right)\right] \quad ...(ix)$$

Here $$\delta = \frac{-(100)^2\beta}{(\alpha + \beta\, 100)}$$

The value of $\delta$ for pure platinum is approximately equal to 1.5.

**Advantage.** The platinum resistance thermometer can be used with great accuracy to measure temperatures ranging from – 200°C to 1200°C. Its accuracy is 0.1°C. Once the platinum resistance thermometer has been standardised with a constant volume hydrogen thermometer, it can be used as a standard thermometer.

Its main disadvantage is that the adjustment of the Callender and Griffith's bridge takes a long time. Moreover, the wire may not attain the temperature of the bath in a short time.

## 13.17 Seebeck Effect

In 1821, Seebeck found that a current flows in a circuit consisting of two dissimilar metals when one junction is heated while the other junction is kept cold. This was a remarkable experiment because *no cell* was used in the circuit. He connected a plate of bismuth between copper wires connected to a galvanometer [Fig.13.7 (*i*)]. He found that if one of the junctions was heated while the other was kept cold, then a current flowed through the galvanometer. He repeated his experiment by taking a thermocouple of Fe and Cu [Fig. 13.7 (*ii*)].

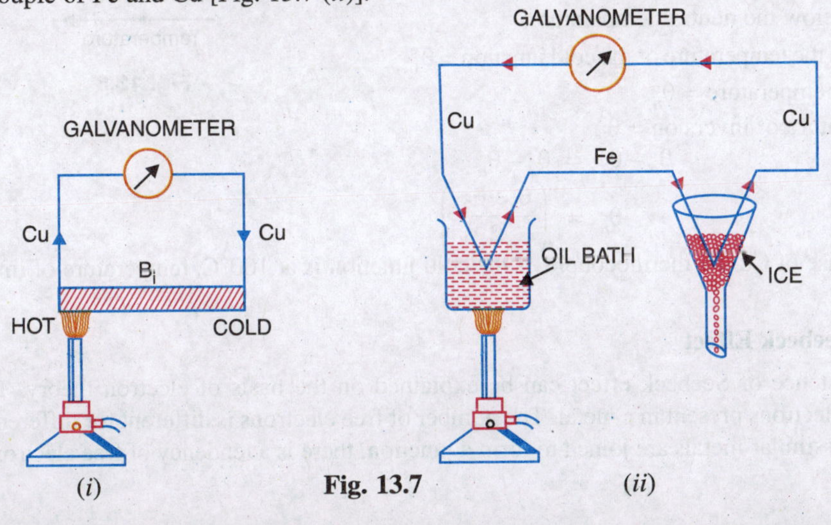


**Fig. 13.7**

If both the junctions are at 0°C, there is no deflection in the galvanometer. When one of the junctions is kept at 0°C, *i.e.*, at the temperature of melting point of ice and the other junction is heated gradually, current flows in the circuit. It was found that current flows from copper to iron at the hot junction and iron to copper at the cold junction. The current increases until the hot junction is at a temperature of 270°C. If the heating is continued beyond 270°C, the current decreases and finally at 540°C the current is zero.

Seeback Effect

It was discovered by *Cumming* in 1823 that, on increasing the temperature of the hot junction beyond 540°C, the direction of the current is reversed. It flows from iron to copper through the hot junction and copper to iron through the cold junction.

The current produced in this way without the use of a cell or a battery is known as *thermo-electric current* and this branch of electricity is known as thermo-electricity. The effect is known as *Seebeck effect.*

The temperature of the hot junction at which maximum current flows in a circuit is known as *neutral temperature* for that couple. The neutral temperature for a given thermocouple is fixed and remains constant whatever may be the temperature of the cold junction. The E.M.F. produced in this way is called *thermo-E.M.F.*

If a graph is plotted between the temperature of the hot junction and the thermo-E.M.F., the cold junction being kept at 0°C the graph is a parabolic curve. The thermo-E.M.F. $E$ varies with temperature according to $E = at + bt^2$, where $a$ and $b$ are constants. The point $A$ represents the neutral temperature. The point $B$ is the temperature of inversion, beyond which the current is reversed (Fig. 13.8).

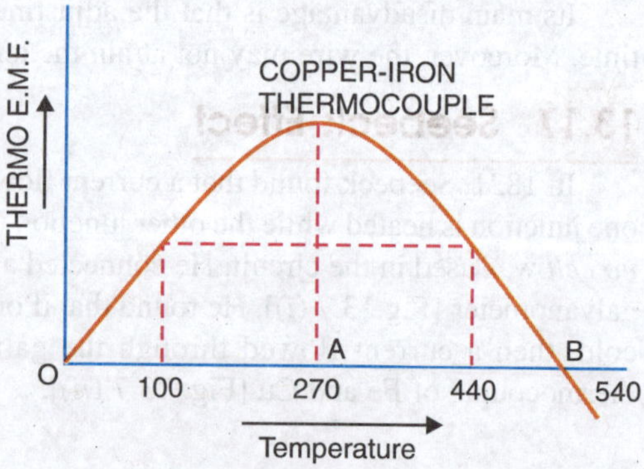

Fig. 13.8

The temperature of inversion is not fixed. It is as much above the neutral temperature as the cold junction is below the neutral temperature.

Suppose the temperature of the cold junction = $\theta_1$

Neutral temperature = $\theta_n$

Temperature of inversion = $\theta_2$

Then, $\theta_n - \theta_1 = \theta_2 - \theta_n$,

or $\theta_n = \left(\dfrac{\theta_1 + \theta_2}{2}\right)$

In the case of Cu–Fe thermocouple, if the cold junction is at 100°C, temperature of inversion = 444°C.

### Causes of Seebeck Effect

The existence of Seebeck effect can be explained on the basis of electron theory. There are always free electrons present in a metal. The number of free electrons is different for different metals. When two dissimilar metals are joined to form a junction, there is a tendency of free electrons of one metal to diffuse into other. The flow continues untill *a contact potential difference* high enough to prevent any further movement of electrons is established. If both the junctions are at the same temperature, the contact potential is the same at both the junctions. But when the junctions are at different temperatures, the contact potential difference at the two junctions are not equal hence a current flows from one junction to the other.

## 13.18 Peltier Effect

Jean Peltier (1785 – 1845) discovered in 1834 a thermo-electric effect which is in converse of Seebeck effect. According to Peltier, '*When a current is passed through a thermocouple whose two junctions are initially at the same temperature, then one of the junctions is heated, while the other is cooled; showing that heat energy is being liberated at one junction* (thereby causing it to be heated) *and absorbed at the other junction* (thereby causing it to be cooled).'

This is explained by Fig. 13.9. Thus, *by the passage of electric current through a thermocouple with junctions initially at the same temperature, difference of temperature is developed between the two junctions.* This effect is known as **Peltier's effect.**

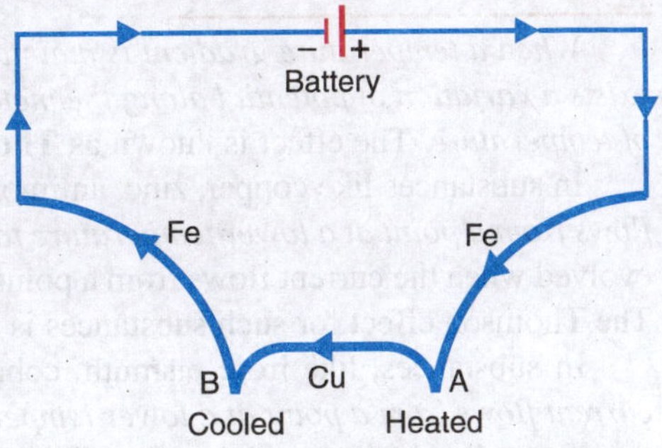

**Fig. 13.9**

For a given pair of metals, the heating or cooling of the junction depends on the direction of the current. If a particular junction is heated by passing the current in one direction, the same is cooled on reversing the direction of current; Thus, the *Peltier effect is reversible.*

### Cause of Peltier Effect

The existence of Peltier effect can be explained on the basis of electron theory. The metals contain free electrons which may be considered as moving about like the molecules of a gas. When dissimilar metals are joined, there will be a flow of electrons from one metal which has a larger number of free electrons per unit volume, to the other having a smaller electron density. The flow continues untill a potential difference high enough to prevent any further movement of electrons is established.

When copper and iron are placed in contact (Fig. 13.9) *the potential of iron is above that of copper.* If a current is passed from a battery as shown, then at B current flows from copper to iron. *i.e.,* from lower potential to higher potentials and hence energy is required for this purpose. This is absorbed from the junction which, therefore, is cooled. At junction A, current flows from iron to copper *i.e.,* from a higher potential to a lower potential and thus energy is given out so that the junction A is heated.

It should be noted that the heating and the cooling of the two junctions in Fig. 13.9, caused by the Peltier effect, provides the conditions necessary for the Seebeck effect of Fig. 13.7. The direction of the resulting Seebeck current is opposite to the battery current. If this thermoelectric current did not oppose, but had been aided by the battery, it would have resulted the heated (cooled) junction further heated (cooled), thereby resulting in a still greater current and so on, without any limit.

## 13.19 Peltier Coefficient ($\pi$)

In S.I. units, *Peltier coefficient is defined as the amount of heat energy in joules absorbed or evolved due to Peltier effect at a junction of two dissimilar metals when one ampere current flows for one second i.e., when a quantity of charge of one coulomb passes through it.* It is denoted by $\pi$.

The value of Peltier coefficient is different for different pairs of metals. For the same pair of metals its value depends upon the temperature of the junction.

If a current of $i$ amperers flows for $t$ seconds through the junction having a Peltier coefficient $\pi$, then

Heat energy absorbed or evolved = $\pi\, it$ Joules ...(*i*)

If V is the contact potential difference at the junction in volts, then

Heat energy absorbed or evolved = Vit Joules ...(*ii*)

Equating (*i*) and (*ii*), we have

$$\pi\, it = Vit$$

or

$$\pi = V \quad ...(iii)$$

Hence *Peltier coefficient (in Joules per coulomb) at a junction is numerically equal to the contact potential difference (in volts).*

## 13.20 Thomson Effect

*When a temperature gradient is maintained between the different parts of the same metal, there exists a variation of potential along the metal i.e., e.m.f. is developed in the metal due to the gradient of temperature.* The effect is known as **Thomson effect.**

In substances like copper, zinc, antimony and cadmium heat energy is absorbed when *a current flows from a point at a lower temperature to a point at higher temperature.* Heat energy is, therefore, evolved when the current flows from a point at a higher temperature to a point at a lower temperature. The Thomson effect for such substances is said to be **positive.**

In substances, like iron, bismuth, cobalt, platinum and nickel heat energy is evolved when a *current flows from a point at a lower temperature to a point at a higher temperature.* Heat energy is, therefore, absorbed when the current flows from a point at a higher temperature to a point at a lower temperature. The Thomson effect for such substances is said to be **negative.**

In lead Thomson effect is zero.

### Cause of Thomson Effect

The existence of Thomson effect can also be explained on the basis of electron theory. The metals contain free electrons which may be considered to be moving about like molecules of a gas. When the temperature of the metal is the same everywhere, the density of the electron gas is also the same at every point. When there is a difference of temperature between two points on the same conductor, there will be a difference in the density of electron gas. This creates a potential difference between two points of a conductor at different temperatures. When electric current flows through the conductor along the direction of potential difference, work is done by the current and heat energy is evolved and when the electric current flows through the conductor in a direction opposite to the direction of potential difference, energy is required, which is absorbed as heat from the conductor.

Thomson effect is also reversible.

## 13.21 Thomson Coefficient (σ)

In S.I. units, *Thomson coefficient is defined as the amount of heat energy in joules absorbed or evolved due to Thomson effect between two points of a conductor which differ in temperature by one kelvin (1K), when one ampere current flows for one second i.e., when a quantity of charge of one coulomb passes through it.* It is denoted by $\sigma$.

Thomson coefficient ($\sigma$) is not constant but varies with temperature.

If a current $i$ ampere flows for $t$ seconds between two points of a conductor having a difference of temperature of 1K and $\sigma$ is its Thomson coefficient, then

Heat energy evolved or absorbed = $\sigma\, it$ Joules. ...(*i*)

If V is the potential difference between the same two points in volts, then

Heat energy absorbed or evolved = $V\, it$ Joules. ...(*ii*)

$$\sigma it = Vit$$

or $$\sigma = V \quad ...(iii)$$

Hence *Thomson coefficient (in Joules per coulomb) is numerically equal to the difference of potential per degree kelvin (in volts).*

*Thomson coefficient is also known as the specific heat of electricity.*

If σ is the Thomson coefficient, $(T_1 - T_2)$ the difference of temperature between two ends of a conductor and a current $i$ is passed through it, then

Heat energy evolved or absorbed due to Thomson effect = $\sigma\, i\, (T_1 - T_2)$.

## 13.22 Thermo-Electric Power $\left(\frac{dE}{dT}\right)$

Thermo-electric power is defined *as the rate of change of thermo-e.m.f. with temperature.*

It we have a thermo-couple with one junction at a temperature TK and the other junction at $(T + dT)$ K and $dE$ is the thermo-e.m.f. developed, then the thermo-electric power at temperature $T$ is given by $\frac{dE}{dT}$.

The thermo-electric power at a particular temperature is given by the slope of the tangent to the $E - T$ curve at that temperature.

In order to determine the variation of thermo-electric power $\frac{dE}{dT}$ with temperature of the hot junction $T$, first of all a graph is plotted between the thermo-e.m.f. $E$ and the temperature of the hot junction $T$. The graph is a parabola, as already shown in Fig. 13.8. Now draw tangents to the parabola at different points. The slopes of these tangents gives the values of thermo-electric power $\frac{dE}{dT}$ at temperature $T$, corresponding to each point selected on the parabola. From these values, a graph $\frac{dE}{dT}$ against $T$ is plotted, This graph showing the relation between thermo-electric power $\frac{dE}{dT}$ and the corresponding absolute temperature $T$ is known as *thermo-electric diagram.* It is always a *straight line* as shown in Fig. 13.10 and is known as **thermo-electric power line.** To construct the thermo-electric diagram for any metal lead is taken as the second metal for the thermo-couple because Thomson coefficient for lead is zero.

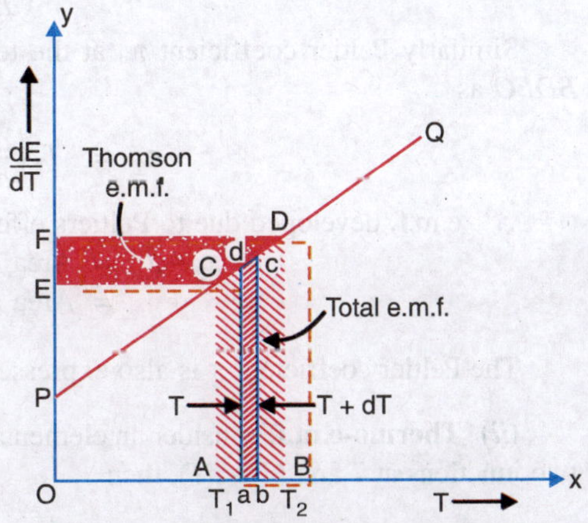


Fig. 13.10

Recalling the graph between the thermo-electric e.m.f. $E$ and the corresponding temperature of the hot junction $T$ (Fig. 13.8). The graph is a parabola. Therefore, when the temperature of the cold junction is kept at 0 °C, the relation between the thermo-electric e.m.f. $E$ and the temperature of the hot junction $T$ is expressed as

$E = aT + bT^2$ (being eqn of parabola) where $a$ and $b$ are constants for the given thermo-couple.

Therefore, the thermo-electric power $\frac{dE}{dT}$ is obtained by differentiating above relation, we get

$$\frac{dE}{dT} = a + 2bT$$

which is the equation of a straight line (like $y = mx + c$). Hence the graph between the *thermoelectric power* $\frac{dE}{dT}$ and the temperature of the hot junction $T$ is a straight line, not passing through the origin as shown in Fig. 13.10.

## Uses of Thermo-Electric Diagram

Fig. 13.10 shows the thermo-electric diagram. Let $PQ$ be the thermo-electric power line for a given metal with respect to lead. Suppose the thermo-couple is maintained with its cold junction at a temperature $T_1$ and the hot junction at a temperature $T_2$. Then $AC$ represents the thermo-electric power at $T_1$ and $BD$ at $T_2$.

**Uses :** The thermo-electric diagram can be used to find the values of

(*i*) Peltier coefficient (*ii*) Thermo e.m.f.
(*iii*) Thomson coefficient (*iv*) Thermo e.m.f. due to a thermo-couple
(*v*) Neutral temperature.

**(*i*) Peltier Coefficient:** The Peltier coefficient $\pi_1$ at the temperature of cold junction $T_1$ is represented by the area $ACEO$ as

$$\pi_1 = T_1\left(\frac{dE}{dT}\right)_{T_1} = OA \times AC = \text{Area } ACEO$$

Similarly Peltier coefficient $\pi_2$ at the temperature of the hot junction $T_2$ is given by the area $BDFO$ as

$$\pi_2 = T_2\left(\frac{dE}{dT}\right)_{T_2} = OB \times BD = \text{Area } BDFO$$

∴ e.m.f. developed due to Peltiers effect is

$$\pi_2 - \pi_1 = \text{Area } BDFO - \text{Area } ACEO$$
$$= \text{Area } ABDFECA \quad ...(i)$$

The Peltier coefficient $\pi$ is also expressed as $\pi = T\frac{dE}{dT}$

**(*ii*) Thermo-e.m.f.** consider an elementary strip $abcd$ of thickness $dT$ having temperature of the two junctions at $T$ and $(T + dT)$, then

$$\text{Area } abcd = \frac{dE}{dT}dT = dE$$

Hence total e.m.f. developed between the temperatures $T_1$ and $T_2$ will be

$$E = \int_{T_1}^{T_2} dE = \text{Sum of the areas of such small strips}$$
$$= \text{Area } ABDC \quad ...(ii)$$

**(*iii*) Thomson coefficient**

The total e.m.f. developed is given by

$$E = (\pi_2 - \pi_1) + \int_{T_1}^{T_2} (\sigma_a - \sigma_b)\,dT$$

$$\therefore \quad \int_{T_1}^{T_2} (\sigma_a - \sigma_b)\,dT = E - (\pi_2 - \pi_1)$$

If the metal $A$ is lead, $\sigma_a = 0$, then

$$\int_{T_1}^{T_2} \sigma_b\,dT = (\pi_2 - \pi_1) - E$$

$= \text{Area } (BDFO - ACEO - ABDC)$

$= \text{Area } CEFD$ ...(*iii*)

Here area *CEFD* represents the e.m.f. produced due to Thomson effect between the temperatures $T_1$ and $T_2$ of two junctions.

In order to find the value of Thomson coefficient at temperature *T*, the area *CEFD* between the temperature $T_1 = T - ½$ and $T_2 = T + ½$ is measured. As $dT = 1$, this area gives the Thomson coefficient at the temperature *T*.

Thomson coefficient is also expressed as

$$\sigma = T \frac{d^2E}{dT^2}.$$

**(*iv*) Thermo-e.m.f. due to thermo-couple.**

Let us consider the case of a thermo-couple consisting of two metals *X* and *Y*. Let *PQ* and *RS* be their thermo-electric power lines respectively as shown in Fig. 13.11. The e.m.f. between the absolute temperatures $T_1$ and $T_2$ for metal *X* is given by the area *ABCD* and for the metal *Y* by the area *ABEF*.

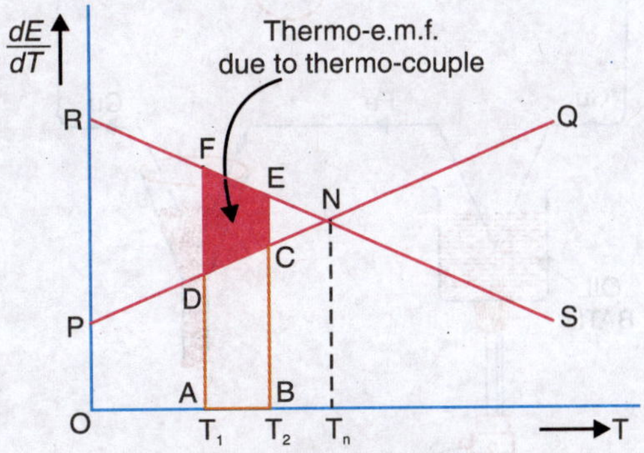


Fig. 13.11

Therefore, e.m.f. for a thermo-couple consisting of metals *X* and *Y* between the same temperatures $T_1$ and $T_2$ will be

$= \text{Area } ABEF - \text{Area } ABCD$

$= \text{Area } DCEF$ ...(*iv*)

**(*v*) Neutral Temperature.** The point *N* where the two thermo-electric power lines intersect gives the neutral temperature ($T_n$) for this thermo-couple formed by two metals *X* and *Y*. At this temperature $\frac{dE}{dT} = 0$.

## 13.23 Thermo-Electric Thermometer

It is based on the principle of Seebeck effect [Refer article 13.17]. Two dissimilar metals are selected. The thermo-E.M.F. produced, depends upon the difference of temperatures between the hot and the cold junctions. For measuring temperatures up to 300°C, copper-constantan thermocouple is used. Iron-nickel is used between 300°C and 600°C and nickel-nichrome is used between 600°C and 1000°C. For temperatures between 1000°C and 1600°C, a thermocouple of platinum and an alloy of platinum and rhodium is used. Beyond 1600°C, up to 2000°C iridium and an alloy of iridium and rubidium is used. For measuring temperatures between 2000°C and 3000°C, tungsten and molybdenum thermocouple is used.

The hot junction is made by welding the junction of two metals and the junction is enclosed in a fine capillary tube. The whole arrangement is enclosed in an outer porcelain tube (Fig. 13.12). The wires 1 and 2 inside the tube are kept separate with the help of mica strips.

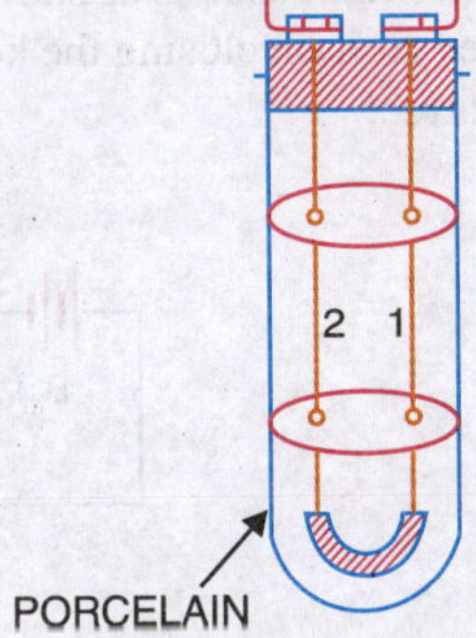


Fig. 13.12

### Measurement of Temperature

The most important part of the experiment for the measurement of temperature is the accuracy with which the thermo-E.M.F. can be measured. Two methods are usually employed, *e.g.*

(*i*) Using a sensitive galvanometer,

(*ii*) Using a potentiometer.

**1. Using a galvanometer :** A sensitive galvanometer is connected with the thermo couple (Fig. 13.13). One of the junctions is kept in melting ice (0°C) and the other junction in an oil bath. The oil bath is heated and thermo-E.M.F. is produced in the circuit. The temperature of the oil bath is determined by using a constant volume hydrogen thermometer or some other sensitive thermometer.

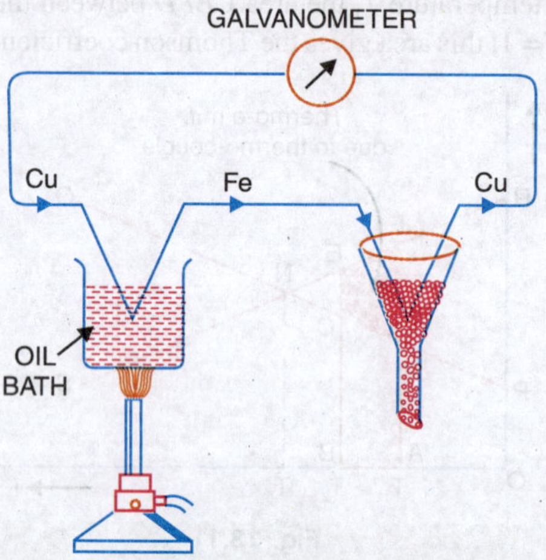


Fig. 13.13

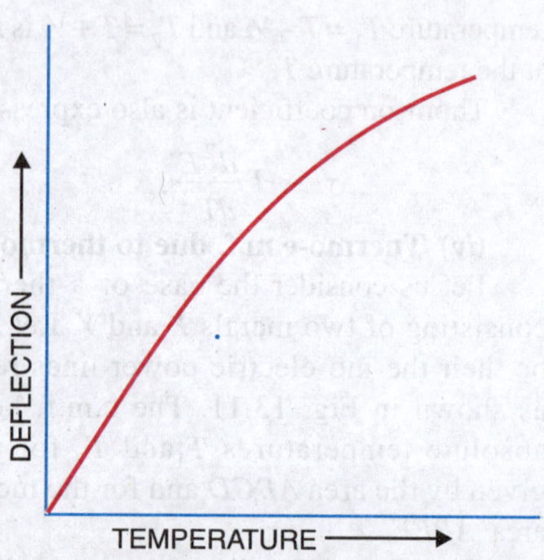


Fig. 13.14

Corresponding to the various temperatures of the hot junction (oil bath) the corresponding deflections in the galvanometer are observed. A graph is plotted between the deflection along the *y*-axis and the temperature of the hot junction along the *x*-axis (Fig. 13.14). Thus for the given thermo-couple the galvanometer is calibrated.

The bath whose temperature is to be measured, is taken. The hot junction of the thermo-couple is immersed in the bath, other junction being kept in melting ice. The deflection produced in the galvanometer is observed. From the graph, the corresponding value of temperature is determined. Moreover, the scale of the galvanometer can be directly marked in °C to show the temperatures directly. Such a thermo-electric thermometer is not very sensitive.

**2. Using a Potentiometer :** The thermo-E.M.F. produced in a thermocouple is of the order of $10^{-5}$ volt. Therefore the potentiometer should be sensitive to measure up to $10^{-6}$ volt. For this purpose the circuit used is as shown in Fig. 13.15. To check the correctness of the connections, first close the key α. If on closing the key *K*, the deflection in the galvanometer decreases, then the connections are correct.

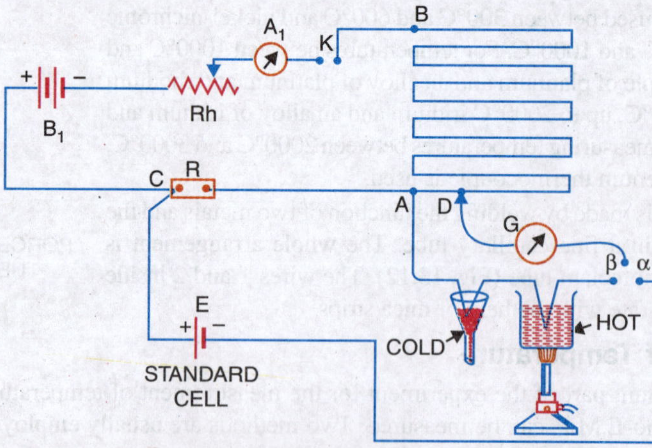


Fig. 13.15

The standard cadmium cell $E$ has an E.M.F. = 1.0183 volts. The potentiometer has ten wires of length one metre each. The potentiometer whose wire has a resistance of one ohm per metre is used. The distance $AD$ = 30 cm. The jockey is kept at $D$.

A resistance of 1018 ohms is introduced in $R$ and the key $\alpha$ is closed. The rheostat $Rh$ is adjusted so that the deflection in the galvanometer is zero. Then, the P.D. across $CD$ = 1.0183 volts.

$$\text{The resistance of } AD = \frac{1 \times 30}{100} = 0.3 \text{ ohm}$$

$$\text{The resistance of } CD = 1018 + 0.3 = 1018.3 \text{ ohm}$$

Fall of potential for one ohm-resistance

$$= \frac{1.0183}{1018.3} = 10^{-3} \text{ volt}$$

The potentiometer wire has a resistance of one ohm per metre. Therefore the fall of potential for one metre length of the potentiometre wire = $10^{-3}$ volt and the fall of potential per mm length of the potentiometer wire

$$= \frac{10^{-3}}{1000} = 10^{-6} \text{ volt per mm.}$$

Thus the potentiometer wire is calibrated.

The key $\alpha$ is taken out and the key $\beta$ is introduced. For the various temperatures of the hot junction, the corresponding balance points are observed. The lengths are measured from the point $A$. Thus the corresponding thermo-E.M.F. produced in the circuits can be calculated. The temperatures of the hot junction are measured with a standard constant volume hydrogen thermometer. A graph is plotted between the thermo-E.M.F. produced and the corresponding temperature. The potentiometer wire can be directly calibrated to measure the temperatures.

The hot junction is kept in the bath of the furnace whose temperature is to be measured and the balance point is found. From the balancing length, the corresponding temperature is either read from the graph or from the chart provided.

For practical purposes charts are provided with each thermo-couple.

**Advantages.** (1) It has a wide range and can measure temperatures from –200°C up to 3000°C. (2) It is useful to measure rapidly varying temperatures. (3) It can be easily constructed and it is cheap. (4) It is useful to measure the temperature of hot furnaces and does not require any calculations, once it has been calibrated.

**Limits and Drawbacks:** This thermometer does not give accurate temperatures over a wide range. Different thermocouples are to be used for different ranges. The potentiometer is to be calibrated for different thermocouples separately. Moreover, the neutral temperature of a thermocouple also limits its range.

## 13.24 Helium Vapour Pressure Thermometer

This thermometer is used to measure temperature up to 0.7 K. The apparatus consists of a bulb $A$ containing liquid helium. This bulb is connected to the manometer limbs $M_1$ and $M_2$ through a connecting tube $C$. The tube $C$ is surrounded by a copper tube $B$ to ensure uniform temperature of the vapour, $R$ is a reservoir containing mercury (Fig. 13.16).

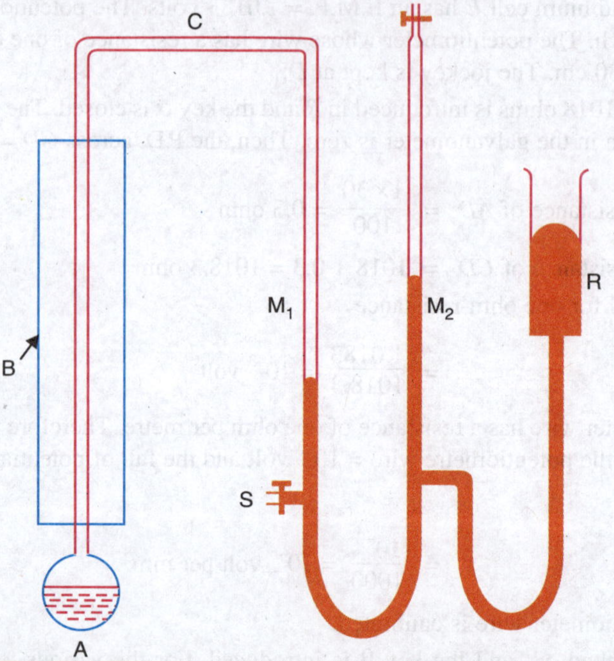


Fig. 13.16

Initially the reservoir $R$ is lowered so that the mercury in the manometers $M_1$ and $M_2$ is below the stop-cock $S$. The tube is connected to an evacuation pump to remove air in the tube $C$ and the bulb $A$. The stop-cock $S$ is closed after evacuation and the bulb $A$ is placed in the bath whose temperature is to be measured. The pressure of saturated helium vapour is measured from the difference in levels of Hg in the limbs $M_1$ and $M_2$. With the help of constant tables, giving the vapour pressure of the liquid at various temperatures, the temperature corresponding to any observed vapour pressure is determined. For lower temperatures, the graph between saturated vapour pressure and temperature is extrapolated.

**TABLE 13.1** **Vapour pressure of helium ($He^4$) at different temperatures**

| *Temp.* *K* | *Pressure* *mm of Hg* |
|---|---|
| 5.00 | 1460 |
| 4.50 | 980 |
| 4.00 | 615 |
| 3.50 | 353 |
| 3.00 | 181 |
| 2.50 | 77 |
| 2.00 | 23.2 |
| 1.50 | 3.6 |
| 1.00 | 0.12 |
| 0.40 | $1.6 \times 10^{-5}$ |
| 0.10 | $3.4 \times 10^{-32}$ |

## 13.25 Standardization and Temperature Scale

The various thermometers depend upon the property of a substance. The temperature scale shown by a thermometer will also depend on the property of the substance selected. Due to this reason, the scale of temperature is arbitrary depending on the property of the substance. The thermometers thus manufactured agree only at the fixed points and not at any other temperature.

Let the magnitude of the property selected be

$x_t, x_{ice}, x_{steam}$ at $t$°C, 0°C and 100°C.

The temperature $t$ can be calculated from the relation

$$\frac{t}{100} = \frac{x_t - x_{ice}}{x_{steam} - x_{ice}} \qquad ...(i)$$

It means that the temperature scale selected and the property of the substance chosen vary uniformly. But it has been found that the properties of a substance *e.g.*, expansion with rise in temperature in mercury thermometer, change of resistance with temperature in a platinum resistance thermometer etc. do not vary uniformly at all temperatures. Due to this reason, all the thermometers do not show the same temperature readings at one fixed temperature. Suppose the temperature of the liquid is measured by (*i*) a mercury thermometer (*ii*) a constant volume hydrogen thermometer and (*iii*) a platinum resistance thermometer at the same time. All these thermometers will not show the same reading. The readings may be 40.11°C, 40°C, and 40.36°C. Thus it is found that the temperature shown by the thermometer depends upon the nature of the thermometer used. However, these thermometers will show the same temperatures at fixed points *e.g.* melting point of ice and boiling point of water.

**TABLE 13.2 Comparison of different thermometers**

| *Constant volume hydrogen thermometer* | *Pt. resistance thermometer* | *Mercury thermometer* |
|---|---|---|
| 0°C | 0°C | 0°C |
| 20°C | 20.24°C | 20.09°C |
| 60°C | 60.36°C | 60.09°C |
| 100°C | 100°C | 100° |

In 1887, an international committee of scientists suggested that a scale of temperature shown by constant volume hydrogen thermometer be taken as a standard scale, 0°C being that of melting ice and 100°C being the temperature of steam at normal pressure. But hydrogen also is not a standard or a perfect gas.

Later on it was thought that temperature scale suggested by Kelvin be used as a standard scale because it does not depend upon property of a substance. The Kelvin scale of temperature (K) [work scale] agrees with the absolute gas scale.

As it appears difficult to realize the Kelvin scale in practice it is suggested in 1933 that an international scale of temperature be adopted. The melting point of ice and boiling point of pure water at normal pressure etc. are given the fixed values and the temperatures between these points are calculated by a specific formula. The international scale of temperature is the nearest practical approach to the Kelvin scale of temperature.

### Fixed Points

**1. Oxygen point.** Temperature of equilibrium between liquid oxygen and gaseous oxygen at normal pressure = – 182.97°C.

**2. Ice point.** Temperature of equilibrium between ice and air saturated water at normal pressure = 0.000°C.

**3. Steam point.** Temperature of equilibrium between liquid water and its vapour at normal pressure = 100.000°C.

**4. Sulphur point.** Temperature of equilibrium between liquid sulphur and its vapour at normal pressure = 444.60°C.

**5. Silver point.** Temperature of equilibrium between solid silver and liquid silver at normal pressure = 960.8°C.

**6. Gold point.** Temperature of equilibrium between solid gold and liquid gold at normal pressure = 1063°C.

The scale is divided into four parts for the purpose of determining the temperature between two fixed points.

(*a*) **From – 190°C to 0°C.** The platinum resistance thermometer is used and the relation is

$$R_t = R_0\,[1 + \alpha t + \beta t^2 - \gamma\,(t - 100)\,t^3]$$

(*b*) **From 0°C to 660°C.** The platinum resistance thermometer is used and the relation is

$$R_t = R_0\,(1 + \alpha t + \beta t^2)$$

(*c*) **From 660°C to 1063°C.** The thermocouple of platinum and an alloy of platinum-rhodium is used. One junction is kept in melting ice. The thermo-E.M.F. produced is measured and the relation used is

$$E = \alpha + \beta t + \gamma t^2$$

(*d*) **Beyond 1063°C.** The ratio of the intensities of monochromatic heat radiation emitted by the black body at $t$°C and at 1063°C is determined and the temperature is calculated using Planck's radiation law.

### Absolute Zero and Ice Point

For approximate purposes, the value of absolute zero or zero degree kelvin is taken as – 273°C. But near about absolute zero its correct determination is a necessity. For accurate work the value of absolute zero is taken as – 273.16°C. Similarly the temperature of melting point of ice was suggested by the International Advisory Committee of Thermometry in 1948, as 273.15 K. But in 1954, the International Committee fixed 273.16 K (triple point of water) as ice point. Thus for accurate purposes,

Ice point = 0°C = 273.16 K

Steam point = 100°C = 373.16 K

Zero degree Kelvin = – 273.16°C

## 13.26 Low Temperature Measurement

The various thermometers used to measure low temperatures are discussed below:

**1.** *Liquid Thermometers.* With alcohol thermometer temperatures up to – 100°C can be measured. With mercury in glass, temperatures up to – 30°C can be measured. The accuracy is about 0.1°C. These are convenient and their response is quick.

**2.** *Gas Thermometers.* Constant volume hydrogen thermometer can be used to measure up to – 250°C and with constant volume, helium thermometer temperatures up to – 268°C can be determined. These thermometers are quite accurate but they are quite bulky.

**3.** *Resistance Thermometers.* With platinum resistance thermometers temperatures up to – 190°C can be measured accurately. Its accuracy is 0.01°C. It is quite accurate but its response is slow.

**4.** *Thermo-electric Thermometers.* With copper-constantan thermocouple or platinum and silver thermocouple, temperatures up to –250°C can be measured. Its accuracy is 0.05°C.

**5.** *Vapour Pressure Thermometers.* These can be used to measure up to – 268°C. Below – 268°C, helium vapour pressure thermometer is used. It can be used up to – 272°C.

**6.** *Magnetic Thermometer.* Near about the absolute zero temperature, magnetic thermometers are used. They are based upon the principle of change in susceptibility with temperature according to Curie's law.

Magnetic Thermometor

## 13.27 High Temperature Measurement

The various thermometers used to measure high temperatures are given below:

**1.** *Liquid Thermometers.* Mercury in glass can measure up to 300°C. If the space above mercury is filled with some inert gas like nitrogen or helium, it can be used up to 600°C.

**2.** *Gas Thermometers.* Constant volume hydrogen thermometer having a platinum bulb, can be used to measure up to 500°C. With a porcelain bulb temperatures up to 1100°C can be measured. Beyond 1100°C, using nitrogen in place of hydrogen, temperatures up to 1500°C can be measured. They are quite bulky for use.

**3.** *Resistance Thermometers.* The platinum thermometer can be used up to 1200°C when properly calibrated. Its accuracy is 0.1°C. It is a slow measuring instrument.

**4.** *Thermo-electric Thermometers.* For measuring up to 300°C copper-constantan thermocouple is used. Iron-nickel thermocouple is used between 300°C and 600°C. Nickel-nichrome thermocouple is used between 600°C and 1000°C. A thermocouple of platinum and rhodium is used between 1000°C and 1600°C. Between 1600°C and 2000°C a thermocouple of iridium and an alloy of iridium and rubidium is used. For measuring between 2000°C and 3000°C, tungsten and molybdenum thermocouple is used.

**5.** *Pyrometers.* For measuring the temperature of furnaces and the sun, optical pyrometers are used.

## NUMERICAL EXAMPLES

### EXAMPLE 13.9

*The resistance of the platinum wire of a platinum resistance thermometer at the ice point is 5* ohms *and at the steam point 5.93 ohms. The pressure exerted by the gas in a constant volume gas thermometer is (i) 100 cm of Hg at ice point (ii) 136.6 cm of Hg at the steam point. When both the thermometers are inserted in a hot bath the resistance of the platinum wire is 5.795 ohms and the pressure of the gas is 131.11 cm of Hg. Calculate Celsius temperature of the both (i) on the platinum scale and (ii) on the gas scale.*

**SOLUTION.**

(*i*) On the platinum scale

$$R_0 = 5 \text{ ohms}$$
$$R_{100} = 5.23 \text{ ohms}$$

$$R_t = 5.795 \text{ ohms}$$

$$t_p = \left(\frac{R_t - R_0}{R_{100} - R_0}\right) \times 100$$

$$= \left(\frac{5.795 - 5}{5.93 - 5}\right) \times 100 = \mathbf{85.48°C}$$

(*ii*) On the gas scale

$$P_0 = 100 \text{ cm Hg}$$

$$P_{100} = 136.6 \text{ cm Hg}$$

$$P_t = 131.11 \text{ cm Hg}$$

$$t = \left(\frac{P_t - P_0}{P_{100} - P_0}\right) \times 100$$

$$= \left(\frac{131.11 - 100}{136.6 - 100}\right) \times 100 = \mathbf{85°C}$$

**EXAMPLE 13.10**

*If the platinum temperature, corresponding to 50°C on the gas scale is 50.25°, what will be the temperature on the platinum scale corresponding to 150°C on the gas scale?*

**SOLUTION.** Let θ be the temperature on the gas scale and *t* on the platinum scale.

(*i*) Let
$$\theta - t = \delta\left[\left(\frac{\theta}{100}\right)^2 - \left(\frac{\theta}{100}\right)\right]$$

$$\theta = 50° \; ; \; t = 50.25$$

$$\therefore \quad 50 - 50.25 = \delta\left[\left(\frac{50}{100}\right)^2 - \left(\frac{50}{100}\right)\right] \quad \text{...(i)}$$

(*ii*)
$$\theta = 150° \; ; \; t = x$$

$$\therefore \quad 150 - x = \delta\left[\left(\frac{150}{100}\right)^2 - \left(\frac{150}{100}\right)\right] \quad \text{...(ii)}$$

Dividing (*ii*) by (*i*) and simplifying

$$x = \mathbf{149.25°C}$$

**EXAMPLE 13.11**

*The resistance of a platinum wire at 0°C, 100°C and 444.6°C is found to be 5.5, 7.5 and 14.5 ohms respectively. The resistance of a wire at a temperature t°C is given by the equation*

$$R_t = R_0 (1 + \alpha t + \beta t^2)$$

*Find the values of α and β.*

**SOLUTION.** Here

$$R_0 = 5.5 \text{ ohms}$$

$$R_{100} = 7.5 \text{ ohms}$$

$$R_{444.6} = 14.5 \text{ ohms}$$

$$R_t = R_0 (1 + \alpha t + \beta t^2)$$

$$\therefore \quad R_{100} = R_0 [1 + \alpha (100) + \beta (100)^2] \quad \text{...(i)}$$

$$R_{444.6} = R_0 [(1 + \alpha (444.6) + \beta (444.6)^2] \quad ...(ii)$$

∴ Substituting the values in equations (*i*) and (*ii*)

$$7.5 = 5.5 [1 + 100\alpha + (100)^2 \beta] \quad ...(iii)$$

$$14.5 = 5.5 [1 + 444.6\alpha + (444.6)^2 \beta] \quad ...(iv)$$

Simplifying equations (*iii*) and (*iv*)

$$\alpha + 100 \beta = \frac{2}{5.5 \times 100} \quad ...(v)$$

$$\alpha + 444.6 \beta = \frac{9}{5.5 \times 444.6} \quad ...(vi)$$

Subtracting equation (*v*) from (*vi*)

$$344.6 \beta = \frac{900 - 2 \times 444.6}{5.5 \times 444.6 \times 100}$$

∴ $$\beta = \mathbf{1.281 \times 10^{-7}/°C}$$

Substituting the value of β in equation (*v*)

$$\alpha + 100 (1.281 \times 10^{-7}) = \frac{2}{5.50 \times 100}$$

$$\alpha = 363.6 \times 10^{-5} - 1.281 \times 10^{-5}$$

$$\alpha = 362.319 \times 10^{-5}$$

$$\alpha = \mathbf{3.62 \times 10^{-3}/°C}$$

**EXAMPLE 13.12**

*The e.m.f. of a thermo-couple, one junction of which is kept at 0°C is given by*

$$E = at + bt^2$$

*Determine the neutral temperature, temperature of inversion and the Peltier and Thomson coefficients.* **(Kanpur 2001; Rohilkhand 2002 Improvement)**

**SOLUTION.** The e.m.f. of a thermo-couple is given by

$$E = at + bt^2 \quad ...(i)$$

Here *t* is in degree celcius. Let us convert it into degree Kelvin by $T = t + 273$ or $t = T - 273$. Substituting, we get

$$E = a (T - 273) + b (T - 273)^2 \quad ...(ii)$$

Differentiating w.r. to *T*, we get

$$\frac{dE}{dT} = a + 2b (T - 273) \quad ...(iii)$$

Again differentiating w.r. to *T*, we get

$$\frac{d^2E}{dT^2} = 2b \quad ...(iv)$$

(*i*) At the neutral temperature $T_n$,

$$T = T_n \text{ and } \frac{dE}{dT} = 0$$

Substituting in (*iii*), we get

$$0 = a + 2b (t_n - 273)$$

or $$T_n = \left(273 - \frac{a}{2b}\right) \text{K} = -\left(\frac{a}{2b}\right) °\text{C}$$

(*ii*) If $T_i$ is the temperature of inversion, then

$$T_n = \frac{273+T_i}{2}$$

or

$$T_i = 2\,T_n - 273 = 2\left(273-\frac{a}{2b}\right) - 273$$

$$= \left(273-\frac{a}{b}\right)\text{K} = -\left(\frac{a}{b}\right)^{\circ}\text{C}$$

(*iii*) The Peltier coefficient (π)

$$\pi = T\frac{dE}{dT}$$

$$= T\,[a + 2b\,(T-273)]$$

$$= (t + 273)\,[a + 2b\,(t)]$$

$$= \mathbf{(t + 273)\,(a + 2bt)}$$

(*iv*) The Thomson coefficient (σ)

$$\sigma = T\frac{d^2E}{dT^2}$$

$$= T\,(2b)$$

$$= \mathbf{2b\,(t + 273)}$$

## EXERCISE–XIII

**13.1** Write an essay on the measurement of high and low temperatures.

**13.2** State with reasons the type of thermometer which you consider most suitable for use at temperatures (*a*) – 250°C (*b*) 700°C and (*c*) 2000°C. Indicate briefly the methods of their use.

**13.3** Describe a platinum resistance thermometer. Explain how it works with the help of Callendar and Griffith's bridge. How does the platinum temperature of a body differ from its true temperature?

**13.4** What do you understand by the absolute scale of temperature? Is the negative temperature possible on this scale? **(Punjab 2002)**

**13.5** Describe the Callendar and Griffith's method of determining the temperature coefficient of platinum. In what respects it is superior to Carey Foster's method? What do you understand by (*i*) platinum leads (*ii*) compensating arm (*iii*) non-inductive winding of the platinum wire? **(Nagpur 2001, Calcutta 2001)**

**13.6** Give an account of the thermo-electric thermometry and discuss the range, sensitivity and usefulness of some important thermocouples.

**13.7** Describe carefully the methods for measuring low temperatures in the range – 100°C to – 273°C. Explain the concept of absolute zero of temperature.

**13.8** Describe a resistance thermometer. Explain how it is used to measure temperatures accurately. Discuss its advantages over a thermo-electric thermometer.

**13.9** Describe and explain how the platinum resistance thermometer can be used to measure temperatures accurately in the region 200°C to 500°C. **(Rajasthan 2001)**

**13.10** Give the construction of a thermo-electric thermo-meter and compare its performance with a standard gas thermometer. What precautions should be taken in measuring temperatures with a thermo-electric thermometer?

**13.11** What is meant by a scale of temperature? On what does the definition of any particular scale depend?

**13.12** Give a detailed account of the experiment you would perform to determine the temperature of boiling on the platinum resistance scale of temperature. By how much does the temperature measured *by* this thermometer differ from that measured by gas thermometer? **(Osmania 2000)**

**13.13** State briefly the principles underlying the working of the following thermometers:
(*i*) Resistance thermometer
(*ii*) Thermo-electric thermometer.

**13.14** Describe Callendar and Griffith's bridge for measuring the resistance of a platinum resistance thermometer at various temperatures.

**13.15** Describe a platinum resistance thermometer. How would you calibrate and use it for measuring the temperature of a body? Mention the advantages of this thermometer. **(Madras 2002, 03, 05)**

**13.16** On what principle is the working of a platinum resistance thermometer based? Describe Callendar and Griffith's bridge for the accurate measurement of resistance. How is the true temperature deduced from the measured platinum temperature?

**13.17** Describe Seebeck effect and explain the terms neutral temperature and temperature of inversion. **(Osmania 2004; Kanpur 2001; Pune 2000)**

**13.18** Explain Seebeck, Peltier and Thomson effects. What are their causes of existence? Are Peltier and Thomson effects reversible? **(Punjab 2002, 2003; Kanpur 2001, Osmania 2005)**

**13.19** Define Peltier coefficient, Thomson coefficient and thermo-electric power. **(Osmania 2004; Bangalore 2004, Punjab 2001)**

**13.20** Discuss the thermo-electric diagram and give its various uses. **(Osmania 2004, Lucknow 2002)**

**13.21** Explain Seebeck effect and give some practical applications of this effect. **(Rohilkhand 2002; Kanpur 2000)**

**13.22** What do you understand by the neutral temperature of a thermo-couple? Explain.

**13.23** Discuss Seebeck and Peltier effects and explain why thermo-e.m.f. is produced. Discuss the nature of the thermo-e.m.f. curve obtained in the case of copper-constantan thermo-couple when difference of temperature of two junctions of thermo-couple is increased. **(Rohilkhand 2001; Kanpur 2002)**

**13.24** What do you understand by thermo-electric power? Obtain an expression for the total e.m.f. of a thermo-couple. **(Osmania s/2005)**

**13.25** Write short notes on :
(*i*) Callendar's compensated constant pressure air thermometer.
(*ii*) Platinum resistance thermometer.
(*iii*) Thermo-electric thermometer.
(*iv*) Seebeck and Peltier effects **(Rohilkhand 2001)**
(*v*) Seebeck effect
(*vi*) Thermoelectric diagram **(Rohilkhand 2001 Improvement)**
(*vii*) Universal gas constant.

**13.26** The temperature of a furnace is 2000°C. What is this temperature (*i*) on the Rankine scale and (*ii*) on the Kelvin scale? **[Ans. (i) 4092°R ; (ii) 2273 K]**

**13.27** The normal boiling point of liquid hydrogen is – 253°C. What is this temperature (*i*) on the Kelvin scale and (*ii*) on the Rankine scale? **[Ans. (i) 20 K ; (ii) 36.6°R]**

**13.28** At what temperature will the Kelvin scale reading be double the Fahrenheit reading? **[Ans. 176.7°F = 353.4 K]**

**13.29** Find the value of the universal gas constant. **[Ans. R = 8.31 × 107 ergs/per g mol. K]** **(Madras 2004; Punjab 2002)**

**13.30** The resistance of a platinum resistance thermometer at the ice point is 5 ohms and at the steam point is 6.93 ohms. The pressure exerted by the gas in constant volume gas thermometer at the ice point is 100 cm of Hg and at the steam point it is 136.6 cm of Hg. When both the thermometers are placed in a bath, the resistance of a resistance thermometer is 5.795 ohms and the pressure of the gas is 114.9 cm of Hg. Calculate the celsius temperature of the liquid (*i*) on the platinum scale and (*ii*) on the gas scale.
**[Ans. $t_p$ = 41.19°C ; θ = 40.71 °C]**

**13.31** The bulb of the Callendar's compensated constant pressure air thermometer is 800 cm³. When the bulb is immersed in a bath, 200 cm³ of mercury has to be drawn out of the reservoir. Calculate the temperature of the bath on the celsius scale. **[Ans. 33.33°C]**

**13.32** If the platinum temperature corresponding to 60°C on the gas scale is 60.25°C, what will be the temperature on the platinum scale corresponding to 120°C on the gas scale? **[Ans. 119.25°C]**

**13.33** If the platinum temperature corresponding to 60°C on the gas scale is 60.36°C, what is the platinum scale temperature corresponding to 151.7°C on the gas scale? **[Ans. 150°C]**

**13.34** The e.m.f. in microvolt of a thermo-couple, one junction of which is at 0ºC is given by

$$E = 1600t - 4t^2$$

where $t$ºC is the temperature of the hot junction. If $t$ = 27ºC, calculate Peltier and Thomson coefficients.
**[Ans. 0.479 volt, 2.4 × 10⁻⁵ volt/K] (Garhwal 2001)**

**13.35** If the thermo-electric power of iron is (1734 – 4.87$t$) e.m.u./ºC and that of copper is (136 + 0.95$t$) e.m.u./ºC where $t$ is the temperature on celcius scale. Find the e.m.f. of their thermo-couple, the junctions of which are at 0ºC and 100ºC in microvolt.
**[Ans. 1402 microvolt] (Rohilkhand 2002 Improvement)**

**13.36** The thermo-electric power of iron is 1734 microvolt at 0ºC and 1247 microvolt at 100ºC, that of copper is 136 microvolt at 0ºC and 231 microvolt at 100ºC. Calculate the e.m.f. of iron-copper couple between the temperature 0ºC and 100ºC. **[Ans. 130.7 millivolt] (Ranchi 2001)**

## OBJECTIVE QUESTIONS

**13.1** If a current is allowed to pass through a circuit consisting of two dissimilar metals, there is either an evolution or absorption of heat at the junction, depending upon the direction of current. The effect is known as **(Kanpur 2001)**

(*a*) Peltier effect (*b*) Seebeck effect
(*c*) Thomson effect (*d*) Joules effect

**13.2** The specific heat of electricity is also known as **(Rohilkhand 2003)**

(*a*) Hall coefficient (*b*) Seebeck coefficient
(*c*) Peltier coefficient (*d*) Thomson coefficient

**13.3** The relation between the neutral temperature ($T_n$) and the temperature of inversion is given by **(Rohilkhand 2002 Improvement)**

(*a*) $T_i = T_n$ (*b*) $T_i = \frac{T_n}{2}$
(*c*) $T_i = 2T_n$ (*d*) None of these

**13.4** Peltier coefficient is given by

(*a*) $\pi = \frac{1}{T}\frac{dE}{dT}$ (*b*) $\pi = \frac{dE}{dT}$
(*c*) $\pi = T\frac{dE}{dT}$ (*d*) $\pi = T^2\frac{dE}{dT}$

**13.5** Thomson effect is

(*a*) reversible (*b*) irreversible
(*c*) neither reversible or irreversible

**13.6** In copper-iron thermo-couple, the current flows from **(Rohilkhand 2002 Improvement)**

(*a*) copper to iron through hot junction (*b*) copper to iron through cold junction
(*c*) iron to copper through hot junction (*d*) iron to copper through cold junction

**13.7** For lead (Pb), the Thomson coefficient is **(Rohilkhand 2003)**

(*a*) positive (*b*) zero
(*c*) negative (*d*) none of these

**13.8** When two different metals are joined to form a closed circuit an e.m.f. is developed if there is a temperature difference between junctions, it is: **(Rohilkhand 2003)**

(*a*) Peltier effect (*b*) Seebeck effect
(*c*) Thomson effect (*d*) None of the above

**13.9** Which of the following is not reversible

(*a*) Joules effect (*b*) Peltier effect
(*c*) Seebeck effect (*d*) Thomson effect

**13.10** Thermo-electric power in a thermo-couple is equal to the rate of change of its. e.m.f. with **(Rohilkhand 2002 Improvement)**

(*a*) temperature difference (*b*) temperature of cold junction
(*c*) temperature of hot junction (*d*) none of these

**13.11** Seebeck effect may be used as

(*a*) a thermo-electric generator (*b*) thermo-electric refrigerator
(*c*) both (*d*) none

**13.12** Comparing with a liquid, it can be shown that Peltier coefficient $\pi$ is analogous to

(*a*) Specific heat (*b*) Latent heat
(*c*) Entropy (*d*) None of these

## ANSWERS

| | | | |
|---|---|---|---|
| **13.1** (*a*) | **13.2** (*d*) | **13.3** (*c*) | **13.4** (*c*) |
| **13.5** (*a*) | **13.6** (*a*) | **13.7** (*b*) | **13.8** (*b*) |
| **13.9** (*a*) | **13.10** (*c*) | **13.11** (*a*) | **13.12** (*b*) |

# Calorimetry 14

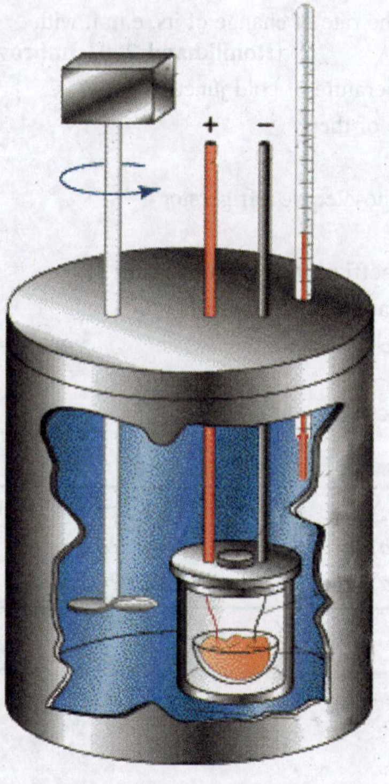


### At a Glance

• Definitions • Regnault's Method of Mixtures • Copper Block Calorimeter • Nernst Vacuum Calorimeter • Newton's Law of Cooling • Sp. Heat of a Liquid-Joule's Electrical Method • Sp. Heat of a Liquid-Callendar and Barne's Continuous Flow Method • Experimental Determination of Heat Capacities • Adiabatic Vacuum Calorimeter • Two Specific Heats of a Gas • $C_V$ of a Gas by Joly's Differential Steam Calorimeter • $C_p$ of a Gas by-Regnault's Method • Continuous Flow Electrical Method • Specific Heat of a Gas at Low Temperatures • Calorific Value of Fuels • Bell Calorimeter • Dulong and Petit's Law • Variation of Specific Heat and Atomic Heat with Temperature • Einstein's Theory of Specific Heat of Solid • Debye's Theory of Specific Heat of Solid

## 14.1 Definitions

When a body is heated, its temperature rises. If 100 g of copper and 100 g of water are heated by similar burners for the same time, the rise in temperature is not the same in the two cases. The rise in temperature depends on the quantity of heat given to the body and the nature of its material. Let $Q$ be the quantity of heat given to a body of mass $m$ and let the rise in temperature be $\theta$.

Then, $Q = mC\theta$

where $C$ is a constant that depends upon the nature of the substance. $C$ is called the specific heat of the substance.

**Calorie.** It is the quantity of heat required to raise the temperature of one gram of water from 14.5°C to 15.5°C. This is the standard unit recommended by International Committee of Pure Physics.

For ordinary purposes, the specific heat of water is taken as 1 but specific heat of water is not 1 at all temperatures. For practical purposes *Calorie* may be defined as the amount of heat required to raise the temperature of 1 gram of water through 1°C.

**Kilogram Calorie.** It is the amount of heat required to raise the temperature of 1 kg of water through 1°C.

1 kg calorie = 1000 calories = 1 kilocalorie

**British Thermal Unit.** It is the amount of heat required to raise the temperature of 1 pound of water through 1°F.

1 *B.T.U.* = 252 calories

**Therm.** It is the amount of heat required to raise the temperature of $10^5$ pounds of water through 1°F.

1 Therm = $2.52 \times 10^7$ calories

**Pound Calorie.** It is the amount of heat required to raise the temperature of 1 pound of water through 1°C.

1 pound calorie = 453.6 calories

It is called centigrade heat unit.

**TABLE 14.1** **Units of Heat**

| *Unit* | *Quantity of water* | *Rise in temp.* | *Relation* |
|---|---|---|---|
| Calorie | 1 g | 1°C | = 1 Calorie |
| Kg Calorie | 1 kg | 1°C | = 1000 Calories |
| BTU | 1 pound | 1°F | = 252 Calories |
| Therm | $10^5$ pounds | 1°F | = $2.52 \times 10^7$ Calories |
| Pound Calorie | 1 pound | 1°C | = 453.6 Calories |

**Specific Heat.** It is defined as the quantity of heat required to raise the temperature of unit mass of a substance through one degree.

Suppose,

Mass of the substance = $m$

Sp. heat of the substance = $C$

Rise in temperature = $\theta$

$$Q = mC\theta$$

or
$$C = \frac{Q}{m\theta}$$

*Unit of specific heat.* In C.G.S. system, the unit of $Q$ is in calories, $m$ is in grams and $\theta$ is in °C. therefore the unit of specific heat will be

$$C = \frac{\text{calorie}}{g\ °C}$$

$$= \text{calorie/g °C}$$

The specific heat of a substance is not constant and it is different at different temperatures.

Ordinarily, the specific heat determined is the mean specific heat. Suppose, $m$ is the mass of the substance, $C$ the mean specific heat and if $Q$ units of heat is required to raise its temperature from $\theta_1$ to $\theta_2$, then the mean specific heat

$$C = \frac{Q}{m(\theta_2 - \theta_1)}$$

For qualitative work, if $dQ$ heat is given to raise the temperature of $mg$ of a substance through $d\theta$.

$$dQ = m\,C\,d\theta$$

$$C = \frac{1}{m} \cdot \frac{dQ}{\theta}$$

**Thermal Capacity.** It is the quantity of heat required to raise the temperature of the whole of the substance through 1°C. Let the mass of the substance be $m$ and its specific heat $C$.

$$\text{Thermal Capacity} = m \times C \times 1$$
$$= m\,C \text{ calories/°C}$$

**Water Equivalent.** It is the amount of water that will absorb the same quantity of heat as the substance for the same rise in temperature. Let the mass of the substance be $m$, specific heat $C$ and rise in temperature $\theta$.

$$Q = m\,C\,\theta$$

If the water equivalent $= w$

$$Q = w \times 1 \times \theta$$

$\therefore$ $$w \times 1 \times \theta = m\,C\,\theta$$

or $$w = m\,C \text{ grams}$$

Water equivalent is numerically equal to the thermal capacity but the unit of water equivalent is grams and that of thermal capacity is calories.

## 14.2 Regnault's Method of Mixtures

**Specific heat of solids.** Regnault was the first (year 1840) to devise the apparatus commonly used in the laboratories to find the specific heat of a solid or liquid employing the *method of mixtures*. In this the given solid is heated to a constant high temperature and then it is quickly transferred into a calorimeter containing water at room temperature. The final temperature of the mixture is noted. From the principle that heat lost by the solid is equal to the heat gained by the calorimeter and contents, the specific heat of the given solid is calculated.

**Theory.** The theory is based on the principle of heat transfer from high temperature to low temperature. The method consists in imparting the quantity of heat to be measured to a certain mass of water contained in a vessel of known thermal capacity and measuring the rise of temperature produced.

If $m_1$ is the mass of a given substance of specific heat $C_1$ and initial temperature $\theta_1$ be plunged into $m_2$ gram of water at temperature $\theta_2$ and $W$ be the thermal capacity (also known as water equivalent) of the calorimeter and $\theta$ be the final temperature of the mixture. Then

Heat lost by the substance,

$$Q_1 = m_1 C_1 (\theta_1 - \theta) \qquad ...(i)$$

and heat gained by water and calorimeter,

$$Q_2 = (m_2 + W)(\theta - \theta_2) \qquad ...(ii)$$

Equating squating eqn. (*i*) and (*ii*), we have

$$m_1C_1(\theta_1 - \theta) = (m_2 + W)(\theta - \theta_2)$$

or

$$C_1 = \frac{(m_2 + W)(\theta - \theta_2)}{m_1(\theta_1 - \theta)} \quad ...(iii)$$

Thus, the specific heat $C_1$ of any unknown substance can be determined by the method of mixtures. Various corrections are, however, necessary, for heat is lost by the system by conduction, convection and radiation. Therefore, for θ, we must put (θ + Δθ), where Δθ is the correction.

**Radiation Correction.** Regnault made use of a more accurate but slightly more elaborate method of *radiation correction*, in the case of solids like rubber, glass etc. A sensitive thermometer capable of measuring 1/10th °C is used. Readings of the mixture are taken at every 1/2 minute internal just after the solid is dropped into the calorimeter. The readings are taken until the calorimeter attains the maximum temperature and then cools down, say by 1°C.

Let θ be the room temperature, $t_1$ be the maximum temperature observed and $n$ the total number of half minute intervals taken to cool down by 1°C. Then the rate of cooling per interval at the mean temperature $\frac{t_1 + (t_1 - 1)}{2}$ °C is $\frac{1}{n}$, while it is zero at θ °C. According to Newton's law, the rate of cooling is proportional to the mean difference of temperature between the cooling body and the surroundings. Hence a straight line graph is drawn with temperature along the *x*-axis and the rate of cooling corresponding to the mean temperature for any interval recorded can be found. The correction for radiation is then made as follows :

**TABLE 14.2**

| *Time Interval* | *Initial Temp.* | *Final Temp.* | *Mean Temp.* | *Rate of Cooling from Graph* | *Correction* | *Corrected Final Temp.* |
|---|---|---|---|---|---|---|
| 0 to 1 | $\theta_0$ | $\theta_1$ | $\frac{(\theta_0 + \theta_1)}{2}$ | $x_1$ | $x_1$ | $\theta_1 + x_1$ |
| 1 to 2 | $\theta_1$ | $\theta_2$ | $\frac{(\theta_1 + \theta_2)}{2}$ | $x_2$ | $x_1 + x_2$ | $\theta_2 + x_1 + x_2$ |
| 2 to 3 and so on. | $\theta_2$ | $\theta_3$ | $\frac{(\theta_2 + \theta_3)}{2}$ | $x_3$ | $x_1 + x_2 + x_3$ | $\theta_3 + x_1 + x_2 + x_3$ |

For the first interval, $x_1$ represents the amount by which the temperature at the end of the interval viz. $\theta_1$ will be less that what it would have been in the absence of loss by conduction, radiation etc. Hence the corrected final temperature for this interval is $\theta_1 + x_1$.

This is how we calculate the corrected final temperatures for 2$^{nd}$, 3$^{rd}$, 4$^{th}$, 5$^{th}$, ....... time intervals. The values in the last column at a certain stage become practically constant giving the maximum temperature the calorimeter and contents would have attained, if there had been no loss of heat.

## 14.3 Copper Block Calorimeter

Specific heat of a solid at high temperature cannot be determined by ordinary apparatus used in the laboratory. Steam or vapours produced would cause considerable loss of heat when a hot solid is mixed with water or a liquid. The result obtained will not be accurate. To overcome this difficulty, *Nernst* and *Lindemann* designed a calorimeter called the *copper block calorimeter.*

It consists of a Dewar flask *A* having a copper block *C* fixed inside it. *T, T* are the terminals of the leads of the thermocouple to measure the temperature of the block *C*. The flask is covered from outside to prevent any loss of heat to the surroundings. Through the tube *B*, a hot solid can be dropped (Fig. 14.1).

The substance whose specific heat is to be determined is heated to a known high temperature and then gently dropped through the tube *B* into the block *C*. The copper block has high thermal conductivity and a uniform final temperature is reached in a short time. The temperature is measured with the help of a calibrated thermocouple, the copper block *C* acting as the hot end. To avoid air currents the tube *B* is closed with an automatic electrical arrangement after the substance has been dropped into the block *C*. The whole apparatus can be maintained at a desired temperature. From the heat gained by copper block and the heat lost by the substance, the specific heat of the substance can be calculated.

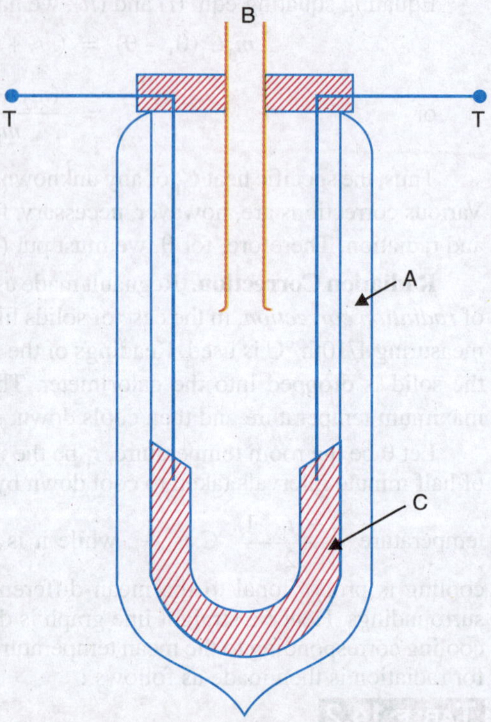

Fig. 14.1

Magnus determined the specific heat of platinum and other metals up to 900°C. Jaeger and his co-workers determined the specific heat of $P_t$, tungsten *Rh*, *Ir* and other metals up to 1600°C. In order to find the specific heat at high temperatures, say between 700°C and 800°C, the experiment is performed twice, first at 700°C and then at 800°C. The mean value of the specific heat is calculated.

## 14.4 Nernst Vacuum Calorimeter

This calorimeter is used to find the specific heat of solids at low temperature. Nernst and Lindemann performed a number of accurate experiments to determine the specific heat of solids at low temperatures.

The Nernst calorimeter consists of an evacuated flask D in which the metal whose specific heat is to be determined, is suspended as shown in Fig. 14.2. *P* and *Q* are two non-conducting loops. The cylinder *B* and the cylindrical plug *A* are made of the same metal. The heading coil C is made of platinum and is wound around, *A*. the leads *LL* are connected to an electrical heating arrangement. The cylinder A fits completely inside the cylinder B. The heating coil C also serves as a platinum resistance thermometer to measure the temperature of the metal.

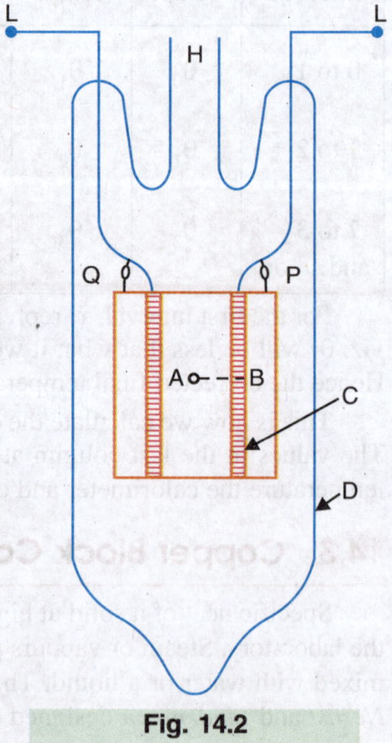

Fig. 14.2

Suppose the resistance of the platinum wire is $R_1$ at temperature $\theta_1$. $R_1$ is found by measuring the current in the wire and the potential difference across the two ends of the wire. Let the potential difference be *E* and the current $I_1$.

$$\therefore \quad R_1 = \frac{E}{I_1} \quad \text{(At temperature } \theta_1\text{)}$$

Similarly resistance $R_2$ at temperature $\theta_2$ is given by

$$R_2 = \frac{E}{I_2} \quad \text{(At temperature } \theta_2\text{)}$$

The platinum wire is initially calibrated and from these values corresponding to $R_1$ and $R_2$ the values $\theta_1$ and $\theta_2$ are found.

The flask $D$ is kept in a bath containing ice, liquid air etc., depending upon the low temperature at which the specific heat of the given metal is to be determined. After the flask $D$ has attained the temperature of the bath, it is evacuated. This initial temperature of the metal is found. Let it be $\theta_1$. Electric current is passed through the coil $C$ for a known time so that the rise in temperature is about 1°C. Suppose the final temperature of the metal is $\theta_2$.

Mass of the metal = $M$ grams

Specific heat = $C$

Rise in temperature = $(\theta_2 - \theta_1)$

Voltmeter reading = $E$ volts

Ammeter reading = $I$ ampere

Time for which current is passed = $t$ seconds

$$\therefore \quad MC(\theta_2 - \theta_1) = \frac{EIt}{4.2}$$

$$C = \frac{EIt}{4.2\,M(\theta_2 - \theta_1)}$$

In this experiment, the heat losses are practically eliminated as the experiment is performed in vacuum. However, a correction is required for the little loss of heat due to radiation.

To determine the specific heat of non-metals, the substance is kept inside a silver vessel. The heating coil is wound round the silver vessel. The experiments is performed in vacuum by suspending the silver vessel inside the flask $D$.

**TABLE 14.3 Specific Heats and Atomic Heats of Solids at 20 °C**

| *Element* | *Atomic Weight* | *Specific Heat* | *Atomic Heat* |
|---|---|---|---|
| Aluminium | 27.0 | 0.212 | 5.724 |
| Boron | 11.0 | 0.307 | 3.377 |
| Carbon | 12.0 | 0.160 | 1.92 |
| Copper | 63.6 | 0.091 | 5.787 |
| Gold | 197.2 | 0.031 | 6.113 |
| Iron | 55.8 | 0.110 | 6.138 |
| Lead | 207.2 | 0.030 | 6.216 |
| Silicon | 28.3 | 0.182 | 5.150 |
| Silver | 107.9 | 0.056 | 6.042 |
| Tin | 118.0 | 0.054 | 6.372 |

## 14.5 Newton's Law of Cooling

Newton's Law of Cooling states that the rate of loss of heat of a body is directly proportional to the difference of temperature of the body and the surroundings. The law holds good only for small difference of temperature. Also, the loss of heat by radiation depends upon the nature of the surface of the body and the area of the exposed surface.

$$\frac{dQ}{dt} \propto (\theta - \theta_0) \quad or \quad -\frac{dQ}{dt} = k\,(\theta - \theta_0)$$

Consider a body of mass *m*, specific heat *C* and at temperature θ. Let $\theta_0$ be the temperature of the surroundings. Suppose, the temperature falls by a small amount *d*θ in time *dt*. Then the amount of heat lost.

$$dQ = mCd\theta$$

∴ Rate of loss of heat

$$\frac{dQ}{dt} = mC\,\frac{d\theta}{dt} \qquad ...(i)$$

From Newton's law of cooling

$$-\frac{dQ}{dt} = k\,(\theta - \theta_0) \qquad ...(ii)$$

where *k* is a constant depending upon the area and the nature of the surface of the body.

From (*i*) and (*ii*)

$$-mC\frac{d\theta}{dt} = k\,(\theta - \theta_0)$$

or

$$\frac{d\theta}{\theta - \theta_0} = -\frac{k}{mC}.\,dt = -\,K.dt \qquad ...(iii)$$

Integrating,

$$\log(\theta - \theta_0) = -\,Kt + c \qquad ...(iv)$$

If a graph is plotted between *t* along the *Z*-axis and log $(\theta - \theta_0)$ along the *Y*-axis, it is a straight line. Hot water is taken in a calorimeter and is placed in a double walled vessel. Temperature of water after regular intervals of time is noted. A graph between log $(\theta - \theta_0)$ and time *t* is plotted (Fig. 14.3 (*i*)). It is a straight line. This verifies Newton's law of cooling.

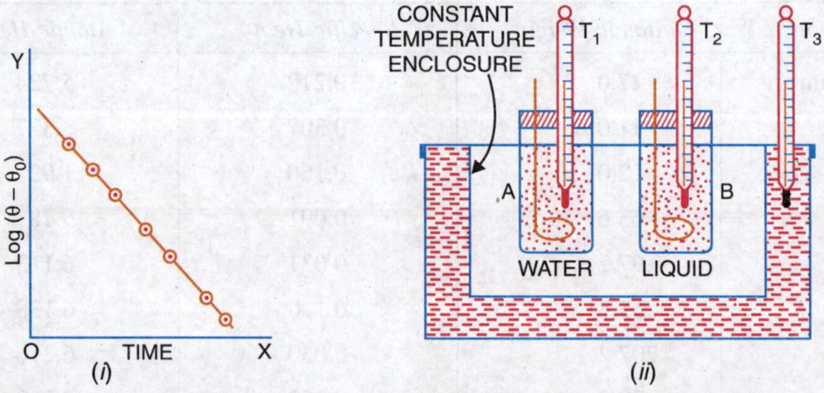


**Fig. 14.3**

**Specific heat of a liquid.** *A* and *B* are two identical calorimeters containing equal volumes of hot water and the hot liquid respectively. The two calorimeters are made of the same material and their outer surfaces are equally polished. The calorimeters are kept inside a constant temperature enclosure. The thermometers $T_1$ and $T_2$ measure the temperature of water and liquid (Fig. 14.3 (*ii*)). The temperature of the two calorimeters are noted after regular intervals of time (say one minute). Graphs are plotted between temperature and time, for water and the liquid (Fig. 14.4).

From equation (*iii*)

$$\int \frac{d\theta}{\theta - \theta_0} = \frac{k}{mC}\int dt$$

In the case of water, suppose

Mass of water $= m$

Water equivalent of the calorimeter $A = w$

Time taken by water to cool from 60°C to 55°C $= t_1$

$$\therefore \quad \int_{60}^{55} \frac{d\theta}{\theta - \theta_0} = -\frac{k}{(m+w)} . t_1 \quad ...(v)$$

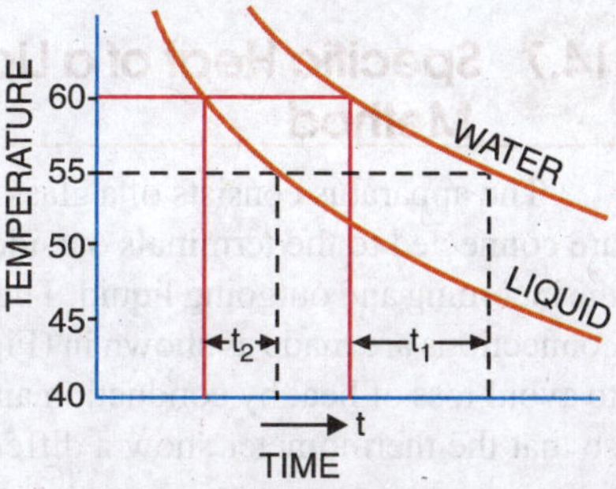

Fig. 14.4

Suppose mass of the liquid $= M$

Water equivalent of the calorimeter $B = w$

Specific heat of the liquid $= C$

Time taken by the liquid to cool from 60°C to 55°C $= t_2$

$$\therefore \quad \int_{60}^{55} \frac{d\theta}{\theta - \theta_0} = -\frac{k}{(MC+w)} . t_2 \quad ...(vi)$$

From equations (*v*) and (*vi*)

$$\frac{MC + w}{t_2} = \frac{(m + w)}{t_1}$$

or
$$C = \frac{(m + w)\, t_2}{Mt_1} - \frac{w}{M} \quad ...(vii)$$

## 14.6 Specific Heat of a Liquid–Joule's Electrical Method

The apparatus consists of a calorimeter in which a heater coil (wire) of resistance *R* is enclosed. The two ends of the wire are connected to the terminals on the lid. The calorimeter is enclosed in a wooden box. The wire R is connected in series with a battery, a key, a rheostat and an ammeter. A voltmeter is connected parallel to the wire (Fig. 14.5).

The liquid whose specific heat is to be determined is taken in the calorimeter. Current is passed through the wire for a known interval of time. The rise in temperature of the calorimeter and the liquid is noted with the help of a thermometer. The current is passed for such a time that the rise in temperature is about 10°C.

Suppose, the mass of the liquid is *M*, specific heat of the liquid is *C* and water equivalent of the calorimeter is *w*.

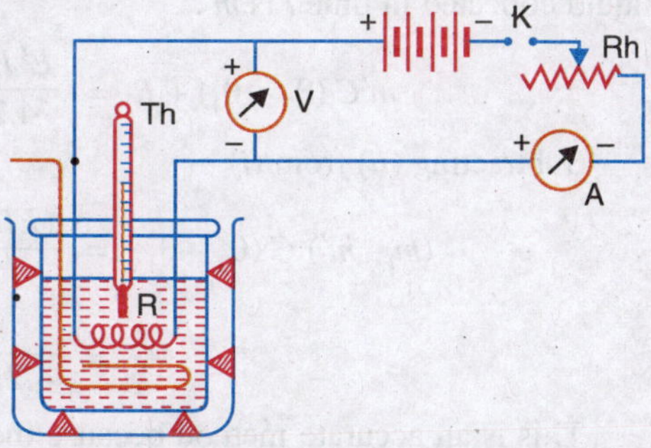

Fig. 14.5

Initial temperature of the liquid $= \theta_1$ °C

Final temperature of the liquid $= \theta_2$ °C

P.D. across the wire $= E$ volts

Current flowing $= I$ amperes

Time $= t$ seconds

Heat produced $= \dfrac{E.I.t}{4.2}$ calories

Heat gained by the liquid and the calorimeter

$$= (MC + w)(\theta_2 - \theta_1)$$

$$\therefore \quad (MC + w)(\theta_2 - \theta_1) = \frac{E.I.t}{4.2}$$

$$C = \frac{E.I.t}{4.2\,(\theta_2 - \theta_1)\, M} - \frac{w}{M}$$

## 14.7 Specific Heat of a Liquid–Callendar and Barnes' Continuous Flow Method

The apparatus consists of a glass tube in which a resistance wire is enclosed. The ends of the wire are connected to the terminals outside the tube. Thermometers $T_1$ and $T_2$ measure the temperature of the incoming and outgoing liquid. The liquid is passed through the tube at a uniform rate. The electrical connections are made as shown in (Fig. 14.6). A vacuum jacket surrounding the glass tube is provided to avoid loss of heat by conduction and convection. Current and the rate of flow of water are adjusted so that the thermometer show a difference of about 10°C.

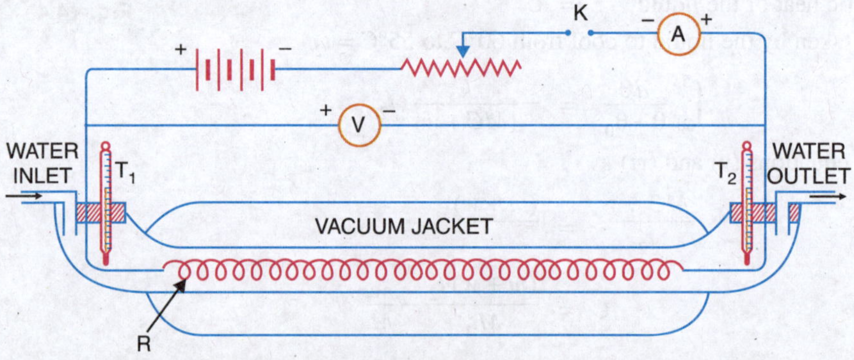


**Fig. 14.6**

Let, $E, I$ and $t$ be the potential difference, current and time for which the liquid is collected. If the mass of liquid collected is $m$ and the difference of temperature between the two thermometers is $(\theta_2 - \theta_1)$, then

$$mC(\theta_2 - \theta_1) + R = \frac{EIt}{4.2} \qquad ...(i)$$

where $R$ is the loss of heat by radiation. The current through the wire is changed. The rate of flow is adjusted so that the difference of temperature $(\theta_2 - \theta_1)$ remains the same. Suppose the amount of liquid collected in time $t$ is $m'$.

$$\therefore \quad m'C(\theta_2 - \theta_1) + R = \frac{E'I't}{4.2} \qquad ...(ii)$$

Subtracting (*ii*) from (*i*)

$$(m - m')\,C(\theta_2 - \theta_1) = \frac{(EI - E'I')t}{4.2}$$

$$C = \frac{(EI - E'I')t}{4.2(m - m')(\theta_2 - \theta_1)}$$

This is an accurate method because the observations are taken under steady state. The thermal capacity of the apparatus does not occur in the calculations. Moreover, the errors due to loss of heat by conduction, convection and radiation have been practically eliminated.

## 14.8 Experimental Determination of Heat Capacities

In modern methods, measurement of heat capacity of substances involves the supply of heat to the system from an electrical source. From the energy received from the source in a given interval of time, a part of it may be used in raising the temperature of the system and the remainder is lost to the surroundings at a lower temperature (Fig. 14.7).

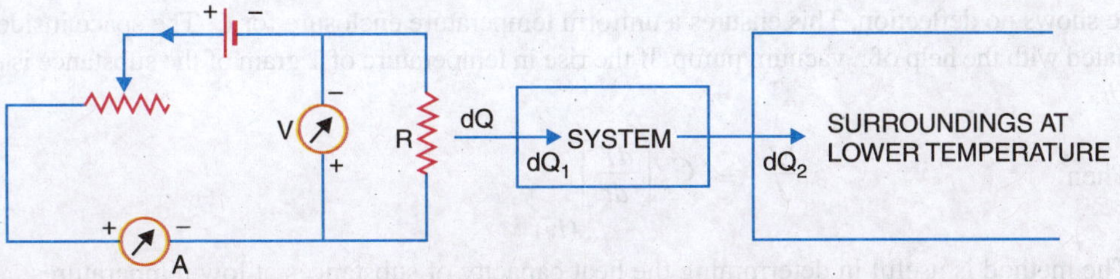


**Fig. 14.7**

Let $dQ$ be the quantity of heat drawn from the electrical source in time $dt$. Here $dQ_1$ is the amount of heat retained by the system to raise its temperature by $d\theta$ and $dQ_2$ is the amount of heat lost to the surroundings.

$$dQ = dQ_1 + dQ_2$$

$$dQ_1 = 1 \times C_p \times dT$$

Here $C_p$ is the specific heat of a substance at constant pressure. If the potential difference across the heater wire is $E$ volts and the current flowing is $I$ amperes, then

$$dQ = \frac{EI.dt}{J}$$

$$\therefore \quad dQ_1 = dQ - dQ_2$$

$$C_p \times dT = \frac{EI.dt}{J} - dQ_2 \quad ...(i)$$

If the system does not lose heat to the surroundings, then

$$dQ_2 = 0$$

$$\therefore \quad C_p \times dT = \frac{EI.dt}{J} \quad ...(ii)$$

$$\frac{EI}{J} = C_p\left[\frac{dT}{dt}\right] \quad ...(iii)$$

To find $\frac{dT}{dt}$, *i.e.*, the rate of rise of temperature of the system with time, a curve is drawn between temperature ($T$) and time ($t$). From the graph $\frac{dT}{dt}$ is found for the temperature at which $C_p$ is to be determined.

## 14.9 Adiabatic Vacuum Calorimeter

This calorimeter was designed by Nernst and later on modified by Simon and Lange. Inside a thin copper vessel $C$, the substance is taken. The vessel $C$ also contains heating coils and a sensitive platinum resistance thermometer to note the temperature. $A$ thermostat $B$ is made of copper and surrounds the vessel $C$. It is also heated electrically. The two junctions of the sensitive thermo-couple are in contact with the inner surface of $B$ and the outer surface of $C$ (Fig. 14.8).

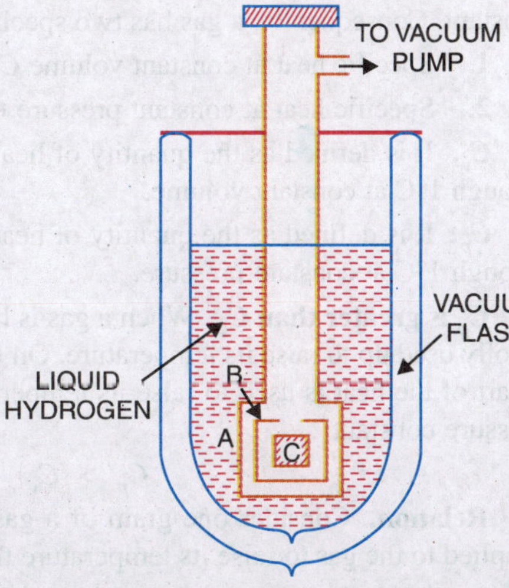


**Fig. 14.8**

The outer vessel $A$ is surrounded by liquid hydrogen

contained in a vacuum flask. The current through the heater coil of $B$ is adjusted so that the thermocouple shows no deflection. This ensures a uniform temperature enclosure for $C$. The space inside $A$ is evacuated with the help of a vacuum pump. If the rise in temperature of 1 gram of the substance is $dT$ in time $dt$,

$$\text{Then} \quad \frac{EI}{J} = C_p\left[\frac{dT}{dt}\right] \quad ...(iv)$$

The method is useful in determining the heat capacity of substances at low temperatures.

To find the value of $C_v$, it is difficult to find its value experimentally at low temperature. Its value is calculated from the equation

$$C_v = C_p - \frac{r}{J} \quad ...(v)$$

## 14.10 Two Specific Heats of a Gas

Consider a gas of mass $m$ at a pressure $P$ and volume $V$. If the gas is compressed, there is rise in temperature. In this case, no heat has been supplied to the gas to raise its temperature.

$\therefore$ Specific heat, $$C = \frac{Q}{m\theta}$$

But $$Q = 0$$

$\therefore$ $$C = 0$$

On the other hand, if heat is supplied to the gas and the gas is allowed to expand such that there is no rise of temperature, then

$$C = \frac{Q}{m\theta}$$

Here $$\theta = 0$$

$\therefore$ $$C = \frac{Q}{m \times 0} = \infty$$

Thus, the specific heat of a gas varies from zero to infinity.

In order to fix the value of the specific heat of a gas, the pressure or volume has to be kept constant. Consequently, a gas has two specific heats.

1. Specific heat at constant volume $C_v$,
2. Specific heat at constant pressure $C_p$.

$C_v$. It is defined as the quantity of heat required to raise the temperature of one gram of a gas through 1°C at constant volume.

$C_p$. It is defined as the quantity of heat required to raise the temperature of one gram of a gas through 1°C at constant pressure.

**$C_p$ is greater than $C_v$.** When a gas is heated at constant volume, the heat supplied to the gas is wholly used up to raise its temperature. On the other hand when a gas is heated at constant pressure, a part of the heat is used to raise its temperature and a part is used to do external work to keep the pressure constant.

$$C_p > C_v$$

**Relation.** Consider one gram of a gas at a pressure $P$, volume $V$ and temperature $T$. Heat is supplied to the gas to raise its temperature through $dT$. As the pressure has to remain constant,

Work done, $$W = P \times A \times x = P \times dV$$

where $dV$ is the change in volume.

From the gas equation

$$PV = rT$$

Differentiating,

$$PdV + VdP = r\,dT$$

But $$dP = 0$$

$\therefore$ $$PdV = r\,dT$$

$\therefore$ Work done in heat units

$$= \frac{r.dT}{J} \text{ calories}$$

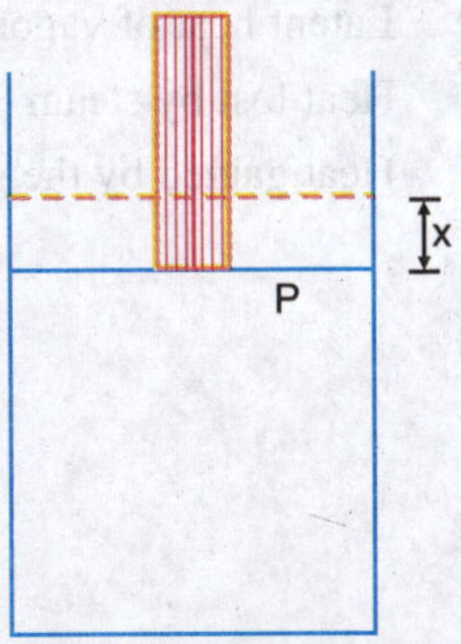


Fig. 14.9

Heat supplied $$= 1 \times C_p \times dT$$

$$= 1 \times C_v \times dT + \frac{r.dT}{J}$$

or $$C_p - C_v = \frac{r}{J}$$

where $r$ is the gas constant for one gram of a gas. If $C_p$ and $C_v$ represent gram molecular specific heats, then

$$C_p - C_v = \frac{R}{J}$$

where $R$ is the universal gas constant.

## 14.11 Specific Heat of a Gas at Constant Volume–Joly's Differential Steam Calorimeter

$C$ is a chamber in which steam can be admitted. $P_1$ and $P_2$ are scale pans suspended from the scale pans $S_1$ and $S_2$ of a balance. $A$ and $B$ are two identical hollow metal spheres. The two spheres are initially evacuated and by placing them in the pans $P_1$ and $P_2$ the balance is counterpoised. In one of the spheres the gas whose $C_v$ is to be determined is admitted at a pressure of about 10 atmospheres and the balance is counterpoised again. Suppose the gas is filled in $B$, the balance will tilt to the left and the extra mass that has to be kept in the pan $S_2$ for counterpoising corresponds to that mass of gas enclosed in $B$ ($M$ grams). Now steam is admitted into the chamber and passed continuously till a constant temperature is reached (temperature of steam). Steam condenses and the mass of steam that has condensed on the pan $P_1$ is more than that condensed on the pan $P_2$ because in the case of pan $P_1$ the enclosed gas also has to be heated from the room temperature to the temperature of steam.

The balance tilts and an extra mass has to be kept in the right hand pan to balance it again. This extra mass corresponds to the extra mass of steam condensed on the pan $P_1$.

The umbrella-shaped vanes $U_1$ and $U_2$ will not allow the steam condensed on the rest of the chamber to fall on the scale pans $P_1$ and $P_2$. The heating coils $H_1$ and $H_2$ will heat the suspension wires to a temperature higher than the temperature of steam and therefore, no steam condenses on these wires.

Suppose,

Mass of the gas $= M$ grams

Sp. heat of gas at constant volume $= C_v$

Initial temperature of the gas $= t_1$°C

Temp. of steam $= t_2$°C

Extra mass of steam condensed on the scale pan $P_1 = m$ grams

Latent heat of vaporisation of water $= L$ cals/g

Heat lost by steam $= mL$

Heat gained by the gas $= MC_v (t_2 - t_1)$

$$MC_v (t_2 - t_1) = mL$$

$$C_v = \frac{mL}{M(t_2 - t_1)} \text{ cals/g °C}$$

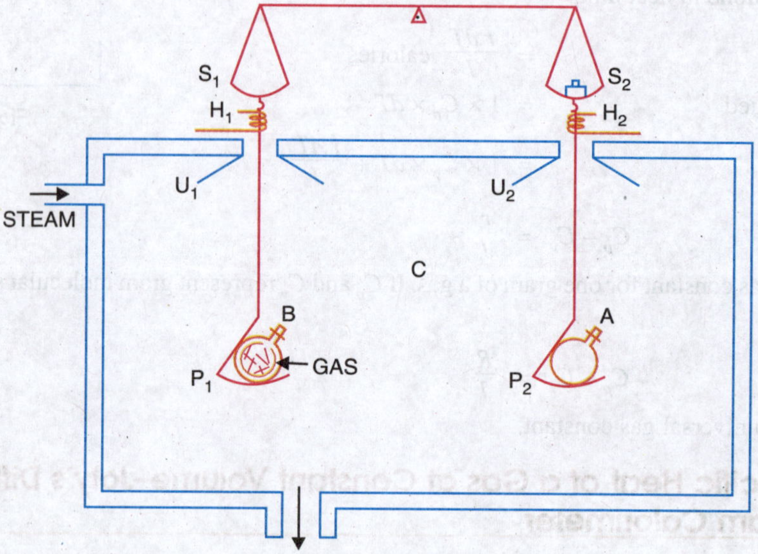


Fig. 14.10

In the second part of the experiment the gas is enclosed in the sphere of pan $P_2$ and the sphere of pan $P_1$ is kept empty and $C_v$ is calculated again. The mean of these two values gives the specific heat of the gas at constant volume.

## 14.12 Specific Heat of a gas at Constant pressure–(Regnault's Method)

The apparatus consists of a reservoir *R* containing the gas at high pressure and at a constant temperature. The pressure of the gas in the reservoir is shown by the pressure gauge. The apparatus is as shown in (Fig. 14.11).

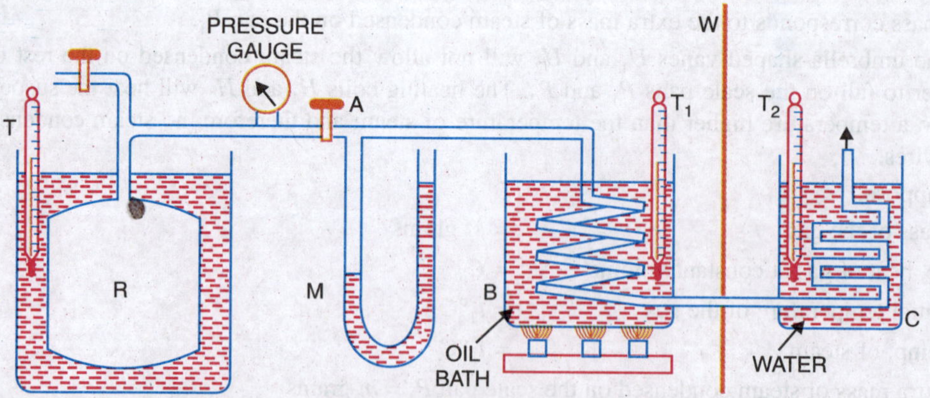


Fig. 14.11

The regulator $A$ allows the gas to flow at a constant pressure through the spiral tubings immersed in the oil bath $B$ and the calorimeter. The pressure of the gas flowing through the spiral tubing is shown by the manometer $M$. Regulator $A$ helps in keeping the level of the liquid in the manometer limbs constant.

Suppose the initial pressure of the gas at any instant in the reservoir is $P_1$, its temperature is $T$, and volume is $V$. The temperature shown by the oil bath is $T_1$ and the calorimeter is at a temperature $T_2$. Gas is allowed to flow for about half an hour. The gas after passing through the oil bath gets heated to temperature $T_1$ and after passing through the calorimeter, gets cooled and gives heat to the calorimeter and its contents. Suppose, the final pressure of the gas in the reservoir is $P_2$, volume is $V$ and the temperature of the calorimeter $C$ and its contents is $T_3$.

**Calculations**

Suppose, mass of the gas flown $= M$

Mass of water in the calorimeter $= m$

Water equivalent of the calorimeter $= w$

Rise of temperatures of calorimeter and its contents

$$= (T_3 - T_2)$$

Mean fall of temperature of the gas

$$= \left(T_1 - \frac{T_2 + T_3}{2}\right)$$

$$\text{Heat gained} = \text{Heat lost}$$

$$(m + w)(T_3 - T_2) = MC_p\left(T_1 - \frac{T_2 + T_3}{2}\right)$$

$$C_p = \frac{(m+w)(T_3 - T_2)}{M\left(T_1 - \dfrac{T_2 + T_3}{2}\right)} \qquad ...(i)$$

To find the mass of the gas ($M$), suppose the density of the gas at NTP = ρ. In the experiment, $V$ cc of the gas at a pressure $(P_1 - P_2)$ and temperature $T$ K has flown through the apparatus. Reducing the volume of the gas to NTP,

$$\frac{(P_1 - P_2)V}{T} = \frac{76\,V_0}{273}$$

or

$$V_0 = \frac{(P_1 - P_2)V \times 273}{76 \times T}$$

∴ Mass

$$M = V_0 \times \rho = \frac{(P_1 - P_2)V \times 273 \times \rho}{76 \times T}$$

Thus, knowing the value of $M$, $C_p$ can be calculated from equation ($i$)

## 14.13 Continuous Flow Electrical Method

The specific heat of a gas at constant pressure by electrical method can be determined by using Callendar and Barnes' continuous flow apparatus. Here $D$ is a special glass vessel. The heating coil is at the axis of the vessel. The incoming gas takes a long zig zag path as indicated by the arrow heads (Fig. 14.12).

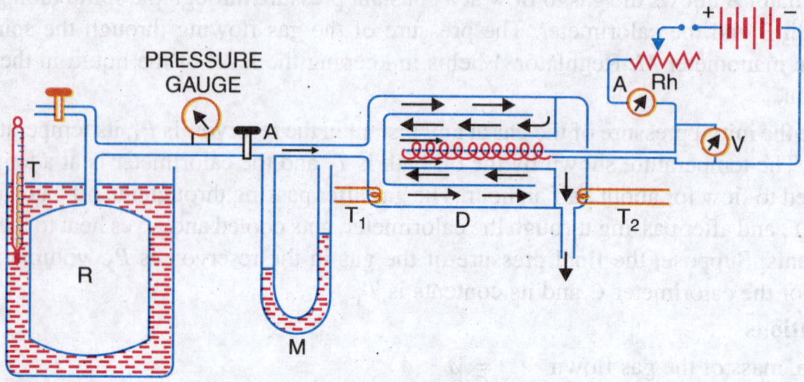


**Fig. 14.12**

The reservoir $R$ contains the gas at high pressure and at a constant temperature. The pressure of the gas is read by the pressure gauge. The regulator $A$ allows the gas to flow at a constant pressure through the vessel $D$. The pressure of the gas flowing through $D$ is indicated by the manometer $M$. The regulator helps in keeping the pressure of the gas constant throughout the experiment. The filament is heated by electric current. The incoming gas is heated due to the heat generated by the filament. The inlet and the outlet temperatures of the gas are measured with the help of the platinum resistance thermometers $T_1$ and $T_2$.

The gas is allowed to flow through the apparatus for some time till the steady state is reached. When the steady state is reached, the thermometers $T_1$ and $T_2$ show constant readings. After the steady state, note the pressure ($P_1$) of the gas in the reservoir. Allow the gas to flow for half an hour and note the final pressure ($P_2$) of the gas in the reservoir. During this half-an-hour, the manometer $M$ and the thermometers $T_1$ and $T_2$ should show constant readings.

### Calculations

Suppose, the mass of the gas flown $= M$

Temperature of incoming gas $= T_1$

Temperature of outgoing gas $= T_2$

Specific heat of the gas at constant pressure $= C_p$

Voltmeter reading $= E$ volts

Ammeter reading $= I$ amperes

Time $= t$ seconds

Heat produced $= \dfrac{EIt}{4.2}$ calories

Heat gained = Heat produced

$$MC_p(T_2 - T_1) = \frac{EIt}{4.2}$$

$$C_p = \frac{EIt}{4.2\,M\,(T_2 - T_1)} \quad \text{...(}i\text{)}$$

To find the mass of the gas ($M$), suppose the density of the gas at N.T.P. = ρ. In the experiment $V$ cc of the gas at a pressure $(P_1 - P_2)$ and temperature $T$ has flown through the apparatus. Here $V$ is the volume of the reservoir $R$. Reducing the volume of the gas to N.T.P.

$$\frac{(P_1 - P_2)V}{T} = \frac{76\,V_0}{273}$$

$$V_0 = \frac{(P_1 - P_2)\,V \times 273}{76\,T}$$

∴ Mass $$M = \rho V_0 = \frac{\rho(P_1 - P_2)\,V \times 273}{76\,T} \text{ grams}$$

Thus knowing the value of $M$, $C_p$ can be calculated from equation ($i$).

The chief advantages of this method are :

**1.** The temperature of every portion of the apparatus remains constant when the steady state is reached.

**2.** The thermal capacity of the flow tube and its contents is eliminated in the calculations.

**3.** The loss of heat due to radiation is minimised due to zig-zag path taken by the gas. The gas flows from outer to the inner region of the tube and absorbs any heat radiated from the inner to the outer portion.

**4.** The temperatures can be measured accurately under steady state.

**5.** The heat produced by electrical arrangement can be calculated accurately.

## 14.14 Specific heat of a gas at low temperatures

The continuous flow method can be used to determine the specific heat of a gas at low temperature. Before the gas is passed into the tube $D$, the gas is initially cooled to a desired low temperature. The experiment is performed in a similar way as discussed in article 14.13.

This method is also used to find the specific heat of a gas at different high pressures. In this case, the gas at high pressure is passed through the tube $D$. Scheel and Heuse employed this method to find the specific heat of various gases up to –180°$C$. Holborn and Jakob used this method to find the specific heat of gases at high pressures.

**TABLE 14.4 Molecular Heats of Gases at 20ºC**

| *Gas* | $C_p$ | $C_v$ |
|---|---|---|
| Hydrogen | 6.87 | 4.88 |
| Nitrogen | 6.95 | 4.96 |
| Oxygen | 7.03 | 5.04 |
| Chlorine | 8.29 | 6.15 |
| Air | 6.95 | 4.96 |
| Carbon dioxide | 8.83 | 6.80 |
| Helium | 4.97 | 2.98 |
| Argon | 4.97 | 2.98 |

The specific heats of hydrogen (for 1 gram) are

$$C_p = 3.435 \text{ and } C_v = 2.44$$

## 14.15 Calorific Value of Fuels

**Definition.** Calorific value of a fuel is defined as the quantity of heat released when a unit quantity of the fuel is completely burnt and the products of combustion are brought to the original temperature.

It is expressed as calories per gram or $B.\ T.\ U.$ per pound. The fuels commonly used are coal, petrol, spirit, diesel oil, wood etc. The calorific value of different fuels can be determined with the help of a bomb calorimeter.

**Bomb Calorimeter.** It consists of a vessel $D$ made of steel or gun metal. A platinum bowl $B$ is suspended inside $D$. The fuel $E$ in the form of powder for a solid or a liquid is taken in $B$. A heating wire is immersed in the fuel and its ends are connected to the terminals $T_1$ and $T_2$. $R$ is a regulating valve for the free supply of oxygen inside $D$. The vessel $D$ is closed with an air tight lid. The whole apparatus is immersed in a calorimeter $C$ containing water. The fuel is ignited by passing current through the heater wire. The heat produced due to the combustion of the fuel is taken by the surrounding water in the calorimeter.

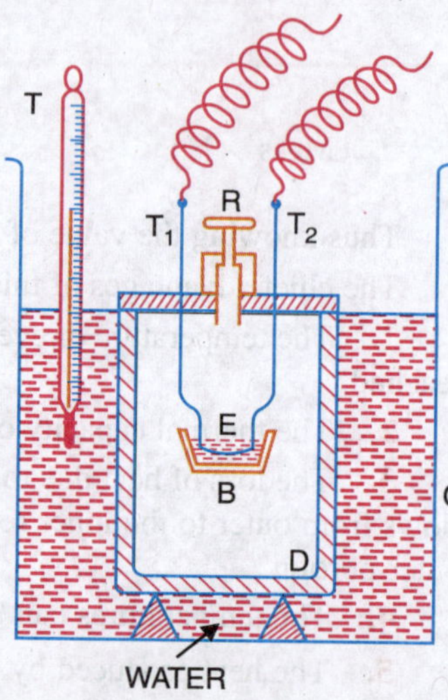


Fig. 14.13

Suppose, the mass of the fuel = $M$

Initial temperature of water = $\theta_1$

Final temperature of water = $\theta_2$

Mass of water = $m$

Water equivalent of the calorimeter and the vessel $D = w$

Heat produced = $(m + w)(\theta_2 - \theta_1)$

$$\therefore \quad \text{Calorific value} = \frac{(m+w)(\theta_2-\theta_1)}{M}.$$

## 14.16 Bell Calorimeter

It is used to find the calorific value of fuels. The apparatus consists of a bell jar $B$ arranged inside a calorimeter containing water at room temperature. In the crucible $C$, the substance is taken in the powdered form. The leads $L$, $L$ are connected to a heating filament dipped inside the fuel in the crucible $C$ (Fig. 14.14). $D$ is a perforated disc through which the burnt gases inside the bell can escape into water. $A$ is a pipe through which oxygen is fed into the combustion chamber.

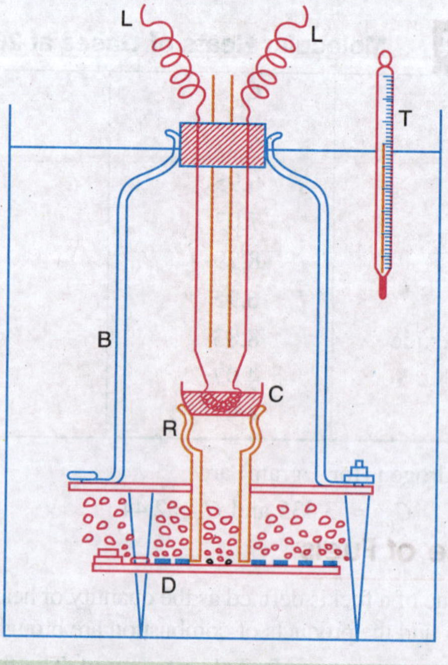


Fig. 14.14

The fuel is ignited by connecting the leads *L*, *L* to a battery. The fuel burns and the burnt hot gases escape into water in the calorimeter, through the disc *D*. When the fuel is completely burnt, the final temperature of water in the calorimeter is noted.

Suppose, the mass of the fuel $= M$

Initial temperature of water $= \theta_1$

Final temperature of water $= \theta_2$

Mass of water $= m$

Water equivalent of the calorimeter $= w$

Heat produced $= (m + w)(\theta_2 - \theta_1)$

Calorific value $= \dfrac{(m+w)(\theta_2 - \theta_1)}{M}$

**TABLE 14.5 Calorific Value of Fuels**

| *Fuel* | *Cal/g* |
|---|---|
| Wood | 2500 |
| Gas coke | 6000 |
| Methylated spirit | 6400 |
| Steam coal | 7500 |
| Anthracite coal | 8800 |
| Heavy Diesel oil | 11350 |
| Petrol | 11100 |
| Paraffin oil | 11200 |

## 14.17 Dulong and Petit's Law

Dulong and Petit, in 1819, studied the specific heat of various elements in a solid state and enunciated a law, called Dulong and Petit's law. According to this law, the *product of the specific heat and the atomic weight i.e. atomic heat of all the elements in the solid state is a constant.* The value of this constant was fixed as 6.4 but it is taken as 6 at present. The exact value is 5.96 which also agrees with the value derived below from the kinetic theory of matter.

The justification of Dulong and Petit's law was obtained from Boltzmann's consideration of the law of equipartition of energy. According to it, the energy associated with one gram atom of a substance for each degree of freedom at temperature *T* is $= \frac{1}{2}RT$. Here *R* is the universal gas constant. If the atom is considered to be vibrating about the mean position, its mean kinetic energy will be equal to its mean potential energy.

For each form of energy there are three degrees of freedom. Therefore, an atom has got six degrees of freedom. Thus the total energy associated with one gram atom of a substance at a temperature *T* is $= 3RT$

$$\therefore \quad E = 3\,RT$$

Therefore, the specific heat at constant volume,

$$C_V = \frac{dE}{dT} = 3R$$

$C_V$ is the atomic heat of the substance

$$C_V = \frac{3\times 8.31\times 10^7}{4.18\times 10^7} \text{ cals/g-atom-}K$$

$$C_V = 5.96 \text{ cals/g-atom-}K$$

**TABLE 14.6 Atomic Heat of Substances at 20ºC**

| *Substance* | *Atomic Weight* | *Specific Heat* | *Atomic Heat* |
|---|---|---|---|
| Aluminium | 27.0 | 0.212 | 5.72 |
| Boron | 10.8 | 0.307 | 3.32 |
| Carbon | 12.0 | 0.160 | 1.92 |
| Copper | 63.6 | 0.091 | 5.79 |
| Gold | 197.2 | 0.031 | 6.11 |
| Iron | 55.8 | 0.110 | 6.12 |
| Lead | 207.2 | 0.030 | 6.21 |
| Silicon | 28.1 | 0.182 | 5.11 |
| Silver | 107.9 | 0.056 | 6.04 |
| Zinc | 65.4 | 0.092 | 6.02 |

Dulong and Petit's law was modified by Woestyn for metallic compounds. According to him, the *molecular heat of a compound* is equal to the sum of the atomic heat of its constituents. Further, it was found by Newmann that the molecular heat of the compounds of similar nature is also constant. The molecular heat in the case of *NaCl*, *AgCl* and *KCl* was found to be nearly 13 and that of $Sb_2O_3$, $Fe_2O_3$, $As_2O_3$ was found to be equal to 26.

## 14.18 Variation of Specific Heat and Atomic Heat with Temperature (Or Departure from Dulong and Petit's Law at Low Temperatures)

It is found that Dulong and Petit's law is not true in the case of carbon, boron and silicon. In the case of these elements the atomic heats at 20°C are 1.92, 3.32 and 5.11. These values differ from the constant

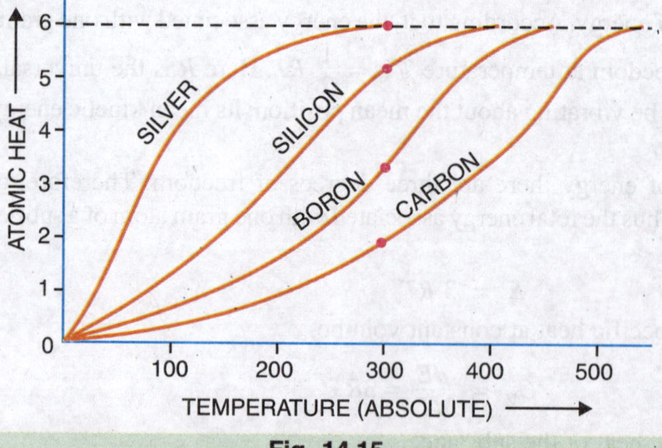


**Fig. 14.15**

value of 6. This variation in atomic heat could not be explained on the basis of kinetic theory of matter. However, it was found by Nernst that the specific heat of a substance decreases with decrease in temperature and at absolute zero the specific heat tends to be zero. Further, he was able to show that the specific heat increases with the rise in temperature and tends to a maximum value. Therefore, the atomic heat of a substance tends to a maximum value of six and decreases with decrease in temperature. In the case of carbon, boron and silicon also, the atomic heat is 6 at high temperatures (Fig. 14.15).

In the case of silver, the atomic heat is 6 at room temperature but it is also less than 6 at lower temperatures.

**TABLE 14.7 Atomic Heats of Silver at Low Temperature**

| *Temperature (K)* | *Atomic heat* |
|---|---|
| 205 | 5.61 |
| 144 | 5.37 |
| 103 | 4.80 |
| 75 | 4.04 |
| 56 | 3.19 |
| 36 | 1.69 |
| 20 | 0.40 |
| 10 | 0.05 |
| 5 | 0.005 |
| 1.4 | 0.00025 |

## 14.19 Einstein's Theory of Specific Heat of Solid

Dulong and Petit's law has been explained by Einstein in 1907 for the first time on the basis of quantum theory of heat radiation. According to quantum theory, heat is radiated in the form of discrete particles called photons. Each particle has energy equal to $h\nu$ where $h$ is the Planck's constant and $\nu$ is the frequency of heat radiation. Einstein proposed his theory under following assumptions :

**Assumptions :**

1. A solid is constituted of atoms. At absolute zero of temperature, these atoms are at rest under the action of mutual attractions and repulsions. The energy of the solid in this state is zero. When the temperature is raised *i.e.*, heated, the atoms are set into vibrations about their mean position of equilibrium under the action of restoring force, and hence executes S.H.M. The frequency of these vibrations is one and the same *i.e.*, natural or characteristic frequency of the particular solid under consideration.
2. Each atom of solid has 3 degrees of freedom like a monatomic gas molecule.
3. The mean energy per degree of freedom is not $kT$ as given by equipartition law but $\frac{h\nu}{e^{h\nu/kT}-1}$ as calculated by Planck by using quantum theory.

**Theory.** Every atom has 3 degrees of freedom, assuming gas to be monatomic.

$\therefore$ the energy of each atom $= \dfrac{3h\nu}{e^{h\nu/kT}-1}$

The energy content of one gram of atom of the solid consisting of $N$ atoms is given by

$$E = \frac{3Nh\nu}{e^{h\nu/kT} - 1}$$

Therefore, the atomic heat at constant volume will be

$$C_V = \frac{dE}{dT} = 3Nh\nu\left[\frac{-1}{(e^{h\nu/kT} - 1)^2}\right]e^{h\nu/kT}\cdot\left(\frac{-h\nu}{kT^2}\right)$$

$$= \frac{3Nh^2\nu^2}{kT^2}\cdot\frac{e^{h\nu/kT}}{(e^{h\nu/kT} - 1)^2}$$

$$= 3Nk\frac{e^{h\nu/kT}}{(e^{h\nu/kT} - 1)^2}\cdot\left(\frac{h\nu}{kT}\right)^2 \quad ...(1)$$

This equation can be conveniently written as

$$C_V = 3R\left[\frac{e^{\theta/T}}{(e^{\theta/T} - 1)^2}\left(\frac{\theta}{T}\right)^2\right] \quad ...(2)$$

where $Nk = R$, gas constant for a gram atom

and $\theta = \frac{h\nu}{k}$

The value of $\nu$ giving good fit for a particular solid is represented by $\nu_E$ and is called the **Einstein's frequency for that solid**. Similarly $\theta$ corresponding to $\nu_E$ is represented by $\theta_E$ and is called **Einstein's temperature**. Here $\theta$ has the dimensions of temperature.

Equation (1) is the Einstein's equation for the atomic heat of the solid at constant volume. From equation (2), it is clear that the atomic heat is a function of temperature. The experimentally observed variation of atomic heat of solids are as follows :

1. At high temperatures, the atomic heat approaches the constant value 3R, as given by Dulong and Petit's law.

2. The atomic heat decreases with decreases of temperature and tends to zero at absolute zero of temperature.

3. The experimental curves drawn for atomic heat against temperature show that the curves have the same form for all substances as shown in Fig. 14.16.

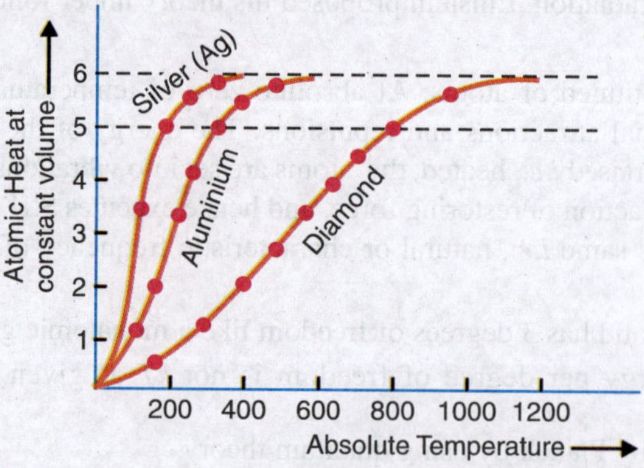


**Fig. 14.16.**

**Discussion :** Let us now explain the above experimentally observed facts on the basis of Einstein's equation (1).

**1. At High Temperature :** At high temperatures, $h\nu/kT$ approaches to very small values so that we may write as

$$(e^{h\nu/kT}-1)^2 = \left[1+\frac{h\nu}{kT}+\frac{1}{2!}\left(\frac{h\nu}{kT}\right)^2+\ldots-1\right]^2$$

$$= \left(\frac{h\nu}{kT}\right)^2$$

Therefore, from equation (1), we have

$$C_V = \frac{3R\left[1+\frac{h\nu}{kT}+\ldots\right]\cdot\left(\frac{h\nu}{kT}\right)^2}{\left(\frac{h\nu}{kT}\right)^2}$$

$$C_V = 3R\left[1+\left(\frac{h\nu}{kT}\right)+\ldots\right] \quad \ldots(3)$$

As $T\to\infty$, $\frac{h\nu}{kT}\to 0$, Hence $C_V \cong 3R$. Thus, this result agrees well with the experiment and Dulong and Petit's law.

**(2) At very low temperatures.** When $T\to 0$, $\frac{h\nu}{kT}$ tends to infinity and hence 1 may be neglected in the expression $(e^{h\nu/kT}-1)^2$. Then, the equation (1) becomes

$$C_V = 3R\frac{e^{h\nu/kT}}{(e^{h\nu/kT})^2}\cdot\left(\frac{h\nu}{kT}\right)^2$$

$$= 3R\frac{1}{e^{h\nu/kT}}\cdot\left(\frac{h\nu}{kT}\right)^2$$

$$= 3R\frac{\left(\frac{h\nu}{kT}\right)^2}{\left[1+\frac{h\nu}{kT}+\frac{1}{2!}\left(\frac{h\nu}{kT}\right)^2+\frac{1}{3!}\left(\frac{h\nu}{kT}\right)^3+\ldots\right]}$$

$$= 3R\frac{1}{\left[\frac{1}{\left(\frac{h\nu}{kT}\right)^2}+\frac{1}{\left(\frac{h\nu}{kT}\right)}+\frac{1}{2!}+\frac{1}{3!}\left(\frac{h\nu}{kT}\right)+\ldots\right]} \quad \ldots(4)$$

As $T\to 0$, $\frac{h\nu}{kT}\to\infty$. Hence $C_V\to 0$ *i.e.*, the atomic heat tends to zero as temperature approaches to absolute zero. This is again in good agreement with experimental results and Dulong and Petit's law.

**3.** According to Einstein's equation (1), the atomic heats of different elements differ only because of different $\nu_E$, the characteristic frequency. At corresponding temperatures, such that the ratio $\nu/T$ is same for all solids, then from equation (1), the atomic heat will also be same for all solids. Thus, the experimental curves have the same form for all substances. This argument also fits with the experimental fact and Dulong and Petit's law.

Hence Einstein's theory predicts the correct values of atomic heat for various elements as observed experimentally. In other words, Einstein's theory is in good agreement with the experimental facts.

### Limitations of Einstein's Theory

(*i*) Although the values of atomic heats agrees well at very high and very low temperatures with the experimental curves, but the variations in between them does not exactly follow. Some elements like copper, aluminium, iron etc. the atomic heat at low temperatures decreases more rapidly than that expected by the Einsteins theory.

(*ii*) The assumptions that the vibrations of all atoms are simple harmonic and they all have one and the same frequency is not valid. In fact, the vibrations of an atom are complex in nature, because it is under the field of force due to large number of neighbouring vibrating atoms.

(*iii*) In Einstein's theory, frequency $v$ and $hv/kT$ have been obtained empirically and cannot be verified from any other independent physical data.

## 14.20. Debye's Theory of Specific Heat of Solids

The discrepancies present in Einstein's theory of specific heat of solids are mostly removed by the theory proposed by Debye as the theory is in excellent agreement with experiment over the whole observable temperature range. Again Debye made some assumptions.

**Assumptions :** Debye's assumptions are :

(*i*) Any solid is capable of vibrating elastically in many different modes, the frequency varying from one mode to another.

(*ii*) The number of modes of vibrations of solids is limited.

(*iii*) When a solid is subjected to continuous elastic vibrations, two kinds of vibrations *i.e.*, (*a*) longitudinal vibrations and (*b*) transverse vibrations are set up.

**Theory :** The number of modes of longitudinal vibrations per unit volume with frequencies between $v$ and $(v + dv)$ is

$$= \frac{4\pi v^2 dv}{C_l^3}$$

where $C_l$ is the velocity of longitudinal vibrations.

Since the transverse vibrations have two independent directions of vibrations *i.e.*, they are equivalent to two waves polarised at right angles to each other.

Therefore, the number of modes of transverse vibrations per unit volume frequencies between $v$ and $(v + dv)$ is

$$= \frac{8\pi v^2 dv}{C_t^3}$$

where $C_t$ is the velocity of transverse vibrations.

∴ The number of independent vibrations per unit volume with frequencies between $v$ and $(v + dv)$ is

$$= 4\pi \left[\frac{1}{C_l^3} + \frac{2}{C_t^3}\right] v^2 dv$$

If $V$ is the volume of one gram-atom of the solid, the number of modes of vibrations for this amount within frequency range $v$ and $(v + dv)$ will be

$$= 4\pi V \left[\frac{1}{C_l^3} + \frac{2}{C_t^3}\right] v^2 dv$$

Thus, the total number of independent modes of vibrations will be

$$= \int_0^{v_m} 4\pi V \left[\frac{1}{C_l^3} + \frac{2}{C_t^3}\right] v^2 dv \qquad ...(i)$$

According to Debye, the upper limit of frequency is not infinite but a definite limit $\nu_m$. This limit chosen so that the total number of possible independent vibrations is equal to the number of vibrations of the separate atoms taken together in the solid. Let there be $N$ atoms in volume $V$, then the possible number of vibrations will be equal to $3N$, because each atom has three degrees of freedom. Therefore, we have

$$\int_0^{\nu_m} 4\pi V\left[\frac{1}{C_l^3}+\frac{2}{C_t^3}\right]\nu^2 d\nu = 3N$$

or $$4\pi V\left[\frac{1}{C_l^3}+\frac{2}{C_t^3}\right]\frac{\nu_m^3}{3} = 3N$$

or $$4\pi V\left[\frac{1}{C_l^3}+\frac{2}{C_t^3}\right] = \frac{9N}{\nu_m^3} \qquad ...(ii)$$

Applying the quantum theory, the energy associated with each degree of freedom is

$$= \frac{h\nu}{e^{h\nu/kT}-1}$$

The thermal energy of 1 gram atom of solid for frequencies between $\nu$ and $(\nu + d\nu)$ is

$$= \frac{h\nu}{e^{h\nu/kT}-1}.4\pi V\left[\frac{1}{C_l^3}+\frac{2}{C_t^3}\right]\nu^2 d\nu$$

$$= \frac{h\nu}{e^{h\nu/kT}-1}.\frac{9N}{\nu_m^3}.\nu^2 d\nu \qquad (\because \text{Eq. } (ii))$$

$$= \frac{9N}{\nu_m^3}.\frac{h\nu^3\, d\nu}{e^{h\nu/kT}-1} \qquad ...(iii)$$

$\therefore$ Total thermal energy for 1 gramatom of solid will be

$$E = \int_0^{\nu_m}\frac{9N}{\nu_m^3}.\frac{h\nu^3 d\nu}{e^{h\nu/kT}-1}$$

$$= \frac{9N}{\nu_m^3}\int_0^{\nu_m}\frac{h\nu^3 d\nu}{e^{h\nu/kT}-1} \qquad ...(iv)$$

Putting $\frac{h\nu}{kT} = x$, then $\nu = \frac{kT}{h}x$, and $d\nu = \frac{kT}{h}dx$.

Let us also introduce a characteristics temperature $\theta$ defined by

$$\theta = \frac{h\nu_m}{k} \qquad ...(v)$$

$$\therefore \qquad \nu_m = \frac{k\theta}{h}$$

Substituting for $\nu_m$ in equation $(iv)$, we get

$$E = \frac{9Nh^3}{k^3\theta^3}\int_0^{\theta/T}\frac{h\,k^3T^3x^3}{(e^x-1)h^3}.\frac{kT}{h}dx$$

or $$= 9Nk\frac{T^4}{\theta^3}\int_0^{\theta/T}\frac{x^3}{(e^x-1)}dx$$

The specific heat of one gram atom at constant volume is given by

$$C_V = \frac{dE}{dT}$$

$$= 9Nk\left[4\left(\frac{T}{\theta}\right)^3\right]\int_0^{\theta/T}\left[\frac{x^3dx}{e^x-1}-\frac{\theta}{T}\cdot\frac{1}{e^{\theta/T}-1}\right]$$

or $$C_V = 3R\left[12\left(\frac{T}{\theta}\right)^3\right]\int_0^{\theta/T}\left[\frac{x^3dx}{e^x-1}-\frac{\theta}{T}\cdot\frac{1}{e^{\theta/T}-1}\right] \quad ...(vi)$$

This is **Debye's equation** for the atomic heat of a monoatomic solid.

It should be noted that the quantity in a square bracket is a function of $\frac{\theta}{T}$ and may be denoted by $f\left(\frac{\theta}{T}\right)$. This function is usually known as **Debye's function.**

Here the parameter $\theta = \frac{h\nu_m}{k}$ varies from element to element and is called **Debye's temperature.**

Using these quantities, from equation (*vi*), the atomic heat (*i.e.*, specific heat of a gram atom) may be expressed as

$$C_V = 3R\,f\left(\frac{\theta}{T}\right) \quad ...(vii)$$

**Discussion**

(*i*) At *high temperatre*, $C_V = 3R$. It means, the atomic heat of all substances tends to a maximum value equal to $3R$.

(*ii*) The atomic heat of substances decreases with decrease of temperature.

(*iii*) The atomic heat tends to zero, as the temperature tends to absolute zero.

(*iv*) *At very low temperatures near absolute zero,* the exponential term $e^{\theta/T}$ becomes large; so that the last term in eqn. (*vi*) can be neglected and the upper limit in the integration may be replaces by infinity. Then, *at low temperatures,* we have

$$C_V = 3R\times12\left(\frac{T}{\theta}\right)^3\int_0^{\infty}\frac{x^3}{e^x-1}dx$$

$$= 3R\times12\left(\frac{T}{\theta}\right)^3\frac{\pi^4}{15}$$

$$= 77.94\times3R\left(\frac{T}{\theta}\right)^3 \quad ...(8)$$

$$\therefore \quad C_V \propto T^3 \quad ...(9)$$

Thus, *at low temperature the atomic heat of solid is directly proportional to the cube of absolute temperature.* This is called Debye's $T^3$ law.

(*v*) The general nature of the curves showing the variation of specific heat with temperature is the same for all the elements *i.e., the behaviour of elements regarding variation of specific heat with temperature is similar.* Any two curves can be made to coincide with each other if temperature scale of one of them is suitably adjusted.

According to Debye, the atomic heat at absolute temperature $T$ is $C_V = 3R.f\left(\frac{\theta}{T}\right)$ where $f\left(\frac{\theta}{T}\right)$ is the function which is same for all elements.

The value of $C_V$ is zero at absolute zero of temperature and tends to $3R$ or 5.96 at high temperature. Hence, we conclude that as per **Debye's theory** a graph plotted atomic heat against function $f\left(\frac{\theta}{T}\right)$ is the same for all elements as shown in Fig. 14.17.

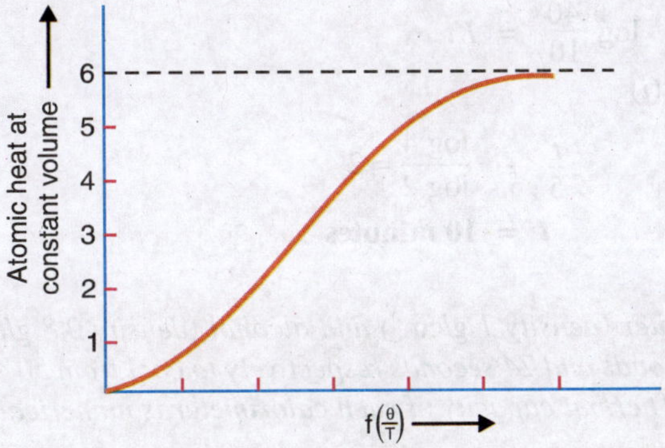


**Fig. 14.17.**

**Shortcomings of Debye's Theory.** The curve obtained by Debye's theory fits well with the experimental curves of all elements. Still Debye's theory has few shortcomings. These are :

(*a*) In the assumptions of Debye's theory, solid is assumed to be continuous. But actually every solid is made up of large number of discrete particles, bound together by a cohesive force.

(*b*) A solid can be assumed to be perfectly isotropic as it is made up of some definite crystals highly prone to some defects.

## NUMERICAL EXAMPLES

### EXMPLE 14.1

*A liquid takes 5 minutes to cool from 80°C to 50°C. How much time will it take to cool from 60°C to 30°C? The temperature of the surrounding is 20°C.* **(Madras 2005 summer)**

**SOLUTION.**

$$\int \frac{d\theta}{\theta - \theta_0} = -K\int dt$$

Given: $\theta_0 = 20°C$

In the first case,

$$\int_{80}^{50} \frac{d\theta}{\theta - \theta_0} = -K \times 5$$

$$\log\left[\theta - \theta_0\right]_{80}^{50} = 5\,K$$

$$\log\left(\frac{50-20}{80-20}\right) = -5\,K$$

$$\log\frac{30}{60} = -5\,K$$

$$\log\frac{60}{30} = 5\,K$$

In the second case, suppose the time taken is $t$ minutes,

$$\int_{60}^{30} \frac{\theta}{\theta-\theta_0} = -K \qquad ...(i)$$

$$\log\left(\frac{30-20}{60-20}\right) = -Kt$$

$$\log\frac{10}{40} = -Kt$$

or
$$\log\frac{40}{10} = Kt \qquad ...(ii)$$

Dividing (*ii*) by (*i*)

$$\frac{t}{5} = \frac{\log 4}{\log 2} = 2$$

$$t = \mathbf{10\ minutes}$$

**EXAMPLE 14.2**

*Equal volumes of water (density 1 g/cm³) and alcohol (density 0.8 g/cm³) when put in similar calorimeters take 100 seconds and 74 seconds respectively to cool from 50 °C to 40 °C. Calculate the specific heat of alcohol. Thermal capacity of each calorimeter is numerically equal to the volume of either liquid.*

**SOLUTION.** Let the volume of either liquid by $V$

Mass of water = $m = V \times 1$

Mass of alcohol = $M \times V \times 0.8$

Water equivalent of each calorimeter = $w = V$

Here
$$t_1 = 100\ s,\ t_2 = 74\ s$$

$$\frac{MC + w}{t_2} = \frac{m + w}{t_1}$$

$$C = \frac{(m+w)\,t_2}{Mt_1} - \frac{w}{M}$$

$$= \frac{(V+V)\,74}{V\times 0.8\times 100} - \frac{V}{V\times 0.8}$$

$$C = \mathbf{0.6\ calorie/g - K}$$

**EXAMPLE 14.3**

*A body cools in 5 minutes from 60 °C to 40 °C. What will be its temperature after the next 5 minutes? Temperature of the surroundings = 10 °C. Assume that the Newton's law of cooling holds good throughout the process.*

**SOLUTION.**
$$\int_{\theta_1}^{\theta_2} \frac{d\theta}{\theta-\theta_0} = -K\int dt = -Kt$$

1. In the first case, $\theta_0 = 10$°C,

$$\theta_1 = 60°\text{C},\ \theta_2 = 40°\text{C},\ t = 5 \text{ minutes}$$

$$\int_{60}^{40} \frac{d\theta}{\theta-\theta_0} = -5K$$

$$\left[\log(\theta-\theta_0)\right]_{60}^{40} = -5K$$

$$\log\left(\frac{40-10}{60-10}\right) = -5K \qquad ...(i)$$

2. In the second case

$$\theta_0 = 10°C$$

$$\theta_1 = 40°C, \theta_2 = x, t = 5 \text{ minutes}$$

$$\int_{40}^{x} \frac{d\theta}{\theta-\theta_0} = -5K$$

$$\log\left(\frac{x-10}{40-10}\right) = -5K \qquad ...(ii)$$

From equations (*i*) and (*ii*)

$$\log\left(\frac{x-10}{40-10}\right) = \log\left(\frac{40-10}{60-10}\right)$$

$$\frac{x-10}{30} = \frac{30}{50}$$

$$x = \mathbf{28°C}$$

## EXAMPLE 14.4

*A liquid takes 4 minutes to cool from 70°C to 50°C. How much time will it take to cool from 50°C to 40°C? The temperature of the surroundings is 25°C. Newton's law of cooling is applicable throughout the process.*

**SOLUTION.**

$$\int_{\theta_1}^{\theta_2} \frac{d\theta}{\theta-\theta_0} = -K\int dt = -Kt$$

1. In the first case,

$$\theta_0 = 25°C$$

$$\theta_1 = 70°C, \theta_2 = 50°C, t = 4 \text{ minutes}$$

$$\int_{70}^{50} \frac{d\theta}{\theta-\theta_0} = -4K$$

$$\log_e\left(\frac{50-25}{70-25}\right) = -4K \qquad ...(i)$$

2. In the second case

$$\theta_0 = 25 °C$$

$$\theta_1 = 50°C, \theta_2 = 40°C, t = ?$$

$$\int_{50}^{40} \frac{d\theta}{\theta-\theta_0} = -Kt$$

$$\log_e\left(\frac{40-25}{50-25}\right) = -Kt \qquad ...(ii)$$

Dividing (*ii*) by (*i*)

$$\frac{t}{4} = \frac{\log_e\left(\frac{15}{25}\right)}{\log_e\left(\frac{25}{45}\right)}$$

$$t = \frac{4 \times 2.3026}{2.3026}\left[\frac{\log_{10}(0.6)}{\log_{10}(0.55)}\right]$$

$$= 4\left[\frac{\log_{10}(0.6)}{\log_{10}(0.55)}\right]$$

$$= 4\left[\frac{\bar{1}.7782}{\bar{1}.7404}\right] = 4\left[\frac{-0.2218}{-0.2596}\right]$$

$$= \frac{8872}{2596}$$

$$= \mathbf{3.418 \text{ min.}}$$

**EXAMPLE 14.5**

*A liquid cools in 6* minutes *from 80°C to 60°C. What will be its temperature after the next 10* minutes*? Temperature of the surroundings is 30°C. Assume that the Newton's law of cooling is applicable throughout the process.*

**SOLUTION.** $\int_{\theta_1}^{\theta_2} \frac{d\theta}{\theta - \theta_0} = -K \int dt = -Kt$

1. In the first case,

$$\theta_0 = 30°\text{C}$$

$$\theta_1 = 80°\text{C}, \theta_2 = 60°\text{C}, t = 6 \text{ min}$$

$$\int_{80}^{60} \frac{d\theta}{\theta - \theta_0} = -6K$$

$$\log\left(\frac{60-30}{80-30}\right) = -6K$$

$$\log 0.6 = -6K \qquad \text{...}(i)$$

2. In the second case,

$$\theta_0 = 30°\text{C}$$

$$\theta_1 = 60°\text{C}, \theta_2 = x, t = 10 \text{ min.}$$

$$\int_{60}^{x} \frac{d\theta}{\theta - \theta_0} = -10K$$

$$\log\left(\frac{x-30}{60-30}\right) = -10K \qquad \text{...}(ii)$$

Dividing (*ii*) by (*i*)

$$\frac{\log\left(\frac{x-30}{60-30}\right)}{\log(0.6)} = \frac{10}{6}$$

$$\log\left(\frac{x-30}{60-30}\right) = \frac{5}{3}\log(0.6)$$

$$\log\left(\frac{x-30}{30}\right) = \log(0.6)^{5/3}$$

$$\frac{x-30}{30} = (0.6)^{5/3}$$

$$\therefore \quad x = \mathbf{42.80\ °C}$$

### EXAMPLE 14.6

*A body initially at 80°C cools to 64°C in 5 minutes and to 52°C in 10 minutes. What will be its temperature after 15 minutes and what is the temperature of the surroundings?*

**SOLUTION.** $\int_{\theta_1}^{\theta_2} \frac{d\theta}{\theta - \theta_0} = -K \int dt = -Kt$

1. In the first case

$$\theta_1 = 80°\text{C}, \theta_2 = 64°\text{C}, t = 5 \text{ min.}$$

$$\int_{80}^{64} \frac{d\theta}{\theta - \theta_0} = -5K$$

$$\log\left(\frac{64 - \theta_0}{80 - \theta_0}\right) = -5K \quad ...(i)$$

2. In the second case,

$$\theta_1 = 64°\text{C}, \theta_2 = 52°\text{C},$$

$$t = 10 - 5 = 5 \text{ min}$$

$$\int_{64}^{52} \frac{d\theta}{\theta - \theta_0} = -5K$$

$$\log\left(\frac{52 - \theta_0}{64 - \theta_0}\right) = -5K \quad ...(ii)$$

From equations (*i*) and (*ii*)

$$\frac{64 - \theta_0}{80 - \theta_0} = \frac{52 - \theta_0}{64 - \theta_0}$$

$$\therefore \quad \theta_0 = \mathbf{16°C}$$

(*iii*) Let the temperature after 15 minutes be $\theta_2$

$$\theta_1 = 52°\text{C}, \theta_0 = 16°\text{C},$$

$$t = 15 - 10 = 5 \text{ minutes}$$

$$\int_{52}^{\theta_2} \frac{d\theta}{\theta - \theta_0} = -5K$$

$$\log\left(\frac{\theta_2 - 16}{52 - 16}\right) = -5K \quad ...(iii)$$

Equating (*ii*) and (*iii*)

$$\frac{\theta_2 - 16}{52 - 16} = \frac{52 - 16}{64 - 16}$$

$$\theta = \mathbf{43°C}$$

### EXAMPLE 14.7

*Find the specific heat of a liquid which takes 2 minutes in cooling from 50°C to 40°C in a vessel in which the same volume of water takes 5 minutes in cooling through the same range of temperature. Mass of water = 100 g, mass of liquid = 85 g, water equivalent of the vessel = 10 g.*

**SOLUTION.** $\frac{MC + w}{t_2} = \frac{m + w}{t_1}$

$$M = 85 \, g$$

$$C = ?$$

$$w = 10\ g$$
$$m = 100\ g$$
$$t_2 = 2 \text{ minutes} = 120\ s$$
$$t_2 = 5 \text{ minutes} = 300\ s$$
$$C = \frac{(m+w)\,t_2}{Mt_1} - \frac{w}{M}$$
$$C = \left[\frac{(100+10)\,120}{85 \times 300}\right] - \frac{10}{85}$$
$$C = \mathbf{0.4 \text{ calories/g-K}}$$

**EXAMPLE 14.8**

*Find the value of the universal gas constant R for one gram molecule of a gas. One gram molecule of a gas at N.T.P. occupies 22400 cm²*

**SOLUTION.**

$$P = 76 \text{ cm of Hg} = 76 \times 13.6 \times 981 \text{ dynes/cm}^2$$
$$V = 22400 \text{ cm}^3$$
$$T = 273\ K$$
$$PV = RT$$
$$R = \frac{PV}{T} = \frac{76 \times 13.6 \times 981 \times 22400}{273}$$
$$R = \mathbf{8.31 \times 10^7 \text{ ergs/mole-K}}$$

**Note.** The value of $R$ is the same for all gases, provided the mass of the gas is one gram molecule.

**EXAMPLE 14.9**

*Calculate the specific heat of air at constant volume, given that specific heat at constant pressure is 0.23, density of air at N.T.P. 1.293 gram/litre and J = 4.2 × 10⁷ ergs/cal.*

**SOLUTION.**

$$C_p = 0.23,\ J = 4.2 \times 10^7 \text{ ergs/cal},\ C_v = ?$$

Density of air at N.T.P. = 1.293 g/litre

Volume of one gram of air at N.T.P.

$$= \frac{1000}{1.293}\ cc$$
$$PV = rT$$
$$r = \frac{PV}{T}$$
$$= \frac{76 \times 13.6 \times 980 \times 1000}{273 \times 1.293}$$
$$C_p - C_v = \frac{r}{J}$$
$$C_v = C_p - \frac{r}{J}$$
$$= 0.23 - \frac{76 \times 13.6 \times 980 \times 1000}{273 \times 1.293 \times 4.2 \times 10^7}$$
$$C_v = 0.23 - 0.0683$$
$$= \mathbf{0.1617}$$

**EXAMPLE 14.10**

*Calculate the difference in the two specific heats of one gram of helium, given that the molecular weight of helium = 4 and the gram molecular volume of helium at N.T.P. = 22.4 litres.*

**SOLUTION.** $$C_p - C_v = \frac{r}{J}$$

Volume of one gram of helium at N.T.P.

$$V = \frac{22400}{4} = 5600\ cm^3$$

$$PV = rT$$

$$r = \frac{PV}{T}$$

$$r = \frac{76 \times 13.6 \times 980 \times 5600}{273}$$

$$C_p - C_v = \frac{r}{J}$$

$$= \frac{76 \times 13.6 \times 980 \times 5600}{273 \times 4.2 \times 10^7}$$

$$= \mathbf{0.4946}$$

**EXAMPLE 14.11**

*An engine consumes 25* litres *of gasoline per hour. The calorific value of gasoline is 6 × $10^6$ calories per* litre. *The output of the engine is 35 kilowatts. Calculate the efficiency of the engine.*

**SOLUTION.** Total heat produced by gasoline in one hour = $25 \times 6 \times 10^6$ cals.

$$\text{Heat produced per second} = \frac{25 \times 6 \times 10^6}{3600} \text{ cals/s}$$

$$\text{Input} = \frac{25 \times 6 \times 10^6 \times 4.2}{3600} \text{ Joules/s, } i.e. \text{ watts}$$

[1 calorie = 4.2 joules]

$$\text{Output} = 35000 \text{ watts}$$

$$\text{Efficiency} = \frac{\text{Useful output}}{\text{Input}} = \frac{35000 \times 3600}{25 \times 6 \times 10^6 \times 4.2}$$

$$= \frac{1}{5}$$

$$\%\ \text{efficiency} = \mathbf{20\%}$$

**EXAMPLE 14.12**

*Calculate Einstein's frequency for a case for which $\theta_E$ = 236 K. Given k = 1.4 × $10^{-23}$ joule/K and h = 6.6 × $10^{-34}$ Joule sec.* **[Kanpur 2000]**

**SOLUTION.** Einstein's frequency $\nu_E$ is expressed by the equation

$$\nu_E = \frac{\theta_E k}{h}$$

$$= \frac{237 \times 1.4 \times 10^{-23}}{6.6 \times 10^{-34}}$$

$$= \mathbf{5.0 \times 10^{12}\ sec^{-1}}.$$

### EXAMPLE 14.13

*For copper $\nu = 2.49 \times 10^{12}$ cycle/sec. (Hertz). Calculate the characteristic temperature $\theta_E$. Given : $h = 6.6 \times 10^{-34}$ joule/s and $k = 1.38 \times 10^{-23}$ joule (K)* **[Rajasthan 2000]**

**SOLUTION.** The characteristic temperature $\theta_E$ is given by

$$\theta_E = \frac{h\nu}{k}$$

$$= \frac{6.6\times10^{-34}\times2.49\times10^{12}}{1.38\times10^{-23}}$$

$$= \mathbf{119.1\ K}$$

### EXAMPLE 14.14

*The specific heat c of a certain substance at very low temperature is found to depend on the absolute temperature T according to the relation $c = AT^3$ where A is a constant.*

*(i) If units are calorie, gram and Kelvin degree, what units must the constant A have ?*

*(ii) How much heat must be added to raise the temperature of a mass m of the material from $T_1$ to $T_2$ ?* **[Purvanchal 2002]**

**SOLUTION.** (*i*) Given, $c = AT^3$

or $A = \dfrac{c}{T^3}$

The units of specific heat $c$ is cal. per gm. per Kelvin and that of $T$ is Kelvin. Substituting, the units of A will be

$$\frac{\text{Cal. per g. per Kelvin}}{(\text{Kelvin})^3} = \text{cal. per g per (kelvin)}^4$$

$$= \mathbf{cal.\ g^{-1}\ K^{-4}}$$

(*ii*) Heat required to raise the temperature of material of mass $m$ through temperature $dT$ is given by

$$dQ = mcdT$$

∴ The heat required to raise its temperature from $T_1$ to $T_2$ will be

$$Q = \int dQ \int_{T_1}^{T_2} mcdT$$

$$= \int_{T_1}^{T_2} mA\,T^3 dT = mA\left[\frac{T^4}{4}\right]_{T_1}^{T_2}$$

$$= \frac{1}{4}mA\left(T_2^4 - T_1^4\right)$$

### EXAMPLE 14.15

*(i) Calculate Einstein's frequency for copper for which $\theta_E = 230$ K. Given : $h = 6.6 \times 10^{-34}$ joule –sec; Boltzmann's constant $K = 1.37 \times 10^{-23}$ joule/K.*

*(ii) Use result (i) to show that the classical theory results $C_V = 3R$ should be valid for copper if $T > 230$ °C.* **[Mumbai 2001]**

**SOLUTION.** (*i*) Einstein's frequency $\nu_E$ in given by

$$\theta_E = \frac{h\nu_E}{k}$$

where $h$ is Planck's constant and $k$ is Boltzmann's constant.

$$\therefore \quad \text{Einstein's frequency, } \nu_E = \frac{\theta_E k}{h} = \frac{230 \times 1.37 \times 10^{-23}}{6.6 \times 10^{-34}} \text{sec}^{-1}$$

$$= \mathbf{4.79 \times 10^{12} \ sec^{-1}}$$

(*ii*) $$C_V = 3R \text{ if } T \gg \frac{h\nu}{k}$$

Here, $$\frac{h\nu}{k} = \frac{6.6 \times 10^{-34} \times 4.79 \times 10^{12}}{1.37 \times 10^{-23}} = \mathbf{230\ K}$$

## EXERCISE–XIV

**14.1** What are the principles of calorimetry ? **[Madras 2004/S]**

**14.2** Describe the continuous flow calorimeter. What are the advantages of the method? How do you determine $C_v$ of a gas by using this method ?

**14.3** Describe fully the working of the continuous flow calorimeter of 'Callendar and Barnes'. What are the advantages of this method? What information has it given regarding the specific heat of water ? **[Madras 2004/W]**

**14.4** Define principal specific heat of a gas. **[Alagappa 2004; Madras 2004]**

**14.5** Describe and explain Joly's steam calorimeter method for finding the specific heat of a gas at constant volume.

**14.6** Point out the difference between the two principal specific heats of a gas and show that for an ideal gas $C_p - C_v = R$ **[Madras 2005/S]**

**14.7** Indicate the manner in which the specific heat of a solid varies with temperature and outline the theory which explains this variation. **[Garhwal 2005]**

**14.8** Discuss Newton's law of cooling. How will you determine the specific heat of a liquid by Newton's law of cooling ? **[M.S. Uni. Tamil Nadu 2007, Madras 2005]**

**14.9** Describe the Regnault's method to find the specific heat of a gas at constant pressure. **[Alagappa 2004]**

**14.10** Discuss the working of a bell calorimeter in determining the calorific value of a fuel.

**14.11** Deduce the relation between the specific heat of a gas at constant pressure and at constant volume. **[Gujarat 2005]**

**14.12** Describe how the specific heat of a gas at constant pressure is determined accurately.

**14.13** Describe Nernst vacuum calorimeter and indicate briefly how it may be used to determine the specific heat at low temperatures.

**14.14** Describe fully the working of a continuous flow electric calorimeter for the measurement of $C_v$ of a gas.

**14.15** Distinguish between the specific heat of a gas at constant pressure and at constant volume. Describe an accurate method to determine the specific heat of a gas at constant pressure.

**14.16** Describe Joly's differential steam calorimeter. How will you use this apparatus for determining the specific heat of a gas at constant volume ? Mention the sources of error.

**14.17** Point out the difference between the two principal specific heats of a gas. Describe an accurate method to determine the specific heat of a gas at constant pressure.

**14.18** State and explain Newton's law of cooling. How will you find the specific heat of a liquid by the method of cooling ?

**14.19** Describe Nernst vacuum calorimeter for determining the specific heat of a good-conducting solid at low temperatures.

**14.20** Describe the nature of variation of specific heat of solids with temperature. **[Rohilkhand 2000, 2001]**

**14.21** Explain Clement and Desorme's method of determining the ratio of two specific heat capacities of a gas. **[Alagappa 2004]**

**14.22** Explain with necessary theory, the experimental determination of $C_V$ by the method of Regnault. Also explain how you will apply correction to the final temperature. **[Madras W/2004]**

**14.23** Give the theory of determining the specific heat capacity of a liquid using the method of mixtures, with cooling correction. **[Madras S/2004]**

**14.24** Define calorific value of fuels. At what temperature water has the minimum value of specific heat? **[Madras 2005/S]**

**14.25** Explain briefly the method of mixtures in the determination of specific heat of a solid. **[Madras 2004/W]**

**14.26** Discuss Dulong and Petit's law and explain the variation of the atomic heat of a substance with temperature. **[Garhwal 2000, Madras 2005]**

**14.27** How do you explain Dulong and Petit's law according to quantum theory. **[Madras 2005, Garhwal 2005]**

**14.28** Explain Einstein's theory of specific heat. **[Garhwal 2002, 2005]**

**14.29** Discuss Debye-Einstein theory regarding atomic heats of solids. Discuss Debye's $T^3$ law. **[Gujarat 2004, 2005, Patna 2005 Garhwal 2005]**

**14.30** State Dulong and Petit's law. Show how the departure from this law at low temperature has been explained by Einstein's theory. **[Mumbai 2005]**

**14.31** State Dulong and Petit's law. How does density of elements depend on specific heat ? **[Madras S/2004]**

**14.32** Obtain an expression for the specific heat of solid according to Einstein's theory. Discuss the success and limitations of this theory. **[Patna 2004, Garhwal 2000, 2001, Agra 2001]**

**14.33** State the assumptions of Einstein's theory to explain the variation of specific heat of a solid with temperature. **[Rohilkhand 2003]**

**14.34** Discuss the variation of specific heat capacity of solids with temperature and give Einstein's theory to explain it. **[Rohilkhand 2000, 2001, Imp.]**

**14.35** Discuss Debye's theory of specific heat of solids as an improvement over the Einstein's theory. **[Agra 2000]**

**14.36** Discuss the main features of Einstein's and Debye's models of specific heat of solids. Derive an expression for the specific heat on the basis of Debye's model. Point out the importance of Debye's temperature.

**14.37** Write short notes on the following :

(*i*) Specific heat of gases **[Kanpur 2000]**

(*ii*) Specific heat of diatomic gases **[Agra 2000]**

(*iii*) Specific heat of liquids

(*iv*) Specific heat of solids

(*v*) Dulong and Petit's law **[Agra 2001]**

(*vi*) Einstein's theory of specific heat of solids

(*vii*) Debye's theory of specific heat of solids **[Kanpur 2001]**

**14.38** A liquid takes 10 minutes to cool from 90ºC to 60ºC. How much time will it take to cool from 70ºC to 40ºC? The temperature of the surroundings is 30ºC. **[Ans. 20 minutes]**

**14.39** A body takes 6 minutes to cool from 80ºC to 50ºC. How much time will it take to cool from 60ºC to 30ºC ? The temperature of the surroundings is 20ºC. **[Ans. 12 minutes]**

**14.40** A body cools from 70°C to 50°C in 6 minutes. What will be its temperature after the next 6 minutes ? The temperature of the surroundings is 15°C. **[Ans. 32.37°C]**

**14.41** A liquid cools in 7 minutes from 60°C to 40°C. What will be its temperature after the next 7 minutes? The temperature of the surroundings is 10°C. **[Ans. 28°C]**

**14.42** A liquid takes 5 minutes to cool from 70°C to 50°C. How much time will it take to cool from 50°C to 35°C ? The temperature of the surroundings in 20°C. **[Ans. 6.785 minutes]**

**14.43** A body takes 3 minutes to cool from 70°C to 60°C. What will be its temperature after the next 6 minutes ? The temperature of the surroundings is 25°C. **[Ans. 46.17°C]**

**14.44** A liquid takes 4 minutes to cool from 65°C to 50°C. What will be its temperature after the next 10 minutes ? The temperature of the surroundings is 35°C. **[Ans. 37.65°C]**

**14.45** A petrol engine consumes 25 kg of petrol per hour. The calorific value of petrol is $11.4 \times 10^6$ cals per kg. The power of the engine is 99.75 kilowatts. Calculate the efficiency of the engine. **[Ans. 30%]**

**14.46** A Nernst calorimeter of lead weighing 396.3 grams and surrounded by a bath at temperature 61 $K$, on being heated electrically for four minutes showed a rise in temperature of 1.219 $K$ under condition of no heat leakage. Calculate the atomic heat of lead at 61 $K$, if the potential drop and average current during the interval of heating were 1.586 volts and 0.1444 ampere respectively. (Atomic weight of lead = 206.4 and $J = 4.18$ joules/calorie.) **[Ans. 5.619]**

**14.47** A copper calorimeter of mass 100$g$ containing 150 $cm^3$ of a liquid of specific heat 0.6 and specific gravity 1.2, is found to cool at the rate of 2°C per minute when its temperature is 50°C above that of the surroundings. If the liquid is emptied out and 150 $cm^3$ of a second liquid of specific heat 0.4 and specific gravity 0.9 are substituted, what will be its rate of cooling, when the temperature is 40°C above that of its surroundings ? (Sp. heat of copper 0.1.) **[Ans. 2.76°C per minute]**

**14.48** The characteristic Debye temperature for diamond is 1860 K and the characteristic Einstein temperature ($h\nu/k$) is 1450 K. Calculate $C_V$ at 20 K from the Einstein and Debye equation. **[Mumbai 2005]**

**14.49** With the help of Einstein Model, find specific heat of solid at high temperature.

Boltzmann constant $k = 1.38 \times 10^{-23}$ Joule/mole °K

Avogadro's number No $= 6.2 \times 10^{26}$ atom/kg. mole **[Gujarat 2005]**

[**Hint :** $C_V = 3R$, since $T > \frac{h\nu}{k}$. Here $R = N_o k$ ]

**14.50** If the frequency of all atoms of solid are equal and Einstein temperature $\theta_E = 100$ K, then calculate the specific heat of the solid at temperature (*i*) 25 K, (*ii*) 500 K **[Ans. :** (*i*) 1.75 (*ii*) 7.5 Cal/g.mole $K^{-1}$**]**

**14.51** Calculate for copper

(*i*) if $\nu_E = 4.8 \times 10^{12}$ Hertz, calculate Einstein's temperature $\theta_E$ for it, and

(*ii*) if Einstein's temperature is $\theta_E = 100$ $K$, calculate Einstein's frequency $\nu_E$.

**[Ans.** (*i*) 229.6 K (*ii*) $2.08 \times 10^{12}$ $sec^{-1}$**]**

## OBJECTIVE QUESTIONS

**14.1** Einstein's theory of specific heat :

(*a*) accepts different frequencies of molecular vibrations

(*b*) accepts same frequency of all molecular vibrations

(*c*) rejects molecular vibrations

(*d*) none of these **(Garhwal 2001)**

**14.2** According to Dulong and Petit's law, the average energy of an atom of a solid at temperature T is :

(*a*) $\frac{1}{2}kT$ (*b*) $kT$

(*c*) $2\ kT$ (*d*) $3\ kT$ **(Kanpur 2001)**

**14.3** According to Dulong and Petit's law, the atomic heat of an element of constant volume :

(*a*) increases with increase of temperature.

(*b*) decreases with increase of temperature

(*c*) becomes zero at absolute zero

(*d*) is constant.

**14.4** At high temperature, the atomic heat of a solid attains a value :

(*a*) Infinite (*b*) $\frac{5}{2}R$

(*c*) 3R (*d*) $\frac{7}{2}R$

**14.5** At very low temperature, the atomic heat of a solid attains a value :

(*a*) zero (*b*) Infinite

(*c*) $\frac{5}{2}R$ (*d*) 3R

**14.6** Einstein's theory of specific heat of solids assumes that the atoms of a solid vibrate simple harmonically (S.H.M.) :

(*a*) like longitudinal waves (*b*) like transverse waves

(*c*) all with the same frequency (*d*) in a complex manner

**14.7** According to Debye's theory, the atomic heat of a solid at low temperature varies as :

(*a*) proportional to its absolute temperature

(*b*) proportional to square of its absolute temperature

(*c*) proportional to cube of its absolute temperature

(*d*) independent of temperature.

## ANSWERS

| | | | |
|---|---|---|---|
| **14.1** (*b*) | **14.2** (*d*) | **14.3** (*d*) | **14.4** (*c*) |
| **14.5** (*a*) | **14.6** (*c*) | **14.7** (*c*) | |

# Transmission of Heat

# 15

### At a Glance

• Introduction • Coefficient of Thermal Conductivity • Rectilinear Flow of Heat along a Bar • Periodic Flow of Heat • Thermal Conductivity Measurements • Propagation of Heat Waves in the Earth's Crust • Ingen-Hausz Experiment • Searle's Method • Lee's Method for Metals • Forbe's Method to find K • Lee's Method for Bad Conductors • Lee's Disc Method for Bad Conductors • Lee's Method for Liquids • Spherical Shell Method (Radial flow of Heat) • Cylindrical Flow of a Heat • Thermal Conductivity of a Rubber • Thermal Conductivity of a glass • Heat Flow Through a Compound Wall • Accretion of Ice on Ponds • Wiedmann-Franz Law • Practical Applications of Conduction of Heat • Davy's Safety Lamp • Convection • Applications of Convection • Central Heating System • Differential Air Thermoscope • Thermopile • Properties of Thermal Radiation • Thermos Flask

## Introduction

Heat can be transferred from one place to the other by three different ways *viz*, conduction, convection and radiation.

**Conduction** is the process in which heat is transmitted from one point to the other through the substance without the actual motion of the particles. When one end of a metal bar is heated, the molecules at the hot end vibrate with higher amplitude (kinetic energy) and transmit the heat energy from one particle to the next and so on. However, the particles remain in their mean positions of equilibrium. This process of conduction is prominent in the case of solids. The property of transmission of heat has been used in Davy's safety lamp. Materials having less conductivity *e.g.* granite, brick walls etc. are used in the construction of a cold storage, furnace of a boiler etc. The space between the two walls of a thermos flask is evacuated because vacuum is a poor conductor of heat. The air enclosed in the woollen fabric helps in protecting us from cold, because air is a poor conductor of heat.

**Convection** is the process in which heat is transmitted from one place to the other by the actual movement of the heated particles. It is prominent in the case of liquids and gases. Land and sea breezes and trade winds are formed due to convection. Convection plays an important part in ventilation, gas filled electric lamps and heating of buildings by hot water circulation.

**Radiation** is the process in which heat is transmitted from one place to the other directly without the necessity of the intervening medium. We get heat radiations directly from the sun without affecting the intervening medium. Heat radiations can pass through vacuum. Their properties are similar to light radiations. Heat radiations also form a part of the electromagnetic spectrum.

## 15.1 Coefficient of Thermal Conductivity

Consider a cube of side $x$ cm and area of each face $A$ sq. cm. The opposite faces of the cube are maintained at temperature $\theta_1$ and $\theta_2$ where $\theta_1 > \theta_2$. Heat is conducted in the direction of the fall of temperature. Quantity of heat conducted across the two opposite faces

$$Q \propto A$$

$$Q \propto (\theta_1 - \theta_2)$$

$$Q \propto t$$

$$\propto \frac{1}{x}$$

$$\therefore \quad Q \propto \frac{A(\theta_1 - \theta_2)}{x}$$

or

$$Q = \frac{KA(\theta_1 - \theta_2)t}{x}$$

Here $K$ is a constant called the coefficient of thermal conductivity of the material of the cube.

If $A = 1$ sq. cm, $(\theta_1 - \theta_2) = 1$°C, $t = 1$ $s$, $x = 1$ cm

then

$$Q = K$$

Therefore, the coefficient of thermal conductivity is defined as the amount of heat flowing in one second across the opposite faces of a cube of side one cm maintained at a difference of temperature of 1 °C.

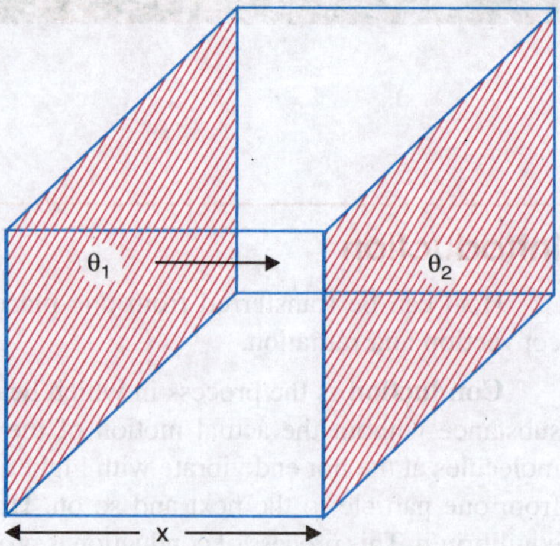


Fig. 15.1

**Temperature gradient.** The quantity $\frac{\theta_1 - \theta_2}{x}$ represents the rate of fall of temperature with respect to distance. The quantity $\frac{d\theta}{dx}$ represents the rate of change of temperature with respect to the distance. As temperature decreases with increase in distance from the hot end, the quantity $\frac{d\theta}{dx}$ is negative and is called the temperature gradient.

$$\therefore \quad Q = -KA\left[\frac{d\theta}{dx}\right]t$$

**Dimensions of *K***

$$K = -\frac{Q}{A\frac{d\theta}{dx}t} = -\frac{Q\,dx}{A\,d\theta\,t}$$

$Q$ represents energy and its dimensions are,

$$[Q] = [ML^2T^{-2}],$$

$$[dx] = [L],\ [A] = [L^2],\ [d\theta] = [\theta],\ [t] = [T]$$

$$[K] = \frac{[ML^2T^{-2}][L]}{[L^2][\theta][T]} = [MLT^{-3}\theta^{-1}]$$

**Thermal Diffusivity.** It is defined as the ratio of thermal conductivity to thermal capacity per unit volume. Let the density of the substance be $\rho$ and its specific heat $S$.

$$h = \frac{K}{\left[\frac{mS}{V}\right]} = \frac{K}{\rho S}$$

$h$ is also called thermometric conductivity.

## 15.2 Rectilinear Flow of Heat along a Bar

### Fourier Equation of Heat Flow

Consider a bar of uniform area of cross-section which is heated at one end. Heat is flowing along the length of the bar. Consider two planes $P_1$ and $P_2$ perpendicular to the length of the bar at distances $x$ and $x + \delta x$ from the hot end. The temperature gradient at the plane $P_1 = \frac{d\theta}{dx}$. Here $\theta$ is the excess of temperature (above the surroundings) of the bar at $P_1$. The excess of temperature

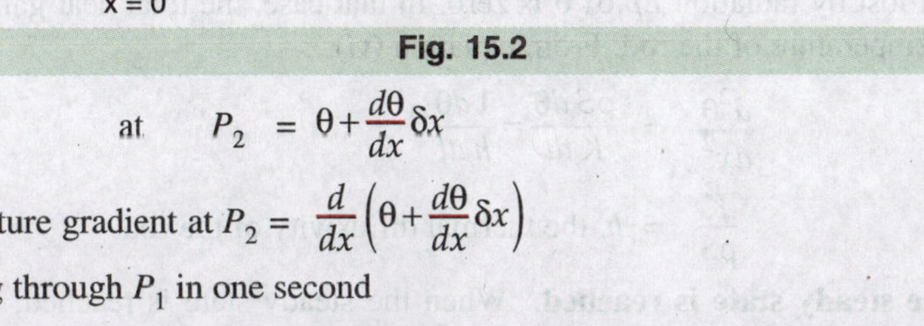


Fig. 15.2

$$\text{at} \quad P_2 = \theta + \frac{d\theta}{dx}\delta x$$

The temperature gradient at $P_2 = \frac{d}{dx}\left(\theta + \frac{d\theta}{dx}\delta x\right)$

Heat flowing through $P_1$ in one second

$$Q_1 = -KA\frac{d\theta}{dx} \qquad ...(i)$$

Heat flowing through $P_2$ in one second

$$Q_2 = -KA\frac{d}{dx}\left(\theta + \frac{d\theta}{dx}\delta x\right) \quad ...(ii)$$

Heat gained per second by the rod between the planes $P_1$ and $P_2$

$$Q = Q_1 - Q_2$$

$$= -KA\frac{d\theta}{dx} + KA\frac{d}{dx}\left(\theta + \frac{d\theta}{dx}\delta x\right)$$

$$Q = KA\frac{d^2\theta}{dx^2}\delta x \quad ...(iii)$$

**Before the steady state is reached.** The quantity of heat $Q$ is used in two ways before the steady state is reached. Partly the heat is used to raise the temperature of the rod and the rest is lost due to radiation. Let the rate of rise of temperature of the bar be $\frac{d\theta}{dt}$. The heat used per second to raise the temperature of the rod

$$= \text{mass} \times \text{specific heat} \times \frac{d\theta}{dt}$$

$$= (A \times \delta x)\rho \times S\frac{d\theta}{dt} \quad ...(iv)$$

Where $A$ is the area of cross-section of the rod, $\rho$ is the density of the material and $S$ is the specific heat.

The heat lost per second due to radiation

$$= Ep\,\delta\,x\,\theta \quad ...(v)$$

where $E$ is the emissive power of the surface, $p$ is the perimeter and $\theta$ is the average excess of temperature of the bar between the planes $P_1$ and $P_2$.

$$Q = A\delta x\rho \times S\frac{d\theta}{dt} + Ep.\delta x.\theta$$

Substituting the value of $Q$ from equation (*iii*)

$$KA\frac{d^2\theta}{dx^2}\delta x = A\delta x\rho\;S\frac{d\theta}{dt} + Ep\delta x\theta$$

$$\frac{d^2\theta}{dx^2} = \frac{\rho S}{K}.\frac{d\theta}{dt} + \frac{Ep}{KA}\theta \quad ...(vi)$$

This is the general equation that represents the rectilinear flow of heat along a bar of uniform area of cross-section, and is known as **Fourier's differential equation.**

## Special Cases

1. **When heat lost by radiation is negligible.** When the rod is completely covered by insulating materials, the heat lost by radiation $Ep.\delta x.\theta$ is zero. In that case, the total heat gained by the rod is used to raise the temperature of the rod. From equation (*vi*)

$$\frac{d^2\theta}{dx^2} = \frac{\rho S\,d\theta}{K\,dt} = \frac{1}{h}\frac{d\theta}{dt} \quad ...(vii)$$

Here, $\frac{K}{\rho S} = h$, the thermal diffusivity of the rod.

2. **After the steady state is reached.** When the steady state is reached, the rod does not require any further heat to raise its temperature. At this stage,

$$\frac{d\theta}{dt} = 0$$

From equation (*vi*)

$$\frac{d^2\theta}{dx^2} = \frac{Ep}{KA}\theta$$

Taking $\frac{Ep}{KA} = \mu^2$

$$\frac{d^2\theta}{dx^2} = \mu^2\theta \quad ...(viii)$$

The general solution of this equation is

$$\theta = Ae^{+\mu x} + Be^{-\mu x} \quad ...(ix)$$

Where *A* and *B* are two unknown constants to be determined from the boundary conditions of the problem.

If the bar is sufficiently long, we can assume that under steady state, no heat is lost from the free end of the bar. This is due to the fact that nearly the whole of the heat energy is lost from the sides as radiation and the free end will be at the temperature of the surroundings.

(*a*) Suppose the bar is of infinite length.

Let the excess of temperature above the surroundings of the hot end be $\theta_0$ and of the other end be zero.

At $x = 0, \theta = \theta_0$

From equation (*ix*)

$$\theta_0 = A + B$$

At $x = \infty, \theta = 0$

From equation (*ix*),

$$0 = Ae^{\infty}$$

But $e^{\infty}$ cannot be zero, therefore

$$A = 0$$

Taking $A = 0$

$$\theta_0 = B$$

Substituting the values of *A* and *B* in equation (*ix*)

$$\theta = \theta_0 e^{-\mu x} \quad ...(x)$$

This equation represents the excess of temperature of a point at a distance *x* from the hot end after the steady state is reached and this represents an exponential curve. The equation is useful in Ingen-Hausz experiment.

(*b*) Suppose the bar is sufficiently long and is of finite length *L*. Then the boundary conditions are

At $x = 0$

$$\theta = \theta_0$$

and $\frac{d\theta}{dx} = 0$ at $x = L$

The values of *A* and *B* in this case are

$$A = \frac{\theta_0}{1 + e^{+2\mu L}}$$

and

$$B = \frac{\theta_0}{1 + e^{-2\mu L}}$$

The solution of equation (*ix*) will be

$$\theta = \theta_0 \left[ \frac{e^{\mu x}}{1+e^{2\mu L}} + \frac{e^{-\mu x}}{1+e^{-2\mu L}} \right]$$

## 15.3 Periodic Flow of Heat

**(Propagation of heat wave in an insulated rod with one end heated sinusoidally).**

So far we have discussed the steady flow of heat along a metallic rod when one end is continuously heated. Let us now examine what happens when one end of metallic bar is alternately heated and cooled *i.e.,* heat is supplied periodically (and not continuously). It is observed that the temperature at different points in the bar also varies periodically.

**Principle:** *Whenever the end is heated, the heat waves flow outward from it and when it is cooled, the heat waves travel towards it and thus two waves are set up in the rod travelling in opposite directions.*

Assuming that there is no loss due to rediation as it is well insulated, the problem becomes simpler. It is just like the propagation of periodic heat only in one direction. Let the periodic variation of temperature be represented by

$$\theta = \theta_0 \cos \omega t \qquad ...(i)$$

where $\theta_0$ is the amplitude and $\omega$ the angular frequency of heat wave or temperature wave along the rod. The variation can be represented by a sinusoidal curve of damped nature as shown in Fig. 15.3

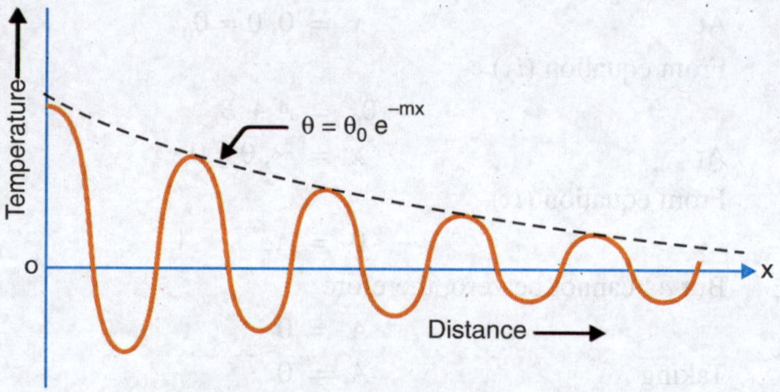


Fig. 15.3

Let the heat flows along positive *x*–axis, the Fourier equation representing unidirectional heat flow along a logged bar is given by

$$\frac{d^2\theta}{dx^2} = \frac{1}{h}\frac{d\theta}{dt} \qquad ...(ii)$$

where $h = \frac{K}{\rho S}$, the thermal diffusivity of the rod.

At $x = 0, \quad \theta = \theta_0 \cos \omega t$

and at $x \to \infty$, $\theta$ do not approach to $\infty$.

Applying these boundary conditions, let us solve the differential equation (*ii*). Let us assume the solution

$$\theta = A\, e^{\alpha x + i\beta t} \qquad ...(iii)$$

where $i = \sqrt{-1}$ and $A$, $\alpha$ and $\beta$ are constants.

Differentiating eqn (*iii*), we get

$$\frac{d\theta}{dt} = Ai\beta\, e^{\alpha x + i\beta t} \qquad ...(iv)$$

and $$\frac{d^2\theta}{dx^2} = A\propto^2 e^{\alpha x + i\beta t}$$

Substituting eqn. (*iv*), in eqn. (*ii*), we get

$$Ai\beta\, e^{\alpha x + i\beta t} = h\, A\alpha^2\, e^{\alpha x + i\beta t}$$

$$\therefore \qquad i\beta = h\,\alpha^2 \qquad \text{or} \qquad \alpha^2 = \frac{i\beta}{h}$$

or
$$\alpha = \pm\sqrt{\left(\frac{i\beta}{h}\right)}$$

Since $(1+i)^2 = 2i$, we have $\sqrt{i} = \frac{1+i}{\sqrt{2}}$

$$\therefore \qquad \alpha = \pm\sqrt{\left(\frac{\beta}{h}\right)} \times \frac{1+i}{\sqrt{2}}$$

$$= \pm\sqrt{\left(\frac{\beta}{2h}\right)} \times (1+i)$$

Substituting in eqn. (*iii*), we have

$$\theta = A\, e^{\pm\sqrt{(\beta/2h)}\,(1+i)x + i\beta t}$$

Since $x \neq \infty$, when $x \to \infty$, we reject +ve sign and hence

$$\theta = A\, e^{-\sqrt{(\beta+2h)}.(1+i)x + i\beta t}$$

$$= A\, e^{-\sqrt{(\beta+2h)}.\,x + i\{\beta t - \sqrt{(\beta+2h)}.\,x\}}$$

This equation can be written in the form

$$\theta = A\, e^{-\sqrt{\beta+2h}.\,x}\left[\cos\left\{\beta t - \sqrt{\left(\frac{\beta}{2h}\right)}\right\}.\,x + i\sin\left\{\beta t - \sqrt{\left(\frac{\beta}{2h}\right)}.\,x\right\}\right] \qquad \ldots(v)$$

At $x = 0$, $\theta$ becomes becomes equal to $A\cos\beta t$. ($\theta = A\cos\beta t$)

But we have assumed, at $x = 0$, $\theta = \theta_0 \cos\omega t$.

Hence, $A = \theta_0$ and constant $\beta = \omega$. Substituting these velues of A and $\beta$ in eqn. (*v*), we get

$$\theta = \theta_0\, e^{-\sqrt{(\cot 2h)}.x}.\cos\left[\omega t - \sqrt{\left(\frac{\omega}{2h}\right)}.x\right] \qquad \ldots(vi)$$

Just to give a general form of wave equation,

We introduce an arbitrary phase constant $\phi$ in the constant term, so that equation (*vi*) now takes the form

$$\theta = \theta_0\, e^{-\sqrt{(\omega/2h)}.x}\cos\left[\left\{\omega t - \sqrt{(\omega/2h)}.x\right\} - \phi\right] \qquad \ldots(vii)$$

This equation represents a **progressive wave,** travelling with a velocity $\sqrt{2\omega h} = \sqrt{\frac{4\pi h}{T}}$ and

$$\text{the amplitude} \qquad = \theta_0\, e^{-\sqrt{(\omega/2h)}.x}$$

$$= \theta_0\, e^{-mx},$$

where $m = \sqrt{\omega/2h}$, and is the **damping factor**.

Substituting for $\omega$ and $h$, the **velocity of heat wave** is given by

$$v = \sqrt{(2\omega h)}$$

$$= \sqrt{2\times\frac{2\pi}{T}\times\frac{K}{\rho S}}$$

$$= \sqrt{\frac{4\pi K}{T\rho S}}$$

But $$v = \frac{\lambda}{T}$$

Therefore, $$\frac{\lambda}{T} = \sqrt{\frac{4\pi K}{T\rho S}}$$

or $$K = \frac{\lambda^2 \rho S}{4\pi T} \qquad \text{...(viii)}$$

## 15.4 Thermal Conductivity Measurement (Angstrom's method)

Angstrom determined the conductivity of a metal bar by using the method of periodic heat flow. He used a long metal bar and enclosed one end of it in a chamber through which steam and cold water at known temperatures could be alternately passed for a definite time, say 12 minutes each. *i.e.,* the end of the bar is heated by steam for 12 minutes and then cooled for 12 minutes by passing cold water through the chamber. Thus, the periodic time of heat flow is $T = 24$ minutes. The process is repeated a number of time so that finally the mean temperature at any point remains constant, *i.e.,* a steady state is reached. Then the maximum temperature at two consecutive points is noted. The distance between these two points is measured by using thermo-couple. From this $\lambda$ can be measured. Knowing $\lambda$ and $T$, the value of $K$, the thermal conductivity can be determined by using the formula given by eqn. (*viii*); the values of $\rho$ and $S$ are known from standard table provided.

## 15.5 Propogation of Heat Waves in the Earth's Crust

This concept is well applicable in case of earth as it rotates about its axis. Its surface is being heated and cooled during daytime and night time alternately. The earth, thus, subjected to a periodic heating and cooling as a result a heat or temperature wave with a period of 24 hours ($T = 24$ hours) is propagated in to the interior of the earth. This wave is called the **dirunal wave.**

The dirunal waves produced are due to earth's rotations around itself. In addition to this, earth rotates around the sun also. Hence earth receives more heat in summer than in the winter. Therefore, this gives rise to another *heat wave* travelling into the interior of the earth. Since the period is of 1 year, this second type is known as **annual wave.**

The conductivity of the earth's crust $K$ can be calculated in a similar manner discussed above and comes out to be

$$K = \frac{\lambda^2 \rho S}{4\pi T} \qquad \text{...(ix)}$$

The periodic variation of temperature ($T$) can be measured by embedding sensitive thermometers in the earth at various depths and thus knowing $\lambda$, $K$ can be calculated using equation (*ix*). Here density $\rho$ and specific heat $S$ is known.

**EXAMPLE 15.1**

*In a periodic flow of heat along an iron bar, the periodic time is 4 minute. If the temperature travels maximum 6 cm in 1 minute, calculate the thermal conductivity of iron. Density of iron = 7.8 gm/cm$^3$, Sp.heat of iron = 0.11 cal/gm °C.*

**SOLUTION.** The velocity of heat wave along the bar is

$$v = \sqrt{2\omega h} = \sqrt{2\times\frac{2\pi}{T}\times\frac{K}{\rho S}} = \sqrt{\frac{4\pi K}{T\rho S}}$$

or $$v^2 = \frac{4\pi K}{T\rho S}$$

Here, $v = 6$ cm/min = 0.1 cm/sec.

$T = 4 \times 60 = 240$ sec

$S = 0.11$ cal/gm °C, $\rho = 7.8$ gm/cm$^3$.

Therefore, the thermal conductivity of the metal bar

$$K = \frac{v^2 T\rho S}{4\pi} = \frac{(0.1)^2 \times 240 \times 7.8 \times 0.11}{4 \times 3.14}$$

$$= \mathbf{0.1639\ cal\ cm^{-1}\ sec^{-1}\ {}^{\circ}C^{-1}}.$$

## 15.6 Ingen-Hausz Experiment

Ingen-Hausz devised an experiment to compare the thermal conductivities of several materials.

Take a box containing hot water or oil. Rods of different materials of the same length and area of cross-section (same size and shape) are taken. Their outer surfaces are electroplated with the same material and are equally polished. This is done such that the emissive power $E$ for all the rods is the same. Rods are fixed at the base of the vessel (Fig. 15.4) and their portions outside the vessel are coated uniformly with wax. Hot water or oil is put into the vessel and the wax starts melting. It is found that the wax melts up to different lengths on different rods.

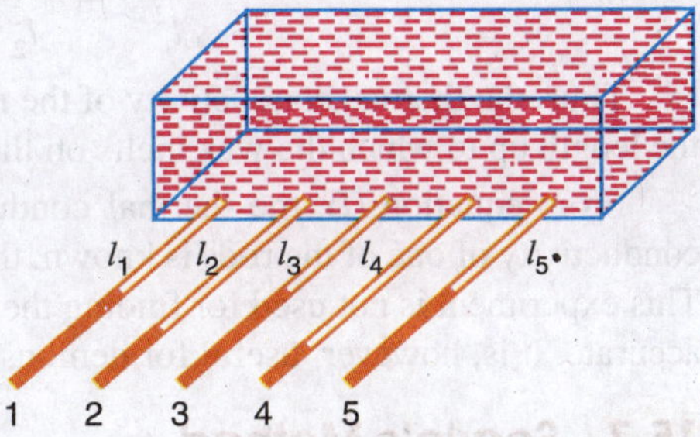


Fig. 15.4

Let the lengths up to which the wax has melted be $l_1, l_2, l_3$..... etc., $\theta_0$ be the excess of temperature of the hot bath above the room temperature and $\theta_m$ the excess of temperature of the melting point of wax above the room temperature.

As, $$\theta = \theta_0 e^{-\mu x}$$

For the first rod,

$$\theta_m = \theta_0 e^{-\mu_1 l_1}$$

For the second rod, $$\theta_m = \theta_0 e^{-\mu_2 l_2}$$

For the third rod,

$$\theta_m = \theta_0 e^{-\mu_3 l_3}$$

..................................

$$\therefore \quad \theta_m = \theta_0 e^{-\mu_1 l_1} = \theta_0 e^{-\mu_2 l_2} = \theta_0 e^{-\mu_3 l_3}$$

$$= \theta_0 e^{-\mu_3 l_3}$$

$$\therefore \quad \mu_1 l_1 = \mu_2 l_2 = \mu_3 l_3 = \text{.........}$$

But $$\mu_1 = \sqrt{\frac{Ep}{K_1 A}}$$

$$\mu_2 = \sqrt{\frac{Ep}{K_2 A}}$$

$$\mu_3 = \sqrt{\frac{Ep}{K_3 A}}$$

Here, $E$, $p$ and $A$ are the same for all the rods

$$\therefore \quad l_1\sqrt{\frac{Ep}{K_1 A}} = l_2\sqrt{\frac{Ep}{K_2 A}} = l_3\sqrt{\frac{Ep}{K_3 A}} = \ldots\ldots\ldots$$

$$\therefore \quad \frac{l_1}{\sqrt{K_1}} = \frac{l_2}{\sqrt{K_2}} = \frac{l_3}{\sqrt{K_3}} = \text{constant}$$

or
$$\frac{K_1}{l_1^2} = \frac{K_2}{l_2^2} = \frac{K_3}{l_3^2} = \text{constant} \qquad \ldots(i)$$

Thus, the thermal conductivity of the material of the rod is directly proportional to the square of the length up to which the wax melts on the rod.

From equation (*i*), the thermal conductivities of any two materials can be compared. If the conductivity of one of the rods is known, the thermal conductivities of other rods can be determined. This experiment is not used for finding the thermal conductivity because the results obtained are not accurate. It is, however, useful for demonstration purposes.

## 15.7 Searle's Method

*XY* is a rod whose coefficient of thermal conductivity is to be determined. The end *X* is enclosed in a steam jacket and steam is continuously passed through it. Heat is conducted along the rod and when the steady state is reached, the thermometers $\theta_1$, $\theta_2$, $\theta_3$ and $\theta_4$ record constant temperatures. The mass of water (*m* grams) collected in a known interval of time (*t* seconds) is noted (Fig. 15.5).

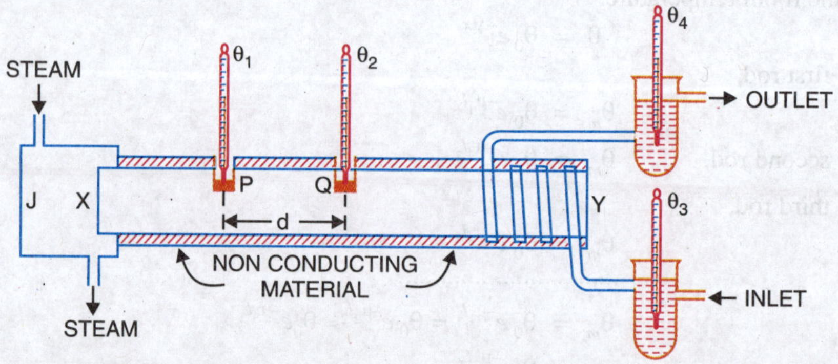


Fig. 15.5

When the steady state is reached, the quantity of heat passing from the section *P* to *Q* is constant and the same quantity of heat flows through the section at *Y*. The rod is well covered with a nonconducting material like, cotton etc. to avoid conduction and convection of heat to the atmosphere.

Quantity of heat that flows from the section at *P* to *Q* in one second,

$$= \frac{KA(\theta_1 - \theta_2)\times 1}{d} \qquad \ldots(i)$$

Amount of heat gained by water in one second

$$= \frac{m(\theta_4 - \theta_3)}{t} \qquad ...(ii)$$

Equating (*i*) and (*ii*)

$$\frac{KA(\theta_1 - \theta_2)}{d} = \frac{m(\theta_4 - \theta_3)}{t}$$

$$K = \frac{md(\theta_4 - \theta_3)}{At(\theta_1 - \theta_2)}$$

Knowing all the quantities, $K$ can be calculated.

## 15.8 Lee's Method for Metals

The thermal conductivity of a metal is different at different temperatures. This method is useful in measuring the thermal conductivity of a metal at various temperatures, preferably at low temperatures.

The apparatus consists of a copper frame $C$ surrounded by a heating coil $H_1$. The specimen rod $AB$ about 10 cm long and 0.5 cm in diameter is fitted inside the frame $C$ (Fig. 15.6). The end $A$ of the rod is heated by a heating coil $H$ and $T_1$ and $T_2$ are two platinum resistance thermometers. These thermometers measure temperatures at the two points distant $d$ apart on the rod $AB$. The end $B$ is fixed in a socket inside the frame $C$. The whole arrangement is surrounded by a Dewar flask $F$.

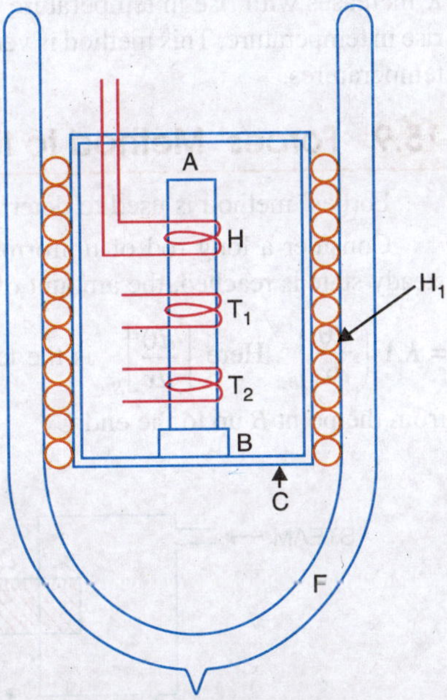


Fig. 15.6

The whole apparatus is immersed in liquid air so that its temperature is below the temperature at which the thermal conductivity is to be measured. The liquid air is removed and with the help of the heating coil $H_1$ the whole apparatus is brought to the desired temperature. The heating coil $H$ is heated by passing a known current through it and the temperatures of $T_1$ and $T_2$ are noted, after they show constant temperature, $\theta_1$ and $\theta_2$ *i.e.,* after the steady state has been reached. Let the current flowing be $I$ amperes and potential difference across the heating coil $H$ be $E$ volts.

$$\text{Heat produced per second} = \frac{EI}{4.2} \text{ calories}$$

$$\therefore \qquad \frac{EI}{4.2} = \frac{KA(\theta_1 - \theta_2)}{d}$$

$$K = \frac{EId}{4.2\,A(\theta_1 - \theta_2)} \qquad ...(i)$$

Hence, the thermal conductivity of the given metal at a desired temperature can be calculated.

**Correction.** When the end $A$ of the rod $AB$ is heated by the heating coil $H$, the copper frame conducts some heat and its temperature rises. This heat does not escape and there is backward flow of heat from the end $B$. Due to this reason, $(\theta_1 - \theta_2)$ will be less than the correct value. Therefore, correction is a necessity. The resistances of $H_1$ and $H$ are made equal. Before performing the main experiment, the same current $I$ is passed through $H_1$, without passing the current through $H$. The difference of temperature $d\theta_2$ between $T_2$ and $T_1$ is noted after the steady state is reached. The observation is repeated after passing the same current $I$ through $H$. Let the difference of temperature be $d\theta_2$. The correction to be applied

$$= \frac{d\theta_1 + d\theta_2}{2}$$

Hence the correct difference in temperature is

$$(\theta_1 - \theta_2) + (d\theta_1 + d\theta_2)/2$$

$$K = \frac{EId}{4.2\, A\left[(\theta_1 - \theta_2) + (d\theta_1 + d\theta_2)/2\right]}$$

Lee found $K$ for various metals between $-170°C$ and $18°C$. For silver, copper, zinc and aluminium $K$ increases with rise in temperature between $-170°C$ and $100°$. For iron and lead, $K$ decreases with rise in temperature. This method is very useful in finding the thermal conductivity of metals at different temperatures.

## 15.9 Forbes' Method to find K

Forbes' method is used to determine **the *absolute*** conductivity of different metals.

Consider a long rod of uniform area of cross-section heated at one end (Fig. 15.7). After the steady state is reached, the amount of heat passing per second across the cross-section at the point $B$ $= KA\left(\frac{d\theta}{dx}\right)_B$. Here $\left[\frac{d\theta}{dx}\right]_B$ is the temperature gradient at $B$. The amount of heat lost by radiation from the point $B$ up to the end $C$

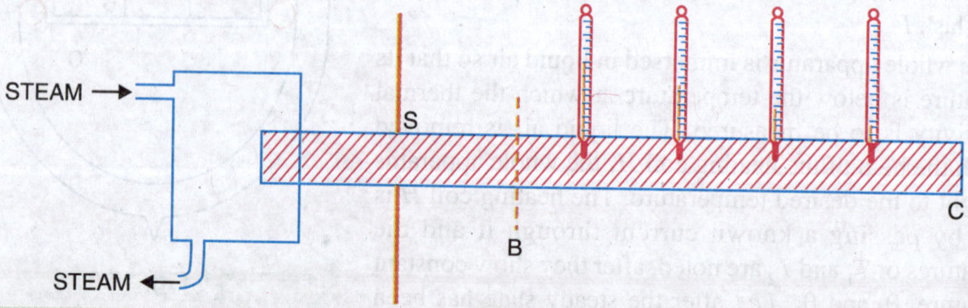


**Fig. 15.7**

$$= \int_B^C (\rho\, A\, dx)\, S\, \frac{d\theta}{dt}$$

$$\therefore \qquad KA\left[\frac{d\theta}{dx}\right]_B = \int_B^C (\rho\, A\, dx)\, S\, \frac{d\theta}{dt}$$

$$K = \frac{\rho S \int_B^C \frac{d\theta}{dt}\, dx}{\left[\frac{d\theta}{dx}\right]_B} \qquad \ldots(i)$$

To find $\int_B^C \frac{d\theta}{dt}\, dx$ and $\left[\frac{d\theta}{dx}\right]_B$ the experiment is divided into two parts.

***(i)* Static Experiment.** A long rod about 2 metres in length is taken and its one end is heated (Fig. 15.7). Forbes heated the rod for about six hours. After the steady state is reached, the thermometers fixed in the rod at various positions are read. A graph is plotted between the temperature (θ) and the distance ($x$) from the hot end. The graph is an exponential curve as shown in (Fig. 15.8). Take the point $B$ and draw a tangent to the curve corresponding to the point $B$.

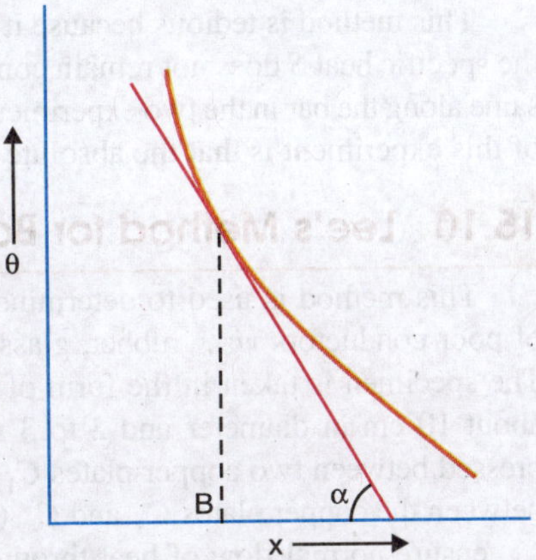


Fig. 15.8

Here $$\left[\frac{d\theta}{dx}\right]_B = \tan\alpha$$

***(ii)* Dynamic Experiment.** The original bar or a small bar of the same material having the same area of cross-section is heated to the temperature of steam. The bar is exposed to the atmosphere and a thermometer is fixed at its middle portion (Fig. 15.9). Note the temperature of the bar after regular intervals of time (say one minute). Draw a graph between temperature and time.

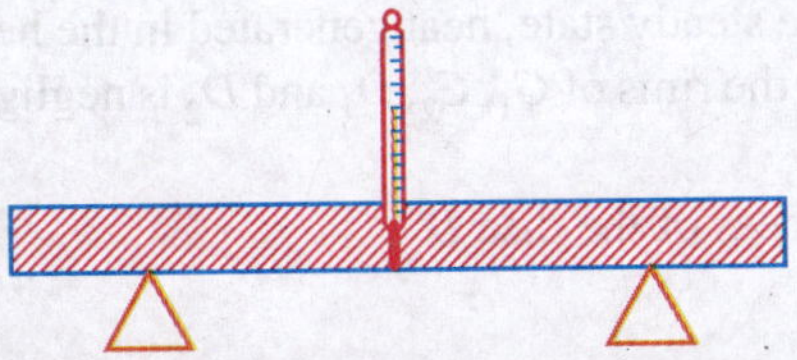

Fig. 15.9

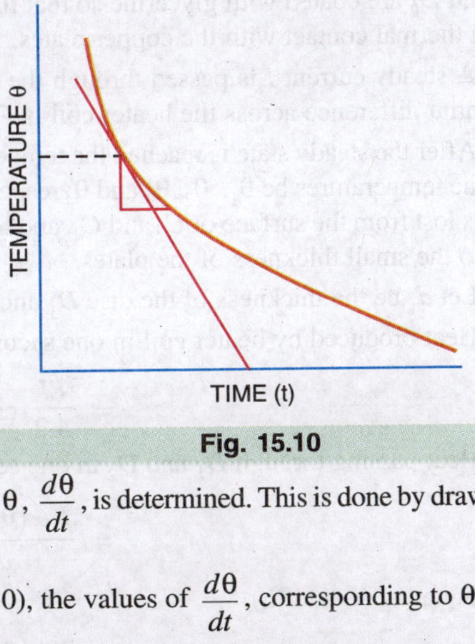


Fig. 15.10

From the graph in (Fig. 15.10), for various values of θ, $\frac{d\theta}{dt}$, is determined. This is done by drawing tangents to the curve at various points on the curve.

From the two graphs given in (Figs. 15.8 and 15.10), the values of $\frac{d\theta}{dt}$, corresponding to θ and $x$ are determined. A graph is plotted between $\frac{d\theta}{dt}$ and $x$. The graph is produced to meet the $x$-axis. Corresponding to the point $B$, a point is located on the graph (Fig. 15.11). The area of the shaded portion = $\int_B^C \frac{d\theta}{dt}\, dx$. The area is measured with the help of the planimeter or from the graph directly.

$$K = \frac{\rho S \int_B^C \frac{d\theta}{dt}\, dx}{\left[\frac{d\theta}{dx}\right]_B}$$

$$K = \frac{\rho S \text{ (area of the shaded portion)}}{\tan\alpha}$$

Hence $K$ can be calculated.

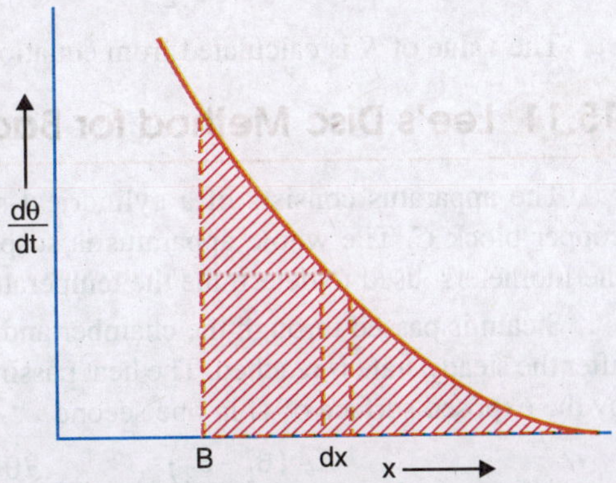


Fig. 15.11

This method is tedious because it takes a long time and three graphs are to be drawn. Moreover, the specific heat $S$ does not remain constant at all temperatures. Also the distribution of heat is not the same along the bar in the two experiments. Therefore, this experiment is not accurate. But the advantage of this experiment is that the absolute conductivity of the material of the rod can be determined.

## 15.10 Lee's Method for Bad Conductors

This method is used to determine the thermal conductivity of poor conductors *viz.,* rubber, glass, ebonite, wood, cork etc. The specimen is taken in the form of two thin discs $D_1$ and $D_2$ about 10 cm in diameter and 2 to 3 mm thick. The disc $D_1$ is pressed between two copper plates $C_1$ and $C_2$ and $D_2$ is pressed between the copper plates $C_3$ and $C_4$ (Fig. 15.12). Discs $C_1$ and $C_3$ ensure normal flow of heat through the experimental plates $D_1$ and $D_2$. $H$ is a heater coil and $T_1$, $T_2$, $T_3$ and $T_4$ are four thermo-couples used to measure temperatures. The surfaces of $D_1$ and $D_2$ are coated with glycerine so that these surfaces make good thermal contact with the copper plates.

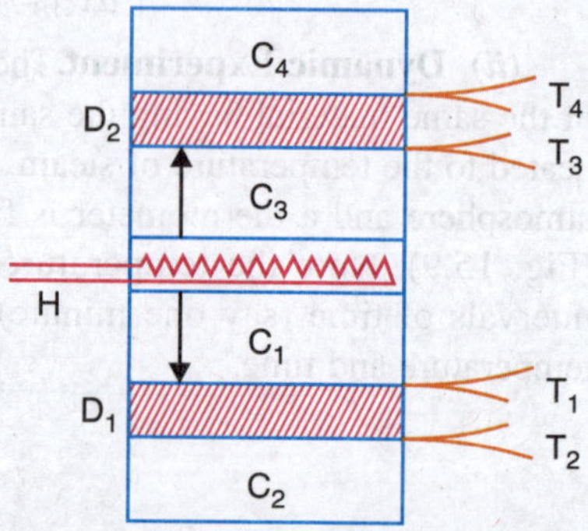


Fig. 15.12

A steady current $I$ is passed through the heater coil $H$. The potential difference across the heater coil is $E$.

After the steady state is reached the temperatures of the thermometers $T_1$, $T_2$, $T_3$ and $T_4$ are noted. Let the temperatures be $\theta_1$, $\theta_2$, $\theta_3$ and $\theta_4$ respectively. In the steady state, heat generated in the heater coil is lost from the surface of $C_2$ and $C_4$, and heat lost from the rims of $C_1$, $C_2$, $D_1$ and $D_2$ is negligible due to the small thickness of the plates.

Let $d_1$ be the thickness of the disc $D_1$ and $d_2$ the thickness of the disc $D_2$.

Heat produced by heater coil in one second

$$= \frac{EI}{4.2} \text{ calories} \qquad ...(i)$$

Heat passing through $D_1$ and $D_2$ in one second

$$= \frac{KA(\theta_1 - \theta_2)}{d_1} + \frac{KA(\theta_3 - \theta_4)}{d_2} \qquad ...(ii)$$

Equating (*i*) and (*ii*)

$$\frac{EI}{4.2} = \frac{KA(\theta_1 - \theta_2)}{d_1} + \frac{KA(\theta_3 - \theta_4)}{d_2} \qquad ...(iii)$$

The value of $K$ is calculated from equation (*iii*).

## 15.11 Lee's Disc Method for Bad Conductors

The apparatus consists of a cylindrical steam chamber $A$, the specimen disc $D$ and brass or copper block $C$. The whole apparatus is suspended from the stand (Fig. 15.13). $T_1$ and $T_2$ are the thermometers used to determine the temperature after the steady state is reached.

Steam is passed through the chamber and the readings of the thermometers $T_1$ and $T_2$ are noted after the steady state is reached. The heat passing through $D$ in one second is equal to the heat radiated by the exposed surface of $C$ in one second.

$$\therefore \quad \frac{KA(\theta_1 - \theta_2)}{d} = ms\frac{d\theta}{dt}\left[\frac{A+S}{2A+S}\right] \qquad ...(i)$$

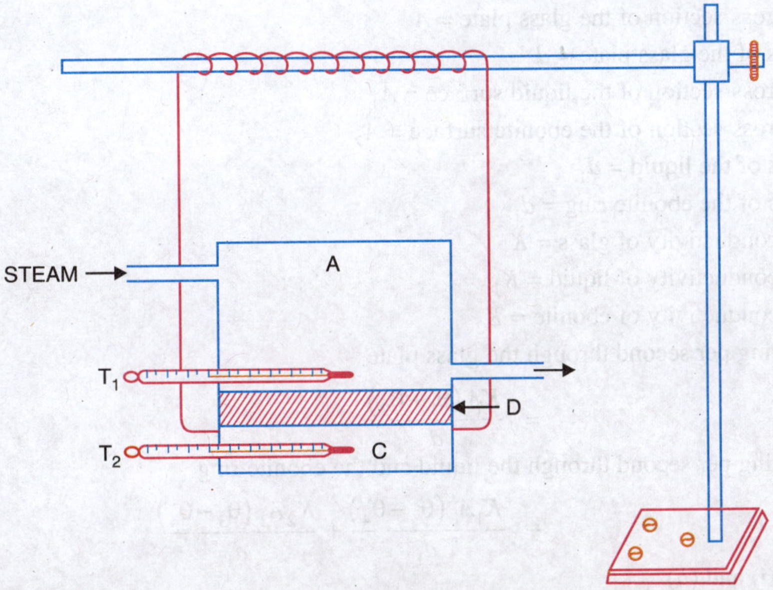


Fig. 15.13

Here $\left[\frac{A+S}{2A+S}\right]$ is the fraction of the total area exposed to the surroundings.

Here $A$ is the area of cross-section of $D$ and $C$. $S$ is the area of the curved surface of $C$, $\frac{d\theta}{dt}$ is the rate of cooling at temperature $\theta_2$, $m$ is the mass and $s$ is the specific heat of $C$.

To find $\frac{d\theta}{dt}$, the disc $D$ is removed and $C$ makes contact with the steam chamber. $C$ is removed when its temperature is about 10ºC higher than $\theta_2$. It is placed over two knife edges and its temperature is observed after equal intervals of time (say one minute). A graph is drawn between temperature and time. From the graph, the value of $\frac{d\theta}{dt}$ at temperature $\theta_2$ is found.

From equation (*i*), $K$ can be calculated.

## 15.12 Lee's Method for Liquids

To determine the thermal conductivity of a liquid, convection current should be determined. To overcome this difficulty, heat should flow from the upper surface to the lower surface of the liquid.

Lee designed an apparatus shown in (Fig. 15.14). $C_1$, $C_2$, $C_3$ and $C_4$ are the copper plates of uniform thickness. $H$ is the heater coil. A glass plate is pressed between $C_2$ and $C_3$. The thermocouples $T_1$, $T_2$, $T_3$ and $T_4$ are used to determine the temperatures. The liquid is contained in a cylindrical tube having an ebonite ring.

A steady current of $I$ amperes is passed through the heater coil $H$. After the steady state is reached, the temperatures shown by the thermocouples are noted. Let these temperatures be $\theta_1$, $\theta_2$, $\theta_3$ and $\theta_4$.

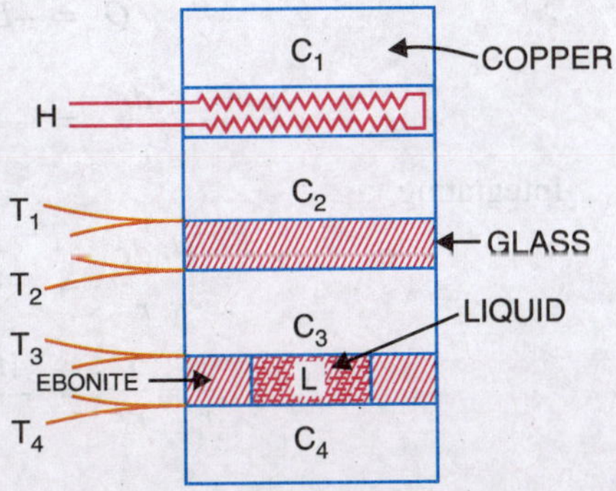


Fig. 15.14

Area of cross-section of the glass plate = $A$
Thickness of the glass plate = $d$
Area of cross-section of the liquid surface = $A_1$
Area of cross-section of the ebonite surface = $A_2$
Thickness of the liquid = $d_1$
Thickness of the ebonite ring = $d_1$
Thermal conductivity of glass = $K$
Thermal conductivity of liquid = $K_1$
Thermal conductivity of ebonite = $K_2$
Heat flowing per second through the glass plate

$$= \frac{KA(\theta_1 - \theta_2)}{d} \quad ...(i)$$

Heat flowing per second through the liquid and the ebonite ring

$$= \frac{K_1A_1(\theta_3 - \theta_4)}{d_1} + \frac{K_2A_2(\theta_3 - \theta_4)}{d_1} \quad ...(ii)$$

Equating (*i*) and (*ii*)

$$\therefore \quad \frac{KA(\theta_1 - \theta_2)}{d} = \frac{K_1A_1(\theta_3 - \theta_4)}{d_1} + \frac{K_2A_2(\theta_3 - \theta_4)}{d_1} \quad ...(iii)$$

The value of $K$, for the liquid can be calculated from equation (*iii*). However, there is small error due to the heat lost by the sides of $C_1$, $C_2$, $C_3$, $C_4$, the glass plate and the ebonite ring.

## 15.13 Spherical Shell Method (Radial Flow of Heat)

Consider two thin spherical shells $A$ and $B$ of radii $r_1$ and $r_2$. The specimen is contained between these two shells $A$ and $B$ (Fig. 15.15). A heating element is placed at the centre of the shells. Heat is conducted through the specimen from the inner to the outer shell. Let the temperatures of the inner and the outer shells be $\theta_1$ and $\theta_2$, after the steady state is reached.

Consider an imaginary shell of radius $r$ and thickness $dr$ having temperatures $\theta$ and $\theta + d\theta$ on its inner and the outer surfaces respectively. The quantity of heat conducted per second through this shell,

$$Q = -KA\frac{d\theta}{dr} \quad ...(i)$$

Here $A = 4\pi r^2$

$$\therefore \quad Q = -K.\, 4\pi r^2 \frac{d\theta}{dr}$$

$$\frac{dr}{r^2} = -\frac{4\pi K}{Q} d\theta$$

Integrating,

$$\int_{r_1}^{r_2} \frac{dr}{r^2} = -\frac{4\pi K}{Q}\int_{\theta_1}^{\theta_2} d\theta$$

$$\frac{1}{r_1} - \frac{1}{r_2} = \frac{4\pi K(\theta_1 - \theta_2)}{Q}$$

$$K = \frac{Q(r_2 - r_1)}{4\pi r_1 r_2(\theta_1 - \theta_2)} \quad ...(ii)$$

$K$ is calculated from equation (*ii*).

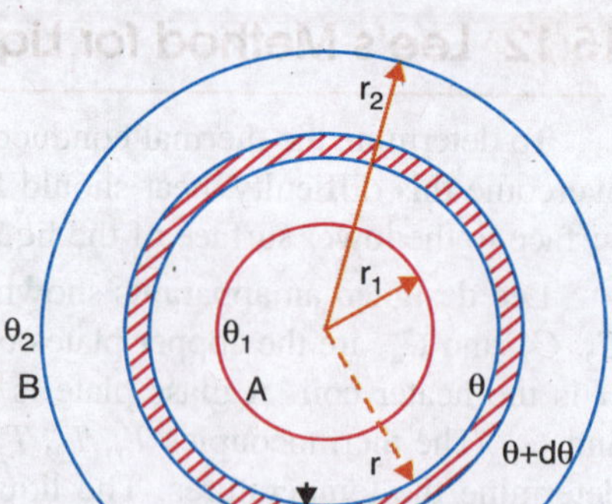

Fig. 15.15

## 15.14 Cylindrical Flow of Heat

Consider a cylindrical tube of length $l$, inner radius $r_1$ and outer radius $r_2$. After the steady state is reached, the temperature on the inner surface is $\theta_1$ and on the outer surface it is $\theta_2$. Here $\theta_1 > \theta_2$. Heat is conducted radially across the wall of the tube. Consider an element of thickness $dr$ and length $l$ at a distance $r$ from the axis (Fig. 15.16).

The quantity of heat flowing per second across the element

$$Q = -KA\frac{d\theta}{dr}$$

But

$$A = 2\pi rl$$

$\therefore$

$$Q = 2\pi Krl\frac{d\theta}{dr}$$

$$Q.\frac{dr}{r} = -2\pi Kl\, d\theta \qquad \text{...}(i)$$

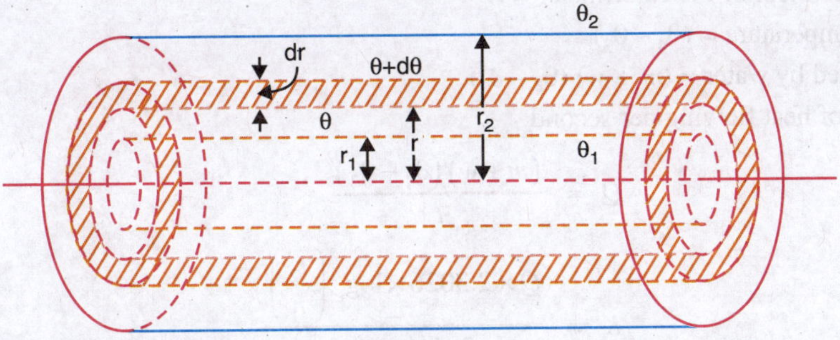


**Fig. 15.16**

$Q$ is constant after the steady state is reached.

Integrating equation ($i$)

$$Q\int_{r_1}^{r_2}\frac{dr}{r} = -2\pi Kl\int_{\theta_1}^{\theta_2} d\theta$$

$$Q\left[\log_e\frac{r_2}{r_1}\right] = -2\pi Kl\,[\theta_2 - \theta_1]$$

$$Q\log_e\frac{r_2}{r_1} = 2\pi kl\,[\theta_1 - \theta_2]$$

$$K = \frac{Q\log_e\dfrac{r_2}{r_1}}{2\pi l\,(\theta_1 - \theta_2)} \qquad \text{...}(ii)$$

or

$$K = \frac{Q\times 2.3026\times\log_{10}\dfrac{r_2}{r_1}}{2\pi l\,(\theta_1 - \theta_2)} \qquad \text{...}(iii)$$

## 15.15 Thermal Conductivity of Rubber

The coefficient of thermal conductivity of a rubber tubing can be determined in the laboratory applying the principle of cylindrical flow of heat.

A known quantity of water is taken in a calorimeter C. A rubber tubing whose inner and outer radii are $r_1$ and $r_2$, is taken and a known length (say 50 cm) of it is immersed in water as shown in (Fig. 15.17). The initial temperature of water is noted. Let it be $\theta_3$. Steam is passed through the rubber tubing for a known time $t$ seconds (say 900 seconds). Let the final temperature of water be $\theta_4$ (after applying radiation correction) and temperature of steam be $\theta_1$. The average temperature on the outer surface of the rubber tubing

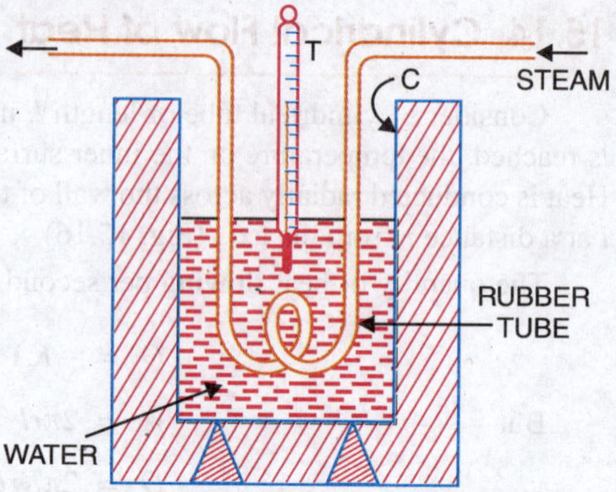

Fig. 15.17

$$\theta_2 = \frac{\theta_3 + \theta_4}{2}$$

Suppose, mass of water = $m$

Water equivalent of the calorimeter = $w$

Rise in temperature = $(\theta_4 - \theta_3)$

Heat gained by water = $(m + w)(\theta_4 - \theta_3)$

Quantity of heat flowing per second

$$Q = \frac{(m+w)(\theta_4 - \theta_3)}{t}$$

But
$$K = \frac{Q \times 2.3026 \times \log_e \left(\frac{r_2}{r_1}\right)}{2\pi l(\theta_1 - \theta_2)}$$

Substituting the values of $\theta_2$ and $Q$

$$K = \frac{(m+w)(\theta_4 - \theta_3) \times 2.3026 \times \log_{10}\left(\frac{r_2}{r_1}\right)}{2\pi l\left(\theta_1 - \frac{\theta_3 + \theta_4}{2}\right)t}$$

Thus, $K$ for rubber can be calculated.

## 15.16 Thermal Conductivity of Glass

A glass tube whose inner and outer radii are $r_1$ and $r_2$ respectively is fixed inside an outer jacket (Fig. 15.18). The apparatus is kept in the slanting position and water is allowed to flow inside the glass tube from lower to the upper end. The two thermometers $T_1$ and $T_2$ are used to determine the temperature of incoming and outgoing water. The rate of flow of water can be adjusted with the help of pinch cock.

Steam is passed inside the outer jacket and the rate of flow of water is adjusted such that the two thermometers show sufficient difference of temperature (say 5 to 10°C). When the two thermometers show constant temperature water flowing through the glass tube for known time $t$ seconds (say 900 seconds) is collected in a beaker. The spiral wire inside the glass tube is kept for making good contact of water with the walls of the tube. Suppose,

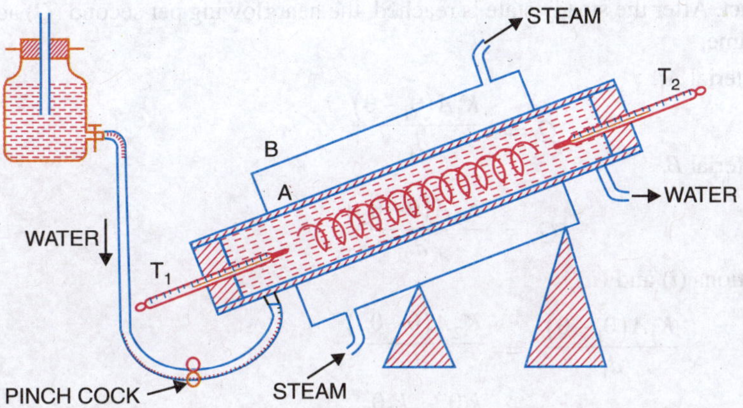


Fig. 15.18

Length of the tube inside the jacket = $l$

Temperature of the steam = $\theta_1$

Temperature of incoming water = $\theta_3$

Temperature of outgoing water = $\theta_4$

Mass of collected water = $m$

Time taken = $t$ seconds.

Quantity of heat gained by water in $t$ seconds = $m(\theta_4 - \theta_3)$

Quantity of heat that flows in one second across the wall of the tube,

$$Q = \frac{m(\theta_4 - \theta_3)}{t}$$

Average temperature of the inside of the glass tube

$$\theta_2 = \frac{\theta_3 + \theta_4}{2}$$

But
$$K = \frac{Q \times 2.3026 \times \log_{10}\left(\frac{r_2}{r_1}\right)}{2\pi l\left(\theta_1 - \frac{\theta_3 + \theta_4}{2}\right)t} \quad ...(i)$$

Substituting the values of $Q$ in equation $(i)$

$$\therefore \quad K = \frac{m(\theta_4 - \theta_3) \times 2.3026 \log_{10}\left(\frac{r_2}{r_1}\right)}{2\pi l\left(\theta_1 - \frac{\theta_3 + \theta_4}{2}\right)t} \quad ...(ii)$$

Hence, $K$ can be determined.

## 15.17 Heat Flow through a Compound Wall

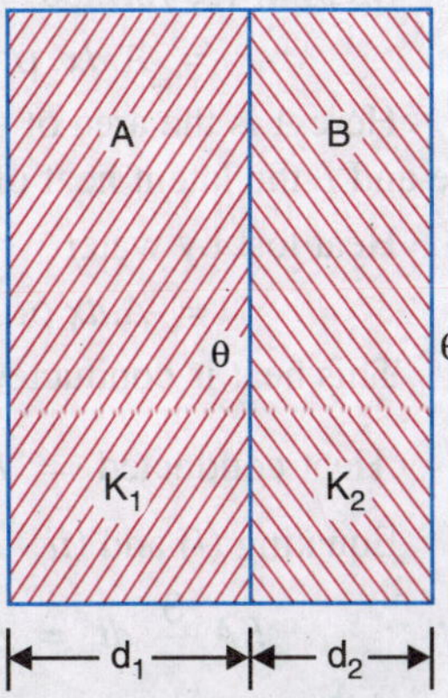


Fig. 15.19

Consider a compound wall (or a slab) made of two materials $A$ and $B$ of thickness $d_1$ and $d_2$ (Fig. 15.19). Let $K_1$ and $K_2$ be the coefficients of thermal conductivity of the two materials. $\theta_1$ and $\theta_2$ are the temperatures of the end faces ($\theta_1 > \theta_2$) and $\theta$ is the temperature of the surface in contact. After the steady state is reached, the heat flowing per second ($Q$) across any cross-section is the same.

For the material $A$,

$$Q = \frac{K_1 A(\theta_1 - \theta)}{d_1} \quad ...(i)$$

For the material $B$,

$$Q = \frac{K_2 A(\theta - \theta_2)}{d_2} \quad ...(ii)$$

From equations (*i*) and (*ii*),

$$\frac{K_1 A(\theta_1 - \theta)}{d_1} = \frac{K_2 A(\theta - \theta_2)}{d_2}$$

$$\theta = \frac{\frac{k_1\theta_1}{d_1} + \frac{k_2\theta_2}{d_2}}{\frac{K_1}{d_1} + \frac{K_2}{d_2}}$$

Substituting the value of $\theta$ in equation (*i*),

$$Q = \frac{A(\theta_1 - \theta_2)}{\frac{d_1}{K_1} + \frac{d_2}{K_2}} \quad ...(iii)$$

In general, for any number of walls or slabs,

$$Q = \frac{A(\theta_1 - \theta_2)}{\sum\left(\frac{d}{K}\right)} \quad ...(iv)$$

## 15.18 Accretion of Ice on Ponds

Consider a layer of ice $x$ cm thick on the surface of a pond. Let the temperature of air over the surface of ice be $-\theta°C$ and that of water below ice $0°C$. Suppose a thickness $dx$ of ice is formed in time $dt$ (Fig. 15.20).

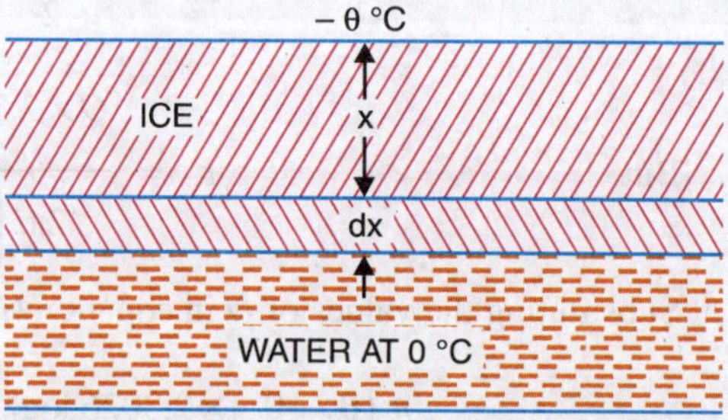


Fig. 15.20

Mass of ice formed

$$= A.dx.\,\rho$$

Here $A$ is the area of the pond, $\rho$ the density of ice and $L$ the latent heat of fusion of ice.

Heat lost by water

$$= A.dx.\rho.\,L \text{ calories} \quad ...(i)$$

This heat is conducted across a layer of ice of thickness $x$ upwards.

$$\text{Heat conducted} = KA.\frac{\theta}{x}\,dt \text{ calories} \quad ...(ii)$$

Equating (*i*) and (*ii*)

$$KA\frac{\theta}{x}\,dt = A\,dx.\rho.\,L$$

$$\frac{dx}{dt} = \frac{K\theta}{\rho L x} \quad ...(iii)$$

$\frac{dx}{dt}$ represents the rate of growth of the thickness of ice.

$$dt = \left(\frac{\rho L}{K.\theta}\right) x dx$$

Total time taken by the layer of ice to increase in thickness by $x$

$$\int dt = \frac{\rho L}{K.\theta} \int x \, . \, dx$$

$$t = \frac{\rho L}{K.\theta} \frac{x^2}{2} + \text{constant}$$

When $t = 0, x = 0$

Hence the constant is zero,

$$\therefore \quad t = \frac{\rho L}{2K.\theta} x^2 \quad \text{...(iv)}$$

## 15.19 Wiedemann-Franz Law

In 1853, Wiedemann and Franz, based on experimental results, put forward a law relating to the thermal conductivity and electrical conductivity of metals. According to this law :

*The ratio of the thermal and electrical conductivities is the same for all metals at the same temperature. Moreover, the ratio is directly proportional to the absolute temperature of the metal.*

Let $K$ and $\sigma$ be thermal and electrical conductivities of a metal at a temperature $T$ degrees absolute,

$$\frac{K}{\sigma} \propto T$$

or

$$\frac{K}{\sigma T} = \text{constant for all metals}$$

This law holds good for a large number of metals between + 100°$C$ and – 100°$C$. At low temperatures the ratio $K/\sigma$ decreases and the value tends to be zero at absolute zero. As the temperature of the metal is decreased, the thermal and electrical conductivities of the metal increase. But the increase in the electrical conductivity is higher and its value tends to infinity at absolute zero.

This corresponds to the super-conducting state of the metal.

Drude explained that thermal and electrical conductivities are due to free electrons in metals. He derived the expression

$$\frac{K}{\sigma T} = \frac{3R^2}{JN^2 e^2} = \text{constant,}$$

where $R$ is the universal gas constant, $N$ is the number of molecules in one gram molecule, $J$ is the Joule's mechanical equivalent of heat and $e$ is the charge on the electron.

**TABLE 15.1** Values of $\frac{K}{\sigma T} \times 10^8$

| *Metal* | *–100°C* | *–50°C* | *0°C* | *18°C* |
|---|---|---|---|---|
| Aluminium | 1.81 | 1.98 | 2.09 | 2.13 |
| Copper | 2.17 | 2.26 | 2.30 | 2.32 |
| Iron | 2.98 | 2.93 | 2.97 | 2.99 |
| Lead | 2.54 | 2.52 | 2.53 | 2.51 |
| Silver | 2.29 | 2.36 | 2.33 | 2.33 |
| Zinc | 2.39 | 2.40 | 2.45 | 2.43 |

## 15.20 Practical Applications of Conduction of Heat

Metals are good conductors of heat and wood, felt, brick, glass, granite, cotton, wool, cork, ebonite, rubber are bad conductors of heat.

1. Sauce pans, hot water buckets, kettles and other utensils are made of metal. They are provided with wooden or ebonite handles so that heat from the utensil is not conducted to the hand.
2. Ice box has a double wall made of tin or iron. The space in between the two walls is packed with cork or felt. This is done because cork and felt are poor conductors of heat and prevent the flow of the outside heat into the box.
3. Thick brick walls are used in the construction of a cold storage. Brick is a bad conductor of heat and does not allow outside heat to flow inside the cold storage.
4. Woollen clothes have fine pores filled with air. Air and wool are bad conductors of heat. Heat from the body does not flow outside to the atmosphere. Therefore, woollen clothes keep the body warm in winter.
5. Two shirts keep the body warmer than a single shirt of the same material and double the thickness. Between the two shirts a fine layer of air acts as a bad conductor and does not allow the heat from the body to flow out to the surroundings.
6. Quilts and bed clothings filled with cotton are used in winter. The air layers in the pores of the cotton are bad conductors of heat. Therefore, the flow of heat to outside is prevented.
7. A steel blade appears colder than a wooden handle in winter. Steel is a good conductor of heat. As soon as a person touches the blade, heat flows from the hand to the blade. Therefore, it appears colder. Since wood is a bad conductor of heat, heat does not flow from the hand to the handle.
8. In cold countries, the windows are provided with double doors. Air in between the two doors forms a non-conducting layer and does not allow heat to flow from inside to outside. In hot countries also, double door windows are used. Heat does not flow from outside to inside because air forms a non-conducting layer in between the two doors.
9. When a stopper fitted tightly to the bottle is to be removed the neck is gently heated. Glass is a poor conductor of heat. Neck expands but heat is not conducted to the stopper. The stopper can be removed easily.

## 15.21 Davy's Safety Lamp

It is also based on the principle of conduction of heat. A wire gauze is placed over a bunsen burner and the gas is lit above the wire gauze (Fig. 15.21). The gas comes out of the wire gauze. A flame appears at the top surface of the wire gauze. The gas below the wire gauze does not get sufficient heat for ignition. The wire gauze conducts away the heat of the flame above it and the temperature at the lower surface does not reach the ignition temperature.

In a Davy's safety lamp, the cylindrical metal gauze of high thermal conductivity surrounds the flame (Fig. 15.22). When this lamp is taken inside a mine, even if explosive gases are present, they do not get ignited because the wire gauze conducts away the heat of the flame. The temperature outside the wire gauze remains lower than the ignition temperature outside the wire gauze remains lower than the ignition temperature of the gases. In the absence of the wire gauze, the gases outside can explode.

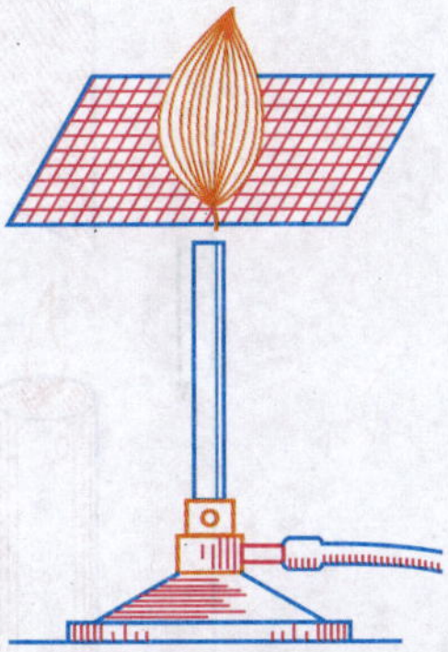

Fig. 15.21

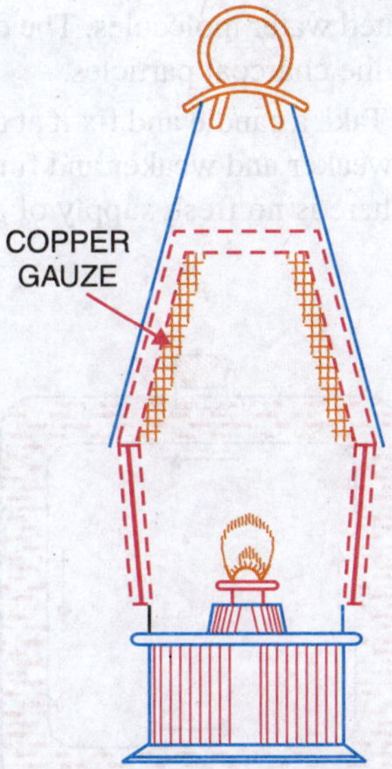


Fig. 15.22

As soon as a miner notices the presence of the explosive gases from the colour of the flames he gives a warning signal. The work is stopped and the mine is thoroughly ventilated.

## 15.22 Convection

**Experiment 1.** Take a flask containing water. Add a large crystal of $KMnO_4$. Heat the flask (Fig. 15.23). Coloured streaks of water rise up and move to the sides. Water at the bottom gets heated and moves up. Water from the sides reaches the bottom, gets heated and rises up. The process continues. This phenomenon is called convection. Every molecule of water comes to the hot point, takes heat and moves up.

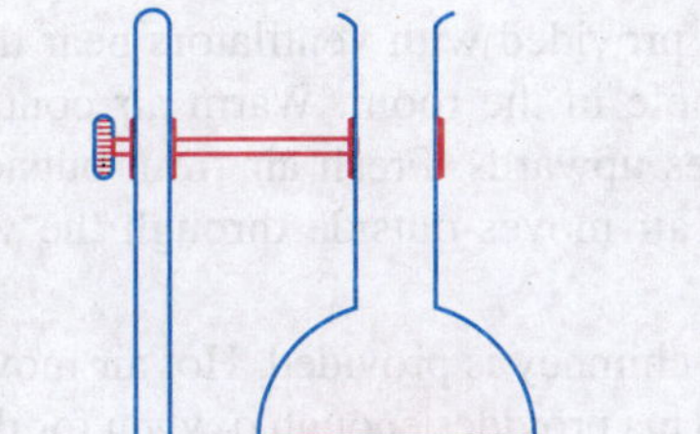

Fig. 15.23

**Experiment 2.** Take a rectangular glass tube and fill it with water (Fig. 15.24). Pour a little charcoal powder in the tube. Heat the side tube gently. The convection currents are set up due to the movement of the heated water molecules. The direction of movement of water molecules is indicated by the movement of the charcoal particles.

**Experiment 3.** Take a candle and fix it at the bottom of a cylinder (Fig. 15.25). Light the candle. The flame becomes weaker and weaker and finally gets extinguished. Air in the cylinder gets heated and is pushed out. There is no fresh supply of air for the burning of the candle.

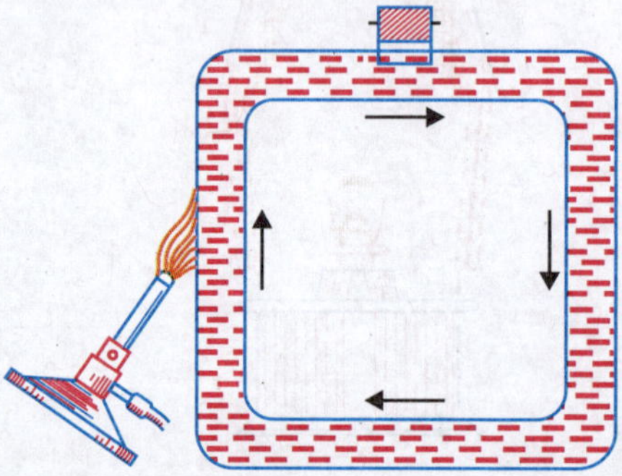

Fig. 15.24

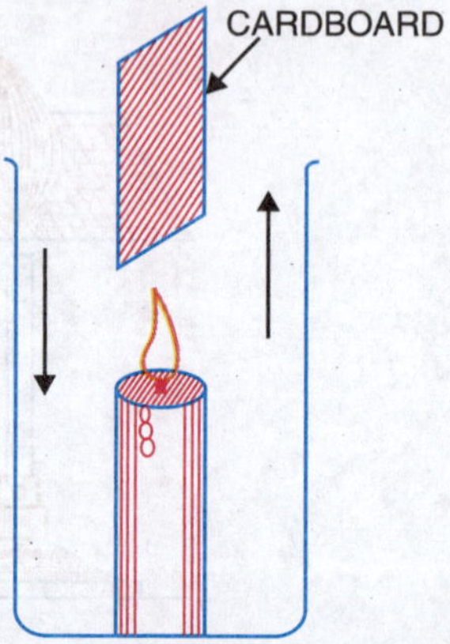


Fig. 15.25

Now, take a cardboard and hold it inside the cylinder. The space above the candle is divided into two parts. Light the candle. The candle continues to burn. Here air above the flame gets heated and moves up through one of the sides of the cardboard. Through the other side of the cardboard fresh air moves in. Convection currents are set up and the candle continues to burn.

In mines, two chimneys are used. Underneath one of the chimneys fire is produced. Through the other chimney fresh air enters into the mines. Convection currents are produced. This provides proper ventilation in the mine.

## 15.23 Applications of Convection

**1. Ventilation.** Rooms are provided with ventilators near the ceiling. Air in the room gets warmer due to respiration of people in the room. Warm air containing more of $CO_2$ and water vapour has less density and moves upwards. Fresh air from outside enters the room through the doors and windows. The impure air moves outside through the ventilators. The phenomenon is continuous.

**2. Chimneys.** In a kitchen, chimney is provided. Hot air moves up the chimney and fresh air enters the fire place from outside. This provides enough oxygen for the continuous combustion of the fuel. Similarly in oil lamps, oil stoves, principle of a chimney is used. When the wick is lighted, air and kerosene oil burn above the wick. Heated air moves up and fresh air from the holes provided at the bottom of the chimney enters the inside. The phenomenon is continuous and the convection currents set up help in supplying enough oxygen for combustion. If holes are not provided, the flame gets extinguished quickly.

In a factory, a high chimney is provided. The rate of circulation of air depends on the difference in pressure between the two ends of the chimney. Higher the chimney, greater is the rate of circulation of air. Due to this reason, sufficient supply of oxygen is provided and the complete combustion of the fuel takes place.

**3. Trade Winds.** The surface of the earth gets heated more at the equator than at the poles. Warm air at the equator moves up and cold air from the poles moves towards the equator. In the northern hemisphere, it is coming from the north and due to the rotation of the earth from west to east, the wind appears to come from north-east. In the southern hemisphere, the wind appears to be from south-west. These winds are called trade winds because they were used by trader for sailing their vessels in ancient days.

Monsoons are also based on the principle of convection.

**4. Ocean Currents.** Water in the ocean at the equator gets heated more than at the poles. Heated water at the equator expands and the level of water rises near the equator. This hot water flows towards the poles. This current is called hot current or gulf streams. The cold water from the poles moves towards the equator beneath the ocean and this current is called the under current.

**5. Land and Sea Breezes.** Near the sea, water becomes less warm than the land during day time. The heated air on the land surface moves up. Air from the surface of the sea moves towards the sea shore. This is called sea breeze. Convection currents are set up. During night when the land becomes cooler than water, the air over the surface of water is warmer and moves upward. Air from the land, moves towards the sea. This is called land breeze.

## 15.24 Central Heating System

In cold countries, the temperature in winter falls below 0ºC. The rooms of a building are kept warm by a central heating system based on the principle of convection.

Water is heated in the boiler at the basement of the building. The rooms are fitted with pipes having radiators fixed on the walls. Hot water from the boiler rises up. It passes through the radiators of different rooms. Radiators get heated and radiated heat to the room. After losing heat to the radiator, water becomes cold and returns back to the boiler as shown in (Fig. 15.26). Hot water also reaches the cold water tank at the top of the building. The hot water flows from the boiler to the cold water tank and the cold water flows from the tank to the boiler. Convection currents are set up and the building is kept warm continuously at a constant temperature.

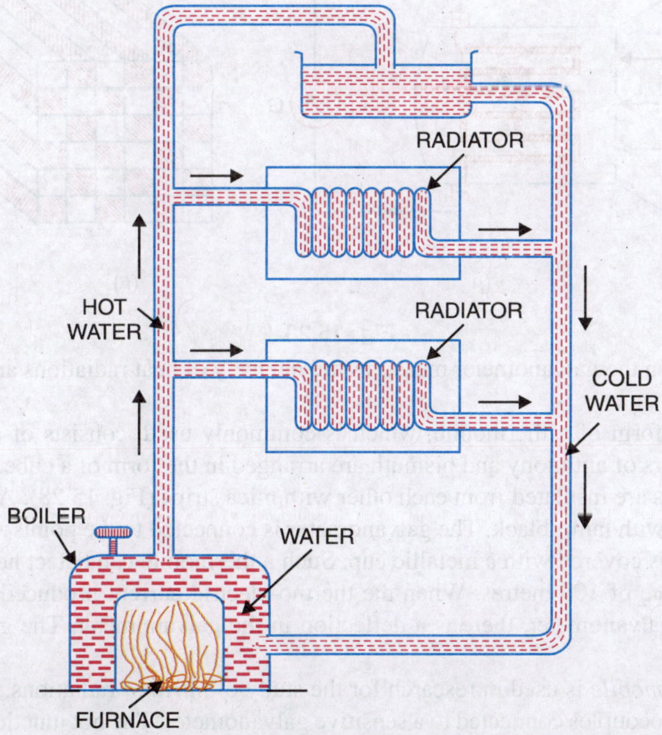


Fig. 15.26

## 15.25 Differential Air Thermoscope

A differential air thermoscope consists of two glass bulbs $A$ and $B$ connected by a narrow glass tubing. The $U$-tube contains sulphuric acid. The bulbs $A$ and $B$ contain air and when they are at the same temperature, there is no difference in level of the liquid in the two limbs. The bulb $A$ is coated with lamp black so that it can completely absorb the whole of the heat radiations falling on it (Fig. 15.27).

When $A$ is exposed to heat radiations, it absorbs heat energy. Air inside the bulb $A$ expands and there is difference in level of the liquid in the two limbs. This thermoscope is very sensitive and can detect heat radiations of even feeble intensity (*i.e.*, from a candle at a large distance).

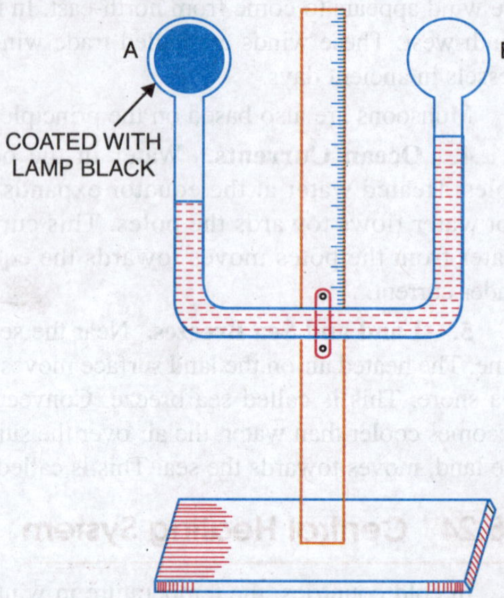


Fig. 15.27

## 15.26 Thermopile

The thermopile originally designed by *Melloni* consisted of a number of thermocouples joined in series. The thermocouples are made of bismuth and antimony. The junctions on one side are coated with lamp black and are exposed to heat radiations. The junctions on the other side are kept cold. The two ends of the thermopile are connected to a sensitive galvanometer (Fig. 15.28).

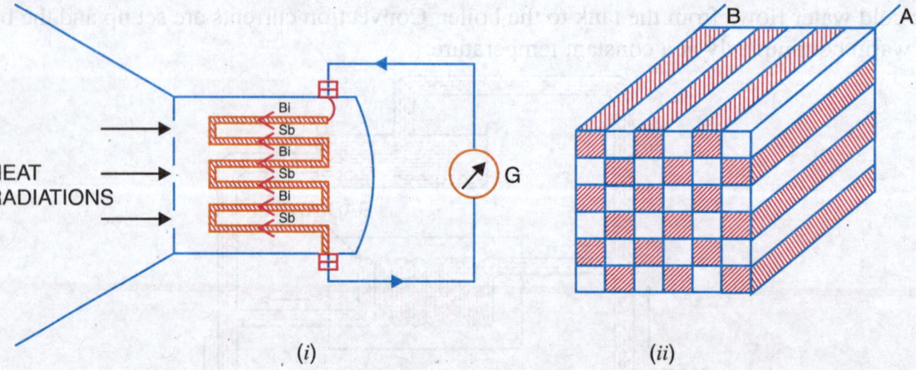


Fig. 15.28

The deflection in the galvanometer measures the intensity of heat radiations and the scale can be calibrated.

The improved form of a thermopile, which is commonly used, consists of a large number of thermopiles. The bars of antimony and bismuth are arranged in the form of a cube. The junctions are soldered but the bars are insulated from each other with mica strips (Fig. 15.28). All the junctions on one side are coated with lamp-black. The galvanometer is connected to the points $A$ and $B$. When not in use, the hot face is covered with a metallic cup. Such a thermopile can detect heat radiations from a candle at a distance of 100 metres. When the thermo-electric current produced in the thermopile flows through the galvanometer, there is a deflection in the galvanometer. The galvanometer scale can be calibrated.

The *linear thermopile* is used in research for the study of *infrared* radiations. It consists of only a few pairs of thermocouples connected to a sensitive galvanometer. The hot junctions are arranged in a line.

The thermopiles can be made sensitive by mounting the thermopile in vacuum and using very short connecting wires. The chief drawback of a thermopile consists in the comparative slowness of its indication and also the long time it takes to return to its zero reading.

## 15.27 Properties of Thermal Radiations

Heat radiations have properties similar to light radiations. They are electromagnetic waves of longer wavelength. The range of wavelengths of heat radiations is from 3 cm to $8 \times 10^{-5}$ cm.

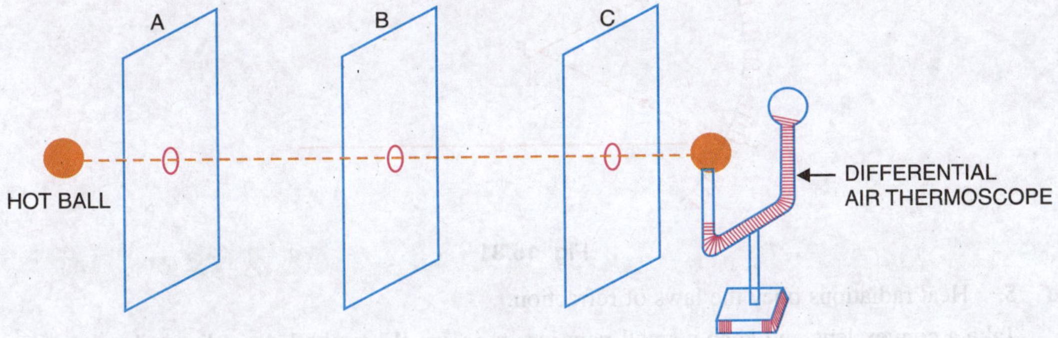


Fig. 15.29

**1.** Heat radiations travel in straight lines.

Take three similar cardboards having holes at the same height. Place a red hot ball in front of the hole of the first cardboard. Bring the blackened bulb of the differential air thermoscope near the hole of the third cardboard (Fig. 15.29). When the hot ball, the three holes and the bulb of the thermoscope are in the same straight line, maximum difference in level is noted. When one of the cardboards is displaced, the thermoscope is not affected. This shows that heat radiations travel in straight lines.

**2.** Heat radiations travel with the velocity of light.

When the sun rises, heat and light are received simultaneously on the earth. During the solar eclipse heat and light are cut off at the same moment. When the eclipse is over, heat and light reach the earth simultaneously. It shows that the heat radiations travel with the velocity of light equal to $3 \times 10^8$ m/s.

**3.** Heat radiations obey inverse square law.

Take a red hot ball. Place the thermopile at a fixed distance from the ball. Note the reading in the galvanometer. Now, place the thermopile at double the distance from the ball. It is found that the deflection of the galvanometer is one-fourth of the deflection in the first case. It shows that heat radiations obey inverse square law.

**4.** Heat radiations obey the laws of reflection.

Take two tubes $A$ and $B$ hinged at $C$ (Fig. 15.30). Keep the metal ball at the mouth of $A$. Adjust the position of $B$ to get the maximum difference in level in the thermoscope. It is found that the difference in level is maximum when the two tubes are equally inclined to the normal at $C$. It means $\angle i = \angle r$.

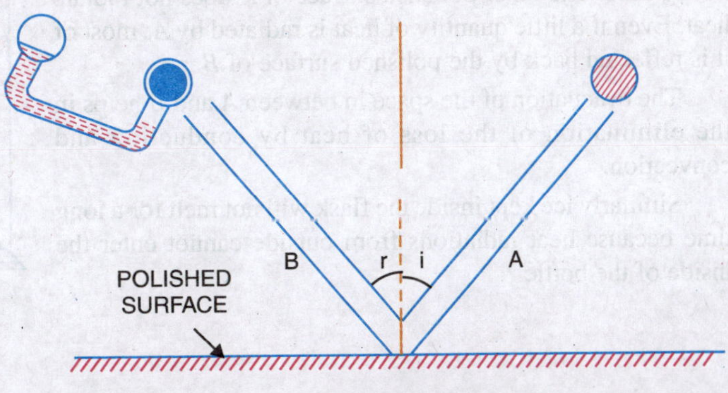


Fig. 15.30

Take a concave metallic reflecting surface. Direct it towards the sun. A small piece of paper placed at the focus of the mirror catches fire. All the heat radiations incident on the mirror after reflection converge at its focus (Fig. 15.31). Room heaters are provided with concave or elliptical reflectors. The heater element is kept at the focus of the mirrors.

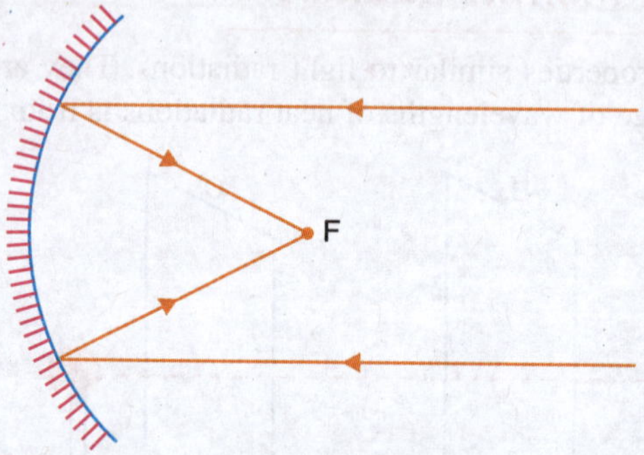


Fig. 15.31

**5.** Heat radiations obey the laws of refraction.

Take a convex lens and keep a small paper at its focus. When the lens is directed towards the sun's rays, the paper catches fire. The parallel heat radiations from the sun after refraction from the lens converge at its focus.

**6.** Heat radiations get diffused when they are incident on a rough and unpolished surface.

**7.** Heat radiations can travel through vacuum.

**8.** Heat radiations do not affect the medium through which they pass. When sun's rays are allowed to pass through a convex lens, they converge at the focus of the lens. The lens does not get heated but a paper at the focus catches fire.

## 15.28 Thermos Flask

A thermos flask consists of a double walled glass bottle. The space in between the two bottles is evacuated and sealed. The inner surface of *B* and the outer surface of *A* are highly polished (*i.e.* silvered) (Fig. 15.32).

When a hot liquid is kept in the bottle, it remains hot for a long time. The outer polished surface of *A* does not radiate heat. Even if a little quantity of heat is radiated by *A*, most of it is reflected back by the polished surface of *B*.

The evacuation of the space in between *A* and *B* helps in the elimination of the loss of heat by conduction and convection.

Similarly ice kept inside the flask will not melt for a long time because heat radiations from outside cannot enter the inside of the bottle.

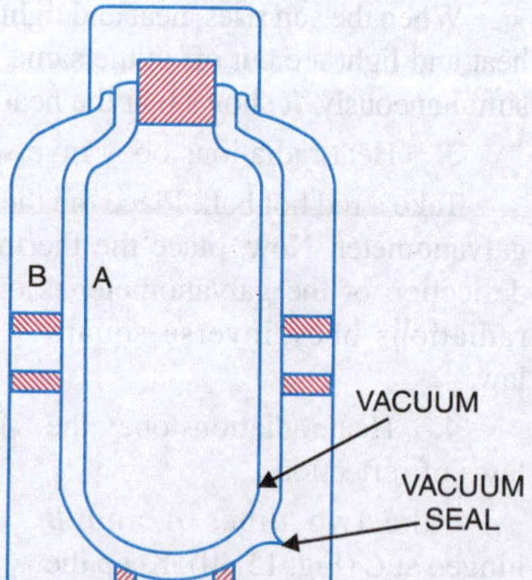


Fig. 15.32

## NUMERICAL EXAMPLES

### EXAMPLE 15.1

*In an Ingen-Hausz experiment, wax melted over 10 cm of copper rod and over 4 cm of iron rod. What is the conductivity of iron when the conductivity of copper is 0.90?*

**SOLUTION.** Here,

$$l_1 = 10 \text{ cm}$$
$$l_2 = 4 \text{ cm}$$
$$K_1 = 0.90$$
$$K_2 = ?$$
$$\frac{K_2}{K_1} = \frac{l_2^2}{l_1^2}$$
$$K_2 = \frac{l_2^2}{l_1^2} \times K_1$$
$$= \frac{16}{100} \times 0.90$$
$$K_2 = \mathbf{0.144}$$

### EXAMPLE 15.2

*The opposite faces of a metal plate of 0.2 cm thickness are at a difference of temperature of 100°C and the area of the plate is 200 sq cm. Find the quantity of heat that will flow through the plate in one minute if $K = 0.2$ CGS units.* **(Purvanchal 2007)**

**SOLUTION.** Here

$$K = 0.2$$
$$A = 200 \text{ sq cm}$$
$$d = 0.2 \text{ cm}$$
$$(\theta_1 - \theta_2) = 100°\text{C}$$
$$t = 60\ s$$
$$Q = \frac{KA(\theta_1 - \theta_2)t}{d}$$
$$= \frac{0.2 \times 200 \times 100 \times 60}{0.2}$$
$$= \mathbf{12 \times 10^5 \text{ cal}}$$

### EXAMPLE 15.3

*A bar of length 30 cm and uniform area of cross-section 5 cm² consists of two halves AB of copper and BC of iron welded together at B. The end A is maintained at 200°C and the end C at 0°C. The sides of the bar are thermally insulated. Find the rate of flow of heat along the bar when the steady state is reached. Thermal conductivity of copper is 0.9 and thermal conductivity of iron is 0.12 CGS units.* **(Calicut 2003, Utkal 2003)**

**SOLUTION.**

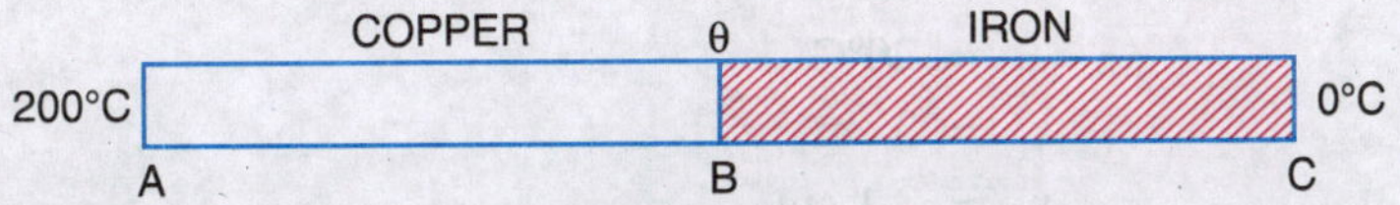


Fig. 15.33

Suppose the temperature at the interface $B$ is $\theta$ after the steady state is reached.

$$\therefore \quad \frac{K_1 A(200-\theta)}{d} = \frac{K_2 A(\theta-0)}{d}$$

$$0.9\,(200-\theta) = 0.12\,\theta$$

$$\theta = 176.5°\text{C}$$

After the steady state is reached, the rate of flow of heat is the same in both the bars.

$$\therefore \quad Q = \frac{K_1 A(\theta_1-\theta)}{d}$$

$$Q = \frac{0.9\times5\times(200-176.5)}{15}$$

or $\quad Q = \mathbf{7.05\ cal/s}$

## EXAMPLE 15.4

*An ice box is built of wood 1.75 cm thick, lined inside with cork 3 cm thick. If the temperature of the inner surface of the cork is 0°C and that of the outer surface of wood is 12°C, what is the temperature of the interface? The thermal conductivity of wood and cork are 0.0006 and 0.00012 CGS units respectively.*

**(Madras 2000)**

**SOLUTION.** Suppose the temperature of the interface is $\theta$ after the steady state is reached

$$\therefore \quad \frac{K_1 A(12-\theta)}{d_1} = \frac{K_2 A(\theta-0)}{d_2}$$

Here $\quad K_1 = 0.0006,\ K_2 = 0.00012$

$$d_1 = 1.75\text{ cm},\ d_2 = 3\text{ cm}$$

$$\frac{0.0006\,(12-\theta)}{1.75} = \frac{0.00012\,(\theta)}{3}$$

$$\theta = \mathbf{10{\cdot}74°C}$$

## EXAMPLE 15.5

*Find the time in which a layer of ice 3 cm thick on the surface of a pond will increase its thickness by 1 mm when the temperature of the surrounding air is – 20 °C.*

**(Purvanchal 2005, Calcutta 2000)**

*Thermal conductivity of ice = 0.005*

*Latent heat of ice = 80 cal/g*

*Density of ice at 0°C = 0.91 g/cm³*

**SOLUTION.** $\quad \int dt = \frac{\rho L}{K\theta}\int x\,dx$

$$t = \frac{\rho L}{2K\theta}\left[x_2^2 - x_1^2\right]$$

Here $\quad \rho = 0.91\text{ g/cm}^3$

$$L = 80\text{ cal/g}$$

$$K = 0.005$$

$$\theta = 20°\text{C}$$

$$x_1 = 3\text{ cm}$$

$$x_2 = 3.1\text{ cm}$$

$$t = \frac{0.91\times80}{2\times0.005\times20}\left[(3.1)^2-(3)^2\right]$$
$$= 222.04 \text{ s}$$
$$= \textbf{3 min 42 s approximately}$$

### EXAMPLE 15.6

*How much time will it take for a layer of ice of thickness 10 cm to increase by 5 cm on the surface of a pond when the temperature of the surroundings is – 10°C ?* **(Purvanchal 2003)**

**SOLUTION.**
$$K = 0.005$$
$$L = 80 \text{ cal/g}$$
$$\rho = 0.90 \text{ g/cm}^3$$

Here
$$\int dt = \frac{\rho L}{K\theta}\int x\,dx$$
$$t = \frac{\rho L}{2K\theta}\left[x_2^2 - x_1^2\right]$$

Here
$$x_1 = 10 \text{ cm} ;\ x_2 = 15 \text{ cm}$$
$$t = \frac{0.90\times80}{2\times0.005\times10}\left[(15)^2-(10)^2\right]$$
$$= 9\times10^4 \text{ s} = \textbf{25 hours}$$

### EXAMPLE 15.7

*Calculate approximately the heat passing per hour through the walls and windows of a room (5 × 5 metres) if the walls are of bricks of thickness 30 cm and have windows of glass 3 mm thick and total area of 5 sq. metre. The temperature of the room is 30°C below that of the outside and the thermal conductivity of bricks and glass is 12 × 10$^{-4}$ and 25 × 10$^{-4}$ C.G.S. units respectively.*

**SOLUTION.** The area of the four walls of the room

$$= 4\times5\times5 \text{ m}^2 = 100 \text{ m}^2 = 100\times10^4 \text{ cm}^2.$$

The total area of the glass windows

$$= 5 \text{ m}^2 = 5\times10^4 \text{ cm}^2$$

Therefore, the area of brick walls = $(100\times10^4 - 5\times10^4)$ cm$^2$ = $95\times10^4$ cm$^2$

The quantity of heat $Q$ passing in time $t$ is given by

$$Q = \frac{KA(\theta_2-\theta_1)t}{x}$$

(*i*) For brick walls : Given K = $12\times10^{-4}$ C.G.S. units, A = $95\times10^4$ cm$^2$

$$\theta_2-\theta_1 = 30°\text{C},\ x = 30 \text{ cm and } t = 1 \text{ hour} = 3600 \text{ sec.}$$

Hence, the quantity of heat passing through the walls is

$$Q_1 = \frac{12\times10^{-4}\times95\times10^4\times30\times3600}{30}$$
$$= \mathbf{4.104\times10^6 \text{ cal.}}$$

(*ii*) For glass windows : Given $K = 25\times10^{-4}$ C.G.S. units, $A = 5\times10^4$cm,$^2$ $\theta_2-\theta_1 = 30°$C

$$x = 3 \text{ mm} = 0.3 \text{ cm and } t = 3600 \text{ sec.}$$

Hence, the quantity of heat passing through the glass windows is

$$Q_2 = \frac{25\times10^{-4}\times5\times10^4\times30\times3600}{0.3}$$

$$= \mathbf{45 \times 10^6\ cal.}$$

Therefore, the total heat passing per hour through the room

$$Q = Q_1 + Q_2$$
$$= (4.104 \times 10^6 + 45 \times 10^6) \text{ cal}$$
$$= \mathbf{49.104 \times 10^6\ cal.}$$

### EXAMPLE 15.8

*Two plates of the same dimensions but of different thermal conductivities $K_1$ and $K_2$ are welded together to form a single plate of double thickness. Show that the equivalent thermal conductivity K is given by*

$$K = \frac{2K_1K_2}{K_1 + K_2}$$

**(Madras 2005, Rohilkhand 2002)**

**SOLUTION.** Le $A$ be the surface area and $x$ the thickness of each plate. $\theta_1$ and $\theta_2$ are the temperatures of the end faces ($\theta_1 > \theta_2$) and $\theta$ is the temperature of the surface in contact. After steady state is reached, the rate of heat flow through both the plates will be the same. Hence

$$\text{Q} = \frac{K_1A(\theta_1 - \theta)}{d} = \frac{K_2A(\theta - \theta_2)}{d} \qquad ...(i)$$

or
$$K_1\theta_1 - K_1\theta = K_2\theta - K_2\theta_2$$

or
$$K_1\theta_1 + K_2\theta_2 = \theta\,(K_1 + K_2)$$

$\therefore$
$$\theta = \frac{K_1\theta_1 + K_2\theta_2}{K_1 + K_2} \qquad ...(ii)$$

Now the thickness of composite plate is 2d and let $K$ be its equivalent thermal conductivity, then the rate of heat flow through it will be

$$Q = \frac{KA(\theta_1 - \theta_2)}{2d} \qquad ...(iii)$$

Equating Eq. (*i*) and Eq. (*iii*), we get

$$\frac{K_1A(\theta_1 - \theta)}{d} = \frac{KA(\theta_1 - \theta_2)}{2d}$$

or
$$K = \frac{2K_1\,(\theta_1 - \theta)}{\theta_1 - \theta_2}$$

$$= \frac{2K_1}{\theta_1 - \theta_2}\left[\theta_1 - \frac{K_1\theta_1 + K_2\theta_2}{K_1 + K_2}\right] \qquad (\because \text{ Eq. } ii)$$

$$= \frac{2K_1}{\theta_1 - \theta_2}\left[\frac{K_1\theta_1 + K_2\theta_1 - K_1\theta_1 - K_2\theta_2}{K_1 + K_2}\right]$$

$$= \frac{2K_1}{\theta_1 - \theta_2}\left[\frac{K_2\,(\theta_1 - \theta_2)}{K_1 + K_2}\right]$$

$$= \frac{\mathbf{2K_1K_2}}{\mathbf{K_1 + K_2}}$$

## EXAMPLE 15.9

*Two thin concentric spherical shells of radii 5 cm and 10 cm respectively have their annular cavity filled with charcoal powder. When energy is supplied at the rate of 10.5 watt to a heater at the centre, a temperature difference of 60° C is set up between the shells. Find the thermal conductivity of charcoal. (J = 4.2 Joule/cal)*

**SOLUTION.** The thermal conductivity of charcoal will be

$$K = \frac{Q\,(r_2 - r_1)}{4\pi r_1 r_2\,(\theta_1 - \theta_2)}$$

Here, heat supplied per sec. is

$$Q = 10.5 \text{ Joule/sec.}$$

$$= \frac{10.5}{J} \text{ cal/sec.}$$

$$= \frac{10.5}{4.2} \text{ cal/sec.}$$

Also given $r_2 = 10$ cm, $r_1 = 5$ cm and $(\theta_1 - \theta_2)$ 60°C

$$\therefore \quad K = \frac{\frac{10.5}{4.2}(10-5)}{4 \times 3.14 \times 10 \times 5 \times 60}$$

$$= \mathbf{3.3 \times 10^{-4} \text{ cal. cm}^{-1} \text{ deg}^{-1} \text{ sec}^{-1}}.$$

## EXERCISE – XV

**15.1** Define coefficient of thermal conductivity. What are its dimensions?

**(Rohilkhand 2002 Imp., Lucknow 2002; Kerala 2001; Meerut 2001, M.S. Uni. Tamil Nadu 2007)**

**15.2** Define temperature gradient and thermal conductivity. Establish equation of heat conduction.

**(Guwahati 2000)**

**15.3** Describe Lee's method to find the coefficient of thermal conductivity of metals. **(Meerut 2003)**

**15.4** Discuss the rectilinear flow of heat along a bar of uniform area of cross-section. Describe expression for thermal conductivity and temperature distribution for a cylindrical rod heated along its axis.

**(Purvanchal 2002, Rohilkhand 2000, Madras 2004, Patna 2005)**

**15.5** Discuss Ingen-Hausz experiment to compare the thermal conductivity of different materials.

**15.6** Discuss in detail Forbes method for finding the coefficient of thermal conductivity of a metal bar.

**15.7** Discuss Lee's disc method for finding the coefficient of thermal conductivity for bad conductors. Can this method be used for good conductors?

**(Madras 2005; Rohilkhand 2003)**

**15.8** Discuss the cylindrical flow of heat. Describe experiments to find the coefficients of thermal conductivity of glass and rubber. Derive the necessary formula. **(Rohilkhand 2000)**

**15.9** Discuss the heat flow through a compound wall.

**15.10** Discuss "Accretion of ice" on ponds.

**15.11** Discuss Wiedemann-Franz law. **(Alagappa 2004)**

**15.12** A thermal conductor in the form of long bar is heated at one end at constant temperature. Discuss the distribution of temperature along the bar before and after the steady state is reached.

**(Rohilkhand 2001)**

**15.13** A metal bar is heated at one end, Derive Fourier's differential equation for rectilinear flow of heat. Obtain an expression for the temperature distribution when the rod is (*a*) open (*b*) covered.

**(Purvanchal 2006, Rohilkhand 2002, 2003)**

**15.14** What is coefficient of thermal conductivity? Describe the radial heat flow through a uniform cylindrical shell, the outer and inner walls of which are maintained at different uniform temperatures.

**15.15** Define thermal conductivity. Describe how the thermal conductivity of glass in the form of a tube can be determined. Derive the necessary formula.

**15.16** Show how the atmospheric pressure changes with height. Calculate the lapse rate.

**(Madras 2001; Madurai Kamraj 2003)**

**15.17** Describe the working of (*i*) disappearing filament pyrometer, (*ii*) total radiation pyrometer and (*iii*) pyrheliometer. **(Meerut 2000)**

**15.18** Describe the cylindrical shell method of determining the thermal conductivity of poor conductors. Derive the formula you will use. **(Alagappa 2004)**

**15.19** Describe Lee's method for determination of the thermal conductivity of a bad conducting material in the form of a disc. **(Rohilkhand 2001, 2003; Meerut 2001, 2003, Alagappa 2004)**

**15.20** Describe a method with necessary theory of determining the thermal conductivity of rubber in the form of a tube.

**15.21** Prove that for an atmosphere in convective equilibrium, the rate of fall of temperature with altitude for dry air is given by the expression

$$\frac{dT}{dh} = \frac{-Mg\,(\gamma-1)}{R\gamma}$$

**(Madras 2001; Alagappa 2000)**

**15.22** Define coefficient of thermal conductivity. Describe Forbes' method for determining the thermal conductivity of a metal rod. Mention some of the limitations of this method. **(Rohilkhand 2001)**

**15.23** Define coefficient of thermal conductivity. What do you understand by a thermal steady state? Explain. **(Purvanchal 2007, Madras 2005)**

**15.24** Write a short note about "Stability of atmosphere". **(Alagappa 2004)**

**15.25** Explain heat conduction and heat radiation with examples. **(Alagappa 2004)**

**15.26** Define thermal conductivity of a substance. Derive the expression for it. **(Alagappa 2004)**

**15.27** Set up the differential equation for one-dimensional flow of heat in a metal bar (taking into account radiation loss) and solve if for the temperature cet the steady state. How will the solution be modified, if radiation loss from the sides of the bar is prevented? **(Culcutta 2003)**

**15.28** The temperature of one end of an insulated rod is being varied sinusoidally with an angulan frequency ω. Assume that the temperature θ is related to *x*, the coordinate at a point on the rod and the time *t* in the following manner

$$\theta = \theta_0\, e^{-\mu x} \sin(kx - \omega t)$$

(here *k* and μ are constants and have their usual meaning)

(*i*) Derive the expressions for *k* and μ.

(*ii*) What is the velocity of temperature wave? **(Calcutta 2003)**

**15.29** Mention the applications of heat radiations?

**15.30** Discuss rectilinear flow of heat along a bar of uniform cross section and show that in the steady state of a metal bar heated at one end.

$$\frac{d^2\theta}{dx^2} = \mu^2\theta$$

where the symbols have their usual meaning.

Hence prove that the length to which the wax melts in a steady state along a wax coated bar is proportional to the square root of the coefficient of thermal conductivity of the material of the bar.

**(Calcutta 2003; Utkal 2003; Guwahati 2000)**

**15.31** Establish the standard Fourier equation for one-dimensional flow of heat when one end of a rod is heated. Deduce an expression for the temperature $\theta$ at a distance $x$ from the hot end when radiation losses are not negligible.

**15.32** Give the theoretical description and Angstrom's method to determine the periodic flow of heat along a long insulated bar. **(Kanpur 2002)**

**15.33** What do you understand by periodic flow of heat? Discuss its application to the heat flow in earth. **(Purvanchal 2006)**

**15.34** Write short notes on the followings.

(*i*) Thermal conductivity
(*ii*) Thermos flask
(*iii*) Conductivity of poor conductors
(*iv*) Forbe's method for coefficient of thermal conductivity of a metal bar **(Rohilkhand 2001)**
(*v*) Ingen-Hausz experiment
(*vi*) Accretion of ice on ponds
(*vii*) Davy's safety lamp **(GND 2000)**
(*viii*) Central heating system
(*ix*) Thermopile

**15.35** Find the time in which a layer of ice 6 cm thick on the surface of a pond will increase its thickness by 2 mm, when the temperature of the surrounding air is –20ºC.

$$K = 0.005, L = 80 \text{ cal/g}, \rho = 0.91 \text{ g/cm}^3$$ **[Ans. 13 min 48.16 s]**

**15.36** How much time will it take for a layer of ice of thickness 8 cm to increase by 4 cm on the surface of a pond when the temperature of the surroundings is – 10ºC.

$$K = 0.005, L = 80 \text{ cal/g}, \rho = 0.90 \text{ g/cm}^3$$ **[Ans. 16 hours]**

**15.37** Thermal conductivity of brass is 4 times of copper. Two bars one of brass and the other of copper of same length and same cross-sectional area are joined together. The free ends of copper and brass are kept at temperatures 0ºC and 100ºC respectively. Find the temperature of the joint in the condition of steady state. Heat lost due to radiation may be neglected.

**[Hint :** $\frac{K_{Bras}(100-\theta)A}{d} = \frac{K_{Copper}(\theta-0)A}{d}$ ] **[Ans. 80ºC] (Meerut 2002)**

## OBJECTIVE QUESTIONS

**15.1** In the Ingen-Hausz experiment the thermal conductivity $K$ and length $L$ of the rod up to which wax melts are related as:

(*a*) $\frac{K}{L}$ = Constant (*b*) $\frac{K^2}{L}$ = Constant

(*c*) $\frac{K}{L^2}$ = Constant (*d*) $KL$ = Constant

**15.2** Under steady state, the temperature of body: **(Rohilkhand 2002 Imp.)**

(*a*) increases with time (*b*) decreases with time

(*c*) does not change with time and is same at all points of the body

(*d*) does not change with time and can be different at different points of the body.

**15.3** If $L$ is the length and $A$ area of cross-section of a rod and $K$ is thermal conductivity of material, then the thermal resistance is given by **(Rohilkhand 2003)**

(*a*) $\frac{KL}{A}$ (*b*) $\frac{A}{KL}$

(*c*) $\frac{KA}{L}$ (*d*) $\frac{L}{KA}$

**15.4** The rate flow of heat through a metal bar of area of cross-section 1 $m^2$ when temperature gradient is 1°C/m under steady state is called: **(Rohilkhand 2002 Improvement)**

(*a*) Thermal resistance (*b*) Thermal conductivity

(*c*) Diffusivity (*d*) Resistivity

**15.5** Thermal conduction in metals takes place by: **(Rohilkhand 2002)**

(*a*) free electrons (*b*) bound electrons

(*c*) vibrations of molecules (*d*) none of the above

**15.6** A single plate of double thickness is made by joining two equal plates of thermal conductivities $K_1$ and $K_2$. The equivalent thermal conductivity of the composite plate will be : **(Rohilkhand 2002, 03)**

(*a*) $\frac{2K_1K_2}{K_1+K_2}$ (*b*) $\frac{K_1K_2}{K_1+K_2}$

(*c*) $\frac{K_1}{K_1+K_2}$ (*d*) $\frac{K_2}{K_1+K_2}$

**15.7** In Searle's method for finding conductivity of metal, the temperature gradient along the bar :

(*a*) is greater near the hot end (*b*) is greater near the cold end

(*c*) is same at all points along the bar (*d*) increases as we go from hot to cold end

**15.8** The S.I. unit of thermal conductivity is :

(*a*) $JS\,m^{-1}\,°C^{-1}$ (*b*) $JS^{-1}\,m^{-1}\,°C^{-1}$

(*c*) $JS^{-1}\,m^{-1}\,°C$ (*d*) $JS^{-1}\,m\,°C^{-1}$

**15.9** Thermal conductivity of bad conductors is measured by : **(Rohilkhand 2003)**

(*a*) Searle's method (*b*) Lee's disc method

(*c*) Callendar and Barne's method (*d*) None of the above

**15.10** Two rods of lengths *L* and 2*L*, thermal conductivities 2*K* and *K* of same cross-sectional area are connected end to end, then the equivalent thermal conductivity of the system is :

(*a*) $\left(\frac{5}{6}\right)$K (*b*) 1.5 K

(*c*) 1.2 K (*d*) $\left(\frac{8}{9}\right)$K

## ANSWERS

| | | | |
|---|---|---|---|
| 15.1 (*d*) | 15.2 (*d*) | 15.3 (*d*) | 15.4 (*b*) |
| 15.5 (*a*) | 15.6 (*a*) | 15.7 (*c*) | 15.8 (*b*) |
| 15.9 (*b*) | 15.10 (*c*) | | |

# Change of State

# 16

### At a Glance

• Change of State • Latent Heat of Fusion • Laws of Fusion • Practical Applications • Effect of Pressure on Freezing Point of Ice • Impurities Lower Freezing Point • Determination of Melting Point of Wax • Determination of Latent Heat of Fusion of Ice • Heat Absorbed in Solution • Vaporization and Condensation • Laws of Ebullition or Boiling • Change in Boiling Point with Pressure • Franklin's Experiment • Latent Heat of Vaporization • Cooling Effect due to Vaporization • Examples of Cooling Caused by Vaporization • Refrigeration • Ammonia Ice Plant • Solid Carbon-dioxide (Dry Ice) • Gas and Vapour • Saturated and Unsaturated Vapour • Vapour Pressure of Liquids • Triple Point • Gibbs' Phase Rule • Weather • Ozone hole and Role of Chlorofluorocarbons (CFCs)

## 16.1 Change of State

A substance can exist in three states *viz.*, solid, liquid and gas. The particular state of a substance depends on its temperature. According to the kinetic theory, the molecules of a substance in the solid state have less degrees of freedom than a substance in the liquid or gaseous state. The molecules are more free to move in the gaseous state. The change of state can be brought about by supplying or withdrawing heat from

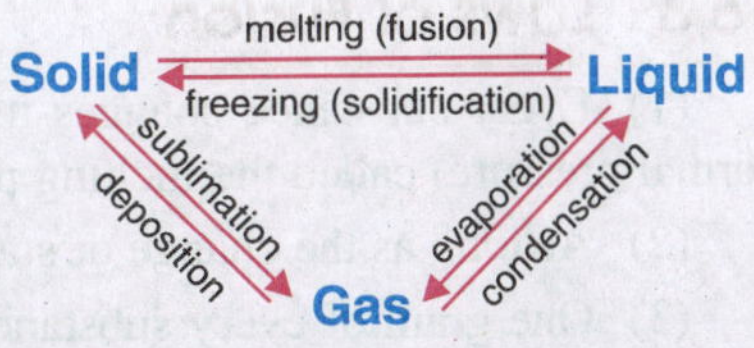

the substance. Ice is the solid state of water. By supplying heat to ice, it can be changed into water. Similarly by supplying heat to water, it can be converted into the gaseous state (steam). The reverse processes occur when heat is withdrawn from steam. This is true for all substances. Even permanent gases like oxygen, nitrogen, hydrogen etc. can be liquefied at low temperatures.

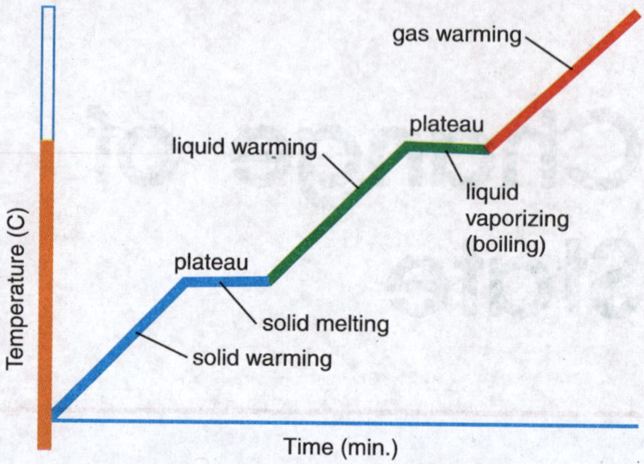


For a given substance, the change of state takes place at a fixed temperature and at a given pressure and *vice versa.* When the substance changes from the solid to the liquid state, the heat supplied at constant temperature (called melting point) is used in overcoming the forces of intermolecular attraction. The mean molecular distance increases and the molecules are more free to move. Similarly, when the substance changes from the liquid to the gaseous state at a fixed temperature (called the boiling point) at standard atmospheric pressure, the heat supplied is used in increasing the mean molecular distance. The molecules become free to move about in the whole space available to them.

## 16.2 Latent Heat of Fusion

Take small pieces of ice in a beaker. Fix a thermometer to note the temperature of ice in the beaker. Heat the beaker slowly. Ice melts but the thermometer does not show any rise in temperature. When the whole ice has melted the temperature of water rises and it is indicated by the thermometer. During the processes of conversion from ice to water (change of state from solid to liquid) the heat supplied is used to change the state of ice from solid state to liquid state. This heat is called latent heat. Latent means hidden *i.e.* which is not indicated by the thermometer.

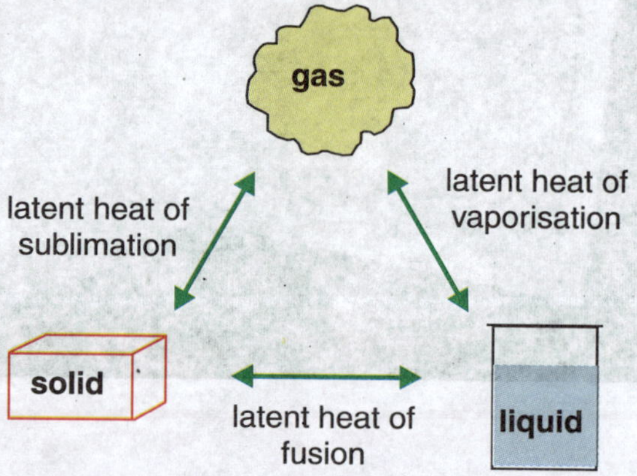


It has been found that one gram of ice takes 80 calories of heat to get itself converted to water. This heat is called latent heat of fusion of ice. Its unit is cals/gram.

Latent heat of fusion of a substance is defined as the amount of heat required to change the state of one gram of a substance from solid to liquid without any change in its temperature.

The melting point of ice is 0ºC at a pressure of 75 cm of Hg.

## 16.3 Laws of Fusion

(1) Every substance changes its state from solid to liquid at a particular temperature (under normal pressure) called the melting point.

(2) As long as the change of state takes place, there is no change in temperature.

(3) One gram of every substance requires a definite quantity of heat for change of state from solid to liquid and it is called the latent heat of fusion of that substance. It is different for different substances.

(4) Some substances show increase in volume on melting *e.g.* wax, ghee etc. while some other substances show decrease in volume on melting *e.g.,* ice.

(5) The melting point of those substances which decrease in volume on melting, is lowered with increase in pressure.

(6) The melting point of those substances which increase in volume on melting, is increased with increase in pressure.

**Note.** (1) *Ice decreases in volume on melting and its melting point is lowered with increase in pressure.*

(2) *The melting point or solidification point is the same for a substance.*

## 16.4 Practical Applications

(1) Aquatic animals live underneath water in ponds and lakes in winter. Ice is formed on the surface of the pond, and it remains on the surface due to its increase in volume and hence lesser density.

(2) Rocks in the mountains break in frosty weather. Water, which may be locked up in the pores of the rock, may freeze into ice and there is increase in volume. This produces cracks in the rocks.

(3) Fertility of the soil increases due to the splitting of the soil in frosty weather. Water freezes into ice and there is increase in volume. This increase in volume of ice disintegrates the soil.

(4) Alloys which expand on solidification are used for better casting. The alloy in the liquid form is put into the cast. When it solidifies, there is increase in volume and a sharp casting is obtained. Type metal expands on solidification.

## 16.5 Effect of Pressure on Freezing Point of Ice–Regelation

Melting point of ice is lowered with increase in pressure. Take a slab of ice. Take a wire and fix two weights (5 kg each) at its two ends. Put the wire over the slab as shown in Fig. 16.1.

The wire passes through the ice slab. Just below the wire, ice melts at a lower temperature due to increase in pressure. When the wire has passed, the water above the wire freezes again. Thus the wire passes through the slab and the slab does not split. This phenomenon of refreezing is called *Regelation.*

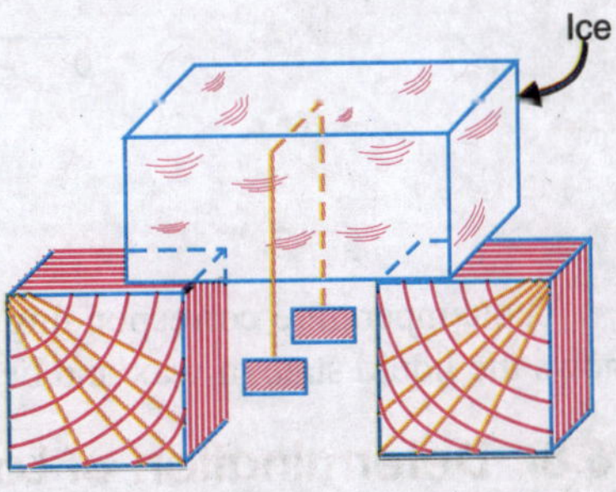


Fig. 16.1

In a similar way it is possible to freeze two pieces of ice by applying pressure and then releasing them.

**Examples.** (1) Water pipes burst in winter in cold countries due to solidification of water into ice. When water is converted into ice there is increase in volume and the pipes are likely to burst.

(2) When wheels of a cart pass over snow, it melts due to increase in pressure exerted by the wheels. When the wheel passes over, the water formed on the wheel solidifies due to regelation. Due to this reason wheels are covered with snow.

(3) Skating is possible on snow due to the formation of a thin layer of water formed below the skates. The water is formed due to the increase of pressure and it acts as a lubricant.

## 16.6 Impurities lower Freezing Point

When salt is mixed with ice, some ice melts taking heat from the salt. The temperature of the mixture decreases. Further, the salt dissolves in the water formed and takes latent heat from the mixture. The temperature of the mixture is decreased further. In this way, with a freezing mixture of salt and ice in the ratio 1 : 3, temperatures as low as – 13°C can be obtained. A freezing mixture consists of powdered ice, common salt and ammonium nitrate.

Similarly when sugar is dissolved in water, the water is cooled.

## 16.7 Determination of Melting Point of Wax

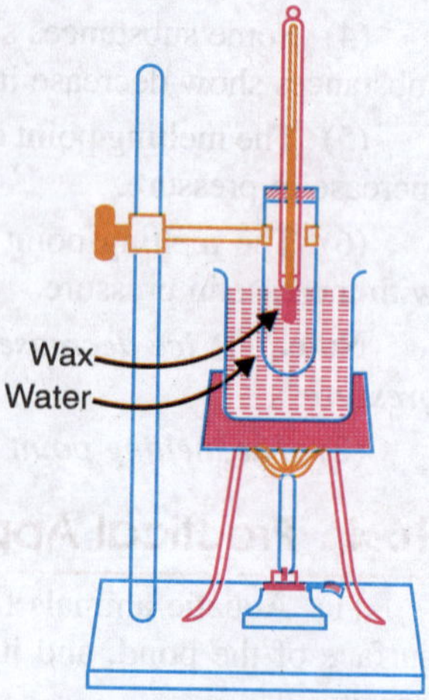

Fig. 16.2

Take sufficient quantity of wax in a test tube and fix a cork along with a thermometer. Adjust the test tube in a beaker containing water (Fig. 16.2). Heat the water to its boiling point. The wax in the test tube melts and is in the liquid state. Extinguish the burner and allow the water and wax to cool. Note the temperature of wax after every one minute till the temperature of wax is about 30°C. Draw a graph between temperature and time. The temperature corresponding to the horizontal line in the graph gives the melting point of wax. At this temperature, liquid wax is converted into the solid wax without change of temperature (Fig. 16.3).

Plot a graph between temperature along the *Y*-axis and time in minutes along the *X*-axis.

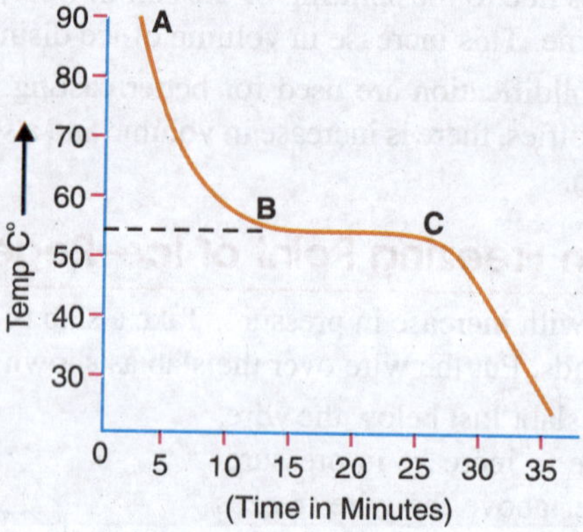

Fig. 16.3

The temperature corresponding to the line *BC* gives the melting point of wax. The portion *AB* shows the liquid state of wax and below *C* the graph represents the solid state.

## 16.8 Determination of Latent Heat of Fusion of Ice

Take a calorimeter. Weigh the calorimeter along with the stirrer. Weigh the calorimeter with stirrer and the lid. Heat some water in a beaker to a temperature of about 10°C higher than the room temperature. Fill the calorimeter to about half, with hot water. Weigh the calorimeter, hot water, stirrer and lid. Place it in a wooden box. Note the temperature of water in the calorimeter. It should be about 4 to 5°C higher than the room temperature. Take small pieces of ice in a blotting paper. Add dry pieces of ice gradually into the calorimeter and stir. Continue adding dry pieces of ice into the calorimeter, till the temperature of water is about 4 to 5°C lower than the room temperature. (This is done to avoid radiation error). Note the final temperature and weigh the whole contents. See that before recording the final temperature the whole of ice has melted.

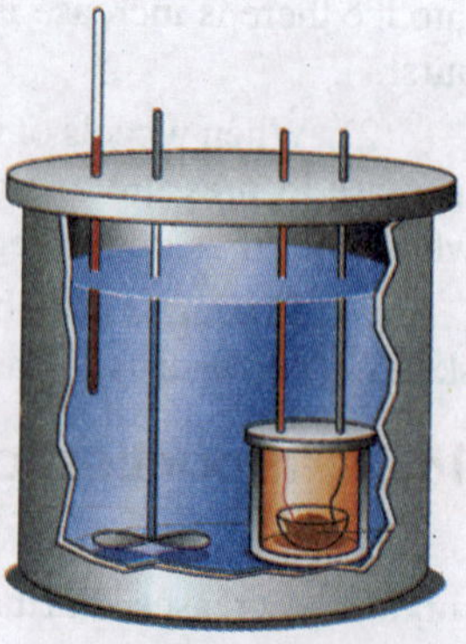
Calorimeter

When one gram of ice melts into water at 0°C, the amount of heat absorbed is called the latent heat of fusion of ice. Suppose $M$ grams of ice is added to $m$ grams of water at $t$°C in a calorimeter whose water equivalent is $W$. The final temperature reached = $T$°C.

Heat lost by water and calorimeter = $(m + W)(t - T)$

$$\therefore \quad \text{Heat gained by ice} = ML + MT$$

$$\therefore \quad \text{Heat gained} = \text{Heat lost}$$

$$ML + MT = (m + W)(t - T)$$

$$L = \frac{(m+W)(t-T)}{M} - T$$

Hence latent heat of fusion of ice can be determined.

## 16.9 Heat Absorbed in Solution

When a solution is formed by dissolving a solute (solid substance) in a solvent (liquid) the following possible changes take place.

(1) The solute gets dissolved in the solvent and takes the necessary latent heat from the solvent. The temperature of the solution decreases. However, in some rarer cases there is absorption of heat.

(2) When a solute is dissolved in a solvent, a chemical reaction may take place. This reaction is generally accompanied by evolution of heat.

(3) When an electrolyte is dissolved in water *e.g.* $CuSO_4$ dissolved in water, ionic dissociation takes place. In this process heat is absorbed.

However, depending on the nature of the substance *i.e.* solute and the solvent, the total effect may be absorption or evolution of heat.

## 16.10 Vaporization and Condensation

The conversion of a substance from the liquid to the vapour state is called *vaporization.* Boiling is the phenomenon of vaporization accompanied by ebullition *i.e.,* violent ebullition of bubbles. However vaporization can take place without boiling. The water kept on an open plate in summer evaporates and there is a change of state. But there is no boiling. Conversion of water to steam is vaporization. The conversion of the substance from the vapour state to the liquid state is called condensation. Conversion of steam into water is called condensation. Rain is caused due to the condensation of water vapour in the clouds.

## 16.11 Laws of Ebullition or Boiling

(1) Every liquid changes its state from liquid to its vapour at a particular temperature (under normal pressure) called the boiling point.

(2) As long as the change of state takes place, there is no change in temperature.

(3) One gram of every liquid requires a definite quantity of heat for change of state from liquid to vapour and it is called the latent heat of vaporization of that liquid. Latent heat is different for different substances.

(4) All liquids show increase in volume on vaporization.

(5) The boiling point of a liquid increases with increase in pressure of a liquid.

(6) The liquid can boil at a lower temperature under reduced pressure.

## 16.12 Change in Boiling Point with Pressure

The effect of pressure on the boiling point (saturation temperature) of a liquid can be studied with the help of the apparatus shown in Fig. 16.4. The temperature at which the change of state from liquid to vapour takes place is called saturation temperature. The saturation temperature increases with increase in pressure. *If the pressure is 1 atmosphere, the saturation temperature is called the boiling point.*

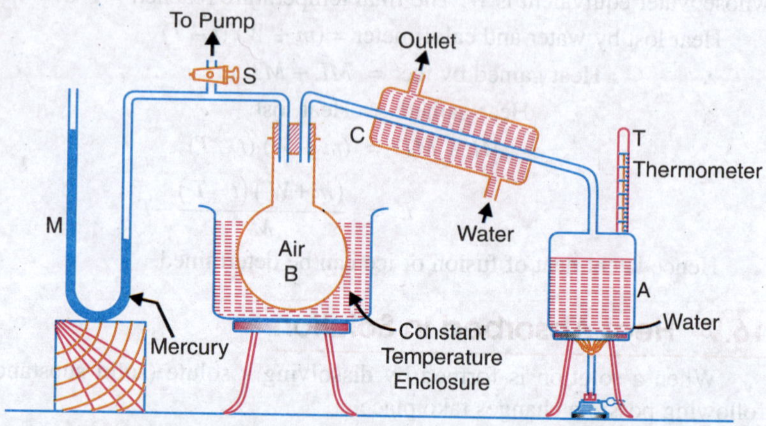


**Fig. 16.4**

The vessel *A* contains the liquid (say water). It is connected to the flask *B* through a tube *C* surrounded by water condenser. To measure the pressure of air over the surface of the liquid in *A*, a manometer is used.

The pressure of air inside *A* and *B* can be changed with the help of the compression pump or evacuation pump.

The stopcock *S* is opened. With the help of the compression pump, air is compressed into *B*. The stopcock is closed. Pressure of air in *A* and *B* is shown by the manometer. It is more than the atmospheric pressure.

The liquid is heated and the temperature at which it boils is recorded by the thermometer. At the boiling point, the temperature of the liquid remains constant. The condenser through which cold water is circulated prevents steam to enter the vessel *B*. The steam is condensed and condensed water flows back to *A*.

In this way at various pressures, the corresponding boiling points of the liquid can be determined. For low pressure, the vacuum pump is used. It is found that the boiling point of a liquid increases with increase in pressure and decreases with decrease in pressure.

The boiling point of water at 76 cm of Hg pressure is 100°C. The boiling point of water at double the normal pressure (*i.e.*, 152 cm of Hg) is about 128°C.

### Applications

*Papin's Digestor and Pressure cookers are* based on the principle that the boiling point of a liquid increases with increase in pressure.

The pressure inside the pressure cooker is much higher than the atmospheric pressure. The steam produced inside the cooker is not allowed to escape and the pressure over the surface of water increases. It means water inside the vessel will be at a temperature higher than the normal boiling point. Vegetables cooked with a pressure cooker are prepared in a short time.

When the pressure increases beyond the desired pressure, the weight placed on the lid lifts up a little and allows the steam to escape slowly and maintains constant pressure inside it. For safety purposes, a safety valve is also provided on the lid. The safety valve on a pressure cooker is a fusible plug. In case, the weight fails to lift, the pressure inside the cooker may increase enormously and may lead to explosion. To prevent this, a fusible plug is provided. It has a small stem made of a material of low melting point. When the pressure inside increases, the saturation temperature also increases and the plug melts. Due to this the steam escapes and the pressure inside the cooker gets lowered.

When a pressure cooker is kept on a stove, on filling it with water to the normal level, vaporization takes place. The water starts evaporating at temperatures lower than the boiling point temperature because of the pressure of air and hence the actual pressure of evaporation is the difference between the atmospheric pressure and the partial pressure of air. As the temperature increases, the rate of evaporation increases and the air is expelled. Gradually, due to the expulsion of air, the vapour pressure of water increases and hence the temperature. After all the air is expelled, boiling takes place. The steam starts rushing out through the opening.

The weight is now placed on the opening. Further, steam that is produced increases the pressure and hence the boiling point is also raised. As soon as the upward force due to the steam pressure is sufficiently high, the weight lifts and allows the steam to escape. This maintains constant pressure and hence constant temperature. If the weight is placed before the expulsion of steam, the partial pressure will not be as high as it ought to be and hence the temperature is not high.

## 16.13 Franklin's Experiment

Take a flask containing water. Heat it till the water begins to boil. When the water is boiling, the air escapes to the atmosphere. After some time, close the mouth of the flask with an air tight rubber cork. Invert the flask and fix it on the retort stand (Fig. 16.5). Pour cold water over the flask. Due to the condensation of steam, the pressure over the surface of water decreases. Water inside the vessel begins to boil. This experiment shows that water boils at a lower temperature under reduced pressure.

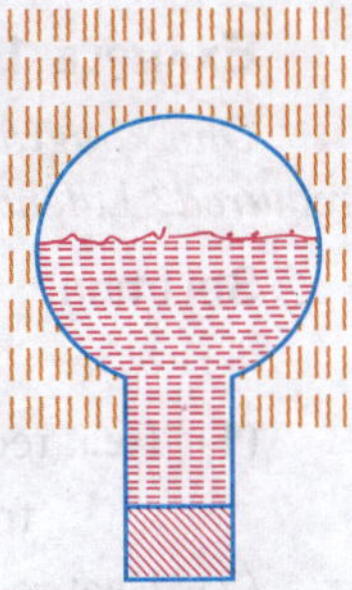

Fig. 16.5

In the laboratory, while preparing steam, the temperature of steam is usually less than 100°C because the atmospheric pressure is less than 76 cm of Hg.

## 16.14 Latent Heat of Vaporization

Latent heat of vaporization of a liquid is the amount of heat required to convert one gram of a liquid into vapour without any change in temperature. For water, latent heat of vaporization is 537 cals/gram at a pressure of 76 cm of Hg. Latent heat of vaporization is different for different substances.

### Determination of latent heat of steam

Take a calorimeter. Weigh the calorimeter along with the stirrer. Weigh the calorimeter along with the stirrer and the lid. Fill the calorimeter to about 2/3 with water and weigh along with the stirrer and the lid. Heat the water in the boiler to produce steam. The boiler is fitted with a steam trap as shown in Fig. 16.6. The whole of the delivery tube is covered with wool. When steam starts coming out of the end of the tube, note the temperature of steam. Note the temperature of water in the calorimeter. Now clean the end of the tube with a blotting paper to remove water drops. Insert the tube quickly into the calorimeter

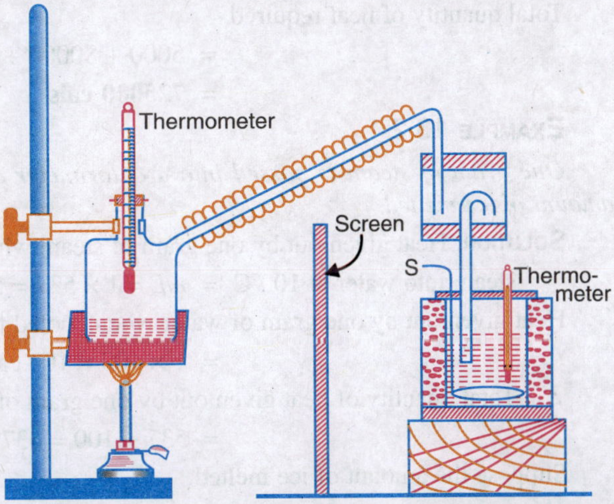


Fig. 16.6

and pass steam. When the temperature of water is raised through about 10ºC, take out the reading carefully. Continue stirring. Note the final temperature. Remove the thermometer carefully and weigh the calorimeter. The calorimeter should be covered with a lid and no water drops should be sticking to the thermometer.

Suppose, $m$ grams of steam is passed into $M$ grams of water at $t_1$°C in a calorimeter of water equivalent $w$. The temperature of steam is $t_2$°C and final temperature is $T$°C.

$$\text{Heat gained by water and calorimeter} = (M + w)(T - t_1)$$

$$\text{Heat lost by steam} = mL + m(t_2 - T)$$

$$\text{Heat gained} = \text{Heat lost}$$

$$(M + w)(T - t_1) = mL + m(t_1 - T)$$

$$L = \frac{(M + w)(T - t_1)}{m} - (t_2 - T)$$

Hence, latent heat of steam can be determined.

## Numerical Examples

### Example 16.1

*One kilogram of ice at – 10ºC is heated until the whole of it evaporates. How much heat is required? Latent heat of fusion of ice = 80 cals per gram and that of steam = 540 cal per gram.*

**Solution.** Mass of ice = 1 kg = 1000 g

Sp heat of ice = 0.5 cal/g-K

(1) Heat required to raise the temperature of ice

from – 10ºC to 0ºC = 1000 × 0.5 × 10 = 5000 cals

(2) Heat required to melt ice at 0ºC

= 1000 × 80 = 80000 cals

(3) Heat required to raise the temperature of water

from 0ºC to 100ºC = 1000 × 100 = 100000 cals

(4) Heat required to convert water into steam at 100ºC

= 1000 × 540 = 540000 cals

Total quantity of heat required

= 5000 + 80000 + 100000 + 540000

= **725000 cals**

### Example 16.2

*One gram of steam is passed into a calorimeter containing 100 grams of ice. Calculate the amount of ice melted.*

**Solution.** Heat given out by one gram of steam when converted from

steam into water at 100ºC = $mL$ = 1 × 537 = 537 cal.

Heat given out by one gram of water when cooled from 100ºC to 0ºC

= 1 × 100 = 100 cal

∴ Total quantity of heat given out by one gram of steam

= 537 + 100 = 637 cal

Suppose the amount of ice melted

= $m$ grams

Latent heat of ice = 80 cal per gram

Heat absorbed by ice = 80 *m* cal

Heat lost by steam = Heat gained by ice

$\therefore$ 637 = 80 *m*

or $m$ = **7.96 g**

**EXAMPLE 16.3**

*What will be the result of adding 52 g of ice to 100 g of water at 40°C?*

**SOLUTION.** Mass of water = 100 g

Temperature of water = 40°C

$\therefore$ Heat lost by water when its temperature falls from

40°C to 0°C = 100 × 40

= **400 cal**

If 52 g of ice melts, heat gained by it should be 52 × 80 = 4160 calories. But water can lose only 4000 calories of heat. Therefore, the whole of ice will not melt.

$\therefore$ Amount of heat gained by ice

= 4000 cal

$$\text{Amount of ice that melts} = \frac{4000}{80} = 50\text{g}$$

Remaining ice = 52 – 50 = 2g

Hence the result will be 2 g of ice and 150 g of water at 0°C.

**EXAMPLE 16.4**

*It takes fifteen minutes for an electric kettle to heat a certain quantity of water from 0°C to the boiling point 100°C. It requires 80 minutes to turn all the water at 100°C into steam. Determine the latent heat of steam. (Assume the radiation losses to be negligible).*

**SOLUTION.** Suppose mass of water = *m*

Heat required to raise its temperature from 0°C to 100°C

= 100*m*

$\therefore$ Heat produced by electric kettle in 15 minutes

= 100 *m*

Heat produced by electric kettle in 1 minute

$$= \frac{100\,m}{15}$$

Heat produced by electric kettle in 80 minutes

$$= \frac{100\,m \times 80}{15}\ \text{cal}$$

Suppose latent heat of steam = *L* cal per gram

Heat required by *m* grams of water to boil off into steam

= *mL* cal

$$\therefore \quad mL = \frac{100\,m \times 80}{15}$$

or $L$ = **533.33 cal/g**

**EXAMPLE 16.5**

*A refrigerator converts 50 g of water at 15°C into ice at – 20°C in one hour. Determine the quantity of heat removed per minute. (Specific heat of ice = 0.5)*

**SOLUTION.** Here $m = 50$ g

Heat removed in cooling water from 15°C to 0°C

$$= 50 \times 1 \times 15 = 750 \text{ cal}$$

Heat removed in converting water into ice at 0°C

$$= 50 \times 80 = 4000 \text{ cal}$$

Heat removed in cooling ice from 0°C to – 20°C

$$= 50 \times 0.5 \times 20 = 500 \text{ cal}$$

Total heat removed in one hour

$$= 750 + 4000 + 500 = 5250 \text{ cal}$$

Heat removed per minute

$$= \frac{5250}{60}$$

$$= \mathbf{87.5 \text{ cal/minute}}$$

## 16.15 Cooling Effect due to Vaporization

If a person pours a little spirit on his hand, he feels a cooling effect. The spirit evaporates and takes latent heat of vaporization from the spirit left behind the hand. Due to this reason one feels a cooling effect

Thus, whenever evaporation of a liquid takes place, the remaining liquid shows a fall in temperature.

The molecules of a liquid are always in a state of random motion. They collide with each other and they may or may not change their respective kinetic energies. Due to the collision a certain molecule may acquire more energy than the energy with which it is held. As soon as this happens the molecule flies off, *i.e.,* it leaves the liquid surface. This is called evaporation. If no external heat is supplied it is obvious that the mean kinetic energy of the molecules decreases and the temperature of the liquid falls. This explains cooling caused by evaporation. If the moving vapour molecules come near to each other at sufficiently low velocities, they experience a greater magnitude of force of attraction. The molecules coalesce and this results in condensation. Rate of evaporation increases with increase in temperature of the liquid and decrease in pressure over the liquid surface.

## 16.16 Examples of Cooling caused by Evaporation

1. In hot weather, we perspire. Water from the body comes out of the fine pores of the skin. Water evaporates and takes heat of vaporization from the body. Due to this reason we feel a cooling sensation.

2. A person wearing wet clothes feels chill. Water evaporates taking the latent heat of vaporization from the body and the person feels chill.

3. Water in an earthen pot remains cool in summer. Water comes out of the pores of the vessel and evaporates. During evaporation, the latent heat is taken from the remaining water. Therefore water remains cool in an earthen vessel.

4. Streets, floors, gardens, lawns etc. are watered in summer. Water evaporates taking the latent heat from them and keeps them cool.

5. Dogs keep their tongues usually out in summer. Water evaporates from the tongue and keeps it cool.

6. *Khas-khas* screens are commonly used in summer. Water is poured on them. Water evaporates and takes heat from them and the air in the room. Due to this reason the room is kept cool.

7. During high fever, a wet cloth soaked in 'Eau de cologne' or ice cold water is kept on the forehead. The liquid evaporates rapidly and takes heat from the head and the body.

8. Hot liquids are poured in a plate to increase the surface area. The rate of evaporation increases and the liquid cools quickly.

9. When a wet muslin piece is wrapped around the bulb of a thermometer, the thermometer records fall of temperature. Water evaporates and takes latent heat from mercury in the bulb.

10. In a packed hall in summer a person feels uneasy. Perspiration from his body does not evaporate. As soon as he comes out of the hall he feels a sudden cooling sensation. Perspiration evaporates rapidly and takes heat from his body.

## 16.17 Refrigeration

The second law of thermodynamics as given by Clausius give the principle of working of a refrigerator and ice plants.

"*It is impossible to make heat flow from a body at a lower temperature to a body at a higher temperature without doing external work on the working substance.*"

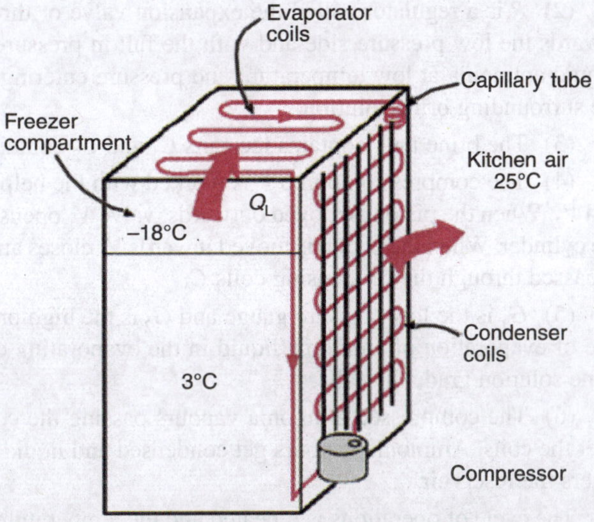


Fig. 16.7. Refrigerator

This part is applicable in the case of ice plants and refrigerators. Heat itself cannot flow from a body at a lower temperature to a body at a higher temperature. But, it is possible, if some external work is done on the working substance. Take the case of an ammonia ice plant. Ammonia is the working substance. Liquid ammonia at low pressure takes heat from the brine solution in the brine tank and is converted into low pressure vapour. External work is done to compress the ammonia vapour to high pressure. This ammonia at high pressure is passed through coils over which water at room temperature is poured. Ammonia vapour gives heat to water at room temperature and gets itself converted into liquid again. This high pressure liquid ammonia is throttled to low pressure liquid ammonia. In the whole process ammonia (the working substance) takes heat from brine solution (at a lower temperature) and gives heat to water at room temperature (at a higher temperature). This is possible only due to the external work done on ammonia by the piston in compressing it. The only work of electricity in the ammonia ice plant is to move the piston to do external work on ammonia. If the external work is not done, no ice plant or refrigerator (Fig. 16.7) will work. Hence, it is possible to make heat flow from a body at a lower temperature to a body at a higher temperature by doing external work on the working substance.

## 16.18 Ammonia Ice Plant

Ammonia ice plant is commonly used to manufacture ice. It is based on the principle that evaporation causes cooling. Moreover in this plant, heat flows from a body at a lower temperature to a body at a higher temperature by doing external work on the working substance. The schematic sketch of an ammonia ice plant is shown in Fig. 16.8.

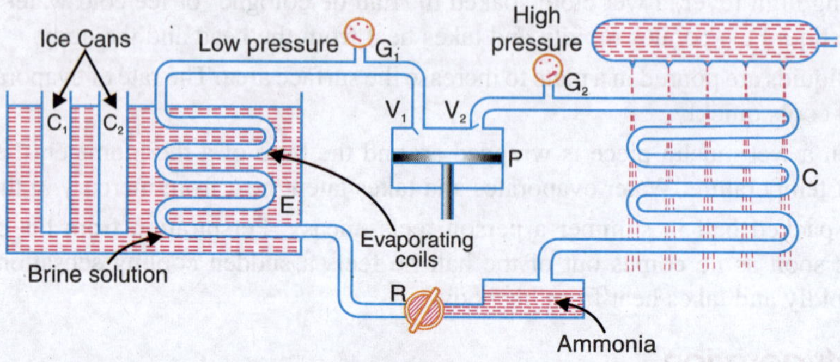


**Fig. 16.8**

(1) $T$ is a reservoir containing ammonia liquid.

(2) $R$ is a regulator called the expansion valve or throttle valve. High pressure liquid expands towards the low pressure side and with the fall in pressure there is a fall in saturation temperature. Liquid ammonia at low temperature and pressure entering the coils evaporates and takes heat from the surrounding brine solution.

(3) The brine tank contains ice cans $C_1$ and $C_2$ containing pure water.

(4) The compression pump $P$ is worked with the help of an electric motor. It has two valves $V_1$ and $V_2$. When the piston is moved outwards, valve $V_1$ opens and $V_2$ is closed. Ammonia vapour enters the cylinder. When the piston is moved inwards $V_1$ closes and $V_2$ opens. Compressed ammonia vapour is passed through the condensing coils $C$.

(5) $G_1$ is the low pressure gauge and $G_2$ is the high pressure gauge. When the pump works, the rate of evaporation of ammonia liquid in the evaporating coils increases and the temperature of the brine solution gradually falls.

(6) The compressed ammonia vapours passing the condensing coils give heat to water falling over the coils. Ammonia vapours get condensed and liquid ammonia is formed. The liquid ammonia enters the reservoir.

The cycle of operations is repeated and the temperature of brine solution in the tank decreases to – 8ºC. Water in the cans freezes to ice.

(*i*) If the pump stops working, then the cyclic process does not take place and the temperature of the brine tank does not fall.

(*ii*) The function of electricity is to move the piston to and fro with the help of the electric motor and to compress the vapours.

(*iii*) Ammonia is the working substance. It takes heat from the brine solution (at low temperature) and transfers heat to the water falling on the condensing coils at the room temperature. It means heat is transferred from a body at a lower temperature to a body at a higher temperature. Ordinarily it is not possible. But by doing external work (with the help of the pump) on the working substance this transfer of heat is possible and this is in accordance with the second law of thermodynamics.

## 16.19 Solid Carbon dioxide – (Dry Ice)

A cylinder containing liquid $CO_2$ is taken. A muslin bag is tied to the nozzle. The cylinder is placed in a slanting position (Fig. 16.9). The valve $V$ is opened and the liquid $CO_2$ enters the muslin bag. The liquid $CO_2$ evaporates rapidly and takes the latent heat of vaporization from the remaining liquid in the bag. Liquid $CO_2$ solidifies and solid $CO_2$ is formed in the muslin bag.

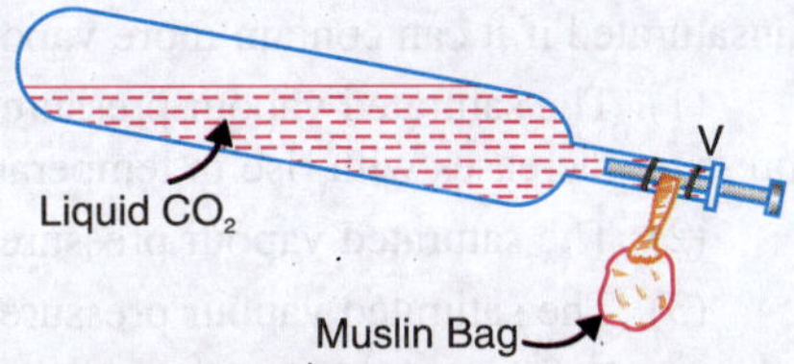

Fig. 16.9

The temperature of solid $CO_2$ is – 78°C and it is also called *dry ice*. It is commonly used by doctors for dressing the wounds. It serves as refrigerant when mixed with ether. Solid $CO_2$ does not melt when exposed to air. It directly evaporates and forms its vapours.

## 16.20 Gas and Vapour

Gas and vapour states are the two distant stages of the same continuous phenomenon. These two states have distinct boundary which is governed by a particular temperature called the critical temperature. Above the critical temperature the substance is in the gaseous state and below the critical temperature it is in the vapour state.

A gas cannot be liquefied by mere application of pressure, howsoever high the pressure may be. A vapour can be liquefied by applying pressure. Therefore to liquefy a gas, first it has to be cooled to a temperature lower than its critical temperature. It means it is converted from a gaseous state to the vapour state. Then a high pressure is applied and the vapours get liquefied.

Critical temperature of a gas is that temperature above which the gas cannot be liquefied by mere application of pressure, howsoever large the pressure may be.

**TABLE 16.1 Critical Temperature**

| *Substance* | *Critical temperature °C* |
|---|---|
| Water | 374 |
| Ammonia | 132 |
| Carbon dioxide | 31 |
| Oxygen | –119 |
| Hydrogen | –240 |
| Helium | –268 |

## 16.21 Saturated and Unsaturated Vapours

Take three clean barometer tubes *A*, *B* and *C* as shown in Fig. 16.10. The level of mercury in each tube is up to the same height. The space above mercury in all the tubes is Torricellian vacuum.

With the help of a bent pipette introduce water into the tube *B*. Water rises to the top and evaporates. Level of mercury in the tube *B* falls. The space above mercury in *B* contains water vapours. There is no water left on the surface of mercury. These vapours are unsaturated and the pressure exerted by the vapours is *h*.

Similarly continue introducing water in the tube *C*. The level of mercury gradually falls. A stage is reached when no more water introduced in the tube *C* evaporates (Fig. 16.10). It means the space above mercury in *C* is completely saturated with water vapours. It cannot contain more vapours. The difference in the levels of mercury in *A* and *C* is *H* and it represents the saturated vapour pressure of water at the room temperature.

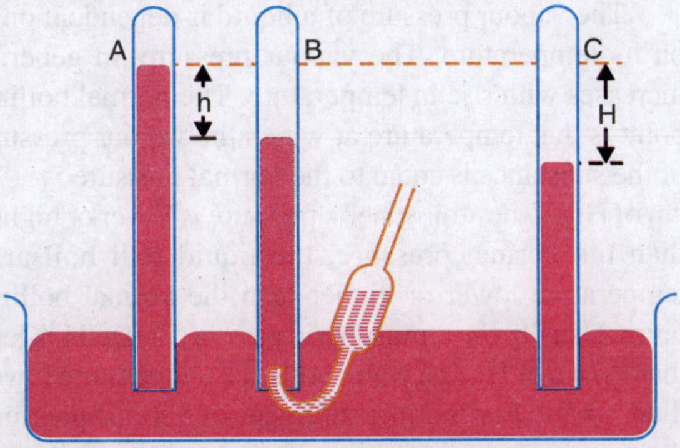

Fig. 16.10

A space is said to be saturated with water vapours if it cannot contain more vapours. It is unsaturated if it can contain more vapours.

(1) The saturated vapour pressure of a liquid depends on its temperature. The saturated vapour pressure increases with rise in temperature and decreases with fall in temperature.

(2) The saturated vapour pressure depends on the nature of the substance.

(3) The saturated vapour pressure does not depend on the volume occupied by the vapours.

(4) The saturated vapour of a liquid is independent of the presence of the vapours of other liquids present. The vapours should not have any chemical action.

(5) The total pressure exerted by the vapours of all substances is equal to the sum of the pressure exerted by the vapours of individual substances,

$$P = P_1 + P_2 + P_3 + \ldots$$

(6) The saturated vapours do not obey the gas laws whereas unsaturated vapours obey the gas laws.

In Fig. 16.11, the curve *AB* shows that with decrease in volume the pressure increases. From *A* to *B* the vapour is unsaturated and obeys Boyle's law.

At *B* the vapour is saturated and exerts maximum pressure. With decrease in volume beyond *B*, the pressure is constant. From *B* to *C* it represents the change from the saturated vapour to the liquid state.

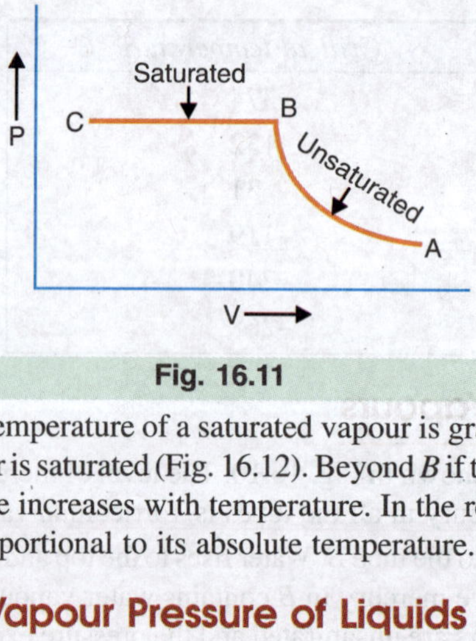


Fig. 16.11

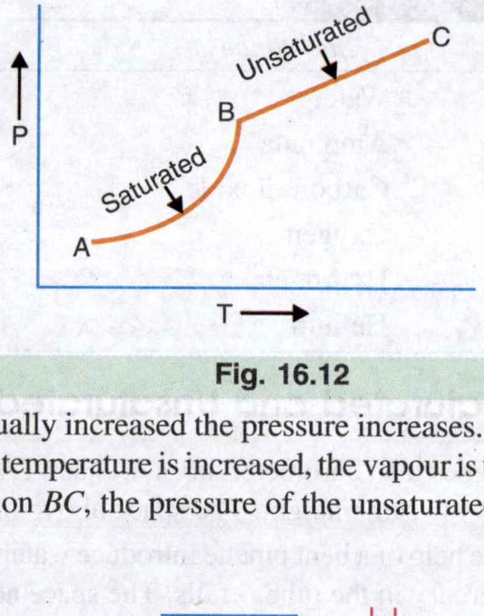


Fig. 16.12

If the temperature of a saturated vapour is gradually increased the pressure increases. From *A* to *B* the vapour is saturated (Fig. 16.12). Beyond *B* if the temperature is increased, the vapour is unsaturated and pressure increases with temperature. In the region *BC*, the pressure of the unsaturated vapour is directly proportional to its absolute temperature.

## 16.22 Vapour Pressure of Liquids

The vapour pressure of a liquid is dependent only on its temperature. The vapour pressure, in general, increases with rise in temperature. The normal boiling point is that temperature at which the vapour pressure of the substance is equal to the normal pressure *i.e.*, 76 cm of Hg. If the atmospheric pressure is lower or higher than the normal pressure, the liquid will boil at a temperature lower or higher than the normal boiling point. Usually the atmospheric pressure in Delhi is less than 76 cm of Hg and water boils at a temperature lower than 100°C. For the measurement of vapour pressure the apparatus shown in Fig. 16.13 is used.

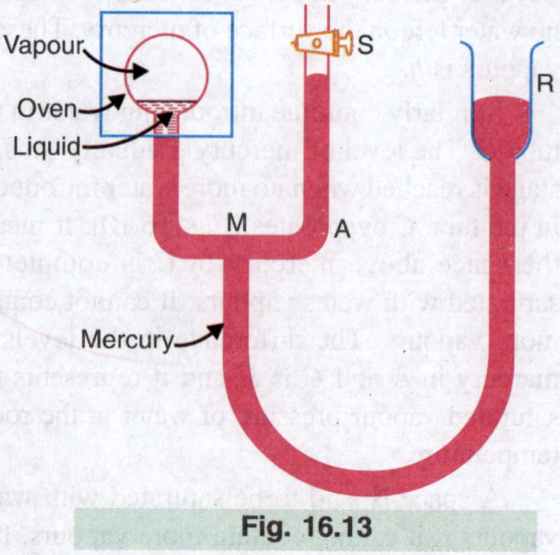


Fig. 16.13

The bulb *B* contains the vapour in contact with its

liquid. $M$ is a manometer limb and $R$ is a reservoir containing mercury. At high temperatures, the stop-cock $S$ is kept open and the tube is filled with a gas at high pressure. Adjust the reservoir $R$ such that the liquid appears in the tube inside the oven. If the pressure of the gas in the limb $A$ is $P_0$ and the difference in levels of mercury in the manometer limbs is $h$, then vapour pressure of the liquid at the temperature $T$ is given by

$$P = P_0 + h \qquad ...(i)$$

In this way the vapour pressure of the liquid at any temperature can be determined.

For low vapour pressure measurement (*i.e*,. at low temperatures) the reservoir $R$ is raised so that the level of mercury in the limb $A$ is above the stop-cock $S$. The stop-cock $S$ is closed. The vessel $B$ is surrounded by the low temperature bath and the reservoir $R$ is adjusted so that the liquid in the tube is within the bath. The difference in the levels of mercury in the manometer directly measures the vapour pressure. In this case $P_0 = 0$

$$\therefore \qquad P = h \qquad ...(ii)$$

As the temperature of the liquid is lowered, a state is reached when the liquid begins to solidify. The temperature and pressure correspond to the *triple point*. At the triple point the substance exists as a solid, liquid and vapour simultaneously.

## 16.23 Triple Point

(*i*) The melting point of ice decreases with increase in pressure. It means that ice melts at a temperature lower than 0°C at a pressure higher than the normal pressure. The curve $AB$ (Fig. 16.14) represents the relation between pressure and temperature. The curve $AB$ is called the ice line. The substance will exist in the solid state (ice) to the left of the curve $AB$ and in the liquid state to the right of the curve $AB$. The curve $AB$ represents the equilibrium between the liquid and the solid states.

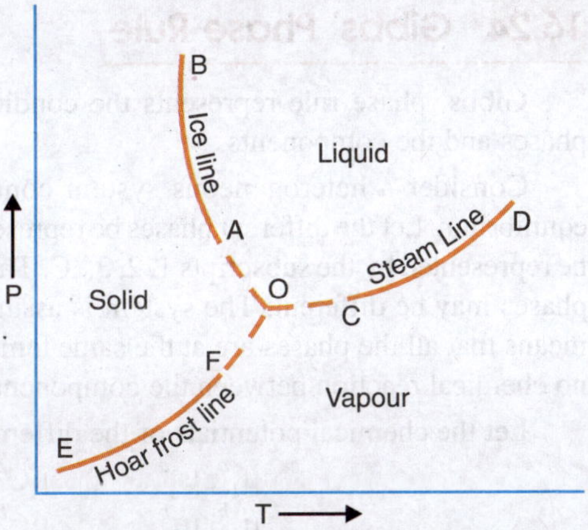


Fig. 16.14

(*ii*) The boiling point of water increases with increase in pressure. The curve $CD$ represents the relation between pressure and temperature and it is called the steam line. Above the curve $CD$, the substance is in the liquid state and below the curve $CD$ it is in the vapour state. The curve $CD$ represents the equilibrium between the liquid and the vapour states.

(*iii*) The curve $EF$ represents equilibrium between the solid and the vapour states of a substance and is called the *Hoar Frost* line. Above the curve $EF$, the substance is in the solid state and below the curve $EF$ it is in the vapour state.

These three curves should meet at a point $O$ as shown in Fig. 16.14. Suppose these curves do not meet at a point as shown in Fig. 16.15 and they enclose an area $ACF$. It means, according to the ice line, the substance must be only in the solid state in the shaded portion. According to the steam line, the substance must be only in the liquid state and according to the hoar frost line the substance must be only in the vapour state.

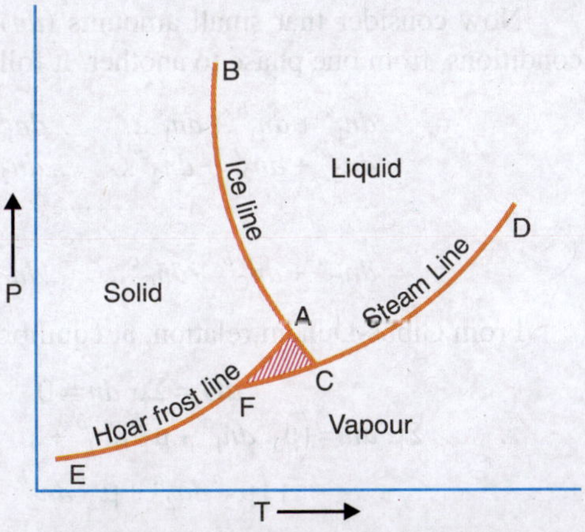


Fig. 16.15

Thus in the shaded portion, the substance must exist simultaneously in the solid, liquid and the vapour states. It is not possible. Therefore, the three curves must meet at a single point *O*, called the **triple point.**

For water, the temperature and pressure corresponding to the triple point are 0.0075 °C and 4.58 mm of Hg respectively. The triple point of water is not fixed but is different for different *allotropic forms of ice.*

**TABLE 16.2 Triple Points**

| | *T (K)* | *Pressure (mm of Hg)* |
|---|---|---|
| $H_2$ | 13.84 | 52.8 |
| $N_2$ | 63.18 | 94.0 |
| $O_2$ | 54.36 | 1.14 |
| $CO_2$ | 216.65 | 3.880 |
| $H_2O$ | 273.1675 | 4.58 |
| I | 387 | 90.0 |

## 16.24 Gibbs' Phase Rule

Gibbs' phase rule represents the condition of equilibrium between the number of coexisting phases and the components.

Consider a heterogeneous system consisting of *P* phases and containing *C* components in equilibrium. Let the different phases be represented by the superscripts *a, b, c...p* and let the components be represented by the subscripts 1, 2, 3,..C. The concentration of any of the components in the different phases may be different. The system is assumed to be in mechanical and thermal equilibrium. This means that all the phases are at the same temperature and pressure. It is further assumed that there is no chemical reaction between the components in the system.

Let the chemical potentials of the different components in the *P*-phases be written as

$$\begin{array}{ll} \mu_1^a, \mu_2^a \ldots & \ldots \mu_C^a \\ \mu_1^b, \mu_2^b \ldots & \ldots \mu_C^b \\ \ldots \quad \ldots & \ldots \\ \ldots \quad \ldots & \ldots \\ \mu_1^p, \mu_2^p \ldots & \ldots \mu_C^p \end{array}$$

Now consider that small amounts (*dn*) of the components are transferred under equilibrium conditions, from one phase to another. It follows that

$$\left.\begin{array}{l} dn_1^a + dn_1^b + dn_1^c \ldots \quad \ldots dn_1^p = 0 \\ dn_2^a + dn_2^b + dn_2^c \ldots \quad \ldots dn_2^p = 0 \\ \ldots \quad \ldots \quad \ldots \quad \ldots \\ \ldots \quad \ldots \quad \ldots \quad \ldots \\ dn_C^a + dn_C^b + dn_C^c \ldots \quad \ldots dn_C^p = 0 \end{array}\right\} \quad \ldots(i)$$

From Gibbs-Duhem relation, at equilibrium, the change in Gibbs' potential

$$dG = \Sigma\mu \, dn = 0$$

$$\left.\begin{array}{rl} \therefore \quad \Sigma\mu \, dn = & \left(\mu_1^a dn_1^a + \mu_1^b dn_1^b + \ldots \quad \ldots \mu_1^p dn_1^p\right) \\ & + \left(\mu_2^a dn_2^a + \mu_2^b dn_2^b + \ldots \quad \ldots \mu_2^p dn_2^p\right) \\ & \ldots \quad \ldots \quad \ldots \quad \ldots \\ & \ldots \quad \ldots \quad \ldots \quad \ldots \\ & + \left(\mu_C^a dn_C^a + \mu_C^b dn_C^b + \ldots \mu_C^p dn_C^p\right) = 0 \end{array}\right\} \quad \ldots(ii)$$

Multiplying the set of equations in (*i*) by *C* different multipliers,

$$\lambda_1, \lambda_2 \ldots \quad \ldots \lambda_C,$$

$$\left.\begin{array}{llll} \lambda_1 dn_1{}^a + \lambda_1 dn_1^b + & \ldots + \lambda_1 dn_1{}^p = 0 \\ \lambda_2 dn_2{}^a + \lambda_2 dn_1^b + & \ldots + \lambda_2 dn_2{}^p = 0 \\ \ldots \quad \ldots & \ldots \quad \ldots \\ \ldots \quad \ldots & \ldots \quad \ldots \\ \lambda_C dn_C{}^a + \lambda_C dn_C^b + & \ldots \lambda_C dn_C{}^p = 0 \end{array}\right\} \qquad \ldots(iii)$$

Adding equations (*ii*) and (*iii*) and equating the coefficients of *dn* terms to zero, we get

$$\left.\begin{array}{l} -\lambda_1 = \mu_1{}^a = \mu_1{}^b = \ldots \quad \ldots = \mu_1{}^p \\ -\lambda_2 = \mu_2{}^a = \mu_2{}^b = \ldots \quad \ldots = \mu_2{}^p \\ \ldots \quad \ldots \quad \ldots \\ \ldots \quad \ldots \quad \ldots \\ -\lambda_C = \mu_C{}^a = \mu_C{}^b = \ldots \quad \ldots = \mu_C{}^p \end{array}\right\} \qquad \ldots(iv)$$

The set of equations in (*iv*) are the equations of phase equilibrium. It means that at equilibrium, the chemical potentials of a component in all its phases must be equal. However chemical potentials are different for different components.

In every phase there are *C* different components having *C* different concentrations. But for a closed system if concentrations of $(C - 1)$ components are arbitrarily chosen, the concentration of the last component is automatically fixed. Therefore, there are $(C - 1)$ composition variables in each phase. For *P* phases the number of composition variables would be

$$P(C - 1)$$

In addition to this value, there are two other variables *viz*., pressure and temperature. Thus the total number of variables will be.

$$= P(C - 1) + 2 \qquad \ldots(v)$$

Further it has been shown that if chemical potential of any one of the *P* phases is defined then the chemical potential of the remaining $(P - 1)$ phases is fixed. Therefore for *C* components the invariant composition variables will be

$$= C(P - 1) \qquad \ldots(vi)$$

Therefore the number of independent variables of a system will be given by

$$F = P(C - 1) + 2 - C(P - 1) \qquad \ldots(vii)$$

Here *F* represents the number of degrees of freedom.

From equation (*vii*)

$$F = C - P + 2 \qquad \ldots(viii)$$

or

$$P + F = C + 2 \qquad \ldots(ix)$$

Water can exist in three states, *viz*., solid, liquid and vapour. These are called the three phases of water. A component is defined as a distinct chemical substance different from others. Ice, water and steam refer to the same component *viz*., water.

In Fig. 16.14, above the curve *CD*, water exists only in the liquid state. In the space *BOD*, the pressure or temperature can be changed independently without any change in the phase of a substance. Thus a substance has two degrees of freedom (temperature and pressure) and only one phase (liquid).

Along the curve *OD*, the substance has two phases (liquid and vapour) but only one degree of freedom (temperature or pressure).

If the pressure is changed the temperature also changes along the curve *OD*.

At the triple point the substance has three phases (solid, liquid and vapour) and no degree of freedom.

According to Gibbs' phase rule, the sum of the number of phases and the number of degrees of freedom is equal to the number of components plus two.

Let $P$ represent the number of phases, $F$ the number of degrees of freedom and $C$ the number of components. Then

$$P + F = C + 2$$

(*i*) For the liquid state, (region *BOD*)

$$P = 1, F = 2 \text{ and } C = 1$$
$$1 + 2 = 1 + 2$$

(*ii*) For the liquid-vapour state (curve *OD*)

$$P = 2, F = 1, C = 1$$
$$2 + 1 = 1 + 2$$

(*iii*) For the solid-liquid-vapour state (triple point)

$$P = 3, F = 0, C = 1$$
$$3 + 0 = 1 + 2$$

The phase rule has practical applications when the components involved are more than one, *e.g.*, freezing mixtures (ice and salt).

## 16.25 Weather

Dew, rain, clouds, fog, mist, hails, snow, hoar frost etc. are due to the condensation of water vapours present under different conditions in the atmosphere.

**Dew.** Dew is formed on grass and leaves in the early hours of the morning. The temperature of the atmosphere during the night gradually falls. In the early morning if the temperature of the atmosphere is less than the dew point temperature, formation of dew takes place. The saturated vapours when they come in contact with grass or leaves condense in the form of fine drops of water.

If the night is cloudy, the temperature of the earth does not fall and dew is not formed. A clear sky free from wind is helpful in the formation of dew.

**Hoar Frost.** If the temperature of the atmosphere falls below 0°C, the dew formed on the leaves and grass freezes and it is called hoar frost.

**Fog and Mist.** When water vapours present in the atmosphere condense on dust particles in the atmosphere, fog is formed. There is no basic difference between fog and mist. Thick mist is called fog. A clear cold night is favourable for the formation of fog and mist.

**Clouds and Rain.** When the fog is formed at high altitudes it is called a cloud. When small drops of water in the cloud combine together, bigger water drops are formed and fall in the form of rain.

**Snow and Hail.** When the cloud suddenly passes through a very cold region, the water drops freeze and fall in the form of snow and hail. Snowfall is common in hill stations because of very low temperature in winter. Hails are caused due to the freezing of rain drops in the atmosphere. Hails are hard and of different sizes whereas snow is quite soft.

## 16.26 Ozone Hole and Role of Chlorofluorocarbons (CFCs)

Sources of radiation (both ionizing and non-ionizing) are natural and artificial. The cosmic rays (for us, Sun is the major source) originate from the outer space. The harmful effects of cosmic radiation on the human race, animals and plants was negligible due to the absorption of these hazardous radiation by the ozonosphere and thereby protecting the earth. But today, due to industrialization and excess use of fertilizers, the situation has become grave. The depletion of the ozone in the atmosphere and formation of 'ozone holes' in the ozonosphere has offered a direct access to these radiation towards the earth. The people living nearer to the equatorial region, the countries like Australia, South Africa, have witnessed the alarming number of cases of cancers, cataracts and mutagenesis.

The terrestrial sunlight consists of visible light, infrared, and ultraviolet (UV) radiation. The visible light and warmth inducing infrared radiation have been essential to human evolution and continued existence on the earth. A stratospheric Ozone, though it is a minor constituent of the Earth's atmosphere stands essential for sustaining of both plant and mammalian on the Earth.

Ozone is restricted to a thin layer in the Earth's atmosphere (such that if all the atmospheric ozone was collected at the Earth's surface, it would form a layer only 3mm thick), the ozonosphere, with a maximum concentration between 20 and 26 km above the earth's surface. It is formed through the combination of atomic oxygen (O) and molecular oxygen ($O_2$). Atomic oxygen (O) is formed by the photo-dissociation of ($O_2$) at around 100 km by solar radiation with $\lambda < 175$ nm, the free oxygen atoms may then combine with oxygen molecules in a three-body collision (to conserve energy) to form ozone.

$$O + O_2 + M \rightarrow O_3 + M \quad \text{...}(i)$$

where M is any atom/molecule (*e.g.*, $O_2$,$N_2$) capable of absorbing the excess energy liberated in the exothermic chemical reaction. This reaction is fast under most atmospheric conditions and therefore the reaction of producing atomic oxygen (O) is the limiting step in the formation of the terrestrial ozone layer. At high altitudes, photon fluxes are high, but $O_2$ concentrations are low, while at low altitudes the reverse is true. Consequently a maxima in the rate of ozone formation (and hence ozone concentrations) is reached at intermediate altitudes, where both photon fluxes and $O_2$ concentrations are significant. This is that region known as stratosphere or ozonosphere.

## THE 'OZONE HOLE'

Ozone is naturally destroyed by photo-dissociation as

$$h\upsilon + O_3 = O_2 + O \quad \text{...}(ii)$$

and chemical reactions producing

$$O_3 + O = 2O_2 \quad \text{...}(iii)$$

These processes being a net result of complex catalytic chemical reactions, either natural or more recently man made. Natural catalytic destruction occurs through the presence of OH radicals formed by photo-dissociation of water vapour in the atmosphere.

$$\begin{array}{ll} & OH + O_3 = HO_2 + O_2 \\ & HO_2 + O = OH + O_2 \\ \hline \text{Net} & O + O_3 = 2O_2 \end{array} \quad \text{...}(iv)$$

in which OH radical survives to convert further ozone in molecular oxygen.

In 1974, Molina and Rowland, (Noble Prize for chemistry awarded in 1995) suggested that the uncontrolled release of chlorofluorocarbons (CFCs) into the terrestrial atmosphere would lead to the catalytic destruction of ozone. Although chemically stable within the troposphere, in the stratosphere the CFCs can be photolysed and release Cl radicals which can then destroy ozone in the catalytic cycle

$$\begin{array}{ll} & Cl + O_3 = ClO + O_2 \\ \text{Net} & ClO + O = Cl + O_2 \\ \hline & O + O_3 = 2O_2 \end{array} \quad \text{...}(v)$$

This mechanism is far more destructive than those naturally occurring mechanisms involving OH radical, as one Cl radical may destroy several thousand ozone molecules before it itself is removed from the ozone chemical cycle. Molina and Rowland suggested even trace (ppm) amounts of any CFC stratosphere could have a dramatic effect upon stratospheric ozone concentration. In 1985 Joe Farmer, Brian Gardiner, and Jonathan Shankin of the British Antarctic Survey discovered a sharp reduction in ozone concentrations above Antarctica; this was soon termed the 'ozone hole'.

Another catalytic destructive agent for ozone is NO. Attention is now being drawn to the effect of other halogen containing species (*e.g.*, bromine derivatives) and the increased emissions of $N_2O$ due to a rise in the use of nitrates in agriculture, in turn leading to an increase in NO within the stratosphere. Other chemical pollutants may yet be found to play a role in ozone depletion. The continued development of industry and introduction of new chemical production schemes make it likely that we will continue unwillingly, to alter the atmosphere's natural balance.

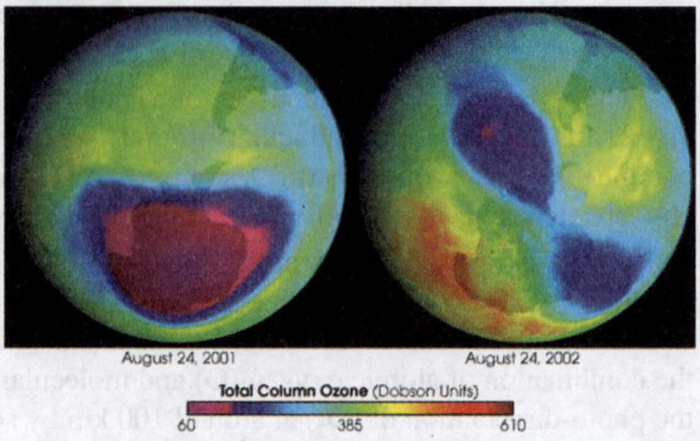


**Antarctic Ozone Hole**

## NUMERICAL EXAMPLES

### EXAMPLE 16.6

*If the temperature of air is 16.5°C and dew point is 6.5°C, find the % relative humidity of air. S.V.P. at 6°, 7°, 16° and 17° are 7.05, 7.51, 13.62 and 14.42 mm respectively.*

**SOLUTION.**

$$\text{Dew point} = 6.5°\text{C}$$

$$\text{Room temperature} = 16.5°\text{C}$$

$$S.V.P. \text{ at } 6.5°\text{C} = \frac{S.V.P. \text{ at } 6° + S.V.P \text{ at } 7°}{2}$$

$$p = \frac{7.05+7.51}{2} = 7.28 \text{ mm}$$

$$S.V.P. \text{ at } 16.5°\text{C} = \frac{S.V.P. \text{ at } 16 + S.V.P \text{ at } 17}{2}$$

$$P = \frac{13.65+14.42}{2} = 14.02 \text{ mm}$$

$$R.H. = \frac{p}{P}\times 100$$

$$= \frac{7.28}{14.02}\times 100 = \mathbf{51.9\%}$$

### EXAMPLE 16.7

*Calculate the dew point when the relative humidity of air at 20°C, is 52%. Given the S.V.P. of water at 20°C, 10°C and 9°C are 17.5,9.2, 8.6 mm of Hg respectively.*

**SOLUTION.**

$$R.H. = \frac{S.V.P. \text{ at dew point}}{S.V.P. \text{ at room temperature}} = \frac{p}{P}$$

$$0.52 = \frac{p}{17.5}$$

$$\therefore \quad p = 9.1 \text{ mm}$$

$$\text{For 1°C difference} = 9.2 - 8.6 = 0.6 \text{ mm}$$

$\therefore$ Dew point temperature

$$= 9 + \frac{0.5}{0.6} = \mathbf{9.83°C}$$

## EXERCISE – XVI

**16.1** What do you understand by change of state?

**16.2** What is triple point? Show that the steam line, the ice line and the Hoar Frost line meet at a single point. **(Kerala 2001; Guwahati 2000)**

**16.3** Discuss Gibbs' phase rule. What do you understand by phase, degree of freedom and component?

**16.4** What is the difference between a gas and a vapour?

**16.5** Distinguish between saturated and unsaturated vapours.

**16.6** Explain briefly the following :

(*i*) Dew (*ii*) Hoar Frost (*iii*) Fog (*iv*) Mist (*v*) Clouds (*vi*) Snow (*vii*) Hail (*viii*) Rain

**16.7** Derive Gibbs' phase rule and discuss the equilibrium of a saturated solution and the solid of the dissolved substance.

**16.8** On a certain day the dew point is 18.6°C and the room temperature is 23.7°C. The saturated vapour pressures of water at 18°C, 19°C, 23°C, and 24°C are 15.46 mm, 16.46 mm, 21.02 mm and 22.32 mm of Hg respectively. Calculate the relative humidity. **[Ans. 73.23%]**

**16.9** Find the result of mixing 10 g of ice at – 10°C with 10 g of water at 10°C. Sp. heat of ice = 0.5 and latent heat of ice = 80 cal/g. **[Ans. A mixture of 9.375 g of ice and 10.625 g of water at 0°C]**

**16.10** What will be the result of mixing 400 g of copper chips at 500°C with 500 g of melting ice? **[Ans. 250 g of ice melts. Final temperature = 0°C]**

**16.11** Calculate how much steam from water boiling at 100°C will just melt 200 g of wax at 15°C. (Melting point of wax = 55°C, Sp. heat of wax = 0.7m Latent heat of fusion of wax = 35 cal per g) **[Ans. 21.54 g]**

**16.12** Explain the principle of refrigeration. **(Osmania 2005)**

**16.13** Calculate the temperature for triple point of water in absence of air. Given that the vapour pressure at 0°C and 1°C are 4.60 and 4.94 mm. respectively, and the lowering of melting point per atmosphere is 0.0072 °C. **[Ans. 0.007156 °C]**

# Air Conditioning Systems

# 17

### At a Glance

• Introduction • Factors Affecting Comfort Air Conditioning • Air Conditioning system • Equipment used in Air Conditioning Systems • Classification of Air Conditioning Systems • Criteria of Comfort Air Conditioning • Summer Air Conditioning System • Winter Air Conditioning System • Year Round Air Conditioning System • Unitary and Central Air Conditioning System • Industrial Air Conditioning System

## Introduction

Modern civilization is showing an increasing fascination towards the air conditioning machines. This demand is increasing day by day to fulfil the 'human comfort'. According to American Society of Heating, Refrigeration and Air Conditioning Engineers (ASHRAE), human comfort is defined as that condition of mind which expresses satisfaction with the thermal environment. Hence the subject has a broad sense which includes the conditioning of air for industrial purposes, life-style of a *hi-fi* society, food processing, storage of food and other materials, computerized and sophisticated instrumentation, auditorium, cinema hall etc.

## 17.1 Factors Affecting Comfort Air Conditioning

The following four factors for comfort air conditioning are basically taken into account. These are:

1. *Temperature of air.* In air conditioning, the control of temperature means the maintenance of any desired temperature within an enclosed space even though the temperature of the outside air is above or below the desired room temperature. This is accomplished either by the addition or removal of heat from the enclosed space as and when demanded. It may be noted that a human being feels comfortable when the air is at 21° *C* with 56% relative humidity.

2. *Humidity of air.* The control of humidity of air means the increasing or decreasing of moisture contents of air during summer or winter respectively in order to produce comfortable and healthy conditions. The control of humidity is not only necessary for human comfort but it also increases the efficiency of the workers. In general, for summer air conditioning, the relative humidity should not be less than 60% whereas for winter air conditioning, it should not be more than 40%.

3. *Purity of air.* It is an important factor for the comfort of a human body. It has been noticed that people do not feel comfortable when breathing contaminated air, even if it is within acceptable temperature and humidity ranges. It is thus obvious that proper filtration, cleaning and purification of air is essential to keep it free from dust and other impurities.

4. *Motion of air.* The motion or circulation of air is another important factor which should be controlled, in order to keep constant temperature throughout the conditioned space. It is, therefore, necessary that there should be equi-distribution of air throughout the space to be air conditioned.

## 17.2 Air Conditioning System

The system that effectively controls the above four conditions to produce the desired effects upon occupants of the space for human comfort, is known as air conditioning system.

## 17.3 Equipments used in Air Conditioning System

The main equipments or parts used in air conditioning system are as under:

1. *Circulation fan.* The main function of this fan is to move air to and from the room.

2. *Air conditioning units.* It is a unit, which consists of cooling and dehumidifying processes for summer air conditioning or heating and humidification processes for winter air conditioning.

3. *Supply duct.* It directs the conditioned air from the circulating fan to the space to be air conditioned at proper point.

4. *Supply outlets.* These are grills, which distribute the conditioned air evenly in the room.

5. *Return outlets.* These are the openings in a room surface which allow the room air to enter the return duct.

6. *Filters.* The main function of the

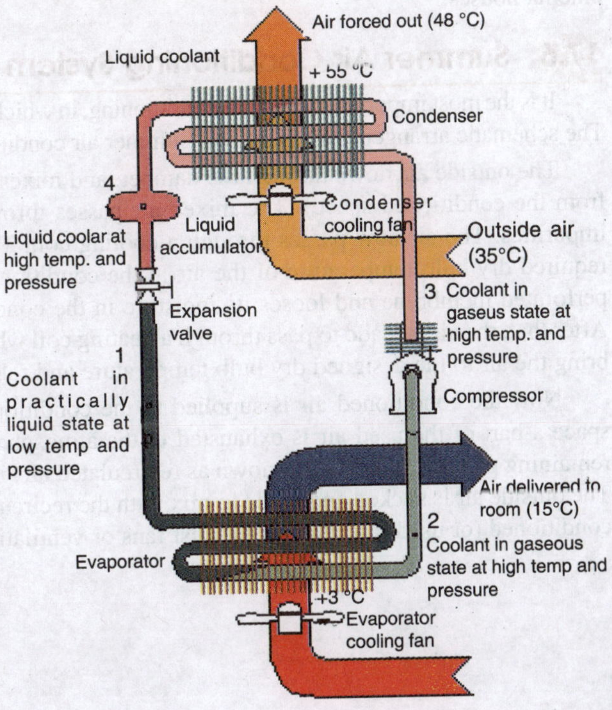

filters is to remove duct, dirt and other harmful bacteria from the air.

## 17.4 Classification of Air Conditioning Systems

The air conditioning systems may be classified broadly into three types. These are;

1. *According to the purpose*
   (*a*) Comfort air conditioning system, and
   (*b*) Industrial air conditioning system.
2. *According to season of the year*
   (*a*) Winter air conditioning system,
   (*b*) Summer air conditioning system, and
   (*c*) Year-round air conditioning system.
3. *According to the arrangement of equipment*
   (*a*) Unitary air conditioning system, and
   (*b*) Central air conditioning system.

## 17.5 Criteria of Comfort Air Conditioning

In comfort air conditioning, the air is brought to the required dry bulb temperature and relative humidity. If sufficient data of the required condition is not given, then it is assumed to be 21° *C* day bulb temperature and 50% relative humidity. The sensible heat factor is, generally, kept as following:

For residence or private office = 0.9
For restaurant or busy office = 0.8
Auditorium or cinema hall = 0.7
Ball room dance hall etc. = 0.6

The comfort air conditioning may be adopted for small cabins, office halls or big halls like cinema houses.

## 17.6 Summer Air Conditioning System

It is the most important type of air conditioning, in which the air is cooled and generally dehumidified. The schematic arrangement of a typical summer air conditioning system is shown in (Fig. 17.1).

The outside air flows through the damper, and mixes up with recirculated air (which is obtained from the conditioned space). The mixed air passes through a filter to remove dirt, dust and other impurities. The air now passes through a cooling coil. The coil has a temperature much below the required dry bulb temperature of the air in the conditioned space. The cooled air passes through a perforated membrane and looses its moisture in the condensed form which is collected in a sump. After that, the air is made to pass through a heating coil which heats up the air slightly. This is done to bring the air to the designed dry bulb temperature and relative humidity.

Now the conditioned air is supplied to the conditioned space by a fan. From the conditioned space a part of the used air is exhausted to the atmosphere by the exhaust fans or ventilators. The remaining part of the used air (known as recirculated air) is again conditioned, as shown in Fig. 17.1. The outside air is sucked and made to mix with the recirculated air in order to make up for the loss of conditioned (or used) air through exhaust fans or ventilation from the conditioned space.

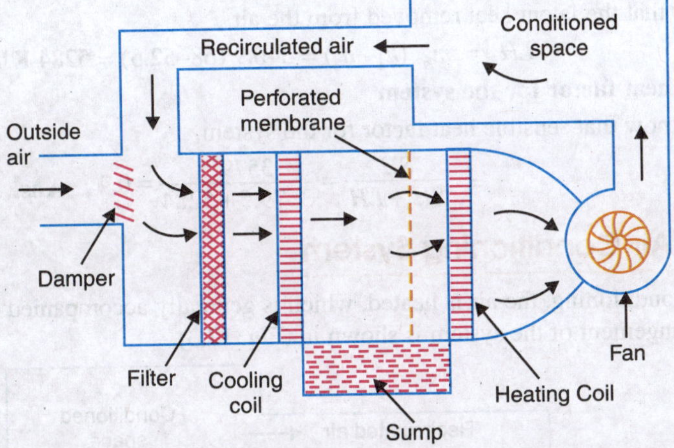


Fig. 17.1 Summer air conditioning system.

### EXAMPLE 17.1

*A theatre of 1200 seating capacity is to be air conditioned for summer conditions with the following data:*

| | |
|---|---|
| *Outdoor conditions* | *30° C DBT and 55% RH* |
| *Required conditions* | *20° C DBT and 60% RH* |
| *Amount of air supplied* | *0.25 m³/min/person.* |

*Find the sensible heat, latent heat removed from the air per minute and sensible heat factor for the system.*

**SOLUTION.** Given : Seating capacity = 1200 person: $t_{d1} = 30^\circ C$ ; $\phi_1 = 55\%$; $t_{d2} = 22^\circ C$ ; $\phi_2 = 60\%$; $v_1 = 0.25\ m^3$/min/person = 0.25 × 1200 = 300 $m^3$/min

#### (i) Sensible heat removed from the air

First of all, mark the initial condition of air at 30° *C* dry bulb temperature and 55% relative humidity on the psychrometric chart as point 1, as shown in (Fig. 17.2.) Now mark the final condition of air at 20° *C* dry bulb temperature and 60% relative humidity on the chart as point 2. Now locate point 3 on the chart by drawing horizontal line through point 2 and vertical line through point 1. From the psychrometric chart, we find that enthalpy of air at point 1,

$$h_1 = 68 \text{ KJ/ kg of dry air}$$

Similarly $\quad h_2 = 42$ KJ/ kg of dry air

and $\quad h_3 = 52.5$ KJ/ kg of dry air

Moreover, specific volume of air at point 1,

$$v_{s1} = 0.88 \text{ m}^3/\text{kg}$$

∴ Mass of air cooled per minute

$$m_a = \frac{v_1}{v_{s1}} = \frac{300}{0.88} = 340.9\ kg$$

and sensible heat removed from the air,

$$SH = m_a(h_3 - h_2) = 340.9$$

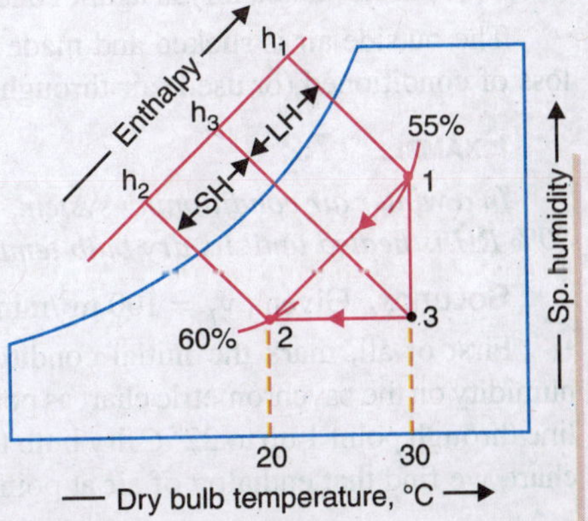


Fig. 17.2

(52.5 – 42) = 3579.5 KJ/min **Ans.**

***(ii)* Latent heat removed from the air**

We know that the latent heat removed from the air,

$$LH = m_a\ (h_1 - h_3) = 340.9\ (68\text{–}52.5) = 5284 \text{ KJ/min} \quad \textbf{Ans.}$$

***(iii)* Sensible heat factor for the system**

We also know that sensible heat factor for the system,

$$= \frac{SH}{SH + LH} = \frac{3579.5}{3579.5 + 5284} = 0.4 \quad \textbf{Ans.}$$

## 17.7 Winter Air Conditioning System

In winter air conditioning, the air is heated, which is generally accompanied by humidification. The schematic arrangement of the system is shown in (Fig. 17.3).

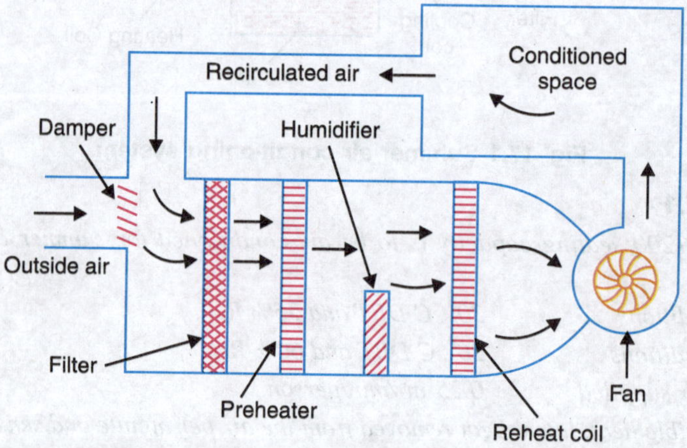


**Fig. 17.3 Winter air conditioning system**

The outside air flows through a damper and mixes up with the recirculated air (which is obtained from the conditioned space). The mixed air passes through a filter to remove dirt, dust and other impurities. The air now passes through a preheat coil in order to prevent the possible freezing of water and to control the evaporation of water in the humidifier. After that, the air is made to pass through a reheat coil to bring the air to the designed dry bulb temperature. Now, the conditioned air is supplied to the conditioned space by a fan. From the conditioned space, a part of the used air is exhausted to the atmosphere by the exhaust fans or ventilators. The remaining part of the used air (known as recirculated air) is again conditioned, as shown in (Fig. 17.3).

The outside air is sucked and made to mix with the recirculated air, in order to make up for the loss of conditioned (or used) air through exhaust fans or ventilation from the conditioned space.

**EXAMPLE 17.2**

*In a winter air conditioning system, 100 m³ of air per minute at 15° C dry bulb temperature and 80% RH is heated until its dry bulb temperature is 22° C. Find heat added to the air per minute.*

**SOLUTION.** Given : $v_1 = 100 \text{ m}^3/\text{min}$; $t_{d1} = 15^\circ C$ ; $\phi_1 = 80\%$; $t_{d_2} = 22^\circ C$

First of all, mark the initial condition of air at 15° *C* dry bulb temperature and 80% relative humidity on the psychrometric chart as point 1. Now mark final condition of air by drawing a horizontal line through point 1 up to 22° *C* dry bulb temperature, as shown in (Fig. 17.4) From the psychrometric chart, we find that enthalpy of air at point 1,

$$h_1 = 36.5 \text{ KJ/ kg of dry air}$$

Similarly, $$h_2 = 43.5 \text{ KJ/ kg of dry air}$$

Moreover, specific volume of air at point 1,

$$v_{s1} = 0.827 \text{ m}^3\text{/kg}$$

Mass of air heated per minute

$$m_a = \frac{v_1}{v_{s1}} = \frac{100}{0.827} = 120.9 \ kg$$

and heat added to the air per minute $m_a(h_2 - h_1) = 120.9$ $(43.5 - 36.5) = 846.3$ KJ **Ans.**

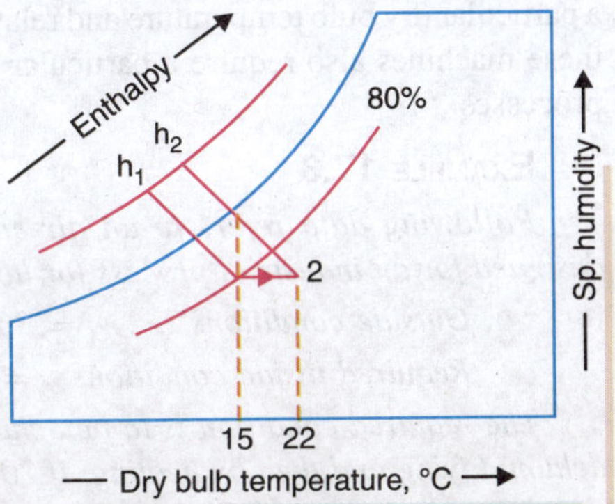


Fig. 17.4

## 17.8 Year Round Air Conditioning System

A year-round air conditioning system should have equipment for both the winter and summer air conditioning. There can be so many combinations for providing year-round air conditioning. The arrangement of the combination should be such that one part is working in winter and the other in summer. For example, heating and humidification is needed in winter, whereas cooling and dehumidification is needed in summer.

## 17.9 Unitary and Central Air Conditioning System

### (*i*) Unitary Air Conditioning System.

In this system, factory assembled air conditioners are installed in or adjacent to the space to be conditioned. The unitary air conditioning systems are of the following two types :

**1.** *Window units.* These are self-contained units of small capacity of 1 tonne to 3 tonnes, and are mounted in a window or through the wall. They are employed to condition the air of one room only. If the room is bigger in size, then two or more units are installed.

**2.** *Vertical packed units.* These are also self-contained units of bigger capacity of 5 to 10 tonnes and are installed adjacent to the space to be conditioned. This is very useful for conditioning the air of a restaurant, bank or small office.

Central Air Conditioning System

The unitary air conditioning system may be adopted for winter, summer or year-round air conditioning.

### (*ii*) Central Air Conditioning System.

This is the most important type of air conditioning system, which is adopted when the cooling capacity required is 25 tonnes or more. The central air conditioning system is also adopted when the air flow is more than 300 m$^3$/min or different zones in a building are to be conditioned.

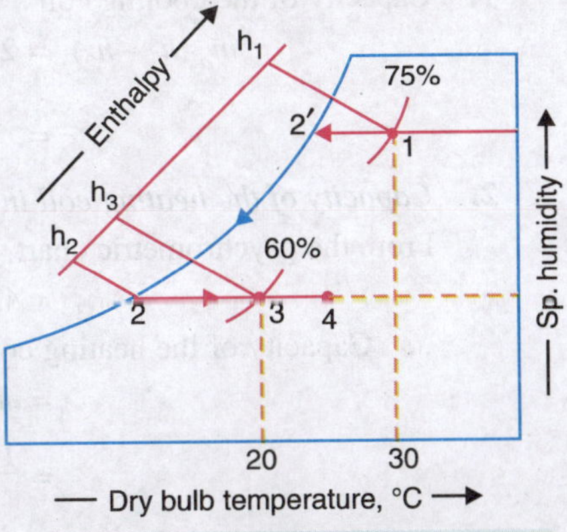


Fig. 17.5

## 17.10 Industrial Air Conditioning System

It is an important system of air conditioning these days in which the inside dry bulb temperature and relative humidity of the air is kept constant for proper working of the machines. Some of the sophisticated electronic and other machines need a particular dry bulb temperature and relative humidity. Sometimes, these machines also require a particular method of psychrometric processes.

Industrial Air Conditioning System

### EXAMPLE 17.3

*Following data refers to an air conditioning system to be designed for an industrial process for hot and wet climate:*

*Outside conditions = 30° C DBT and 75% RH*

*Required inside conditions = 20° C DBT and 60% RH*

*The required condition is to be achieved first by cooling and dehumidifying and then by heating. If 20 m³ of air is absorbed by the plant every minute, find: 1. capacity of the cooling coil in tonne of refrigeration ; 2. capacity of the heating coil in kW ; and 3. amount of water removed per hour.*

**SOLUTION.** Given: $t_{d1} = 30°C$ ; $\phi_1 = 75\%$; $t_{d2} = 20°C$ ; $\phi_2 = 60\%$; $v_1 = 20\ m^3/min$

**1.** ***Capacity of the cooling coil in tonne of refrigeration***

First of all, mark the initial condition of air at 30°C dry bulb temperature and 75% relative humidity, on the psychrometric chart as point 1, as shown in (Fig. 17.5). Then mark the final condition of air at 20°C dry bulb temperature and 60% relative humidity on the chart as point 3.

Now locate the points 2′ and 2 on the saturation curve by drawing horizontal lines through points 1 and 3, as shown in (Fig. 17.5). On the chart, the process 1–2′–2 represents the cooling and dehumidifying process and 2-3 represents the heating process. From the psychrometric chart, we find that the specific volume of air at point 1,

$$v_{s1} = 0.886\ m^3/kg \text{ of dry air}$$

Enthalpy of air at point 1,

$$h_1 = 81.8\ kJ/kg \text{ of dry air}$$

and enthalpy of air at point 2,

$$h_2 = 34.2\ kJ/kg \text{ of dry air}$$

We know that mass of air absorbed by the plant,

$$m_a = \frac{v_1}{v_{s1}} = \frac{20}{0.886} = 22.6\ kg / min$$

∴ Capacity of the cooling coil

$$m_a (h_1 - h_2) = 22.6\ (81.8 - 34.2) = 1075.76\ kJ/min$$

$$= \frac{1075.76}{210} = 5.1\ TR \quad \textbf{Ans.}$$

**2.** ***Capacity of the heating coil in kW***

From the psychrometric chart, we find that enthalpy of air at point 3,

$$h_3 = 42.6\ kJ/kg \text{ of dry air}$$

∴ Capacity of the heating coil

$$= m_a (h_3 - h_2) = 22.6\ (42.6 - 34.2) = 189.84\ kJ / min$$

$$= \frac{189.84}{60} = 3.16\ kW \quad \textbf{Ans.}$$

**3.** *Amount of water removed per hour*

From the psychrometric chart, we find that specific humidity of air at point 1,

$$W_1 = 0.0202 \text{ kg/kg of dry air}$$

and specific humidity of air at point 2,

$$W_2 = 0.0088 \text{ kg/kg of dry air}$$

∴ Amount of water removed per hour

$$= m_a(W_1 - W_2) = 22.6\,(0.0202 - 0.0088) = 0.258 \text{ kg / min}$$

$$= 15.48 \text{ kg}/h \quad \textbf{Ans.}$$

## EXERCISE–XVII

**17.1** Mention the factors affecting comfort air conditioning.

**17.2** Define air conditioning system. What are the equipments used?

**17.3** Give the classification of air conditioning systems. Mention criteria of comfort air conditioning.

**17.4** Explain summer air conditioning system.

An office hall of 70 persons capacity is to be provided with summer air conditioning plant with the following data:

Outside conditions = 35° CDBT and 26° WBT

Inside conditions = 20° CDBT and 45% RH

Volume of air supplied = 0.3 $m^3$/min/person

Find the sensible heat factor for the plant. **[Ans. 0.35]**

**17.5** Explain winter air conditioning system.

**17.6** A cinema hall of seating capacity 1500 persons has been provided with an air conditioned plant with the following data :

Outdoor conditions = 40° CDBT and 20% C WBT

Required conditions = 20° CDBT and 60% RH

Amount of air supplied = 0.3 $m^3$/min/person

If the required condition is achieved first by adiabatic humidifying and then cooling, find the capacity of cooling coil and capacity of the humidifier. **[Ans. 37.6 TR ; 54.2 kg/h]**

**17.7** Write short notes on

(*i*) Human comfort

(*ii*) Winter air conditioning system.

(*iii*) Summer air conditioning system.

(*iv*) Year round air conditioning system.

(*v*) Industrial air conditioning system.

**17.8** Explain what do you mean by year round air conditioning ?

**17.9** Describe unitary and central air conditioning system.

**17.10** Explain industrial air conditioning system.

**17.11** Write a note on Air conditioning. **[Madras 2005/S]**

# Global Warming 18

## At a Glance

• Introduction • Consequences of Global Warming • The Green-house Effect • Types of Greenhouse Gases • Evidence of Global Warming • Possible Causes of Global Warming • Effects of Global Warming • Efforts to Control Global Warming • National and International Programs • Our Contribution In reduction of Global Warming • Conclusions

## Introduction

In early, January 1996, meteorologists in New York city, Washington (D.C.) reported that the earth's average surface temperature in 1995 was the highest of any previous year recorded. Various research centers in Great Britain also noticed similar results for 1995 i.e. the temperature 14.84 $^0$C (58.72 $^0$F), the highest since records were kept first in 1856. Gradual rise in average temperature of atmosphere is considered as the most relevant evidence of global warming trend. Global warming refers to an increase in the earth's average surface air temperature. Global warming and cooling in themselves are not necessarily bad, since the earth has gone through cycles of temperature change many times in 4.65 billion years of its history, without any noticeable change in its average temperature. However, human activities are adding to its natural level of warming, resulting into the rapid increase in its average temperature. The global warming usually means a fast, unnatural increase that is enough to cause the expected climate conditions to change rapidly and often catastrophically.

The conceivable effects of global warming are wide-ranging and possibly severe. The consequences of global climate change may hamper the mammalian and plants life on the earth due to increase in the average temperature of the atmosphere, oceans and landmasses. The chief cause of this warming is thought to be burning of fossil fuels, such as coal, oil, and natural gas, which releases into the atmospheric carbon dioxide and other substances known as greenhouse gases. As the atmosphere becomes richer in these gases, it becomes a better insulator, retaining more of the heat provided to the earth by the sun. Growing of excess of green house gases in earth's atmosphere, our earth planet is miserably facing the problem of global warming. Today, human community at large is of the opinion that we experimenting with our planet earth, under the name of industrialization and globalization and are forcing it into a danger zone.

## 18.1 Consequences of Global Warming

The average surface temperature of earth is about 15 °C (59 °F). Over the last century, this average has risen by about 0.6 Celsius degrees (1 Fahrenheit degree). Scientists predict further warming of 1.4 to 5.8 Celsius degrees (2.5 to 10.4 Fahrenheit degrees) by the year 2100. This temperature rise is expected to melt polar ice-caps and glaciers as well as warm the oceans, all of which will expand ocean volume and raise sea level by an estimated 9 to 100 cm (4 to 40 inches), flooding some coastal regions and even entire islands. Some regions in warmer climate will receive more rainfall than before, but soil will dry out faster between storms, causing soil desiccation. This soil desiccation may damage food crops, thereby disrupting food supplies in some part of the world. Plants and animal species will shift their ranges toward the poles or to higher elevations seeking cooler temperatures and species that cannot do so, may be ruined and hence become distinct.

## 18.2 The Green-house Effect

Energy reaching the earth is mostly in the form of visible light and heat which lights and warms the earth and is necessary for sustaining mammalian and plants life (refer article 8.36). Fig.18.1 depicts the role of green house effect in heating the earth's atmosphere.

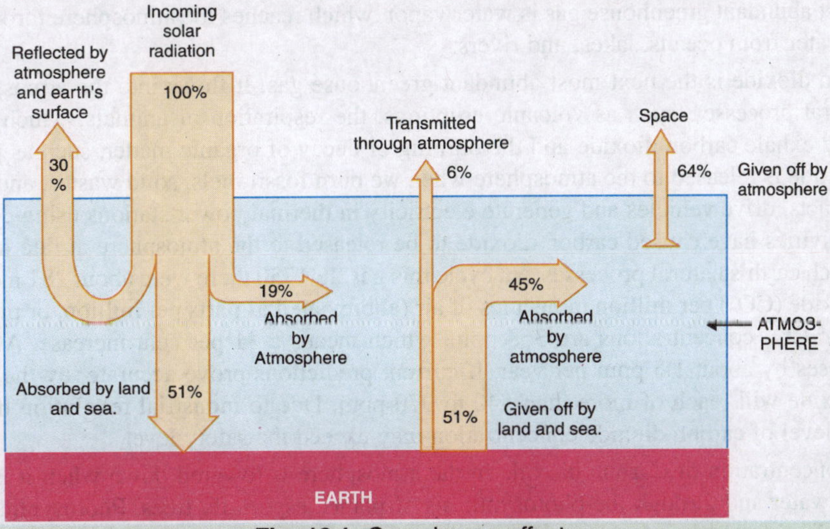


**Fig. 18.1.** Greenhouse effect

Visible light has a wavelength range 400 nm to 700 nm which is shorter than infra-red (i.e. heat) radiation, both being electromagnetic waves. Out of the total energy projected towards the earth, particles in the upper atmosphere reflect a small amount of this energy off into the space and most of the energy floods onto the earth surface. When this sunlight energy strikes the surface of earth, the light energy changes into heat i.e. long-wave infra-red radiation. As a consequence of this, most of the light energy, in turn, is released by the earth on reflection as long-wave infra-red radiation and add further to the warming of earth atmosphere.

Due to the presence of many layers of water vapor and other molecules in the atmosphere, its conductivity becomes poor and therefore heat is transferred out through the upper atmosphere slowly. When the global temperature is stable, the atmosphere eventually transfers all this energy back into the space; but a portion remains trapped in earth's atmosphere. Certain gases in the atmosphere including water vapor, carbon dioxide and methane provide the trap. The continuous process of absorbing sunlight and reflecting infra-red waves by the earth, these gases conserve heat as the glass does in a greenhouse and thus these gases are known as **greenhouse gases**. The Swedish scientist Svante Arrhenius in 1896, first described this natural process, called it as the "**greenhouse effect**". In fact, Arrhenius estimated that the natural greenhouse effect of the atmosphere was responsible for the average global temperature and is favorable for sustaining life on the earth. In other words, all life on earth relies on this greenhouse effect- without it, the planet would have been colder by about 33 Celsius degrees (59 Fahrenheit degrees), and ice would have covered earth from N-pole to S-pole. This is the advantage of greenhouse effect in order to maintain the average temperature on the earth's surface. Thus, the greenhouse effect works as a gift offered to mankind by the nature.

However, as the concentration of these greenhouse gases in the atmosphere increases, more heat energy remains trapped in the blanket of atmosphere. The lag of heat transfer raises the total amount of heat present on the earth's surface and thus the surface temperature goes on increasing constantly. The rise of concentration of these greenhouse gases tampers the greenhouse balance and progresses in the anti-balancing direction- towards continual warming and resulting into "**global warming**". This inculcates the detrimental implications on human society and life on earth.

## 18.3 Types of Greenhouse Gases

Greenhouse gases occur naturally in the environment and also result from human activities. By far the most abundant greenhouse gas is water vapor, which reaches the atmosphere through evaporation of water from oceans, lakes, and rivers.

Carbon dioxide is the next most abundant greenhouse gas. It flows into the atmosphere from many natural processes, such as volcanic eruptions, the respiration of animals, which breathe in oxygen and exhale carbon dioxide and the burning or decay of organic matter, such as plants. The carbon dioxide is released to the atmosphere when we burn fossil fuels, solid wastes, and wood and wood products, drive vehicles and generate electricity in thermal power stations using coal as fuel. Human activities have caused carbon dioxide to be released to the atmosphere at rate much faster than at which earth's natural processes can cycle this gas. In 1750 there were about 281 molecules of carbon dioxide ($CO_2$) per million molecules of air (abbreviated as parts per million, or ppm). Today atmospheric $CO_2$ concentrations are 368 ppm, which means a 31 per cent increase. Atmospheric $CO_2$ increases by about 1.5 ppm per year. If current predictions prove accurate, by the year 2100 carbon dioxide will reach of more than 540 to 970 ppm. Due to industrial revolution through the world, the level of carbon dioxide concentration may exceed the safety level.

The concentration of Carbon dioxide in the atmosphere is lowered down when it is absorbed into ocean water and through the photosynthesis of plants, especially trees. Photosynthesis breaks up carbon dioxide, releasing oxygen into the atmosphere and incorporating the carbon into new plant tissue. At the same time, the number of trees available to absorb carbon dioxide through photosynthesis has been greatly reduced by massive deforestation, particularly clearing land for agricultural activities in tropical regions and rapid growth in industrialization throughout the world.

Methane is an even more effective insulator, trapping over 20 times more heat than does the same amount of carbon dioxide. Methane is emitted during the production and transport of coal, natural gas and oil. Methane also comes from decomposing organic waste in landfills, and it is released from certain animals, especially cows, as a byproduct of digestion. Since the beginning of the Industrial Revolution in the mid-1700s, the amount of methane in the atmosphere has more than doubled.

Nitrous oxide is another powerful insulating gas released primarily by burning fossil fuels and by plowing farm soils. Nitrous oxide traps about 300 times more heat than does the same amount of carbon dioxide. The concentration of nitrous oxide in the atmosphere has increased 17 per cent over pre-industrial levels.

In addition, greenhouse gases are produced in many manufacturing processes. Perfluorinated compounds result from the smelting of aluminium. Hydrofluorocarbons are formed during the manufacture of many products, including the foams used in insulation, furniture and car seats. Refrigerators built in some developing nations still use chlorofluorocarbons as coolants. In addition of their ability to trapping heat, they destroy the earth's high-altitude ozone layer, forming "ozone hole". Scientists are of growing concern about these hazardous gases produced that pose an environmental catastrophe.

## 18.4 Evidence of Global Warming

Global average temperatures strongly suggest that global warming has actually begun to take place. This is not, however, a simple trend to document. Temporary confusing factors can mask on overall trend in average temperatures. For example, when Mount Pinatubo erupted in the philippines in 1991, the volcano released enormous quantities of dust and vapor into the upper atmosphere. The material ejected in the eruption blocked enough sunlight to depress global average temperature for two years. These kinds of events only temporarily obscure an overall trend that reappears afterward as though the masking factor had never occurred. Inevitable variations in local climate, such as the recent snowstorm in the Eastern United States, attract attention but are not meaningful in the overall average.

Data for global temperature calculation have been collected for a too short period to reveal its confirmation, whether the undeniable warming trend since 1856 is human –induced warming or part of some natural cycle. In many senses this is an impractical debate because it cannot be settled completely until there are, say, 1000 years for continuous data collection would be too late to take action. In this context, a global body of thousands of scientists, in 1995 issued a consensus statement concluding that human –induced global warming had probably begun. These scientists, chosen by their governments as the best in their fields, made up a United Nations (UN) advisory body known as the Intergovernmental Panel on Climate Change (IPCC).

Although the question of the human contribution to the present warming trend cannot yet be resolved definitively, some longer-term sets of data exist that can provide clues about past temperature changes and their relation to concurrent greenhouse gas concentrations. Glacial ice can store this information. Glaciers are formed when the accumulating weight of annual increments of snow turns lower layers of the snow into ice. In the process, tiny bubbles of air are trapped that are essentially micro-samples of past atmospheric composition. These air bubbles can be sampled by taking an ice core. Because the ratio of oxygen isotopes in the air varies predictably with temperature, oxygen isotopes in the samples provide an accurate indication of what the temperature was when the bubble formed, and modern analysis techniques can reveal $CO_2$ or $CH_4$ concentrations from the same atmosphere. A number of ice cores have been taken; the oldest is from the Russian Antarctic Station Vostoc. A major problem with making any prediction about climate change is that the situation is not static. Computer models suggests that a global average temperature increase of 1.5 to 3.5 °C (2 to 7 °F) with a doubling of $CO_2$.

## 18.5 Possible Causes of Global Warming

In fact, it is very difficult to assign the exact causes of global warming because the earth's climatic system is complex and some factors that affects climate may vary in cycles over time. For instance, the activity of the Sun varies cyclically in every 11 years. Sunspots, transient cool spots on the face of the Sun, that are associated with bursts of solar energy, waxing and waning with a period of 11 -year period.

In another cyclic pattern, periodic shifts in Pacific Ocean currents create the EI Nino effect. EI Nino occurs when warm water in the western Pacific Ocean flows unusually far eastward toward South America. The effect dramatically alters rainfall patterns from Peru to California, from Borneo to northeast Brazil, and even in Mediterranean Europe.

Other mechanisms of global climate give some indication of the complexity of the overall system. The spinning of the planet and the change of sunlight as earth traces its path around the Sun generate air and ocean currents that transfer heat from one part of the planet to another. Another cycle of heat transfer in the global climatic system occurs when heated air rising off the equatorial regions moves pole ward, cools, and descends.

The curve rebounds later in the year as deciduous trees drop their leaves, annual plants die back, and decomposition releases carbon from the dead plant matter. The same process takes place in the Southern Hemisphere. Most carbon in this carbon cycle is continuously moving through air, water, microorganisms, plants and animals. As tiny fraction of the carbon stops moving through the cycle - for instance, by burial in a landslide or swamp and is preserved in sediments. Given enough time this carbon would become fossil fuel of the future. In addition, seashells store carbon in the form of calcium carbonate, which may eventually become limestone or marble.

Human activity has disturbed this natural cycle of carbon absorption and emission. This disruption causes atmospheric $CO_2$ to increase. Massive deforestation, particularly in tropical region and from a variety of causes, together with other forms of biomass burning, is probably releasing more than a gigatons of carbon in the atmosphere annually (compared with 5.5 gigatons for industrial $CO_2$). A variety of other processes including photosynthesis of plants, removes some of this $CO_2$.

## 18.6 Effects of Global Warming

As mentioned earlier, scientists use elaborate computer models of temperature, precipitation patterns, and atmosphere circulation to study global warming. Based on these models, scientists have made several predictions about how global warming will affect weather, sea levels, coastlines, agriculture, wildlife, and human health.

**(i) Weather**

Scientists predict that during global warming, the northern regions of the Northern Hemisphere will heat up more than other areas of the planet, northern and mountain glaciers will shrink, and less ice will float on northern oceans. Regions that now experience light winter snows may receive no snow at all. In temperate mountains, snowlines will be higher and snowpacks will melt earlier. Growing seasons will be longer in some areas. Winter and night-time temperatures will tend to rise more than summer and daytime ones.

The warmed world will be generally more humid as a result of more water evaporating from the oceans. Scientists are not sure whether a more humid atmosphere will encourage or discourage further warming. One the one hand, water vapor is a greenhouse gas, and its increased presence should add to the insulating effect. On the other hand, more vapor in the atmosphere will produce more clouds, which reflect sunlight back into space, which should slow the warming process. Greater humidity will increase rainfall, on average, about 1 per cent for each Fahrenheit degree of warming. (Rainfall over the continents has already increased by about 1 per cent in the last 100 years.) Storms are expected to be more frequent and more intense. However, water will also evaporate more rapidly from soil, causing it to dry out faster between rains. Some regions might actually become drier than before. Winds will blow harder and perhaps in different patterns. Hurricanes, which gain their force from the evaporation of water, are likely to be more severe. Against the background of warming, some very cold periods will still occur. Weather patterns are expected to be less predictable and more extreme.

### (ii) Sea Levels

As the atmosphere warms, the surface layer of the ocean warms as well, expanding in volume and thus raising sea level. Warming will also melt much glacier ice, especially around Greenland (Fig. 18.2). Greenland glaciers are melting into the water twice as fast as previously believed, resulting into further swelling of the sea. Sea levels worldwide rose 10 to 25 cm (4 to 10 inch) during the 20th century, and IPCC scientists predict a further rise of 9 to 88 cm (4 to 35 inch) in the 21st century (Fig. 18.3).

**Fig. 18.2. Ice under fire : Antarctica.**

**Fig. 18.3. Glacier recession around the world.**

Sea level changes will make life complicated in many coastal regions. A 100-cm (40 inch) rise could submerge 6 per cent of the Netherlands, 17.5 per cent of Bangladesh, and most or all of many islands. Erosion of cliffs, beaches, and dunes will increase (Fig. 18 4). Storm surges, in which winds locally pile up water and raise the sea, will become more frequent and damaging. As the sea invades the mouths of rivers, flooding from runoff will also increase upstream. This pushes the boundaries of life (Fig. 18.5). Wealthier countries will spend huge amounts of money to protect their shorelines, while poor countries may simply evacuate low-lying coastal regions.

**Fig. 18.4. Light House coast is coming closer.**

**Fig. 18.5. Pushing boundaries of life:Antarctica.**

Even the modest rise in sea level will greatly change coastal ecosystems. A 50 cm (20 inch) rise will submerge about half of the present coastal wetlands of the United States. New marshes will form in many places, but now where urban areas and developed landscapes block the way. This sea-level rise will cover much of the Florida Everglades.

**(iii) Agriculture**

A warmed globe will probably produce as much food as before, but not necessarily in the same places. Southern Canada, for example, may benefit from more rainfall and a longer growing season. At the same time, the semiarid tropical farmlands in some parts of Africa may become further impoverished. Desert farm regions that bring in irrigation water from distant mountains may suffer if the winter snowpack, which functions as a natural reservoir, melts before the peak growing months. Crops and woodlands may also be afflicted by more insect and plant diseases.

**(iv) Animals and Plants**

Animals and plants will find it difficult to escape from or adjust to the effects of warming because humans occupy so much land. Under global warming, animals will tend to migrate toward the poles and up mountainsides toward higher elevations (Fig. 18.6). Plants will shift their ranges, seeking new areas as old habits grow too warm. In many places, however, human development will prevent this shift. Species that find cities or farmlands blocking their may north or south may die out. Some types of forests, unable to propagate toward the poles fast enough, may disappear.

**Fig.18.6. Peoples shifting toward higher elevations.**

**(v) Human Health**

In a warmer world, scientists predict that more people will get sick or die from heat stress, due less to hotter days than to warmer nights (giving the sufferers less relief). Diseases now found in the tropics, transmitted by mosquitoes and other animal hosts, will widen their range as these animals hosts move into regions formerly too cold for them. Today 45% of the world's people live where they might get bitten by a mosquito carrying the parasite that causes malaria; that percentage may increase to 60%, if temperatures rise. Other tropical diseases may spread similarly, including dengue fever, yellow fever, and encephalitis. Scientists also predict rising incidence of allergies and respiratory diseases as warmer air grows more charged with pollutants, mold spores, and pollens.

## 18.7 Efforts To Control Global Warming

There are two major approaches to slowing the buildup of greenhouse gases. The first is to keep carbon dioxide out of the atmosphere by storing the gas or its carbon component somewhere else, a strategy called carbon sequestration. The second major approach is to reduce the production of greenhouse gases.

**Carbon Sequestration:**

The simplest way to sequester carbon is to preserve trees and to plant more. Trees, especially young and fast-growing ones, soak up a great deal of carbon dioxide, break it down in photosynthesis, and store the carbon in new wood in the formation of its fresh tissues. Worldwide, forests are being cut down at an alarming rate, particularly in tropics. In many areas, there is little regrowth as land loses fertility or is changed to other uses, such as farming or building housing developments. Reforestation could overcome these losses and counter part of the greenhouse buildup.

Many companies and governments all over the world, particularly in the United States, India, Norway, Brazil, Malaysia, Russia, and Australia have initiated reforestation projects by teaching local residents about tree-farming practices. The trees planted are expected to absorb up to 50 million tons of carbon dioxide over 40 years.

Carbon dioxide can also be sequestered directly. Carbon dioxide has traditionally been injected into oil well to force more petroleum out of the ground or seafloor. Now it is being injected simply to isolate it underground in oil fields, coal beds, or aquifers.

**Reduction in Production of Greenhouse Gases:**

Scientists are looking forward for alternative energy sources that will produce small or even no trace of carbon dioxide. At one natural gas drilling platform off the coast of Norway, carbon dioxide brought to the surface with the natural gas is captured and injected into an aquifer from which it cannot escape. The same process can be used to store carbon dioxide released by a power plant, factory, or any large stationary source. Deep ocean waters could also absorb a great deal of carbon dioxide. The feasibility and environmental effects of both these options are now under study by international teams.

In an encouraging trend, energy use around the world has slowly shifted away from fuels that release a great deal of carbon dioxide toward fuels that release somewhat less of this heat-trapping gas. Wood was the first major source of energy used by humans. With the dawn of the Industrial Revolution in 18$^{th}$ century, coal became the dominant energy source. By the mid-19$^{th}$ century oil had replaced coal in dominance, fueling the internal combustion engines that were eventually used in automobiles. By the 20$^{th}$ century, natural gas began to be used worldwide for heating and lighting. In this progression, combustion of natural gas releases less carbon dioxide than oil, which in turn releases less of the gas than do either coal or wood.

Nuclear energy, though controversial for reasons of safety and the high coasts of nuclear waste disposal, releases no carbon dioxide at all. Solar power, wind power, and hydrogen fuel cells also emit no greenhouse gases. Someday these alternative energy sources may prove to be practical, low-pollution energy sources, although progress today in this direction is slow.

## 18.8 National and International Programs

Enormous contributions can be made by energy conservation and energy efficiency. For a number of years, the United State's economy grew without any increase in energy use, mostly through improvements in efficiency. Power companies have found it in their economic interest to encourage energy conservation by their customers because it is cheaper than building new power plants.

Alternative energy sources-hydroelectric, geothermal, solar, and wind power-are all making important contributions as well. Individual solar panels are now a practical substitute for massive power lines in rural electrification. Nuclear energy is also a possible substitute but tends to be very expensive, involves a waste disposal problem, and, for certain kinds of reactors, has a nuclear proliferation potential as well. Future alternative energy sources, such as hydrogen fuel cells, may be closer to practically than most people currently realize.

Another short- to medium-term way to address the problem of increasing atmospheric $CO_2$ is to take advantage of the ability of plants and other organisms to remove carbon from the air and incorporate it into cellulose in their tissues or into other compounds. For example, the reforestation of degraded nonarable land could remove carbon from air. It has been calculated that there is enough such land in the world to remove a billion tons of $CO_2$ annually for the first 30 years of rapid growth of new trees. Reforestation, together with eliminating deforestation and burning in the tropics, could reduce the annual net increase of $CO_2$ emissions from 3.5 gigatons to 1.5 gigatons.

Many of these efforts, however, lack of fund availability. In addition, there is an urgent need for strategies to reduce the potential impact of greenhouse gas emissions that result from economic development in developing nations. One important way to achieve progress in these areas is through activities implemented jointly under the UN Climate Convention produced at the 1992 "Earth Summit" in Rio de Janeiro, Brazil. Under joint implementation, a company in one country that undertakes an activity that will increase $CO_2$ emissions can offset its environment impact using a number of different strategies. The company might complete a reforestation project in another country; or help a company in another country release less $CO_2$, perhaps by providing energy-efficient technology; or protect a so-called carbon pool, such as an existing forest. Even though a system of international credits is yet to be agreed upon, some companies and some countries, such as Costa Rica, are doing pioneering work.

In the meantime, the UN through the Climate Convention meets periodically to try to move toward the goal of reducing $CO_2$ emissions. It is not an easy task. The current goal is for industrialized nations to return to 1990 emissions levels by the year 2000; but without major changes in technology this could be very hard to attain.

Union of Concerned Scientists (U.C.S.) and economists in U.S.A. have given a call to nation's leaders and appealed them to swiftly establish and implement policies to bring about deep reductions in "U.S. heat-trapping emissions". Accordingly, President Barack Obama and his new Environmental Protection Agency (EPA) Administrator Lisa Jackson have moved to put $CO_2$ and other greenhouse gases under regulation by the "Clean Air Act". Recently, on April 17, 2009, EPA issued a proposal finding that these gases endanger human health and well-being.

**Nuclear Weapons & Global Security for Clean Air**

Under the Clean Air Act, scientists and policy experts in U.S.A. work to reduce some of the biggest security threats facing the world today, including the risks posed by nuclear weapons, nuclear terrorism, and space weapons. Recent success in Nuclear weapons and global security includes a consensus agenda signed by leading U.S.A. -based prominent physicists, including 23 Nobel Prize winners, supported the recommendations on an eventual ban on all nuclear weapons. The program will be worked out with scientists around the globe to increase international understanding of these issues and to foster and strengthen efforts to increase international security.

## 18.9 Our Contribution In Reduction of Global Warming

**(General Code of Conduct)**

We release greenhouse gases as a result of using energy to drive, using electricity to light and heat your home, and through other activities that support our quality of life like growing food, raising livestock and throwing away garbage. Greenhouse gas emissions can be reduced through simple measures like changing light bulbs and properly inflating your tires. The following 22 steps can be easily exercised as general code of conduct that will not only reduce our greenhouse gas emissions, but also reduce air pollution, increase the nation's energy independence and save money. For simplicity, the general code of conduct is subdivided into four categories, depending upon where you are i.e., at home, at office, at school or college, and on the road.

• **At Home**

Making a few small changes in your home and yard can lead to big reductions of greenhouse gas emissions and save money. Explore our list of nine simple steps you can take around the house and yard to reduce greenhouse gas emissions:

**1. Change 5 lights**

Replace at least your 5 most frequently used conventional incandescent bulbs (that emit heat radiation) by CFL lamps that will help the environment and will save money on energy bills. When we save the electricity means we put less load on thermal power generator, less coal burn and ultimately less amount of $CO_2$ poured into the atmosphere. This simple action would prevent greenhouse gases equivalent to the emissions from nearly 10 million cars.

**2. Look for qualified products**

When buying new products, always look for qualified products (Energy Star in U.S.A.) including lighting, home electronic appliances, heating and cooling equipment and appliances.

**3. Properly tune heating and cooling equipment**

Simple steps like cleaning air filters regularly and having your heating and cooling equipment tuned annually by a licensed contractor can save energy and increase comfort at home, and at the same time reduce greenhouse gas emissions.

**4. Use of green power**

Green power is environmentally friendly electricity that is generated from renewable energy sources such as wind and the sun. There are two ways to use green power: you can buy green power or you can modify your house to generate your own green power by installing wind mill and solar panels at your home.

**5. Reduce, reuse and recycle**

If there is an access of recycling, recycle your newspapers, beverage containers, paper and other goods. Use products in containers that can be recycled and items that can be repaired or reused. Reducing, reusing and recycling in your home help to conserve energy and reduces pollution and greenhouse gases.

**6. Be green in your yard**

Try to maintain the lawn or garden green that will benefit the environment. Use a push mower for cutting grass, instead a gas or electric mower that consumes no fossil fuels and emits no greenhouse gases. Composting your food and yard waste reduces the amount of garbage that you send to landfills and reduces greenhouse gas emissions.

**7. Use water efficiently**

Saving water around the home is essential. Municipal water systems require a lot of energy to purify and distribute water to households and saving water can lower greenhouse gas emissions. Be smart when irrigating your lawn or landscape; only water when needed and do it during the coolest part of the day, early morning is best. Turn the water off while shaving or brushing teeth. Do not use your toilet as a waste basket - water is wasted with each flush.

**8. Spread the message**

Spread the message by telling your family and friends that use as much small electricity as possible as it reduces pressure on Thermal Power Generators (less $CO_2$ -will be poured into the atmosphere). High energy efficiency of their home and office appliances is good for the environment because it lowers greenhouse gas emissions and air pollution.

• **At Office**

Business and home offices use a significant amount of electricity for heating and cooling, lighting and operating equipment. Here are three easy ways to reduce greenhouse gas emissions and helps to clean the air:

**9. Manage office equipment**

Office equipment and electronic appliances use energy even when idle or stand-by. To save energy and reduce greenhouse gas emissions at work, always activate the power management features on your computer and monitor, unplug laptop power cords when not in use and turn off equipment and lights at the end of the day. Consider using a power strip that can be turned off when you are not using your computers, printers, wireless routers and other electronic appliances. Use only qualified products (Energy Star in U.S.A.) including lighting and electronic appliances that will consume minimum electric power.

**10. Use less energy for your commute**

Switch to public transportation, carpooling, biking, telecommuting and other innovative ways to save energy and reduce greenhouse gas emissions on your way to and from work. If you do drive, ensure that the fuel efficiency of your vehicle is high.

**11. Reduce, reuse and recycle**

You can reduce, reuse and recycle office paper, newspapers, beverage containers, electronic equipment and batteries at the office from resource extraction, manufacturing, and disposal. You can reduce, reuse and recycle by using two-sided printing and copying; buying supplies made with recycled content; and recycling used printer cartridges.

• **At School and College**

Students, educators and school administrators can all play a key role in reducing greenhouse gas emissions. Here is a directory of some education and action planning resources:

**12. Bring science in life**

Explore the climate change through video films and make the students to watch the climate animations and their impact that brings science in life. The administrator should provide games that help students, their parents and their teachers learn about both the science of climate change and what actions they can take to reduce greenhouse gas emissions.

**13. Get involved at your school, college and university activities**

College students can play an important role in reducing greenhouse gas emissions at their colleges or universities by reducing their emissions from energy they use in dorm rooms. Students can also work with school administrators to increase energy efficiency on campus, reduce their school's greenhouse gas emissions by using green power, create a campus climate action plan, or develop an inventory of their school's greenhouse gas emissions.

**14. Teach students about climate change and ecosystems**

Use toolkit for formal and informal education of the climate change and wildlife to learn about the science of climate change and its potential effects on our nation's wildlife and their habitats so as to preserve the ecosystem.

**15. Engage middle school students in estimating emissions**

Enhance critical thinking skills by framing their syllabus so as to make them efficient to estimate the emission of greenhouse disturbing gases.

**16. Reduce, reuse and recycle**

You can ask students to reduce, reuse and recycle the school or classroom papers, newspapers, beverage containers, electronic equipment and batteries. You can reduce, reuse and recycle by using two-sided printing and copying; buying supplies made with recycled content and recycling used printer cartridges. You can reuse by donating your old electronic instruments to the needy schools and organizations.

- **On The Road**

The burning of fuels releases carbon dioxide ($CO_2$) into the atmosphere and contributes to climate change, but these emissions can be reduced by improving your car's fuel efficiency. You can take the following actions to reduce your greenhouse gas emissions, reduce the nation's dependence on oil and save money.

**17. Buy smart: Efficient vehicle**

Before buying a new or used vehicle (or even before renting a vehicle), check out its efficiency EPA's Green Vehicle Guide and fuel economy guide (standard in USA). These resources provide information about the emissions and fuel economy performance of different vehicles. The Green Vehicle Guide provides detailed information on emissions (including Air Pollution and Greenhouse Gas scores for each model) and the Fuel Economy Guide focuses on fuel efficiency. There are a wide range of cleaner, more fuel-efficient vehicles available in the market today that produce fewer greenhouse gas emissions.

**18. Drive smart:**

Many factors affect the fuel economy of your car. To improve fuel economy and reduce greenhouse gas emissions, go easy on the brakes and avoid hard accelerations, reduce time spent idling and unload unnecessary items in your trunk to reduce weight.

**19. Tune your ride**

A well-maintained car is more fuel-efficient, produces fewer greenhouse gas emissions, is more reliable and is safer. Keep your car well tuned, follow the manufacturer's maintenance schedule, and use the recommended grade of motor oil. Also check and replace your vehicle's air filter regularly.

**20. Check your tire pressure**

Check your tire pressure regularly. Under-inflation increases tire wear, reduces your fuel economy by up to 3 percent and leads to higher greenhouse gas emissions and releases of air pollutants.

**21. Give your car a break**

Use public transportation whenever possible to avoid using your car. Leaving your car at home just two days a week will reduce greenhouse gas emissions by an average of 1,600 pounds per year.

**22. Use renewable fuels**

Both E85 and bio-diesel are renewable fuels that can reduce greenhouse gas emissions from your vehicle. E85 is a fuel blend containing 85% ethanol that can be used in certain vehicles called Flex Fuel Vehicles (FFVs). Bio-diesel is a renewable fuel made from agricultural resources such as vegetable oils.

## 18.10 Conclusions

Global warming is one of the most serious challenges facing us today. To protect the health and economic well-being of current and future generations, we must reduce our emissions of heat-trapping gases by using the technology, know-how, and practical solutions already at our disposal. Efforts are being made through out the world including developed and developing countries to reduce global warming so as to give clean air to the future generation. The time has come now that every one of us should contribute to the code of conduct to produce fewer greenhouse gas emissions.

At the end, it is common sense to do something to prevent global warming. This is not a backyard experiment that humankind can walk away from. We are experimenting with the entire planet, and this is where we live.

## EXERCISE–XVIII

**18.1** What is greenhouse effect? Explain the role of greenhouse effect in heating the earth's atmosphere.

**18.2** Explain global warming in context with greenhouse effect. What are the consequences of global warming.

**18.3** What are greenhouse gases? Explain their role in producing global warming.

**18.4** Estimate the evidence of global warming.

**18.5** Explain the possible causes of global warming. What is the role of carbon dioxide gas.

**18.5** Explain the effect of global warming on the following:

(*i*) Weather

(*ii*) Sea levels

(*iii*) Agriculture

(*iv*) Animals and plants

(*v*) Human health

**18.7** Enunciate the efforts employed to control global warming

**18.8** What is EPA. Where it is established? Explain its role in controlling global warming

**18.9** Explain how a common man can contribute his efforts in controlling global warming when he is at home, at office, at school and on road.

**18.10** 'Global warming is one of the most serious challenges facing us today', Explain.

# APPENDIX – I
# PHYSICAL CONSTANTS
## COMMON 'SI' DERIVED QUANTITIES

| *Quantity* | *Symbol* | *Unit* | *Symbol* | |
|---|---|---|---|---|
| **Supplementary** | | | | |
| Plane angle | $\phi, \theta$ | radian | rad | |
| Solid angle | $\omega, \Omega$ | steradian | sr | |
| **Derived** | | | | |
| Area | $A$ | square metre | | $m^2$ |
| Volume | $V$ | cubic metre | | $m^3$ |
| Frequency | $f, v$ | hertz | Hz | $s^{-1}$ |
| Density | $d, \rho$ | kilogram per cubic metre | | kg $m^{-3}$ |
| Concentration | $c$ | mole per cubic metre | | mol $m^{-3}$ |
| Velocity | $v$ | metre per second | | m $s^{-1}$ |
| Angular velocity | $\omega$ | radian per second | | rad $s^{-1}$ |
| Acceleration | $a$ | metre per second squared | | m $s^{-2}$ |
| Angular acceleration | $\alpha$ | radian per second squared | | rad $s^{-2}$ |
| Force | $F$ | newton | N | m kg $s^{-2}$ |
| Pressure, stress | $P, p$ | pascal | Pa | N $m^{-2}$ |
| Viscosity (dynamic) | $\eta$ | pascal-second | | pa s |
| Viscosity (kinematic) | | metre squared per second | | $m^2$ $s^{-1}$ |
| Energy, work | $E, W$ | joule | J | N m |
| Power, radiant flux | $P$ | watt | W | J $s^{-1}$ |
| Quantity of electricity | $q$ | coulomb | C | A s |
| Potential difference, electromotive force | $V$ | volt | V | W $A^{-1}$ |
| Electric field strength | $E$ | volt per metre | | V $m^{-1}$ |
| Electric resistance | $R$ | ohm | $\Omega$ | V $A^{-1}$ |
| Electric conductance | $G$ | siemens | S | $\Omega^{-1}$ |
| Capacitance | $C$ | farad | F | C $V^{-1}$ |
| Magnetic flux | $\Phi$ | weber | Wb | V s |
| Magnetic flux density | $B$ | tesla | T | Wb $m^{-1}$ |
| Inductance | $L$ | henry | H | $\Omega$ s |
| Magnetic field strength | $H$ | ampere per metre | | A $m^{-1}$ |
| Magnetomotive force | $M$ | ampere | A | |
| Wave number* | $v$ | 1 per metre | | $m^{-1}$ |
| Activity (radioactive) | $A$ | 1 per second | | $s^{-1}$ |
| Luminous flux | $F$ | lumen | lm | cd sr |
| Luminance | $l$ | candela per square metre | | cd $m^{-2}$ |
| Illuminance | $E$ | lux | lx | lm $m^{-2}$ |
| Heat flux density, irradiance | | watt per square metre | | W $m^{-2}$ |
| Heat capacity, entropy | $s$ | joule per kelvin | | J $K^{-1}$ |
| Specific heat capacity, specific entropy | $S$ | joule per kilogram kelvin | | J $kg^{-1}$ $K^{-1}$ |
| Thermal conductivity | $K$ | watt per metre kelvin | | W $m^{-1}K^{-1}$ |
| Molar energy | | joule per mole | | J $mol^{-1}$ |
| Molar entropy, molar heat capacity | $C$ | joule per mole kelvin | | J $mol^{-1}K^{-1}$ |

* Wave number in the infrared are still often expressed in $cm^{-1}$.

## TABLE OF FUNDAMENTAL CONSTANTS

| | Symbol | SI Unit | Standard error (parts in $10^6$) |
|---|---|---|---|
| **Principal constants** | | | |
| Velocity of light | c | $2.997\ 925 \times 10^8$ m s$^{-1}$ | 1 |
| Planck's constant | h | $6.626\ 2 \times 10^{-34}$ J s | 8 |
| Planck's constant (h/2π) | $\hbar$ | $1.054\ 592 \times 10^{-34}$ J s | 8 |
| Charge of electron | e | $1.602\ 192 \times 10^{-19}$ C | 4.5 |
| Mass of electron | $m_e$ | $9.109\ 56 \times 10^{-31}$ kg | 6 |
| Mass of electron | $Nm_e$ | $5.485\ 93 \times 10^{-4}$ amu | 6 |
| Avogadro's constant (number of atoms in 12 kg of $^{12}$C) | N | $6.022\ 17 \times 10^{26}$ kmol$^{-1}$ | 7 |
| Mass of any atom ÷ atomic weight | $N^{-1}$ | $1.660\ 53 \times 10^{-27}$ kg amu$^{-1}$ | 7 |
| Faraday constant of electrolysis | F (= Ne) | $9.648\ 67 \times 10^7$ C kmol$^{-1}$ | 6 |
| Constant of gravitation | G | $6.673 \times 10^{-11}$ N m$^2$ kg$^{-2}$ | 500 |
| **Spectroscopy and atoms** | | | |
| Planck's constant | h | $4.135\ 71 \times 10^{-15}$ eV s | 4 |
| Planck's constant (h/2π) | $\hbar$ | $6.582\ 18 \times 10^{-16}$ eVs | 4 |
| Charge/mass ratio of electron | $e/m_e$ | $1.758\ 803 \times 10^{11}$ C kg$^{-1}$ | 3 |
| Fine structure constant | $\alpha$ | $7.297\ 35 \times 10^{-3}$ | 1.5 |
| Fine structure constant, reciprocal | $\alpha^{-1}$ | 137.036 0 | 1.5 |
| Scale of the fine structure | $\alpha^2 R_\alpha$ | $5.843\ 66 \times 10^2$ m$^{-1}$ | 3 |
| Rydberg's constant (fixed nucleus) | $R_\alpha$ | $1.097\ 373\ 1 \times 10^7$ m$^{-1}$ | 0.1 |
| Rydberg's constant (hydrogen H) | $R_H$ | $1.096\ 775\ 8 \times 10^7$ m$^{-1}$ | 0.1 |
| Bohr radius $(4\pi/\mu_0 c^2)\ \hbar^2/m_e e^2$ | $a_0$ | $5.291\ 771 \times 10^{-11}$ m | 1.5 |
| Compton wavelength of electron | $\lambda_c$ | $2.426\ 310 \times 10^{-12}$ m | 3 |
| Classical 'radius' of electron $(\mu_0 c^2/4\pi)e^2/mc^2$ | $r_e$ | $2.817\ 94 \times 10^{-15}$ m | 5 |
| Thomson cross-section $8\pi r_c^2/3$ | $\sigma_T$ | $6.652\ 45 \times 10^{-29}$ m | 10 |
| Zeeman effect, $\mu_B/hc$ | | 46.686 0 m$^{-1}$ T$^{-1}$ | 3 |
| Bohr magneton e $\hbar/2m_e$ | $\mu_B$ | $9.274\ 10 \times 10^{-24}$ J T$^{-1}$ | 7 |
| Nuclear magneton e $\hbar/2m_p$ | $\mu_N$ | $5.050\ 95 \times 10^{-27}$ J T$^{-1}$ | 10 |
| Ratio of masses proton/electron | $m_p/m_e$ | $1.836\ 11 \times 10^3$ | 6 |
| Gyromagnetic ratio of proton | $\gamma_p = \mu_p/\frac{1}{2}\hbar$ | $2.675\ 196 \times 10^8$ s$^{-1}$ T$^{-1}$ | 3 |
| Band-spectra constant | $Nh/8\pi^2 c$ | $1.685\ 80 \times 10^{-17}$ amu m | 7 |
| **Conversion factors for Mass, Energy and Wavelength** | | | |
| **Energy and mass** | | | |
| Electron volt | | $1.602\ 192 \times 10^{-19}$ J | 4.5 |
| Atomic mass unit in MeV | | 931.481 | 6 |
| 1 MeV in a.m.u | | $1.073\ 559 \times 10^{-3}$ | 6 |
| Rest-mass of electron | | 0.511 004 MeV | 3 |
| Joules per kmol for 1 eV per molecule | | $9.648\ 67 \times 10^7$ | 6 |

| | | | |
|---|---|---|---|
| **Frequency, wavelength and energy** | | | |
| Quantum energy ÷ wave number / Energy × wavelength | | $1.986\ 48 \times 10^{-25}$ J m | 8 |
| | | $1.239\ 854 \times 10^{-6}$ eV m | 4 |
| Wave number ÷ energy | | $8.065\ 46 \times 10^{5}$ $eV^{-1}$ $m^{-1}$ | 4 |
| Quantum energy ÷ frequency | | $4.135\ 71 \times 10^{-15}$ eV $Hz^{-1}$ | 4 |
| Frequency ÷ energy | | $2.417\ 966 \times 10^{14}$ Hz $eV^{-1}$ | 4 |
| **Thermal constants** | | | |
| Gas constant | $R_o$ | $8.314\ 3 \times 10^{3}$ J $K^{-1}$ $kmol^{-1}$ | 42 |
| Boltzmann's constant $R_o/N$ | k | $1.380\ 62 \times 10^{-23}$ J $K^{-1}$ | 43 |
| Boltzmann's constant | | $8.617\ 1 \times 10^{-5}$ eV $K^{-1}$ | 42 |
| Energy kT for T = 273.15 K | | 0.023 538 eV | 42 |
| Stefan's constant $2\pi^5k^4/15c^2h^3$ | | $5.669\ 6 \times 10^{-8}$ W $m^{-2}$ $K^{-1}$ | 170 |
| Constants in Planck's formula (power flow) | $2\pi hc^2$ | $3.741\ 84 \times 10^{-16}$ W $m^2$ | 8 |
| Constants in Planck's formula (energy density) | $8\pi hc$ | $4.992\ 58\ 10^{-24}$ J m | 8 |
| Constants in Planck's formula | hc/k | $1.438\ 83 \times 10^{-2}$ m K | 43 |

**Note:** Magnetic moments are defined so that mechanical energy = **μ.B**; the unit is 1 $JT^{-1}$ = 1J $Wb^{-1}$ $m^2$ = 1 A $m^2$ (1 tesla = 1 weber $m^{-2}$ = $10^4$ gauss).

## TEMPERATURE AND HEAT

### EMISSIVITY CORRECTIONS (°C) FOR THE TOTAL RADIATION PYROMETER*

| Observed temp./°C | Total emissivity 0.1 | 0.2 | 0.3 | 0.4 | 0.5 | 0.6 | 0.7 | 0.8 | 0.9 |
|---|---|---|---|---|---|---|---|---|---|
| 200 | 337 | 211 | 148 | 107 | 78 | 56 | 38 | 23 | 11 |
| 400 | 513 | 326 | 230 | 168 | 123 | 89 | 61 | 37 | 17 |
| 600 | 675 | 429 | 304 | 223 | 163 | 118 | 80 | 49 | 23 |
| 800 | 833 | 530 | 375 | 275 | 202 | 145 | 100 | 61 | 28 |
| 1000 | 989 | 630 | 446 | 327 | 240 | 173 | 118 | 73 | 34 |
| 1200 | 1146 | 729 | 517 | 379 | 278 | 200 | 137 | 84 | 39 |
| 1400 | 1302 | 828 | 587 | 430 | 316 | 228 | 156 | 96 | 45 |
| 1600 | 1457 | 928 | 658 | 482 | 354 | 255 | 175 | 107 | 50 |
| 1800 | 1613 | 1027 | 728 | 534 | 392 | 282 | 193 | 119 | 55 |

* The value of $T_0$ is barely significant in the temperature range given in the table, but has been taken as 298.15 K (25°C).

## COEFFICIENTS OF LINEAR EXPANSION OF SOLIDS

Over small ranges of temperature, the variation of length ($l$) with temperature ($t$) of most solid substances is represented by the equation $l_t = l_{to}\{1 + \alpha(t - t_o)\}$, where α is the mean coefficient of linear expansion over the temperature range $(t - t_o)$. The values of α below apply to normal ambient temperatures unless otherwise stated.

| *Substance* | $\frac{\alpha}{10^{-6}/^0 C}$ | *Substance* | $\frac{\alpha}{10^{-6}/^0 C}$ |
|---|---|---|---|
| **Metals** | | Gold | 14 |
| Aluminium | 23 | Gold 0 to 500 ℃ | 15 |
| Aluminium 0 to 600 ℃ | ~29 | Iridium | 6.5 |
| Antimony | ~11 | Iridium 0 to 1750 ℃ | 9 |
| Antimony ǁ axis | ~16 | Iron (electrolytic) | 11.7 |
| Antimony ⊥ axis | ~8 | Iron (electrolytic) 0 to 700 ℃ | 15 |
| Bismuth | ~13 | Iron (wrought) | 12 |
| Bismuth ǁ axis | ~16 | Lead | 29 |
| Bismuth ⊥ axis | ~12 | Lead 0 to 300 ℃ | 32 |
| Cadmium | ~30 | Magnesium | 25 |
| Cadmium ǁ axis | ~52 | Magnesium 0 to 400 ℃ | 29 |
| Cadmium ⊥ axis | ~20 | Nickel | 12.8 |
| Chromium | ~7 | Nickel 0 to 1000 ℃ | 18 |
| Chromium 0 to 900 ℃ | ~11 | Palladium | ~11 |
| Cobalt | ~12 | Palladium 0 to 1000 ℃ | ~14 |
| Cobalt 25 to 350 ℃ | ~14 | Platinum | 8.9 |
| Copper | 16.7 | Platinum at 800 ℃ | ~11 |
| Copper 0 to 1000 ℃ | 20 | Rhodium | 8.4 |
| Silver | 19 | 36% Ni at about 20 ℃ | 0-1.5 |
| Silver 0 to 900 ℃ | –22 | 36% Ni at about 500 ℃ | 9 |
| Tantalum | 6.5 | 43% Ni | 7.9 |
| Tantalum 0 to 400 ℃ | 6.6 | 58% Ni | 11.4 |
| Thallium | ~28 | Phosphor bronze | 17 |
| Tin | ~21 | Platinum-iridium, 90 Pt. 10 Ir | 8.7 |
| Tin ǁ axis | ~30 | Speculum metal, ~68 Cu, 32 Sn | ~19 |
| Tin ⊥ axis | ~16 | Stainless steel, typical, | |
| Titanium | ~9 | 18 Cr, 8 Ni | ~16 |
| Tungsten | 4.5 | 12% Cr, 20 to 100 ℃ | ~10 |
| Tungsten 600 to 1400 ℃ | ~6 | Steel, carbon, typical | ~11 |
| Tungsten 1400 to 2200 ℃ | ~7 | Stellite, ~80 Co, 20 Cr | |
| Vanadium | ~8 | 20 to 100 ℃ | 14 |
| Zinc (cast) | ~30 | 20 to 600 ℃ | 16 |
| Zinc (crystal); ǁ axis | ~60 | Y-alloy | 22 |
| Zinc (crystal); ⊥ axis | ~13 | **Miscellaneous** | |
| Zinc (rolled sheet; ǁ rolling) | ~31 | Alundum $Al_2O_3$ at 100 ° C | ~8 |
| Zinc (rolled sheet; ⊥ rolling) | ~20 | Building materials, etc. | |
| **Alloys** | | Brick | 3-10 |
| Aluminium bronze, ~92 Cu, | | Cement and concrete | 10-14 |
| 7 AI, 0.4 Zn | 16-17 | Granite | 5-10 |
| Brass, ~68 Cu, 32 Zn | 18-19 | Limestone | 4-10 |

| *Substance* | $\frac{\alpha}{10^{-6}/^0C}$ | *Substance* | $\frac{\alpha}{10^{-6}/^0C}$ |
|---|---|---|---|
| Yellow brass (Muntz metal), | | Marble | 3-15 |
| ~60 Cu, 40 Zn, 20 to 100°C | 19-23 | Portland stone | ~3 |
| Bronze (gunmetal, bell metal), | | Sandstone | 5-12 |
| ~80 Cu, 20 Sn | 17-18 | Slate | 6-12 |
| Constantan, ~60 Cu, 40 Ni | 15-17 | Fluorspar, $CaF_2$ | 19 |
| Cupro-nickel (Monel metal), | ~14 | Glass | |
| ~63 Ni, 30 Cu, with about | | soft soda 14% $Na_2O$ | ~8.5 |
| 2% each of Fe, Mn and Pb | | hard potash 20% $K_2O$ | ~10 |
| between 500 and 600°C | ~19 | flint, lead, 46% PbO | ~8 |
| Delta metal, 60 Cu, 36 Zn, | | borate | 3-7 |
| 2 Sn, 2 Fe | ~20 | pyrex | 3 |
| Duralumin | 23 | hysil (chance) | 3.3 |
| ~95 Al, 4 Cu, 20 to 500 °C | 27 | Porcelain, insulators | 3-6 |
| German silver (nickel silver) | ~18 | Porcelain, refractory | 2-5 |
| Magnalium, ~90 Al, 10 Mg | ~23 | Porcelain, laboratory | 3-4 |
| Nichrome, ~90 Ni, 10 Cr | 13 | Quartz, ‖ axis | 8 |
| 87 Ni, 9 Cr, 1.5 Fe | 12.5 | Quartz, ⊥ axis | 14 |
| 85 Ni, 10 Cr, 3 Fe | 13 | Silica, fused | 0.4 |
| Nickel-steels (including Invar) | | Silica, fused 0 to 1000 °C | 0.5 |
| 10% Ni | 13 | Woods, typical, along grain | ~3-5 |
| | | Woods, typical, across grain | ~35-60 |

## COEFFICIENTS OF CUBICAL EXPANSION OF LIQUIDS

The following table gives values of $\gamma = (1/V)\ (dV/dt)$, where $t = 20°C$. The expansion coefficient generally increases with rising temperature.

| *Substance* | $\frac{\gamma}{10^{-5}/^0C}$ | *Substance* | $\frac{\gamma}{10^{-5}/^0C}$ |
|---|---|---|---|
| Acetic acid | 107 | Glycerol | 47 |
| Acetone | 143 | Mercury | 18.1 |
| Alcohol, methyl | 119 | Methyl iodide | 120 |
| Alcohol, ethyl | 108 | n-Pentane | 155 |
| Aniline | 85 | Phenol | 79 |
| Benzene | 121 | Toluene | 107 |
| Carbon bisulphide | 119 | Turpentine | 96 |
| Carbon tetrachloride | 122 | m-Xylene | 99 |
| Chloroform | 127 | Water | 21 |
| Ether, ethyl | 163 | $H_2SO_4$ 100% | 56 |
| Ethyl bromide | 140 | | |

## THERMAL CONDUCTIVITIES OF METALLIC ELEMENTS : UNIT IS W $m^{-1}$ $k^{-1}$

| Metal | Temperature/K | | | | |
|---|---|---|---|---|---|
| | 173.2 | 273.2 | 373.2 | 573.2 | 973.2 |
| Aluminium | 241 | 236 | 240 | 233 | 92 |
| Antimony | 33 | 25.5 | 22 | 19 | 27 |
| Beryllium | 367 | 218 | 168 | 129 | 93 |
| Bismuth | 11 | 8.2 | 7.2 | 13 | 17 |
| Cadmium | 100 | 97 | 95 | 89 | 45 |
| Caesium | 37 | 36 | 20 | 20.6 | 17.7 |
| Cerium | 8 | 11 | 13 | 16 | — |
| Chromium | 120 | 96.5 | 92 | 82 | 66 |
| Cobalt | 130 | 105 | 89 | 69 | 53 |
| Copper | 420 | 403 | 395 | 381 | 354 |
| Dysprosium | 9 | 10.5 | — | — | — |
| Erbium | 14 | 15 | — | — | — |
| Gadolinium | 12 | 10 | — | — | — |
| Gallium | 43 | 41 | 33 | 45 | — |
| Gold | 324 | 319 | 313 | 299 | 272 |
| Hafnium | 25 | 23 | 22 | 21 | 21 |
| Holmium | 14 | 16 | 17 | — | — |
| Indium | 92 | 84 | 76 | 42 | — |
| Iridium | 156 | 147 | 145 | 139 | — |
| Iron | 99 | 83.5 | 72 | 56 | 34 |
| Lanthanum | 12 | 13 | 14.5 | — | — |
| Lead | 37 | 36 | 34 | 32 | 21 |
| Lithium | 94 | 86 | 82 | 47 | 59 |
| Lutetium | 18 | 17 | — | — | — |
| Manganese | 160 | 157 | 154 | 150 | — |
| Manganese | 7 | 8 | — | — | — |
| Mercury | 29.5 | 7.8 | 9.4 | 11.7 | — |
| Molybdenum | 145 | 139 | 135 | 127 | 113 |
| Nickel | 113 | 94 | 83 | 67 | 71 |
| Niobium | 53 | 53 | 55 | 58 | 64 |
| Osmium | 93 | 88 | 87 | 87 | — |
| Palladium | 72 | 72 | 73 | 79 | 93 |
| Platinum | 73 | 72 | 72 | 73 | 78 |
| Plutonium | 4 | 6 | 8 | — | — |
| Potassium | 105 | 104 | 53 | 45 | 32 |
| Praesodymium | 9.9 | 12 | 13.4 | — | — |
| Rhenium | 52 | 49 | 47 | 44 | 45 |
| Rhodium | 156 | 151 | 147 | 137 | — |

| Metal | Temperature/K | | | | |
|---|---|---|---|---|---|
| | 173.2 | 273.2 | 373.2 | 573.2 | 973.2 |
| Rubidium | 59 | 58 | 32 | 29 | 22 |
| Ruthenium | 123 | 117 | 115 | 108 | 98 |
| Samarium | 10 | 13 | 13 | 14 | — |
| Scandium | 15 | 16 | — | — | — |
| Silver | 432 | 428 | 422 | 407 | 377 |
| Sodium | 141 | 142 | 88 | 78 | 60 |
| Tantalum | 58 | 57 | 58 | 58.5 | 60 |
| Technetium | — | 51 | 50 | 50 | — |
| Terbium | 11 | 10.5 | — | — | — |
| Thallium | 51 | 47 | 44 | — | — |
| Thorium | 55 | 54 | 54 | 56 | 58 |
| Thulium | 16 | 17 | — | — | — |
| Tin | 76 | 68 | 63 | 32 | 40 |
| Titanium | 26 | 22 | 21 | 19 | 21 |
| Tungsten | 188 | 177 | 163 | 139 | 119 |
| Uranium | 24 | 27 | 29 | 33 | 43 |
| Vanadium | 32 | 31 | 31 | 33 | 38 |
| Yttrium | 16.5 | 17 | — | — | — |
| Zinc | 117 | 117 | 112 | 104 | 66 |
| Zirconium | 26 | 23 | 22 | 21 | 23 |

## SPECIFIC HEAT CAPACITY OF WATER : UNIT IS J $g^{-1}$ $^{O}C^{-1}$

| °C | 0 | 1 | 2 | 3 | 4 | 5 | 6 | 7 | 8 | 9 |
|---|---|---|---|---|---|---|---|---|---|---|
| 0 | 4.2174 | 4.2138 | 4.2104 | 4.2074 | 4.2045 | 4.2019 | 4.1996 | 4.1974 | 4.1954 | 4.1936 |
| 10 | 4.1919 | 4.1904 | 4.1890 | 4.1877 | 4.1866 | 4.1855 | 4.1846 | 4.1837 | 4.1829 | 4.1822 |
| 20 | 4.1816 | 4.1810 | 4.1805 | 4.1801 | 4.1797 | 4.1793 | 4.1790 | 4.1787 | 4.1785 | 4.1783 |
| 30 | 4.1782 | 4.1781 | 4.1780 | 4.1780 | 4.1779 | 4.1779 | 4.1780 | 4.1780 | 4.1781 | 4.1782 |
| 40 | 4.1783 | 4.1784 | 4.1786 | 4.1788 | 4.1789 | 4.1792 | 4.1794 | 4.1796 | 4.1799 | 4.1801 |
| 50 | 4.1804 | 4.1807 | 4.1811 | 4.1814 | 4.1817 | 4.1821 | 4.1825 | 4.1829 | 4.1833 | 4.1837 |
| 60 | 4.1841 | 4.1846 | 4.1850 | 4.1855 | 4.1860 | 4.1865 | 4.1871 | 4.1876 | 4.1882 | 4.1887 |
| 70 | 4.1893 | 4.1899 | 4.1905 | 4.1912 | 4.1918 | 4.1925 | 4.1932 | 4.1939 | 4.1946 | 4.1954 |
| 80 | 4.1961 | 4.1969 | 4.1977 | 4.1985 | 4.1994 | 4.2002 | 4.2011 | 4.2020 | 4.2029 | 4.2039 |
| 90 | 4.2048 | 4.2058 | 4.2068 | 4.2078 | 4.2089 | 4.2100 | 4.2111 | 4.2122 | 4.2133 | 4.2145 |

## SPECIFIC HEAT CAPACITIES OF VARIOUS SUBSTANCES

| *Substance* | *Temp./°C* | *Sp. ht. $Jg^{1}$ °$C^{1}$* | *Substance* | *Temp./°C* | *Sp. ht. $Jg^{1}$ °$C^{1}$* |
|---|---|---|---|---|---|
| **Alloys** | | | **Miscellaneous** | | |
| Brass, red | 0 | 0.377 | Air (dry), $C_p$ | 20 | 1.006 |
| Brass, yellow | 0 | 0.368 | ,, | 100 | 1.011 |
| Eureka (constantan) | 18 | 0.410 | ,, | 500 | 1.092 |
| German silver | 0-100 | 0.398 | ,, | 1000 | 1.192 |
| Solder Sn 54% + Pb 46% | | 0.176 | ,, | –100 | 1.008 |
| Steel, carbon | 50-100 | 0.48 | ,, (100 atm) | –80 | 1.902 |
| Steel, stainless | 50-100 | 0.51 | Asbestos | 20-100 | 0.84 |
| **Liquids** | | | Basalt | 20-100 | 0.84-1.00 |
| Alcohol, amyl | 18 | 2.30 | Ebonite | 20-100 | 1.38 |
| Alcohol, ethyl | 0 | 2.29 | Fluorspar, $CaF_2$ | 30 | 0.88 |
| Alcohol, ethyl | 40 | 2.71 | Glass. crown | 10-50 | 0.67 |
| Alcohol, methyl | 12 | 2.52 | Glass. flint | 10-50 | 0.50 |
| Aniline | 15 | 2.15 | Glass. Pyrex | 26 | 0.78 |
| Benzene | 10 | 1.42 | Granite | 20-100 | 0.80-0.84 |
| Benzene | 40 | 1.77 | Ice | –250 | 0.15 |
| Brine | 20 | 2.89 | ,, | –160 | 1.0 |
| Brine | 15 | 3.01 | ,, | –21 to –1 | 2.0-2.1 |
| Ether. ethyl | 18 | 2.34 | Indiarubber | 15-100 | 1.13-2.01 |
| Glycerine | 18-50 | 2.43 | Marble, white | 18 | 0.88-0.92 |
| Oil, castor | 20 | 2.13 | NaCI | + 10 | 0.88 |
| Oil, linseed | 20 | 1.84 | KCI | 277 | 0741 |
| Oil, olive | 7 | 1.97 | Paraffin wax | 0-20 | 2.9 |
| Oil, paraffin | 20-60 | 2.13-2.26 | Porcelain | 15-1000 | 1.07 |
| Oil, rape | 20 | 2.04 | Quartz, $SiO_2$ | 0 | 0.73 |
| Oil, sperm | 20 | 2.06 | Quartz | 350 | 1.17 |
| Sea-water | 17 | 3.93 | Sand | 20-100 | 0.80 |
| Toluene | 18 | 1.67 | Silica (fused) | 15-200 | 0.84 |
| Turpentine | 18 | 1.76 | ,, | 15-800 | 1.04 |

## RATIO OF SPECIFIC HEAT CAPACITIES γ

| Gas | Temp./°C | γ | Gas | Temp./°C | γ |
|---|---|---|---|---|---|
| **Monatomic gases** | | | | | |
| Helium | 0 | 1.63 | Air (dry) | 500 | 1.357 |
| Argon | 0 | 1.667 | Air (dry) | 900 | 1.32 |
| Neon | 19 | 1.642 | Air (dry) (200 atm) | 0 | 1.828 |
| Krypton | 19 | 1.689 | | –79.3 | 2.333 |
| Xenon | 19 | 1.666 | Hydrogen | 4-17 | 1.407/8 |
| Mercury vapour | 310 | 1.666 | Nitrogen | 20 | 1.401 |
| **Diatomic** | | | Oxygen | 5-14 | 1.400 |
| Air (dry) | –79.3 | 1.405 | Carbon monoxide | 1800 | 1.297 |
| Air (dry) | 0-17 | 1.40/2 | Nitric oxid | — | 1.394 |
| **Triatomic** | | | Acetylene, $C_2H_2$ | — | 1.26 |
| Ozone | — | 1.29 | Ethylene, $C_2H_4$ | — | 1.264 |
| Water vapour | 100 | 1.334 | Benzene $C_6H_6$ | 20 | 1.40 |
| Carbon dioxide | 4-11 | 1.300 | Benzene $C_6H_6$ | 99.7 | 1.105 |
| Carbon dioxide | 300 | 1.22 | Chloroform $CHC1_3$ | 24-42 | 1.110 |
| Carbon dioxide | 500 | 1.20 | | 99.8 | 1.150 |
| Ammonia, $NH_3$ | — | 1.336 | $CCI_4$ | — | 1.130 |
| Nitrous oxide, $N_2O$ | — | 1.324 | Me. alcohol | 99.7 | 1.256 |
| Nitrogen peroxide $N_2O_4$ | 20 | 1.172 | Me. bromide | — | 1.274 |
| Nitrogen peroxide $NO_2$ | 150 | 1.31 | Me. chloride | 19-30 | 1.279 |
| $H_2S$ | — | 1.340 | Me. iodide | — | 1.286 |
| $CS_2$ | — | 1.239 | Et. alcohol | 53 | 1.133 |
| Sulphur dioxide | 16-34 | 1.26 | Et. alcohol | 99.8 | 1.134 |
| | 500 | 1.2 | Et. bromide | — | 1.188 |
| **Polyatomic gases** | | | Et. chloride | 22.7 | 1.187 |
| Methane, $CH_4$ | — | 1.313 | Et. ether | 12-20 | 1.024 |
| Ethane, $C_2H_6$ | — | 1.22 | Et. ether | 99.7 | 1.112 |
| Propane, $C_3H_8$ | — | 1.130 | Acetic acid | 136.5 | 1.147 |

## [A] EMISSION SPECTRA (PROMINENT LINES)

Wavelengths in vacuum in Angstrom Unit (A = $10^{-10}$m). Fraunhoffer names are also given. Strong lines are written in bold letters.

| *Argon* | *Cadmium* | *Helium* | *Hydrogen* | *Mercury* | *Neon* | *Sodium* |
|---|---|---|---|---|---|---|
| 3949.0 | 4678.16 | **3888.65** | 3970 | 4046.56 (H) | 5330.9 | **5889.96($D_1$)** |
| 4044.4 | **4799.92** | **4026.19** | 4101.74 ($H_\delta$) | 4077.81 | 5341.1 | **5895.93($D_2$)** |
| 4158.6 | **5085.82** | 4387.93 | 4340.46 ($H_\delta$) | 4339.2 | 5400.7 | 6154.2 |
| 4191.0 | **6438.47** | 4437.55 | | 4347.5 | 5852.5 | 6160.8 |
| 4198.3 | | **4471.48** | 4861.33 ($H_\beta$) | **4358.35** | 5881.9 | |
| 4200.7 | | 4713.14 | **6562.82 ($H_\alpha$)** | 4916.04 | 5944.8 | |
| 4259.4 | | 4921.93 | | 4959.7 | 6143.1 | |
| 4266.3 | | **5015.67** | | **5460.74** | 6266.5 | |
| 4272.2 | | **5875.62** | | **5769.59** | 6383.0 | |
| 4300.1 | | 6678.15 | | **5790.65** | 6402.4 | |
| 4345.2 | | 7055.2 | | 6152 | 6506.5 | |
| 4596.1 | | | | 6232 | 6598.9 | |
| 4628.4 | | | | 6908 | 7245.2 | |
| 4702.3 | | | | | | |

## [B] FLAME SPECTRA

| *Salt of Sodium* | *Potassium* | *Lithium* | *Strontium* |
|---|---|---|---|
| 5895.93 $D_1$ | 7699.0 | 6707.8 | 7070.15 |
| 5889.96 $D_2$ | 7664.9 | | 4607.34 |
| | 4047.2 | | 4415.51 |
| | 4044.2 | | 4077.71 |

## NORMAL SPECTRAL EMISSIVITIES

| *Material* | *Emissivity Wavelength* | | | *T/K* |
|---|---|---|---|---|
| | *0.65 μm* | *1.0 μm* | *5.0 μm* | |
| Aluminium (polished surface) | (0.1 at 2 μm) | | 0.05 | 500 |
| Aluminium (rough surface) | (0.3 at 2 μm) | | 0.25 | 500 |
| Aluminium oxide | 0.1 | 0.06 | 0.39 | 1200 |
| (recrystallized alumina) | 0.15 | 0.07 | 0.43 | 1400 |
| | 0.25 | 0.1 | 0.46 | 1600 |
| | 0.4 | 0.2 | 0.6 | 1800 |
| Beryllia* | 0.5 | 0.35 | 0.8 | 1100 |
| Chromium | 0.35 | — | — | 1550 |
| Cobalt | 0.35 | 0.25 | — | 1300 |
| Gold | 0.15 | 0.05 | 0.03 | 1000 |
| Hafnium | 0.45 | — | — | 1200 |
| Iridium | 0.3 | 0.23 | 0.1 | 1500 |
| Iron | 0.35 | 0.3 | 0.15 | 1400 |
| Stainless steel | 0.33 | 0.3 | 0.2 | 1200 |
| „ „ (oxidized) | 0.8 | 0.8 | 0.7 | 1200 |
| Kanthal A (oxidized) | 0.85 | 0.85 | 0.75 | 1300 |
| Molybdenum | 0.4 | 0.3 | 0.15 | 2000 |
| Magnesia* | 0.25 | 0.2 | 0.37 | 1400 |
| Nickel | 0.45 | 0.35 | 0.15 | 1100 |
| Nickel (oxidized) | 0.88 | 0.84 | 0.75 | 1300 |
| Nickel/20% chromium | 0.4 | — | — | 1200 |
| „ „ (oxidized) | 0.9 | (0.85) | (0.8) | 1200 |
| Niobium | 0.4 | 0.32 | 0.2 | 2000 |
| Osmium | 0.43 | — | — | 1500 |
| Palladium | 0.35 | — | — | 1500 |
| Platinum | 0.35 | 0.25 | 0.08 | 1200 |
| Platinum 13% rhodium | 0.28 | — | — | 1100 |
| Rhenium | 0.4 | 0.35 | 0.2 | 1400 |
| Rhodium | 0.2 | — | — | 1400 |
| Ruthenium | 0.34 | — | — | 1500 |
| Silicon | 0.4 | 0.25 | 0.25 | 1500 |
| Silver | 0.05 | 0.05 | 0.05 | 1000 |
| Tantalum | 0.41 | 0.3 | 0.18 | 2200 |
| Thoria* | 0.4 | 0.35 | 0.37 | 1400 |
| Titanium | 0.6 | 0.3 | 0.18 | 1000 |
| Tungsten | 0.43 | 0.38 | 0.12 | 2400 |
| Uranium | 0.3 | — | — | 1100 |
| Vanadium | — | 0.65 | 0.28 | 800 |
| Zirconium | 0.5 | 0.45 | 0.3 | 1600 |
| Zirconia* | 0.4 | 0.2 | 0.45 | 1400 |

* These materials have an emissivity which increases with temperature in a way similar to that of aluminium oxide.

## DIELECTRIC CONSTANTS AND DIELECTRIC STRENGTHS OF MATERIALS

| *Material* | *Dielectric const. K* | *Dielectric strength (volt/m).* |
|---|---|---|
| Vacuum | 1.0000 | ∞ |
| Air | 1.00059 | $3 \times 10^6$ |
| Bakelite | 4.9 | $24 \times 10^6$ |
| Fused quartz | 3.78 | $8 \times 10^6$ |
| Pyrex glass | 5.6 | $14 \times 10^6$ |
| Polystyrene | 2.56 | $24 \times 10^6$ |
| Teflon | 2.1 | $60 \times 10^6$ |
| Neoprene rubber | 6.7 | $12 \times 10^6$ |
| Nylon | 3.4 | $14 \times 10^6$ |
| Paper | 3.7 | $16 \times 10^6$ |
| Strontium Titanate | 2.33 | $8 \times 10^6$ |
| Water | 80 | — |
| Silicone oil | 2.5 | $15 \times 10^6$ |

## MAGNETIC SUSCEPTIBILITIES OF SOME PARAMAGNETIC & DIAMAGNETIC MATERIALS

| *Material* | $\chi_m$ | *Materials* | $\chi_m$ |
|---|---|---|---|
| Aluminium | $2.3 \times 10^{-5}$ | Silver | $-2.86 \times 10^{-5}$ |
| Bismuth | $-1.66 \times 10^{-5}$ | Sodium | $-0.24 \times 10^{-5}$ |
| Copper | $-0.98 \times 10^{-5}$ | Tungsten | $6.8 \times 10^{-5}$ |
| Gold | $-3.6 \times 10^{-5}$ | Water | $-0.88 \times 10^{-5}$ |
| Lead | $-1.7 \times 10^{-5}$ | Hydrogen (1 Atm) | $-9.9 \times 10^{-9}$ |
| Magnesium | $1.2 \times 10^{-5}$ | Oxygen (1 Atm) | $2.1 \times 10^{-6}$ |

## LINEAR ABSORPTION COEFFICIENTS FOR GAMMA RAYS

| *Photon Energy/MeV* | *Linear Absorption coefficient/cm$^{-1}$* | | | |
|---|---|---|---|---|
| | *Water* | *Aluminium* | *Iron* | *Lead* |
| 0.1 | 0.167 | 0.432 | 2.69 | 59.8 |
| 0.15 | 0.149 | 0.359 | 1.43 | 20.8 |
| 0.2 | 0.139 | 0.324 | 1.08 | 10.1 |
| 0.3 | 0.118 | 0.278 | 0.833 | 3.79 |
| 0.4 | 0.106 | 0.249 | 0.722 | 2.35 |
| 0.5 | 0.0967 | 0.227 | 0.651 | 1.64 |
| 0.6 | 0.0894 | 0.210 | 0.598 | 1.29 |
| 0.8 | 0.0786 | 0.184 | 0.525 | 0.946 |
| 1.0 | 0.0706 | 0.166 | 0.468 | 0.772 |
| 1.5 | 0.0576 | 0.135 | 0.380 | 0.581 |
| 2.0 | 0.0493 | 0.116 | 0.332 | 0.510 |
| 3.0 | 0.0396 | 0.0953 | 0.282 | 0.463 |
| 5.0 | 0.0302 | 0.0767 | 0.247 | 0.486 |
| 10.0 | 0.0221 | 0.0626 | 0.236 | 0.568 |

# APPENDIX – II

## NOBEL LAUREATES IN PHYSICS

| Year | Nobel Laureate | | Recognized for |
|---|---|---|---|
| 1901 | Wilhelm Konrad Rontgen | 1845-1923 | discovery of X-rays |
| 1902 | Hendrik Antoon Lorentz | 1853-1928 | their researches into the influence of magnetism |
| | Peiter Zeeman | 1865-1943 | upon radiation phenomena |
| 1903 | Antoine Henri Becquerel | 1852-1908 | his discovery of spontaneous radioactivity; their joint researches on the radiation |
| | Pierre Curie | 1859-1906 | phenomena |
| | Marie Sklodowskan Curie | 1867-1934 | discovered by Prof. Henri Becquerel |
| 1904 | John William Strutt, (Lord Rayleigh) | 1842-1919 | investigations of the densities of the most important gases and his discovery of argon |
| 1905 | Philipp Eduard Anton von Lenard | 1862-1947 | his work on cathode rays |
| 1906 | Joseph John Thomson | 1856-1940 | his theoretical and experimental investigations on the conduction of electricity by gases |
| 1907 | Albert Abraham Michelson | 1852-1931 | his optical precision instruments and metrological investigations carried out with their aid |
| 1908 | Gabriel Lippmann | 1845-1921 | his method of reproducing colors photographically based on the phenomena of interference |
| 1909 | Guglielmo Marconi | 1874-1937 | their contribution to the development of |
| | Karl Ferdinand Braun | 1850-1918 | wireless telegraphy |
| 1910 | Johannes Diderik Van der Waals | 1837-1932 | his work on the equation of state of gases and liquids |
| 1911 | Wilhelm Wien | 1864-1928 | his discoveries regarding the laws governing the radiation of heat |
| 1912 | Nils Gustaf Dalen | 1869-1937 | his invention of automatic regulators for use in conjunction with gas accumulators for illuminating lighthouses and buoys |
| 1913 | Heike Kamerlingh Onnes | 1853-1926 | his investigations of the properties of matter at low temperatures which led, inter alia to the production of liquid helium. |
| 1914 | Max Theodor Felix von Laue | 1879-1960 | his discovery of the diffraction of Rontgen rays by crystals |
| 1915 | William Henry Bragg | 1862-1942 | their services in the analysis of crystal struc |
| | William Lawrence Bragg | 1890-1971 | ture by means of X-rays |
| 1917 | Charles Glover Barkla | 1877-1944 | his discovery of the characteristic X-rays of the elements |
| 1918 | Max Planck | 1858-1947 | his discovery of energy quanta |
| 1919 | Johannes Stark | 1874-1957 | his discovery of the Doppler effect in canal rays and the splitting of spectral lines in electric fields |

| | | | |
|---|---|---|---|
| 1920 | Charles Edouard Guillaume | 1861-1938 | the service he has rendered to precision measurements in Physics by his discovery of anomalies in nickel steel alloys |
| 1921 | Albert Einstein | 1879-1955 | his services to Theoretical Physics, and especially for his discovery of the laws of the photoelectric effect |
| 1922 | Niels Henrik David Bohr | 1885-1962 | the investigation of the structure of atoms, and of the radiation emanating from them |
| 1923 | Robert Andrews Millikan | 1868-1953 | his work on the elementary charge of electricity and on the photoelectric effect. |
| 1924 | Karl Manne Georg Siegbahn | 1888-1979 | his discovery and research in the field of X-ray spectroscopy |
| 1925 | James Franck<br>Gustav Hertz | 1882-1964<br>1887-1975 | their discovery of the laws governing the impact of an electron upon an atom |
| 1926 | Jean Baptiste Perrin | 1870-1942 | his work on the discontinuous structure of matter, and especially for his discovery of sedimentation equilibrium |
| 1927 | Arthu Holly Compton<br>Charles Thomson Rees Wilson | 1892-1962<br>1869-1959 | his discovery of the effect named after him, his method of making the paths of electrically charged particles visible by condensation of vapour |
| 1928 | Owen Willans Richardson | 1879-1959 | his work on the thermionic phenomenon, and especially for the discovery of the law named after him |
| 1929 | Prince Louis Victor de Broglie | 1892-1987 | his discovery of the wave nature of electrons |
| 1930 | Sir Chandrasekhara Venkata Raman | 1888-1970 | his work on the scattering of light and for the discovery of the effect named after him |
| 1932 | Werner Heisenberg | 1901-1976 | the creation of quantum mechanics, the application of which has, among other things, led to the discovery of the allotropic forms of hydrogen |
| 1933 | Erwin Schrödinger<br>Pal Adrien Maurice Dirac | 1887-1961<br>1902-1984 | the discovery of new productive forms of atomic theory |
| 1934 | (no award) | | |
| 1935 | James Chadwick | 1891-1974 | his discovery of the neutron |
| 1936 | Victor Franz Hess<br>Carl David Anderson | 1883-1964<br>1905-1991 | the discovery of cosmic radiation<br>his discovery of the position |
| 1937 | Clinton Joseph Davisson<br>George Paget Thomson | 1881-1958<br>1892-1975 | their experimental discovery of the diffraction of electron by crystals |
| 1938 | Enrico Fermi | 1901-1954 | his demonstrations of the existence of new radioactive elements produced by neutron irradiation, and for his related discovery of nuclear reactions brought about by slow neutrons |

| | | | |
|---|---|---|---|
| 1939 | Ernest Orlando Lawrence | 1901-1958 | the invention and development of the cyclotron and for results obtained with it, especially for artificial radioactive elements |
| 1943 | Otto Stern | 1888-1969 | his contribution to the development of the molecular rays method and his discovery of the magnetic moment of the proton |
| 1944 | Isidor Issac Rabi | 1898-1988 | his resonance method for recording the magnetic properties of atomic nuclei |
| 1945 | Wolfgang Pauli | 1900-1958 | his discovery of the Exclusion Principle (Pauli Principle) |
| 1946 | Percy Williams Bridgman | 1882-1961 | the invention of an apparatus to produce extremely high pressure, and for the discoveries he made therewith in the field of high-pressure physics |
| 1947 | Sir Edward Victor Appleton | 1892-1965 | his investigations of the physics of the upper atmosphere, especially for the discovery of the so-called Appleton layer |
| 1948 | Patrick Maynard Stuart Blackett | 1897-1974 | his development of the Wilson cloud chamber method, and his discoveries therewith in nuclear physics and cosmic radiation |
| 1949 | Hideki Yukawa | 1907-1981 | his prediction of the existence of mesons on the basis of theoretical work on nuclear forces |
| 1950 | Cecil Frank Powell | 1903-1969 | his development of the photographic method of studying nuclear processes and his discoveries regarding mesons made with this method |
| 1951 | Sir John Douglas Cockcroft | 1897-1967 | their pioneer work on the transmutation of atomic nuclei by artificially accelerated atomic particles |
| | Ernest Thomas Sinton Walton | 1903- | |
| 1952 | Felix Bloch | 1905-1983 | their development of new methods for nuclear magnetic precision measurements and discoveries in connection therewith |
| | Edward Mills Purcell | 1912- | |
| 1953 | Frits Zernike | 1888-1966 | his demonstration of the phase contrast method, especially for his invention of the phase contrast microscope |
| 1954 | Max Born | 1882-1970 | his fundamental research in quantum mechanics especially for his statistical interpretation of the wave function |
| | Walther Bothe | 1891-1958 | the coincidence method and his discoveries made therewith |
| 1955 | Willis Eugene Lamb | b. 1913 | his discoveries concerning the fine structure of the hydrogen spectrum |
| | Polykarp Kusch | b. 1911 | his precision determination of the magnetic moment of the electron |
| 1956 | William Bradford Shockley | 1910-1989 | their researches on semiconductors and their discovery of the transistor effect |
| | John Bardeen | 1908-1991 | |
| | Walter Houser Brattain | 1902-1987 | |

| | | | |
|---|---|---|---|
| 1957 | Chen Ning Yang<br>Tsung Dao Lee | b. 1922<br>b. 1926 | their penetrating investigation of the parity laws which has led to important discoveries regarding elementary particles |
| 1958 | Pavel Alekseyevich Cerenkov<br>Ilya Mikhaylovich Frank<br>Igor Yevgenyevich Tamm | b. 1904<br>1908-1990<br>1895-1971 | the discovery and interpretation of the Cerenkov effect |
| 1959 | Emillio Gino Segre<br>Own Chamberlain | 1905-1990<br>b. 1920 | their discovery of the antiproton |
| 1960 | Donald Arthur Glaser | b. 1926 | the invention of the bubble chamber |
| 1961 | Robert Hofstadter | 1915-1990 | his pioneering studies of electron scattering in atomic nuclei and for his thereby achieved discoveries concerning the structure of the nucleons |
| | Rudolf Ludwig Mossbauer | b. 1929 | his research concerning the resonance absorption of γ rays and his discovery in this connection of the effect which bears his name |
| 1962 | Lev Davidovich Landau | 1908-1968 | his pioneering theories of condensed matter, especially liquid helium |
| 1963 | Eugene Paul Wigner | b. 1902 | his contributions to the theory of the atomic nucleus and the elementary particles, particularly through the discovery and application of fundamental symmetry principles |
| | Maria Goeppert Mayer<br>J. Hans D. Jensen | 1906-1972<br>1907-1973 | their discoveries concerning nuclear shell structure |
| 1964 | Charles Hard Townes<br>Nikolay Gennadiyevich Basov<br>Alekanndr Mikhazlovich Prokhorov | b. 1915<br>b. 1922<br>b. 1916 | fundamental work in the field of quantum electronics which has led to the constructi - on of oscillators and amplifiers based on the maser-laser principle |
| 1965 | Schinichiro Tomonaga<br>Julian Schwinger<br>Richard Phillips Feynman | 1906-1979<br>b. 1918<br>1918-1988 | their fundamental work in quantum electrodynamics, with profound consequences for the physics of elementary particles |
| 1966 | Alfred Kastler | 1902-1984 | the discovery and development of optical methods for studying Hertzian resonance in atoms |
| 1967 | Hans Albrecht Bethe | b. 1906 | his contributions to the theory of nuclear reactions, especially his discoveries concerning the energy production in stars |
| 1968 | Luis Walter alvarez | 1911-1988 | his decisive contribution to elementary particle physics, in particular the discovery of a large number of resonance states, made possible through his development of the technique of using the hydrogen bubble chamber and data analysis |

| | | | |
|---|---|---|---|
| 1969 | Murray Gell-Mann | b. 1929 | his contributions and discoveries concerning the classification of elementary particles and their interactions. |
| 1970 | Hannes Olof Gosta Allven | b. 1908 | fundamental work and discoveries in magnetohydro-dynamics with fruitful applications in different parts on plasma physics |
| | Louis Eugene Felix Neel | b. 1904 | fundamental work and discoveries concerning antiferromagnetism and ferrimagnetism which have led to important applications in solid-state physics |
| 1971 | Dennis Gabor | 1900-1979 | his discovery of the principles of holography |
| 1972 | John Bardeen<br>Leon Neil Cooper<br>John Robert Schrieffer | 1908-1991<br>b. 1930<br>b. 1931 | their development of a theory of superconductivity |
| 1973 | Leo Esaki | b. 1925 | his discovery of tunneling in semiconductors; |
| | Jvar Giaever | b. 1929 | his discovery of tunneling in superconductors; |
| | Brian David Josephson | b. 1940 | his theoretical prediction of the properties of a supercurrent through a tunnel barrier |
| 1974 | Anthony Hewish | b. 1924 | the discovery of pulsars |
| | Sir Martin Ryle | 1918-1984 | his pioneering work in radio astronomy |
| 1975 | Aage Bohr<br>Ben Mottelson<br>Leo James Rainwater | b. 1922<br>b. 1926<br>1917-1986 | the discovery of the connection between collective motion and particle motion and the development of the theory of the structure of the atomic nucleus based on this connection |
| 1976 | Burton Richter<br>Samuel Chao Chung Ting | b. 1931<br>b. 1936 | their (independent) discovery of an important fundamental particle |
| 1977 | Phillip Warren Anderson<br>Sir Nevill Francis Mott<br>John Hasbrouck Van Vleck | b. 1923<br>b. 1905<br>1899-1980 | their fundamental theoretical investigations of the electronic structure of magnetic and disordered system |
| 1978 | Pyotr Leonidovich Kapitsa | 1894-1984 | his basic inventions and discoveries in low temperature physics |
| | Arno Allan Penzias<br>Robert Woodrow Wilson | b. 1926<br>b. 1936 | their discovery of cosmic microwave background radiation |
| 1979 | Sheldon Lee Glashow<br>Abdus Salam<br>Steven Weinberg | b. 1932<br>b. 1926<br>b. 1933 | their unified model of the action of the weak and electromagnetic forces and for their prediction of the existence of neutral currents. |
| 1980 | James Watson Cronin<br>Val L. Fitch | b. 1931<br>b. 1923 | the discovery of violations of fundamental symmetry principles in the decay of neutral K mesons |
| 1981 | Nicolaas Bloembergen<br>Arthur Leonard Schawlow | b. 1920<br>b. 1921 | their contribution to the development of laser spectroscopy; |
| | Kai M. Siegbahn | b. 1918 | His contribution to high resolution electron microscopy. |
| 1982 | Kenneth Geddes Wilson | b. 1936 | his method of analyzing the critical phenomena inherent in the changes of matter under the influence of pressure and temperature |

| | | | |
|---|---|---|---|
| 1983 | Subrahmanyan Chandrasekhar | b. 1910 | his theoretical studies of the structure and evolution of stars |
| | William A. Fowler | b. 1911 | his studies of the formation of chemical elements in the universe |
| 1984 | Carlo Rubbia<br>Simon van der Meer | b. 1934<br>b. 1925 | their decisive contributions to the large project, which led to the discovery of the field particles W and Z, communicators of the weak interaction |
| 1985 | Klaus von Klitzing | b. 1943 | his discovery of the quantized Hall resistance |
| 1986 | Ernst August Friedrich Ruska | 1906-1988 | his invention of the electron microscope |
| | Gerd Binning<br>Heinrich Rohrer | b. 1947<br>b. 1933 | their invention of the scanning-tunneling electron microscope |
| 1987 | Kar Alex Muller<br>Johannes Georg Bednorz | b. 1927<br>b. 1950 | their discovery of a new class of superconductors |
| 1988 | Leon Max Lederman<br>Malvin Schwartz<br>Jack Steinberger | b. 1922<br>b. 1932<br>b. 1921 | the neutrino-beam method and the demonstration of the doublet structure of the leptons through the discovery of the muon neutrino |
| 1989 | Norman Foster Ramsey, Jr. | b. 1915 | his invention of the separated-oscillatory-field method and its use in the hydrogen maser and other atomic clocks |
| | Hans Georg Dehmelt<br>Wilhelm Paul | b. 1922<br>b. 1913 | their development of the ion trap technique |
| 1990 | Jerome I, Friedman<br>Henry W. Kendall<br>Richard E. Taylor | b. 1930<br>b. 1926<br>b. 1929 | their experiment which led to breakthrough in our understanding of matter |
| 1991 | Pierre-Gilles de Gennes | b. 1932 | his work on polymers and liquid crystals |
| 1992 | Georges Charpak | | |
| 1993 | Joseph H. Taylor<br>Russel A. Hulse | | discovery of binary pulsar |
| 1994 | Betran N. Brockhouse<br>Clifford G. Shull | | their development of neutron scattering technique |
| 1995 | Martin L. Perl<br>Frederick Reines | | |
| 1996 | David Lee<br>Robert Richardson<br>Douglas Osheroff | | discovery of superfluidity in $^3$He isotope of helium. |
| 1997 | Steven Chu<br>Claude Cohen-Tannoudji<br>William D. Phillips | | development of methods to cool and trap atoms with laser light |
| 1998 | Robert C. Laughlin<br>Horst L. Stoermer<br>Daniel C. Tsui | | discovery of how subatomic particles can behave like a fluid. |

| | | | |
|---|---|---|---|
| 1999 | Gerardus Hooft<br>Martinus J. G. Veltman | | theoretical work on the structure and motion of subatomic particles. |
| 2000 | Zhores I. Alfernov<br>Herbert Kroemer and<br>Jack Kilby | | development of technology used in satellite communications and cellular phones.<br>invention and development of the ICs, and invention of the pocket calculator. |
| 2001 | Erick A. Cornell<br>Carl E. Wieman<br>Wolfgang Ketterle | | the achievement of Bose-Einstein condensation in dilute gases of alkali atoms and for early fundamental studies of the properties of the condensates. |
| 2002 | Raymond Davis Jr.<br>Masatoshi Koshiba;<br>Riccardo Giacconi | | their research into cosmic neutrinos and; his construction of instruments needed to investigate cosmic X-rays radiation which is absorbed in the Earth's atmosphere |
| 2003 | Alexei A. Abrikosov<br>Vitaly L. Gineburg<br>Anthony J. Leggett | | their work in quantum physics concerning superconductivity and superfluidity. |
| 2004 | David Gross<br>David Politzer<br>Frank Wilczek | | a unified description of all forces of Nature – from the tiniest distances within the atomic nucleus to the vast distances of the universe. |
| 2005 | Roy Glouber<br>John Hall<br>Theodor Haensch | | their contributions to the development of Laser based precision spectroscopy including the optical frequency conbtechnique. |
| 2006 | John C. Mather<br>George F. Smoot | b. 1946<br>b. 1945 | for the discovery of the black body form and anisotropy of the cosmic microwave backgroud radiation. |
| 2007 | Albert Fert<br>Peter Grunberg | | Giant Magnetoresistance. |
| 2008 | Yoichiro Nambu | | Mechanism of spontaneous broken symmetry in subatomic physics. |
| | Makoto Kobayashi<br>Toshihide Maskawa | | The origin of the broken symmetry which predicts the existence of at least three families of quarks in nature. |

# INDEX

## A



## B



## C

## D



## E



## F

## U



## V



## W



## Y



## Z

# NOTES

# NOTES

# NOTES